WebAssign

W9-BCQ-535

OVER ONE BILLION PROBLEMS GRADED

The most widely used online homework management system, with more than 2.2 million users and over one billion problems graded, **WebAssign®** is the premier course solution for instructors and students. Backed by superior reliability, support, and a proven track record, **WebAssign** and **Cengage Learning** boast 99.99% uptime and unparalleled customer service and training.

For Instructors:

- A simple, user-friendly interface
- Two-step course creation
- An *Intuitive Question Editor* allows questions to be edited or added without learning a programming language.
- Automatic grading of online homework assignments.

For Students:

For students, there are problems from the textbook.

For more information and to view sample assignments, please visit: **www.webassign.net/cengage**

BROOKS/COLE
CENGAGE Learning™

PHYSICS 2010–2011 Edition

Vern J. Ostdiek, Benedictine College
Donald J. Bord, University of Michigan

Publisher: Mary Finch

Developmental Editor: Jamie Bryant, B-books, Ltd.

Associate Developmental Editor: Brandi Kirksey

Product Development Manager, 4LTR Press: Steven E. Joos

Project Manager, 4LTR Press: Clara Goosman

Editorial Assistant: Joshua Duncan

Brand Executive Marketing Manager, 4LTR Press: Robin Lucas

Marketing Director: Patrick Leow

Marketing Manager: Nicole Mollica

Marketing Coordinator: Kevin Carroll

Marketing Communications Manager: Belinda Krohmer

Production Director: Amy McGuire, B-books, Ltd.

Senior Content Project Manager: Cathy Brooks

Senior Media Editor: Rebecca Berardy-Schwartz

Manufacturing Coordinator: Miranda Klapper

Production Service: B-books, Ltd.

Senior Art Director: Cate Rickard Barr

Cover Design: Studio Montage

Cover Image: © Bol/Begsteiger / PHOTOTAKE

Photography Manager: Deanna Ettinger

Photo Researchers: Jamie Bryant, B-books, Ltd., Sam Marshall, and Terri Wright

© 2011 Brooks/Cole, a part of Cengage Learning

ALL RIGHTS RESERVED. No part of this work covered by the copyright herein may be reproduced, transmitted, stored or used in any form or by any means graphic, electronic, or mechanical, including but not limited to photocopying, recording, scanning, digitizing, taping, Web distribution, information networks, or information storage and retrieval systems, except as permitted under Section 107 or 108 of the 1976 United States Copyright Act, without the prior written permission of the publisher.

For product information and technology assistance, contact us at
Cengage Learning Academic Resource Center, 1-800-423-0563

For permission to use material from this text or product,
submit all requests online at **www.cengage.com/permissions**
Further permissions questions can be emailed to
permissionrequest@cengage.com

© 2011 Cengage Learning. All Rights Reserved.

Library of Congress Control Number: 2009943791

SE ISBN-13: 978-0-538-73539-1
SE ISBN-10: 0-538-73539-2

Brooks/Cole Cengage Learning
20 Davis Drive
Belmont, CA 94002-3098
USA

Cengage Learning products are represented in Canada by Nelson Education, Ltd.

For your course and learning solutions, visit **academic.cengage.com**
Purchase any of our products at your local college store or at our preferred online store **www.CengageBrain.com**

Printed in the United States of America
1 2 3 4 5 6 7 13 12 11 10

Brief Contents

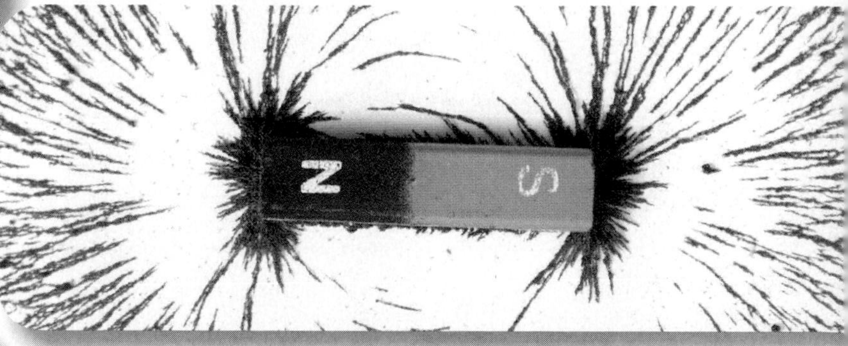

Contents

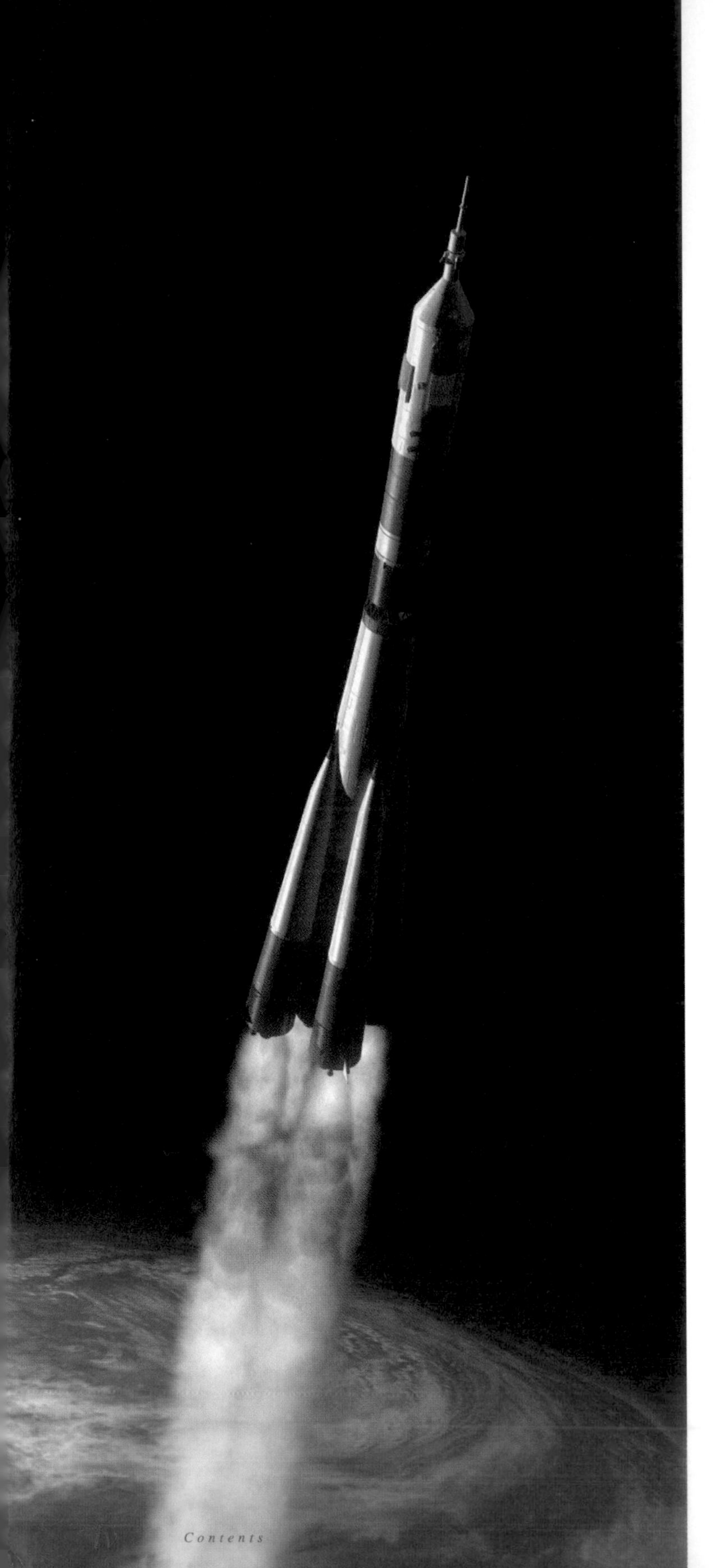

Contents

Contents

Contents

Dedication
This book is dedicated to the memory of Vern Ostdiek—friend, colleague, and co-author—whose passion for physics and for teaching infuses every page.
Vern J. Ostdiek
1953–2008
© roberta casaliggi/iStockphoto.com

Prologue

P.1 Why Learn Physics?

The answer is easy for those majoring in physics, engineering, or other sciences: Physics will provide them with important tools for their academic and professional lives. The technology that our modern society relies on comes from applying the discoveries of physics and other sciences. From designing safe, efficient passenger jets to producing sophisticated, inexpensive computers and cell phones, engineers apply physics every day.

Landing astronauts on the Moon and returning them safely to Earth, one of the greatest feats of the twentieth century, is a good example of physics applied on many levels. The machines involved—from powerful rockets to on-board computers—were designed, developed, and tested by people who knew a lot about physics. The planning of the orbits and the timing of the rocket firings to change orbits involved scientists with a keen understanding of basic physics, like gravity and the "laws of motion." Often, the payoff is not so tangible or immediate as a successful Moon landing. Behind great technological advances are years or even decades of basic research into the properties of matter.

For instance, take a portable CD player. A laser reads data off a spinning disc, integrated circuit chips inside "translate" the digital data into electrical signals, tiny magnets in the headphones help convert those signals into sound, and a liquid crystal display provides information for the user. If you could send this device back

NASA / © Mike Bentley/iStockphoto.com

It took a lot of physics to get astronaut Neil Armstrong to the Moon. The Moon landing celebrated its 40th anniversary in July 2009.

in time to when your grandparents were children, it would astound the greatest physicists and electrical engineers of the day. But even at that time, scientists were studying the properties of semiconductors (the raw material for lasers and integrated circuit chips) and liquid crystals.

For you and others like you, taking perhaps just one physics course in your life, the usefulness of physics is probably not a big reason for studying it. We will see that with even one course, you can use physics to determine, for example, how large a raft has to be to support you or whether using a toaster and a hair dryer at the same time will trip a circuit breaker. But you are not going to make a living with your understanding of physics, nor will you be using it (knowingly) every day. So why should *you* study physics? There are both aesthetic and practical reasons for learning physics. Seeing the order that exists in Nature and understanding that it follows from a relatively small number of "rules" can be fascinating—similar to learning the inspirations behind a musician's or an artist's work. Learning how common devices operate gives you a better understanding of how to use them and may reduce any frustration you have with them. An elementary knowledge of physics also helps you make more informed decisions regarding important issues facing you, your community, your nation, and the world. As you progress through this book, keep track of events through the news

© Matthew T Tourtellott/Shutterstock / © iStockphoto.com

© Popperfoto/Getty Images

"In a century that will be remembered foremost for its science and technology—in particular for our ability to understand and then harness the forces of the atom and universe—one person clearly stands out as both the greatest mind and paramount icon of our age: The kindly, absent-minded professor whose wild halo of hair, piercing eyes, engaging humanity, and extraordinary brilliance made his face a symbol and his name a synonym for genius, Albert Einstein." (*Time* magazine.)

In 1999, *Time* magazine, known for naming its annual "person of the year," set out to choose its "person of the century." A daunting task it was, considering the events of the turbulent twentieth century and the inevitable criticism that would come from those favoring someone not named. Would it be a world leader who shaped significant spans of the century—for good or bad—such as Franklin D. Roosevelt (a runner-up for person of the century) or Joseph Stalin (1939 and 1942 person of the year)? Perhaps a military figure like Dwight D. Eisenhower (1944 person of the year) or a spiritual leader like Pope John Paul II (1994 person of the year)? Would it be a champion of peace or justice such as Mahatma Gandhi (the other runner-up) or Martin Luther King, Jr. (1963 person of the year)? In the end, the selection was someone with enormous name recognition, but whose work most people would confess ignorance about, the physicist Albert Einstein.

Einstein was chosen because he symbolized the great strides made during the 1900s in deciphering and harnessing fundamental aspects of the material universe. But his style, his manner, and his allure had something to do with it as well. At times, he dominated physics with spectacular results, in a way that is reminiscent of certain "athletes of the 20th century" (Aaron, Ali, Gretzky, Jordan, Montana, Navratilova, Nicklaus . . .). Like them, only his last name is needed to identify him. In 1905, while working as a civil servant far removed from the great centers of physics research, Einstein had three scientific papers—about three different subjects—published in a German physics journal. They were so extraordinary that any one would likely have led to his receiving the Nobel Prize in physics—then, as now, the highest award in the field.

Had *Time* magazine been in business during previous centuries, the editors might well have honored other physicists in the same way, perhaps Galileo or Newton for the seventeenth century or Maxwell for the nineteenth. Such is the high regard that Western civilization has for the field of physics and those who excel in it. Partly, it is the impact their discoveries often have on our lives, by way of technological gadgets or civilization-threatening weaponry. But often it is the intellectual resonance we have with the revolutionary insights they give us about the universe (it is the Earth that moves around the Sun, it is matter being converted into energy that makes the Sun shine . . .).

© iStockphoto.com

media of your choice. You may be surprised at just how often physics is in the news—directly or indirectly.

If you start this excursion into the world of physics with a sense of curiosity and a thirst for knowledge, you won't be disappointed. And you will have the two most important characteristics needed to make the endeavor both easy and successful. Learning how sunlight and raindrops combine to make a rainbow will deepen your appreciation of its beauty. Knowing about centripetal force will help you understand why ice or gravel on a curved road is dangerous. Learning the basics of nuclear physics will help you understand the danger of radon gas and the promise of nuclear fusion. Knowing the principles behind stereo speakers, aircraft altimeters, refrigerators, lasers, microwave ovens, guitars, and Polaroid sunglasses will give you a better appreciation for how these devices do what they do.

Throughout this book, we encourage the reader to be inquisitive. Just memorizing definitions and facts doesn't lead you to a real understanding of a subject, any more than memorizing a manual on playing soccer means you can jump into a game and do well. You have to practice, try things, think of situations and how events would evolve, and so on. So it is with physics. Being able to recite Newton's third law of motion is good, but understanding what it means and how it works in the real world is what's really important.

Our goal with this book is to encourage you to learn through inquiry. Ask yourself questions. Experiment. Try things out. Make the concepts real. To help you in this process, we've made sure that many of the questions at the ends of the chapters are inquiry-based. Once you get used to this method of learning, you will find that you will learn the material faster and more deeply than before.

Welcome to the world of physics. You are embarking on an introduction to a field that continues to fascinate people in all walks of life. Tell a friend or family member that you are now studying physics. That will quite likely impress them in a way that most other subjects would not. Whether it should do so is an interesting question. We hope that when you finish this endeavor, you will answer yes.

P.2 What Is Physics?

Because physics is one of the basic sciences, it is important to first have an idea of just what science is. Science is the *process* of seeking and applying knowledge about our universe. Science also refers to the *body of knowledge* about the universe that has been amassed by humankind. Pursuing knowledge for its own sake is pure or basic science; developing ways to use this knowledge is applied science. Astronomy is mostly a pure science, while engineering fields are applied science. The material we present in this book is a combination of fundamental concepts that we believe are important to know for their own sake and of examples of the many ways that these concepts are applied in the world around us.

Physics is not as easy to define as some areas of science like biology, the study of living organisms. If you ask a dozen physicists to define the term, you are not likely to get two answers exactly alike. One suitable definition is that physics is the study of the fundamental structures and interactions in the physical universe. In this book, you will find much about the structures of things like atoms and nuclei, along with close looks at how things interact by way of gravity, electricity, magnetism, and so on. Within physics, there is a wide range of divisions. Table P.1 lists some of the common areas based on one measure of research activity. There is a lot of overlap between the divisions,

Table P.1 Some Commonly Identified Divisions of Physics, Ranked Roughly by Number of Doctorates Earned Each Year (Based on information from the American Institute of Physics)

Area	Topics of Investigation
1. Condensed Matter	structures and properties of solids and liquids
2. Astronomy and Astrophysics	stars, galaxies, evolution of the universe
3. Particles and Fields	fundamental particles and fields, high-energy accelerators
4. Nuclear Physics	nuclei, nuclear matter, quarks and gluons
5. Atomic and Molecular	atoms, molecules
6. Optics and Photonics	light, laser technology
7. Biophysics	physics of biological phenomena
8. Materials Science	applications of condensed-matter physics
9. Applied Physics	device development, technological innovation
10. Atmospheric and Space Physics	meteorology, Sun, planets

and some of them are clearly allied with other sciences like biology and chemistry.

The field of physics is divided differently when the basics are being taught to beginners. The topics presented to students in their first exposure to physics are usually ordered according to their historical development (study of motion first, elementary particles and cosmology last). This ordering also approximates the ranking of areas by our everyday experience with them. We've all watched people in motion and things collide, but few people encounter the idea of quarks before taking a physics class—even though we and all of the objects we deal with are mainly composed of quarks.

The vast majority of students who take an introductory course in physics are not majoring in it. Most of those who do earn a degree in physics find employment in business, industry, government, or education. Data compiled by the American Institute of Physics for 2005–2006 indicate that the employer distribution for individuals receiving bachelor's, masters, and doctoral degrees is remarkably similar with about 50 percent of each group finding work in the private sector, another quarter in educational institutions (high schools, colleges, and universities), and some 10 percent or so in government facilities (national laboratories and the military). In addition to the expected occupations like researcher and teacher, people with physics degrees also have job titles like engineer, manager, computer scientist, and technician. Often, physicists are hired not so much for their knowledge of physics as for their experience with problem solving and advanced technology.

P.3 How Is Physics Done?

So how does one "do" physics, or science in general? How did humankind come by this mountain of scientific knowledge that has been amassed over the ages? A blueprint exists for scientific investigation that makes an interesting starting point for answering these questions. It is at best an oversimplification of how scientists operate. Perhaps we should regard it as a game plan that is frequently modified when the action starts. It is called the **scientific method**. One version of it goes something like this: Careful *observation* of a phenomenon induces an investigator to question its cause. A *hypothesis* is formed that purports to explain the observation. The scientist devises an *experiment* that will test this hypothesis, hoping to show that it is correct—at least in one case—or that it is incorrect. The outcome of the experiment often raises more questions that lead to a *modification* of the hypothesis and further experimentation. Eventually, an accepted hypothesis that has been verified by different experiments can be elevated to a *theory* or a *law*. The term that is used—*theory* or *law*—is not particularly important in physics: Physicists hold Newton's second *law* of motion and Einstein's special *theory* of relativity in roughly the same regard in terms of their validity and their importance.

One nice thing about the scientific method is that it is a logical procedure that is practiced by nearly everyone from time to time (see Figure P.1). Let's say that you get into your car and find that it won't start (observation). You speculate that maybe the battery is dead (hypothesis). To see if this is true, you turn on the radio or the lights to see if they work (experiment). If they don't, you may look for someone to give you a jump start. If they do, you probably guess that something else, such as the starter, is causing the problem. Clearly, a good mechanic must be proficient at this way of investigating things. Healthcare professionals making a medical diagnosis use similar procedures. And they are useful to students as well.

© Helene Rogers/Alamy

Figure P.1
The basics of the scientific method, with an example.

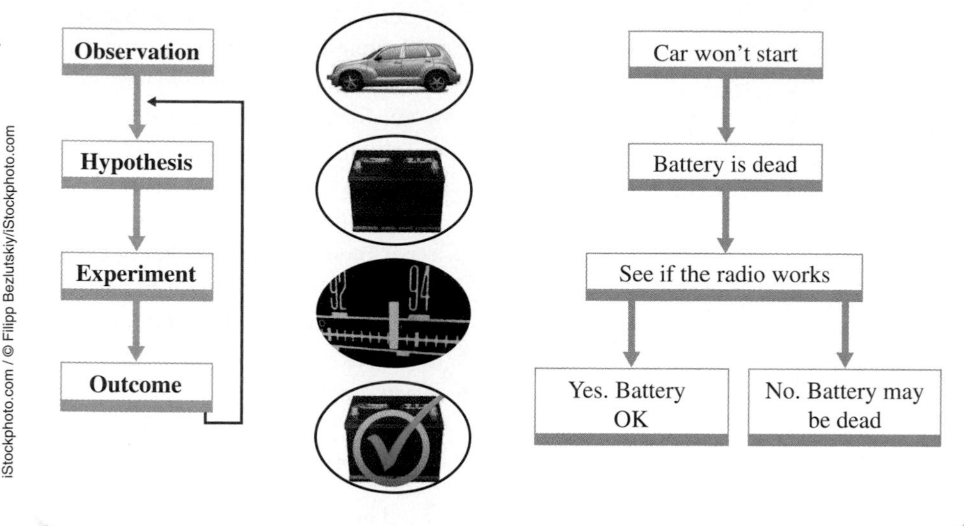

© iStockphoto.com / © Don Nichols/iStockphoto.com / © charles taylor/iStockphoto.com / © Filipp Bezlutskiy/iStockphoto.com

One of the architects of the scientific method was Galileo Galilei (1564–1642). He thought a great deal about how science should be done and applied his ideas to his study of motion. Galileo believed that science had to have a strong logical basis that included precise definitions of terms and a mathematical structure with which to express relationships. He introduced the use of controlled experiments, which he applied with great success in his studies of how objects fall. By ingenious experimental design, he overcame the limitations of the crude timing devices that existed and measured the acceleration of falling bodies. Galileo was a shrewd observer of natural phenomena, from the swinging of a pendulum to the moons orbiting Jupiter. By drawing logical conclusions from what he saw, he demonstrated that rules could be used to predict and explain natural phenomena that had long seemed mysterious or magical.

The scientific method is important, particularly in the day-to-day process of filling in the details about a phenomenon being studied. But it is not the whole story. Even a brief look at the history of physics reveals that there is no simple recipe that scientists have followed to lead them to breakthroughs. Some great discoveries were made by traditional physicists working in labs, proceeding in a "scientific method" kind of way. However, unplanned events often come into play, such as accidents (Galvani discovered electric currents while performing biology experiments) or luck (Becquerel stumbled upon nuclear radiation because of a string of cloudy days in Paris). Sometimes "thought experiments" were required because the technology of the time didn't allow "real" experiments. Newton predicted artificial satellites, and Einstein unlocked relativity this way. Often, it was hobbyists, not professional scientists, who made significant discoveries (e.g., the statesman Benjamin Franklin and schoolteacher Georg Simon Ohm). Sometimes, it is scientists correctly interpreting the results of others who failed to "connect the dots" themselves (Lise Meitner and her nephew Otto Frisch identified nuclear fission that way). The point is that there are no "hard and fast" rules for making scientific discoveries, and this is no less true in physics than it is in chemistry, biology, astronomy, or any other scientific discipline.

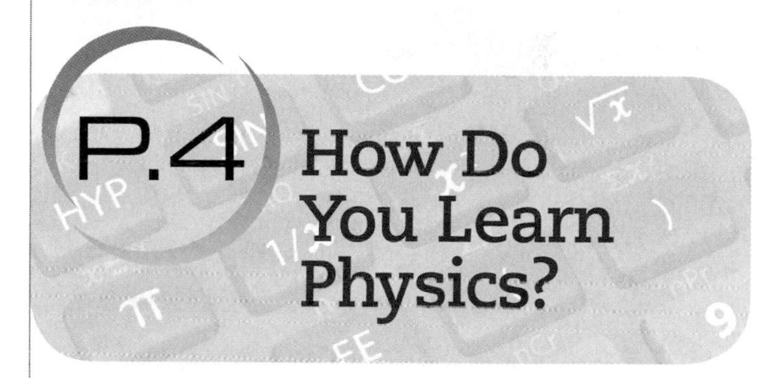

P.4 How Do You Learn Physics?

The goal of learning physics and any other science is to gain a better understanding of the universe and the things in it. We generally focus attention on only a small segment of the universe at one time, so that the structural complexities and interactions within it are manageable. We call this a *system.* Some examples of systems that we will talk about are the nucleus of an atom, the atom itself, a collection of atoms inside a laser, air circulating in a room, a rock moving near the Earth's surface, and the Earth with satellites in orbit about it (see Figure P.2).

The kinds of things a person might want to know about a system include: (1) its structure or configuration; (2) what is going on in it and why; and (3) what will happen in it in the future. The first step is often relatively easy—identifying the objects in the system. Protons, electrons, chromium atoms, heated air, a rock, and the Moon are some of the things in the systems mentioned above. Often, the items in a physical system are already familiar

Figure P.2

Combination of six figures representing different systems we will be examining. The scale of the different parts varies greatly, from smaller than can be seen with a microscope to thousands of miles.

© Nicholas Monu/iStockphoto.com / © Slavoljub Pantelic/iStockphoto.com

to us. A host of other intangible things in a system must also be identified and labeled before real physics can begin. We must define things like the *speed* of the rock, the *density* of the air, the *energy* of the atoms in the laser tube, and the *angular momentum* of a satellite to understand what is going on in a system and what its future evolution will be. We will call these things and others like them **physical quantities**. Most will be unfamiliar to you unless you have studied physics before. Together with the named objects, they form what can be called the *vocabulary of physics.* There are hundreds of physical quantities in regular use in the various fields of physics, but for our purposes in this book, we will need only a fraction of these.

Physics seeks to discover the basic ways in which things interact. Laws and principles express relationships that exist between physical quantities. For example, the law of fluid pressure expresses how the pressure at some location in a fluid depends on the weight of the fluid above. This law can be used to find the water pressure on a submerged submarine or the air pressure on a person's chest. These "rules" are used to understand the interactions in a system and to predict how the system will change with time in the future. The laws and principles themselves were formed after repeated, careful observations of countless systems by scientists throughout history. They withstood the test of time and repeated experimentation before being elevated to this status. You might regard physics as the continued search for, and the application of, basic rules that govern the interactions in the universe.

The process of learning physics has two main thrusts:

1. The need to develop an understanding of the different physical quantities used in each area (establish a vocabulary)
2. The need to grasp the significance of the laws and principles that express the relationships among these physical quantities.

Visualizing Relationships

Memorizing the definitions and laws is only a first step—but that alone won't do it. See, do, think, interact, visualize. Get involved in the physical world. That's how you learn physics. Everywhere you look, you can find examples of physical theories in action. This book includes dozens of worked out examples based on real world situations to help you connect the discussions in each chapter with how they play out in the real world.

Another tool we use to help you visualize the relationships in physics is the *concept map.* Concept maps were developed in the 1960s and are used in a wide range of fields in a variety of ways. A concept map presents an overview of issues, examples, concepts, and skills in the form of a set of interconnected *propositions.* Two or more concepts joined by linking words or phrases make a proposition. The meaning of any particular concept is the sum of all the links that contain the concept. To "read" a concept map, start at the top with the most general concepts, and work your way down to the more specific items and examples at the bottom. Concept Map P.1 is one example. It is used to show some of the connections

Concept Map P.1

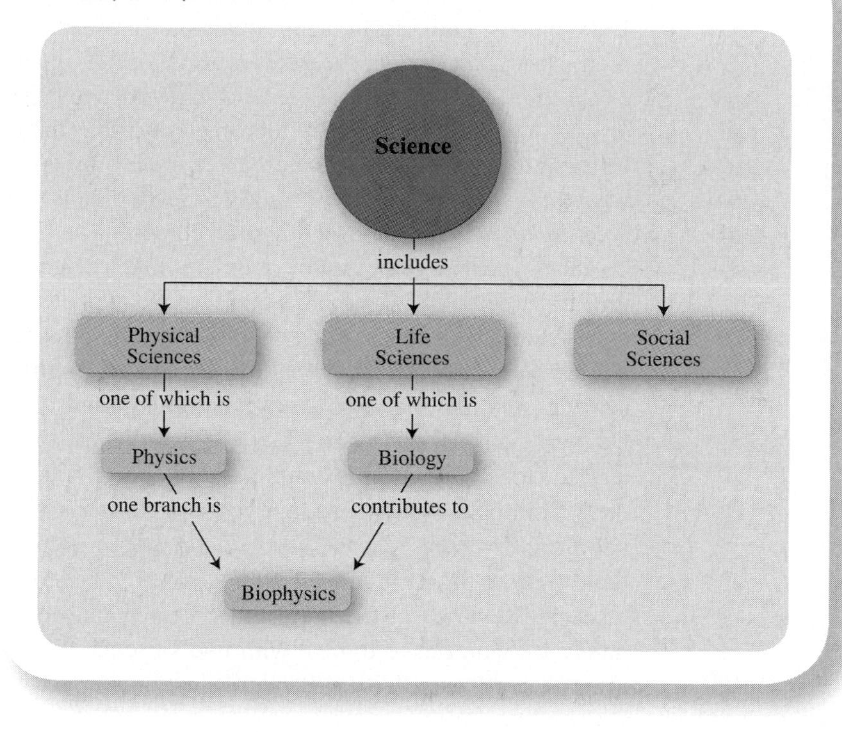

between the general concept, science, and one branch of physics, biophysics. This particular concept map could easily be expanded by, for example, showing all of the social sciences or all of the branches of physics.

The Review Cards at the back of your book contain concept maps designed as summaries to help you organize the ideas, facts, and applications of physics. You should understand that there are many possible maps that could be constructed from a given set of concepts. The maps drawn on the cards represent one way of organizing and understanding a particular set of concepts.

One of the main reasons physics has been so successful is that it harnesses the power of mathematics in useful ways. Many of the most important relationships involving physical quantities are best expressed mathematically. Predictions about the future conditions in a system usually involve math. An essential part of learning physics is developing an understanding of, and an appreciation for, this powerful side of physics. The good news for beginners is that the simplest of mathematics—what most of us learned before age 16 or so—is all that is needed for this purpose.

Physical Quantities and Measurement

To be useful in physics, physical quantities must satisfy some conditions. A physical quantity must be *unambiguous*, its meaning clear and universally accepted. Understanding the meaning of a term involves more than just memorizing the words in its definition. Words can only describe something. To understand a concept, you must go beyond words. For example, the simple definition "force is a push or a pull on a body" does not really convey our complete conceptual understanding of what is meant by force. Many physical quantities (speed, pressure, power, density, and others) can be defined by an equation. Mathematical statements tend to be more precise than ones in words, making the meanings of these terms clearer.

Observation yields *qualitative* information about a system. Measurement yields *quantitative* information, which is central in any science that strives for exactness. Consequently, physical quantities must be measurable, directly or indirectly. One must be able to assign a numerical value that represents the amount of a quantity that is present. It is easy to visualize a measurement of distance, area, or even speed, but other quantities, like pressure, voltage, energy, or power, are a bit more abstract. Each of these can still be measured in prescribed ways. They would be useless if this were not so.

The basic act of measuring is one of comparison. To measure the height of a person, for instance, one would compare the distance from the floor to the top of the person's head against some chosen standard length, such as a foot or a meter (Figure P.3). The height of the person is the number of units 1 foot or 1 meter long (including fractions) that have to be put together to equal that distance. The **unit of measure** is the standard used in the measurement—the foot or meter in this case. A complete measurement of a physical quantity, then, consists of a number and a unit of measure. For example, a person's height might be expressed as

$$\text{height} = 5.75 \text{ feet} \qquad \text{or} \qquad h = 5.75 \text{ ft}$$

Physical quantities will appear in these boxes throughout the book.

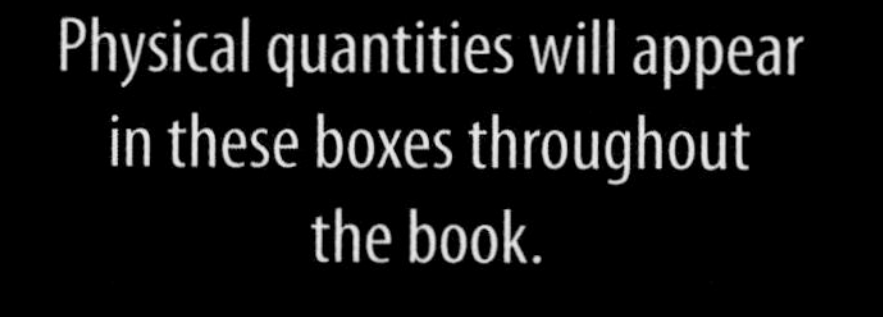

Figure P.3

Measurement is an act of comparison. A person's height is measured by comparison with the length of a chosen standard. In this case, height equals five 1-foot lengths plus a segment 0.75 feet long. The same person's height is also equal to 1 meter plus 0.75 meter.

© Jeff DeVries/iStockphoto.com

Here h represents the quantity (height) and ft the unit of measure (feet). The same height in meters is

$$h = 1.75 \text{ m}$$

So, when we introduce a physical quantity into our physics vocabulary—another "tool," so to speak—we must specify more than just a verbal definition. We should also give a mathematical definition (if possible), relate it to other familiar physical quantities, and include the appropriate units of measure.

In the world today, there are two common systems of measure. The United States uses the **English system**, and the rest of the world, for the most part, uses the **metric system**. An attempt has been made in the United States to switch completely to the metric system, but so far it has not succeeded. The metric system has been used by scientists for quite some time, and we will use it a great deal in this book. It is a convenient system to use because the different units for each physical quantity are related by powers of 10. For example, a kilometer equals 1,000 meters, and a millimeter equals 0.001 meter. The prefix itself designates the power of 10. *Kilo-* means 1,000, *centi-* means 0.01 or $\frac{1}{100}$, and *milli-* means 0.001 or $\frac{1}{1000}$. A *kilo*meter, then, is 1,000 meters. Table P.2 illustrates the common metric prefixes. You may not know what an *ampere* is, but you should see immediately that a *milliampere* is one thousandth of an ampere.

Having to use two systems of units is like living near the border between two countries and having to deal with two systems of currency. Most people who grew up in the United States have a better feel for the size of English-system units like feet, miles per hour, and pounds than for metric-system units like meters, kilometers per hour, and newtons. Often, the examples in this book will use units from both systems, so that you can compare them and develop a sense of the sizes of the metric units. A table relating the units in the two systems is included on a special Review Card at the back of your book. Fortunately we won't have to deal with two systems of units after we reach electricity (Chapter 7).

Table P.2 Common Metric Prefixes and Their Equivalents

1 *centi*meter = 0.01 meters	1 meter = 100 centimeters
1 *milli*meter = 0.001 meters	1 meter = 1,000 millimeters
1 *kilo*meter = 1,000 meters	1 meter = 0.001 kilometers
EXAMPLES	
189 centimeters = 1.89 meters	72.39 meters = 7,239 centimeters
25 millimeters = 0.025 meters	0.24 meters = 240 millimeters
7.68 kilometers = 7,680 meters	23.4 meters = 0.0234 kilometers

*intro

A prologue is an introductory development. This prologue is an introduction to the field of physics, our approach to teaching it, and how to get started learning it. The groundwork has been laid, and we are now ready to proceed.

© Linda & Colin McKie/iStockphoto.com

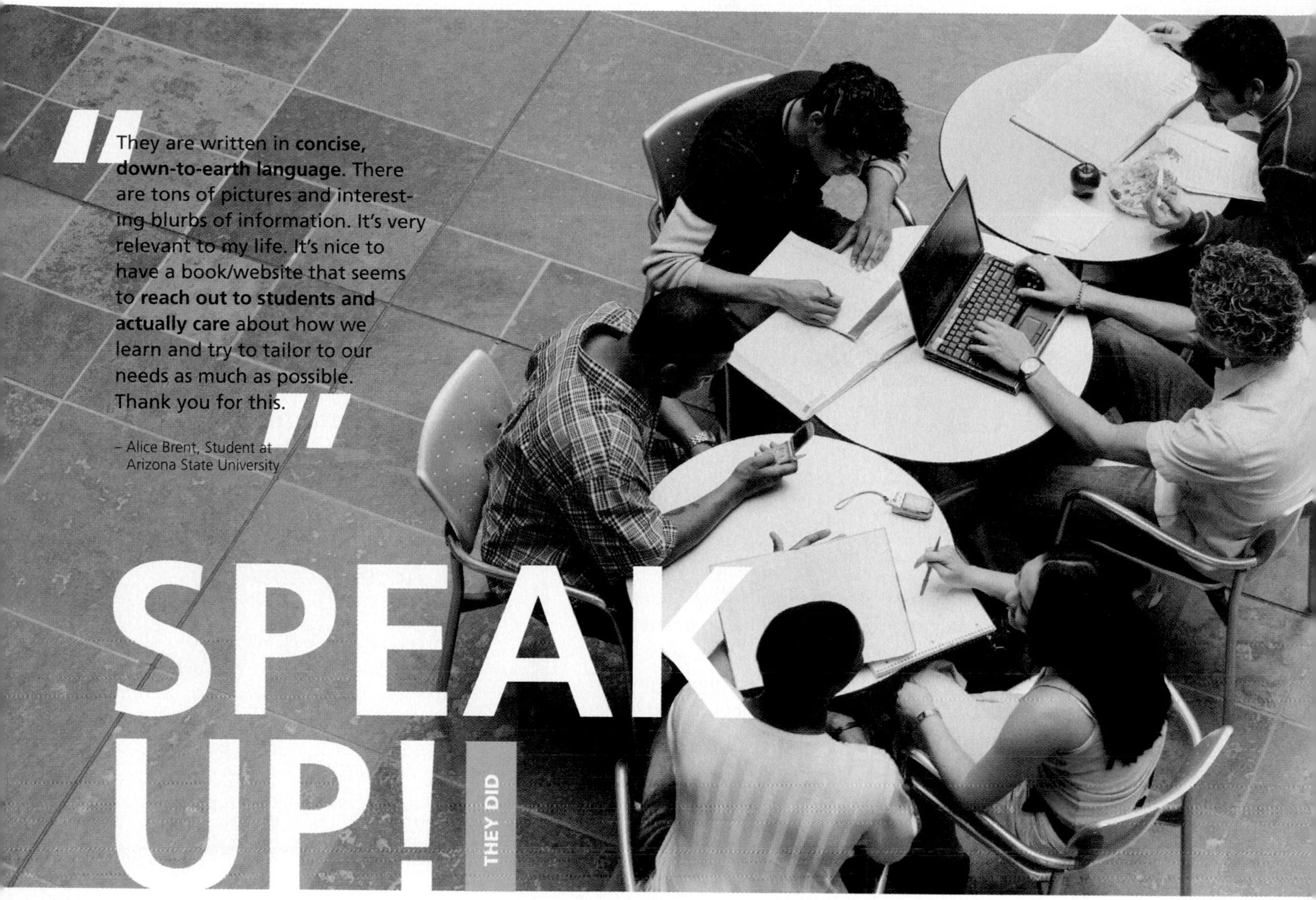

"They are written in **concise, down-to-earth language.** There are tons of pictures and interesting blurbs of information. It's very relevant to my life. It's nice to have a book/website that seems to **reach out to students and actually care** about how we learn and try to tailor to our needs as much as possible. Thank you for this."

– Alice Brent, Student at Arizona State University

SPEAK UP!

THEY DID

PHYSICS was built on a simple principle: to create a new teaching and learning solution that reflects the way today's faculty teach and the way you learn.

Through conversations, focus groups, surveys, and interviews, we collected data that drove the creation of the current version of PHYSICS that you are using today. But it doesn't stop there – in order to make PHYSICS an even better learning experience, we'd like you to SPEAK UP and tell us how PHYSICS worked for you.

What did you like about it?
What would you change?
Are there additional ideas you have that would help us build a better product for next semester's physics students?

At **4ltrpress.cengage.com/physics** you'll find all of the resources you need to succeed in physics – **simulations, interactive quizzes, flashcards,** and more.

Speak Up! Go to **4ltrpress.cengage.com/physics.**

The Study of Motion

sections

1.1 Fundamental Physical Quantities

Acceleration, pure and simple. That's the point of drag racing. Increase the speed of the dragster as quickly as possible so it will travel the one-quarter of a mile (about 400 meters) faster than your opponent. The elite machines have engines that are much more powerful than those on the largest airplanes in World War II (about 5,000 horsepower). They go from 0 to over 300 mph in less than 5 seconds. At the end of their brief run, another critical feat must be accomplished: They have to be brought to a stop safely. How do you slow a vehicle that is going much faster than a passenger jet when it lands? You deploy a parachute.

Imagine driving such a machine: Five seconds of forward acceleration pressing you to the back of your seat, followed by many more seconds of acceleration in the opposite direction—backward—that strains the straps holding you in. And for much of that time you are going hundreds of miles per hour. Drag racing is one of the most dramatic examples of human beings deliberately accelerating themselves. The faint of heart need not apply for this job.

Our main task involves a closer look at motion—what it is, how one quantifies it, what the simplest kinds of motion are, and so on. Fortunately for beginners, most of the concepts and terms are familiar because motion is such an important part of our everyday world. We have a sense of how

Two types of acceleration are involved in drag racing, speeding up and slowing down. Is there another type of acceleration? Can a car accelerate without changing its speed? Read on to find out.

© Dynamic Graphics/Jupiterimages

fast 60 miles per hour (or 97 kilometers per hour) is and how to compute speed if we know the distance traveled and the time elapsed. The unit of measure pretty much tells us how to do that: "Miles per hour" means divide the distance in miles by the time in hours. But there are some important subtleties that must be examined, and our basic ideas about velocity and acceleration need to be expanded.

In physics, there are three basic aspects of the material universe that we must describe and quantify in various ways: *space*, *time*, and *matter.* All physical quantities used in this textbook involve measurements (or combinations of measurements) of space, time, and the properties of matter. The units of measure of all of these quantities can be traced back to the units of measure of distance, time, and two properties of matter called *mass* and *charge.* We will not deal with charge until Chapter 7.

Distance, time, and mass, known as **fundamental physical quantities**, are such basic concepts that it is difficult to define them, particularly time. **Distance** represents a measure of space in one dimension. Length, width, and height are examples of distance measurements.

© Martin Taylor/iStockphoto.com / © iStockphoto.com

Physical Quantity	Metric Units	English Units
Distance d (or l, w, h)	meter (m)	foot (ft)
	millimeter (mm)	inch (in.)
	kilometer (km)	mile (mi)
	centimeter (cm)	

© Marie-france Bélanger/iStockphoto.com / © Brian Hagiwara/
Brand X Pictures/Jupiterimages

This same format will be used for all physical quantities that have several common units.

© iStockphoto.com

The table above lists the common distance units of measure, in both the metric system and the English system, and their abbreviations.

Why are there so many different units in each system? Generally, it is easier to use a unit that fits the scale of the system being considered. The meter is good for measuring the size of a house, the millimeter for measuring the size of a coin, and the kilometer for measuring the distance between cities. Eighty kilometers is the same as 80,000 meters, but the former measurement is easier to conceptualize and to use in calculations. Similarly, it might be correct to say that a coin is 0.000019 kilometers in diameter, but 19 millimeters is a more convenient measure (see Figure 1.1).

The sizes of all of the distance units, including the English units, are defined in relation to the meter. In this book, we will use meters most often in our examples. You should try to get used to distance measurements expressed in meters and have an idea, for example, of how long 25 and 0.2 meters are. Table 1.1 shows some representative distances expressed in metric and in English units.*

It is a rather simple matter to convert a distance expressed in, say, meters to a distance expressed in another unit. For example, you might solve a problem and find that the answer is, "The sailboat travels 23 meters in 10 seconds." Just how far is 23 meters? Think of any numerical measurement as a number multiplied by a unit of measure. In this case, 23 meters equals 23 *times* 1 meter. Then the 1 meter can be replaced by the corresponding number of feet—a conversion factor found on a tear-out card at the end of your book.

$$23 \text{ meters} = 23 \times 1 \text{ meter}$$

But 1 meter = 3.28 feet; therefore:

$$23 \text{ meters} = 23 \times 1 \text{ meter} = 23 \times 3.28 \text{ feet}$$

$$23 \text{ meters} = 75.44 \text{ feet}$$

Two other physical quantities that are closely related to distance are *area* and *volume*. Area commonly refers to the size of a surface, such as the floor in a room or the outer skin of a basketball. The concept of area can apply to surfaces that are not flat and to "empty," two-dimensional spaces such as holes and open windows (see Figure 1.2). Area is a much more general idea than "length times width," an equation you may have learned that applies only to rectangles. The area of something is just the number of squares 1 inch by 1 inch (or 1 meter by 1 meter, or 1 mile by 1 mile, etc.) that would have to be added or placed together to cover it.

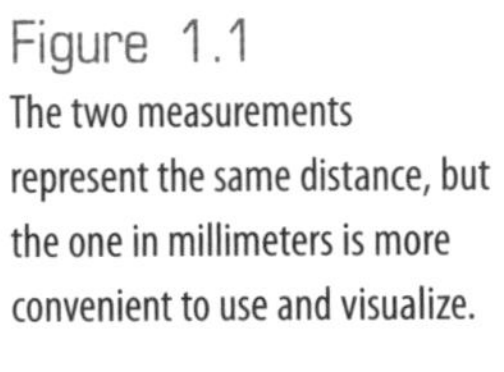

Figure 1.1
The two measurements represent the same distance, but the one in millimeters is more convenient to use and visualize.

© iStockphoto.com

Table 1.1 Some Representative Sizes and Distances

Size/Distance	Metric	English
Size of a nucleus	1×10^{-14} m	4×10^{-13} in.
Size of an atom	1×10^{-10} m	4×10^{-9} in.
Size of a red blood cell	8×10^{-6} m	3×10^{-4} in.
Typical height of a person	1.75 m	5.75 ft
Tallest building	818 m	2,685 ft
Diameter of Earth	1.27×10^{7} m	7,920 miles
Earth-Sun distance	1.5×10^{11} m	9.3×10^{7} miles
Size of our galaxy	9×10^{20} m	6×10^{17} miles

* We will use scientific notation occasionally. If you're rusty on this, use your Math Review Cards at the back of the book to brush up.

Figure 1.2
Every surface, whether it is flat or curved, has an area. The area of this rectangle and the area of the surface of the basketball are equal.

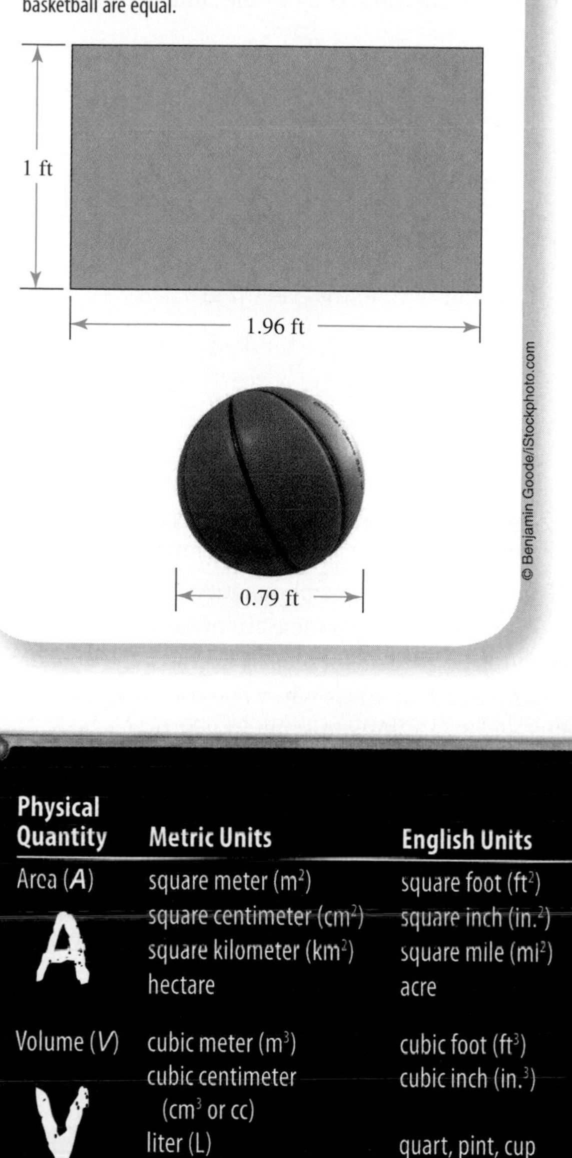

© Benjamin Goode/iStockphoto.com

Physical Quantity	Metric Units	English Units
Area (*A*)	square meter (m^2)	square foot (ft^2)
	square centimeter (cm^2)	square inch ($in.^2$)
	square kilometer (km^2)	square mile (mi^2)
	hectare	acre
Volume (*V*)	cubic meter (m^3)	cubic foot (ft^3)
	cubic centimeter (cm^3 or cc)	cubic inch ($in.^3$)
	liter (L)	quart, pint, cup
	milliliter (mL)	teaspoon, tablespoon
Time (*T*)	second (s)	second (s)
	minute (min)	minute (min)
	hour (h)	hour (h)

© Stefan Klein/iStockphoto.com / © Greg Nicholas/iStockphoto.com

In a similar manner, the volume of a solid is the number of cubes 1 inch or 1 centimeter on a side needed to fill the space it occupies. For common geometric shapes such as a rectangular box or a sphere, there are simple equations to compute the volume. We use volume measures nearly every time we buy food in a supermarket.

Area and volume are examples of physical quantities that are based on other physical quantities—in this case, just distance. Their units of measure are called *derived units* because they are derived from more basic units (1 square meter = 1 meter × 1 meter = 1 meter^2). Until we reach Chapter 7, all of the physical quantities will have units that are derived from units of distance, time, mass, or a combination of these.

The measure of **time** is based on periodic phenomena—processes that repeat over and over at a regular rate. The rotation of the Earth was originally used to establish the universal unit of time, the *second.* The time it takes for one rotation was set equal to 86,400 seconds (24 × 60 × 60). Both the metric system and the English system use the same units for time.

Clocks measure time by using some process that repeats. Many mechanical clocks use a swinging pendulum. The time it takes to swing back and forth is always the same. This is used to control the speed of a mechanism that turns the hands on the clock face. Mechanical wristwatches and stopwatches use an oscillating balance wheel for the same purpose. Quartz electric clocks and digital watches use regular vibrations of an electrically stimulated crystal made of quartz.

The first step in designing a clock is to determine exactly how much time it takes for one cycle of the oscillation. If it were 2 seconds for a pendulum, for example, the clock would then be designed so that the second hand would rotate once during 30 oscillations. The **period** of oscillation in that case is 2 seconds.

DEFINITION

Period *The time for one complete cycle of a process that repeats. It is abbreviated T, and the units are seconds, minutes, and so forth.*

T

But there is another way to look at this. The clock designer must determine how many cycles must take place before 1 second (or 1 minute or 1 hour) elapses. In

© iStockphoto.com

our example, one-half of a cycle takes place during each second. This is called the **frequency** of the oscillation.

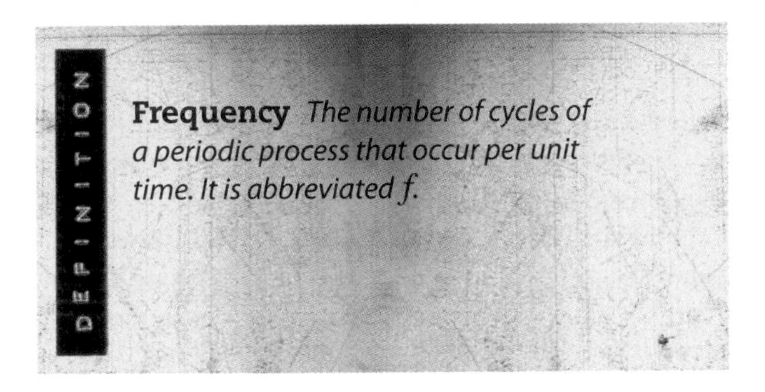
DEFINITION

Frequency *The number of cycles of a periodic process that occur per unit time. It is abbreviated f.*

The standard unit of frequency is the hertz (Hz), which equals 1 cycle per second.

$$1\ \text{Hz} = 1/\text{s} = 1\ \text{s}^{-1}$$

The frequency of AM radio stations is expressed in kilohertz (kHz) and those of FM stations in megahertz (MHz). *Mega-* is the metric prefix signifying 1 million. So, 91.5 MHz equals 91,500,000 Hz. Another metric prefix, *giga-*, is also commonly used with hertz. For example, the "speed" of a computer's processor—actually, the frequency of the electrical signal it uses—might be 2 gigahertz, which is 2 billion (2,000,000,000) hertz.

The relationship between the period of a cyclic phenomenon and its frequency is simple: The period equals 1 divided by the frequency, and vice versa.

$$\text{period} = \frac{1}{\text{frequency}}$$

$$\text{T} = \frac{1}{f}$$

Also:

$$f = \frac{1}{T}$$

EXAMPLE 1.1

A mechanical stopwatch uses a balance wheel that rotates back and forth 10 times in 2 seconds. What is the frequency of the balance wheel?

$$\text{frequency} = \text{number of cycles per time}$$

$$f = \frac{10\ \text{cycles}}{2\ \text{s}}$$

$$= 5\ \text{Hz}$$

What is the period of the balance wheel?

$$\text{period} = \text{time for one cycle}$$

$$= 1\ \text{divided by the frequency}$$

$$T = \frac{1}{f} = \frac{1}{5}\ \text{Hz}$$

$$= 0.2\ \text{s}$$

The balance wheel oscillates 300 times each minute.

The third basic physical quantity is mass. The **mass** of an object is basically a measure of how much matter is contained in it. (This statement illustrates the sort of circular definition that arises when one tries to define fundamental concepts.) We know intuitively that a large body, such as a locomotive, has a large mass because it is composed of a great deal of material. Mass is also a measure of what we sometimes refer to in everyday speech as *inertia.* The larger the mass of an object, the greater its inertia and the more difficult it is to speed up or slow down.

Mass is not in common use in the English system; note the unfamiliar unit, the slug. Weight, a quantity that is related to, but is *not* the same as mass, is used

© James Allred/iStockphoto.com

Physical Quantity	Metric Units	English Units
Mass (m)	kilogram (kg) gram (g)	slug

instead. We can contrast the two ideas: When you try to lift a shopping cart, you are experiencing its weight. When you try to speed it up or slow it down, you are experiencing its mass (see Figure 1.3). We will take another look at mass and weight in Chapter 2.

Figure 1.3
(a) When changing the speed of something, you experience its mass. (b) When lifting something, you must overcome its weight.

1.2 Speed and Velocity

A key concept to use when quantifying motion is **speed**.

DEFINITION

Speed *Rate of movement. Rate of change of distance from a reference point. The distance traveled divided by the time elapsed.*

Physical Quantity	Metric Units	English Units
Speed (v)	meter per second (m/s) kilometer per hour (km/h)	foot per second (ft/s) mile per hour (mph)

A couple of aspects of speed are worth highlighting. First, speed is *relative*. A person running on the deck of a ship cruising at 20 mph might have a speed of 8 mph relative to the ship, but the speed relative to the water or a pier would be 28 mph (if headed toward the front of the ship). If we use the ship as the reference point, the speed is 8 mph. With the water or a pier as the reference point, the speed is 28 mph (Figure 1.4). If you are traveling at 55 mph on a highway and a car passes you going 60 mph, its speed is 5 mph relative to your car. Most of the time speed is measured relative to the surface of the Earth; this will be the case in this book, unless stated otherwise.

Second, it is important to distinguish between *average speed* and *instantaneous speed.* An object's average speed is the total distance it travels during some period of time divided by the time that elapses:

$$\text{average speed} = \frac{\text{total distance}}{\text{total elapsed time}}$$

If a 1,500-mile airline flight lasts 3 hours, the airplane's average speed is 500 mph. Of course, the airplane's speed changes during the 3 hours, so its speed is not 500 mph at each instant. (You can average 75% on four exams without actually getting 75% on each one.) Similarly, a sprinter who runs the 100-meter dash

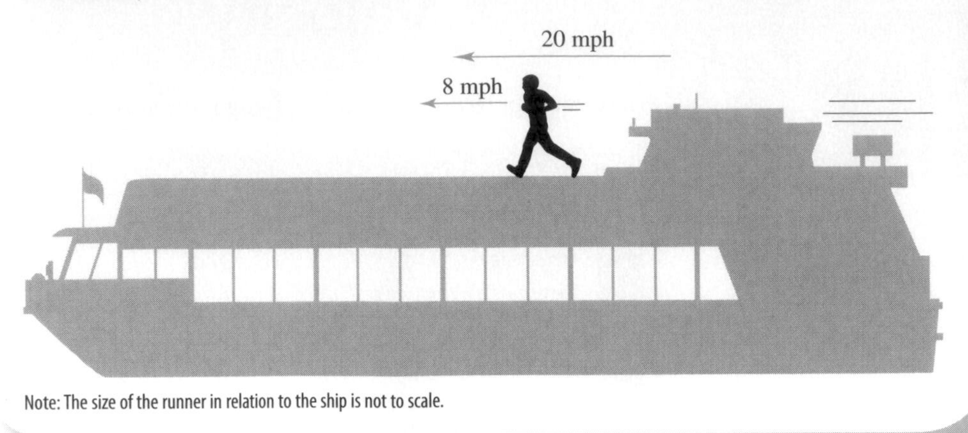

Figure 1.4

Speed is relative. The speed of a person running on a ship is 8 mph relative to the deck. If the ship's speed is 20 mph, the person's speed relative to the pier is either 28 mph (if headed toward the front of the ship) or 12 mph (if headed toward the rear).

Note: The size of the runner in relation to the ship is not to scale.

in 10 seconds has an average speed of 10 m/s but is not traveling with that speed at each moment. This gives rise to the concept of instantaneous speed—the speed that an object has at an instant in time. A car's speedometer actually gives instantaneous speed. When it shows 55 mph, it means that, if the car traveled with exactly that speed for 1 hour, it would go 55 miles. How can one determine the instantaneous speed of something that is not equipped with a speedometer? Instantaneous speed can't be measured exactly using the basic meaning of speed (distance traveled divided by the time it takes) because an "instant" implies that *zero* time elapses. But we can get a good estimate of an object's instantaneous speed by timing how long it takes to travel a very short distance:

$$\text{instantaneous speed} = \frac{\text{very short distance}}{\text{very short time}}$$

For example, given sophisticated equipment, we might measure how long it takes a car to travel 1 meter. If that time is found to be 0.05 seconds (not an "instant" but a very short time), a good estimate of the car's instantaneous speed is

$$\text{instantaneous speed} = \frac{1\ \text{m}}{0.05\ \text{s}} = 20\ \text{m/s} = 44.8\ \text{mph}$$

In drag racing, the maximum instantaneous speed of a dragster is estimated by timing how long it takes to travel the final 60 feet (about 18 meters) of the quarter-mile race. This is typically less than 0.15 seconds.

Global positioning system (GPS) receivers used by pilots, drivers, hikers, and others employ radio signals from satellites to determine location. They can compute the device's approximate instantaneous speed by computing how far it travels in a short time and then dividing by that time. One common GPS receiver updates its position every second, so it can estimate instantaneous speed using 1 second as its "short time." A typical bicycle speedometer uses a sensor on the front wheel. The device can time how long it takes the wheel to make one rotation and estimate the instantaneous speed by using the circumference of the wheel (typically around 2.1 meters) as the "short distance."

The conversion from meters per second to miles per hour is done in the same manner that was used with distance in Section 1.1. From the Table of Conversion Factors on the card at the end of your book, 1 m/s = 2.24 mph; so

$$20\ \text{m/s} = 20 \times 1\ \text{m/s} = 20 \times 2.24\ \text{mph} = 44.8\ \text{mph}$$

Actually, this is the car's average speed during the time span. In normal situations, a car's speed will increase or decrease by at most 1 mph during a twentieth of a second, so this is a pretty good estimate. Our answer is within ±1 mph of the true value. During a collision, a car's speed can change a great deal during 0.05 seconds, so one would have to somehow use a much shorter time interval to calculate instantaneous speed in this case. An automaker might use high-speed videotapes of a crash test to measure much shorter periods and corresponding distances. In general, "very short time" means an interval during which the object's speed won't change by an amount greater than the desired error limit of the estimate. The concept of instantaneous speed is what is important here, more so than the technical details of how it is measured in different situations.

In some cases, an object may have been traveling for a while and already moved some distance before we start taking measurements to determine its speed (average or instantaneous). Perhaps we want to measure a sprinter's average speed during the last part of a race. Then the distance and time that we use would be the values at the end of the segment being timed (the *final* values) minus the values at the beginning of the segment (the *initial* values)—that is, the *changes* in distance and time. The general expression for speed is

$$\text{speed} = \frac{\text{change in distance}}{\text{change in time}} = \frac{d_{\text{final}} - d_{\text{initial}}}{t_{\text{final}} - t_{\text{initial}}}$$

$$v = \frac{\Delta d}{\Delta t}$$

(The symbol Δ is the Greek letter delta and is used to represent a "change in" a physical quantity.) This equation can represent both average and instantaneous speed. When Δt is the total elapsed time for a trip, v is the average speed. When Δt is a very short time, then v is the instantaneous speed.

EXAMPLE 1.2

An analysis of a videotape of Olympic gold-medal winner Florence Griffith Joyner (1959–1998) running a 100-meter dash might yield the data shown in the table below. Compute her average speed for the race and estimate her peak instantaneous speed. For the entire race, this would be as follows:

$$\text{average speed} = v = \frac{\Delta d}{\Delta t} = \frac{d_{\text{final}} - d_{\text{initial}}}{t_{\text{final}} - t_{\text{initial}}}$$

$$v = \frac{100 \text{ m} - 0 \text{ m}}{10.50 \text{ s} - 0 \text{ s}} = \frac{100 \text{ m}}{10.50 \text{ s}}$$

$$= 9.52 \text{ m/s} \ (= 21.3 \text{ mph})$$

d (meters)	t (seconds)
0	0
60	6.85
70	7.76
80	8.67
90	9.58
100	10.50

The equally spaced times between 60 and 90 meters indicate that her speed was constant. Therefore, we can use any segment from this part of the race to compute her instantaneous speed. For the segment between 80 and 90 meters, this would be:

$$v = \frac{\Delta d}{\Delta t} = \frac{d_{\text{final}} - d_{\text{initial}}}{t_{\text{final}} - t_{\text{initial}}}$$

$$= \frac{90 \text{ m} - 80 \text{ m}}{9.58 \text{ s} - 8.67 \text{ s}} = \frac{10 \text{ m}}{0.91 \text{ s}}$$

$$= 11.0 \text{ m/s} \ (= 24.6 \text{ mph})$$

A speedometer displays instantaneous speed.

© iStockphoto.com

The speed of a car being driven around in a city changes quite often. The instantaneous speed may vary from 0 mph (at stoplights) to 45 mph. The average speed for the trip is the total distance divided by the total time, maybe 20 mph.

When the speed of an object is constant, the average speed and the instantaneous speed are the same. In this case, we can express the relationship between the distance traveled and the time that has elapsed as follows:

$$d = vt \quad \text{(when speed is constant)}$$

This is an example of what is called a *proportionality:* We say that d is proportional to t (abbreviated $d \propto t$). If the time is doubled, the distance is doubled. The constant speed v is called the *constant of proportionality.* We will encounter many examples in which one physical quantity is proportional to another.

Most drivers are accustomed to speed measured in miles per hour or kilometers per hour. This is most convenient when talking about travel times for distances greater than a few miles. Often, it is more enlightening to use feet per second. For example, a car going 65 mph travels 130 miles in 2 hours (using the preceding equation). But for a potential accident, it is relevant to consider how far the car will travel in a few seconds. Since 65 mph equals 95.6 feet per second, if a driver takes 2 seconds to decide how to avoid an accident, the car will have traveled 191 feet—more than 10 car lengths.

Try to develop a feel for the different speed units, particularly meters per second. You might keep in mind this comparison:

$$65 \text{ mph} = 95.6 \text{ ft/s} = 29.1 \text{ m/s}$$

Here is one last example of doing conversions. From the Table of Conversion Factors, 1 mph = 0.447 m/s:

$$65 \text{ mph} = 65 \times 1 \text{ mph}$$

$$= 65 \times 0.447 \text{ m/s} = 29.1 \text{ m/s}$$

Apparently, the universe was created with an absolute speed limit—the speed of light in empty space. This speed is represented by the letter c. Nothing has ever been observed traveling faster than c. The value of c and some other speeds are included in Table 1.2.

Table 1.2 Some Speeds of Interest

Description	Metric	English
Speed of light, *c* (in vacuum)	3×10^8 m/s	186,000 miles/second
Speed of sound (in air, room temperature)	344 m/s	771 mph
Highest instantaneous speeds:		
Running (cheetah)	28 m/s	63 mph
Swimming (sailfish)	30.4 m/s	68 mph
Flying—level (merganser)	36 m/s	80 mph
Flying—dive (peregrine falcon)	97 m/s	217 mph
Humans (approximate):		
Swimming	2.5 m/s	5.6 mph
Running	12 m/s	27 mph
Ice skating	14 m/s	31 mph

An important aspect of motion is *direction*. We will see that changing the direction of a body's motion can produce effects that are equivalent to changing the speed. **Velocity** is a physical quantity that incorporates both ideas.

DEFINITION

Velocity *Directed motion. Speed in a particular direction (same units as speed).*

The speed of a ship might be 10 m/s, while its velocity might be 10 m/s east. Whenever a moving body changes direction, such as a car going around a curve or someone walking around a corner, the velocity changes even if the speed does not. Being told that an airplane flies for 2 hours from a certain place and averages 100 mph does not tell you enough to determine where it is. Knowing that it travels in a straight line due north would allow you to pinpoint its location.

A speedometer alone gives the instantaneous speed of a vehicle. A speedometer used in conjunction with a compass would give velocity: speed and the direction of motion. In a car traveling along a winding road, the movement of the compass needle indicates that the velocity is changing even if the speed is not.

Velocity is an example of a physical quantity called a **vector**. Vectors have both a numerical size (magnitude) and a direction associated with them. Quantities that do not have a direction are called **scalars**. Speed by itself is a scalar. Only when the direction of motion is included do we have the vector velocity. Similarly, we can define the vector *displacement* as distance in a specific direction. For the airplane referred to earlier, the *distance* it travels in 2 hours is 200 miles. Its actual location can be determined only from its *displacement*—200 miles due north, for example. The basic equation for speed, $v = \Delta d/\Delta t$, is also the equation for velocity (that's why v is used) with d representing a vector displacement.

A distant lightning strike provides a good demonstration of the difference between the speed of sound and the speed of light.

© Clint Spencer/iStockphoto.com / © iStockphoto.com

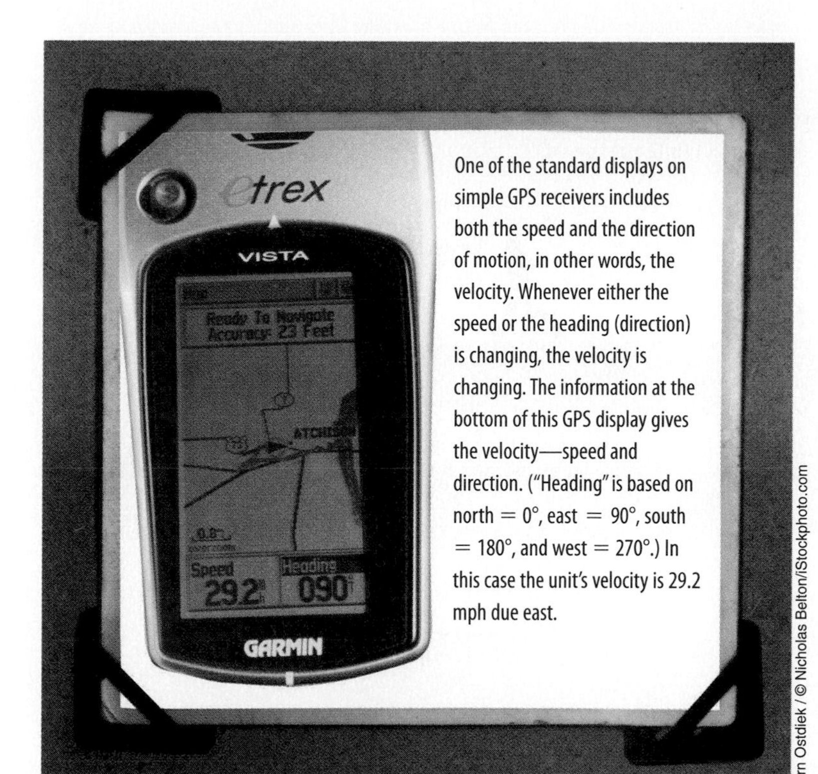

One of the standard displays on simple GPS receivers includes both the speed and the direction of motion, in other words, the velocity. Whenever either the speed or the heading (direction) is changing, the velocity is changing. The information at the bottom of this GPS display gives the velocity—speed and direction. ("Heading" is based on north = 0°, east = 90°, south = 180°, and west = 270°.) In this case the unit's velocity is 29.2 mph due east.

© Vern Ostdiek / © Nicholas Belton/iStockphoto.com

We can classify most physical quantities as scalars or vectors. Time, mass, and volume are all scalars because there is no direction associated with them.

Vectors are represented by arrows in drawings, the length of the arrow being proportional to the size or magnitude of the vector (see Figure 1.5). If a car is traveling twice as fast as a pedestrian, then the arrow representing the car's velocity is twice as long as that representing the pedestrian's velocity.

When an object can move forward or backward along a line, its velocity is positive when it is going in one direction and negative when it is going in the opposite direction. When going forward in a car, you might give the velocity as +10 m/s (positive). If the car stops and then goes backward, the velocity is negative, −5 m/s, for example. The velocity of a person on a swing is positive, then negative, then positive, and so on. Which direction is associated with positive velocity is somewhat arbitrary. When dribbling a basketball, you could say its velocity is positive when it is moving downward and negative when it is moving upward, or vice versa. Speed does not become negative when the velocity does. The negative sign is associated with the direction of motion.

When dealing with a system in which the direction of motion can change, even if forward and backward are the only possible directions, it is better to use velocity than speed because the + or − sign indicates the direction of motion. In a situation in which the direction of motion does not change, like a falling object, the terms *speed* and *velocity* are often used interchangeably. However, in the remainder of this text, *velocity* will be used in all situations in which the direction of motion can be important. In cases where an object's direction of motion doesn't change, we will use the direction of the object's initial motion as the positive direction (unless stated otherwise).

Figure 1.5
Vector quantities can be represented by arrows. Arrow length indicates the vector's size, and arrow direction shows the vector's direction. In both figures, velocity is represented with an arrow. The car's velocity is much greater than the pedestrian's, so the arrow representing the car's velocity is longer. Because both objects are moving in the same direction, their velocity vectors (arrows) are parallel.

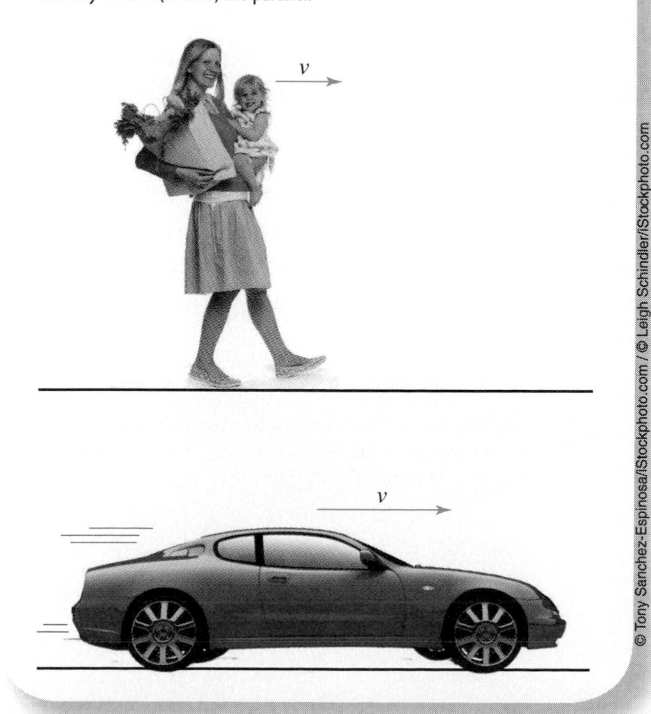

© Tony Sanchez-Espinosa/iStockphoto.com / © Leigh Schindler/iStockphoto.com

Vector Addition

Sometimes a moving body has two velocities at the same time. The runner on the deck of the ship in Figure 1.4 has a velocity relative to the ship and a velocity because the ship itself is moving. A bird flying on a windy day has a velocity relative to the air and a velocity because the air carrying the bird is moving relative to the ground. The velocity of the runner relative to the water or that of the bird relative to the ground is found by adding the two velocities together to give the *net*, or *resultant*, *velocity*. Let's consider how two velocities (or two vectors of any kind) are combined in *vector addition*.

Figure 1.6

(a) The resultant velocity of a runner on the deck of a ship is found by adding the runner's velocity and the ship's velocity. The result is 28 mph forward. (b) Using the same procedure when the runner is headed toward the rear of the ship, the resultant velocity is 12 mph forward.

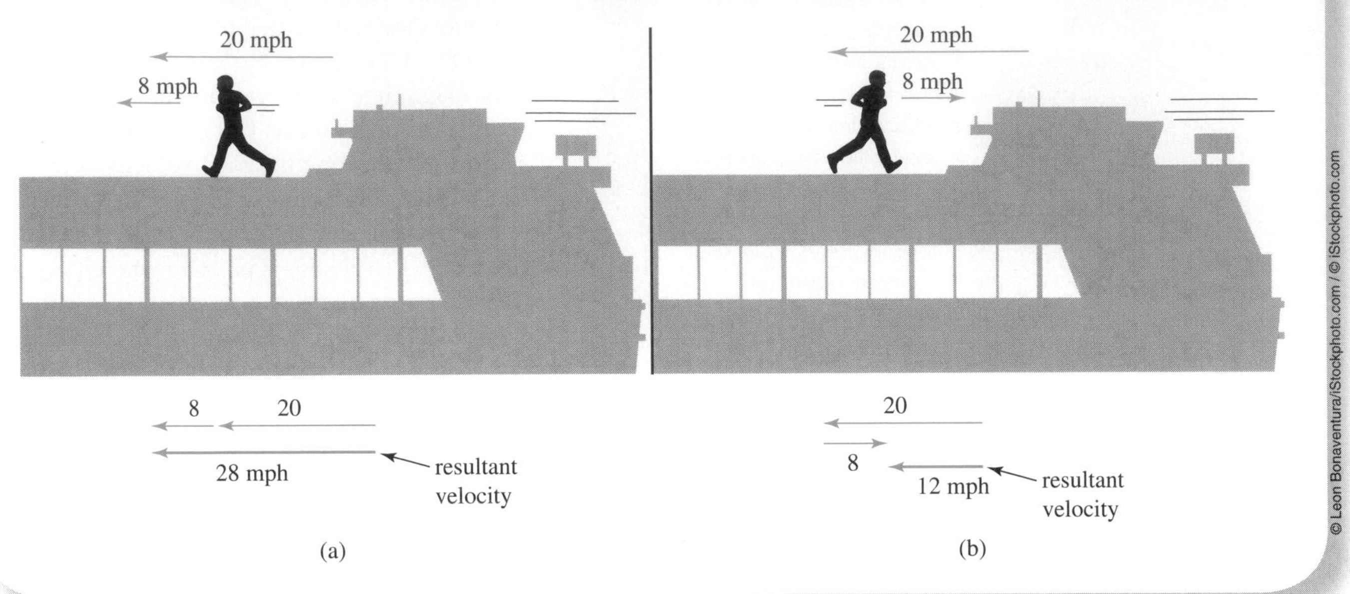

When adding two velocities, you represent each as an arrow with its length proportional to the magnitude of the velocity—the speed. For the runner on the ship, the arrow representing the ship's velocity is $2\frac{1}{2}$ times as long as the arrow representing the runner's velocity because the two speeds are 20 mph and 8 mph, respectively. Each arrow can be moved around for convenience, provided its length and its direction are not altered. Any such change would make it a different vector. The procedure for adding two vectors is as follows.

> Two vectors are added by representing them as arrows and then positioning one arrow so its tip is at the tail of the other. A new arrow drawn from the tail of the first arrow to the tip of the second is the arrow representing the resultant vector—the sum of the two vectors.

Figure 1.6 shows this for the runner on the deck of the ship. In Figure 1.6a, the runner is running forward in the direction of the ship's motion, so the two arrows are parallel. When the arrows are positioned "tip to tail," the resultant velocity vector is parallel to the others, and its magnitude—the speed—is 28 mph (8 mph + 20 mph). In Figure 1.6b, the runner is running toward the rear of the ship, so the arrows are in opposite directions. The resultant velocity is again parallel to the ship's velocity, but its magnitude is 12 mph (20 mph − 8 mph).

Vector addition is done the same way when the two vectors are not along the same line. Figure 1.7a shows a bird with velocity 8 m/s north in the air while the air itself has velocity 6 m/s east. The bird's velocity observed by someone on the ground, Figure 1.7b, is the sum of these two velocities. This we determine by placing the two arrows representing the velocities tip to tail as before and drawing an arrow from the tail of the first to the tip of the second (Figure 1.7c and d). The direction of the resultant velocity is toward the northeast. Watch for this when you see a bird flying on a windy day: Often the direction the bird is moving is not the same as the direction its body is pointed.

What about the magnitude of the resultant velocity? It is not simply 8 + 6 or 8 − 6, because the two velocities are not parallel. With the numbers chosen for this example, the magnitude of the resultant velocity—the bird's speed—is 10 m/s. If you draw the two original arrows with correct relative lengths and then measure the length of the resultant arrow, it will be $\frac{5}{4}$ times the length of the arrow representing the 8 m/s vector. Then 8 m/s times $\frac{5}{4}$ equals 10 m/s.*

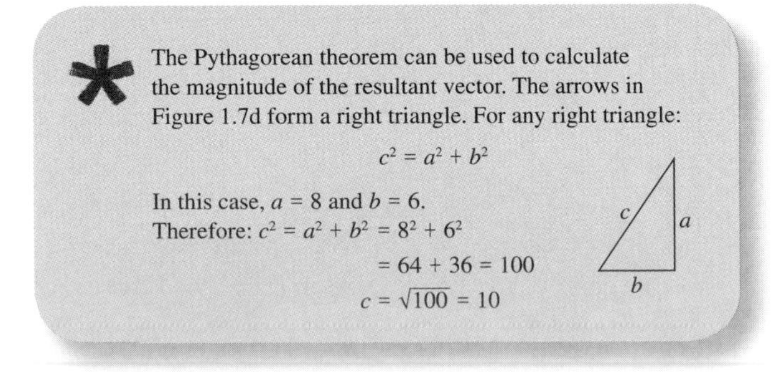

* The Pythagorean theorem can be used to calculate the magnitude of the resultant vector. The arrows in Figure 1.7d form a right triangle. For any right triangle:

$$c^2 = a^2 + b^2$$

In this case, $a = 8$ and $b = 6$.

Therefore: $c^2 = a^2 + b^2 = 8^2 + 6^2$

$= 64 + 36 = 100$

$c = \sqrt{100} = 10$

Figure 1.7
The velocity of a bird relative to the ground (b) is the vector sum of its velocity relative to the air and the velocity of the air (wind). (c) and (d) show that the vectors can be added two different ways, but the resultant is the same vector.

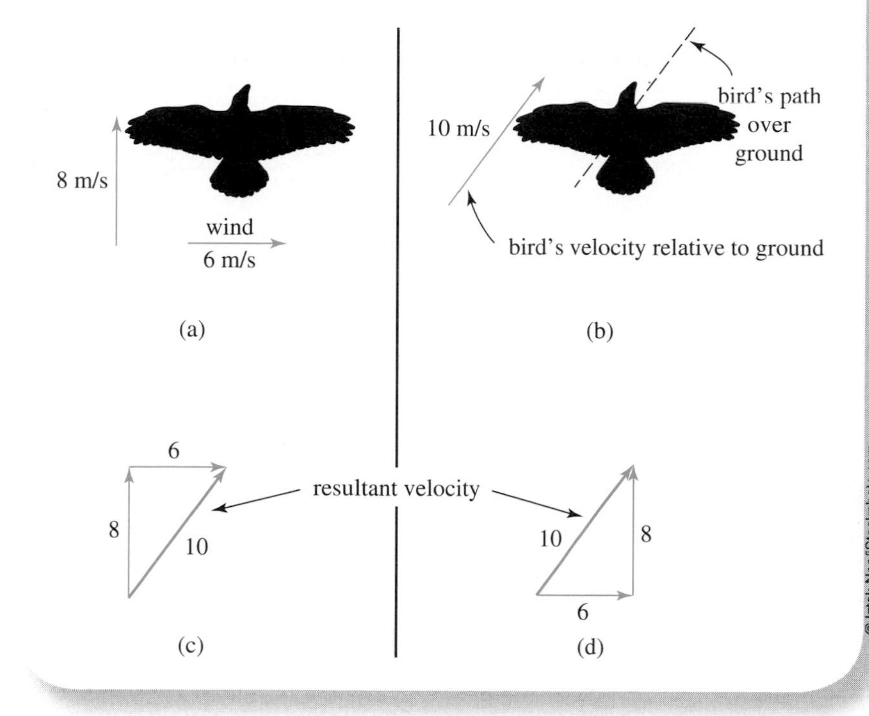

© Iztok Noc/iStockphoto.com

Velocity involves more than just speed. It's also about direction.

Vector addition is performed in the same manner, no matter what the directions of the vectors. Figure 1.8 shows two other examples of a bird flying with different wind directions. The magnitudes of the resultants are best determined by measuring the lengths of the arrows. There are many other situations in which a body's net velocity is the sum of two (or more) velocities (for example, a swimmer or boat crossing a river). Displacement vectors are added in the same fashion. If you walk 10 meters south, then 10 meters west, your net displacement is 14.1 meters southwest.

The process of vector addition can be "turned around." Any vector can be thought of as the sum of two other vectors, called *components* of the vector. When we observe the bird's single velocity in Figure 1.7b, we would likely realize that the bird has two velocities that have been added. Even when a moving body only has one "true" velocity, it may be convenient to think of it as two velocities that have been added together. For example, a soccer player running southeast across a field can be thought of as going south with one velocity and east with another velocity at the same time. A car going down a long hill has one velocity component that is horizontal and another that is vertical (downward).

Figure 1.8
Other examples of vector addition. The bird has the same speed and direction in the air, but the wind direction is different.

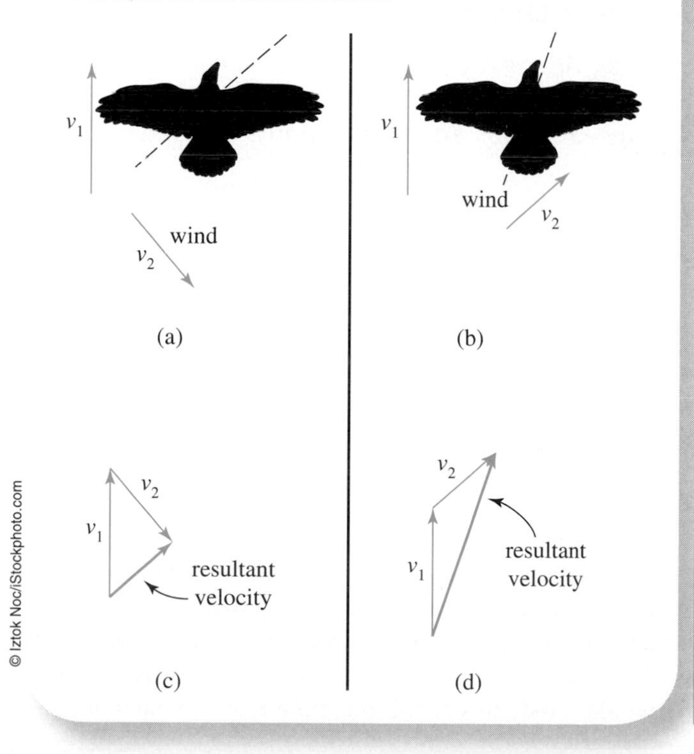

© Iztok Noc/iStockphoto.com

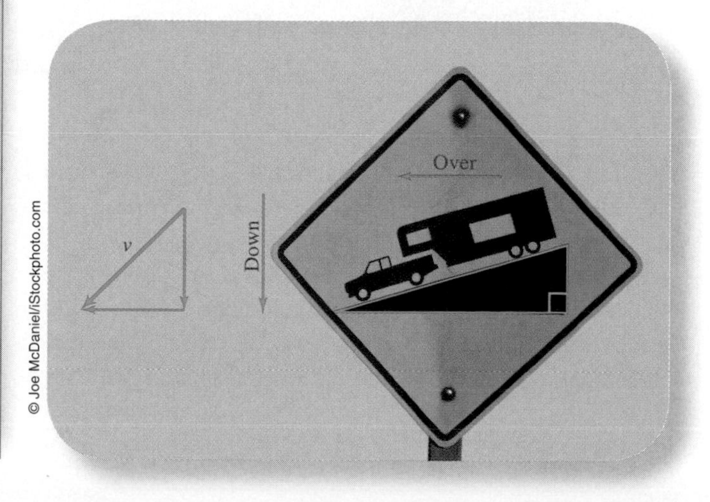

© Joe McDaniel/iStockphoto.com

1.3 Acceleration

The physical world around us is filled with motion. But think about this for a moment: Cars, bicycles, pedestrians, airplanes, trains, and other vehicles all change their speed or direction often. They start, stop, speed up, slow down, and make turns. The velocity of the wind usually changes from moment to moment. Even the Earth as it moves around the Sun is constantly changing its direction of motion and its speed, though not by much as reckoned on a daily basis. The main thrust of Chapter 2 is to show how the change in velocity of an object is related to the force acting on it. For these reasons, a very important concept in physics is **acceleration**.

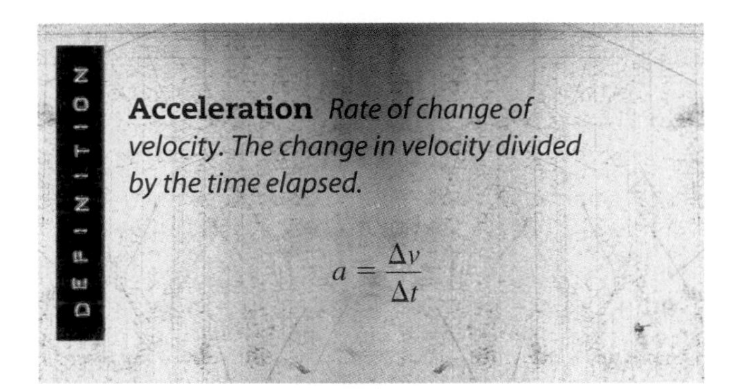

Physical Quantity	Metric Units	English Units
Acceleration (a)	meter per second^2 (m/s^2)	foot per second^2 (ft/s^2) mph per second (mph/s)

Whenever something is speeding up or slowing down, it is undergoing acceleration. As you travel in a car, anytime the speedometer's reading is changing, the car is accelerating. Acceleration is a vector quantity, which means it has both magnitude and direction. Note that the relationship between acceleration and velocity is the same as the relationship between velocity and displacement. Acceleration indicates how rapidly velocity is changing, and velocity indicates how rapidly displacement is changing.

EXAMPLE 1.3

A car accelerates from 20 to 25 m/s in 4 seconds as it passes a truck (Figure 1.9). What is its acceleration?

Since the direction of motion is constant, the change in velocity is just the change in speed—the later speed minus the earlier speed.

$$a = \frac{\Delta v}{\Delta t} = \frac{\text{final speed} - \text{initial speed}}{\Delta t}$$

$$= \frac{25 \text{ m/s} - 20 \text{ m/s}}{4 \text{ s}}$$

$$= \frac{5 \text{ m/s}}{4 \text{ s}}$$

$$= 1.25 \text{ m/s}^2$$

This means that the car's speed increases 1.25 m/s during each second.

When something is slowing down (its speed is decreasing), it is undergoing acceleration. In everyday speech, the word *acceleration* is usually applied when speed is increasing, *deceleration* is used when speed is decreasing, and a change in direction of motion is not referred to as an acceleration. In physics, one word, *acceleration*, describes all three cases because each represents a change in velocity.

EXAMPLE 1.4

After a race, a runner, traveling in a fixed direction, takes 5 seconds to come to a stop from a speed of 9 m/s. The acceleration is as follows:

$$a = \frac{\Delta v}{\Delta t} = \frac{0 \text{ m/s} - 9 \text{ m/s}}{5 \text{ s}}$$

$$= \frac{-9 \text{ m/s}}{5 \text{ s}}$$

$$= -1.8 \text{ m/s}^2$$

The minus sign means the acceleration and the velocity (which is positive) are in opposite directions.

Perhaps the most important example of accelerated motion is that of an object "falling freely" near the Earth's surface. By an object falling freely, we mean that only the force of gravity is acting on it and we can ignore things like air resistance. A rock falling a few meters would satisfy this condition, but a feather would not.

Figure 1.9
While passing a truck, a car increases its speed from 20 to 25 m/s.

© Luis Carlos Torres/iStockphoto.com

© iStockphoto.com

Freely falling bodies move with a constant downward acceleration. The magnitude of this acceleration is represented by the letter g.

$$g = 9.8\ \text{m/s}^2 \quad \text{(acceleration due to gravity)*}$$
$$= 32\ \text{ft/s}^2 = 22\ \text{mph/s}$$

The downward velocity of a falling rock increases 22 mph each second that it falls. The letter g is often used as a unit of measure of acceleration. An acceleration of 19.6 m/s^2 equals 2 g. (Some representative accelerations are given in Table 1.3.)

Table 1.3 Some Accelerations of Interest

Description	Acceleration	
Freely falling body (on the Moon)	1.6 m/s^2	0.16 g
Freely falling body (on the Earth)	9.8 m/s^2	1 g
Space shuttle (maximum)	29 m/s^2	3 g
Drag racing—car (average for $\frac{1}{4}$ mile)	32 m/s^2	3.3 g
Highest (sustained) survived by human	245 m/s^2	25 g
Clothes—spin cycle in a typical washing machine	400 m/s^2	41 g
Tread of a typical car tire at 65 mph	2,800 m/s^2	285 g
Click beetle jumping	3,920 m/s^2	400 g
Bullet in a high-powered rifle	2,000,000 m/s^2	200,000 g
Projectile in an electromagnetic "rail gun"	1.96×10^9 m/s^2	$2 \times 10^8\ g$

Another way to think of acceleration is in terms of the changes in the arrows drawn to represent velocity. When a car accelerates from one speed to a higher speed, the *length* of the arrow used to represent the velocity *increases* (see Figure 1.10). If we place the two arrows side by side, the change in velocity can be represented by a third arrow, Δv, drawn from the tip of the first arrow to the tip of the second. The later velocity equals the initial velocity plus the change in velocity. The original arrow plus the arrow representing Δv equals the later arrow. The acceleration vector is this change in velocity divided by the time. Note that the acceleration vector points forward. Increasing the velocity requires a forward acceleration.

* In section 2.8, we will learn precisely why the acceleration due to gravity on Earth assumes this value when we study Newton's law of Universal Gravitation.

Centripetal Acceleration

The concept of acceleration includes changes in *direction* of motion as well as changes in speed. A car going around a curve and a billiard ball bouncing off a cushion are accelerated, even if their speeds do not change.

We can once again use arrows to indicate the change in velocity. Figure 1.11 shows a car at two different times as it goes around a curve. If we place the two arrows representing the car's velocity next to each other, we see that the direction of the arrow has changed. The change in velocity, Δv, is represented by the arrow drawn from the tip of v_1 to the tip of v_2. In other words, the original arrow plus the arrow representing Δv equals the later arrow, just like in straight-line acceleration (Figure 1.10).

Note the direction of the change in velocity Δv: It is pointed toward the center of the curve. Because the acceleration is in the same direction as Δv, it too is directed toward the center. For this reason, the acceleration of an object moving in a circular path is called **centripetal acceleration** (for "center-seeking"). The centripetal acceleration is always perpendicular to the object's velocity—directed either to its right or to its left.

Figure 1.10

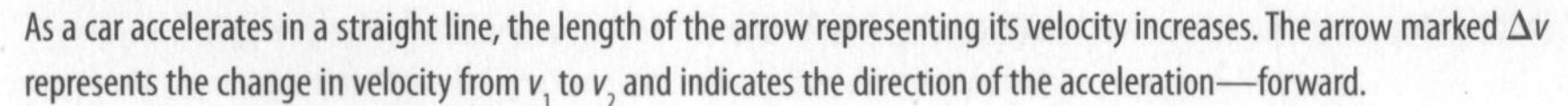

As a car accelerates in a straight line, the length of the arrow representing its velocity increases. The arrow marked Δv represents the change in velocity from v_1 to v_2 and indicates the direction of the acceleration—forward.

© Carina Lochner/iStockphoto.com

Figure 1.11

As a car, traveling at constant speed, rounds a curve, it undergoes centripetal acceleration. Here the arrow representing the car's velocity changes direction. Because the change in velocity, Δv, is directed toward the center of the curve, so is the acceleration.

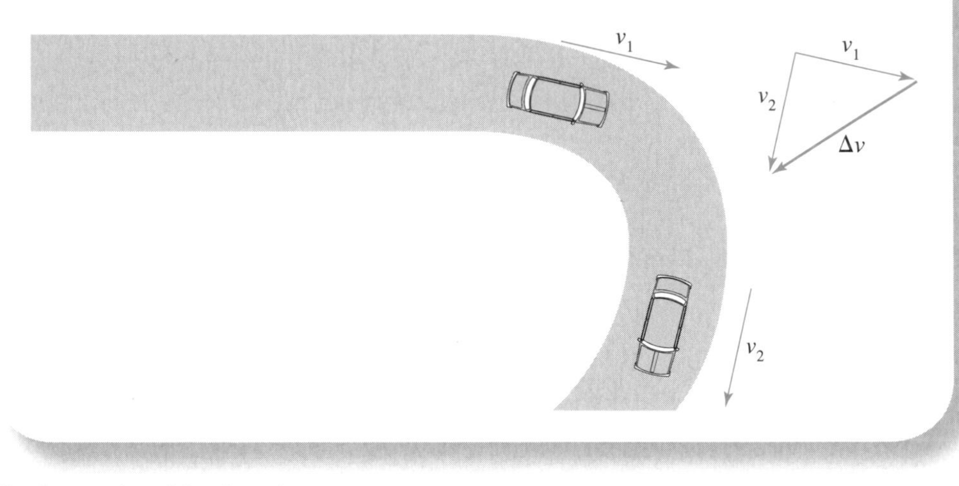

So we know the direction of the acceleration of a body moving in a circular path, but what about its magnitude? The faster the body is moving, the more rapidly the direction of its velocity is changing. Consequently, the magnitude of the centripetal acceleration depends on the speed v. It also depends on the radius r of the curve (Figure 1.12). A larger radius means the path is not as sharply curved, so the velocity changes more slowly and the acceleration is smaller. The actual equation for the size of the acceleration is

$$a = \frac{v^2}{r} \qquad \text{(centripetal acceleration)}$$

We can also describe the relationship between a, v, and r by saying that the *acceleration is proportional to the square of the speed:*

$$a \propto v^2$$

Figure 1.12

An object moving along a circular path is accelerated because its direction of motion, and therefore its velocity, are changing. Its centripetal acceleration equals the square of its speed, divided by the radius of its path. Doubling its speed would quadruple its acceleration. Doubling the radius of its path would halve the acceleration.

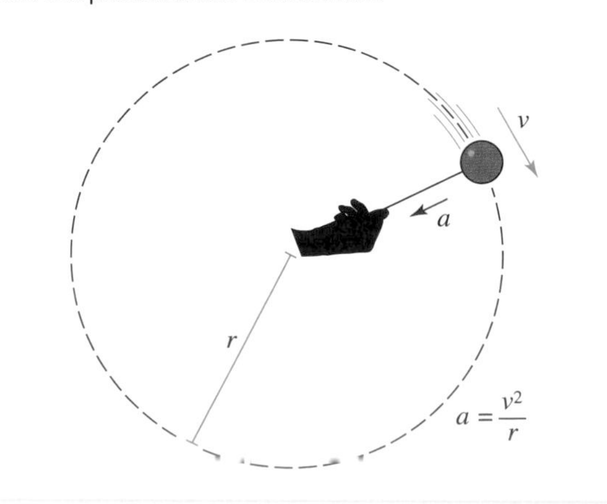

© iStockphoto.com

and the *acceleration is inversely proportional* to the *radius r:*

$$a \propto \frac{1}{r}$$

This means that when the speed is doubled, the acceleration becomes four times as large. If the radius is doubled, the acceleration becomes one-half as large. We will encounter these two relationships again.

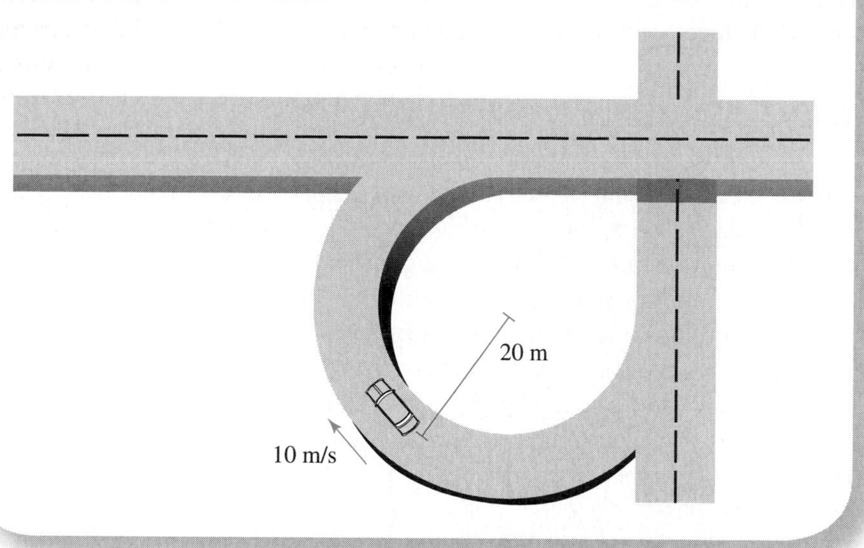

Figure 1.13
Car on a cloverleaf with a 20-meter radius.

EXAMPLE 1.5

Let's estimate the acceleration of a car as it goes around a curve. The radius of a segment of a typical cloverleaf is 20 meters, and a car might take the curve with a constant speed of 10 m/s (about 22 mph; see Figure 1.13).

Because the motion is circular,

$$a = \frac{v^2}{r} = \frac{(10 \text{ m/s})^2}{20 \text{ m}}$$

$$= \frac{100 \text{ m}^2/\text{s}^2}{20 \text{ m}} = 5 \text{ m/s}^2$$

If the car could go 20 m/s and stay on the road (it could not), its acceleration would be four times as large—20 m/s^2 or about 2 *g*.*

Many people have difficulty accepting the idea of centripetal acceleration, or they don't see how changing the direction of motion of a body is the same kind of thing as changing its speed. But some common experiences show that it is. Let's say that you are riding in a bus and that a book is resting on the slick seat next to you. The book will slide over the seat in response to the bus's acceleration. As the bus speeds up, the book slides backward. As the bus slows down, the book slides forward. In both cases, the reaction of the book is to move in the *direction opposite* the bus's acceleration. What happens when the bus goes around a curve? The book slides toward the *outside* of the curve, showing that the bus is accelerating toward the *inside* of the curve—a centripetal acceleration.

The cornering ability of a car is often measured by the maximum centripetal acceleration it can have when it rounds a curve. Automotive magazines often give the "cornering acceleration" or "lateral acceleration" of a car they are evaluating. A typical value for a sports car is 0.85 *g*, which is equal to 8.33 m/s^2.

* The speed, heading, and time displayed on a GPS receiver can be used to compute the centripetal acceleration of a vehicle going around a curve. The equation is different because the radius is not used. If v = speed in mph, Δhdg = change in heading in degrees, and Δt = time elapsed in seconds, then the acceleration in mph/s is:

$$a = v \times \frac{\Delta hdg}{\Delta t} \times 0.0175$$

(The "0.0175" converts degrees into proper units.) For example, if the heading changes from 30° to 90° in 10 s, and the car is going 65 mph, then Δhdg= 60° and a = 6.8 mph/s = 0.31 *g*. (This all works without a GPS receiver if the change in heading is known—for example, a 90° turn.)

1.4 Simple Types of Motion

Now let's take a look at some simple types of motion. In each example, we'll consider a single body moving in a particular way. Our goal is to show how distance and speed depend on time. Take note of the different ways that the relationships can be shown or expressed.

Zero Velocity

The simplest situation is one in which *no motion* occurs: a single body sitting at a fixed position in space. We characterize this system by saying that the distance of the object, from whatever fixed reference point we choose, is constant. Nothing is changing. This means that the object's velocity and acceleration are both zero.

Constant Velocity

The next simplest case is *uniform motion.* Here the body moves with a constant velocity—a constant speed in a fixed direction. An automobile traveling on a straight, flat highway at a constant speed is a good example. A hockey puck sliding over smooth ice almost fits into this category, because it slows down only slightly due to friction. Note that the acceleration equals zero. That much should be obvious. The interesting relationship here is between distance and time. We can express that relationship in four different ways: with words, mathematics, tables, or graphs.

Let's take the example of a runner traveling at a steady pace of 7 m/s. If you are standing on the side of the road, how does the distance from you to the runner change with time, after the runner has gone past you? In words, the distance increases 7 meters each second: Two seconds after the runner passes by, the distance would be 14 meters (see Figure 1.14). Stated mathematically, the distance in meters equals the time multiplied by the velocity, 7 m/s. The time is measured in seconds, starting just as the runner passes you. The shorthand way of writing this is the mathematical equation

$$d = 7t \quad (d \text{ in meters, } t \text{ in seconds})$$

This is an example of the general equation that we saw earlier with $v = 7$ m/s:

$$d = vt$$

The same information can also be put in a table of values of time and distance (see the table below). These values all satisfy the equation $d = 7t$. The table shows the values of distance (d) at certain times (t). You could make the table longer or shorter by using different time increments.

Time (s)	Distance (m)
0	0
1	7
2	14
3	21
4	28

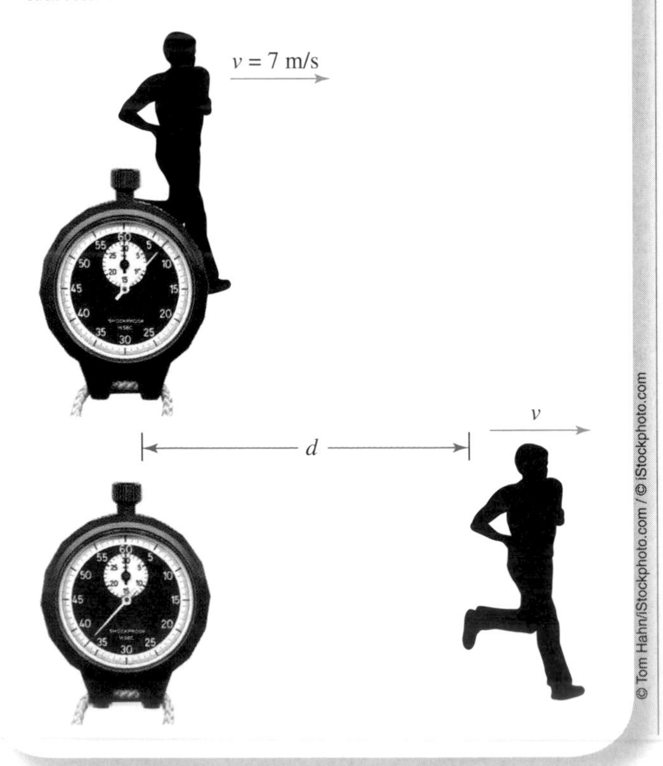

Figure 1.14
The velocity of a runner is 7 m/s. This means that the runner's distance from a fixed point (you standing behind the runner) increases 7 meters each second.

The fourth way to show the relationship between distance and time is to graph the values in the table (see Figure 1.15a). The usual practice is to graph distance versus time, which puts distance on the vertical axis. Note that the data points lie on a straight line. Remember this simple rule: When the speed is constant, the graph of distance versus time is a straight line. In general, when one quantity is proportional to another, the graph of the two quantities is a straight line.

An important feature of this graph is its *slope*. The slope of a graph is a measure of its steepness. In particular, the slope is equal to the *rise* between two points on the line divided by the *run* between the points. This is illustrated in Figure 1.15b. The rise is a distance, Δd, and the run is a time interval, Δt. So the slope equals Δd divided by Δt, which is also the object's velocity. The slope of a distance-versus-time graph equals the velocity.

The graph for a faster-moving body, a racehorse for instance, would be steeper—it would have a larger slope. The graph of d versus t for a slower object (a person walking) would have a smaller slope (see Figure 1.16). When an object is standing still (when it has no motion), the graph of d versus t is a flat line parallel with the horizontal axis. The slope is zero because the velocity is zero.

Figure 1.15
(a) Graph of distance versus time when velocity is a constant 7 m/s. (b) Same graph with the slope indicated.

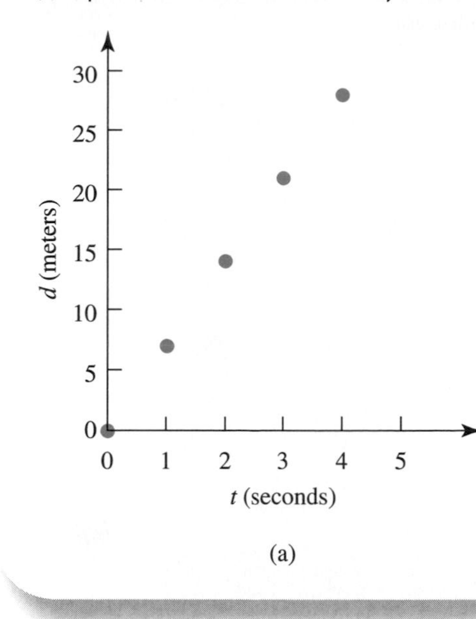

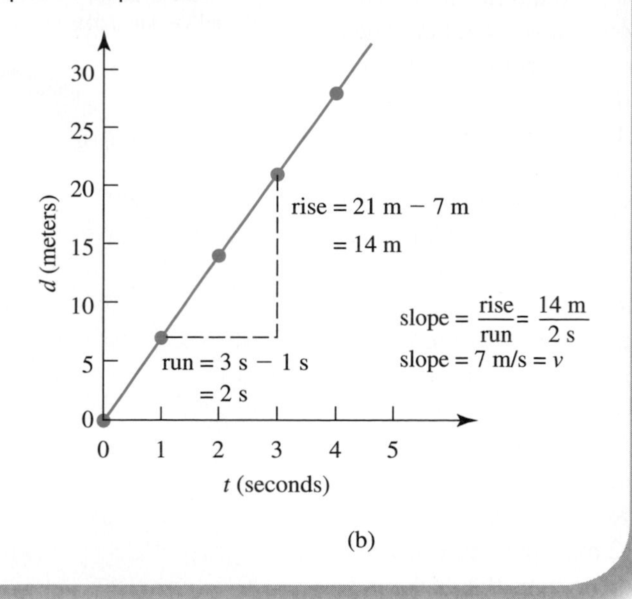

Figure 1.16
The slope of a distance-versus-time graph equals the velocity. The graph for a body with a higher velocity (the horse) is steeper—that is, it has a larger slope. When the velocity is very low, the graph is nearly a horizontal line.

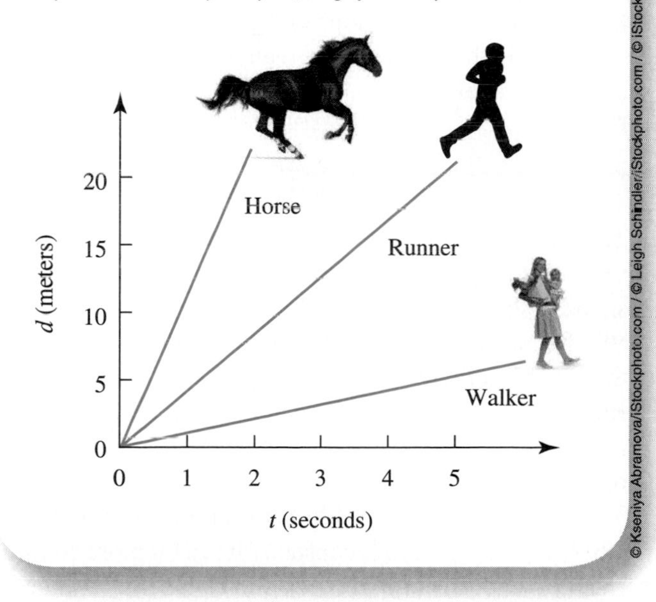

© Kseniya Abramova/iStockphoto.com / © Leigh Schindler/iStockphoto.com / © iStockphoto.com

Figure 1.17
Graph of distance versus time for a car with varying velocity. At point *A* on the graph, the slope starts to increase as the car accelerates. At *B*, its velocity is constant. The slope decreases to zero at point *C* when the car is stopped. At *D*, the car is backing up, so its velocity is negative.

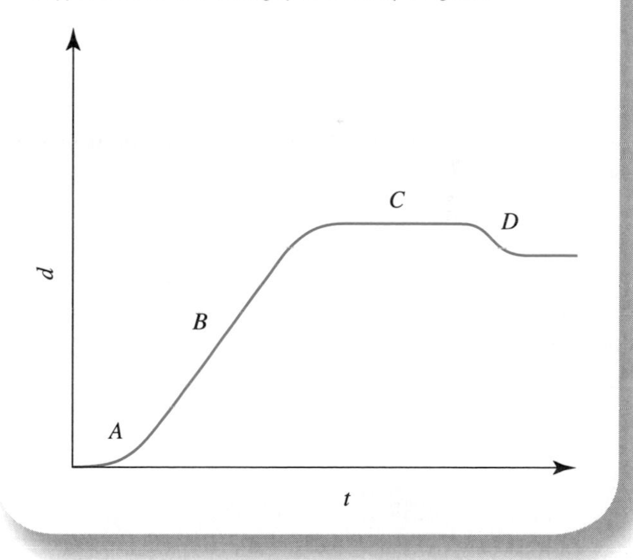

Even when the velocity is not constant, the slope of a *d* versus *t* graph is still equal to the velocity. In this case, the graph is not a straight line because, as the slope changes (a result of the changing velocity), the graph curves or bends. The graph in Figure 1.17 represents the motion of a car that starts from a stop sign, drives down a street, and then stops and backs into a parking place. When the car is stopped, the graph is flat. The distance is not changing and the velocity is zero. When the car is backing up, the graph is slanted downward. The distance is decreasing, and the velocity is negative.

Constant Acceleration

The next example of simple motion is *constant acceleration in a straight line.* This means that the object's velocity is changing at a fixed rate. A freely falling body is the best example. A ball rolling down a straight, inclined plane is another. Often, cars, runners, bicycles, trains, and aircraft have nearly constant acceleration when they are speeding up or slowing down.

Let's use free fall as our example. Assume that a heavy rock is dropped from the top of a building and that we can measure the instantaneous velocity of the rock and the distance that it has fallen at any time we choose (see Figure 1.18). The rock falls with an acceleration equal to g. This is $9.8\ m/s^2$, which is the same as 22 mph per second. First, consider how the rock's velocity changes. The velocity increases 9.8 m/s, or 22 mph, each second. This means that the rock's velocity equals the time (in seconds) multiplied by 9.8 (for m/s) or 22 (for mph). Mathematically,

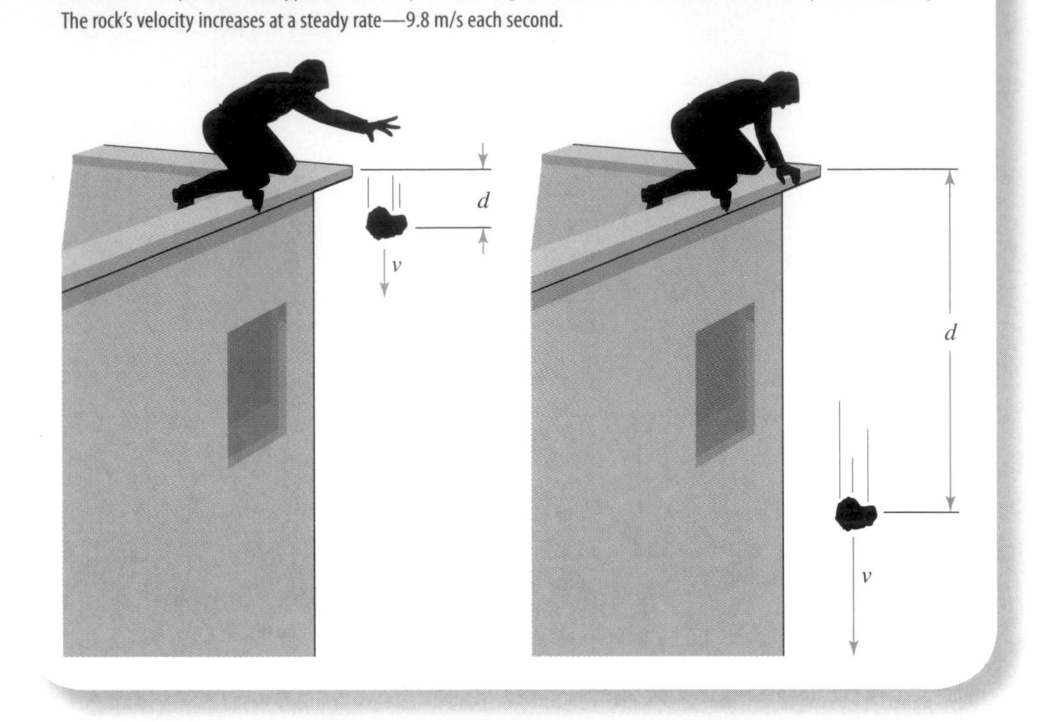

Figure 1.18

A rock falls freely after it is dropped from the top of a building. The distance d is measured from the top of the building. The rock's velocity increases at a steady rate—9.8 m/s each second.

$$v = 9.8t \quad (v \text{ in m/s, } t \text{ in seconds})$$

$$v = 22t \quad (v \text{ in mph, } t \text{ in seconds})$$

The general form of the equation that applies to any object that starts from rest and has a constant acceleration a is

$$v = at \quad (\text{when acceleration is constant})$$

So in constant acceleration, the velocity is proportional to the time. The proportionality constant is the acceleration a. A table of values for this example is shown below.

Time	Velocity	
(s)	(m/s)	(mph)
0	0	0
1	9.8	22
2	19.6	44
3	29.4	66
4	39.2	88

The graph of velocity versus time is a straight line (see Figure 1.19). The slope of a graph of velocity versus time equals the acceleration. This is because the rise is a change in velocity, Δv, and the run is a change in time Δt; so

$$\text{slope} = \frac{\Delta v}{\Delta t} = a$$

The corresponding graph for a body with a smaller acceleration, say, a ball rolling down a ramp, would have a smaller slope.

The similarity between the graphs in Figure 1.19 and the graphs in Figure 1.15 for uniform motion is obvious. But keep in mind that the graph of *distance* versus time is a straight line for uniform motion. Here it is the graph of *velocity* versus time.

What is the relationship between distance and time when the acceleration is constant? It is a bit more complicated, as expected. Figure 1.20 shows that the distance a falling body travels during each successive time interval grows larger as it falls. Since the velocity is continually changing, the distance equals the average velocity times the time. What is the average velocity? The object starts with velocity equal to zero, and after accelerating for a time t, its velocity is at. Its average velocity is:

$$\text{average velocity} = \frac{0 + at}{2} = \frac{1}{2}at$$

Figure 1.19

(a) Graph of velocity versus time for a freely falling body. (b) The same graph with the slope indicated.

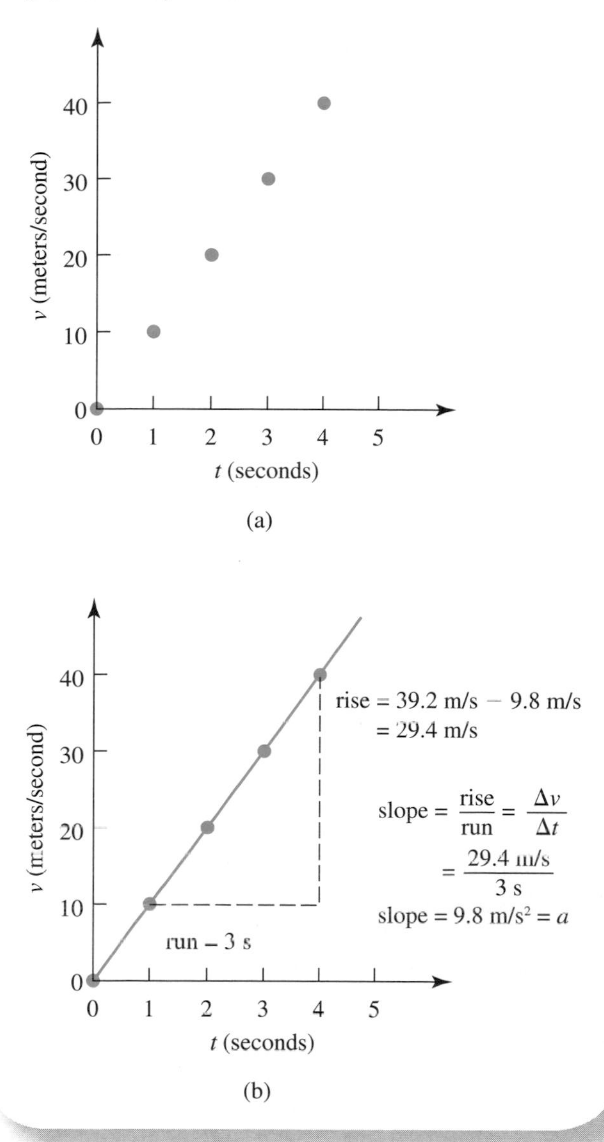

Figure 1.20

A falling ball photographed with a flashing strobe light. Each image shows where the ball was at the instant when the light flashed. The images are close together near the top because the ball is moving more slowly at first. As it falls, it picks up speed and thus travels farther between flashes.

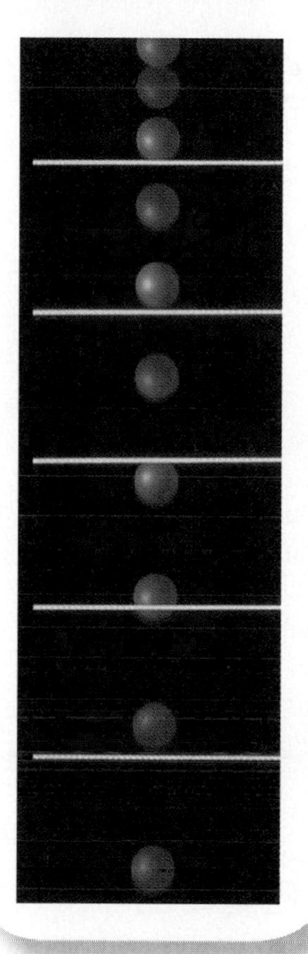

© Richard Megna/Fundamental Photographs

(If you take two quizzes and get 0 on the first and 8 on the second, your average grade is 4—one-half of 0 plus 8.)

The distance traveled is this average velocity times the time:

$$d = \text{average velocity} \times t = \frac{1}{2}at \times t$$

$$d = \frac{1}{2}at^2 \qquad \text{(when acceleration is constant)}$$

In the case of a falling body, the acceleration is 9.8 m/s^2; therefore,

$$d = \frac{1}{2}at^2 = \frac{1}{2} \times 9.8 \times t^2$$

$$= 4.9t^2 \qquad (d \text{ in meters, } t \text{ in seconds})$$

So in the case of constant acceleration, the distance is proportional to the square of the time. The constant of proportionality is one-half the acceleration.*

A table of distance values for a falling body is shown below. This distance increases rapidly. The graph of distance versus time curves upward (see Figure 1.21). This is because the velocity of the body is increasing with time, and the slope of this graph equals the velocity. Table 1.4 and your Review Card summarizes these three simple types of motion.

Time (s)	Distance (m)
0	0
1	4.9
2	19.6
3	44.1
4	78.4

Rarely does the acceleration of an object stay constant for long. As a falling body picks up speed, air resistance causes its acceleration to decrease (more on this in Section 2.6). When a car is accelerated from a stop, its acceleration usually decreases, particularly when the transmission is shifted into a higher gear. Figure 1.22 shows the velocity of a car as it accelerates from 0 to 80 mph. Note that acceleration steadily

* For an object that is already moving with a velocity v_{initial} and then undergoes constant acceleration, the average velocity after a time t is

$$v_{\text{average}} = \frac{v_{\text{initial}} + (v_{\text{initial}} + at)}{2} = v_{\text{initial}} + \frac{1}{2}at$$

Therefore,

$$d = v_{\text{initial}}\,t + \frac{1}{2}at^2$$

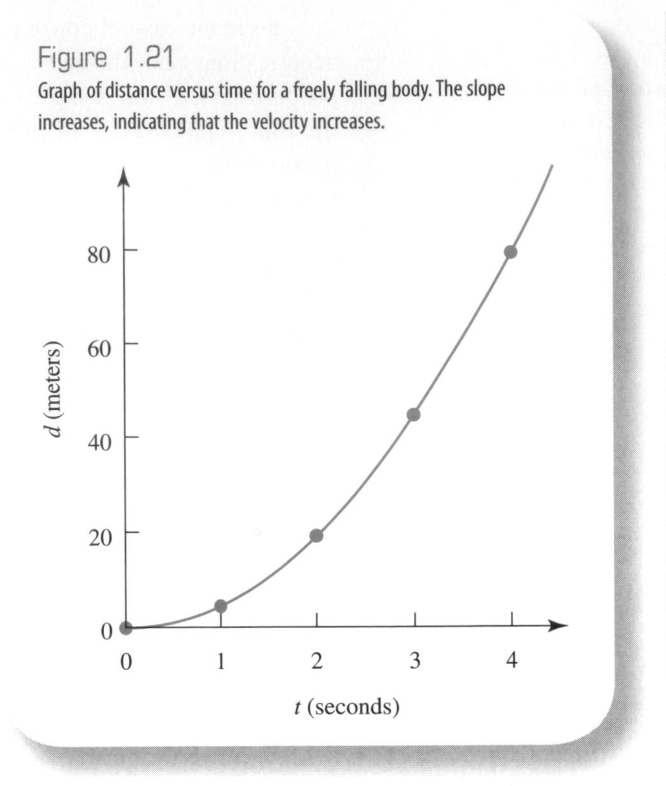

Figure 1.21
Graph of distance versus time for a freely falling body. The slope increases, indicating that the velocity increases.

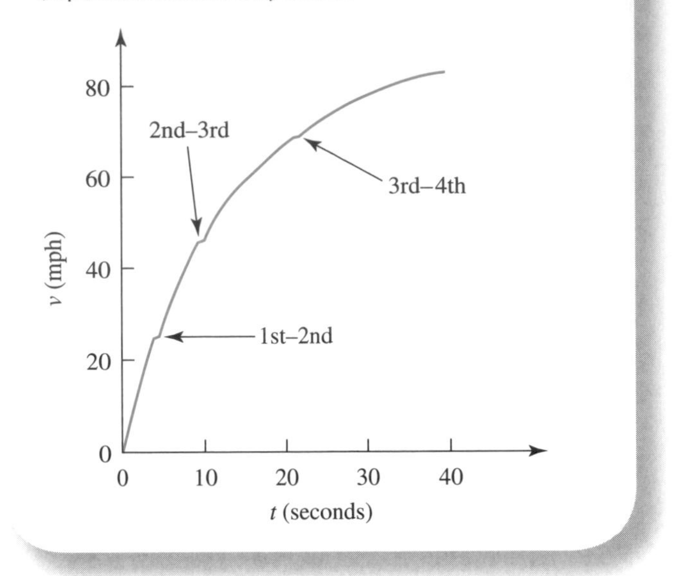

Figure 1.22
Graph of a car's velocity versus time as the car is accelerated. The notches in the curve occur at transmission shifts when the engine is momentarily disconnected from the drivetrain. Between the shifts, the acceleration (slope) decreases as the velocity increases.

decreases (the slope gets smaller). During the short time the transmission is shifted, the acceleration is zero.

While graphs may be a good way to visualize the relationships between physical quantities, mathematics is the most precise way, and it is not limited to two (or sometimes three) variables, as are graphs. However, math is inherently abstract and also somewhat like a language.

In this book, we will "translate" most of the mathematics we use into statements and also express many of the relationships graphically. But we will also retain some of the "original" language, mathematics, and show that even an understanding of high school math allows you to solve physics problems. The application of physics in our society by engineers and others requires mathematics. The Golden Gate Bridge could not have been designed and built through the use of words only.

That said, when you see a graph, the first thing you should do is take careful note of the quantities that are plotted. You should notice that some of the graphs for the examples of motion are straight lines. When the graph shows *distance* versus time, a straight line means the *velocity* is constant. But when it shows *velocity* versus time, a straight line means that the *acceleration* is constant. Even though the shapes of the two graphs are similar, because the quantities plotted are different for each graph, they represent very different situations. To a business executive, a rising graph showing profits will inspire joy, while a rising graph showing expenses will cause concern.

The most important thing to look for in the shape of a graph is a trend. Is the slope always positive or always negative—is the graph continually going up or going

Table 1.4 Summary of Examples of Motion

Type of Motion	Behavior of Physical Quantities	Equations
Stationary object	Distance constant	$d = \text{constant}$
	Speed zero	$v = 0$
	Acceleration zero	$a = 0$
Uniform motion†	Distance proportional to time	$d = vt$
	Velocity constant	$v = \text{constant}$
	Acceleration zero	$a = 0$
Uniform acceleration† (from rest)	Distance proportional to time squared	$d = \frac{1}{2}at^2$
	Velocity proportional to time	$v = at$
	Acceleration constant	$a = \text{constant}$

† Distance measured from object's initial location.

Figure 1.23

(a) Graph of the position of a fist versus time during a karate blow. Contact is made at about 6 milliseconds. (b) Graph of the velocity of the fist versus time. At contact, the fist accelerates rapidly. [From S. R. Wilke, R. E. McNair, M. S. Feld. "The Physics of Karate." *American Journal of Physics* 51 (September 1983): 783–790.]

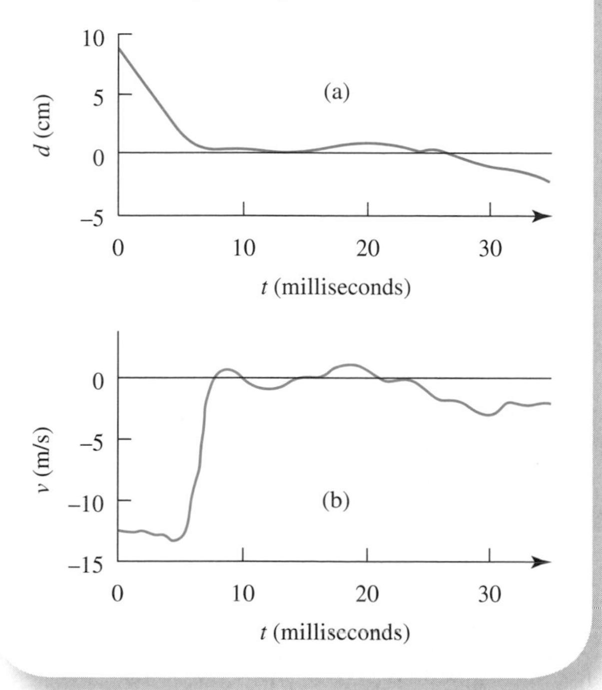

down? If the slope changes, consider what that signifies. If it is a graph of velocity versus time, for example, choose a particular time, and note whether the velocity is increasing, decreasing, or remaining the same.

Let's look at an application of the kinds of graphs we've been using. During a karate demonstration, a concrete block is broken by a person's fist. Figure 1.23a shows the distance of the hand above the block versus time, measured using high-speed photography. The fist travels downward until it contacts the block, at about 6 milliseconds. This causes a large acceleration as the fist is brought to a sudden stop. Figure 1.23b shows the velocity of the fist, just the slope of the distance versus time graph at each time. Contact with the concrete is indicated by the steep part of the graph as the velocity goes to zero. If we take the slope of this segment of the velocity graph, we find that the magnitude of the acceleration of the fist at that moment is about 3,500 m/s^2, or 360 g (ouch!). What happened at about 25 milliseconds?

questions&problems

Questions

(▶ Indicates a review question, which means it requires only a basic understanding of the material to answer. The others involve integrating or extending the concepts presented thus far.)

1. Two rectangular shaped rugs are on display in a showroom. If one rug is twice as long as the other, does this necessarily imply that its area is also twice as large as that of the second? Explain.
2. ▶ Explain what a "derived unit" of measure is.
3. A pendulum clock is taken to a repair shop. Its pendulum is replaced by a shorter one that oscillates with a smaller period than the original. What effect, if any, does this have on how the clock runs?
4. ▶ What are the "basic" or "fundamental" physical quantities? Why are they called that?
5. Many countries that formerly used the English system of measure have converted to the metric system. Why is the metric system simpler to use, once you are familiar with it?
6. A wind is blowing from the north (the air is moving towards the south). When a person is walking towards the north, is the relative speed of the wind that the person senses greater than, the same as, or less than the speed the person senses when not walking? How about when the person is walking towards the south?
7. Scenes in films or television programs sometimes show people jumping off moving trains and having unpleasant encounters with the ground. If someone is on a moving flat-bed train car and wishes to jump off, how could the person use the concept of relative speed to make a safer dismount?
8. ▶ List the physical quantities identified in this chapter. From which of the fundamental physical quantities is each derived? Which of them are vectors, and which are scalars?
9. ▶ What is the distinction between speed and velocity? Describe a situation in which an object's speed is constant but its velocity is not.

10. ▶What is "vector addition," and how is it done?
11. Can the resultant of two velocities have zero magnitude? If so, give an example.
12. A swimmer heads for the opposite bank of a swiftly flowing river. Make a sketch showing the swimmer's two velocities and the resultant velocity.
13. A basketball player shoots a free throw. Make a sketch showing the basketball's velocity just after the ball leaves the player's hands. Draw in two components of this velocity, one horizontal and one vertical. Repeat the sketch for the instant just before the ball reaches the basket. What is different?
14. ▶What is the relationship between velocity and acceleration?
15. ▶How does the velocity of a freely falling body change with time? How does the distance it has fallen change? How about the acceleration?
16. Using concepts and physical quantities discussed in this chapter, explain why it is usually safe for a person standing on the seat of a chair to jump horizontally and land on the floor, but not for a person standing on the roof of a tall building to jump horizontally and land on the ground.
17. ▶What is centripetal acceleration? What is the direction of the centripetal acceleration of a car going around a curve?
18. During 200-meter and 400-meter races, runners must stay in lanes as they go around a curved part of the track. If runners in two different lanes have exactly the same speed, will they also have exactly the same centripetal acceleration as they go around a curve? Explain.
19. An insect is able to cling to the side of a car's tire when the car is going 5 mph. How much harder is it for the insect to hold on when the speed is 10 mph?
20. As a car goes around a curve, the driver increases its speed. This means the car has two accelerations. What are the directions of these two accelerations?
21. The following are speeds and headings displayed on a GPS receiver. (Heading gives the direction of motion based on: north = 0°, east = 90°, south = 180°, etc.) In each case, indicate whether the receiver was accelerating during the time between the displays, and if it was, describe in what way the receiver was accelerating.
    a) Initially: 60 mph, 70°. Five seconds later: 50 mph, 70°.
    b) Initially: 50 mph, 70°. Five seconds later: 70 mph, 70°.
    c) Initially: 60 mph, 70°. Five seconds later: 60 mph, 90°.
22. In Figure 1.12, arrows show the directions of the velocity and the acceleration of a ball moving in a circle. Make a similar sketch showing these directions for a car (a) speeding up from a stop sign and (b) slowing down as it approaches a stop sign.
23. ▶If a ball is thrown straight up into the air, what is its acceleration as it moves upward? What is its acceleration when it reaches its highest point and is stopped at an instant?
24. ▶What does the slope of a distance-versus-time graph represent physically?
25. Sketch a graph of velocity versus time for the motion illustrated in Figure 1.17. Indicate what the car's acceleration is at different times.

Problems

1. A yacht is 20 m long. Express this length in feet.
2. Express your height (a) in meters and (b) in centimeters.
3. A convenient time unit for short time intervals is the *millisecond*. Express 0.0452 s in milliseconds.
4. One mile is equal to 1,609 m. Express this distance in kilometers and in centimeters.
5. A hypnotist's watch hanging from a chain swings back and forth every 0.8 s. What is the frequency of its oscillation?
6. The quartz crystal used in an electric watch vibrates with a frequency of 32,768 Hz. What is the period of the crystal's motion?
7. A passenger jet flies from one airport to another 1,200 miles away in 2.5 h. Find its average speed.
8. A runner in a marathon passes the 5-mile mark at 1 o'clock and the 20-mile mark at 3 o'clock. What is the runner's average speed during this time period?
9. In Figure 1.8, assume that $v_1 = 8$ m/s and $v_2 = 6$ m/s. Use a ruler to estimate the magnitudes of the resultant velocities in (c) and (d).
10. On a day when the wind is blowing toward the south at 3 m/s, a runner jogs west at 4 m/s. What is the velocity (speed and direction) of the air relative to the runner?
11. How far does a car going 25 m/s travel in 5 s? How far would a jet going 250 m/s travel in 5 s?
12. A long-distance runner has an average speed of 4 m/s during a race. How far does the runner travel in 20 min?
13. Draw an accurate graph showing the distance versus time for the car in Problem 11. What is the slope?
14. The graph in Figure 1.24 shows the distance versus time for an elevator as it moves up and down in a building. Compute the elevator's velocity at the times marked a, b, and c.

Figure 1.24

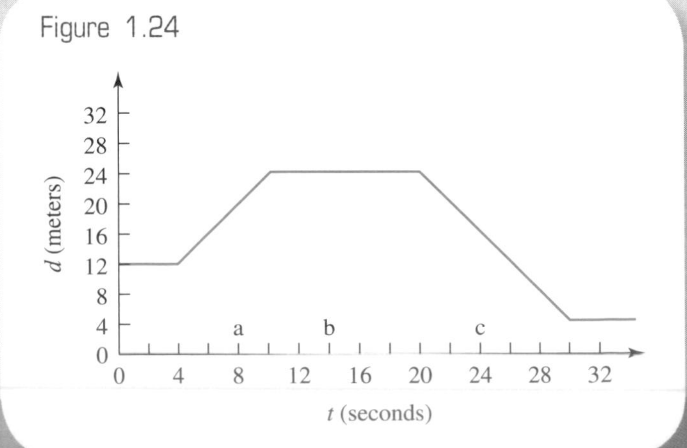

15. A high-performance sports car can go from 0 to 100 mph (44.7 m/s) in 7.9 s.
    a) What is the car's average acceleration?
    b) The same car can come to a complete stop from 30 m/s in 3.2 s. What is its average acceleration?
16. As a baseball is being thrown, it goes from 0 to 40 m/s in 0.15 s.
    a) What is the acceleration of the baseball?
    b) What is the acceleration in *g*'s?
17. A child attaches a rubber ball to a string and whirls it around in a horizontal circle overhead. If the string is 0.5 m long, and the ball's speed is 10 m/s, what is the ball's centripetal acceleration?
18. A child sits on the edge of a spinning merry-go-round that has a radius of 1.5 m. The child's speed is 2 m/s. What is the child's acceleration?
19. A runner is going 10 m/s around a curved section of track that has a radius of 35 m. What is the runner's acceleration?
20. During a NASCAR race a car goes 50 m/s around a curved section of track that has a radius of 250 m. What is the car's acceleration?
21. A rocket accelerates from rest at a rate of 60 m/s².
    a) What is its speed after it accelerates for 40 s?
    b) How long does it take to reach a speed of 7,500 m/s?
22. A train, initially stationary, has a constant acceleration of 0.5 m/s².
    a) What is its speed after 15 s?
    b) What would be the total time it would take to reach a speed of 25 m/s?
23. a) Draw an accurate graph of the speed versus time for the train in Problem 22.
    b) Draw an accurate graph of the distance versus time for the train in Problem 22.
24. Draw an accurate graph of the velocity versus time for the elevator in Problem 14.
25. A skydiver jumps out of a helicopter and falls freely for 3 s before opening the parachute.
    a) What is the skydiver's downward velocity when the parachute opens?
    b) How far below the helicopter is the skydiver when the parachute opens?
26. A rock is dropped off the side of a bridge and hits the water below 2 s later.
    a) What was the rock's velocity when it hit the water?
    b) What was the rock's average velocity as it fell?
    c) What is the height of the bridge above the water?
27. The roller coaster in Figure 1.25 starts at the top of a straight track that is inclined 30° with the horizontal. This causes it to accelerate at a rate of 4.9 m/s² (1/2 *g*).
    a) What is the roller coaster's speed after 3 s?
    b) How far does it travel during that time?

Figure 1.25

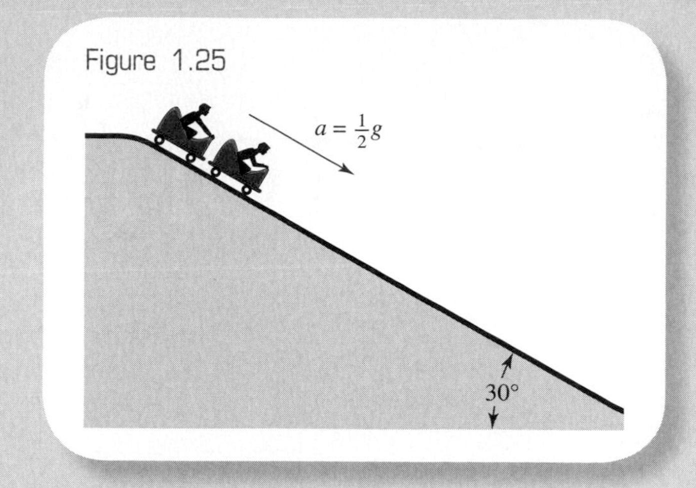

28. During takeoff, an airplane goes from 0 to 50 m/s in 8 s.
    a) What is its acceleration?
    b) How fast is it going after 5 s?
    c) How far has it traveled by the time it reaches 50 m/s?
29. The graph in Figure 1.26 shows the velocity versus time for a bullet as it is fired from a gun, travels a short distance, and enters a block of wood. Compute the acceleration at the times marked a, b, and c.

Figure 1.26

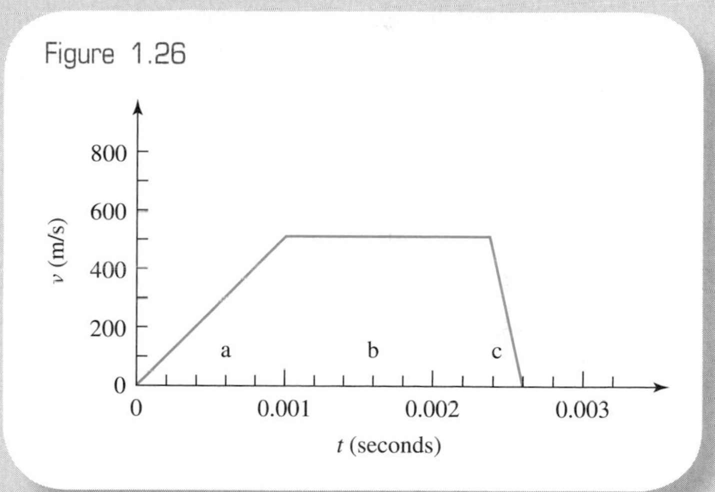

30. A bungee jumper falls for 1.3 s before the bungee cord begins to stretch. Until the jumper has bounced back up to this level, the bungee causes the jumper to have an average acceleration upward of 4 m/s².
    a) How fast is the jumper going when the bungee cord begins to stretch?
    b) How far below the diving platform is the jumper at that moment?
    c) How long after the bungee cord begins to stretch does the jumper reach the low point of the drop?
    d) How far below the diving platform is the jumper at the instant the speed is zero?
31. A drag-racing car goes from 0 to 300 mph in 5 s. What is its average acceleration in *g*'s?

Newton's Laws

sections

2.1 Force

Since the 1960s, robotic space probes have been sent throughout the solar system to examine planets and their moons, asteroids, comets, and other objects of scientific interest. The most conspicuous bodies not yet visited are Pluto and Charon. But that will change in 2015. The *New Horizons* spacecraft is barreling through the solar system, traveling faster than any previous probe, on its way to a flyby of Pluto and Charon. Launched in January 2006, *New Horizons* passed by Jupiter in late February 2007, not so much to examine this often-visited giant planet but to gain even more speed by way of a "gravity assist" from it.

The scientific instruments on *New Horizons* and the rockets that sent it on its journey are products of modern technology. But the principles behind interplanetary travel are rooted in laws established by Sir Isaac Newton more than 300 years ago. His third law of motion explains how a vehicle can accelerate in empty space where there is nothing to "push" against. His second law makes it possible to figure out how long to fire a rocket engine to change a spacecraft's velocity by a desired amount. His law of universal gravitation is the tool for taking into account the effect of the Earth, the Sun, and other planets on a spacecraft's path. As if that weren't enough, Newton also invented calculus, the branch of mathematics essential for performing the latter two tasks.

Sir Isaac Newton (1642–1727) was an English scholar who made many fundamental discoveries in both physics and mathematics. Newton is often regarded as the father of modern physical science, and for this reason, it is difficult to overestimate his importance in the development of today's civilization. The dominant scientific, industrial, and technological advancements of the last two centuries were triggered in part by Newton's work.

In formulating his mechanics, Newton began with Galileo's ideas about motion and then sought systematic rules that govern motion and, more importantly, *changes in motion.* The key concept in Newtonian mechanics is **force**.

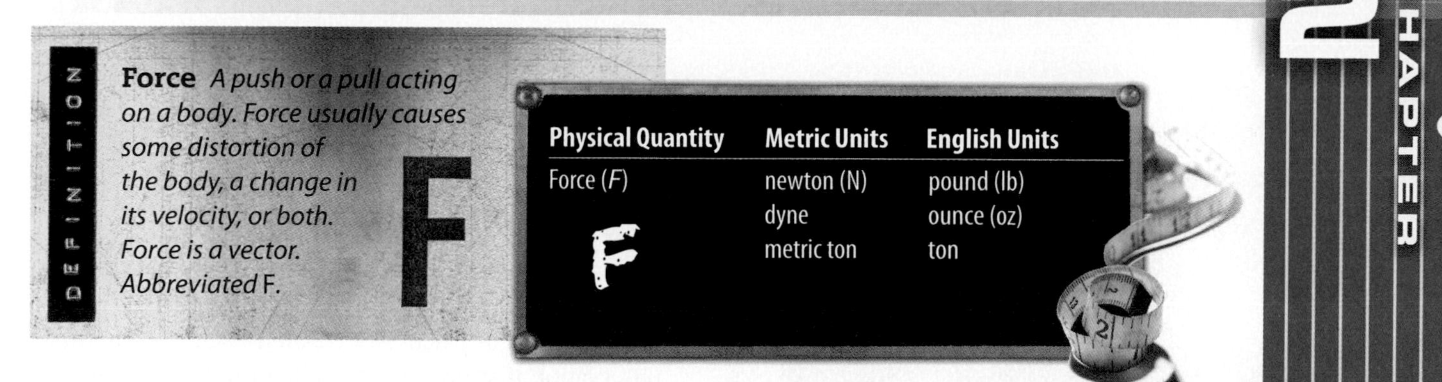

DEFINITION

Force *A push or a pull acting on a body. Force usually causes some distortion of the body, a change in its velocity, or both. Force is a vector. Abbreviated* F.

Physical Quantity	Metric Units	English Units
Force (*F*)	newton (N)	pound (lb)
	dyne	ounce (oz)
	metric ton	ton

The conversion factors between the primary units of force in the two measurement systems are

$$1\ \text{N} = 0.225\ \text{lb} \qquad 1\ \text{lb} = 4.45\ \text{N}$$

This means that a force of 150 pounds is equal to a force of 668 newtons.

The distortion caused by a force is often obvious, such as the compression of a sofa cushion when you sit on it. Sometimes it cannot be observed without help. For example, high-speed photography

© Julie de Leseleuc/iStockphoto.com / © Bill Noll/iStockphoto.com

Figure 2.1

Force on the golf ball distorts it and accelerates it.

© Edward Kinsman/Photo Researchers, Inc.

reveals that a golf ball is flattened as a club hits it (Figure 2.1). Spring scales, like the produce scales in supermarkets, use distortion to measure forces. The greater the force acting to stretch or compress a spring, the greater the distortion (Figure 2.2).

Force is a bit difficult to define and is sometimes regarded as a fundamental quantity like time and distance. The English-system units should indicate to you that the idea of force is quite common. The words *push*, *shove*, *lift*, *pull*, and *yank* are some everyday synonyms that we use to describe force. The concepts of force and energy are probably the two most ubiquitous and useful in all of physics. Newton's laws of motion are simple, direct statements about forces in general and their relationship to motion.

Let's consider some examples of force to show how versatile the concept is. We exert forces on objects in dozens of everyday situations, such as in pushing or pulling a door open, lifting a box, pulling out a drawer, and throwing a ball. Many machines we use are designed to exert forces. Cranes, hoists, jacks, and vises are all examples of such machines. Automobiles and other propelled vehicles function by creating forces that act on them. If this last statement seems a bit odd to you, come back to it after you have read about Newton's third law of motion in Section 2.7.

The most common force in our lives is **weight**. Most of us measure the weight of our bodies regularly. Many things that we buy, such as flour, cat food, and nails, are sold by weight.

The direction of the force of gravity is what determines the directions up and down. We are so accustomed to living with this force that pulls everything toward the Earth's surface that we often forget that it is a force (Figure 2.3). A child's simple observation that things naturally "fall down" leads one to Aristotle's idea of motion toward a natural place. But Newton made the important observation that objects are pulled toward the Earth by a force in much the same way that a sled is pulled by a person.

The weight of an object depends on two things: the amount of matter comprising the object (its mass) and whatever external, mass-containing agents with which the object is interacting gravitationally. The second point means that a body's weight depends on where it is. The weight of an object on the Moon is about one-sixth the weight it would have on Earth because the gravitational pull of the Moon is less than the Earth's. Even on the Earth, the weight of an object varies slightly with location. A 190-pound person would weigh about 1 pound less at the equator than at the North or South Pole.

Another important, common force is **friction**. Friction is at work when a chair slides across a floor, when brakes keep a car from rolling down a hill, when air resistance slows

Figure 2.2

(a) A force of 5 newtons stretches the spring 9 centimeters. (b) Doubling the force on the spring doubles the distance it is stretched. The scale itself has a spring inside and makes use of this principle.

© Vern Ostdiek / © Nicholas Belton/iStockphoto.com

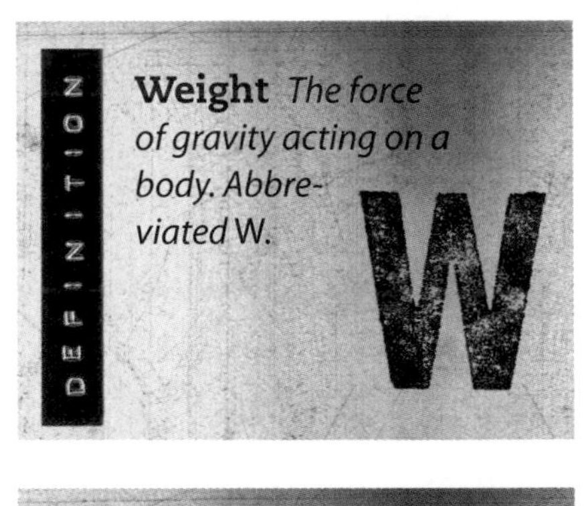

DEFINITION **Weight** *The force of gravity acting on a body. Abbreviated W.*

DEFINITION **Friction** *A force of resistance to relative motion between two bodies or substances in physical contact.*

Figure 2.3
Weight (W) is the downward force of gravity. This force acts on objects whether they are stationary, moving horizontally, or moving vertically. (Arrows are not to scale.)

down a baseball, and when a boat glides over the water. Frictional forces arise at the surfaces or boundaries of the materials involved. We can distinguish two types of friction, static and kinetic. When there is no relative motion between two objects, the friction that acts is **static friction**. If you try to push a refrigerator across a floor, you must first overcome the force of static friction between it and the floor. A person who is walking or running relies on the static friction between his or her shoes and the ground to maintain their motion. Even though the person's body is moving relative to the ground, there is no relative motion between the shoes and the ground when they are in contact. It is difficult to walk on ice because the force of static friction is reduced, often leading to slipping (and falling). Even a moving car relies on the static friction between its tire surfaces and the pavement. As long as the tires are not skidding or spinning, there is no relative motion between them and the pavement where they are in contact with one another.

A block of wood resting on an inclined plane is a simple system that relies on static friction (Figure 2.4). If the plane is horizontal, there is no friction. When the plane is tilted a small amount, friction acts to keep the block from sliding down. Here the force of friction opposes the component of the force of gravity—weight—which acts parallel to the plane. The steeper the plane is, the greater the force of static friction needed to keep the block from sliding down. When the angle of the plane reaches a certain value, the block will begin to move down. The component of the weight parallel to the plane will have exceeded the maximum force of static friction. Notice that the force of static friction between two surfaces can have any value between zero and some characteristic maximum.

Kinetic friction acts when there is relative motion between two substances in contact, such as an aircraft moving through the air, a fish swimming underwater, or a tire skidding on pavement. In Figure 2.4d, kinetic friction acts on the block of wood when it is sliding down the inclined plane. The force of kinetic friction that acts between two solids is usually less than the maximum static friction that can act. This is why a car can be stopped more quickly when its tires are not skidding.

The effects of kinetic friction are often undesirable. A car that is traveling on a flat road at a constant speed consumes fuel mainly because it must act against the forces of kinetic friction—air resistance acting on the car's

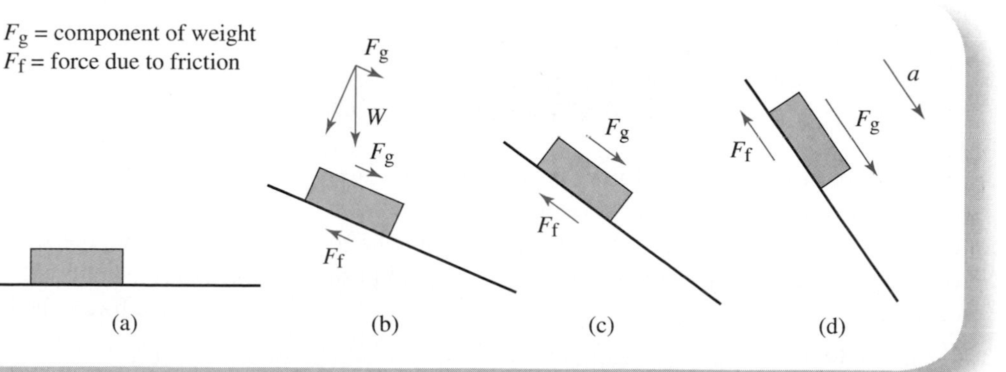

Figure 2.4
Static friction opposes gravity, which acts to pull the block down the incline. As the plane is tilted, both forces (F_g and F_f) increase until the maximum possible force of static friction is reached (c). Further tilting makes F_g larger than F_f, and the block accelerates down the incline (d).

Figure 2.5

Bicycle brakes use kinetic friction as the pads rub against the rim of the wheel.

exterior and friction between various moving parts in the axles, transmission, and engine. This also applies to aircraft and ships. Brakes represent a useful application of kinetic friction. On most bicycles, the brakes consist of pads that rub against the wheel rims (Figure 2.5). When the brakes are applied, the force of kinetic friction between the pads and the rim slows the bicycle.

Often it is difficult to include the effects of friction in mathematical models of physical systems. Consequently, we will find it helpful to consider situations in which frictional forces are small enough to be ignored. It is easier to analyze the motion of a falling rock than that of a falling feather. We need to understand these simple systems before we can incorporate the complexities of friction.

2.2 Newton's First Law of Motion

As we have just seen, many forces act in everyday situations. Newton's laws make no reference to specific types of forces. They are general statements that apply whether a force is caused by gravity, friction, or a direct push or pull.

LAWS

Newton's First Law of Motion *An object will remain at rest or in motion with constant velocity unless acted on by a net external force.*

This sentence requires some explanation. An *external* force is one that is caused by some agent outside the object or system in question. Weight is an external force because it is caused by something outside an object (for example, the Earth). If your car stalls, you cannot move it by sitting in the driver's seat and pushing on the windshield. This would be an internal force. You must exit the car and push from the outside. The *net* force is the vector sum of the external forces acting on the body. If one person pushes forward on your car while another pushes backward with equal effort, the net force is zero (Figure 2.6a). If two forces act in the same direction, the net force is the sum of the two (Figure 2.6b). If two forces act in different directions, the net force is found by adding the two vectors together (Figure 2.7).

At first glance, Newton's first law does not seem to be profound. Obviously, an object will remain stationary unless a net force causes it to move. But the law also states that anything that is already moving will not speed up, slow down, or change direction unless a net force acts on it. This means that the states of "no motion" and of "uniform motion" are equivalent as far as forces are concerned. That is, Newton's first law implies that a force is required only to *change* the state of motion. This may seem to run counter to your intuition, but that is because you rarely see a moving object with no net force acting on it. A car traveling at a constant velocity has zero net force acting on it: The various forces acting all cancel each out.

As you throw a ball, your hand exerts a net force that is in the same direction as the ball's velocity. The ball's speed increases because the force acts in the same direction as its motion. When you catch the ball, the force is opposite to the direction of the ball's velocity. Here the force slows down the ball.

Another important point implied by the first law is that a force is required to change the direction of motion of an object. Velocity, a vector, includes direction. Constant velocity implies constant direction as well as constant speed. To make a moving object change its direction of motion, a net external force must act on it. To deflect a moving soccer ball, a player must exert a sideways force on it with the foot or head (Figure 2.8). Without such a force, the ball will not change direction.

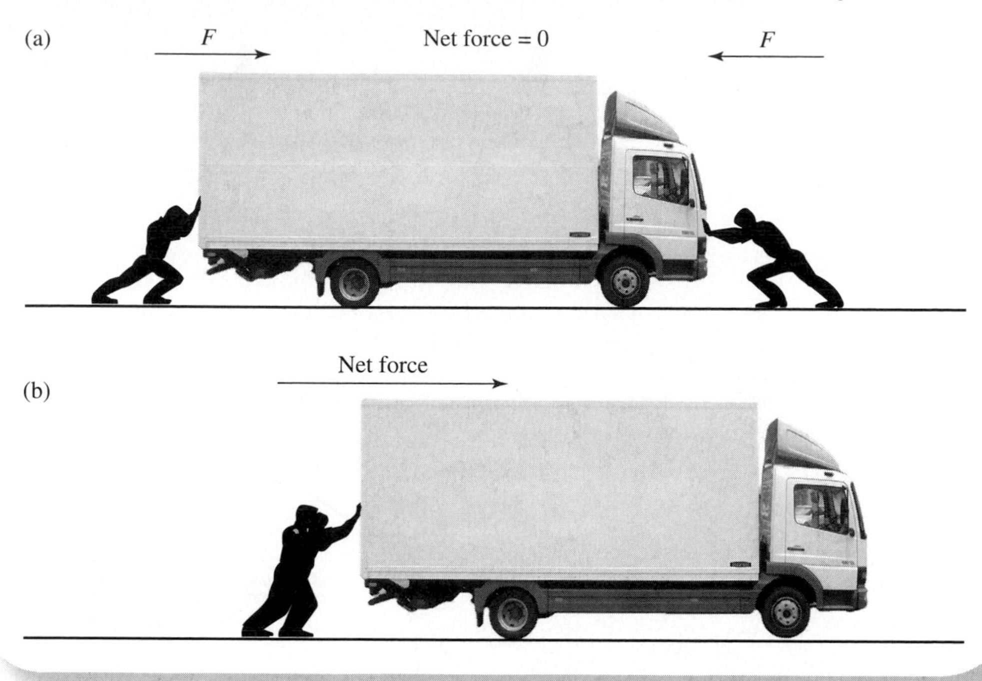

Figure 2.6
(a) Equal forces in opposite directions produce a net force equal to zero. (b) Forces in the same direction add together.

Figure 2.7
Two sailors push on a boat in different directions. The net force is the vector sum of the two applied forces.

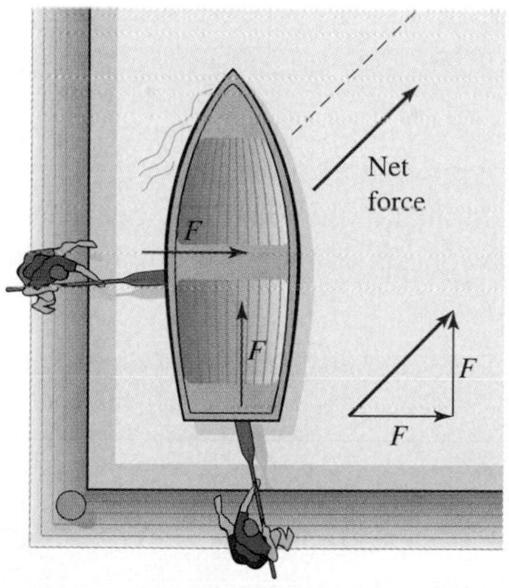

The implication of Newton's first law is that a net external force must act on an object to speed it up, slow it down, or change its direction of motion. Because each of these is an acceleration of the object, the first law tells us that a force is required to produce any acceleration. The centripetal acceleration inherent in circular motion requires a force; it is called the **centripetal force**. Like centripetal acceleration, the centripetal force on an object is directed toward the center of the object's path. As a car goes around a curve, the centripetal force that acts on it is a sideways frictional force between the tires and the road. Without this friction, as is nearly the case on ice, there is no centripetal force, and the car will go in a straight line—off the curve.

Imagine a rubber ball tied to a string and whirled around in a horizontal circle overhead. The ball "wants" to move in a straight line (by the first law) but is prevented from doing so by the centripetal force acting through the string. If the string breaks, the path of the ball will be a straight line in the direction the ball was moving at the instant the string broke (Figure 2.9). You may be tempted to think that the ball would move in a direct line away from the center of its previous circular motion. But that is not the case. If you ever see a hammer

Figure 2.8
Force is needed to change the direction of motion.

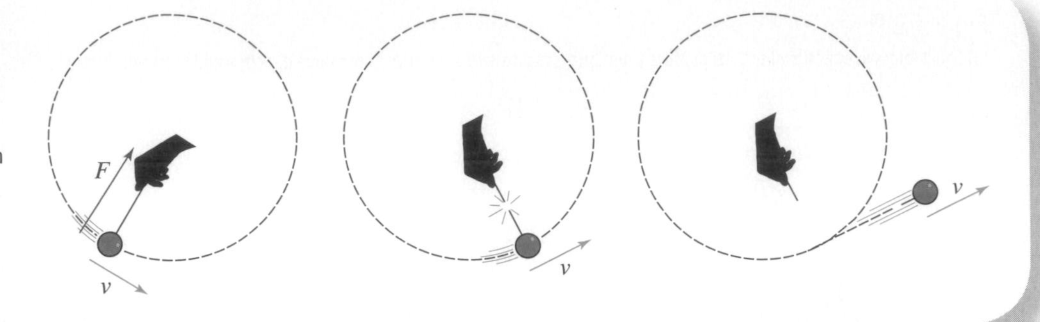

Figure 2.9
Centripetal force must act on any object to keep it moving on a circular course. If the force is removed, the object will move at the same speed in a straight line.

thrower, a discus thrower, or a catapult in action, notice the path of the projectile after it is released. It moves along a line tangent to its original circular path.

There are other examples of this effect. Children (or even physics professors) riding on a spinning merry-go-round must hold on because their bodies "want" to travel in a straight line, not in a circle. Similarly, clothes are partially dried during the spin cycle in an automatic washer because the water droplets tend to travel in straight lines and move through the holes in the side of the tub. The clothes are "pulled away" from the water as they spin. The same effect could be used to create an "artificial gravity" on a space station by making it spin (Figure 2.10). (The classic movie *2001: A Space Odyssey* showed this well.) You may have tried out a "spinning room" at an amusement park; it uses the same principle.

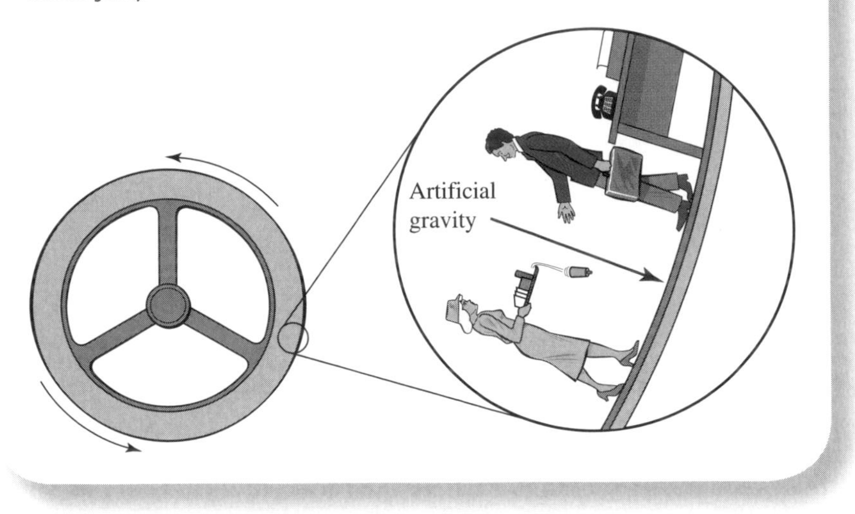

Figure 2.10
Model of a rotating space station. The circular paths of the occupants would make them experience artificial gravity.

2.3 Mass

Newton's second law of motion states the exact relationship between the net force acting on an object and its acceleration. But before stating the second law, we will take another look at mass. Imagine the effect of a small net force acting on a car (see Figure 2.11). The resulting acceleration would be quite small. The same net force would cause a much larger acceleration if it acted on a shopping cart.

The property of matter that makes it resist acceleration is often referred to as *inertia.* A car is more difficult to accelerate than a shopping cart because it has more inertia. The concept of inertia is embodied in the physical quantity *mass.*

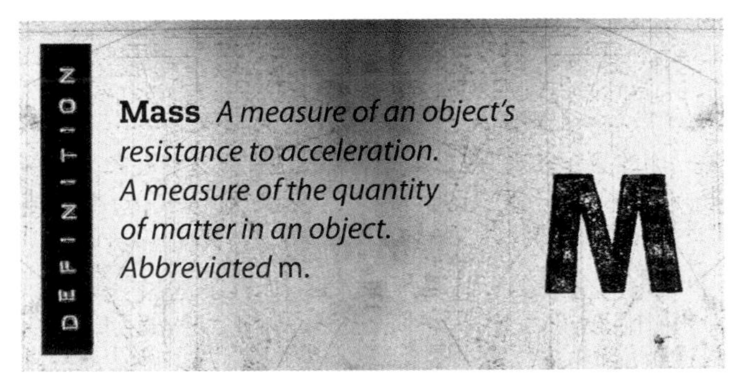

The concept of mass is not in everyday use in the English system. For example, flour is purchased by weight (pounds), not by mass (slugs). In the metric system, the situation is the reverse. Mass is commonly used

Figure 2.11
The effect of a net force on an object depends on the object's mass. If the mass is very large, the acceleration of the object will be small. Conversely, if the mass is small, the acceleration will be large.

instead of weight. Flour is purchased by the kilogram, not by the newton.*

The actual mass of a given object depends on its size (volume) and its composition. A rock has more mass than a pebble because of its size; it has more mass than an otherwise identical piece of Styrofoam because of its composition. The rock is harder to accelerate in the sense that to achieve a given acceleration a, larger force is required. Fundamentally, the mass of a body is determined mainly by the total numbers of subatomic particles that comprise it (protons, neutrons, and electrons).

Mass is *not* the same as weight. Mass and weight are related to each other in that the weight of an object is proportional to its mass. But weight is a force that arises from the gravitational interaction. Mass is an intrinsic property of matter that does not depend on any external phenomenon. One major cause of confusion is that users of both the English system and the metric system often incorrectly lump together the two concepts in everyday use. Weight is often incorrectly used in place of mass in countries using the English system, and mass is often incorrectly used in place of weight in countries using the metric system. It usually doesn't matter since they are proportional to each other, but if sometime in the future people routinely travel to the Moon or to other places where the acceleration due to gravity is different from that on the Earth, then it will.

If a hammer has a mass of 1 kilogram, that will not change if it is taken into space aboard a spacecraft and then to the Moon's surface. Its mass is 1 kilogram wherever it is. The hammer's weight, however, varies with the location, because weight depends on gravity (see Figure 2.12). Its weight on Earth is 9.8 newtons, its weight would appear to be zero in a spacecraft orbiting the Earth, and its weight would be only 1.6 newtons on the Moon's surface. Though "weightless" in the spacecraft, the hammer is not massless and would still resist acceleration as it does on Earth.

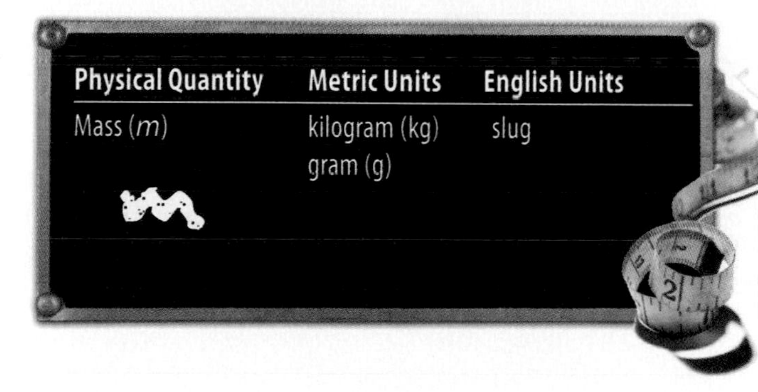

Physical Quantity	Metric Units	English Units
Mass (m)	kilogram (kg) gram (g)	slug

Figure 2.12
The hammer's weight depends on where it is, but its mass is always the same—1 kilogram.

* To help you get a feel for the size of the kilogram, we offer the following "mixed" conversion:

1 kilogram *weighs* 2.2 pounds (on Earth)

One kilogram *does not equal* 2.2 pounds! On Earth, anything with that mass has a *weight* of 9.8 newtons—which happens to equal 2.2 pounds.

Another way of approaching mass and weight is to think of them as two different characteristics of matter. We might call these aspects *inertial* and *gravitational.* Mass is a measure of the inertial property of matter—how difficult it is to change its velocity. Weight illustrates the gravitational aspect of matter: any object experiences a pull by the Earth, the Moon, or any other body near it.

2.4 Newton's Second Law of Motion

We are now ready to state Newton's second law of motion. This law is our most important tool for applying mechanics in the real world.

LAWS

Newton's Second Law of Motion *An object is accelerated whenever a net external force acts on it. The net force equals the object's mass times its acceleration.**

$$F = ma$$

This law expresses the exact relationship between force and acceleration. For a given body, a larger force will cause a proportionally larger acceleration (see Figure 2.13). Force and acceleration both are vectors. The direction of the acceleration of an object is the same as the direction of the net force acting on it.

EXAMPLE 2.1

An airplane with a mass of 2,000 kilograms is observed to be accelerating at a rate of 4 m/s². What is the net force acting on it?

$$F = ma = 2{,}000 \text{ kg} \times 4 \text{ m/s}^2$$

$$= 8{,}000 \text{ N}$$

Another way of stating Newton's second law is that an object's acceleration is equal to the net force acting on the object divided by its mass.

$$a = \frac{F}{m}$$

This means that a large force acting on a large mass can result in the same acceleration as a small force acting on a small mass. A force of 8,000 newtons acting on a 2,000-kilogram airplane causes the same acceleration as a force of 4 newtons acting on a 1-kilogram toy.

The following example illustrates what can be done with the mechanics that we have learned so far.

EXAMPLE 2.2

An automobile manufacturer decides to build a car that can accelerate uniformly from 0 to 60 mph in 10 s (see Figure 2.14). In metric units, this is from 0 to 27 m/s. The car's mass is to be about 1,000 kilograms. What is the force required?

First, we must determine the acceleration and then use Newton's second law to find the force. As we did in Chapter 1:

$$a = \frac{\Delta v}{\Delta t} = \frac{27 \text{ m/s} - 0 \text{ m/s}}{10 \text{ s}}$$

$$= 2.7 \text{ m/s}^2$$

Figure 2.13
The acceleration of an object is proportional to the net force. Tripling the force on an object triples its acceleration.

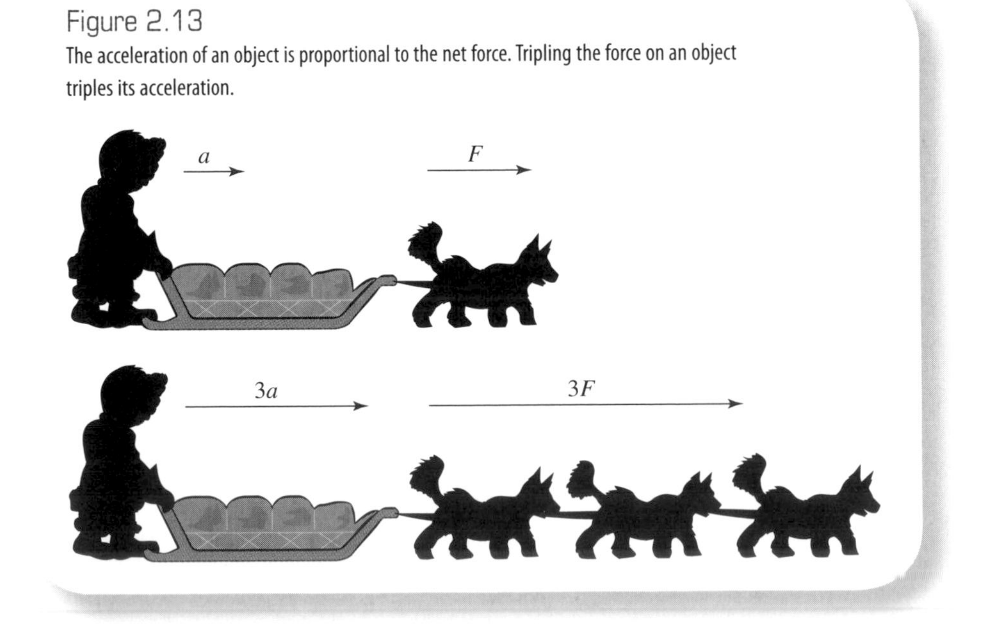

* The unit of measure of force, the newton, is established by this law. One newton is the force required to give a 1-kilogram mass an acceleration of 1 m/s². In other words:

1 newton = 1 kilogram meter/second²

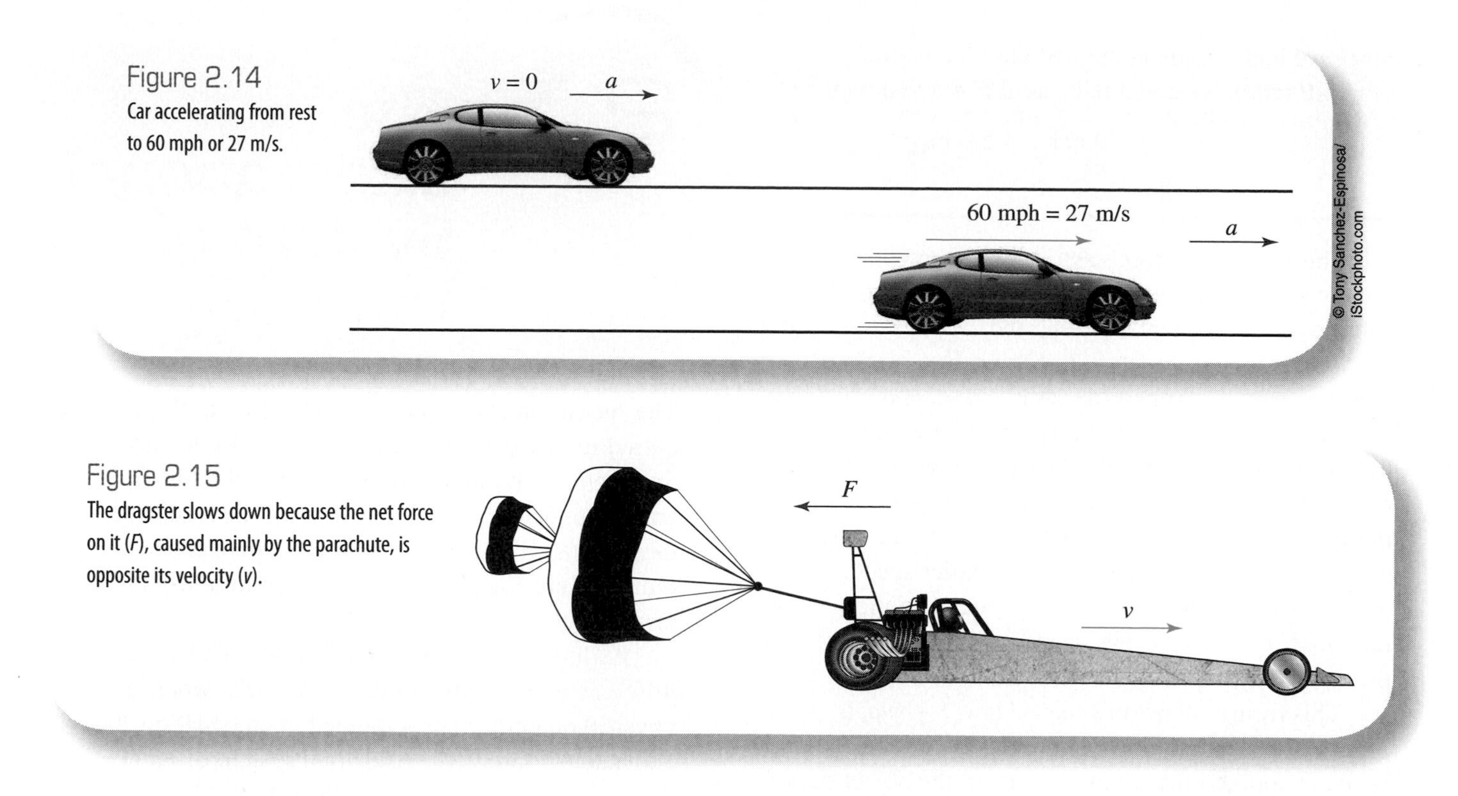

Figure 2.14
Car accelerating from rest to 60 mph or 27 m/s.

Figure 2.15
The dragster slows down because the net force on it (F), caused mainly by the parachute, is opposite its velocity (v).

So the force needed to cause this acceleration is

$$F = ma = 1{,}000 \text{ kg} \times 2.7 \text{ m/s}^2$$
$$= 2{,}700 \text{ N}$$

(This is equal to 2,700 × 0.225 lb = 607.5 lb.) In the next chapter, we will use this information to determine the size (power output) of the car's engine.

Newton's second law establishes the relationship between mass and weight. Recall that any freely falling body has an acceleration equal to g (9.8 m/s^2). By the second law, the size of the gravitational force needed to cause this acceleration is

$$F = ma = mg$$

We call this force the object's weight. So in this case the expression "force is equal to mass times acceleration" translates to "weight is equal to mass times acceleration due to gravity."

$$F = ma \rightarrow W = mg$$

On Earth, where the acceleration due to gravity is 9.8 m/s^2, the weight of an object is

$$W = m \times 9.8 \qquad \text{(on Earth, } m \text{ in kg, } W \text{ in N)}$$

The weight of a 2-kilogram brick is

$$W = 2 \text{ kg} \times 9.8 \text{ m/s}^2 = 19.6 \text{ N}$$

On the Moon, the acceleration due to gravity is 1.6 m/s^2. Therefore,

$$W = m \times 1.6 \qquad \text{(on the Moon, } m \text{ in kg, } W \text{ in N)}$$

Forces that cause deceleration or centripetal acceleration also follow the second law. To slow down a moving object, a force must act in the direction opposite the object's velocity (see Figure 2.15). A force acting sideways on a moving body causes a centripetal acceleration. The object moves along a circular path as long as the net force remains perpendicular to the velocity. The size of the centripetal force required is

$$F = ma$$

and since

$$a = \frac{v^2}{r}$$

$$F = \frac{mv^2}{r} \qquad \text{(centripetal force)}$$

EXAMPLE 2.3

In Example 1.5, we computed the centripetal acceleration of a car going 10 m/s around a curve with a radius of 20 meters. If the car's mass is 1,000 kilograms, what is the centripetal force that acts on it?

$$F = \frac{mv^2}{r} = \frac{1{,}000 \text{ kg} \times (10 \text{ m/s})^2}{20 \text{ m}}$$
$$= 5{,}000 \text{ N} = 1{,}120 \text{ lb}$$

Since we had already computed the acceleration ($a = 5.0 \text{ m/s}^2$), we could have used $F = ma$ directly.

$$F = ma = 1{,}000 \text{ kg} \times 5.0 \text{ m/s}^2$$
$$= 5{,}000 \text{ N}$$

The centripetal force acting on a car going around a flat curve is supplied by the friction between the tires and the road. Note that the faster the car goes, the greater the force required to keep it moving in a circle. If two identical cars go around the same curve, and one is going two times as fast as the other, the faster car needs four times the centripetal force (see Figure 2.16). The required centripetal force is inversely proportional to the radius of the curve. A tighter curve (one with a smaller radius) requires a larger force or a smaller speed. If a car goes around a curve too fast, the force of friction will be too small to maintain the needed centripetal force, and the car will go off the outside of the curve.

This form of Newton's second law, $F = ma$, is not the original version, but it is the most useful one for our purposes. It applies only when the mass of the object doesn't change. This is almost always the case, but there are important exceptions. For example, the mass of a rocket decreases rapidly as it consumes its fuel. Consequently, its acceleration will increase even if the net force on the rocket doesn't change. Mathematics more advanced than we will use is needed to deal with such cases.

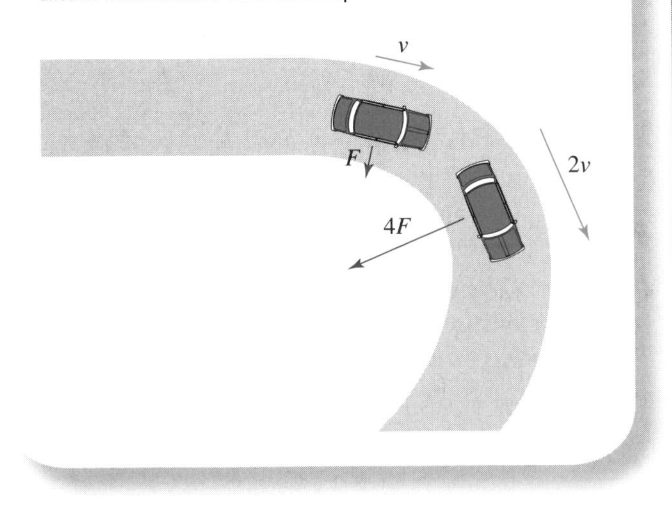

Figure 2.16
The centripetal force necessary to keep a car on a curved road is supplied by static friction between the tires and the road. The size of the force is proportional to the square of the speed. One car going twice as fast as another would need four times the centripetal force.

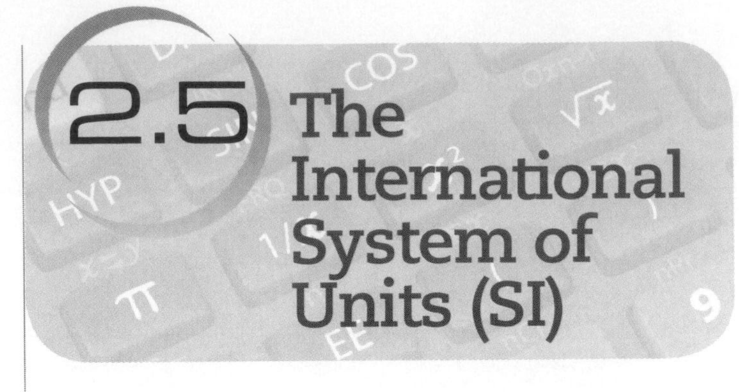

2.5 The International System of Units (SI)

The metric unit of force—the newton—is the force required to cause a 1-kilogram mass to accelerate 1 m/s². So when a mass in kilograms is multiplied by an acceleration in meters per second squared, the result is a force in newtons. If grams or centimeters per second squared (cm/s²) were used instead, the force unit would not be newtons.

At this point, you may be getting confused by the different units of measure, particularly when different physical quantities are combined in an equation. To help alleviate this problem, a separate system of units within the metric system was established. It is called the **international system**, or **SI**, after the French phrase *Système International d'Unités.* This system associates only one unit of measure with each physical quantity (Table 2.1). Each unit is chosen so that the system is internally consistent, that is, when two or more physical quantities are combined in an equation, the result will be in SI units if the original physical quantities are in SI units. For

Table 2.1 SI Units (Partial List)

Physical Quantity	SI Unit
Distance (d)	meter (m)
Area (A)	square meter (m²)
Volume (V)	cubic meter (m³)
Time (t)	second (s)
Frequency (f)	hertz (Hz)
Speed and velocity (v)	meter per second (m/s)
Acceleration (a)	meter per second squared (m/s²)
Force (F) and weight (W)	newton (N)
Mass (m)	kilogram (kg)

example, the SI units of distance and time are the meter and the second, respectively. Consequently, the SI unit of speed is the meter per second. Table 2.1 gives the SI units of the physical quantities that we have used so far.

You may notice that, for the most part, we have been using SI units in our examples. From now on, we will use SI units exclusively when working in the metric system.

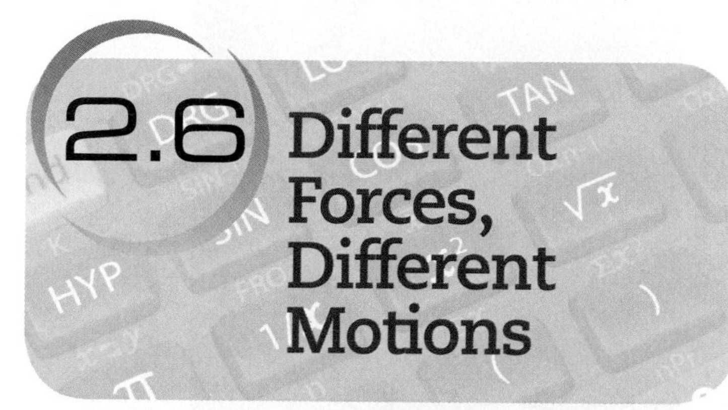

2.6 Different Forces, Different Motions

In Chapter 1, we considered three simple states of motion: zero velocity, constant velocity, and constant acceleration. Now, let's see how force is involved in these cases and then look at some other types of motion. Here the particular cause of the force is not important. A constant net force acting on a moving object has the same effect whether it is due to gravity, friction, or someone's pushing the object. What interests us now is the relationship between the force—both its magnitude and direction—and the motion. Our concerns are the direction of the force compared to the object's velocity, whether the force is constant, and the way in which the force varies if it is not constant.

The acceleration of a stationary object or of one moving with uniform motion (constant velocity) is zero. By the second law, this means that the net force is also zero (see Figure 2.17). It is that simple. When the net force on a body is zero, we say that it is in *equilibrium*, whether it is stationary or moving with constant velocity.

Figure 2.17
Whether a boat is moving with constant velocity or sitting at rest, the net force in both cases is zero.

In uniform acceleration, the second law implies that a constant force acts to cause the constant acceleration. A steady net force acting in a fixed direction will make an object move with a constant acceleration. A freely falling body is a good example: The force (weight) is constant and always acts in a downward direction.

A net force that acts opposite to the direction of motion of an object will cause it to slow down. If the force continues to act, the object will come to a stop at an instant and then accelerate in the direction of the force, opposite its original direction of motion. This is what happens when you throw a ball straight up into the air (see Figure 2.18). It is accelerating downward the whole time.

Figure 2.18
When a ball is thrown straight up, its acceleration is 1 g downward. This causes it to slow down on its way upward, stop at an instant, and then speed up on its way downward.

© Geoff Kuchera/iStockphoto.com

Figure 2.19

(a) The path of a projectile is shown with its velocity vector at selected moments. At the very top its velocity is horizontal. (b) The same projectile with the vertical and horizontal components of the velocity shown separately. The horizontal velocity component stays constant, while the vertical component decreases, and then increases downward, because of the constant downward acceleration due to gravity. (c) A strobe view of three "projectiles" being juggled.

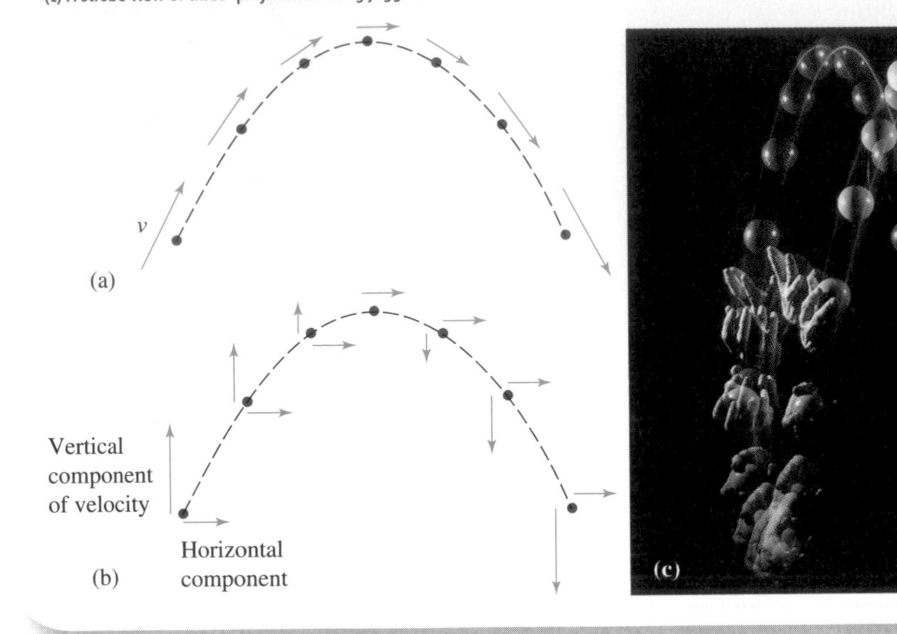

© Richard Megna/Fundamental Photographs

While on the way up, its speed decreases until it reaches its highest point. There it has zero speed at an instant and then it falls with increasing speed. Its acceleration is always g, even at the instant that it is stopped.

What happens when an object is thrown upward at an angle to the Earth's surface? This is one example of a classic mechanics problem—projectile motion. The motion is a composite of horizontal and vertical motions. The path that a projectile takes has a characteristic arc shape that is nicely illustrated by the stream of water from a water fountain. This shape, an important one in mathematics, is called a *parabola.*

The key to understanding projectile motion is to realize that the vertical force of gravity has no effect on the horizontal motion. An object initially moving horizontally has the same downward acceleration g as an object that is simply dropped. So, ignoring air resistance, a projectile moves horizontally with constant speed and vertically with constant downward acceleration (Figure 2.19). As an object moves along its path, the vertical component of its velocity decreases to zero (at the highest point of the arc) and then increases downward. The horizontal component of its velocity stays constant. Over level ground, a projectile will travel farthest if it starts at an angle of 45° to the ground. When the force of air resistance is large enough to affect the motion, such as when a softball is thrown very far, the maximum range occurs at smaller angles.

Simple Harmonic Motion

Here is a simple situation with a force that is not constant. Imagine an object, say, a block, attached to the end of a spring (Figure 2.20). When it is not moving, the block is in equilibrium, and the net force on it is zero. If you lift the block up a bit and then release it, it will experience a net force downward, toward its original (rest) position. The reverse occurs if you pull it down a bit below its rest position and release it. Then the net force on it will be up, again toward its original position. Such a force is called a *restoring force* because it acts to restore the system to the original configuration. In this case, the net force is proportional to the displacement from the rest position. The farther the block is displaced, the

The stream of water from a water fountain shows the shape of the path of projectiles—a parabola.

© John Frink/iStockphoto.com

Figure 2.20

The net force that acts on a block hanging from a spring depends on the displacement of the block from its rest (equilibrium) position (a). If raised and then released, the block experiences a net force downward (b). If the block is pulled downward and then released (c), the net force is upward.

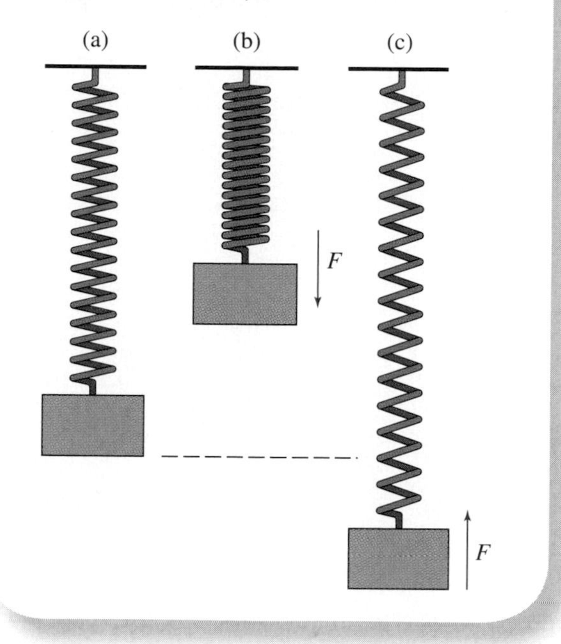

Figure 2.21

Graphs of *d* versus *t* and *v* versus *t* for simple harmonic motion. Notice that the velocity is zero when the distance is largest, and vice versa. At the high and low points of the motion, the mass is not moving at an instant. As it passes through the rest position, it has its highest speed.

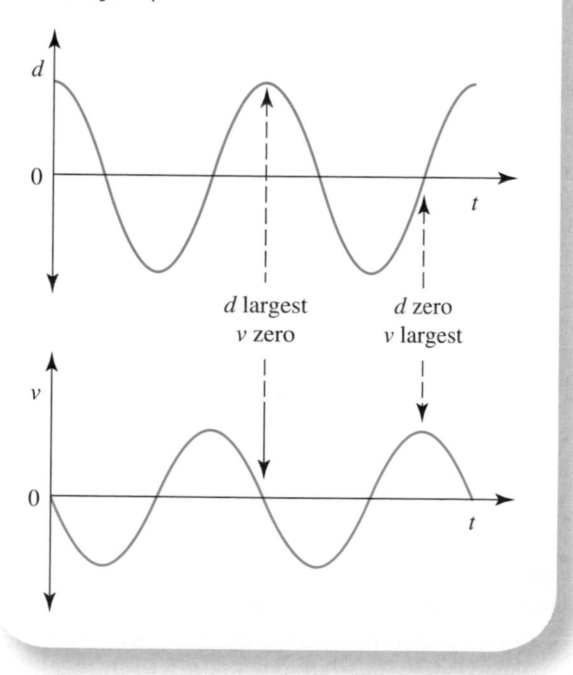

greater the force acting to move it back. (Figure 2.2 also shows this.)

What kind of motion does this force cause? If an object is pulled down and then released, the upward force will cause it to accelerate. It will pick up speed as it moves upward, but the force and acceleration will decrease as it nears the rest position. When it reaches that point, the force is zero, and it stops accelerating but continues its upward motion (Newton's first law). Once it has moved past the rest position, the force is downward. The object will slow down, stop at an instant, and then gain speed downward. This process is repeated over and over: The object oscillates up and down.

This type of motion, which is very important in physics, is called **simple harmonic motion**. It occurs in many other systems—a pendulum swinging through a small angle, a cork bobbing up and down in water, a car with very bad shock absorbers, and air molecules vibrating with the sound from a tuning fork. In fact, simple harmonic motion is involved in all kinds of waves, as we shall see.

The graphs of distance versus time and velocity versus time show the characteristic oscillation (Figure 2.21). Both graphs have what is known as a *sinusoidal* shape. This shape arises often in physics, and we will see it again when we talk about waves (Chapter 6).

Simple harmonic motion is cyclical motion with a constant frequency. In our example, the frequency of oscillation depends only on the mass of the object and the strength of the spring.* The same mass on a stronger spring will have a higher frequency. For a given spring, a larger mass will have a smaller (lower) frequency.

Falling Body with Air Resistance

The force of air resistance that acts on things moving through the air, like a thrown baseball or a falling skydiver, is one example of kinetic friction. This force is in the opposite direction of the object's velocity and

* For the mathematically inclined—the strength of a spring is found by measuring the force F needed to stretch it a distance d. Then one computes:

$$k = \frac{F}{d}$$

where k is called the *spring constant*. A strong (stiff) spring has a large k, and a weak spring has a small k. A mass m attached to the spring would oscillate with a frequency given by:

$$f = \frac{1}{2\pi} \times \sqrt{\frac{k}{m}} = 0.159 \times \sqrt{\frac{k}{m}}$$

Figure 2.22

Successive views of a falling body affected by air resistance (ar). The upward force of air resistance increases as the object's speed increases. When this force is large enough to offset the downward weight (W), the net force (F_{net}) is zero (far right). The object's speed is constant and is called the terminal speed (v_t).

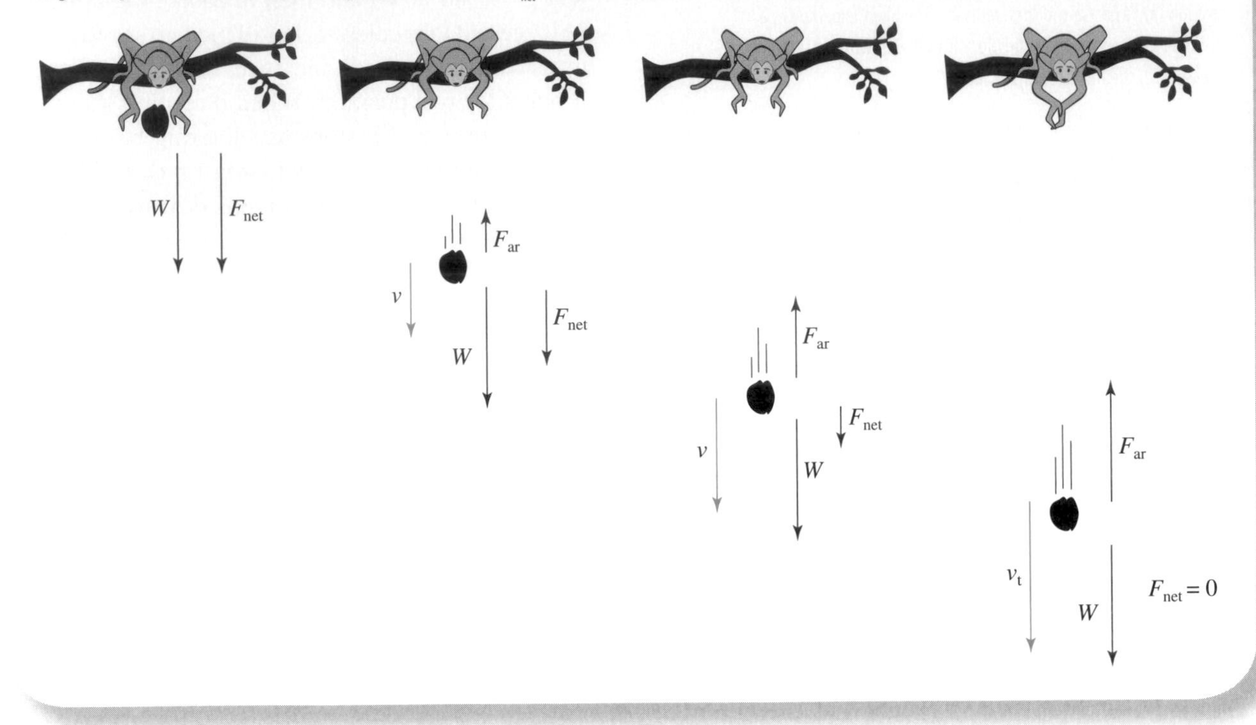

will cause the object to slow down if no other force opposes it. The faster an object goes, the larger the force of air resistance. There is no simple equation for the size of the force of air resistance. For things like baseballs, bicyclists, cars, and aircraft, the force of air resistance is approximately proportional to the square of the object's speed relative to the air. For example, on a day with no wind the force of air resistance on a car going 60 mph is about four times as large as when it is going 30 mph.

Figure 2.23

Graph of speed versus time for a falling body with air resistance. At first the speed increases rapidly as in free fall (compare to Fig. 1.19b). But the increasing force of air resistance gradually reduces the acceleration to zero, at which point the speed is constant—the terminal speed v_t.

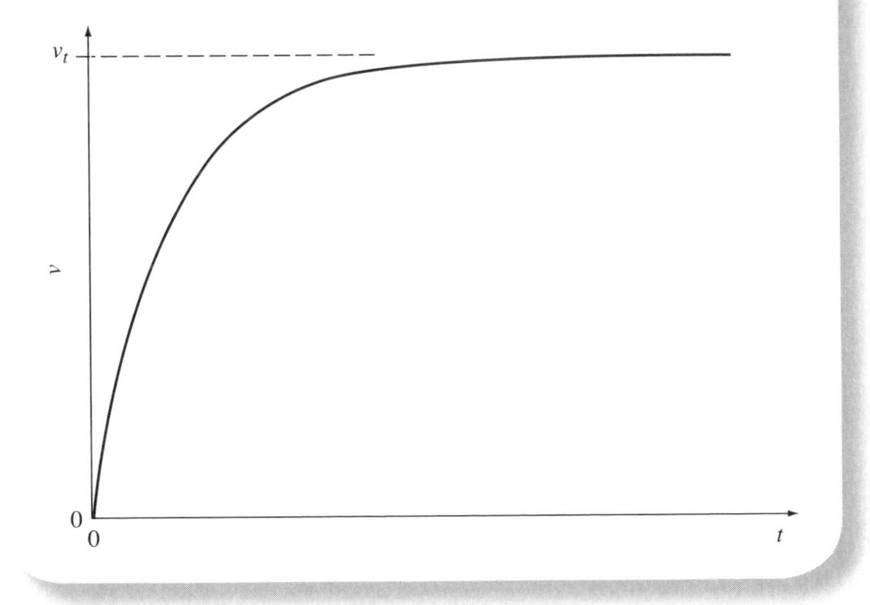

Without air resistance the constant force on a falling body (its weight) gives it a constant acceleration g. Its speed increases steadily until it hits the ground. But as a body falls through the air, the force of air resistance grows as the speed increases and eventually affects the motion (Figure 2.22). This increasing force acts opposite to the downward force of gravity, so the net force decreases. This continues as the body gains speed until the force of air resistance acting upward equals the weight acting downward. At this point, the net force is zero, so the speed stays constant from then on. This speed is called the *terminal speed* of the body. Figure 2.23 shows a graph of speed versus time for this motion.

Rocks and other dense objects have large terminal speeds and may fall for many seconds before air resistance affects their motion appreciably. Feathers, dandelion seeds, and balloons take less than a second to reach their terminal speeds, which are quite low. If a skydiver jumps out of a hovering helicopter or balloon, he or she will fall for about 2 or 3 s before the force of air resistance starts to have a major effect. The terminal speed depends on the skydiver's size and orientation when falling, but it is typically around 120 mph (about 54 m/s).* Then, when the parachute opens, the increased air resistance slows the skydiver to a much lower terminal speed—maybe 10 mph.

In the spring-and-block example in Figure 2.20, the net force depends on the object's distance from its equilibrium position. When air resistance affects the motion of a falling body, the net force depends on the body's speed. These examples illustrate how interconnected the different physical quantities are in real physical systems. Things can get quite complicated: Imagine a spring and mass suspended under water. Here the net force would depend on both the block's position and its speed. See if you can describe how the block would move. Table 2.2 summarizes the types of forces that we have considered thus far.

In all of the examples in this section, the position of the object can be predicted for any instant of time in the future through the use of Newton's laws of motion and appropriate mathematical techniques—provided we have accurate information about the initial position and velocity of the object and the forces that affect its motion. This ability to predict the position of the object is the main reason Newton's laws are so important. It allows us to send spacecraft to planets billions of miles away and to predict what a roller coaster will do before it is built. But discoveries in the last century show that we can't always predict the future configurations of systems. In Chapter 10, we describe how the science of quantum mechanics, the essential tool for dealing with systems on the scale of atoms or smaller, tells us that such arbitrarily accurate predictions cannot be made because it is impossible to know precisely both the position and velocity of particles such as electrons. The study of chaos has revealed that even in some relatively simple systems there is inherent randomness: The future configuration at some instant in time cannot be predicted no matter how accurately we know the forces, initial positions, and initial velocities. However, the mechanics based on Newton's laws remain one of the most valuable tools for applying physics in the world around us.

2.7 Newton's Third Law of Motion

Newton's third law of motion is a statement about the nature of forces in general. It is a simple law that adds an important perspective to the understanding of forces.

LAWS

Newton's Third Law of Motion *Forces always come in pairs: When one object exerts a force on a second object, the second exerts an equal and oppositely directed force on the first.*

Table 2.2 Summary of Examples of Forces

Nature of Net Force	Description of Motion
Zero net force	Constant velocity: stationary or motion in straight line with constant speed
Constant net force	Constant acceleration
Force parallel to velocity	Motion in a straight line with increasing speed
Force opposite to velocity	Motion in a straight line with decreasing speed
Force perpendicular to velocity	Motion in a circle: radius depends on speed and force
Restoring force proportional to displacement	Simple harmonic motion (oscillation)
Net force decreases as speed increases	Acceleration decreases: velocity reaches a constant value

* Terminal speed also depends on how thin the air is. In 1960, Captain Joseph Kittinger jumped from a balloon at 100,000 feet and may have fallen faster than the speed of sound! As he descended into denser air he slowed down—even before he opened his parachute.

If object A causes a force on object B, object B exerts an equal force in the opposite direction on A.

$$F_{\text{B on A}} = -F_{\text{A on B}}$$

If you push against a wall with your hand, the wall exerts an equal force back on your hand (see Figure 2.24). A book resting on a table exerts a downward force on the table equal to the weight of the book. The table exerts an upward force on the book also equal to the book's weight. The Earth pulls down on you with a force that is called your weight. Consequently, you exert an equal but upward force on the Earth. When you dive off a diving board, the Earth's gravitational force accelerates you downward. At the same time, your equal and opposite pull upward on the Earth accelerates it toward you. But the Earth's mass is about one hundred thousand billion billion (10^{23}) times your mass, so its acceleration is negligible. Yours, by contrast, is easily perceptible.

Figure 2.24

If one body exerts a force on a second body, the second exerts an equal and opposite force on the first.

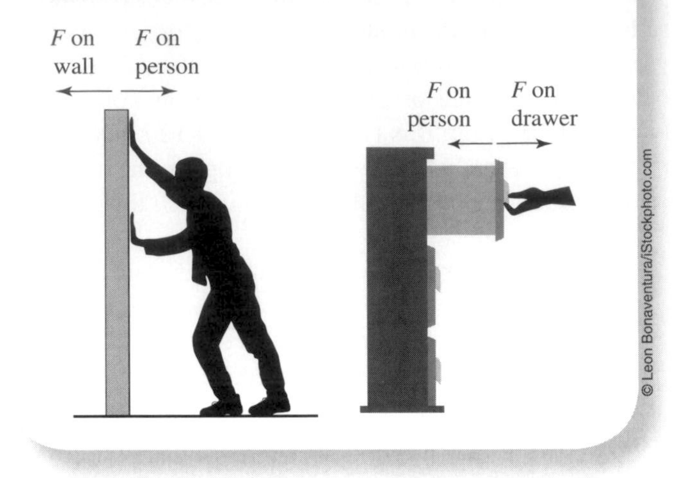

The third law gives a new insight into what actually happens in many physical systems. If you are on inline skates and push hard against a wall, you accelerate backward (see Figure 2.25). Think about this for a moment: A *forward* force by your foot makes you accelerate in the *opposite direction.* In reality, the equal and opposite force of the wall on your foot causes the acceleration. If you stand in the middle of the roller rink, you can't use your foot to move yourself because you need to push against something. Similarly, when a car speeds up, the engine causes the tires to push backward on the road. It is the road's equal and opposite force on the tires that causes the car to accelerate forward. The same thing happens when the brakes slow the vehicle. The recoil of a gun is caused by the third law: The large force accelerating the bullet produces an equal and opposite force on the gun—the "kick."

Figure 2.26 illustrates Newton's third law. A small cart is fitted with a spring-loaded plunger that can be pushed into

Figure 2.25

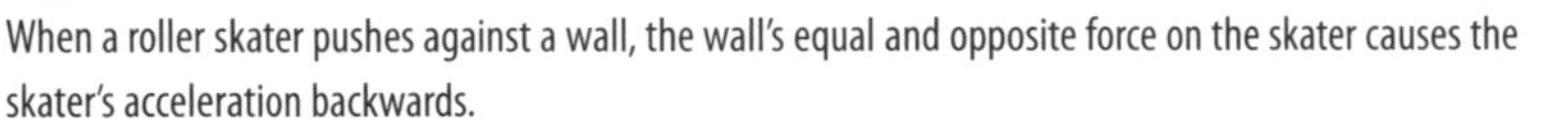

When a roller skater pushes against a wall, the wall's equal and opposite force on the skater causes the skater's acceleration backwards.

Figure 2.26

A small cart is fitted with a spring-loaded plunger and trigger. The cart is accelerated only if the plunger can exert a force on something else (b and c); then the equal and opposite force on the cart causes the acceleration.

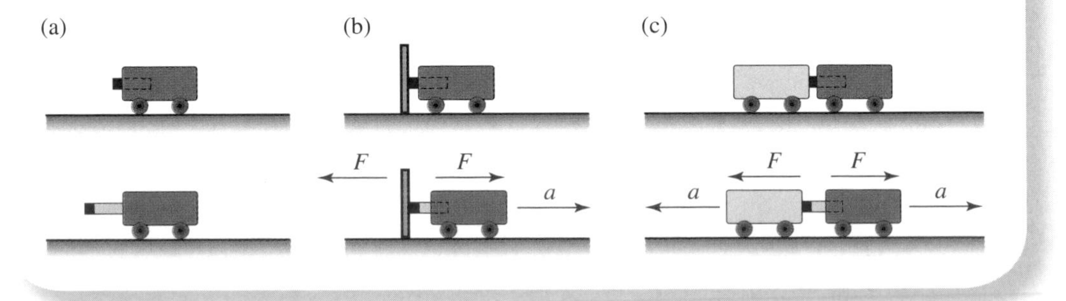

Figure 2.27

(a) Author Vern Ostdiek having fun with Newton's third law. (b) A wing on a flying aircraft deflects the air downward. This downward force on the air causes an equal and opposite upward force on the wing.

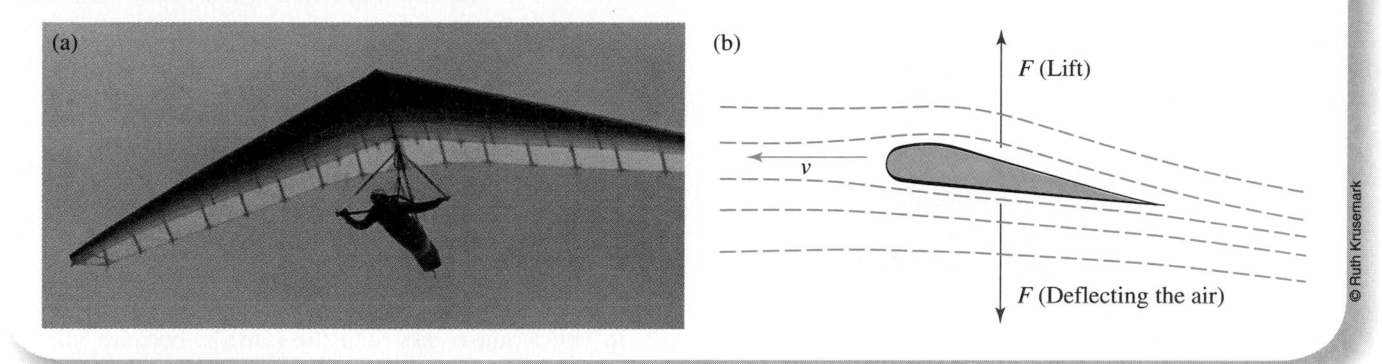

the cart and retained. When the spring is released, the plunger moves out. If there is nothing for the plunger to push against (Figure 2.26a), the cart does not move afterwards. The plunger can exert no force, so there is no equal and opposite force to accelerate the cart. If the cart is next to a wall when the spring is released (Figure 2.26b), the force on the wall results in an opposite force on the cart, and the cart accelerates away from the wall. If a second cart is next to the plunger (Figure 2.26c), both carts feel the same force in opposite directions, and they accelerate away from each other. In Chapter 3, we will show that the ratio of the speeds of the carts depends on the ratio of their masses. If one has twice the mass of the other, it will have half the speed.

Rockets and jet aircraft are propelled by ejecting combustion gases at a high speed. The engine exerts a force that ejects the gases, and the gases exert an equal and opposite force on the rocket or jet. Unlike cars, they do not need to push against anything to be propelled.

Birds, airplanes, and gliders can fly because of the upward force exerted on their wings as they move through the air (Figure 2.27a). As the air flows under and over the wing, it is deflected (forced) downward. This results in an equal and opposite upward force on the wing, called *lift* (see Figure 2.27b). Propellers on airplanes, helicopters, boats, and ships employ the same basic idea. They pull or push the air or water in one direction, resulting in a force on the propeller in the opposite direction. Propellers are useless in a vacuum.

Whenever an object is accelerating, it exerts an equal and opposite force on whatever is accelerating it. You have probably noticed this when riding in a car, bus, or airplane as it accelerates: The seat back exerts a forward force on you that causes you to accelerate. Your body pushes back on the seat with an equal and opposite force. It seems that there is some force "pulling" you back against the seat. This is not a real force, just a reaction of your mass to acceleration. The same effect is observed when an object is undergoing a centripetal acceleration. As a car or bus goes around a curve, you seem to be "pulled" to the side. Again, it is just a reaction to a net force causing an acceleration.

Like Newton's first law, the third law is mainly conceptual rather than mathematical. It emphasizes that forces arise only during interactions between two or more things. By thinking in terms of pairs of forces, we can more easily distinguish the causes and effects in such interactions.

2.8 The Law of Universal Gravitation

Newton's fourth major contribution to the study of mechanics is not a law of motion but a law relating to gravity. Newton made an important intellectual leap: He realized that the force that pulls objects toward the Earth's surface also holds the Moon in its orbit. Moreover, he claimed that every object exerts an attractive force on every other object. This concept is called *universal gravitation:* Gravity acts everywhere and on all things. The force that the Earth exerts on objects near its surface—weight—is just one example of universal gravitation.

What determines the size of the gravitational force that acts between two bodies? Newton used his deep understanding of mathematics and mechanics along with information about the orbits of the Moon and the planets to reason what the force law must be. The third law of motion states that when two bodies exert forces on one another, the forces are equal in size. Since the Earth's gravitational force depends on an object's mass, Newton reasoned, the force on each object in a pair is proportional to each object's mass (see Figure 2.28). If the mass of either object is doubled, the sizes of the forces on both are doubled.

By thinking in terms of pairs of forces, we can more easily distinguish the causes and effects in such interactions.

The size of the gravitational force between two bodies must also depend on the distance between them. The Sun is much more massive than the Earth, but its force on you is much less than the Earth's, because you are much closer to the Earth. For geometric reasons, Newton felt that the size of the gravitational force is inversely proportional to the square of the distance between the bodies. But he needed proof—proof that he got from examining the Moon's orbit.

Long before Newton's time, astronomers had measured the radius of the Moon's orbit and, knowing this and the period of its motion, had determined its orbital speed. So Newton could calculate the Moon's (centripetal) acceleration, v^2/r, which turned out to be about $g/3{,}600$. In other words, the Moon's acceleration is about 3,600 times smaller than that of

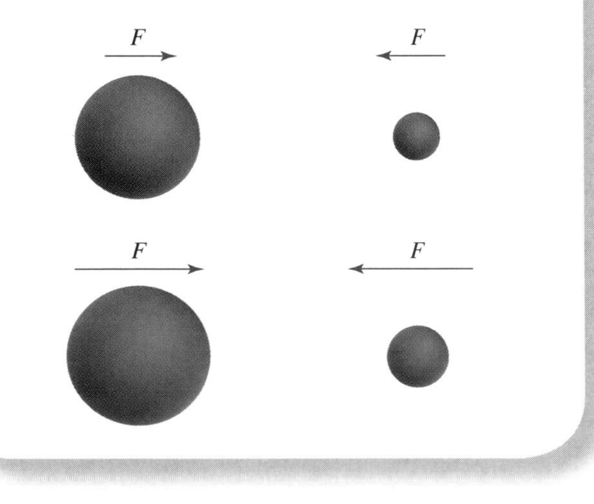

Figure 2.28
The equal and opposite gravitational forces that act on any pair of objects depend on the masses of both objects. If either object is replaced by another with a larger mass, the two forces are proportionally larger.

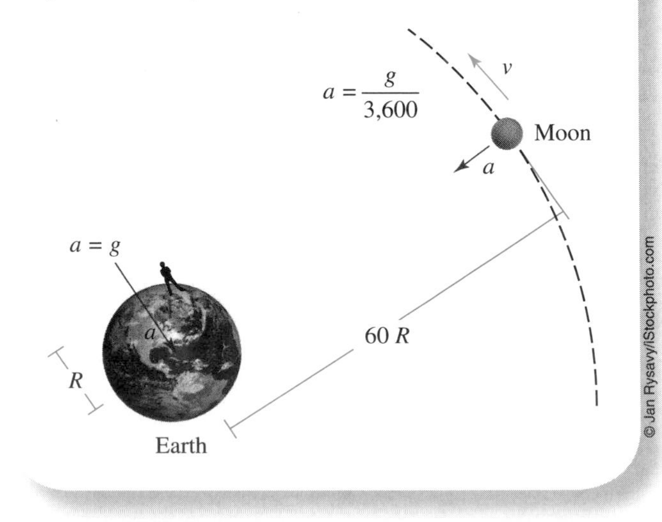

Figure 2.29
The gravitational force that the Earth exerts on other objects is inversely proportional to the square of the distance from the Earth's center to the object's center. That is why the Moon's centripetal acceleration is much less than g. (Not drawn to scale.)

an object falling freely near the Earth. Newton also realized that the Moon is about 60 times farther from the Earth's center than an object on the Earth's surface. The acceleration of the Moon is 60 squared, or 3,600, times smaller because it is 60 times as far away (see Figure 2.29). Since acceleration is proportional to force, Newton had his proof that the gravitational force is inversely proportional to the square of the distance between the centers of the two objects.

LAWS

Newton's Law of Universal Gravitation *Every object exerts a gravitational pull on every other object. The force is proportional to the masses of both objects and inversely proportional to the square of the distance between their centers.*

$$F \propto \frac{m_1 m_2}{d^2}$$

where m_1 *and* m_2 *are the masses of the two objects, and* d *is the distance between their centers.*

This force acts between the Earth and the Moon, the Earth and you, two rocks in a desert—any pair of objects. Your weight depends on the distance between you and the Earth's center. If you were twice as far from the Earth's center (8,000 miles instead of 4,000 miles), you would weigh one-fourth as much (see Figure 2.30). If you were three times as far, you would weigh one-ninth as much.

In 1798, the English physicist Henry Cavendish performed precise measurements of the actual gravitational forces acting between masses. Cavendish used a delicate torsion balance (Figure 2.31). Two masses were balanced on a beam that was suspended from a thin wire attached to the beam's midpoint. Two large masses were placed next to the smaller ones on the beam. The gravitational forces on the smaller masses were strong enough to rotate the beam slightly and twist the wire. Cavendish used the amount of twist to measure the size of the gravitational force. His results showed that the force between two 1-kilogram masses 1 meter apart would be 6.67×10^{-11} newtons. With this result, Newton's law of universal gravitation can be expressed as an equation:

$$F = \frac{6.67 \times 10^{-11}\, m_1 m_2}{d^2} \quad \text{(SI units)}$$

The constant of proportionality in the equation is called the gravitational constant G.

$$G = 6.67 \times 10^{-11}\ \text{N-m}^2/\text{kg}^2$$

Since G is such a small number, the gravitational force between objects is usually quite small. For example, the force between two persons, each with mass equal to 70 kilograms, when they are 1 m apart is only 0.00000033 newtons, or 0.0000012 ounces. The Earth's gravitational force on us (and our force on the Earth) is so large because the Earth's mass is huge.

In fact, we can use the law of universal gravitation to compute the mass of the entire Earth. First compute the gravitational force exerted on a body of mass m by the Earth—the body's weight—using the equation $W = mg$. This force can also be calculated using:

$$F = \frac{GmM}{R^2}$$

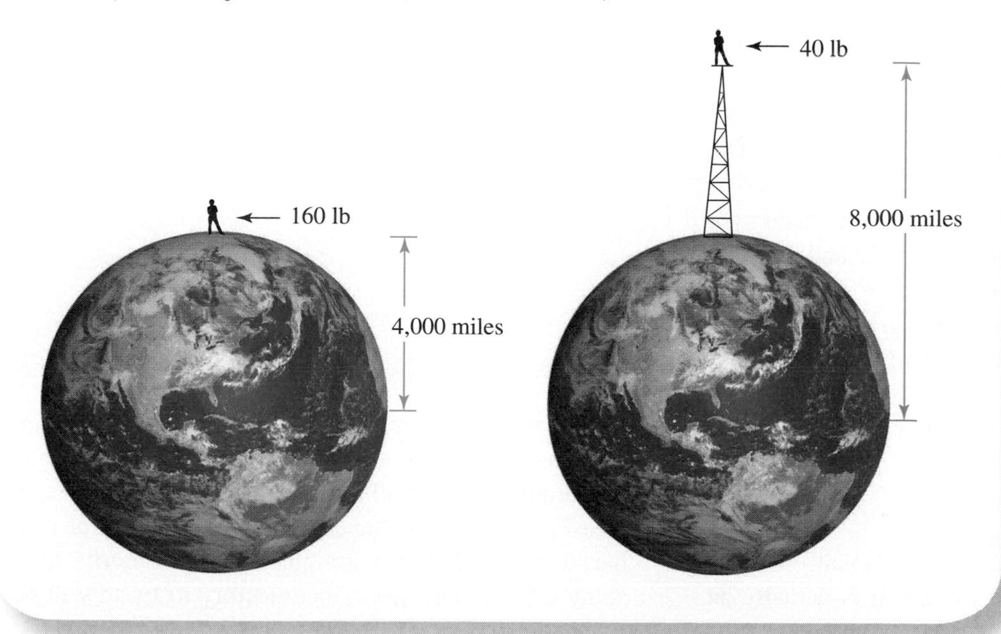

Figure 2.30
On a tower 4,000 miles high, you would be twice as far from the Earth's center as you are when standing on its surface. On the tower, you would weigh one-fourth as much. (Person not drawn to scale.)

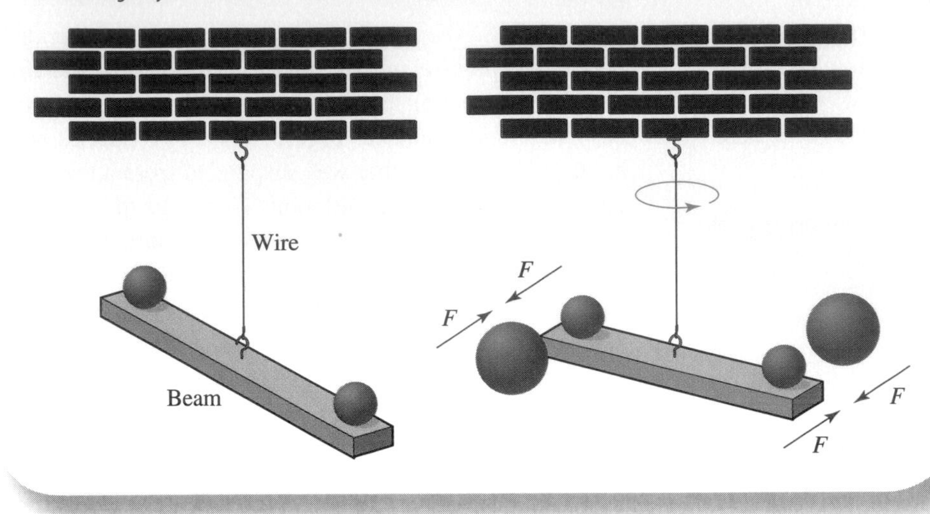

Figure 2.31
Simplified sketch of the torsion balance used by Cavendish to measure the gravitational force between masses. The two large objects exert forces on the two small ones, causing the suspending wire to twist.

Table 2.3 Acceleration due to Gravity in our Solar System

Location	Acceleration due to Gravity
Earth	1.0 *g*
Sun	27.9 *g*
Moon	0.16 *g*
Mercury	0.38 *g*
Venus	0.88 *g*
Mars	0.39 *g*
Jupiter	2.65 *g*

Here M represents the mass of the Earth, and R is the radius of the Earth, the distance the body is from the Earth's center. The value of R, first measured by the Greeks over 2,000 years ago, is about 6.4×10^6 meters. Since these two forces are equal to each other:

$$W = F$$

$$mg = \frac{GmM}{R^2}$$

Canceling the m:

$$g = \frac{GM}{R^2}$$

The values of g, G, and R can be inserted and the resulting equation solved for M, the mass of the Earth. The result: $M = 6 \times 10^{24}$ kg.

The acceleration due to gravity on the Moon is not the same as it is on the Earth. Likewise, each of the other planets has its own "g." This variation exists simply because the masses and the radii are all different. The values of the acceleration due to gravity have been computed for the Moon, the Sun, and the planets, using the preceding equation and each one's mass M and radius R (see Table 2.3 for some of these).

Newton used his law of universal gravitation to explain a variety of phenomena that had been mysteries before. These included the theoretical basis for orbital motion, the motion of comets, and the cause of tides.

Orbits

Newton used an elegant "thought experiment" to illustrate that orbital motion about the Earth is actually an extension of projectile motion. Imagine that a cannon is placed at the top of a very high mountain and that it can shoot a cannonball horizontally at any desired speed (see Figure 2.32). If the cannonball just rolls out of the barrel, it will fall in a straight line to the Earth. Given some small initial speed, its trajectory to Earth will be a

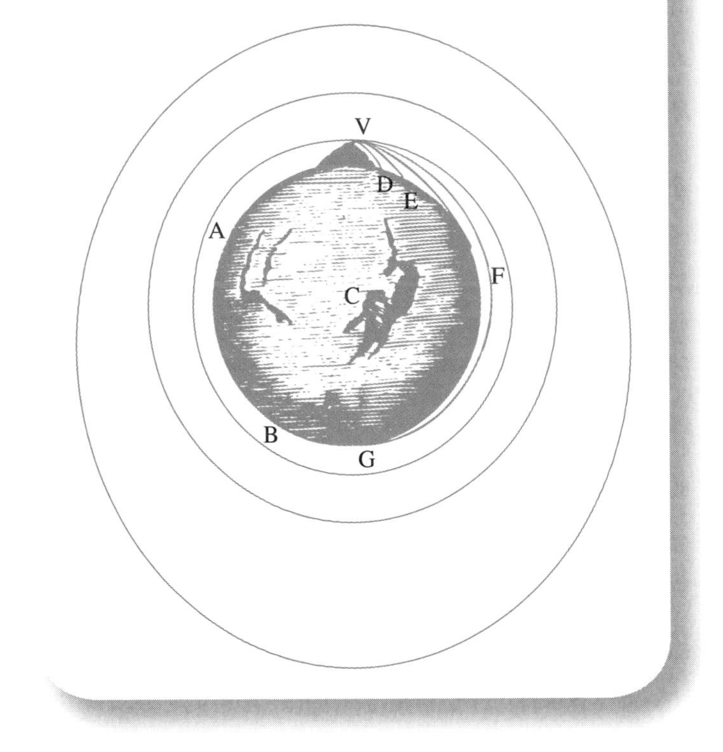

Figure 2.32
A re-creation of the original drawing of Newton's cannon "thought experiment." The cannon is placed on a very high mountaintop (V). The path of the cannonball depends on its initial speed.

parabola (path D in Figure 2.32). But if its speed is increased more, the ball travels farther and farther around the Earth before it hits the ground (paths E, F, and G). If it were possible to shoot a cannonball with a high enough speed, it would travel in a full circle around the Earth and hit the cannon in the rear. It would be in orbit about the Earth, just as the Moon is. An object in orbit is continually "falling" toward the Earth. On an even higher mountain, one could place the cannonball in an orbit with a larger radius.

Realistically, this could not be done because of the force of air resistance. To orbit the Earth, an object must be above (outside) the Earth's atmosphere. But the idea is a handy way of showing the connection between a terrestrial phenomenon—projectile motion—and the motion of the Moon about the Earth. It also shows what Newton predicted—that we can put satellites into orbit.

It is quite easy to estimate how fast something has to move to stay in orbit. When an object is moving in a circle around the Earth, the centripetal force on it is the Earth's gravitational pull. For a satellite orbiting just above the Earth's surface, the gravitational force is approximately equal to the satellite's weight when it is on the Earth's surface—mg. Also, the radius of the satellite's orbit is approximately equal to the radius of the Earth, R (see Figure 2.33). The required speed is found by equating the centripetal force to the gravitational force. The centripetal force is mv^2/R, where R is the radius of the Earth. So

$$\frac{mv^2}{R} = mg$$
$$v^2 = gR = 9.8 \text{ m/s}^2 \times (6.4 \times 10^6 \text{ m})$$
$$= 63{,}000{,}000 \text{ m}^2/\text{s}^2$$
$$v = 7{,}900 \text{ m/s}$$

Newton's mathematical analysis of orbital motion is not restricted to Earth orbits. It also applies to the motions of the planets around the Sun and to the moons in orbit about the other planets. His results allowed astronomers to calculate the orbits of celestial objects with higher accuracy and with fewer observational data than before. He showed that comets, such as Halley's Comet, are moving in orbits around the Sun. The orbits of most of them are flattened ellipses with the Sun near one end, at a point called the *focus* of the ellipse (Figure 2.34). This is why Halley's Comet is only near enough to the Earth to be seen every 76 years. It spends most of its time far from the Sun and the Earth.

The orbits of all of the planets, Earth included, are actually ellipses, although they are much closer to being circular than the orbit of Halley's Comet. The Sun is at one focus of these orbits. Note in Figure 2.34 that the elliptical nature of the orbit of the minor (or dwarf) planet Pluto causes it to spend part of the time inside the orbit of Neptune.

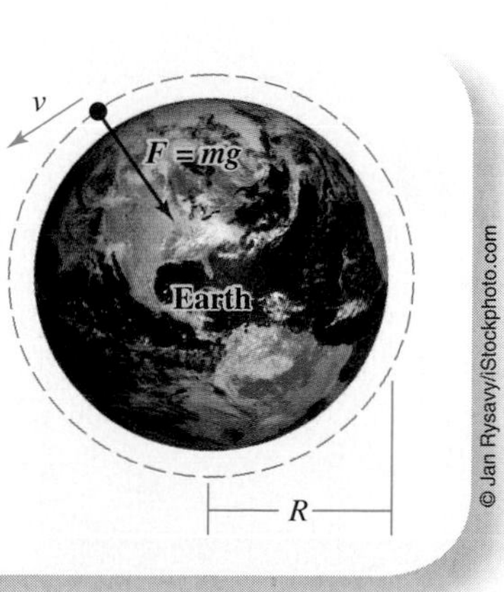

Figure 2.33
Satellite in a low, circular orbit about the Earth. The gravitational (centripetal) force on the satellite is about equal to the satellite's weight when on the Earth, *mg*, and the radius of the orbit is about equal to the Earth's radius, *R*.

Gravitational Field

Gravitation is an example of action at a distance. Objects exert forces on each other, even though they may be far apart and there is no matter between them to transmit the forces. The other forces we have talked about involve direct contact between things. Gravitation does not.

Figure 2.34
Orbit of Halley's Comet around the Sun. The comet is seen from Earth only when it is in that part of its orbit near the Sun. The Earth's orbit is the smallest circle.

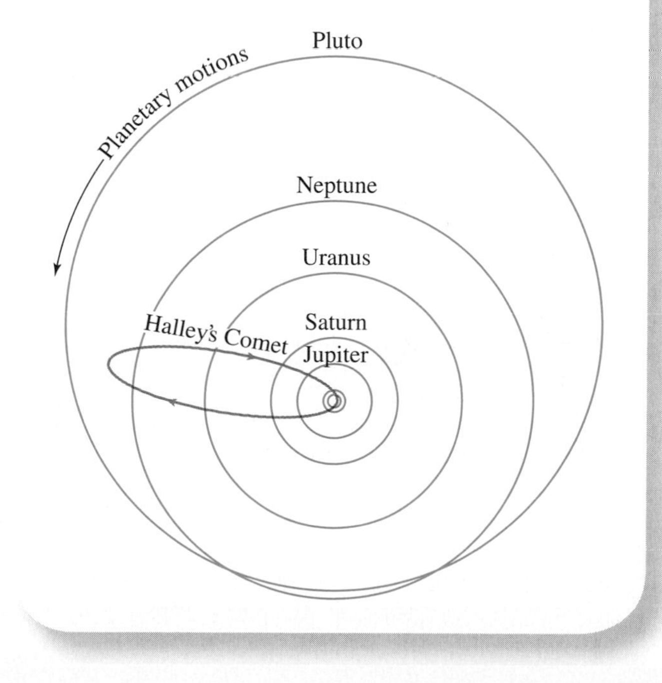

How is force possible without contact? One way to get better insight into this situation is to use the concept of a *field.* Imagine an object situated in space. The matter in the object causes an effect or a disturbance in the space around it. We call this a *gravitational field.* This field extends out in all directions but becomes weaker at greater distances from the object. In this model, the field itself causes a force to act on any other object. It plays the role of an invisible agent for the gravitational force. Whenever a second body is in the field of the first, it experiences a gravitational force. But the field is present even when there is no other object around to experience its effect.

We might call this a "force field" because it causes forces on other bodies. But do not imagine it to be the kind of "invisible wall" one sees in science-fiction movies. One way to represent the shape of the gravitational field around an object is by drawing arrows at different points in space. They show the magnitude and the direction of the force that would act on anything placed at each point (Figure 2.35a). These arrows are long near the object and short farther away because the gravitational force decreases as the distance increases. Another way to represent the gravitational field is to connect the arrows, making "field lines." The direction of the field line at any point in space again shows the direction of the force that would act on an object placed there. The strength of the gravitational field is represented by the spacing of the lines: The lines are farther apart where the field is weaker (Figure 2.35b).

All of the forces in nature can be traced to four fundamental forces. The gravitational force is one of them. The others are the electromagnetic force, responsible for electric and magnetic effects; the strong nuclear force; and the weak nuclear force. These will be discussed in later chapters. But what about friction and other forces involving direct contact? Most of these can be traced back to the electromagnetic force. This force determines the sizes and shapes of atoms and molecules. For example, a stretched spring pulls back because of the electrical forces between the atoms in it.

Tides

For the most part, tides are the result of gravitational forces exerted on the Earth by the Moon. To understand this, consider Figure 2.36. Points A, B, and C all lie on a line through the center of the Earth and the center of the Moon. According to Newton's law of universal gravitation, Earth material at A experiences a stronger attractive force toward the Moon than does material at C because it is closer to the Moon. Similarly, material at C experiences a greater attractive force toward the Moon than does material at B. Thus, the material at A is pulled away from material at C, while the material at C is in turn pulled away from the material at B. The net effect is to separate these three points. Thus the shape of the Earth is elongated or stretched by the gravitational pull of the Moon along a line connecting the two bodies.

This view is one that might be seen by an observer outside the Earth-Moon system looking in. One might ask what an observer at point C sees. Such an observer would notice that point A is pulled toward the Moon and away from C. The observer would also see that point B, relative to C, seems to be pushed away from C, in the direction away from the Moon.

We can redraw our earlier figure now indicating how an Earthbound observer at C views the forces exerted by the Moon at points A and B (Figure 2.37). Notice that

Figure 2.35
Two ways of showing the gravitational field around an object: (a) with arrows and (b) with field lines.

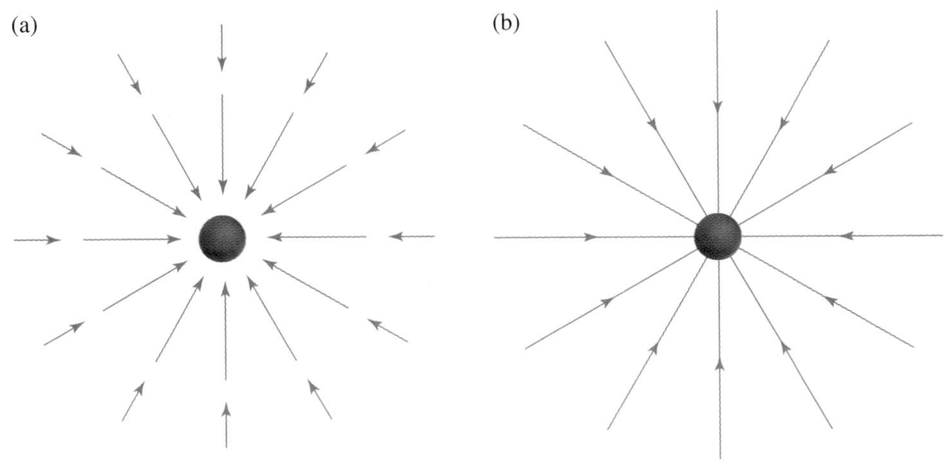

(a)

(b)

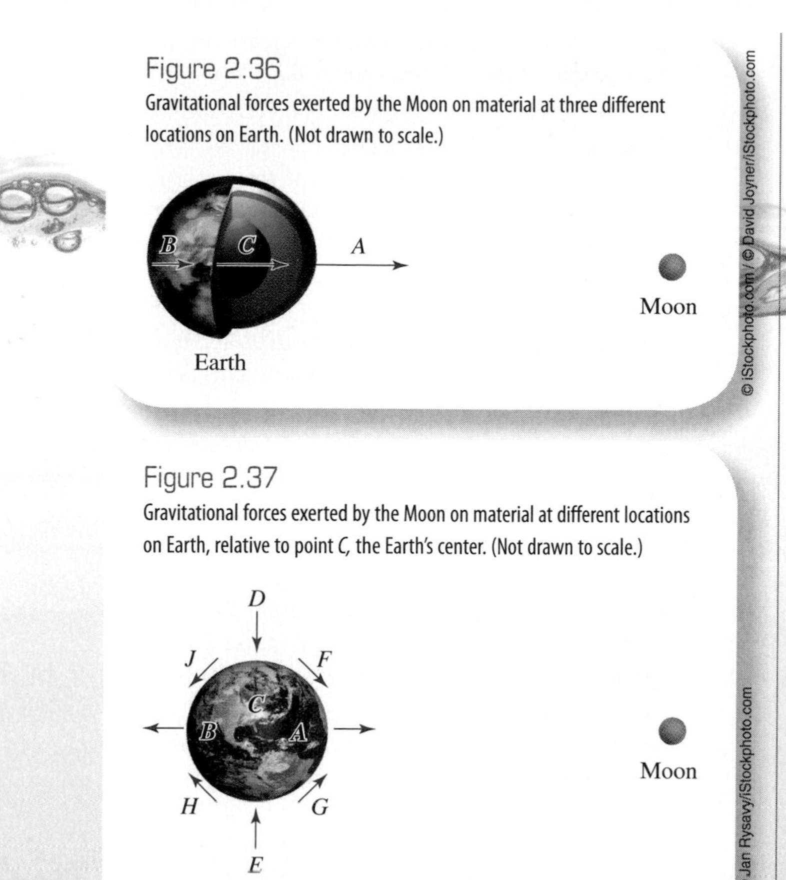

Figure 2.36
Gravitational forces exerted by the Moon on material at three different locations on Earth. (Not drawn to scale.)

© iStockphoto.com / © David Joyner/iStockphoto.com

Figure 2.37
Gravitational forces exerted by the Moon on material at different locations on Earth, relative to point C, the Earth's center. (Not drawn to scale.)

© Jan Rysavy/iStockphoto.com

the same stretching or elongation of the Earth along the direction of the Moon occurs as before, but now the symmetry of the situation as seen from C is manifested.

But what has this to do with tides? Suppose we were to perform this same kind of analysis for points D, E, F, G, H, and J. We would find that, relative to C, forces along the directions of the arrows shown in Figure 2.37 would exist because of the gravitational presence of the Moon. Now imagine the Earth to be covered with an initially uniform depth of water. How would this fluid move in response to these forces? Again, applying the laws of mechanics, we are led to the following, perhaps somewhat startling, results. (1) Water at points D and E would weigh slightly more than water elsewhere because, in addition to the Earth's own gravitational pull toward C, there is a small component of force toward C produced by the Moon. (2) Conversely, water at points A and B would weigh slightly less than water elsewhere because the Moon exerts small forces in directions opposite to the Earth's own inward gravitational attraction toward C. These forces work to counteract gravity and, hence, to reduce the weight of the water. (3) Water at points F, G, H, and J would experience a force parallel to the Earth's surface and would begin to flow in the directions of the arrows at each point, much as water flows down any inclined surface in response to gravity. This flow causes the water to pile up at opposite sides of the Earth along the line joining the Earth to the Moon. These "piles" of water are called the tidal bulges, and as the Earth rotates, points on its surface move into and then out of the bulges, resulting in the high and low tides observed at intervals of roughly 6 hours (Figure 2.38).

Many complications to this simple picture cause the behavior of real tides to deviate somewhat from that predicted by this model. In particular, we have ignored such things as tidal effects caused by the Sun, the influence of the Earth's rotation on the motion of the tidal bulges, and the effects on the local heights of tides caused by the Earth's rugged, uneven surface. The first of these complications leads to such phenomena as the lower-than-average *neap tides*, which occur when the Moon is in either the first- or third-quarter phase (and hence at right angles to the Sun in the sky), and the higher-than-average *spring tides*, which happen when the Moon is in either the new or the full phase (and hence in line with the Sun in the sky). Despite having ignored these problems, we are still able to understand the basic characteristics of tides in terms of this simple Newtonian model.

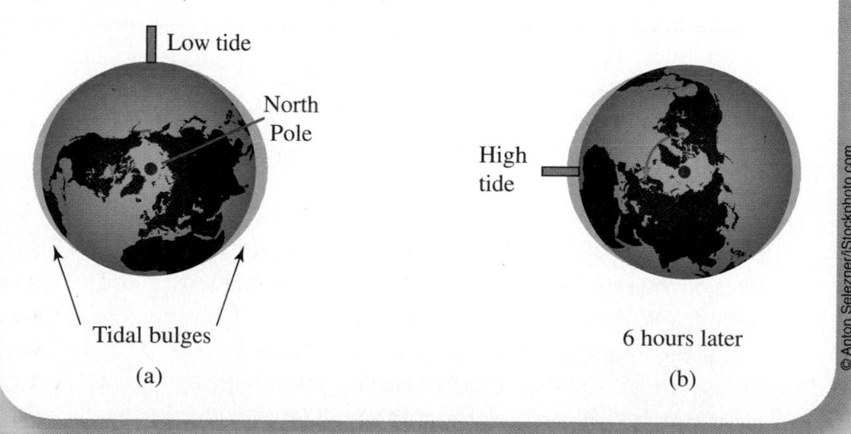

Figure 2.38
(a) The "tidal bulges" in the oceans caused by the gravitational pull of the Moon. (b) As the Earth rotates, places on its surface move from low tide to high tide, back to low tide, and so on. (Not drawn to scale.)

© Anton Seleznev/iStockphoto.com

questions&problems

Questions

(▶ Indicates a review question, which means it requires only a basic understanding of the material to answer. The others involve integrating or extending the concepts presented thus far.)

1. ▶ What is force? Identify several of the forces that are currently acting on or around you.
2. ▶ What is weight? Under what circumstances might something be weightless?
3. A person places a book on the roof of a car and drives off without remembering to remove it. As the book and the car move down a street at a steady speed, there are two horizontal forces acting on the book. What are they?
4. ▶ What are the two types of friction? Can both types act on the same object at the same time?
5. A person places a hand on a closed book resting on a table, and then presses downward while pushing outward. Either the book slides across the table or the hand slides across the book. What determines which of these alternatives occur? Which type(s) of friction is (are) involved?
6. ▶ What do we mean by "external" force? Why can only external forces (as opposed to internal ones) cause objects to move?
7. At one moment in a football game, player A exerts a force to the east on player B. At the same time, a teammate of A exerts the same-sized force to the south on player B. In what direction is B likely to go because of these forces?
8. ▶ How does an object move when it is subject to a steady centripetal force? How does it move if that force suddenly disappears?
9. A person is riding on a train while watching the display on a GPS unit (refer back to page 11). The person notices that both the "speed" and the "heading" readings are not changing. What can the person conclude about the net force acting on the train car?
10. ▶ Discuss the distinction between mass and weight.
11. ▶Two astronauts in an orbiting space station "play catch" (throw a ball back and forth to each other). Compared to playing catch on Earth, what effect, if any, does the "weightless" environment have on the process of accelerating (throwing and catching) the ball?
12. An "extreme" roller coaster is moving along its track. During a brief period, the track exerts a downward force on the cars. Describe what is happening. (What is the shape of the track at this point?)
13. Single-engine airplanes usually have their propeller at the front. Boats and ships usually have their propeller(s) at the rear. From the perspective of Newton's second law of motion, is this significant?
14. As a rocket ascends, its acceleration increases even though the net force on it stays constant. Why?
15. ▶ What is the international system of units (SI)?
16. An archer aims an arrow exactly horizontal over a flat field and shoots it. At the same instant, the archer's watchband breaks and the watch falls to the ground. Does the watch hit the ground before, at the same time as, or after the arrow hits the ground? Defend your answer.
17. ▶ Describe the variation of the net force on and the acceleration of a mass on a spring as it executes simple harmonic motion.
18. ▶ Explain how the change in the force of air resistance on a falling body causes it to eventually reach a terminal speed.
19. The terminal speed of a ping-pong ball is about 20 mph. From the top of a building a ping-pong ball is thrown *downward* with an initial speed of 50 mph. Describe what happens to the ball's speed as it moves downward from the moment it is thrown to the moment when it hits the ground.
20. At least two forces are acting on you right now. What are these forces? On what is the equal and opposite force to each of these acting?
21. As any car travels with constant velocity on a straight, flat section of highway, the road still exerts both a vertical (upward or downward) force and a horizontal (forward or backward) force on the car. Which direction is each of these forces?
22. How is Newton's third law of motion involved when you jump straight upward?
23. Jane and John are both on roller skates and are facing each other. First Jane pushes John with her hands and they move apart. Later they get together, and John pushes Jane equally hard with his hands and they move apart. Do they move any differently in the two cases? Why or why not?
24. What would be the gravitational force on you if you were at a point in space a distance R (the Earth's radius) above the Earth's surface?
25. ▶ Describe how the magnitude of the gravitational force between objects was first measured.
26. If suddenly the value of G, the gravitational constant, increased to a billion times its actual value, what sorts of things would happen?
27. The first "Lunar Olympics" is to be held on the Moon inside a huge dome. Of the usual Olympic events—track and field, swimming, gymnastics, and so on—which would be drastically affected by the Moon's gravity? In which events would Earth-based records be broken? In which events would the performances be no better—or perhaps worse—than on the Earth?
28. ▶ In the broadest terms, what causes tides?

29. The Sun exerts much larger forces on the Earth and its oceans than the Moon does, yet the tides caused by the Sun are much lower than those caused by the Moon. Why?
30. We have studied four different laws authored by Sir Isaac Newton. For each of the following, indicate which law is best for the task described.
    a) Calculating the net force on a car as it slows down
    b) Calculating the force exerted on a satellite by the Earth
    c) Showing the mathematical relationship between mass and weight
    d) Explaining the direction that a rubber stopper takes after the string that was keeping it moving in a circle overhead is cut
    e) Explaining why a gun recoils when it is fired
    f) Explaining why a wing on an airplane is lifted upward as it moves through the air

Problems

1. Express your weight in newtons. From this, determine your mass in kilograms.
2. A child weighs 300 N. What is the child's mass?
3. Suppose an airline allows a maximum of 30 kg for each suitcase a passenger brings along.
    a) What is the weight in newtons of a 30-kg suitcase?
    b) What is the weight in pounds?
4. The mass of a certain elephant is 1,130 kg.
    a) Find the elephant's weight in newtons.
    b) Find its weight in pounds.
5. The mass of a subway car and passengers is 40,000 kg. If its acceleration as it leaves a station is 0.9 m/s^2, what is the net force acting on it?
6. A motorcycle and rider have a total mass equal to 300 kg. The rider applies the brakes, causing the motorcycle to accelerate at a rate of -5 m/s^2. What is the net force on the motorcycle?
7. As a 2-kg ball rolls down a ramp, the net force on it is 10 N. What is the acceleration?
8. In an experiment performed in a space station, a force of 60 N causes an object to have an acceleration equal to 4 m/s^2. What is the object's mass?
9. The engines in a supertanker carrying crude oil produce a net force of 20,000,000 N on the ship. If the resulting acceleration is 0.1 m/s^2, what is the ship's mass?
10. A person stands on a scale inside an elevator at rest (Figure 2.39). The scale reads 800 N.
    a) What is the person's mass?
    b) The elevator accelerates upward momentarily at the rate of 2 m/s^2. What does the scale read then?
    c) The elevator then moves with a steady speed of 5 m/s. What does the scale read?
11. A jet aircraft with a mass of 4,500 kg has an engine that exerts a force (thrust) equal to 60,000 N.
    a) What is the jet's acceleration when it takes off?
    b) What is the jet's speed after it accelerates for 8 s?
    c) How far does the jet travel during the 8 s?
12. At the end of Section 1.4, we mentioned that the maximum acceleration of a fist during a particular karate blow was measured to be about 3,500 m/s^2 (Figure 1.23).
    a) If the mass of the fist was approximately 0.7 kg, what was the maximum force?
    b) What was the maximum force on the concrete block?
13. A sprinter with a mass of 80 kg accelerates from 0 m/s to 9 m/s in 3 s.
    a) What is the runner's acceleration?
    b) What is the net force on the runner?
    c) How far does the sprinter go during the 3 s?
14. As a baseball is being caught, its speed goes from 30 to 0 m/s in about 0.005 s. Its mass is 0.145 kg.
    a) What is the baseball's acceleration in m/s^2 and in *g*'s?
    b) What is the size of the force acting on it?
15. On aircraft carriers, catapults are used to accelerate jet aircraft to flight speeds in a short distance. One such catapult takes a 18,000-kg jet from 0 to 70 m/s in 2.5 s.
    a) What is the acceleration of the jet (in m/s^2 and *g*'s)?
    b) How far does the jet travel while it is accelerating?
    c) How large is the force that the catapult must exert on the jet?
16. At the end of an amusement park ride, it is desirable to bring a gondola to a stop without having the acceleration exceed 2 *g*. If the total mass of the gondola and its occupants is 2,000 kg, what is the maximum allowed braking force?

17. An airplane is built to withstand a maximum acceleration of 6 *g*. If its mass is 1,200 kg, what size force would cause this acceleration?

18. Under certain conditions, the human body can safely withstand an acceleration of 10 *g*.

    a) What net force would have to act on someone with mass of 50 kg to cause this acceleration?

    b) Find the weight of such a person in pounds, then convert the answer to (a) to pounds.

19. A race car rounds a curve at 60 m/s. The radius of the curve is 400 m, and the car's mass is 600 kg.

    a) What is the car's (centripetal) acceleration? What is it in *g*'s?

    b) What is the centripetal force acting on the car?

20. A hang glider and its pilot have a total mass equal to 120 kg. While executing a 360° turn, the glider moves in a circle with an 8-m radius. The glider's speed is 10 m/s.

    a) What is the net force on the hang glider?

    b) What is the acceleration?

21. A 0.1-kg ball is attached to a string and whirled around in a horizontal circle overhead. The string breaks if the force on it exceeds 60 N. What is the maximum speed the ball can have when the radius of the circle is 1 m?

22. On a highway curve with radius 50 m, the maximum force of static friction (centripetal force) that can act on a 1,000-kg car going around the curve is 8,000 N. What speed limit should be posted for the curve so that cars can negotiate it safely?

23. A centripetal force of 200 N acts on a 1,000-kg satellite moving with a speed of 5,000 m/s in a circular orbit around a planet. What is the radius of its orbit?

24. As a spacecraft approaches a planet, the rocket engines on it are fired (turned on) to slow it down so it will go into orbit around the planet. The spacecraft's mass is 2,000 kg and the thrust (force) of the rocket engines is 400 N. If its speed must be decreased by 1,000 m/s, how long must the engines be fired? (Ignore the change in the mass as the fuel is burned.)

25. A space probe is launched from the Earth, headed for deep space. At a distance of 10,000 miles from the Earth's center, the gravitational force on it is 600 lb. What is the size of the force when it is at each of the following distances from the Earth's center?

    a) 20,000 miles

    b) 30,000 miles

    c) 100,000 miles

***remember**

There are more quizzes and study tools online at 4ltrpress.cengage.com/physics.

Discover your **PHYSICS** online experience at **4ltrpress.cengage.com/physics**.

You'll find everything you need to succeed in your class.

- Interactive Quizzes
- Interactive and Printable Flashcards
- Simulations
- Glossary
- And more

4ltrpress.cengage.com/physics

Energy and Conservation Laws

sections

© Mark Evans/iStockphoto.com

3.1 Conservation Laws

Newton's laws of motion, in particular the second law, govern the instantaneous behavior of a system. They relate the forces that are acting at any instant in time to the resulting changes in motion. **Conservation laws** involve a different approach to mechanics, more of a "before-and-after" look at systems. A conservation law states that the total amount of a certain physical quantity present in a system stays constant (is conserved). For example, we might state the following conservation law.

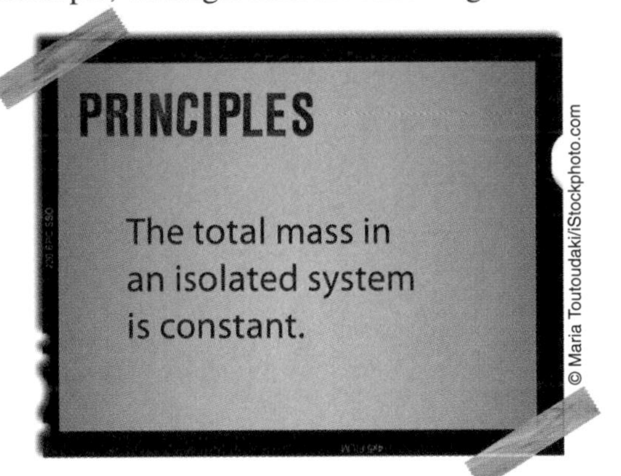
PRINCIPLES

The total mass in an isolated system is constant.

© Maria Toutoudaki/iStockphoto.com

An "isolated system" in this case means that no matter enters or leaves the system. The law states that the total mass of all the objects in a system doesn't change regardless of the kinds of interactions that go on in it.

A simple example of this rather obvious conservation law is given by aerial refueling. One aircraft, called a tanker, pumps fuel into a second aircraft while both are in flight. If we ignore the small amount of fuel that both aircraft consume during the refueling, this can be regarded as an isolated system. The law of conservation of mass tells us that the total mass of both aircraft remains constant. So if the receiving aircraft gains 2,000 kilograms of fuel, we automatically know that the tanker loses 2,000 kilograms of fuel. If one tanker refuels several aircraft in a formation, we include all of the aircraft in the system. The total mass of fuel dispensed by the tanker equals the total mass of fuel gained by the other aircraft. This fact may be useful. Let's say that the fuel gauge on one of the receiving aircraft is faulty. To determine how much fuel that aircraft was given, the crews use the conservation of mass: The total mass of fuel unloaded from the tanker minus the total mass of fuel given to the other aircraft in the formation equals the mass of fuel given to the aircraft in question.

The preceding example illustrates how we can use a conservation law. Without knowing the details about an interaction (for example, the actual rate at which fuel is transferred between the aircraft), we can still extract quantitative information by simply comparing the total amount of mass before and after. Three more conservation laws are presented in this chapter. Although they are a bit less intuitive and a bit more complicated to use than the law of conservation of mass, they are applied in the same way.

Refueling midair

© Getty Images / © iStockphoto.com

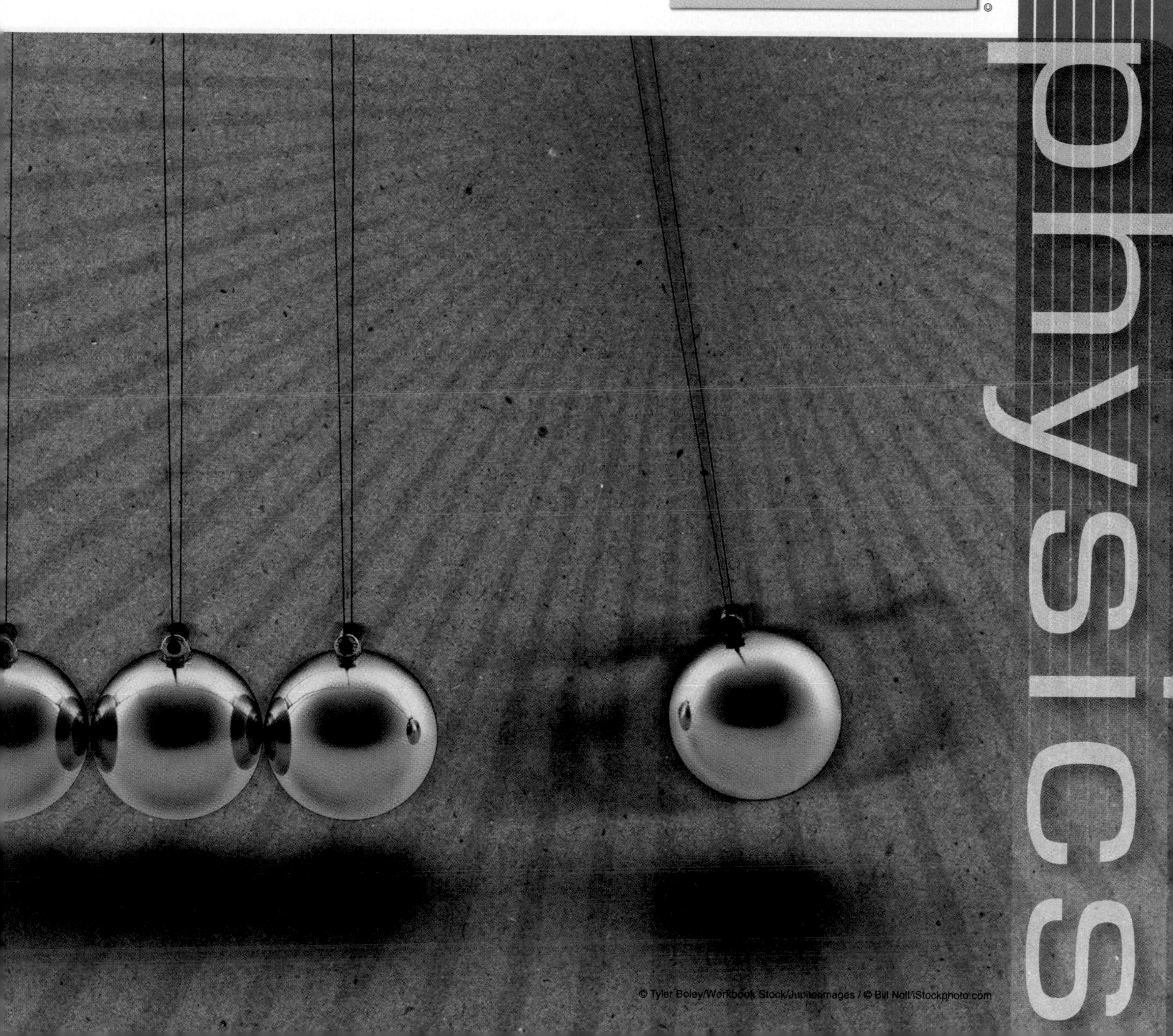

© Tyler Boley/Workbook Stock/Jupiterimages / © Bill Noll/iStockphoto.com

3.2 Linear Momentum

The conservation law for **linear momentum** follows directly from Newton's laws of motion, and we will consider it first.

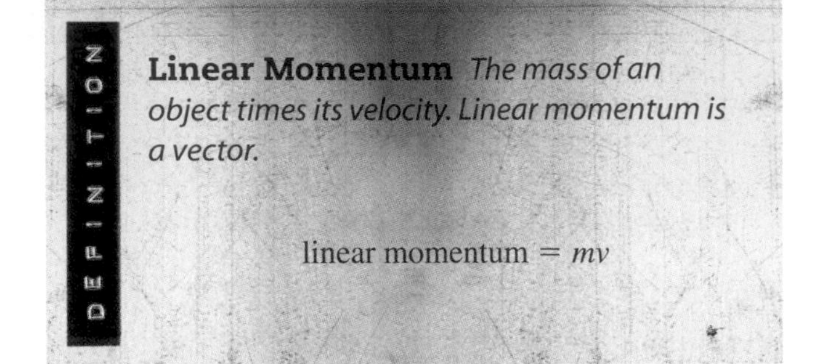

DEFINITION

Linear Momentum *The mass of an object times its velocity. Linear momentum is a vector.*

$$\text{linear momentum} = mv$$

Linear momentum is often referred to simply as *momentum.* It is a vector quantity (since velocity is a vector), and its SI unit of measure is the kilogram-meter/second (kg-m/s).

Linear momentum incorporates both mass and motion. Anything that is stationary has zero momentum. The faster a body moves, the larger its momentum. A heavy object moving with a certain velocity has more momentum than a light object moving with the same velocity (Figure 3.1). For example, the momentum of a bicycle and rider with a total mass of 80 kilograms and a speed of 10 m/s is

$$mv = 80 \text{ kg} \times 10 \text{ m/s} = 800 \text{ kg-m/s}$$

The linear momentum of a 1,200-kilogram car with the same speed is 12,000 kg-m/s. The bicycle and rider would have to be going 150 m/s (not likely) to have this same momentum.

Newton originally stated his second law of motion using linear momentum. In particular, we can restate this law as follows.

LAWS

Newton's Second Law of Motion (alternate form) *The net external force acting on an object equals the rate of change of its linear momentum.*

$$\text{force} = \frac{\text{change in momentum}}{\text{change in time}}$$

$$F = \frac{\Delta(mv)}{\Delta t}$$

To change an object's linear momentum, a net force must act on it. The larger the force, the faster the momentum will change. If the mass of the object stays constant, which is true in most cases, this equation is equivalent to the first form of Newton's second law (Section 2.4) because

$$\frac{\Delta(mv)}{\Delta t} = m\frac{\Delta v}{\Delta t} = ma \quad \text{(if mass is constant)}$$

Either form of the second law could be used for cars, airplanes, baseballs, and so on. For rockets and similar things with changing mass, only the alternate form should be used.

We can get yet another useful form of the second law by multiplying both sides by Δt to obtain

$$\Delta (mv) = F\,\Delta t$$

The quantity on the right side is called the *impulse.* The same change in momentum can result from a small force acting for a long time or a large force acting for a short

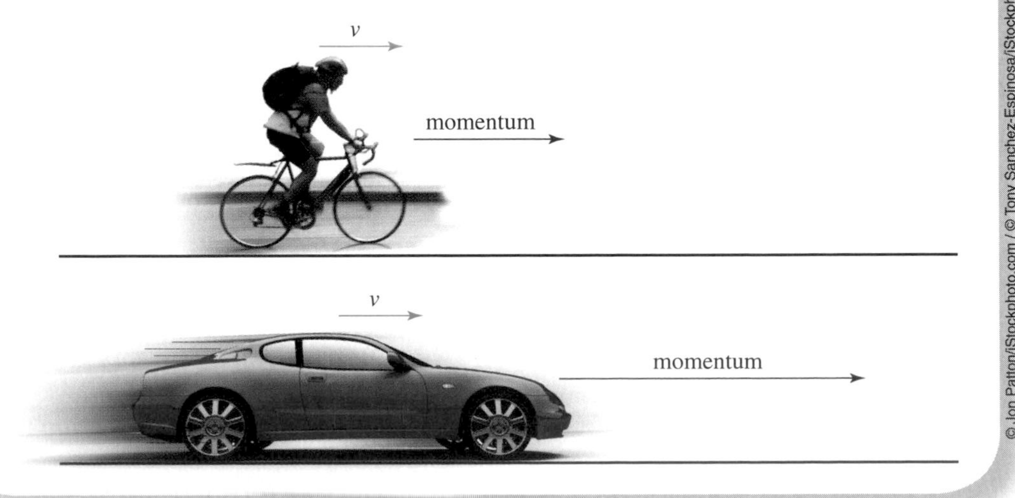

Figure 3.1
Linear momentum depends on the mass and the velocity. If a car and a bicycle have the same velocity, the car has a larger momentum because it has a larger mass.

time. This equation is useful for analyzing what goes on during impacts in sports that use balls and clubs.

When you throw a tennis ball, a small force acts on it for a relatively long period of time (as your hand moves through the air). When the ball is served at the same speed with a racquet, a large force acts for a short period of time. In either case the result is a change in momentum of the ball, $\Delta(mv)$. One reason to have good follow-through on a shot is to prolong the time of contact, Δt, between the ball and the racquet. This leads to a greater change in momentum, so the ball will leave the racquet with a higher speed.

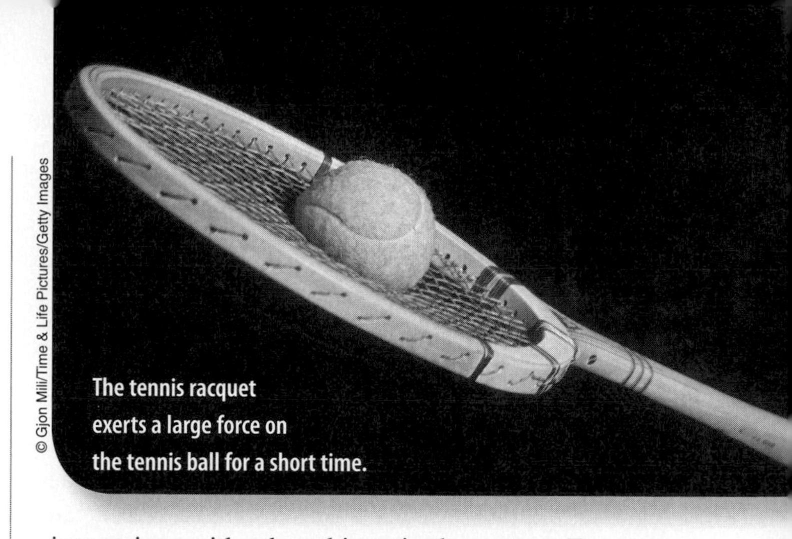
The tennis racquet exerts a large force on the tennis ball for a short time.

© Gjon Mili/Time & Life Pictures/Getty Images

EXAMPLE 3.1

Let's estimate the average force on a tennis ball as it is served. The ball's mass is 0.06 kilograms, and it leaves the racquet with a speed of, say, 40 m/s (90 mph). High-speed photographs indicate that the contact time is about 5 milliseconds (0.005 s).

Since the ball starts with zero speed (and hence, zero momentum), its change in momentum is

$$\begin{aligned}\Delta(mv) &= \text{momentum afterwards}\\ &= 0.06\text{ kg} \times 40\text{ m/s}\\ &= 2.4\text{ kg-m/s}\end{aligned}$$

So the average force is

$$\begin{aligned}F &= \frac{\Delta(mv)}{\Delta t} = \frac{2.4\text{ kg-m/s}}{0.005\text{ s}}\\ &= 480\text{ N} = 108\text{ lb}\end{aligned}$$

Our main application of the idea of linear momentum is based on the following conservation law.

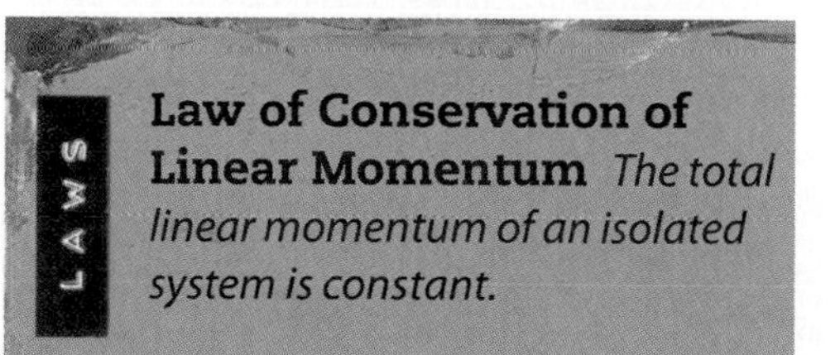

LAWS

Law of Conservation of Linear Momentum *The total linear momentum of an isolated system is constant.*

For this law, an isolated system means that there are *no outside forces* causing changes in the linear momenta of the objects inside the system. The momentum of an object can change only because of interactions with other objects in the system. For example, once the cue ball on a pool table has been shot (given some momentum), the pool table and the balls form an isolated system. If the cue ball collides with another ball initially at rest, its momentum is changed (decreased) (see Figure 3.2). This change occurs because of an interaction with another object in the system. The momentum of the ball with which it collides is increased, although the total linear momentum of the system remains the same. If someone put a hand on the table and stopped the cue ball, the system would no longer be isolated, and we couldn't assume that the total linear momentum of the system was constant.

The most important use of the law of conservation of linear momentum is in the analysis of collisions. Two billiard balls colliding, a traffic accident, and two skaters running into each other are some familiar examples of

Figure 3.2
Shown below are before (top) and after (bottom) views of a collision. The cue ball loses momentum, and the eight ball gains momentum. The total momentum of the two balls is the same before and after. The momentum vector before the collision equals the sum of the two momentum vectors after, as shown by the arrows at the bottom of the figure.

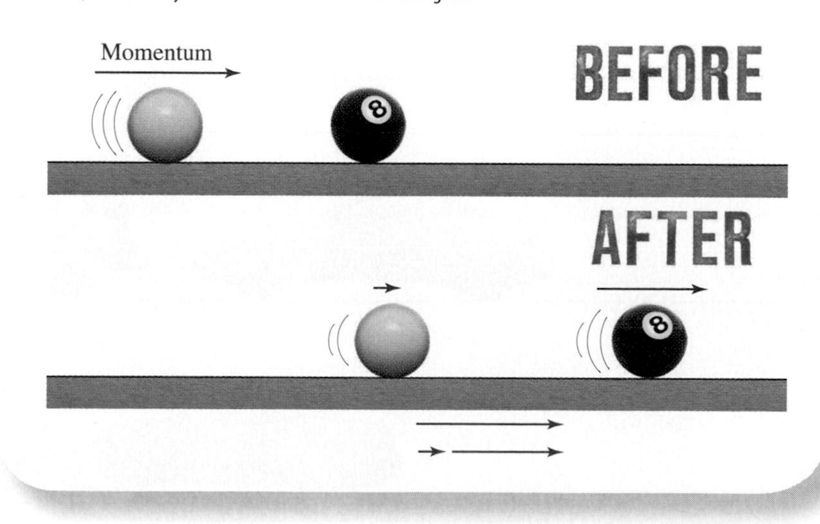

© Simon Askham/iStockphoto.com / © Glen Jones/iStockphoto.com / © iStockphoto.com

collisions. We will limit our examples to collisions involving only two objects that are moving in one dimension (along a line).

During any collision, the objects exert forces of equal magnitude and opposite direction on each other that cause them to accelerate (in opposite directions). These forces are usually quite large and are often due to direct contact between the bodies, as is the case with billiard balls and automobiles. "Action-at-a-distance" forces between objects that don't actually come into contact, like the gravitational pull between a spacecraft and a planet, can also be involved in collisions. The following statement is the key to applying the law of conservation of linear momentum to a collision:

LAWS

The total linear momentum of the objects in the system before the collision is the same as the total linear momentum after the collision.

$$\text{total } mv \text{ before} = \text{total } mv \text{ after}$$

EXAMPLE 3.2

We can use the law of conservation of linear momentum to analyze a simple automobile collision. A 1,000-kilogram automobile (car 1) runs into the rear of a stopped car (car 2) that has a mass of 1,500 kilograms. Immediately after the collision, the cars are hooked together, and their speed is estimated to have been 4 m/s (Figure 3.3). What was the speed of car 1 just before the collision?

Our conservation law tells us that the total linear momentum of the system will be constant:

$$\text{total } (mv)_{\text{before}} = \text{total } (mv)_{\text{after}}$$

Before the collision, only car 1 is moving. The total linear momentum before the collision is

$$(mv)_{\text{before}} = m_1 \times v_{\text{before}} = (1{,}000 \text{ kg}) \times v_{\text{before}}$$

After the collision, both cars are moving as one body. The total momentum after the collision is

$$\begin{aligned}(mv)_{\text{after}} &= (m_1 + m_2) \times v_{\text{after}} \\ &= (1{,}000 \text{ kg} + 1{,}500 \text{ kg}) \times 4 \text{ m/s} \\ &= 2{,}500 \text{ kg} \times 4 \text{ m/s} = 10{,}000 \text{ kg-m/s}\end{aligned}$$

These two linear momenta are equal. Therefore,

$$\begin{aligned}1{,}000 \text{ kg} \times v_{\text{before}} &= 10{,}000 \text{ kg-m/s} \\ v_{\text{before}} &= \frac{10{,}000 \text{ kg-m/s}}{1{,}000 \text{ kg}} \\ &= 10 \text{ m/s}\end{aligned}$$

This type of analysis is routinely used to reconstruct traffic accidents. For example, it can be used to determine whether a vehicle was exceeding the speed limit just before a collision. In this problem the car was going 10 m/s, or about 22 mph. If the accident happened in a 15-mph speed zone, the driver of car 1 would have been speeding.

The speed of a bullet or a thrown object can be measured similarly. A bullet is fired into, and becomes embedded in, a block of wood hanging from a string

Figure 3.3

A 1,000-kilogram car collides with a 1,500-kilogram car that is stationary. Afterward, the two cars are hooked together and move with a speed of 4 m/s. The conservation of linear momentum allows us to determine that the speed of the first car was 10 m/s.

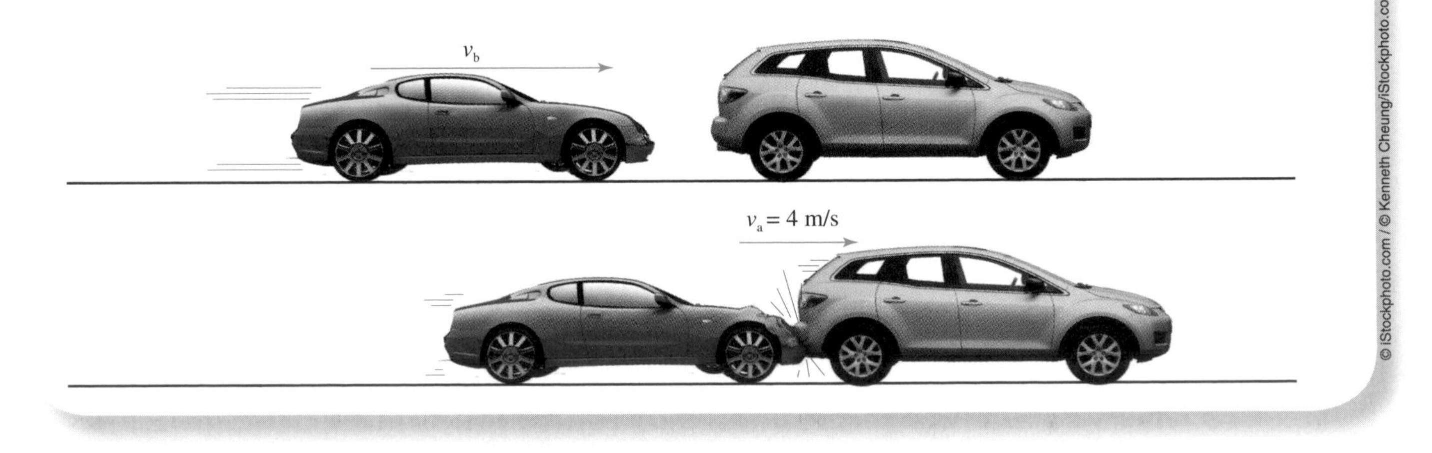

© iStockphoto.com / © Kenneth Cheung/iStockphoto.com

(Figure 3.4). If one measures the masses of the bullet and the wood block and the speed of the block immediately afterward, the initial speed of the bullet can be found by using the conservation of linear momentum. You can measure the speed of your "fast ball" by throwing a lump of sticky clay at a hanging block and proceeding in the same way. The key in both cases is measuring the speed of the block of wood after the collision. One can do this most easily by using the law of conservation of energy. We will show how in Section 3.5.

In Section 2.7, we described a simple experiment in which we used two carts (Figure 2.26c). One cart has a spring-loaded plunger that pushes on the other cart, causing both carts to be accelerated. By the law of conservation of linear momentum,

$$(mv)_{\text{before}} = (mv)_{\text{after}}$$

Before the spring is released, neither cart is moving, so the momentum is zero:

$$(mv)_{\text{before}} = 0$$

Therefore, the total linear momentum afterward is zero. But since both carts are moving, this momentum equals:

$$(mv)_{\text{after}} = 0 = (mv)_1 + (mv)_2$$

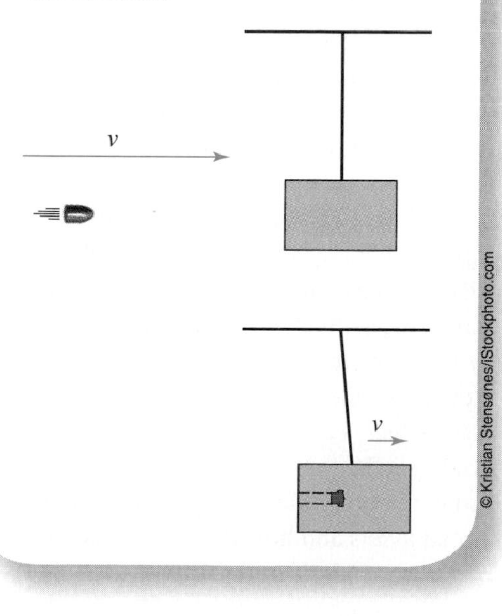

© Kristian Stensønes/iStockphoto.com

Figure 3.4
A bullet becomes embedded in a block of wood. If the speed of the block and the masses of the block and the bullet are measured, the initial speed of the bullet can be computed using the conservation of linear momentum.

So:

$$(mv)_1 = -(mv)_2$$

The momentum of one of the carts is negative. This has to be the case, because they are moving in opposite directions. (Remember: linear momentum is a vector.) Since $(mv)_1 = m_1 \times v_1$:

$$m_1 \times v_1 = -m_2 \times v_2$$

$$v_1 = -\frac{m_2}{m_1} \times v_2$$

$$\frac{v_1}{v_2} = -\frac{m_2}{m_1}$$

So we have substantiated the statement made in Section 2.7: The ratio of the speeds of the two carts is the inverse of the ratio of their masses. If the mass of cart 2 is 3 kilograms and the mass of cart 1 is 1 kilogram, the ratio is 3 to 1. Cart 1 moves away with three times the speed of cart 2 (Figure 3.5). Note that this gives us only the ratio of the speeds, not the actual speed of each cart, which would depend on the strength of the spring. With a weak spring, the speeds might be 1 m/s and 3 m/s. With a stronger spring, the speeds might be 2.5 m/s and 7.5 m/s.

We can use linear momentum conservation to get a different view of some of the situations described in Section 2.7. There we used forces and Newton's third law of motion. When a gun is fired, the bullet acquires momentum in one direction, and the gun gains equal momentum in the opposite direction—the "kick" of the gun against the shoulder or hand. When a hockey player hits

Figure 3.5
The mass of the cart on the right, cart 2, is three times the mass of the cart on the left, cart 1. After the spring is released, the speed of the lighter cart is three times the speed of the more massive cart.

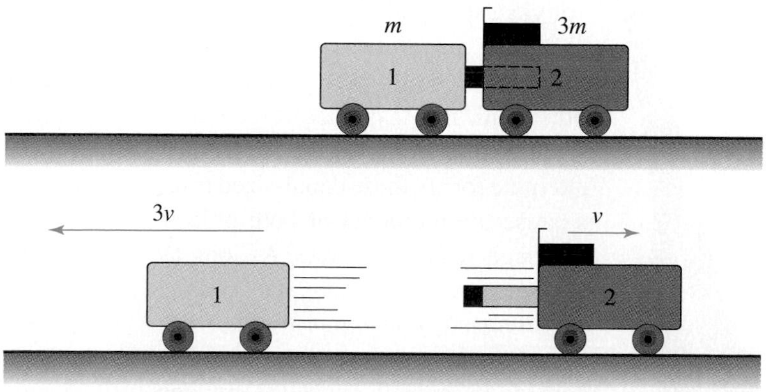

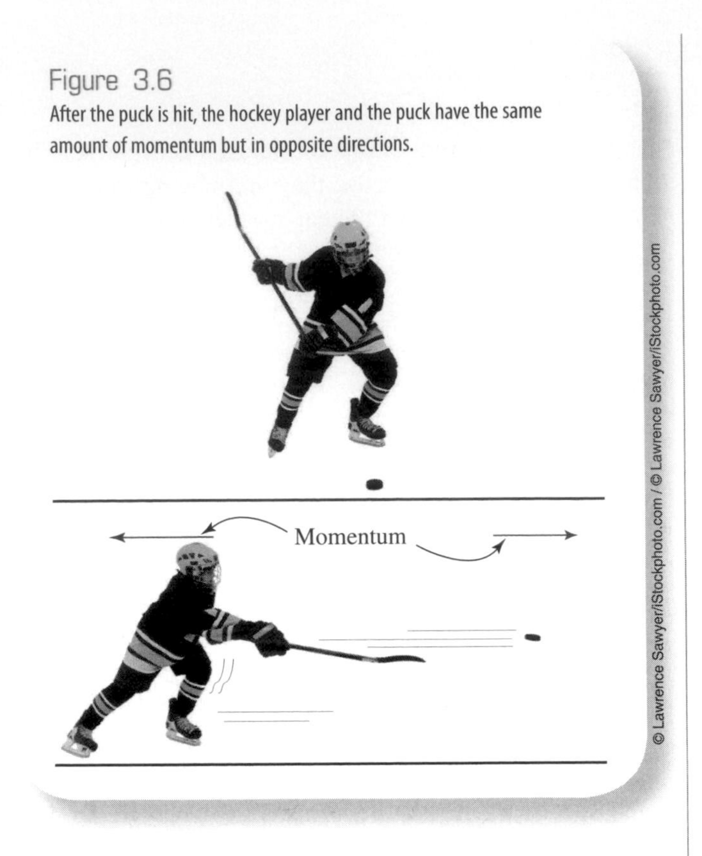

Figure 3.6
After the puck is hit, the hockey player and the puck have the same amount of momentum but in opposite directions.

© Lawrence Sawyer/iStockphoto.com / © Lawrence Sawyer/iStockphoto.com

a puck forward, he or she will move backward with equal momentum (Figure 3.6). Rockets and jets give momenta to the ejected exhaust gases. They in turn gain momentum in the forward direction. For a rocket with no external forces on it, the increase in momentum during each second will depend on how much gas is ejected—which equals the mass of fuel burned—and on the speed of the gas.

Why is linear momentum conserved? We can use Newton's second and third laws of motion to answer this question. When two objects exert forces on each other by colliding or via a spring plunger, the forces are equal in magnitude and opposite in direction. The objects push on each other with the same size force but in opposite directions. By the second law (alternate form), these equal-sized forces cause the momenta of both objects to change at the same rate. As long as the objects are interacting, *they change one another's momentum by the same amount but in opposite directions.* The momentum gained (or lost) by one object is exactly offset by the momentum lost (or gained) by the other. The total linear momentum is not changed. In Example 3.2, the 1,000-kilogram car is slowed from 10 m/s before the collision to 4 m/s after the collision (Figure 3.3). Its momentum is decreased by 6,000 kg-m/s (1,000 kilograms × 10 m/s − 1,000 kilograms × 4 m/s). The 1,500-kilogram car goes from 0 m/s before to 4 m/s after. So its momentum is increased by 6,000 kg-m/s (1,500 kilograms × 4 m/s).

© iStockphoto.com

These examples illustrate the usefulness of conservation laws. The approach is different from that used with Newton's second law in Chapter 2, where it was necessary to know the size of the force that acts on the object at each moment to determine the object's velocity. With the law of conservation of linear momentum, we do not have to know the details of the interactions—how large the forces are and how long they act. All we need is some information about the system before and after the interaction. Note that in Example 3.2 we used information from after the collision to determine the speed of the car *before* the collision. In the example with the carts (Figure 3.5), we determined the ratio of the speeds *after* the interaction.

3.3 Work: The Key to Energy

The law of conservation of energy is arguably the most important of the conservation laws. Not only is it useful for solving problems, it is also a powerful theoretical statement that can be used to understand widely diverse phenomena and to show what hypothetical processes are or are not possible. As we mentioned earlier, the concept of **energy** is one of the most important in physics. This is because energy takes many forms and is involved in all physical processes. One could say that every interaction in our universe involves a transfer of energy or a transformation of energy from one form to another.

We can compare the concept of energy to that of financial assets, which can take the form of cash, real estate, material goods, or investments, among other things. The study of economics is in part a study of these forms of financial assets and how they are transferred and transformed. Much of physics deals with the forms of energy and the transformations that occur during interactions.

When first encountered, the concept of energy is a bit difficult to understand because there is no simple way to define it. As an aid, we will first introduce **work**, a physical quantity that is quite basic and that gives us a good foundation for understanding energy.

The idea of work in physics arises naturally when one considers *simple machines* like the lever and the inclined plane when used in situations with negligible friction. Let's say that you use a lever to raise a heavy rock (Figure 3.7). If you place the fulcrum close to the rock, you find that a small, downward force on your end results in a larger, upward force on the rock. However, the distance your end moves as you push it down is correspondingly larger than the distance the rock is raised. By measuring the forces and distances, we find that dividing the larger force by the smaller force gives the same number (or ratio) as dividing the larger distance by the smaller distance. In particular:

$$\frac{F \text{ on left end}}{F \text{ on right end}} = \frac{d \text{ right end moves}}{d \text{ left end moves}}$$

We can multiply both sides by *F on right end* and *d left end moves* and get the following result:

$$(F \text{ on left}) \times (d \text{ left moves}) = (F \text{ on right}) \times (d \text{ right moves})$$

$$F_{\text{left}}\, d_{\text{left}} = F_{\text{right}}\, d_{\text{right}}$$

In other words, even though the two forces and the two distances are different, the quantity *force times distance* has the same value for both ends of the lever. We might say that raising the rock is a fixed task. One can perform the task by lifting the rock directly or by using a lever. In the former case, the force is large, equal to the rock's weight, but the distance moved is small. When using the lever, the force you exert is smaller, but the distance through which your hands move is larger. The quantity *Fd* is the same, regardless of which way the task is accomplished.

We reach the same conclusion when considering an inclined plane. Let's say that a barrel must be placed on a loading dock (Figure 3.8). Lifting the barrel directly requires a large force acting through a small distance,

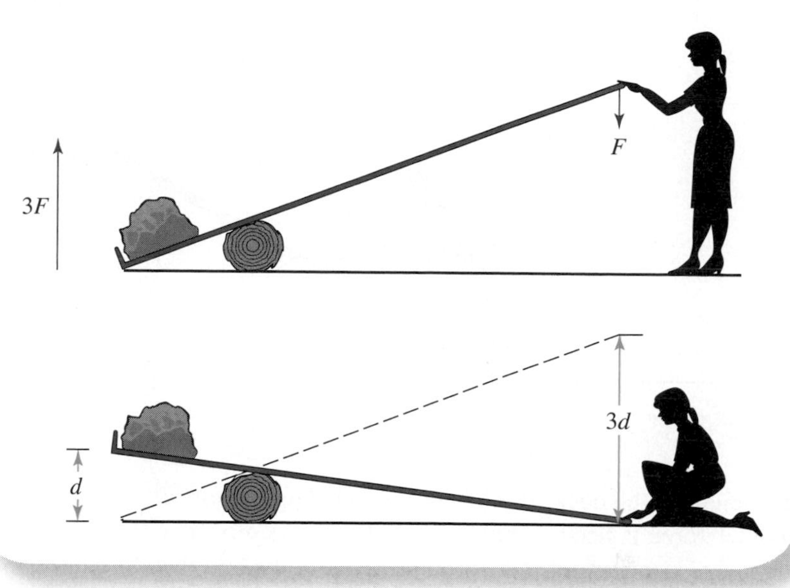

Figure 3.7

When a lever is used to raise a rock, a small force on the right end results in a larger force on the left end. But the right end moves a greater distance than the left end. The force multiplied by the distance moved *is the same* for both ends.

© Jaimie D. Travis/iStockphoto.com

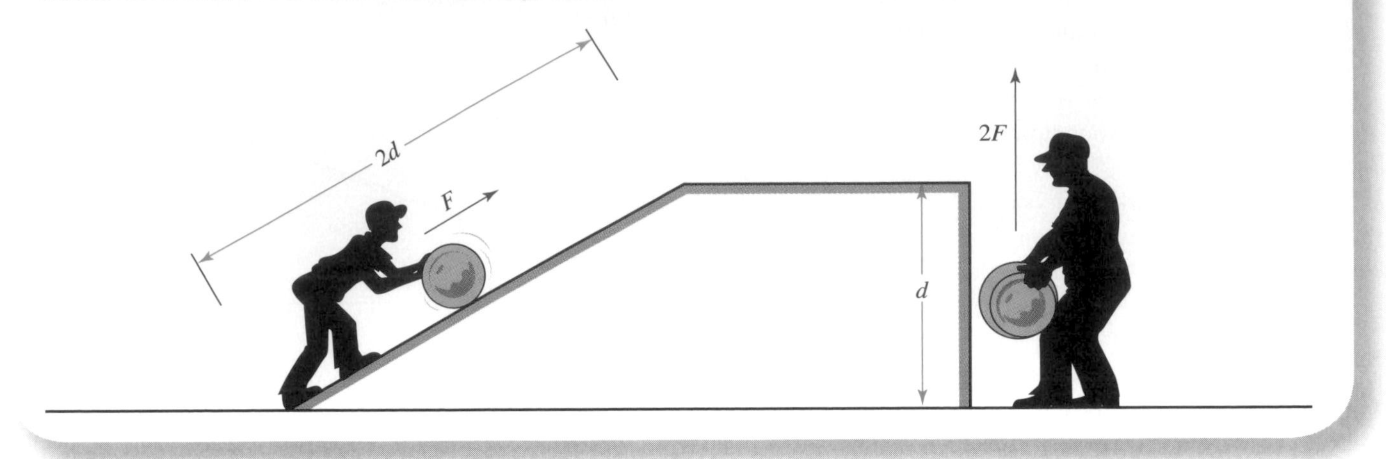

Figure 3.8
Two identical barrels need to be placed on the loading dock. One is lifted directly, requiring a large force. The other is rolled up the ramp, an inclined plane. A smaller force is needed, but it must act on the barrel over a longer distance.

the height of the dock. If the barrel is rolled up a ramp, a smaller force is needed, but the barrel must be moved a greater distance. Again, the product of the force and the distance moved is the same for the two methods.

$$(F \text{ lifting}) \times \text{height} = (F \text{ rolling}) \times (\text{ramp length})$$

$$F_{\text{lifting}} d_{\text{lifting}} = F_{\text{rolling}} d_{\text{rolling}}$$

The quantity *force times distance* is obviously a useful way of measuring the "size" of a task. It is called *work.*

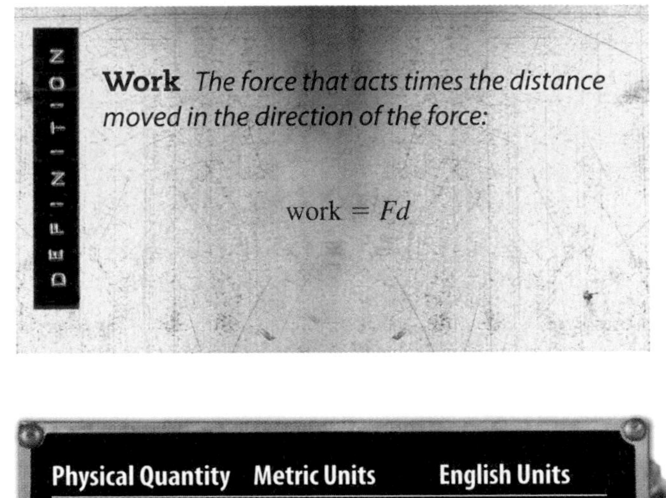
DEFINITION

Work *The force that acts times the distance moved in the direction of the force:*

$$\text{work} = Fd$$

Physical Quantity	Metric Units	English Units
Work	joule (J) [SI unit]	foot-pound (ft-lb)
	erg	British thermal unit (Btu)
	calorie (cal)	
	kilowatt-hour (kWh)	

Since force is a vector, work equals the distance moved times the component of the force parallel to the motion. Work itself is not a vector. There is no direction associated with work. Work can be positive or negative however. Whenever an object moves and there is a force acting on the object in the same or opposite direction that it moves, work is done. When the force and the motion are in the same direction, the work is positive. When they are in opposite directions, the work is negative.

EXAMPLE 3.3

Because of friction, a constant force of 100 newtons is needed to uniformly slide a box across a room (Figure 3.9). If the box moves 3 meters, how much work is done?

$$\begin{aligned} \text{work} &= Fd \\ &= 100 \text{ N} \times 3 \text{ m} \\ &= 300 \text{ N-m} \end{aligned}$$

The unit in the answer is the newton-meter (N-m). This is called the *joule*.

$$1 \text{ joule} = 1 \text{ newton-meter} = 1 \text{ newton} \times 1 \text{ meter}$$

$$1 \text{ J} = 1 \text{ N-m}$$

The joule is a derived unit of measure, and it is the SI unit of work and energy. Just remember that if you use only SI units for force, distance, and so on, your unit for work and energy will always be the joule (J).

EXAMPLE 3.4

Let's say that the barrel in Figure 3.8 has a mass of 30 kilograms and that the height of the dock is 1.2 meters. How much work would you do when lifting the barrel?

$$\text{work} = Fd$$

Assuming the barrel is raised with constant speed, the applied force equals the weight of the barrel, mg.

$$F = W = mg = 30 \text{ kg} \times 9.8 \text{ m/s}^2$$
$$= 294 \text{ N}$$

Since the applied force is parallel to the barrel's motion:

$$\text{work} = Fd = Wd$$
$$= 294 \text{ N} \times 1.2 \text{ m}$$
$$= 353 \text{ J}$$

The work you do when rolling the barrel up the ramp would be the same. The force would be smaller, but the distance would be larger.

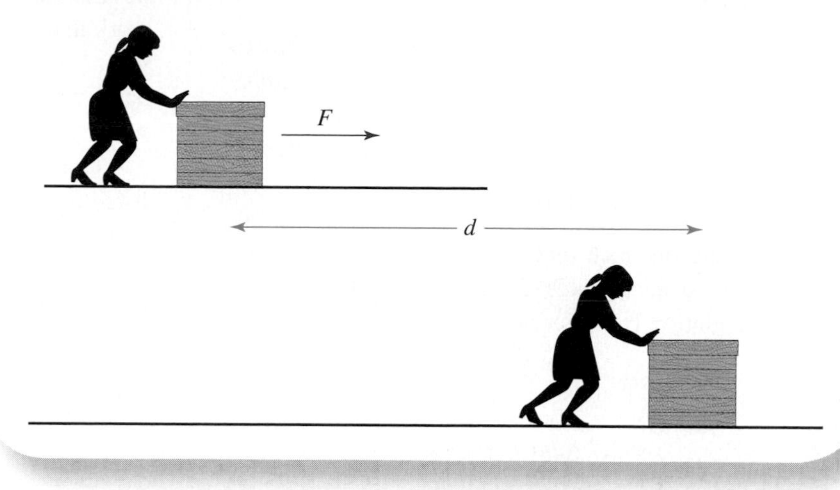

Figure 3.9
Pushing on a box with a force of 100 newtons causes it to slide over the floor. If the box moves 3 meters, you do 300 joules of work.

It is important to note that no work is done on an object if the force is perpendicular to the object's displacement or motion. When you simply carry a box across a room, your force on the box is vertical, whereas the displacement of the box is horizontal. Hence, you do no work on the box (Figure 3.10a).

Uniform circular motion is another situation in which a force acts on a moving body but no work is done. Recall that a centripetal force must act on anything to keep it moving uniformly along a circular path (Figure 3.10b). This force is always toward the center of the circle and perpendicular to the object's velocity at each instant. Therefore, the force does not do work on the object.

Work can be done in a circular motion by a force that is not a centripetal force. When you turn the crank on a pencil sharpener or a fishing reel, for instance, you exert a force on the handle that is in the same direction as the handle's motion. Hence, you do work on the handle.

Work is done on an object when it is accelerated in a straight line. The following example shows how the amount of work can be

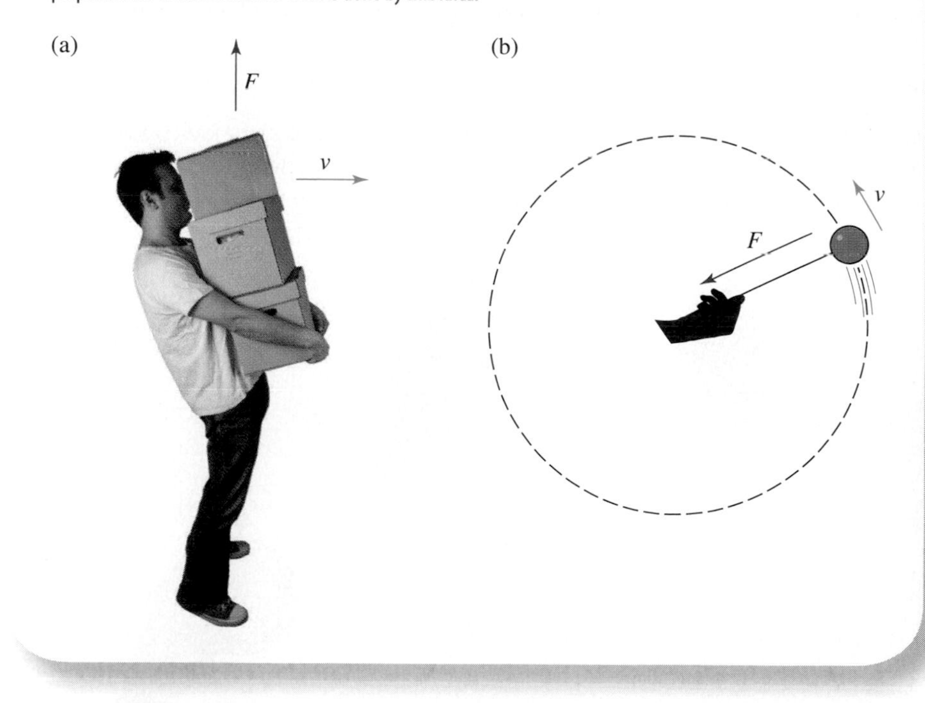

Figure 3.10
(a) When you carry a box across a room, the force on the box is perpendicular to the direction the box moves. No work is done on the box by the force holding it up. (b) The string exerts a force on the ball that is perpendicular to its motion. No work is done by this force.

computed. (In the next section, we will see that there is an easier way to do this.)

EXAMPLE 3.5

In Example 2.2, we used Newton's second law to compute the force needed to accelerate a 1,000-kilogram car from 0 to 27 m/s in 10 seconds. Our answer was $F = 2{,}700$ newtons. How much work is done?

$$\text{work} = Fd$$

To find the distance that the car travels, we use the fact that the car accelerates at 2.7 m/s² for 10 seconds. Using the equation from Section 1.4:

$$\begin{aligned} d &= \frac{1}{2}at^2 \\ &= \frac{1}{2} \times 2.7 \text{ m/s}^2 \times (10 \text{ s})^2 \\ &= 1.35 \text{ m/s}^2 \times 100 \text{ s}^2 = 135 \text{ m} \end{aligned}$$

So the work done is

$$\begin{aligned} \text{work} &= Fd \\ &= 2{,}700 \text{ N} \times 135 \text{ m} \\ &= 364{,}500 \text{ J} \end{aligned}$$

We have seen that work is done (a) when a force acts to move something against the force of gravity (Figures 3.7 and 3.8), (b) when a force acts to move something against friction (Figure 3.9), and (c) when a force accelerates an object. There are many other possibilities. When a force distorts something, work is done. For example, to compress or stretch a spring, a force must act on it. This force acts through a distance in the same direction as the force, so work is done.

Work is also done when a force causes something to slow down. When you catch a ball, your hand exerts a force on the ball. As the ball slows down, it pushes your hand back with a force of equal magnitude and opposite direction (see Figure 3.11). In this case, the ball does work on your hand. Your hand does negative work on the ball. The amount of work that the ball does on your hand is equal to the amount of work that was originally done on the ball to accelerate it (ignoring the effect of air resistance). If you allow your hand to move back as you catch the ball, the force of the ball on you will be less than if you try to keep your hand stationary. The work that the ball will do on your hand is the same either way. Since work = force × distance, the force on your hand will be smaller if the distance that the ball moves while you catch it is larger.

One last example: When an object falls freely, the force of gravity does work on it. As in the previous example with the car, this work goes to accelerate the object. In particular, if a body falls a distance d, the work done on it by the force of gravity is

$$\text{work} = Fd$$

But

$$F = W = mg$$

So:

$$\text{work} = Wd = mgd$$

The work that the force of gravity does on an object as it falls is equal to the work that was done to lift the object the same distance. When something is lifted, we say that work is done *against* the force of gravity. The movement is in the opposite direction of the gravitational force. When something falls, work is done *by* the force of gravity. The movement is in the same direction as the gravitational force.

In summary, work is done by a force whenever the point of application moves in the direction of the force. Work is done against the force whenever the point of

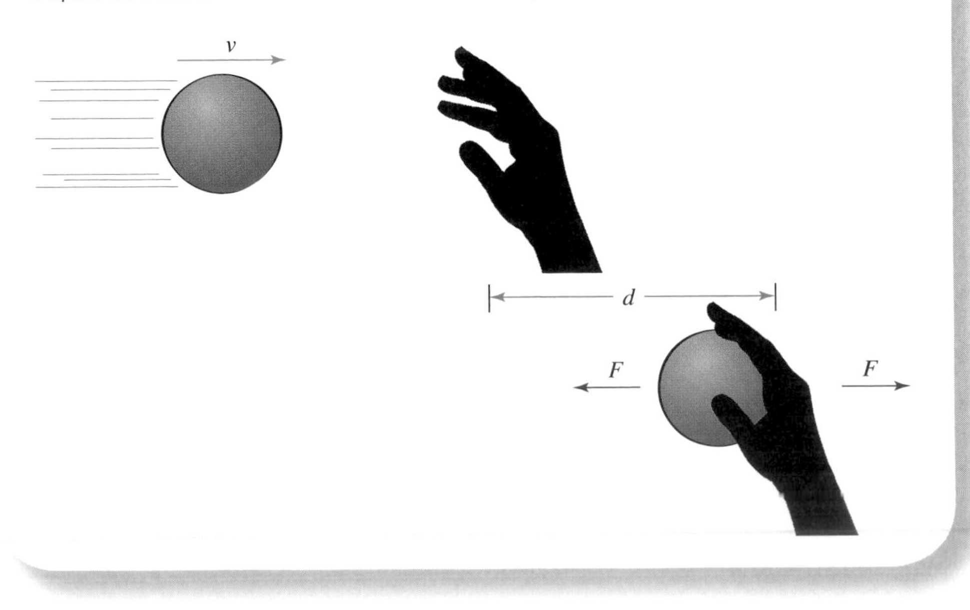

Figure 3.11
As you catch a ball, your hand exerts a force on the ball. By Newton's third law, the ball exerts a force of equal magnitude but opposite direction on your hand. This force does work on you as you slow down the ball. The work done is equal to the work that was done to accelerate the ball in the first place.

application moves opposite the direction of the force. Forces always come in pairs that are equal in magnitude and opposite in direction (Newton's third law of motion). Consequently, when work is done by one force, this work is being done against the other force in the pair.

3.4 Energy

In Section 3.3, we saw that work can often be "recovered." The work that is done when a ball is thrown is equal to the work done by the ball when it is caught. The work done when lifting an object against the force of gravity is equal to the work done by the force of gravity when the object falls. When work is done on something, it gains **energy**. This energy can then be used to do other work later. A bowler gives energy to her ball upon releasing it at the top of the alley. This energy is given up to do work when the ball scatters the pins at the opposite end.

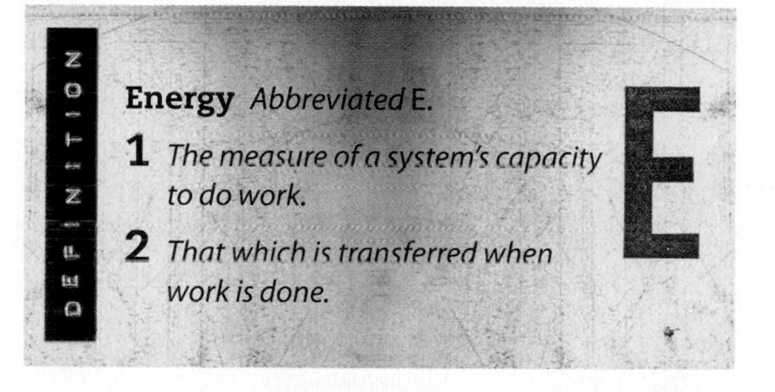

The units of energy are the same as the units of work. Like work, energy is a scalar.

The more work done on something, the more energy it gains, and the more work it can do in return. We might say that energy is "stored work." To be able to do work, an object must have energy. When you throw a ball, you are transferring some energy from you to the ball. The ball can then do work. In Figure 3.10, no energy is transferred because no work is done.

There are many forms of energy corresponding to the many ways in which work can be done. Some of the more common forms of energy are chemical, electrical, nuclear, and gravitational, as well as the energy associated with heat, sound, light, and other forms of radiation. Anything possessing any of these forms of energy is capable of doing work.

In mechanics, there are two main forms of energy, which we can classify under the single heading of *mechanical energy.* Anything that has energy because of its motion or because of its position or configuration has mechanical energy. We refer to the former as **kinetic energy** and to the latter as **potential energy**.

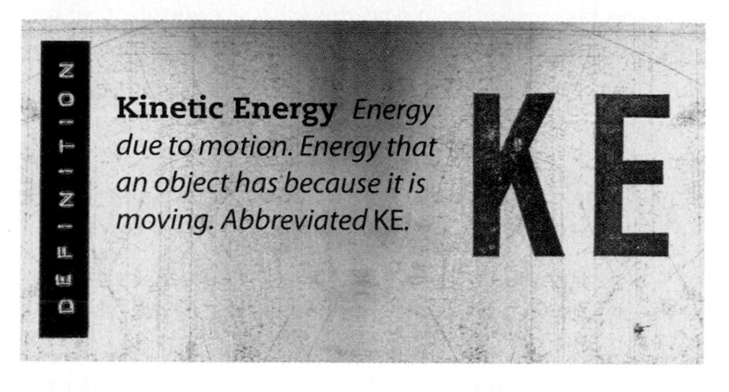

Anything that is moving has kinetic energy. The simplest example is an object moving in a straight line. The amount of kinetic energy an object has depends on its mass and speed. In particular:

$$KE = \frac{1}{2}mv^2 \quad \text{(kinetic energy)}$$

The kinetic energy that an object has is equal to the work done when accelerating the object from rest.* So another way to determine the amount of work done when accelerating an object is to compute its kinetic energy. This calculation also shows that the work done depends only on the object's final speed and not on how rapidly or slowly it was accelerated.

EXAMPLE 3.6

In Example 3.5, we computed the work that is done on a 1,000-kilogram car as it accelerates from 0 to 27 m/s. The car's kinetic energy when it is traveling 27 m/s is

$$KE = \frac{1}{2}mv^2 = \frac{1}{2} \times 1{,}000 \text{ kg} \times (27 \text{ m/s})^2$$
$$= 500 \text{ kg} \times 729 \text{ m}^2/\text{s}^2$$
$$= 364{,}500 \text{ J}$$

This is the same as the work done as the car accelerates. The car can do 364,500 joules of work because of its motion.

---

*For the mathematically inclined, we can compute the work done to accelerate something from rest in the case of constant acceleration.

$$\text{work} = Fd$$

Now we use what we learned in Chapters 1 and 2. The force needed is $F = ma$, and the distance traveled is $d = \frac{1}{2}at^2$. So:

$$\text{work} = Fd = ma \times \tfrac{1}{2}at^2 = \tfrac{1}{2}m \times a^2 \times t^2 = \tfrac{1}{2}m \times (at)^2$$

But at equals the speed v that the body acquires in t seconds. So:

$$\text{work} = \tfrac{1}{2}mv^2 = KE$$

The kinetic energy of a moving body is proportional to the square of its speed. If one car is going twice as fast as a second, identical car, the faster one has four times the kinetic energy. It takes four times as much work to stop the faster car. Note that the kinetic energy of an object can never be negative. (Why? Because m is always positive and v^2 is positive even when v is negative.)

Since speed is relative, kinetic energy is also relative. A runner on a moving ship has KE relative to the ship and a different KE relative to a buoy anchored in the water.

Another way that an object can have kinetic energy is by rotating. To make something spin, work must be done on it. A dancer performing a pirouette has kinetic energy. A spinning top has kinetic energy, as do the Earth, Moon, Sun, and other astronomical objects as they spin about their axes. The amount of kinetic energy that a spinning object has depends on its mass, its rotation rate, and the way its mass is distributed. Many common objects, from simple children's toys to sophisticated hybrid vehicles, employ rotating mechanical elements as a means to "store" energy.

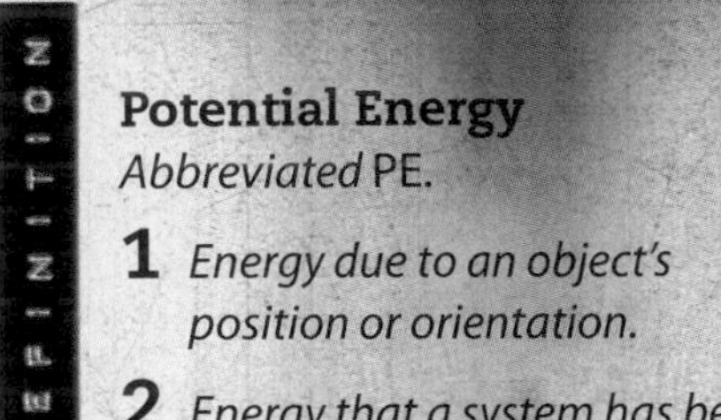

DEFINITION

Potential Energy
Abbreviated PE.

1 *Energy due to an object's position or orientation.*

2 *Energy that a system has because of its configuration.*

PE

© Paul A. Souders / Corbis

A spinning dancer has kinetic energy, even though she stays in one place.

A flywheel stores rotational kinetic energy.

Wind-up toy

© iStockphoto.com / © Vern Ostdiek

The amount of potential energy that a system acquires is equal to the work done to put it into a given configuration. When an object is lifted, it is given potential energy. It can use this energy to do work. For example, when the weights on a cuckoo clock or any other gravity-powered clock are raised, they are given potential energy. As they slowly fall, they do work in operating the clock.

In Section 3.3, we computed the work done when an object is lifted. This work is equal to the potential energy that the object is given. Since the work is done against the force of gravity, it is called **gravitational potential energy**.

$$PE = \text{work done} = \text{weight} \times \text{distance raised}$$
$$= Wd = mgd$$
$$PE = mgd \quad \text{(gravitational potential energy)}$$

Gravitational potential energy is one of the most common types of potential energy, and it is often referred to simply as *potential energy*. Potential energy is a relative quantity because elevation can be measured relative to different levels.

EXAMPLE 3.7

A 3-kilogram brick is lifted to a height of 0.5 meters above a table (Figure 3.12). Its potential energy relative to the table is

$$PE = mgd$$
$$PE = 3\text{ kg} \times 9.8\text{ m/s}^2 \times 0.5\text{ m}$$
$$= 14.7\text{ J} \quad \text{(relative to the table)}$$

But the tabletop itself may be 1 meter above the floor. The brick's height above the floor is 1.5 meters, and so its potential energy relative to the floor is

$$PE = mgd$$
$$= 3\text{ kg} \times 9.8\text{ m/s}^2 \times 1.5\text{ m}$$
$$= 44.1\text{ J} \quad \text{(relative to the floor)}$$

A person sitting in a chair has gravitational potential energy relative to the floor, to the basement of the building, and to the level of the oceans. Usually, some convenient reference level is chosen for determining potential energies. In a room it is handy to use the floor as the reference level for measuring heights and, consequently, potential energies.

An object's potential energy is negative when it is *below* the chosen reference level. Often the reference level is chosen so that negative potential energies signify that an object cannot "escape." For example, it is reasonable to measure potential energy relative to the ground level on flat areas outdoors. Anything on the ground has zero potential energy, and anything above the ground has positive potential energy. If there is a hole in the ground, any object in the hole will have negative potential energy (Figure 3.13). Any object that has zero or positive potential energy is free to move about if it has any kinetic energy. For example, the ball in Figure 3.13 can roll about horizontally on the surface of the ground. Objects with negative potential energy are confined to the hole and cannot escape. They must be given enough energy initially to get out of the hole before they can move freely.

Springs and rubber bands can possess another type of potential energy, **elastic potential energy**. Work must be done on a spring to stretch or compress it, giving the spring potential energy. This "stored energy" can then be used to do work. The actual amount of potential energy a spring has depends on two things: how much it is stretched or compressed and how strong it is. In

Figure 3.12
The potential energy (*PE*) of an object depends on how its distance (*d*) above some reference level is measured. The brick has 14.7 joules of potential energy relative to the table and 44.1 joules of potential energy relative to the floor. In both cases, the potential energy equals the work done to raise the brick from that level.

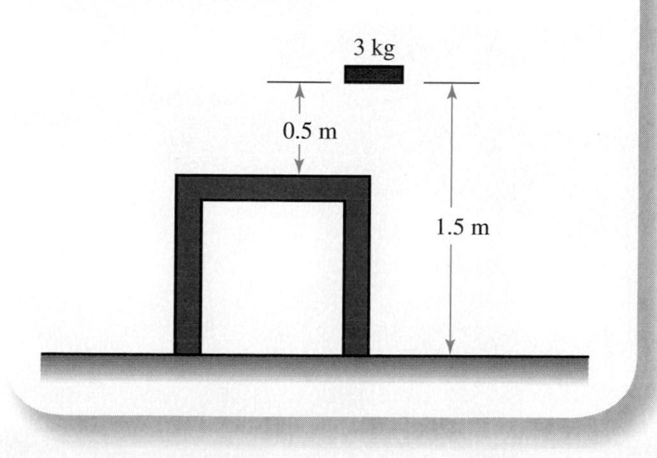

Figure 3.13
The potential energy of golf ball A is positive relative to the ground. The potential energy of B is zero, and that of C is negative because it is below ground level. Balls A and B can move horizontally while C is restricted to the hole.

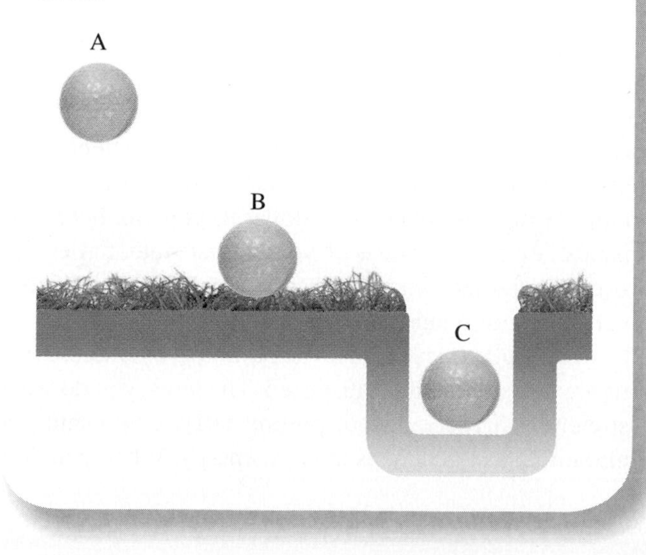

© iStockphoto.com

Figure 3.5, the combined kinetic energies of the carts after the spring is released equal the original elastic potential energy of the spring. A stronger spring would possess more potential energy and would give the carts more *KE*, so they would go faster.

Many devices use elastic potential energy. Toy dart guns have a spring inside that is compressed by the shooter. When the trigger is pulled, the spring is released and does work on the dart to accelerate it. The potential energy of the spring is converted to kinetic energy of the dart. The bow and arrow operate the same way, with the bow acting as a spring. Windup devices such as clocks, toys, and music boxes use energy stored in springs to operate the mechanism. Usually, the spring is in a spiral shape, but the principle is the same. Rubber bands provide lightweight energy storage in some toy airplanes.

Another form of energy important in mechanical systems is **internal energy**. Internal energy, heat, and temperature are discussed formally in Chapter 5. Basically, the internal energy of a substance is the total energy of all the atoms and molecules in the substance. To raise something's temperature, to melt a solid, or to boil a liquid all require increasing the internal energy of the substance. Internal energy decreases when a substance's temperature decreases, when a liquid freezes, or when a gas condenses.

Internal energy is involved whenever there is kinetic friction. In Figure 3.9, the work done on the box as it is pushed across the floor is converted into internal energy because of the friction between the box and the floor. The result is that the temperatures of the floor and the box are raised (although not by much). As a car or a bicycle brakes to a stop, its kinetic energy is converted into internal energy in the brakes. Automobile disc brakes can become red hot under extreme braking. Meteors (shooting stars) are a spectacular example of kinetic energy being converted into internal energy. When they enter the atmosphere at very high speed, air resistance heats them enough to glow and melt. Gravitational potential energy can also be converted into internal energy. A box slowly sliding down a ramp has its gravitational potential energy converted into internal energy by friction. If you climb a rope and then slide down it, you can burn your hands severely as some of your initial potential energy is converted into internal energy because of the friction between your hands and the rope.

Internal energy can also be produced by internal friction when something is distorted. The work you do when stretching a rubber band, pulling taffy, or crushing an aluminum can generates internal energy. When you drop something like a book and it doesn't bounce, most of the energy the object had is converted into internal energy on impact.

Internal energy arising from friction, unlike kinetic energy and potential energy, usually cannot be recovered. Work done to lift a box and give it potential energy can be recovered as work or some other form of energy. Work done to slide a box across a floor becomes internal energy that is "lost" (made unavailable). In every mechanical process, some energy is converted into internal energy. It has been estimated that in the United States the annual financial losses associated with overcoming friction are several hundred billion dollars.

The archer's work bending the bow gives it elastic potential energy.

© Volodymyr Vyshnivetskyy/iStockphoto.com

In summary, work always results in a transfer of energy from one thing to another, in a transformation of energy from one form to another, or both. In our earlier analogy in which we compared energy to financial assets, work plays the role of a transaction such as buying, selling, earning, or trading. These transactions can be used to increase or decrease the net worth of an individual or to convert one form of asset into another. Work done *on* a system increases its energy. Work done *by* the system decreases the energy of the system. Work done *within* the system results in one form of energy being changed into another.

© John R. Foster/Photo Researchers, Inc.

A meteoroid leaves a glowing trail in the night sky. Air resistance causes its kinetic energy to be converted into internal energy.

3.5 The Conservation of Energy

In the preceding section, we described several situations in which one form of energy was converted into another. These included a dart gun (potential energy in a spring converted to kinetic energy of the dart), a car braking (kinetic energy converted into internal energy), and a box sliding down an inclined plane (gravitational potential energy becoming internal energy). There are countless situations involving many other forms of energy transfer or conversion.

Many devices in common use are simply energy converters. Some examples are listed in Table 3.1. Some of these devices involve more than one conversion. In a hydroelectric dam (see Figure 3.14), the potential energy of the water behind the dam is converted into kinetic energy of the water. The moving water then hits a turbine (propeller), which gives it kinetic energy of rotation. The rotating turbine turns a generator, which converts the kinetic energy into electrical energy.

Internal energy is an intermediate form of energy in both a car engine and a nuclear power plant. In a car engine, the chemical energy in fuel is first converted into internal energy as the fuel burns (explodes). The heated

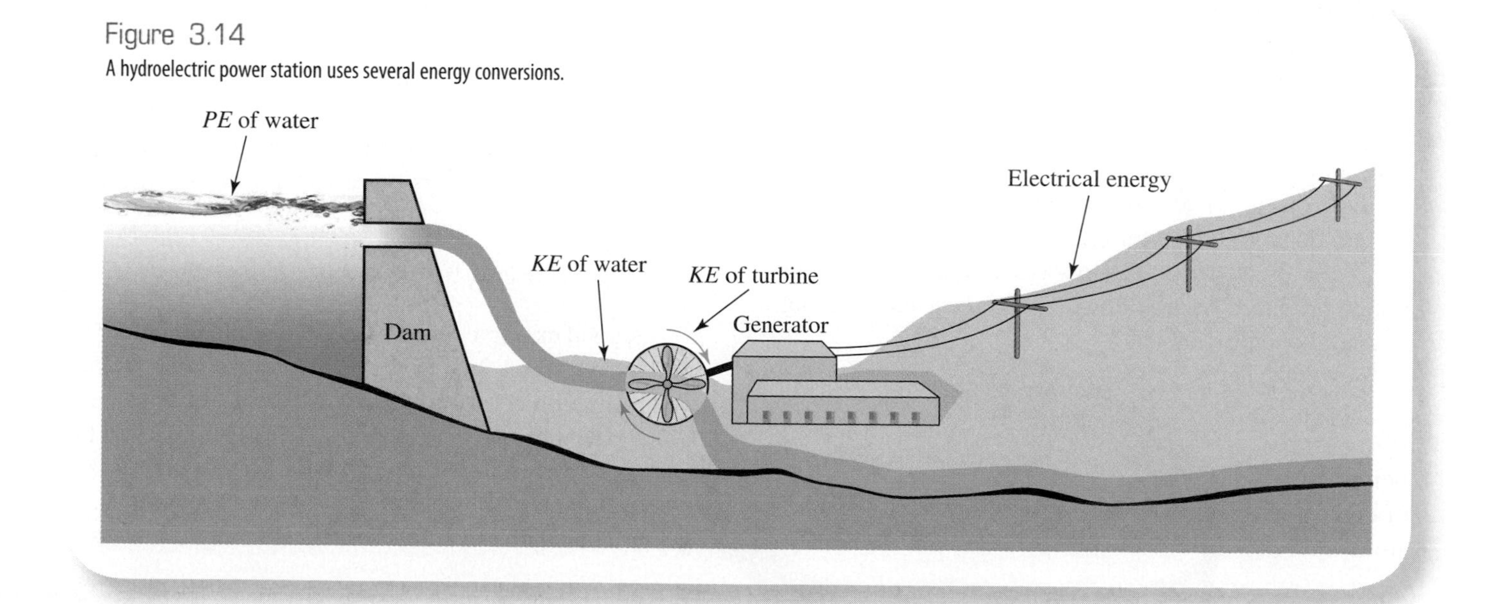

Figure 3.14
A hydroelectric power station uses several energy conversions.

gases expand and push against pistons (or rotors in rotary engines). These in turn make a crankshaft and flywheel rotate to ultimately turn the drive shaft and wheels. In a nuclear power plant, nuclear energy is converted into internal energy in the reactor core. This internal energy is used to boil water into steam. The steam is used to turn a turbine, which turns a generator that produces electrical energy.

Internal energy is also a "wasted" by-product in all of the devices in Table 3.1. Any device that has moving parts has some kinetic friction. Some of the input energy is converted into internal energy by this friction. The generator, electric motor, car engine, and the electrical power plants all have unavoidable friction. In some of the devices, internal energy is produced because of the nature of the process. Over 95% of the electrical energy used by an incandescent light bulb is converted to internal energy, not usable light. Over 60% of the available energy in coal-fired and nuclear power plants goes to unused internal energy. We will investigate this further in later chapters.

Table 3.1 **Some Energy Converters**

Device	Energy Conversion
Light bulb	Electrical energy to radiant energy
Car engine	Chemical energy (in fuel) to kinetic energy
Battery	Chemical energy to electrical energy
Elevator	Electrical energy to gravitational potential energy
Generator	Kinetic energy to electrical energy
Electric motor	Electrical energy to kinetic energy
Solar cell	Radiant energy (light) to electrical energy
Flute	Kinetic energy (of air) to acoustic energy (of sound wave)
Nuclear power plant	Nuclear energy to electrical energy
Hydroelectric dam	Gravitational potential energy (of water) to electrical energy

Even though there are many different forms of energy and countless devices that involve energy conversions, the following law always holds.

LAWS

Law of Conservation of Energy *Energy cannot be created or destroyed, only converted from one form to another. The total energy in an isolated system is constant.*

For a system to be isolated, energy cannot leave or enter it. For an isolated mechanical system, work cannot be done on the system by an outside force, nor can the system do work on anything outside of it.

This law means that energy is a commodity that cannot be produced from nothing or disappear into nothing. If work is being done or a form of energy is "appearing," then energy is being used or converted somewhere.

Energy cannot be produced from nothing or disappear into nothing.

Unlike money, which you can counterfeit or burn, you cannot manufacture or eliminate energy.

The law of conservation of energy is both a practical tool and a theoretical tool. It can be used to solve problems, notably in mechanics, and it is a necessary condition that proposed theoretical models must satisfy. As an example of the latter, a theoretical astrophysicist may develop a model that explains how stars convert nuclear energy into heat and radiation. A first test of the validity of the model is whether or not energy is conserved.

We now illustrate the practical usefulness of the law by considering some mechanical systems. In each case, we assume that friction is negligible, so we do not have to take into account any conversion of potential energy or kinetic energy into internal energy.

The basic approach in using the law of conservation of energy is the same as that used with the corresponding law for linear momentum. If there is a conversion of energy in the system from one form to another, the total energy before the conversion equals the total energy after the conversion.

$$\text{total energy before} = \text{total energy after}$$

A good example that we have encountered before (in Sections 1.4 and 2.6) is the motion of a freely falling body. If an object is raised to a height d, it has gravitational potential energy. If it is released, the object will fall and convert its potential energy into kinetic energy. The conversion process is continuous. As the object falls, its potential energy decreases because its height decreases,

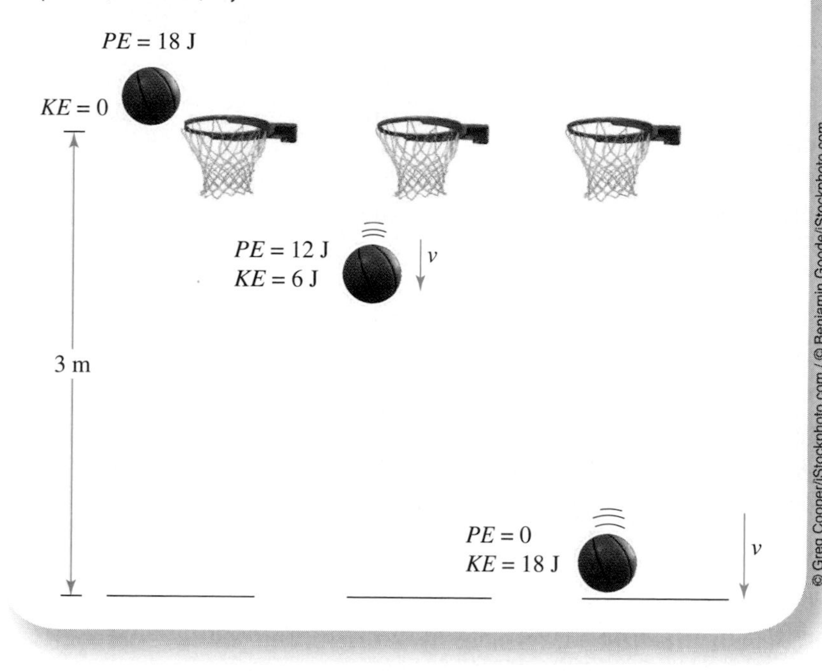

Figure 3.15
A basketball rolls off the rim and falls to the floor. Initially, it has potential energy only. As it falls, its potential energy decreases as its kinetic energy increases. Just before it hits the floor, it has kinetic energy only. At each point as it falls, its total energy, kinetic plus potential, is the same, 18 joules.

© Greg Cooper/iStockphoto.com / © Benjamin Goode/iStockphoto.com

while its kinetic energy increases because its speed increases. If it is falling freely, there is no air resistance, and the only two forms of energy are kinetic and gravitational potential energy. Energy conservation then means that the sum of the kinetic energy and the potential energy of the object is always the same (see Figure 3.15).

$$E = KE + PE = \text{constant}$$

We can use this fact to show how the object's speed just before it hits the floor depends on the height *d*. We do this by looking at the object's energy just as it is dropped and then just before it hits. At the instant it is released, its kinetic energy is zero, because its speed is zero, and its potential energy is *mgd*.

$$E = KE + PE = 0 + PE$$

$$= PE = mgd \quad \text{(just released)}$$

When it reaches the floor, at the instant before impact, its potential energy is zero. So:

$$E = KE + PE = KE + 0$$

$$= KE = \frac{1}{2}mv^2 \quad \text{(just before impact)}$$

These two quantities are equal since the amount of the object's energy has not changed, only the form. The potential energy the object had when it was released equals the kinetic energy it has just before impact.

$$\frac{1}{2}mv^2 = mgd$$

Dividing both sides by m and multiplying by 2, we get:

$$v^2 = 2gd$$

$$v = \sqrt{2gd} \quad \text{(speed after falling distance } d\text{)}$$

EXAMPLE 3.8

In 2003, a man went over Horseshoe Falls, part of Niagara Falls, and survived. He is thought to be the first person to do so without the aid of any safety devices. The height of the falls is about 50 meters. Estimate the speed of the man when he hit the water at the bottom of the falls.

Assuming that the air resistance is too small to affect the motion appreciably, we can use the preceding result:

$$v = \sqrt{2gd} = \sqrt{2 \times 9.8 \text{ m/s}^2 \times 50 \text{ m}}$$

$$v = \sqrt{980 \text{ m}^2/\text{s}^2}$$

$$= 31.3 \text{ m/s} \quad \text{(about 70 mph)}$$

The speed of a freely falling body does not depend on its mass, only on how far it has fallen *(d)* and the acceleration due to gravity. Table 3.2 shows the speed of an object after it has fallen various distances. In Section 1.4, we illustrated the relationships between speed and time and between distance and time. This completes the picture.

We have the reverse situation when an object is thrown or projected straight up. The object starts with kinetic energy, rises until all of that has been converted into potential energy, and then falls (Figure 2.18). Conservation of energy tells us that the kinetic energy that it begins with equals its potential energy at the highest point. This results in the same equation relating the initial speed v to the maximum height reached, *d:*

$$v^2 = 2gd$$

which we can rewrite as:

$$d = \frac{v^2}{2g}$$

To compute how high something will go when thrown straight upward, insert its initial speed for v in this equation. We can also use Table 3.2 "backward": A ball thrown upward at 12 mph will reach a height of 5 feet.

Now let's consider a similar problem. A roller coaster starts from rest at a height d above the ground. It rolls without friction or air resistance down a hill (see Figure 3.16). What is its speed when it reaches the bottom?

Again, the only forms of energy are gravitational potential energy and kinetic energy, since we assume there is no friction. The total energy, which is kinetic energy plus potential energy, is constant. The kinetic energy of the roller coaster at the bottom of the hill must equal its potential energy at the top.

$$KE \text{ (at bottom)} = PE \text{ (at top)}$$

$$\frac{1}{2}mv^2 = mgd$$

This is the same result obtained for a freely falling body. The speed at the bottom is consequently given by the same equation:

$$v = \sqrt{2gd}$$

It would have been impossible to solve this problem using only the tools from Chapter 2. To use Newton's second law, $F = ma$, one needs to know the net force that acts at every instant. The net force driving the roller coaster along its path varies as the slope of the hill changes, so using Newton's second law makes this a very complicated problem. The principle of energy conservation allows us to solve a problem easily that we could not have solved before. In the process, we also come up with the following general result: The law of conservation of energy tells us that for an object affected by gravity but not friction, the speed that it has at a distance *(d)* below its starting point is given by the preceding equation *regardless of the path it takes.* The speed of a roller coaster that rolls down a hill is the same as that of an object that falls vertically from the same height. The roller coaster does take more time to build up that speed (its acceleration is smaller) however, and the freely falling body therefore reaches the ground sooner than does a roller coaster car.

The motion of a pendulum involves the continuous conversion of gravitational potential energy into kinetic energy and back again. Let's say that a child on a swing is pulled back (and up; see Figure 3.17). The child then has gravitational potential energy because of his or her height above the rest position of the swing. When released, the child swings downward, and

Table 3.2 Speed Versus Distance for a Freely Falling Body

A. SI Units		B. English Units	
Distance d (m)	Speed v (m/s)	Distance d (ft)	Speed v (mph)
0	0	0	0
1	4.4	1	5.5
2	6.3	2	7.7
3	7.7	3	9.5
4	8.9	4	11
5	9.9	5	12
10	14	10	17
20	20	20	24
100	44	100	55

Note: Parts A and B are independent. The distances and speeds in A and B are not equivalent.

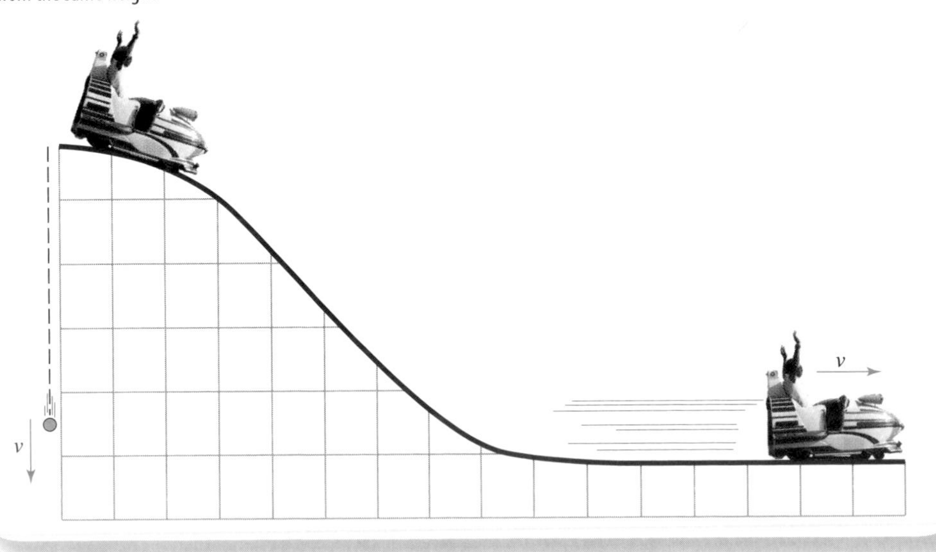

Figure 3.16
As a roller coaster travels down a hill, its potential energy is converted into kinetic energy. If there is no friction, its kinetic energy at the bottom equals its potential energy at the top. Its speed at the bottom is the same as that of an object dropped from the same height.

© Simon Bremner/Getty Images

the potential energy is converted into kinetic energy. At the lowest point in the arc, the child has only kinetic energy, which equals the original potential energy. The child then swings upward and converts the kinetic energy back into potential energy. This continues until the swing stops at a point nearly level with the starting point. This process is repeated over and over as the child swings. Air resistance takes away some of the kinetic energy. If the child is pushed each time, the work done puts energy into the system and counteracts the effect of the air resistance. Without air resistance or any other friction, the child would not have to be pushed each time and would continue swinging indefinitely.

Figure 3.17

As a child swings back and forth, gravitational potential energy is continually converted into kinetic energy and back again. The potential energy at the highest (turning) points equals the kinetic energy at the lowest point.

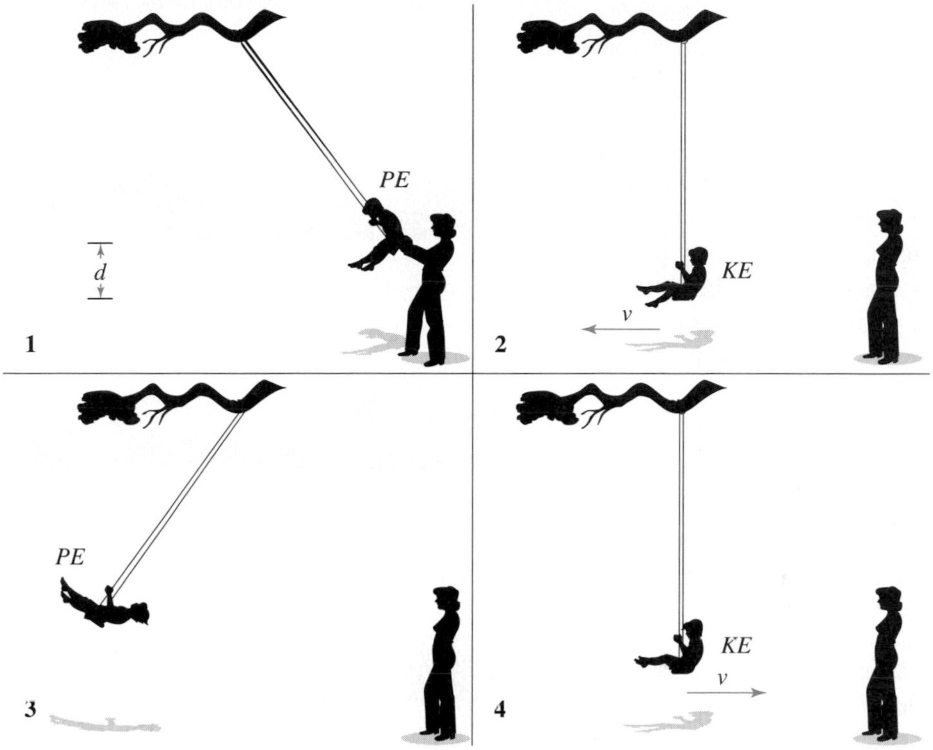

The maximum height that a pendulum reaches (at the turning points) depends on its total energy. The more energy a pendulum has, the higher the turning points. In Section 3.2, we described a way to measure the speed of a bullet or a thrown object (Figure 3.4). The law of conservation of linear momentum is used to relate the speed of the bullet before the collision to the speed of the block (and bullet) afterward. If the wood block is hanging from a string, the kinetic energy it gets from the impact causes it to swing up like a pendulum. The more energy it gets, the higher it will swing. We can determine the speed of the block after impact by measuring how high the block swings. The potential energy of the block (and bullet) at the high point of the swing equals the kinetic energy of the block (and bullet) right after impact. This results in the same equation relating the speed at the low point to the height reached.

$$v = \sqrt{2gd}$$

In Section 2.6, we discussed the motion of an object hanging from a spring (Figure 2.20). This motion also consists of a continual conversion of potential energy into kinetic energy and back again as the spring is alternately compressed and extended.

Figure 3.18 shows someone chipping a ball at a golf course. Since the ball rests in a small valley below ground level, its potential energy is negative relative to the level ground. When the ball is not moving, its total energy is negative because its kinetic energy is zero and its potential energy is negative. The golfer gives the ball kinetic energy by hitting it with a club. A weak stroke gives it enough energy to roll back and forth but not to "escape" from the hole (Figure 3.18a). The ball's total energy is larger but still negative. In Figure 3.18b, the golfer gives the ball more energy by hitting it harder. Because its total energy is still negative, the ball again oscillates back and forth, although reaching a higher point on each side before turning around. In Figure 3.18c, the golfer hits the ball just hard enough for the ball to roll out of the valley and stop once it is out. The ball is given just enough kinetic energy to make its total energy equal to zero, and it "escapes" from the little valley. If the ball were hit even harder, it would escape and have excess kinetic energy—it would continue to roll on the level ground.

This is the principle behind rocking a car when it is stuck. If a tire is in a hole, it is best to make the car oscillate back and forth. By giving it some energy during each cycle, by pushing or by using the engine, one can often give the car enough energy to leave the hole.

There are many analogous systems in physics in which an object is bound unless its total energy is equal to or greater than some value. A satellite in orbit around the Earth is an important example. The satellite's motion from one side of the Earth to the other and back is similar to the motion of the golf ball in the valley. If it is given enough energy, the satellite will escape from the Earth and move away, much like the golf ball. The minimum speed that will give a satellite enough energy to leave the Earth is called the *escape velocity*, approximately 11,200 m/s or 25,000 mph.

When water boils, the individual water molecules are given sufficient energy to break free from the liquid (Chapter 5). Sparks and lightning occur only after electrons are given enough energy to break free from their atoms (Chapter 7). The transition of a system from a bound state to a free state is quite common in physics and is not limited to mechanical systems.

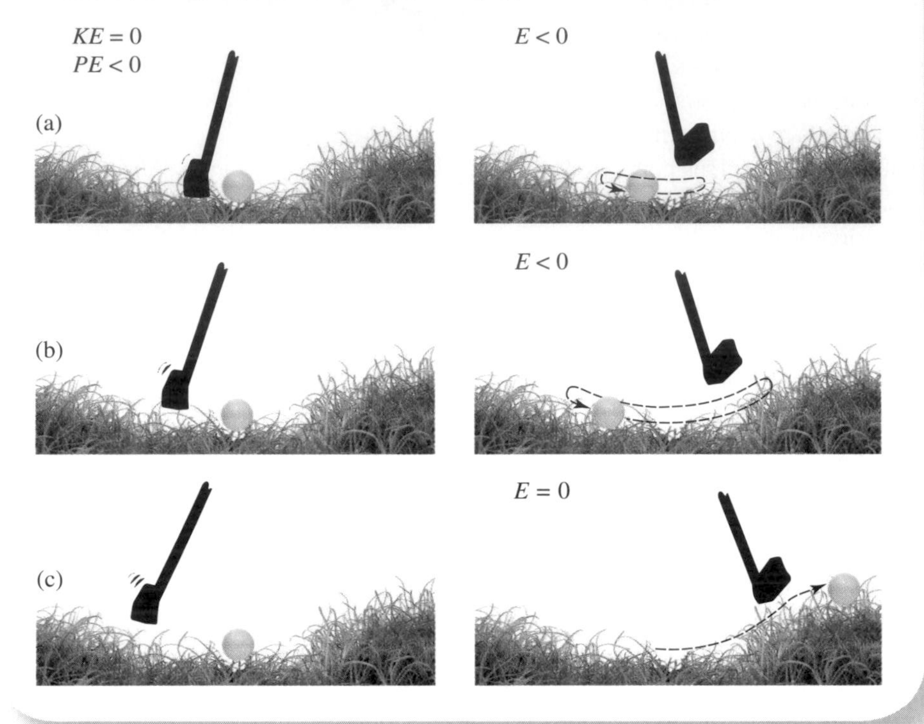

Figure 3.18

A golf ball at rest in the small valley has negative potential energy. Hitting the golf ball gives it kinetic energy, but it oscillates inside the valley if its total energy is negative, (a) and (b). If the golf ball is given enough kinetic energy to make its total energy zero, it rolls out of the valley and stops (c).

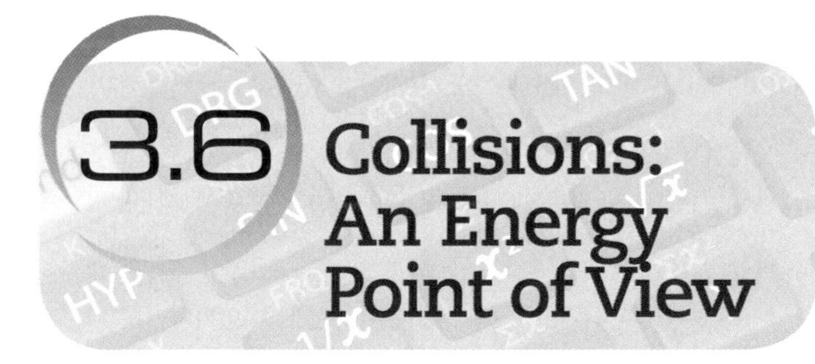

3.6 Collisions: An Energy Point of View

Earlier in this chapter we pointed out that the main "tool" for studying all collisions is the law of conservation of linear momentum (Section 3.2). In this section, we look at collisions from an energy standpoint. In some collisions, the only form of energy involved, before and after, is kinetic energy. In other collisions, forms of energy like potential energy and internal energy play a role. Collisions can be classified as follows.

DEFINITION

Elastic Collision *A collision in which the total kinetic energy of the colliding bodies after the collision equals the total kinetic energy before the collision.*

Inelastic Collision *A collision in which the total kinetic energy of the colliding bodies after the collision* is not equal to *the total kinetic energy before. The total kinetic energy after can be greater than, or less than, the total kinetic energy before the collision.*

In an elastic collision, kinetic energy is conserved. The total energy is *always* conserved in both types of collisions, but in elastic collisions no energy conversions take place that make the total kinetic energy after different from the total kinetic energy before.

Figure 3.19 illustrates examples of these two types of collisions. Two equal-mass carts traveling with the same speed but in opposite directions collide. In both collisions, the total linear momentum before the collision equals the total momentum after. (This total is equal to zero. Why?) In Figure 3.19a, the carts bounce apart because of a spring attached to one of them. After the collision, each cart has the same speed it had before, but it is going in the opposite direction. Consequently, the total kinetic energy of the two carts is the same after the collision as it was before. This is an elastic collision.

Figure 3.19b is an example of an inelastic collision. This time the two carts stick together (because of putty on one of them) and stop. The total kinetic energy after the collision is zero in this case. The automobile collision analyzed in Example 3.2 (Figure 3.3) is also an inelastic collision. To see this, use the information from that example to calculate the total kinetic energy before and after the collision.

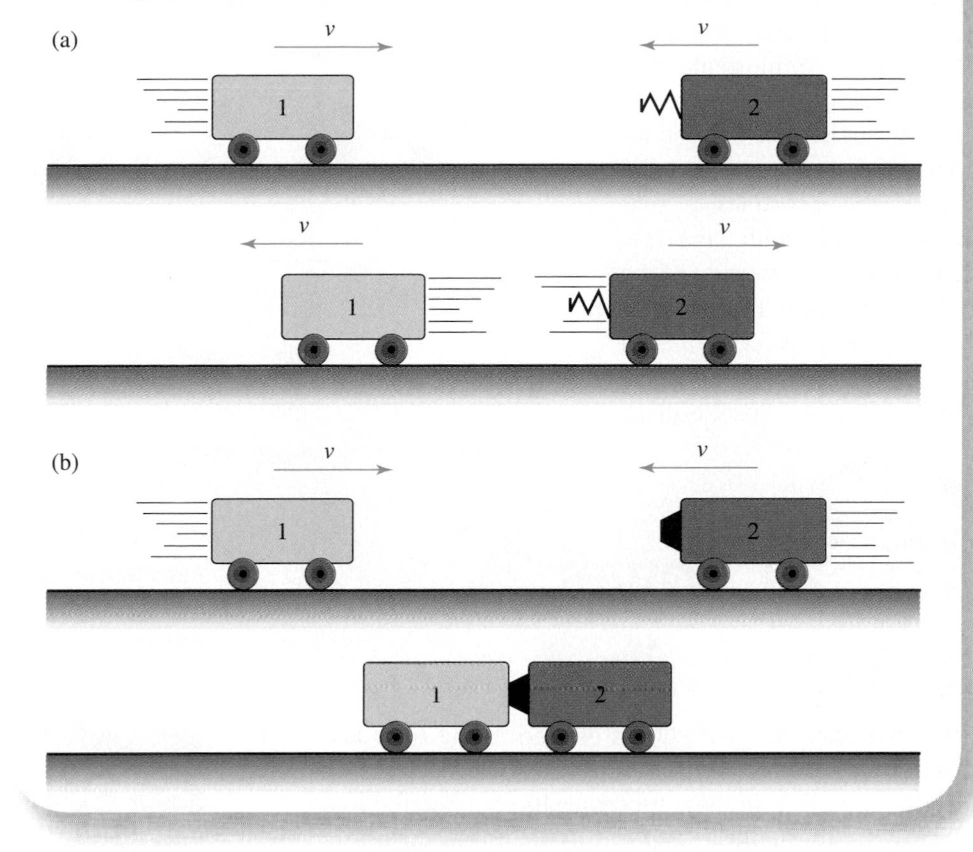

Figure 3.19

(a) Two carts with the same mass and speed collide head on and bounce apart. The total kinetic energy of the carts is the same before and after the collision. (b) This time the two carts stick together after the collision. In this case, the kinetic energy after the collision is zero.

EXAMPLE 3.9

Recall the automobile collision analyzed in Example 3.2 (Figure 3.3). Compare the amounts of kinetic energy in the system before and after the collision.

The kinetic energy before the collision was

$$KE_{\text{before}} = \frac{1}{2} \times 1{,}000 \text{ kg} \times (10 \text{ m/s})^2$$

$$= 50{,}000 \text{ J}$$

The kinetic energy after the collision was

$$KE_{\text{after}} = \frac{1}{2} \times 2{,}500 \text{ kg} \times (4 \text{ m/s})^2$$

$$= 20{,}000 \text{ J}$$

So 30,000 joules (60%) of the kinetic energy before the collision was converted into other forms of energy.

In these two examples of inelastic collisions, part or all of the original kinetic energy of the colliding bodies is converted into other forms of energy, mostly internal energy, but also some sound (the "crash" that we would hear). In Figure 3.19b, *all* of the kinetic energy is converted into other forms of energy.

In some collisions, the total kinetic energy after the collision is greater than the total kinetic energy before the collision. If our old friend the cart, with its plunger pushed in ("loaded"), is struck by a second cart, the plunger will be released by the shock (Figure 3.20). The plunger's potential energy is transferred to both carts as kinetic energy. Therefore, the total kinetic energy after the collision is greater than the total kinetic energy before the collision. This is an *inelastic* collision. The stored energy is released by the collision.

The use of collisions is an invaluable tool in studying the structure and properties of atoms and nuclei. Much of the information in Chapters 10, 11, and 12 was gleaned from the careful analysis of countless collisions. Linear accelerators, cyclotrons, betatrons, and other devices produce high-speed collisions between atoms, nuclei, and subatomic particles. The collisions are recorded and analyzed using the law of conservation of linear momentum and other principles. If a collision is inelastic, the amount of kinetic energy "lost" or "gained" in the collision is useful for determining the properties of the colliding particles.

Collisions are also responsible for other phenomena such as gas pressure and the conduction of heat. Air molecules colliding with the inner surface of a balloon keep the balloon inflated. When you touch a piece of ice, the molecules in your finger collide with, and lose energy to, the molecules of the ice. This lowers the temperature of your finger. (See related article on the slingshot effect for more on elastic collisions.)

Figure 3.20

Cart 2 has energy stored in its spring-loaded plunger. When this cart is struck by cart 1, this potential energy is converted into kinetic energy, which is then shared by both carts. The total kinetic energy after the collision is greater than the total kinetic energy before the collision.

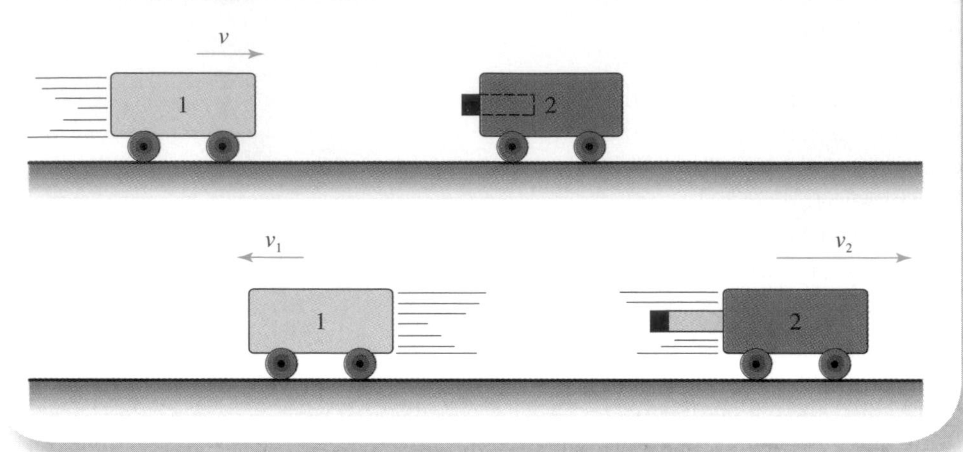

3.7 Power

We have seen many examples of work being done and energy being transformed into other forms. The amount of time involved in these processes has not entered into the discussion until now. Let's say that a ton of bricks needs to be loaded from the ground onto a truck (see Figure 3.21). This might be done in one of two ways. First, a person could lift the bricks one at a time and place them on the truck. This might take the person an hour. Second, a forklift could be used to load the bricks all at once. This might take only 10 seconds. In both cases, the same amount of work is done. The force on each brick (its weight) times the distance it is moved (the height of the truck bed) is the same whether the bricks are loaded one at a time or all at once. The work done, Fd, is the same in both cases, but the **power** is different.

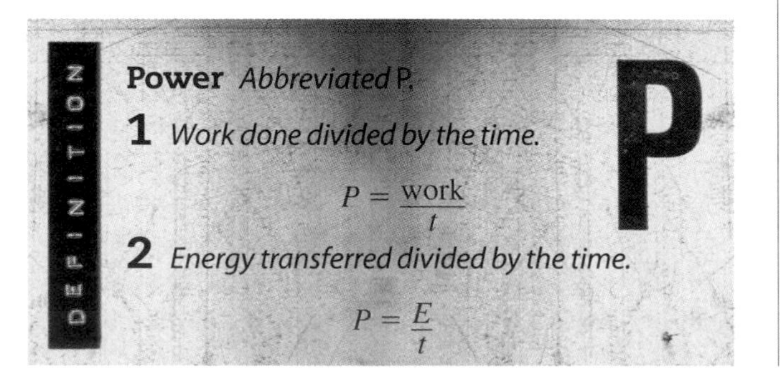

Figure 3.21

A ton of bricks is loaded onto a truck in two different ways. The work done is the same, but since the forklift does the job much faster, the power is much greater.

In SI units, the weight of a ton of bricks is 8,900 newtons. If the height of the truck bed is 1.2 meters, then the work done is

$$\text{work} = Fd = 8{,}900 \text{ N} \times 1.2 \text{ m} = 10{,}680 \text{ J}$$

If the forklift does this work in 10 seconds, the power is

$$P = \frac{\text{work}}{t} = \frac{10{,}680 \text{ J}}{10 \text{ s}}$$

$$= 1{,}068 \text{ J/s} = 1{,}068 \text{ W}$$

© Sergii Tsololo/iStockphoto.com / © Tricia Shay Photography/Getty Images

SLINGSHOT EFFECT

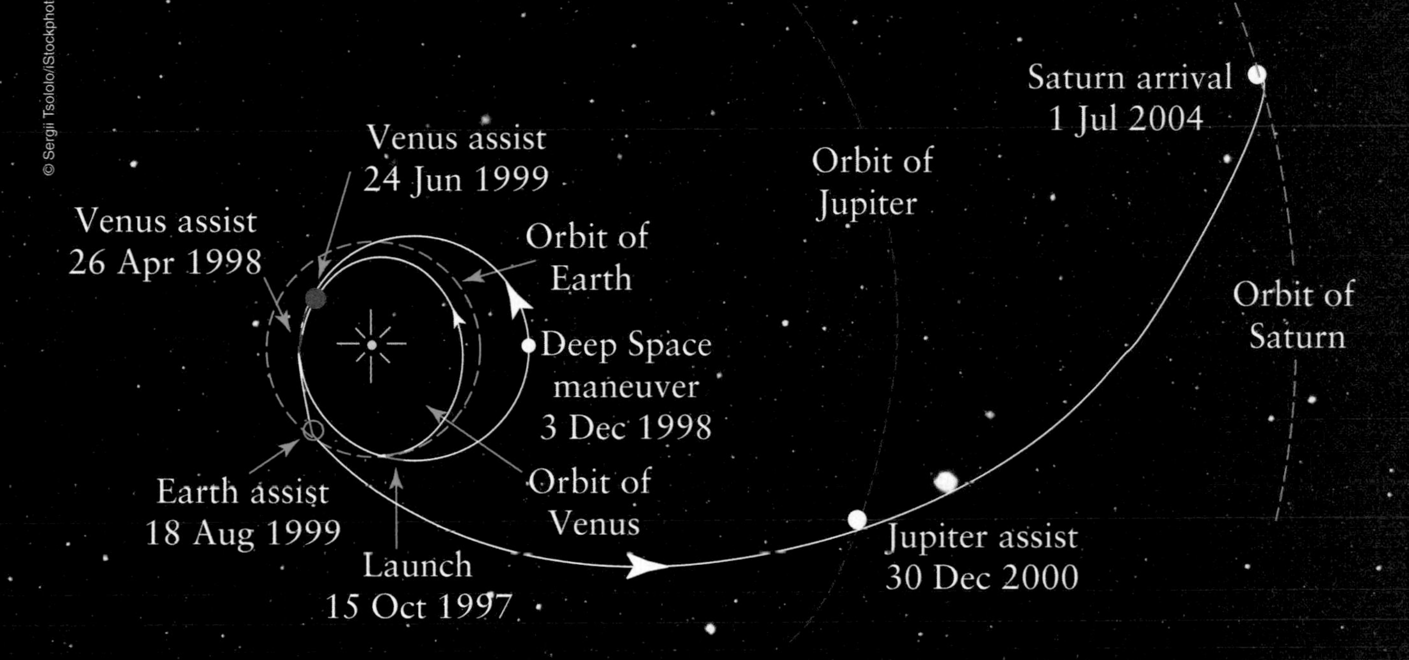

Elastic collisions involving the force of gravity (and no physical contact) are used in space exploration. Called the *slingshot effect* or *gravity assist*, the technique involves having a spacecraft overtake a planet and pass it on its side away from the Earth. The spacecraft gains kinetic energy from the planet, just as the eight ball gains kinetic energy in the collision depicted in Figure 3.2. The planet loses kinetic energy, but the planet is so huge that its decrease in speed is imperceptible. This effect is like that of a volleyball bouncing off the front of a moving locomotive. Space probes sent to the outer part of the solar system (beyond Jupiter) have relied on gravity assists. The *Voyager 2* spacecraft used gravity assists from Jupiter, Saturn, and Uranus on its journey to Neptune and beyond. (Its speed was increased by 10 miles per second by the Jupiter gravity assist.)

Two recent missions exploited gravity assists from the inner planets, thereby allowing heavy spacecraft to be started on their journeys using relatively small rockets. On the way to its 1995 arrival at Jupiter, the *Galileo* spacecraft used two gravity assists from Earth and one from Venus. The 6-ton *Cassini* spacecraft used two gravity assists from Venus, one from Earth, and one from Jupiter to reach Saturn in 2004.

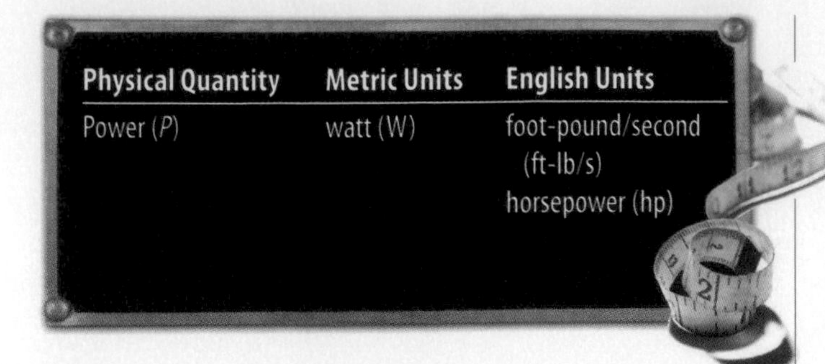

Physical Quantity	Metric Units	English Units
Power (*P*)	watt (W)	foot-pound/second (ft-lb/s) horsepower (hp)

The unit joule per second (J/s) is defined to be the watt (W), the SI unit of power.

$$1 \text{ watt} = \frac{1 \text{ joule}}{1 \text{ second}}$$
$$1 \text{ W} = 1 \text{ J/s}$$

The watt should be familiar to you because it is commonly used to measure the power consumption of electrical devices. A 60-watt light bulb uses electrical energy at the rate of 60 joules each second. A 1,600-watt hair dryer uses 1,600 joules of energy each second.

Horsepower is the most commonly used unit of power in the English system of units. Automobile engines, lawn mowers, and many other motorized devices are rated in horsepower. The basic power unit, pound-foot per second, is the unit of work, foot-pound, divided by the unit of time, second. The conversion factors are

$$1 \text{ hp} = 550 \text{ lb-ft/s} = 746 \text{ W}$$

A device that could raise 550 pounds a distance of 1 foot in 1 second would output 1 horsepower. Raising 110 pounds a distance of 5 feet in 1 second would also require 1 horsepower.

The relationship between power and work (or energy) is the same as that between speed and distance. Power is the rate of change of work (energy). Speed is the rate of change of distance.

$$P = \frac{\text{work}}{\text{time}} \longleftrightarrow v = \frac{\text{distance}}{\text{time}}$$

A runner and a bicyclist can both travel a distance of 10 miles, but the latter can do it much faster because a bicycle is capable of much higher speeds. A person and a forklift can both do 10,680 joules of work raising the bricks, but the forklift can do it much faster because it has more power.

EXAMPLE 3.10

In Examples 2.2 and 3.5, we computed the acceleration, force, and work for a 1,000-kilogram car that goes from 0 to 27 m/s in 10 seconds. We can now determine the required power output of the engine. The work, 364,500 joules, is done in 10 seconds. Hence the power is

$$P = \frac{\text{work}}{t} = \frac{364{,}500 \text{ J}}{10 \text{ s}}$$
$$= 36{,}450 \text{ W} = 48.9 \text{ hp}$$

The car's kinetic energy when going 27 m/s is also 364,500 joules (see Example 3.6). It takes the engine 10 seconds to give the car this much kinetic energy, so we get the same result using energy divided by time.

Given enough food or fuel, there is usually no limit to how much work a device or a person can do. But there is a limit on how fast the work can be done. The power is

© Ruslan Olinchuk/iStockphoto.com / © Stock Image/Jupiterimages

The *Gossamer Albatross* became the first human-powered aircraft to cross the English Channel (12 June 1979). The pilot turned the propeller using a bicycle-type mechanism. A steady power output of about 250 watts was required for the flight, which lasted 2 hours and 49 minutes.

© Jim Sugar/Corbis / © Shelly Perry/iStockphoto.com

limited. In other words, only so much work can be done each second. The power output can be anything from zero (no work) to some maximum. For example, a 100-horsepower automobile engine can put out from 0 to 100 horsepower. When accelerating as fast as possible, the engine is putting out its maximum power. While cruising down a flat highway, the engine may be putting out only 10–20 horsepower, enough to counteract the effects of air resistance and other frictional forces.

The human body has a maximum power output that varies greatly from person to person. In the act of jumping, an outstanding athlete can develop more than 8,000 watts, but only for a fraction of a second. The same person would have a maximum of less than 800 watts if the power level had to be maintained for an hour. The average person can produce 800 watts or more for a few seconds and perhaps 100–200 watts for an hour or more. When running, the body uses energy to overcome friction and air resistance. In short races, the best runners can maintain a speed of 10 m/s for about 20 seconds. In longer races, the speeds are lower because the power level has to be maintained longer. For a race lasting about 30 minutes, the best average speed is about 6 m/s.

3.8 Rotation and Angular Momentum

Our final conservation law applies to rotational motion, as in a spinning ice skater or a satellite moving in orbit around the Earth. You might say that this law is the rotational analogue or counterpart of the law of conservation of linear momentum.

LAWS

Law of Conservation of Angular Momentum *The total angular momentum of an isolated system is constant.*

For a system to be isolated so that the law of conservation of angular momentum applies, the only net external force that can act on the object must be directed toward or away from the center of the object's motion. The centripetal force required to keep an object moving in a circle fits this condition. A force that acts in any direction other than toward or away from the center of motion produces what is called *torque* (from the Latin word for "twist"). When a spacecraft fires its rocket engines to reenter the atmosphere, the force on it acts opposite to its direction of motion and therefore produces a torque that decreases its angular momentum. Torque is the rotational analogue of force: A net external force changes an object's linear momentum, and a net external torque changes an object's angular momentum.

But what exactly is **angular momentum**? We first introduce angular momentum for the simple case of a body moving in a circle, and then extend it to motion along noncircular paths. Imagine a small object moving along a circular path. In this case, the angular momentum equals the product of the object's mass, its speed, and the radius of its path.

$$\text{angular momentum} = mvr \quad \text{(circular path)}$$

Notice that the angular momentum of the object is also equal to its linear momentum (mv) multiplied by the radius of its circular path.

To illustrate the law of conservation of angular momentum, imagine a ball circling overhead on the end of a string that passes through a tube (Figure 3.22a). (We assume there is no friction or air resistance.) The faster the ball goes, the greater its orbital angular momentum. Using a longer string for the same speed would also make its angular momentum larger. Imagine suddenly shortening the string by pulling downward on the end with your free hand, letting the string slide through the tube. This makes the ball move in a circle with a smaller radius but with a higher speed (Figure 3.22b). Since the force exerted on the ball is directed toward the center of its motion, angular momentum is conserved—it has the same value before and after the change. Because the radius of the ball's path is now smaller, its speed must be higher. If you let the string out so the radius is larger, the ball will slow down, keeping the angular momentum constant.

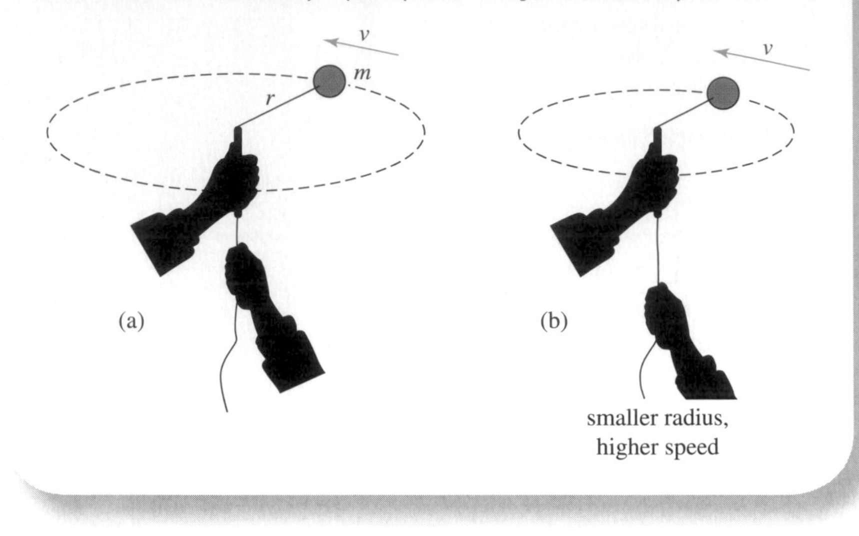

Figure 3.22

(a) The angular momentum of an object moving in a circle equals *m* times *v* times *r*.

(b) If the radius is decreased, the object speeds up so that the angular momentum stays the same.

In the process of pulling the string downward, you do work. This work increases the kinetic energy of the ball.

With caution, we can use this definition of angular momentum—*mvr*—for an object moving in a path other than a circle. Figure 3.23 shows the elliptical orbit of a satellite moving around the Earth. At points *A* and *B*, the satellite's velocity will be perpendicular to a line from the satellite to the Earth's center. At these points the satellite's path is like a short segment of a circle. Consequently, the satellite's angular momentum is *mvr* at those points. At point *B*, *r* is smaller than at point *A*. Because the angular momentum is the same at *A* as at *B*, the satellite's speed is greater at *B*. For example, if the satellite is 13,000 kilometers (about 8,000 miles, twice the Earth's radius) from the Earth's center at point *B* and 26,000 kilometers from the Earth's center at point *A*, its speed at *B* will be two times its speed at *A*. The actual values for the speeds are about 6,400 m/s (about 14,000 mph) when it is closest to the Earth and about 3,200 m/s when it is farthest away.

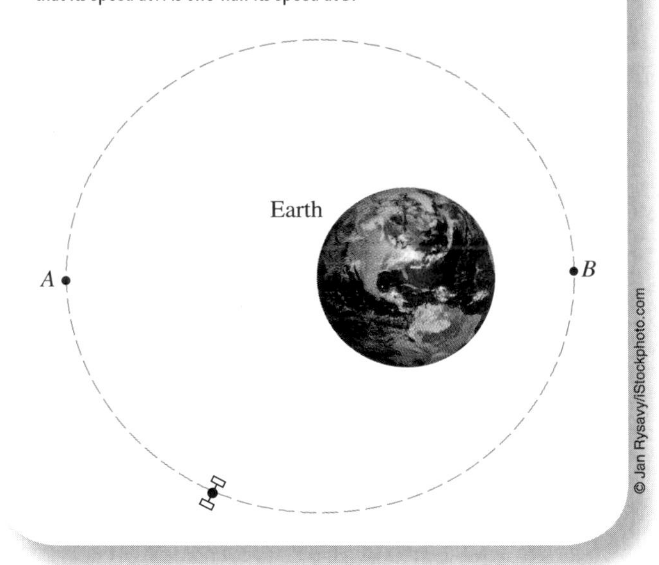

Figure 3.23

A satellite in orbit around the Earth is twice as far from the Earth's center at point *A* as it is at point *B*. Conservation of angular momentum then tells us that its speed at *A* is one-half its speed at *B*.

© Jan Rysavy/iStockphoto.com

An object spinning about an axis, like a top or an ice skater doing a spin, has angular momentum. We can think of each part of the object as moving in a circle and having orbital angular momentum. For example, a spinning skater's hands, arms, shoulders, and other body parts are all moving in circles. The combined angular momentum of the parts of a spinning body remains constant if no torque acts on it. The rate of spinning can be increased or decreased by repositioning parts of the object. For example, the spinning ice skater can start with arms extended outward. When the arms are pulled in closer to the body, the skater spins faster. Each part of the skater's arms has a certain amount of angular momentum as it moves in a circle with a certain radius. Pulling the arms in decreases the radius. As the radius of rotation decreases, the speed of rotation must increase to keep angular momentum constant. Consequently the rest of the skater's body spins faster, so that the total angular momentum remains the same. The reverse happens if the arms are moved out—the skater slows down.

questions&problems

Questions

(▶ Indicates a review question, which means it requires only a basic understanding of the material to answer. The others involve integrating or extending the concepts presented thus far.)

1. ▶What is a conservation law? What is the basic approach taken when using a conservation law?
2. ▶Why is the alternate form of Newton's second law of motion given in this chapter the more general form?
3. Could the linear momentum of a turtle be greater than the linear momentum of a horse? Explain why or why not.
4. An astronaut working with many tools some distance away from a spacecraft is stranded when the "maneuvering unit" malfunctions. How can the astronaut return to the spacecraft by sacrificing some of the tools?
5. ▶For what type of interaction between bodies is the law of conservation of linear momentum most useful?
6. Describe several things you have done today that involved doing work. Are you doing work right now?
7. If we know that a force of 5 N acts on an object while it moves 2 meters, can we calculate how much work was done with no other information? Explain.
8. During a head-on collision of two automobiles, the occupants are decelerated rapidly. Use the idea of work to explain why an air bag that quickly inflates in front of an occupant reduces the likelihood of injury.
9. When climbing a flight of stairs, do you do work on the stairs? Do the stairs do work on you? Explain.
10. People and machines around us do work all the time. But is it possible for things like magnets and the Earth to do work? Explain.
11. How can you give energy to a basketball? Is there more than one way?
12. ▶Identify as many different forms of energy as you can that are around you at this moment.
13. When you throw a ball, the work you do equals the kinetic energy the ball gains. If you do *twice* as much work when throwing the ball, does it go *twice as fast?* Explain.
14. ▶An object has kinetic energy, but it stays in one place. What must it be doing?
15. ▶How can the gravitational potential energy of something be negative?
16. ▶What is elastic potential energy? Is there anything around you now that possesses elastic potential energy?
17. Identify the energy conversions taking place in each of the following situations. Name all of the relevant forms of energy that are involved.
    a) A camper rubbing two sticks together to start a fire.
    b) An arrow shot straight upward, from the moment the bowstring is released by the archer to the moment when the arrow reaches its highest point.
    c) A nail being pounded into a board, from the moment a carpenter starts to swing a hammer to the moment when the nail has been driven some distance into the wood by the blow.
    d) A meteoroid entering the Earth's atmosphere.
18. Solar-powered spotlights have batteries that are charged by solar cells during the day and then operate lights at night. Describe the energy conversions in this entire process, starting with the Sun's nuclear energy and ending with the light from the spotlight being absorbed by the surroundings. Name all of the forms of energy that are involved.
19. Truck drivers approaching a steep hill that they must climb often increase their speed. What good does this do, if any?
20. If you hold a ball at eye level and drop it, it will bounce back, but not to its original height. Identify the energy conversions that take place during the process, and explain why the ball does not reach its original level.
21. A ball is thrown straight upward from the surface of the Moon. Is the maximum height it reaches less than, equal to, or greater than the maximum height reached by a ball thrown upward on the Earth with the same initial speed (no air resistance in both cases)? Explain.
22. ▶Describe the distinction between elastic and inelastic collisions. Give an example of each.
23. Many sports involve collisions, between things—like balls and rackets—and between people—as in football or hockey. Characterize the various sports collisions as elastic or inelastic.
24. Carts A and B stick together whenever they collide. The mass of A is twice the mass of B. How could you roll the carts toward each other in such a way that they would be stopped after the collision? (Assume there is no friction.)
25. Is it possible for one object to gain mechanical energy from another without touching it? Explain.
26. Two cranes are lifting identical steel beams at the same time. One crane is putting out twice as much power as the other. Assuming friction is negligible, what can you conclude is happening to explain this difference?
27. A person runs up several flights of stairs and is exhausted at the top. Later the same person walks up the same stairs and does not feel as tired. Why is this? Ignoring air resistance, does it take more work or energy to run up the stairs than to walk up?
28. ▶How can a satellite's speed decrease without its angular momentum changing?
29. ▶Why do divers executing midair somersaults pull their legs in against their bodies?
30. It is possible for a body to be both spinning and moving in a circle in such a way that its total angular momentum is zero. Describe how this can be.

Problems

1. A sprinter with a mass of 65 kg reaches a speed of 10 m/s during a race. Find the sprinter's linear momentum.
2. Which has the larger linear momentum: a 2,000-kg houseboat going 5 m/s or a 600-kg speedboat going 20 m/s?
3. In Section 2.4, we computed the force needed to accelerate a 1,000-kg car from 0 to 27 m/s in 10 s. Compute the force using the alternate form of Newton's second law. The change in momentum is the car's momentum when going 27 m/s minus its momentum when going 0 m/s.
4. A runner with a mass of 80 kg accelerates from 0 to 9 m/s in 3 s. Find the net force on the runner using the alternate form of Newton's second law.
5. A pitcher throws a 0.5-kg ball of clay at a 6-kg block of wood. The clay sticks to the wood on impact, and their joint velocity afterward is 3 m/s. What was the original speed of the clay?
6. A 3,000-kg truck runs into the rear of a 1,000-kg car that was stationary. The truck and car are locked together after the collision and move with speed 9 m/s. What was the speed of the truck before the collision?
7. A 50-kg boy on roller skates moves with a speed of 5 m/s. He runs into a 40-kg girl on skates. Assuming they cling together after the collision, what is their speed?
8. Two persons on ice skates stand face to face and then push each other away (Figure 3.24). Their masses are 60 and 90 kg. Find the ratio of their speeds immediately afterward. Which person has the higher speed?

Figure 3.24

9. A loaded gun is dropped on a frozen lake. The gun fires, with the bullet going horizontally in one direction and the gun sliding on the ice in the other direction. The bullet's mass is 0.02 kg, and its speed is 300 m/s. If the gun's mass is 1.2 kg, what is its speed?
10. A running back with a mass of 80 kg and a speed of 8 m/s collides with, and is held by, a 120-kg defensive tackle going in the opposite direction. How fast must the tackle be going before the collision for their speed afterward to be zero?
11. A motorist runs out of gas on a level road 200 m from a gas station. The driver pushes the 1,200-kg car to the gas station. If a 150-N force is required to keep the car moving, how much work does the driver do?
12. In Figure 3.7, the rock weighs 100 lb and is lifted 1 ft by the lever.
    a) How much work is done?
    b) The other end of the lever is pushed down 3 ft while lifting the rock. What force had to act on that end?
13. A weight lifter raises a 100-kg barbell to a height of 2.2 m. What is the barbell's potential energy?
14. A microwave antenna with a mass of 80 kg sits atop a tower that is 50 m tall. What is the antenna's potential energy?
15. A personal watercraft and rider have a combined mass of 400 kg. What is their kinetic energy when they are going 15 m/s?
16. As it orbits the Earth, the 11,000-kg Hubble Space Telescope travels at a speed of 7,900 m/s and is 560,000 m above the Earth's surface.
    a) What is its kinetic energy?
    b) What is its potential energy?
17. The kinetic energy of a motorcycle and rider is 60,000 J. If their total mass is 300 kg, what is their speed?
18. In compressing the spring in a toy dart gun, 0.5 J of work is done. When the gun is fired, the spring gives its potential energy to a dart with a mass of 0.02 kg.
    a) What is the dart's kinetic energy as it leaves the gun?
    b) What is the dart's speed?
19. A worker at the top of a 629-m-tall television transmitting tower in North Dakota accidentally drops a heavy tool. If air resistance is negligible, how fast is the tool going just before it hits the ground?
20. A student drops a water balloon out of a dorm window 12 m above the ground. What is its speed when it hits the ground?
21. A child on a swing has a speed of 7.7 m/s at the low point of the arc (Figure 3.25). How high will the swing be at the high point?

Figure 3.25

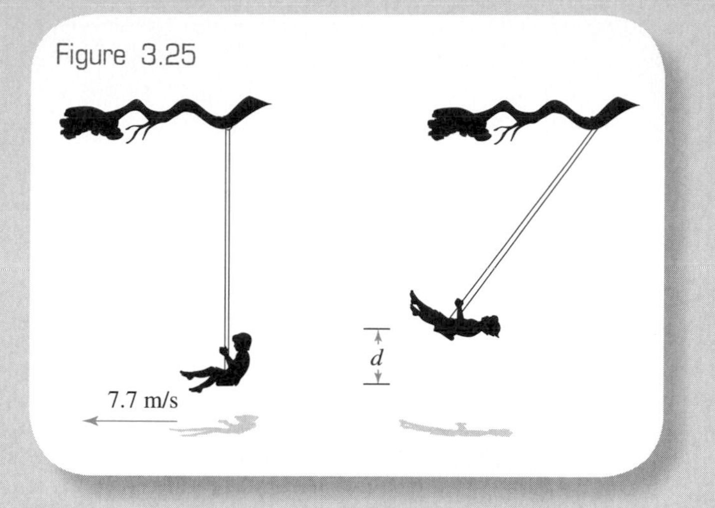

22. The cliff divers at Acapulco, Mexico, jump off a cliff 26.7 m above the ocean. Ignoring air resistance, how fast are the divers going when they hit the water?
23. In the Drop Tower at Bremen, Germany, the 300-kg test chamber falls freely from a height of 110 m.
    a) What is the chamber's potential energy at the top of the tower?
    b) How fast is it going when it reaches the bottom of the tower? (You may want to convert your answer to mph for comparison to highway speeds.)
24. The fastest that a human has run is about 12 m/s.
    a) If a pole vaulter could run this fast and convert all of his or her kinetic energy into gravitational potential energy, how high would he or she go?
    b) Compare this height with the world record in the pole vault.
25. A bicycle and rider going 10 m/s approach a hill. Their total mass is 80 kg.
    a) What is their kinetic energy?
    b) If the rider coasts up the hill without pedaling, at what point will the bicycle come to a stop?
26. In January 2003, an 18-year-old student gained a bit of fame for surviving—with only minor injuries—a remarkable traffic accident. The vehicle he was driving was "clipped" by another one, left the road, and rolled several times. He was thrown upward from the vehicle (he wasn't wearing a seat belt) and ended up dangling from an overhead telephone cable and a ground wire about 8 meters above the ground. Rescuers got him down after 20 minutes. It is estimated that he reached a maximum height of about 10 meters.
    a) How fast was the driver's body going when he was thrown from the vehicle?
    b) If he had not landed in the wires, how fast would he have been going when he hit the ground?
27. The ceiling of an arena is 20 m above the floor. What is the minimum speed that a thrown ball would have to have to reach the ceiling?
28. Compute how much kinetic energy was "lost" in the collision in Problem 6.
29. Compute how much kinetic energy was "lost" in the collision in Problem 7.
30. A 1,000-W motor powers a hoist used to lift cars at a service station.
    a) How much time would it take to raise a 1,500-kg car 2 m?
    b) If it is replaced with a 2,000-W motor, how long would it take?
31. How long does it take a worker producing 200 W of power to do 10,000 J of work?
32. An elevator is able to raise 1,000 kg to a height of 40 m in 15 s.
    a) How much work does the elevator do?
    b) What is the elevator's power output?
33. A professor's little car can climb a hill in 10 s. The top of the hill is 30 m higher than the bottom, and the car's mass is 1,000 kg. What is the power output of the car?
34. In the annual Empire State Building race, contestants run up 1,575 steps to a height of 1,050 ft. In 1983, the winner of this race was Al Waquie, with a time of 11 min and 36 s. Mr. Waquie weighed 108 lb.
    a) How much work was done?
    b) What was the average power output (in ft-lb/s and in hp)?
35. It takes 100 minutes for a middle-aged physics professor to ride his bicycle up the road to Alpe d'Huez in France. The vertical height of the climb is 1,120 m and the combined mass of the rider and bicycle is 85 kg. What is the bicyclist's average power output?

*remember

There are more quizzes and study tools online at 4ltrpress.cengage.com/physics.

Physics of Matter

sections

4.1 Matter: Phases, Forms, and Forces

The subject of this section is matter—anything that has mass and occupies space. The Earth, the water we drink, the air we breathe, our bodies, everything we touch is composed of matter. Obviously, matter exists in different forms that have different physical properties. We can classify matter into four categories: **solid**, **liquid**, **gas**, and **plasma**. These are called the four **phases**, or *states*, of matter. (In physics, *gas* does *not* refer to gasoline, and *plasma* does *not* refer to the liquid part of blood.) Briefly, we can distinguish the four phases as follows.

DEFINITION

Solids *Rigid; retain their shape unless distorted by a force. Examples: rock, wood, plastic, iron.*

Liquids *Flow readily; conform to the shape of a container; have a well-defined boundary (surface); have higher densities than gases. Examples: water, alcohol, gasoline, blood.*

Gases *Flow readily; conform to the shape of a container; do not have a well-defined surface; can be compressed (squeezed into a smaller volume) readily. Examples: air, carbon dioxide, nitrogen, helium.*

Plasmas *Have the properties of gases but also conduct electricity; interact strongly with magnetic fields; commonly exist at higher temperatures. Examples: gases in operating fluorescent, neon, and vapor lights (compact fluorescents); matter in the Sun and stars.*

Nearly all of the matter in our everyday experience appears as solid, liquid, or gas. Traditionally, these have been referred to as the three states of matter, with plasmas being a special "fourth" state of matter. Although plasmas are rare on Earth, most of the visible matter in the universe is in the form of plasmas in stars. In the last 50 years, the study of plasmas has grown to be one of the major subfields of physics because of the interest in nuclear fusion (the topic of Section 11.7). Nuclear fusion is the source of energy for stars, the Sun included. One of the main goals of plasma physics is to artificially produce starlike plasmas in which fusion can occur.

Many substances do not fit easily into one of these categories. Granulated sugar and salt flow readily and take the shape of a container. But they are considered to be solids because a single granule of each fits the description of a solid

Fluorescent lights use glowing plasmas.

© Chris Hill/iStockphoto.com / © iStockphoto.com

physics

© Michael Krinke/iStockphoto.com / © iStockphoto.com

and because both can be crystallized into larger chunks. Tar and molasses do not flow readily, particularly when they are cold, and so they act somewhat like solids. But given time they do flow and take the shape of their container; they are considered liquids.

© Jorge Salcedo/iStockphoto.com / © Maria Toutoudaki/iStockphoto.com

Many substances are composites of matter in two different phases. Styrofoam behaves like a solid but is composed mostly of gas trapped in millions of tiny, rigidly connected bubbles. The water in many rivers carries along tiny, solid particles that will settle if the water is allowed to stand. The mists that fill a shower stall and comprise fog and low clouds consist of millions of small droplets of water mixed in with the air. Apples and potatoes are solid but contain a great deal of liquid.

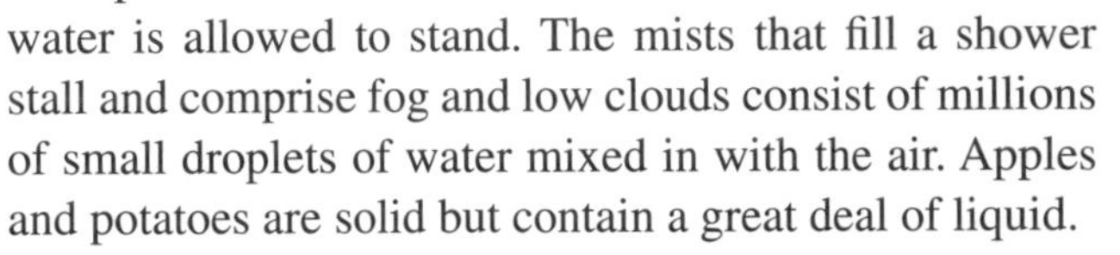

Another factor that complicates our neat classification of matter is that the phase of a given substance can change with temperature and pressure. Water is a good example. Normally a liquid, water becomes a solid (ice) when cooled below 0°C (32°F). Under normal pressure, water becomes a gas (steam) when its temperature is raised above 100°C (212°F). Even at room temperature, water can be made to boil if the air pressure is reduced, however. Propane, carbon dioxide, and many other gases can be forced into the liquid phase at room temperature by increasing the pressure. Most refrigerators and air conditioners depend on the pressure-induced liquification of a gaseous refrigerant for their proper function.

To simplify our discussion in this chapter, we will consider only matter in one "pure" phase: Solid like a rock, liquid like pure water, or gaseous like carbon dioxide. When we say that a substance exists in a particular phase, we mean its phase at normal room temperature and pressure (unless stated otherwise).

The phases of matter refer to the macroscopic (external) form and properties of matter. These in turn are determined by the microscopic (internal) composition of matter. Around 2,500 years ago, some Greek philosophers theorized that all matter as we see and experience it is composed of tiny, indivisible pieces. It turns out that this is true up to a point: Diamond, water, and oxygen are all composed of extremely tiny "building blocks" or units that are the smallest entities that retain the identity of the substance. (But, as we shall see, the building blocks themselves are composed of still smaller particles.) Water is a liquid, and diamond is a solid because of the properties of the small units that comprise each. Thus, we can reclassify all substances by the nature of their intrinsic composition. We will begin with the simplest class of matter and proceed in order of increasing complexity.

The chemical **elements** represent the simplest and purest forms of everyday matter. At this time, scientists have identified 117 different elements and have agreed upon names for 110 of them.* Some of the common substances used in our society are elements. These include hydrogen, helium, carbon, nitrogen, oxygen, neon, gold, iron, mercury, and aluminum. Each element is composed of incredibly small particles called **atoms**. There are 117 different atoms, one for each of the different known elements. Only about 90 of the elements exist naturally on Earth. The others are artificially produced in laboratories. The majority of the elements are quite rare and have names familiar only to chemists and other scientists.

The atom is not an indivisible particle: It has its own internal structure. Every atom has a very dense, compact core called the **nucleus**, which is surrounded by one or more particles called **electrons**. The nucleus itself is composed of two kinds of particles, **protons** and **neutrons**** (Figure 4.1). The protons and electrons have equal but opposite electric charges and attract each other. The electrons are much lighter than protons and neutrons, and they move in orbits with the attraction of the protons supplying the required centripetal force.

Every atom associated with a particular element has the same unique number of protons. For example, atoms that have 2 protons are atoms of helium, those with 8 protons are atoms of oxygen, those with 79 protons are atoms of gold, and so on. The *atomic number* of an element is the number of protons that are in each atom of the element. The atomic number of helium is 2 because every atom of helium has 2 protons in it. For oxygen it is 8, for gold it is 79, and so on. Each element is also given an abbreviation called its *chemical symbol.* Table 4.1 contains the chemical symbols and atomic numbers of

* The discovery of element 122 was reported in 2008, but as of this time, it remains unconfirmed. In addition, it has been proposed that element 112 be named Copernicium (symbol Cp), but this recommendation has yet to be accepted by the International Union of Pure and Applied Chemistry.

** Protons and neutrons are themselves composed of smaller particles called quarks. More on this in Chapter 12.

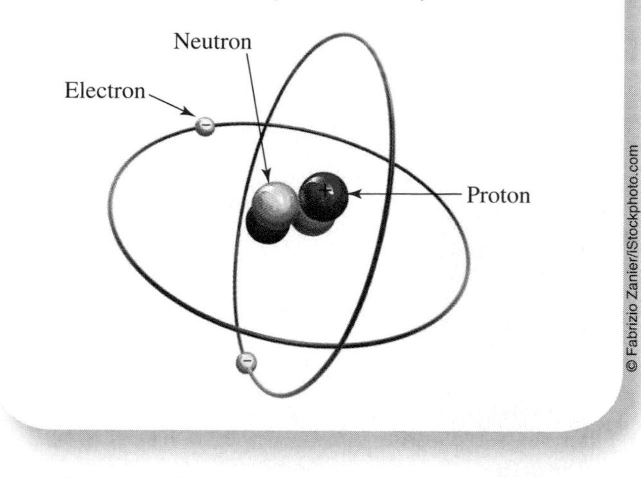

Figure 4.1
Idealized diagram of an atom. An atom consists of electrons in orbit around a compact nucleus, which in turn contains protons and neutrons. All atoms of a particular element have the same number of protons. This is the element's atomic number. (Not drawn to scale.)

some familiar elements. It also includes the phase of each element at room temperature and normal pressure. The periodic table of the elements on the card in the back of the book shows all of the known elements and some of their properties. (We will take a closer look at the structure of atoms in Chapters 7 and 10.)

Chemical **compounds** are the next simplest form of everyday matter. There are millions of different compounds, including many common substances such as water, salt, sugar, and alcohol. Compounds are similar to elements in that each compound also has a unique building block called a **molecule**. Every molecule of a particular compound consists of the same unique combination of two or more atoms held together by electrical forces. For example, each molecule of water consists of two atoms of hydrogen attached to one atom of oxygen (Figure 4.2). Similarly, one atom of carbon attached to one atom of oxygen forms a molecule of carbon monoxide. Each compound can be represented by a "formula"—a shorthand notation showing both the kinds and the numbers of atoms in each of the compound's molecules. You are probably familiar with the formula for water, H_2O. Some others are NaCl (table salt), CO_2 (carbon dioxide), CO (carbon monoxide), $C_{12}H_{22}O_{11}$ (sugar), and C_2H_5OH (ethyl alcohol). The study of how atoms combine to form molecules is a major part of chemistry.

Some elements in the gas phase are also composed of molecules. The oxygen in the air we breathe is an element, but most of the oxygen atoms are paired up to form O_2 molecules. Ozone (O_3) is a rarer form of oxygen that is present in air pollution and also in the ozone layer, centered about 15 miles above the Earth's surface. Each ozone molecule is composed of three oxygen atoms. Nitrogen gas also exists in the form of molecules, N_2. Helium and neon are both gaseous elements whose atoms do not easily form molecules.

Many substances, such as air, stone, and seawater, are composed of two or more different compounds or elements that are physically mixed together. These are

Table 4.1 Some Common Chemical Elements

Element	Symbol	Atomic Number	Phase
Hydrogen	H	1	Gas
Helium	He	2	Gas
Carbon	C	6	Solid
Nitrogen	N	7	Gas
Oxygen	O	8	Gas
Neon	Ne	10	Gas
Sodium	Na	11	Solid
Aluminum	Al	13	Solid
Silicon	Si	14	Solid
Chlorine	Cl	17	Gas
Calcium	Ca	20	Solid
Iron	Fe	26	Solid
Cobalt	Co	27	Solid
Nickel	Ni	28	Solid
Copper	Cu	29	Solid
Zinc	Zn	30	Solid
Silver	Ag	47	Solid
Barium	Ba	56	Solid
Gold	Au	79	Solid
Mercury	Hg	80	Liquid
Lead	Pb	82	Solid
Uranium	U	92	Solid

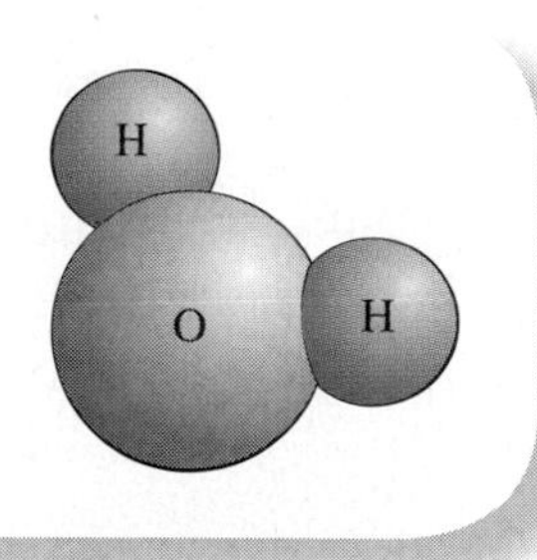

Figure 4.2
Water molecules consist of two hydrogen atoms attached to an oxygen atom. This structure is common to all molecules of water, ice, and steam.

classified as **mixtures** and **solutions**. Air consists of dozens of different gases that are mixed together. Table 4.2 shows the composition of clean, dry air. The air we breathe also contains water vapor and pollutants (like carbon monoxide). The amount of such gases present in the air varies considerably from place to place and from day to day. The air in the Sahara Desert does not have quite the same composition as the air in Los Angeles.

A mixture of two elements is not the same as a compound. If you mix hydrogen and oxygen, you simply have a gaseous mixture. The individual hydrogen atoms and oxygen molecules remain separate. If you ignite the mixture, the hydrogen and oxygen atoms will combine to form water molecules. Energy is released explosively in the form of heat (fire) because the atoms have less total energy when they are bound to each other.

The example of hydrogen, oxygen, and water also illustrates that the properties of a compound (water) are usually quite different from the properties of the constituent elements (hydrogen and oxygen). Sodium (Na) is a solid that reacts violently with water. Chlorine (Cl) is a gas that is used to kill bacteria in drinking water. If either element were ingested alone, it could be fatal. But when sodium and chlorine are combined chemically, the result is table salt (NaCl), a necessary part of our diet (Figure 4.3).

The basic unit of life is the cell, and each cell is composed of many different compounds. Some of these contain billions of atoms in each molecule, the DNA molecule being an important example. The main constituents of such "organic" molecules are hydrogen, carbon, oxygen, and nitrogen. Many other elements are present in smaller amounts. For example, your bones and teeth contain calcium.

So atoms are basic to compounds as well as to elements. It is difficult to imagine how small atoms are and

Table 4.2 Composition of Clean, Dry Air

Gas	Percent Composition (by volume)
Nitrogen (N_2)	78.1
Oxygen (O_2)	20.9
Argon (Ar)	0.93
Carbon dioxide (CO_2)	0.03
Other gases (Ne, He, and so on)	0.04

Figure 4.3

Separately, chlorine and sodium are poisonous. When chemically combined, they form table salt.

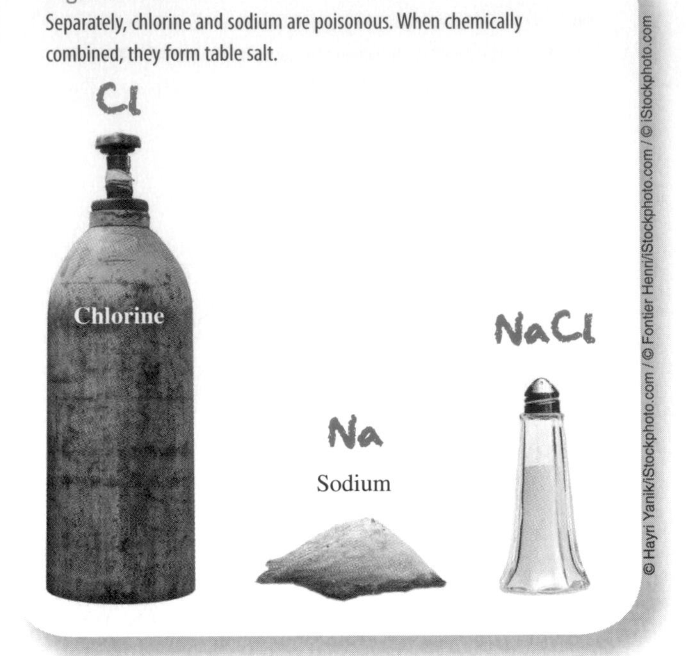

just how many there are in common objects. Atoms are about 1 ten-millionth of a millimeter in diameter. A ball 1 inch in diameter is about midway between the size of an atom and the size of the Earth (Figure 4.4). In other words, if a 1-inch ball were expanded to the size of the Earth, each atom in the ball would expand to about 1 inch in diameter.

Since atoms are so small, huge numbers of them are present in anything large enough to be seen. There are about 100 thousand billion billion (one followed by 23 zeros) atoms in each of your fingernails. Even the smallest particles that can be seen with the naked eye contain far more atoms than there are people on Earth.

Atoms are nearly indestructible—only nuclear reactions affect them—and they are continually recycled. The Earth and everything on it are believed to be composed of the debris of stars that exploded billions of years ago. Processes such as combustion, decay, and growth result in atoms being combined with or dissociated from other atoms. A single carbon atom in your earlobe may once have been part of a dinosaur, a redwood tree, a rose, or Leonardo da Vinci—or all four.

Behavior of Atoms and Molecules

The constituent particles of matter, atoms and molecules, exert electrical forces on each other. ("Static cling" is an example of an electrical force.) The nature of these forces determines the properties of the substance. The forces

Figure 4.4

Think about how many golf balls it would take to fill up a hollow ball the size of the entire Earth. That's about how many atoms there are in each golf ball.

© iStockphoto.com / © Jan Rysavy/iStockphoto.com / © Fabrizio Zanier/iStockphoto.com

between atoms in an element depend on the configuration of the electrons in each atom. In a compound, the size and shape of the molecules, as well as the forces between the molecules, affect its observed form and properties. We can relate the three common phases of matter to the interparticle forces as follows.

DEFINITION

Solids *Attractive forces between particles are very strong; the atoms or molecules are rigidly bound to their neighbors and can only vibrate.*

Liquids *The particles are bound together, though not rigidly; each atom or molecule can move about relative to the others but always remains in contact with other atoms or molecules.*

Gases *Attractive forces between particles are too weak to bind them together; atoms or molecules move about freely with high speed and are generally widely separated; particles are in contact only when they collide.*

A standard model for representing the interparticle forces in a solid consists of each atom connected to its neighbors by a spring. The atoms are free to oscillate like a mass hanging from a spring. (The vibration of the atoms or molecules is related to the temperature, as we shall see in Chapter 5.) Often atoms or molecules in a solid form a regular geometric pattern called a *crystal* (Figure 4.5). Table salt is a crystalline compound in which the sodium and chlorine atoms alternate with each other. In solids that do not have a regular crystal structure, called *amorphous solids*, the atoms or molecules are "piled together" in a random fashion. Glass is a good example of such a solid.

Carbon is very interesting because it is an element with two common crystalline forms (graphite and diamond) that possess very different properties, plus a large number of recently discovered molecular forms. In diamond, each carbon atom is strongly bonded to each of its four nearest neighbors resulting in a crystalline solid that is the hardest known natural material (Figure 4.6a). In graphite, the primary ingredient in the misnamed "lead" in pencils, the

Figure 4.5

(a) The atoms or molecules in a crystalline solid are arranged in a regular three-dimensional pattern, similar to the way rooms are arranged in a large apartment or office building. The atoms or molecules exert forces on each other. (b) A crystal behaves much like an array of particles that are connected to each other by springs. (c) Image showing the geometric arrangement of silicon atoms, produced with a scanning tunneling microscope (STM) at the IBM Thomas J. Watson Research Center.

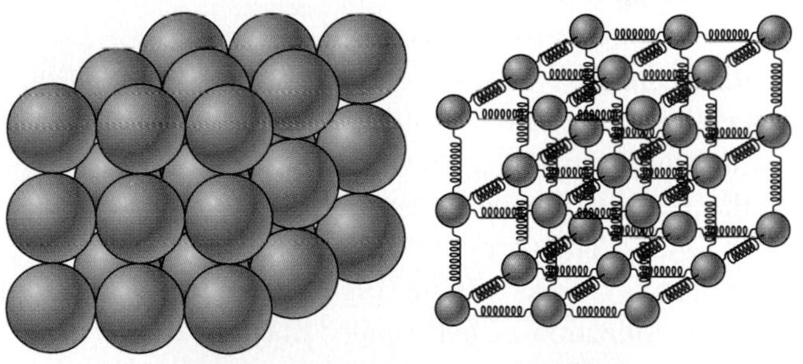

(a) (b)

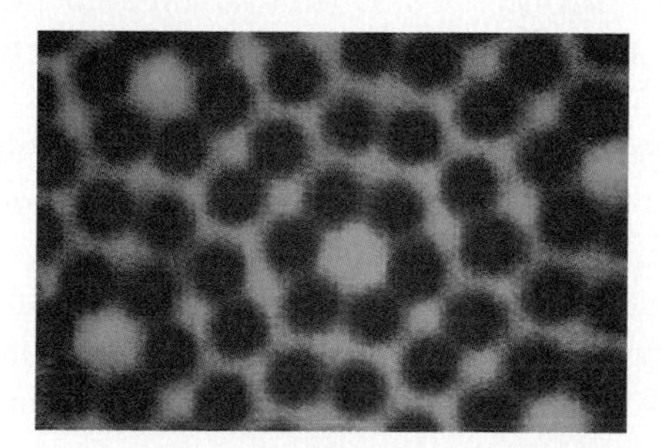

(c)

© Phototake, Inc./Alamy

carbon atoms form sheets, with each atom strongly bonded to its three nearest neighbors in the same layer to form a mesh of hexagons (Figure 4.6b). These sheets can easily be forced to slide relative to each other making graphite an excellent "dry" lubricant. Carbon atoms can also bond together to form large molecules, the most famous being C_{60}, named buckminsterfullerene ("buckyball" for short) after the famous engineer and philosopher R. Buckminster Fuller. The molecule reminded its discoverers of geodesic domes invented by Fuller. The 60 atoms in each buckminsterfullerene molecule arrange themselves into a soccer-ball shape consisting of 12 pentagons and 20 hexagons. The C_{60} molecules can in turn bond together to form a crystal (Figure 4.6c). Dozens of larger and smaller hollow carbon molecules can also form, as well as "carbon nanotubes" consisting of graphitelike sheets of carbon atoms rolled up into tubes. The C_{60} and other molecular carbon forms show promise for useful applications in microelectronics and medicine.

Figure 4.6
Three forms of solid carbon. Each small sphere represents a carbon atom, and the connecting rods represent the force holding pairs of carbon atoms together: (a) diamond; (b) graphite; and (c) C_{60} crystal. Copyright 1993, Henry Hill, Jr.

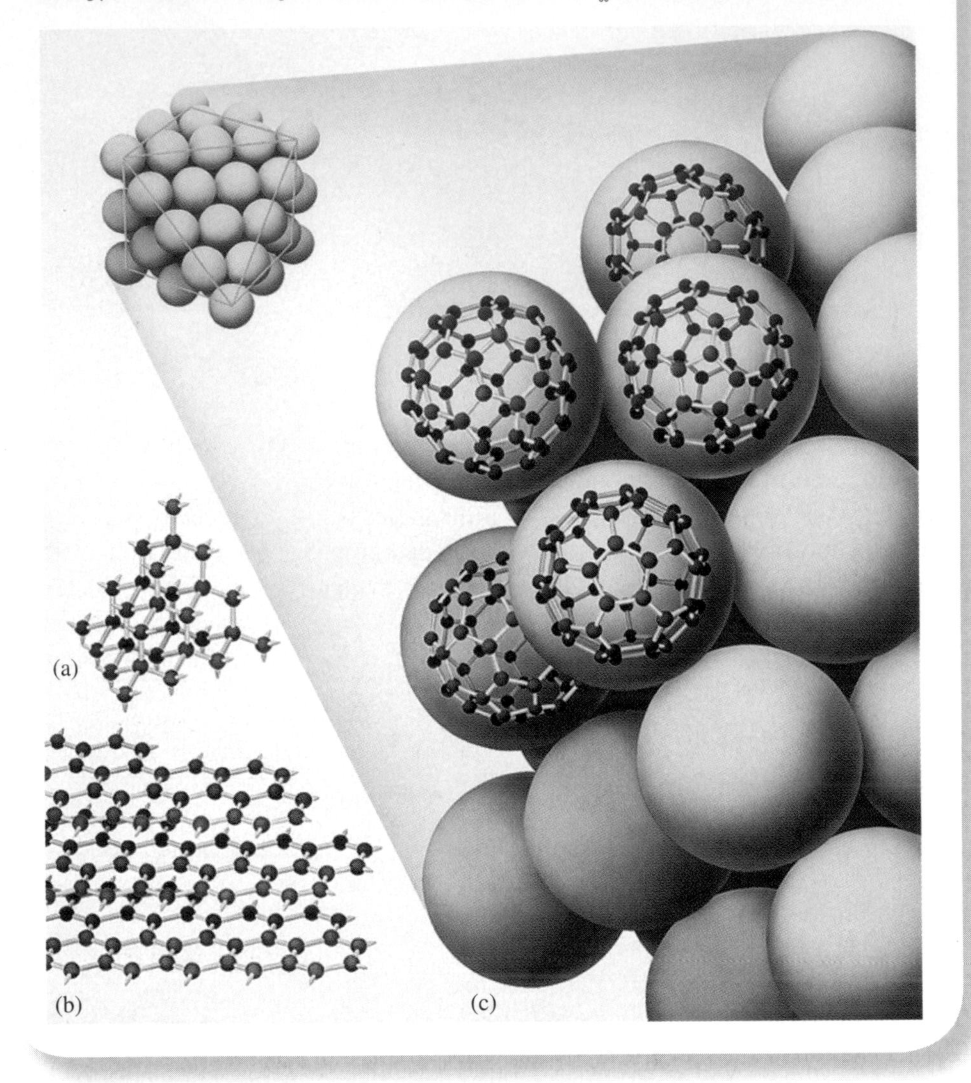

In a liquid, the forces between the particles are not strong enough to bind them together rigidly. The atoms or molecules are free to move around as well as vibrate (Figure 4.7a). This is similar to a collection of ball-shaped magnets clinging to each other. The forces between the particles are responsible for surface tension (the natural attraction between the surface of a liquid and another nearby surface) and the spherical shape of small drops.

Many compounds have an interesting intermediate phase between solid and liquid called the *liquid crystal* phase. The molecules have some mobility, as in a liquid, but they are arranged regularly, as in a solid. Liquid crystal displays (LCDs) in calculators, flat video displays, and watches use electrically induced

Figure 4.7
(a) The atoms or molecules in a liquid remain in contact with each other, but they are free to move about. (b) In a gas, the atoms or molecules are not bound to each other. They move with high speed and interact only when they collide.

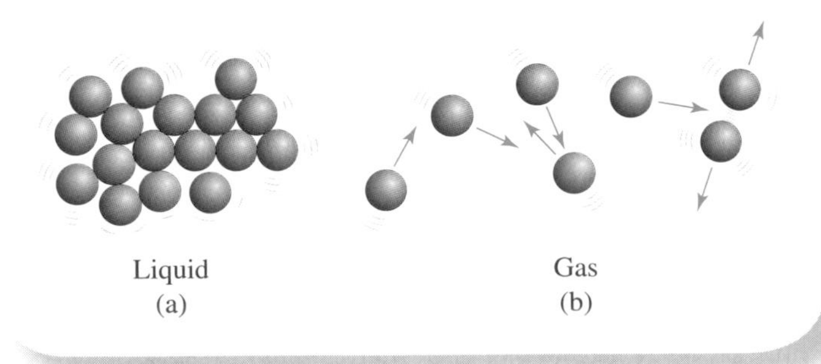

movement of the molecules to change the optical properties of the liquid crystal. (More on this in Section 9.1.)

In a gas, the atoms and molecules are widely separated and move around independently, except when they collide (see Figure 4.7b). Their speeds are surprisingly high: The oxygen molecules you are now breathing have an average speed of over 1,000 mph. As each molecule collides randomly with other molecules, its speed is sometimes increased and other times decreased.

Often, the surface of a solid or a liquid forms a boundary for a gas. The surface of water in a glass, a lake, or an ocean forms a lower boundary for the air above. The inside of the walls of a tire are a boundary for the air inside (Figure 4.8). The high-speed atoms or molecules in the gas exert a force on the surface as their random motions cause them to collide with it. This is the basis for gas pressure, as mentioned in Section 3.6.

Can we extend our study of mechanics to fluid motion? The answer is a qualified yes.

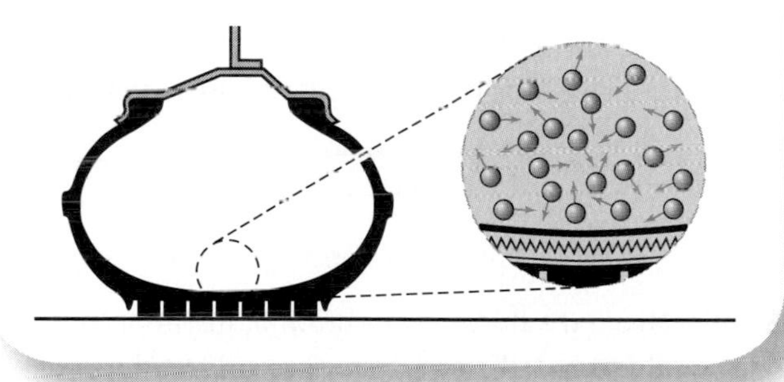
Figure 4.8
The air molecules inside a tire collide with the walls and exert forces on it. If the molecules weren't moving, the tire would be flat.

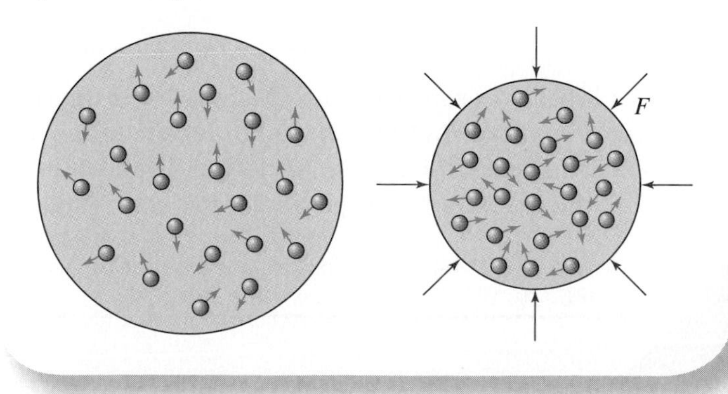

Figure 4.9
Gases can be compressed because the atoms and/or molecules are widely separated. Increasing the force (pressure) on the boundaries of a gas squeezes the particles closer together.

In other words, the weight of a car is supported by the collisions of air molecules with the inner walls of its tires.

At normal temperatures and pressures, the average distance between the centers of the atoms or molecules in a gas is about 10 times that in a solid or liquid (Figure 4.9). There is a great deal of empty space in a gas, and that is why gases can be compressed easily.

4.2 Pressure

Now that we have established some of the basic properties of solids, liquids, and gases, it is time to consider how the mechanics we've presented in the preceding chapters relates to extended matter. We have already seen in Chapters 1, 2, and 3 that extended objects can often be treated as particles (for example, a rock as it falls, satellites and even planets as they move in orbits, trucks during collisions, and so on). Furthermore, the law of conservation of angular momentum (Section 3.8) allows us to examine interesting types of rotational motions of solid bodies. But what about gases and liquids? The fluids in and around us—blood in our veins and arteries, the atmosphere, streams, and oceans—are in constant motion. Can we extend our study of mechanics to fluid motion? The answer is a qualified yes.

Newton's laws and the conservation laws can be applied to fluids using appropriate extensions of physical quantities like mass and force, but the mathematics is much more complicated and beyond the level of this text. (Some of the most powerful computers in the world are used exclusively to solve problems involving fluids, such as the motions of the atmosphere.) So we will limit our study of fluid motion to one (restricted) conservation law (Section 4.7). However, there are several phenomena related to fluids at rest that are important in our daily lives and that can be dealt with using simple mathematics. These phenomena are the principal topics of most of the rest of this chapter. We first introduce the physical quantities **pressure** and **density**, which are extensions of force and mass, respectively. These are two essential quantities for studying fluids both at rest and in motion.

Forces that are exerted by gases, liquids, and solids normally are spread over a surface. The force that the floor exerts on you when you are standing is distributed over the bottoms of your feet. A boat floating on water has an upward force spread over its lower surface. The air around a floating balloon exerts forces on the balloon's surface. In situations like these, the physical quantity pressure is quite useful.

The water exerts an upward force that is spread over the submerged surface of the boat.

© Christopher Pattberg/iStockphoto.com / © Maria Toutoudaki/iStockphoto.com

DEFINITION

Pressure *The force per unit area for a force acting perpendicular to a surface. The perpendicular component of a force acting on a surface divided by the area of the surface.*

$$p = \frac{F}{A}$$

Pressure is a scalar not a vector: There is no direction associated with it.

p is the abbreviation for pressure, and P is the abbreviation for power.

© iStockphoto.com

Physical Quantity	Metric Units	English Units
Pressure (p)	pascal (Pa) (1 Pa = 1 N/m^2)	pound per square foot (lb/ft^2) pound per square inch (psi)
	millimeters of mercury (mm Hg)	inches of mercury (in. Hg)

The standard pressure units consist of a force unit divided by an area unit. The SI unit of pressure is the pascal (Pa), which equals 1 newton per square meter. The English unit for pressure (psi) may be the most familiar to you. Air pressure in tires is commonly measured in psi. For comparison:

$$1 \text{ psi} = 6{,}890 \text{ Pa}$$

This number is so large because a force of 1 pound on each square inch would produce a huge force on 1 square meter—a much larger area. The two pressure units involving mercury will be explained in Section 4.4. Another common unit of pressure is the atmosphere (atm). It equals the average air pressure at sea level. (Later, we will see that the air pressure is lower at higher altitudes.) It is related to the other units as follows:

$$1 \text{ atm} = 1.01 \times 10^5 \text{ Pa} = 14.7 \text{ psi}$$

This unit is like the g used as a unit of acceleration. We are all subject to the pressure of the air around us and to the acceleration due to gravity, so it is natural to compare other pressures and accelerations to them. The old practice of using the length of a person's foot as a unit of distance is another example of taking a unit of measure provided by Nature. Table 4.3 lists some representative pressures.

EXAMPLE 4.1

A 160-pound person stands on the floor. The area of each shoe that is in contact with the floor is 20 square inches. What is the pressure on the floor? Assuming the person's weight is shared equally between the two shoes, the force of one shoe is 80 pounds. So:

$$p = \frac{F}{A} = \frac{80 \text{ lb}}{20 \text{ in.}^2}$$

$$= 4 \text{ psi} \qquad \text{(standing on both feet)}$$

By Newton's third law of motion, the floor exerts an equal and opposite force on the shoes. Thus the pressure of the floor on the shoes is also 4 psi.

If the person stands on one foot instead, that shoe has all of the weight and the pressure is

$$p = \frac{160 \text{ lb}}{20 \text{ in.}^2}$$

$$= 8 \text{ psi} \qquad \text{(standing on one foot)}$$

What if the person put on high-heeled shoes and balanced on the heel of one shoe? The bottom of the heel might measure 0.5 inches by 0.5 inches. The area is then 0.25 square inches. So the pressure in this case would be

$$p = \frac{160 \text{ lb}}{0.25 \text{ in.}^2} = 640 \text{ psi} \qquad \text{(balanced on a narrow heel)}$$

© Michal Koziarski/iStockphoto.com / © iStockphoto.com

Table 4.3 Some Pressures of Interest

Description	p (Pa)	p (psi)	p (atm)
Lowest laboratory pressure	7×10^{-11}	1×10^{-14}	7×10^{-16}
Atmospheric pressure at an altitude of 100 km	0.06	9×10^{-6}	6×10^{-7}
Lowest recorded sea level atmospheric pressure	0.87×10^{5}	12.6	0.86
Average sea level atmospheric pressure	1.01×10^{5}	14.7	1
Highest recorded sea level atmospheric pressure	1.08×10^{5}	15.7	1.07
Inside a tire (typical)	3.1×10^{5}	45	3.1
4,000 m underwater (*Titanic's* remains)	4×10^{7}	5,800	390
Center of the Earth	1.7×10^{11}	2.5×10^{7}	1.7×10^{6}
Highest sustained laboratory pressure	3×10^{11}	4.4×10^{7}	3×10^{6}
Center of the Sun	1.3×10^{14}	1.9×10^{10}	1.3×10^{9}

Example 4.1 shows that the same force causes a much higher pressure when it acts over a smaller area. We can think of pressure as a measure of how "concentrated" a force is.

There are many situations in which a liquid or a gas is under pressure and exerts a force on the walls of its container. The relationship between force and pressure can be used to determine the force on a particular area of the walls. Since pressure equals force divided by area, the pressure times the area equals the force. In other words, the total force on a surface equals the force on each square inch (the pressure) times the total number of square inches (the area).

$$F = pA$$

EXAMPLE 4.2

In the late 1980s, there were several spectacular (and tragic) aircraft mishaps involving rapid loss of air pressure in the passenger cabins. (The causes included failure of a cargo door, outer skin rupture due to corrosion or cracks or both, and small bombs.) Because the cabins are pressurized, there are large outward forces acting on windows, doors, and the aircraft skin.* Let's estimate the sizes of these forces.

The pressure inside a passenger jet cruising at high altitude (about 7,500 meters or 25,000 feet) is about 6 psi (0.41 atmospheres) greater than the pressure outside. What is the outward force on a window measuring 1 foot by 1 foot and on a door measuring 1 meter by 2 meters? The area of the window is

$$A = 1 \text{ ft} \times 1 \text{ ft} = 12 \text{ in.} \times 12 \text{ in.}$$
$$= 144 \text{ in.}^2$$

(The area has to be in square inches because the pressure is in pounds per square *inch.*)

The force on the window is

$$F = pA = 6 \text{ psi} \times 144 \text{ in.}^2$$
$$= 864 \text{ lb}$$

A rather small pressure causes a large force on the window. For the door, we use SI units:

$$p = 6 \text{ psi} = 6 \times 1 \text{ psi} = 6 \times 6{,}890 \text{ Pa}$$
$$= 41{,}340 \text{ Pa}$$
$$A = 1 \text{ m} \times 2 \text{ m} = 2 \text{ m}^2$$

Consequently, the force on the door is

$$F = pA = 41{,}340 \text{ Pa} \times 2\text{m}^2$$
$$= 82{,}680 \text{ N}$$

The force on the door is greater than the weight of 10 small automobiles, or 100 people.

* One may ask why aircraft do not fly at low altitudes so that cabins would not need to be pressurized. There are several reasons why jets fly at 10,000 meters or higher.

1. If something goes wrong and an aircraft starts losing altitude, the pilot has more time to recover when the plane is very high to begin with. More than once an aircraft has been saved after diving over 7,000 meters.
2. Because the air is thinner at high altitudes, the force of air resistance on an aircraft is smaller. Consequently, it can travel faster and use less fuel than when flying at low altitudes.
3. The air is usually much "smoother" at high altitudes. There is much less turbulence, so passengers are more comfortable.

© Nicholas Belton/iStockphoto.com

Pressure is a relative quantity. When you test the air pressure in a tire, you are comparing the pressure of the air inside the tire with the pressure outside the tire. When a tire is flat, there is still air inside it, but the pressure inside is the same as the atmospheric pressure outside.

For example, let us say that the atmospheric pressure is 14 psi. A tire is tested, and the air pressure gauge shows 30 psi. This means that the air pressure inside the tire is 30 psi higher than the air pressure outside the tire. So the actual pressure on the inner walls of the tire is $30 + 14 = 44$ psi. This is sometimes referred to as the *absolute pressure.* The pressure relative to the outside air (30 psi in this case) is then called the *gauge pressure* (Figure 4.10).

We can look at this another way. The pressure inside the tire, 44 psi, causes an outward force of 44 pounds on each square inch of the tire wall. The air pressure outside the tire causes an inward force of 14 pounds on each square inch. The net force due to air pressure on each square inch of the tire is then 30 pounds.

The gauge pressure of the air in a tire changes if the outside air pressure changes. What happens if the car is driven into a large chamber and the air pressure in the chamber is increased from 14 psi to 44 psi? The gauge pressure in the tires will then be zero, and the tires will go flat even though no air has been removed from them. When the air pressure is reduced to 14 psi, the tires will expand again to their normal shape.

In the previous example of the pressurized aircraft, the 6 psi is the gauge pressure. The absolute pressure inside the cabin might be 11 psi, and the air pressure outside might be 5 psi. The atmospheric pressure at 26,000 feet is about 5 psi.

The standard pen-shaped tire-pressure tester nicely illustrates some of the physics that we have considered so far. It consists of a hollow tube (cylinder) fitted with a piston that can slide back and forth in the cylinder (Figure 4.11). Air from the tire enters the cylinder at the left end and pushes on the piston. The right end allows air from the outside to push on the other side of the piston. If the pressure inside the tire is greater than the outside air pressure, there is a net force to the right on the piston. A spring placed behind the piston is compressed by this net force. The greater the net force on the piston, the greater the compression of the spring. A calibrated shaft extends from the right side of the piston and out of the right end. When the piston is pushed to the right, the shaft protrudes from the right end the same distance that the spring is compressed. The length of shaft showing indicates the gauge pressure in the tire, since that is what causes the force on the piston.

We conclude this section with one last important note about pressure. Since gases are compressible, the

Figure 4.10
An absolute pressure inside of 44 psi produces a gauge pressure of 30 psi.

Figure 4.11

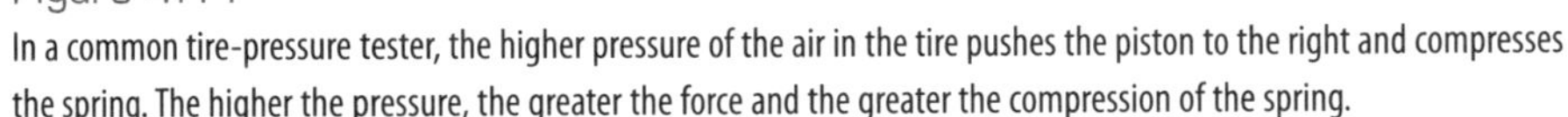
In a common tire-pressure tester, the higher pressure of the air in the tire pushes the piston to the right and compresses the spring. The higher the pressure, the greater the force and the greater the compression of the spring.

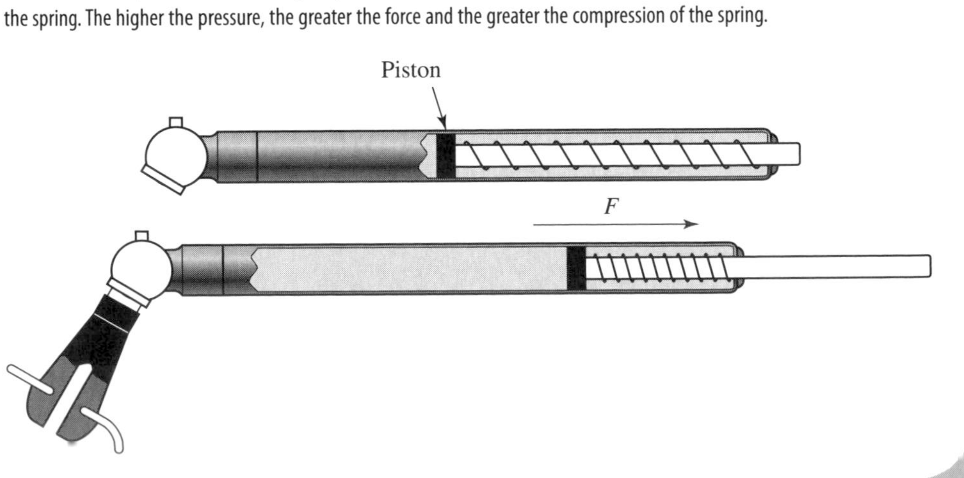

volume of a gas can be changed (Figure 4.9). Whenever the volume of a fixed amount of gas is changed, the pressure in the gas changes also. Increasing the volume of a gas reduces the pressure. Decreasing the volume increases the pressure. To understand why this is the case, recall from the previous section that the collisions of the atoms and molecules in a gas with a surface cause gas pressure. If the volume is decreased, the particles are squeezed together so there are more of them near each square inch of the boundaries. More collisions occur each second, which means more force on each square inch and higher pressure. The opposite happens when the volume is increased.

The temperature of a gas influences the speeds of the atoms or molecules. Consequently, the pressure is also affected by the gas temperature. When the temperature of a given quantity of gas is kept constant, the pressure p is related to the volume V as follows:

$$pV = \text{a constant} \qquad \text{(gas at fixed temperature)}$$

This means that the *volume of a gas is inversely proportional to the pressure.* If the pressure is doubled, the volume is halved. In Chapter 5, we will see precisely what effect temperature has on pressure and volume.

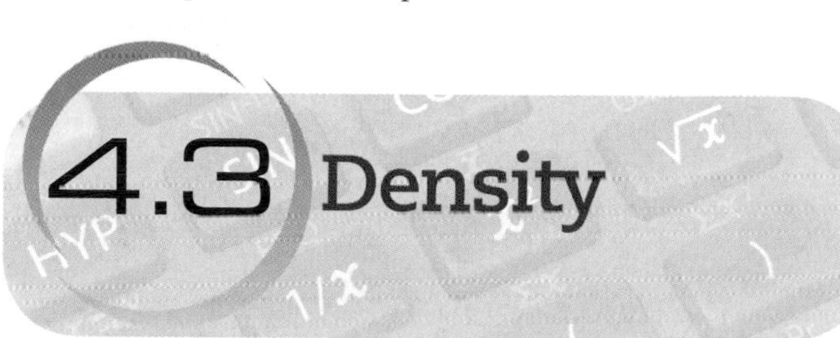

Pressure is an extension of the idea of force. Similarly, **mass density** is an extension of the concept of mass. Just as pressure is a measure of the concentration of force, density is a measure of the concentration of mass.

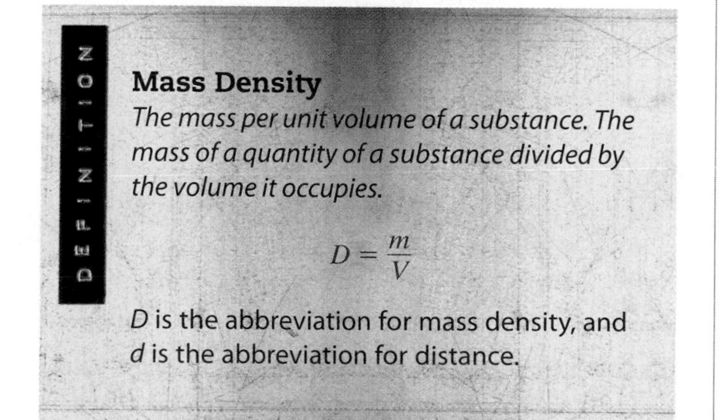

Physical Quantity	Metric Units	English Units
Mass density (D)	kilogram per cubic meter (kg/m^3) gram per cubic centimeter (g/cm^3)	slug per cubic foot

To find the mass density of a substance, one measures the mass of a part or "sample" of it and divides by the volume of that part or sample. It doesn't matter how much is used: The greater the volume, the greater the mass.

EXAMPLE 4.3

The dimensions of a rectangular aquarium are 0.5 meters by 1 meter by 0.5 meters. The mass of the aquarium is 250 kilograms larger when it is full of water than when it is empty (Figure 4.12). What is the density of the water? First, the volume of the water is

$$V = l \times w \times h$$
$$= 1\ \text{m} \times 0.5\ \text{m} \times 0.5\ \text{m}$$
$$= 0.25\ \text{m}^3$$

So:

$$D = \frac{m}{V} = \frac{250\ \text{kg}}{0.25\ \text{m}^3}$$
$$= 1{,}000\ \text{kg/m}^3 \qquad \text{(mass density of water)}$$

If a tank with twice the volume were used, the mass of the water in it would be twice as great, and the density would be the same. The mass density of any amount of pure water is 1,000 kg/m^3.

When the same tank is filled with gasoline, the mass of the gasoline is found to be 170 kilograms. Therefore, the density of gasoline is

$$D = \frac{170\ \text{kg}}{0.25\ \text{m}^3}$$
$$= 680\ \text{kg/m}^3$$

Except for small changes caused by variations in temperature or pressure, the mass density of any pure solid or liquid is constant. It is an identifying trait of that substance. The mass densities of pure gases, on the other hand, vary greatly with changes in temperature or pressure. We have seen that doubling the pressure on a gas

Figure 4.12

The mass of the water in this aquarium is 250 kilograms. Using this and the volume of the aquarium, we can compute the mass density of water.

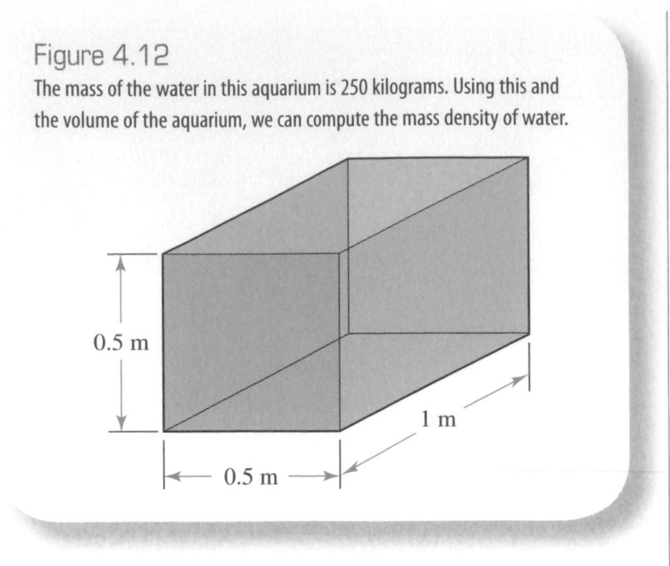

halves the volume. That would double the mass density as well. In Chapter 5, we will describe how changing the temperature of a gas also alters its volume and, therefore, its density. By convention, the tabulated densities of gases are given at the standard temperature and pressure (STP): 0°C temperature and 1 atmosphere pressure.

The mass density of each element or compound is fixed. Water, lead, mercury, salt, oxygen, gold, and so on, all have unique mass densities that have been measured and catalogued. The density of a mixture containing two or more substances depends on the density and percentage of each component. Most metals in common use are alloys, consisting of two or more metallic elements and other elements like carbon. For example, 14-karat gold is only about 58% gold. However, the mass density of a mixture with a particular composition is constant.

Table 4.4 lists the mass densities (column 3) of some common substances. The values for mixtures can vary and so are merely representative. The mass densities of the gases are for standard conditions.

Having a list of the densities of common substances is quite useful for three reasons. First, one can use mass density to *help identify a substance*. For example, one can determine whether or not a gold ring is solid gold by measuring its density and comparing it to the known density of pure gold. Second, density measurement is used routinely to *determine how much of a particular substance is present in a mixture*. The coolant in an automobile radiator is usually a mixture of water and antifreeze. These two liquids have different densities (Table 4.4), so the density of a mixture depends on the ratio of the amount of water to the amount of antifreeze. The higher the density, the greater the antifreeze content and the lower the freezing temperature of the coolant. By simply measuring the coolant density, one can determine the coolant's freezing temperature. If you have donated blood to a blood bank, part of the screening included checking to see whether the hemoglobin content of your blood was high enough. This is done by determining whether the blood's density is greater than an accepted minimum value.

Third, one can *calculate the mass* of something if one knows what its volume is. The mass of a substance equals the volume that it occupies times its mass density.

$$m = V \times D$$

In some cases, it is practical to use another type of density called **weight density**. It is commonly used in the English system of units because weight is in more common use than mass.

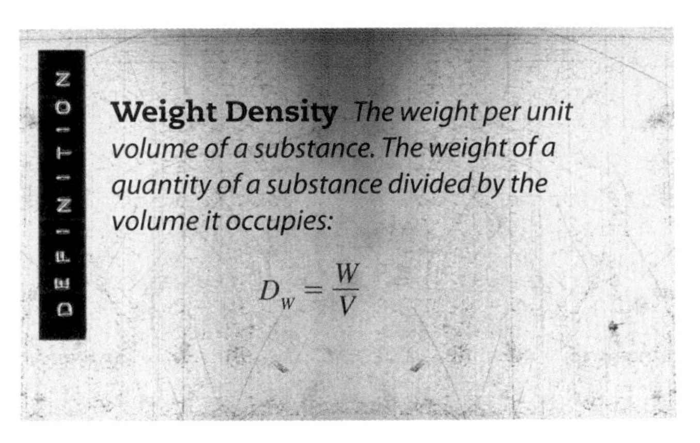

Weight Density *The weight per unit volume of a substance. The weight of a quantity of a substance divided by the volume it occupies:*

$$D_W = \frac{W}{V}$$

The weight density of a substance is equal to the mass density times the acceleration due to gravity. This

EXAMPLE 4.4

The mass of water needed to fill a swimming pool can be computed by measuring the volume of the pool. Let's say a pool is going to be built that will be 10 meters wide, 20 meters long, and 3 meters deep. How much water will it hold? The volume of the pool will be

$$V = l \times w \times h = 20 \text{ m} \times 10 \text{ m} \times 3 \text{ m} = 600 \text{ m}^3$$

$$m = V \times D = 600 \text{ m}^3 \times 1{,}000 \text{ kg/m}^3$$

$$= 600{,}000 \text{ kg}$$

That is a lot of water (about 3,000 bathtubs full).

© David Joyner/iStockphoto.com

Table 4.4 **Densities of Some Common Substances**

Substance	Type*	Mass Density, D (kg/m^3)	Weight Density, D_W (lb/ft^3)	Specific Gravity
Solids				
Juniper wood	m	560	35	0.56
Ice	c	917	57.2	0.917
Ebony wood	m	1,200	75	1.2
Silicon	e	2,300	146	2.3
Concrete	m	2,500	156	2.5
Aluminum	e	2,700	168	2.7
Diamond (Carbon)	e	3,400	210	3.4
Iron	e	7,860	490	7.86
Brass	m	8,500	530	8.5
Nickel	e	8,900	555	8.9
Copper	e	8,930	557	8.93
Silver	e	10,500	655	10.5
Lead	e	11,340	708	11.34
Uranium	e	19,000	1,190	19
Gold	e	19,300	1,200	19.3
Liquids				
Gasoline	m	680	42	0.68
Ethyl alcohol	c	791	49	0.791
Water (pure)	c	1,000	62.4	1.00
Seawater	m	1,030	64.3	1.03
Antifreeze	m	1,100	67	1.1
Sulfuric acid	c	1,830	114	1.83
Mercury	e	13,600	849	13.6
Gases (at 0°C and 1 atm)				
Hydrogen	e	0.09	0.0056	0.00009
Helium	e	0.18	0.011	0.00018
Air	m	1.29	0.08	0.00129
Carbon dioxide	c	1.98	0.12	0.00198
Radon	e	10	0.627	0.010

**Note*:* "e" stands for element, "c" for compound, and "m" for mixture.

Physical Quantity	Metric Units	English Units
Weight density D_W	newton per cubic meter (N/m^3)	pound per cubic foot (lb/ft^3) pound per cubic inch (lb/in.^{3})

is because the weight of a substance is just g times its mass.

$$W = m \times g \rightarrow D_W = D \times g$$

Column 4 in Table 4.4 shows representative weight densities in English units. These can be used in the same ways that mass densities are used. One interesting exercise is to compute the weight of the air in a room. (Often people do not realize that air and other gases *do* have weight.)

EXAMPLE 4.5

A college dormitory room measures 12 feet wide by 16 feet long by 8 feet high. What is the weight of air in it under normal conditions (Figure 4.13)? The weight is related to the volume by the equation:

$$W = D_W \times V$$

But

$$V = l \times w \times h = 16 \text{ ft} \times 12 \text{ ft} \times 8 \text{ ft}$$
$$= 1{,}536 \text{ ft}^3$$

So the weight is

$$W = D_W \times V = 0.08 \text{ lb/ft}^3 \times 1{,}536 \text{ ft}^3$$
$$= 123 \text{ lb}$$

This is for sea-level pressure and 0°C temperature. At normal temperature, 20°C, the weight would be 115 pounds.

Traditionally, the term *density* has referred to mass density for those using the metric system and to weight density for those using the English system. When comparing the densities of different substances, yet another quantity is often used, the **specific gravity**. The specific gravity of a substance is the ratio of its density to the density of water. Column 5 of Table 4.4 shows the specific gravities of the substances. Diamond is 3.4 times as dense as water, so its specific gravity is 3.4. This means that a certain volume of diamond will have 3.4 times the mass and 3.4 times the weight of an equal volume of water.

Water is used as the reference for specific gravities simply because it is such an important and yet common substance in our world. The density of water is another example of a "natural" unit of measure like the atmosphere (atm) and the acceleration of gravity (g). As a matter of fact, the original definition of the unit of mass in the metric system, the gram, was based on the density of water. The gram was defined to be the mass of 1 cubic centimeter of water at 0°C. This is why the density of water is exactly 1,000 kg/m^3.

4.4 Fluid Pressure and Gravity

A fluid is any substance that flows readily. All gases and liquids are fluids, as are plasmas. Granulated solids such as salt or grain can be considered fluids in some situations because they can be poured and made to flow.

Fluids are very important to us: Without air and water, life as we know it would be impossible. In the remainder of this chapter, we will discuss some of the properties of fluids in general. Usually, the statements will apply to both liquids and gases, but the fact that gases are compressible and liquids are not is sometimes important.

We live in a sea of air—the atmosphere. Though we are usually not aware of it, the air exerts pressure on everything in it. This pressure varies with altitude and can cause one's ears to "pop" when riding in a fast elevator. When swimming underwater, the same phenomenon can occur if you go deeper under the surface. In both the atmosphere and underwater, the pressure is caused by the force of gravity. Without gravity, air and water would not be pulled to the Earth. They would simply float off into space, just as we would. The pressure in any fluid increases as you go deeper in it and decreases when you rise through it.

Before considering the exact relationship between depth and pressure, we state two general properties of pressures in fluids. First, fluid pressures act in all directions. When you put a hand under water, the pressure acts not only on the top of your hand, but also on the sides and bottom. Second, the force of gravity causes the pressure in a fluid to vary with depth only, not with horizontal position. In liquids, the pressure depends on the

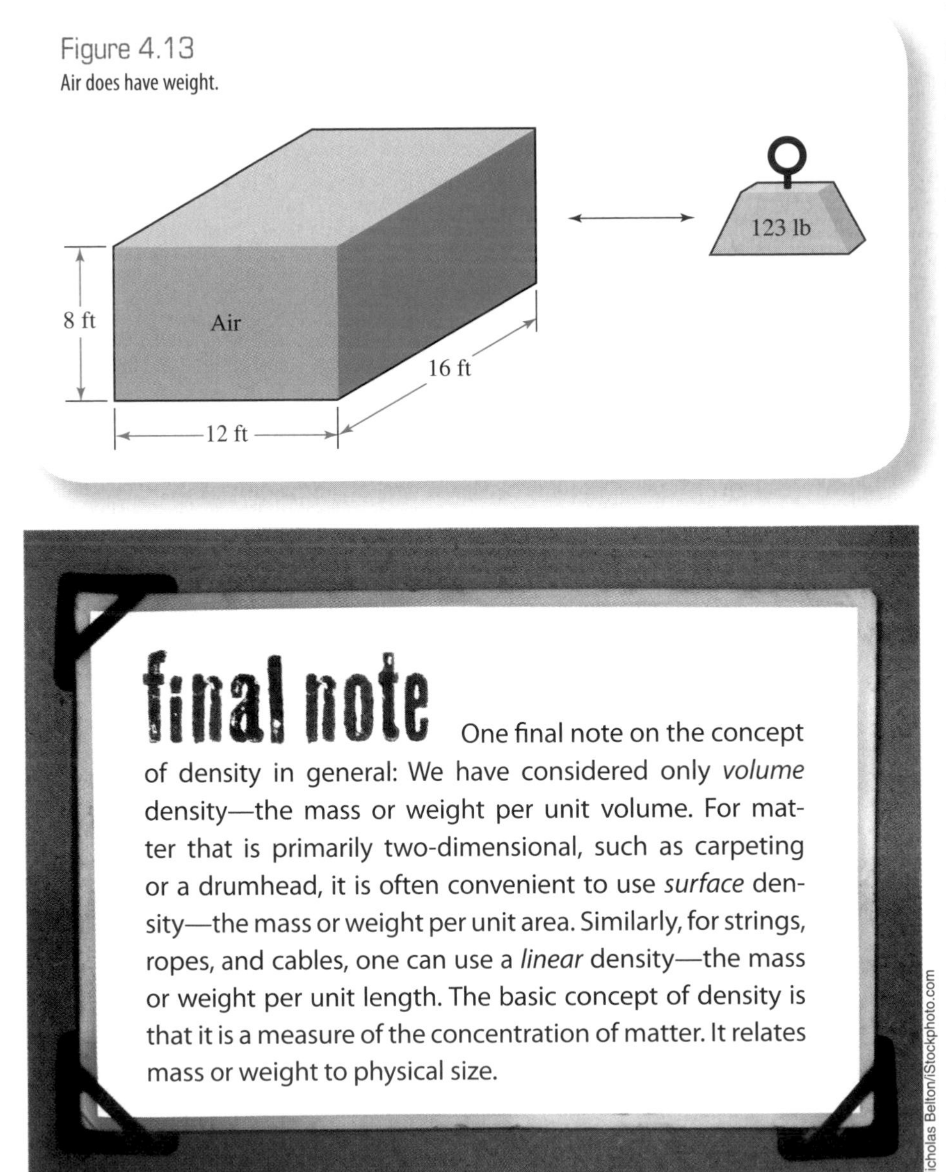

Figure 4.13
Air does have weight.

final note One final note on the concept of density in general: We have considered only *volume* density—the mass or weight per unit volume. For matter that is primarily two-dimensional, such as carpeting or a drumhead, it is often convenient to use *surface* density—the mass or weight per unit area. Similarly, for strings, ropes, and cables, one can use a *linear* density—the mass or weight per unit length. The basic concept of density is that it is a measure of the concentration of matter. It relates mass or weight to physical size.

© Nicholas Belton/iStockphoto.com

vertical distance from the surface and is independent of the shape of the container.

One can illustrate both of these principles by filling a rubber boot with water. When holes are punched in the boot, the pressure causes water to run out (Figure 4.14). Water will run out of holes in the top, sides, and bottom of the toe because the pressure acts in all directions on the inner surface. Water comes out faster from holes that are farther below the surface because the pressure is greater. The speed is the same for all holes that are at the same level, regardless of their orientation (up, down, or sideways) and their location (toe, heel, etc.). This is because the pressure in a particular fluid depends only on the depth, not on the lateral position.

The following law explains how the pressure in a fluid is related to gravity.

LAWS

Law of Fluid Pressure *The (gauge) pressure at any depth in a fluid at rest equals the weight of the fluid in a column extending from that depth to the free-standing surface of the fluid divided by the cross-sectional area of the column.*

This law is as much a prescription for determining the pressure in a fluid as it is a description of what causes pressure in a fluid. For liquids, we can use it to derive the simple relationship between pressure and depth. Let's say a tank is filled with a liquid so that the bottom is some distance h below the liquid's surface. On the bottom, we look at a rectangular area that has length l and width w (Figure 4.15). All of the liquid directly above the rectangle is in a column with dimensions l by w by h. The weight of this liquid pushes down on the rectangle. This causes a pressure on the bottom that is equal to the weight of the liquid in the column divided by the area of the rectangle.

$$p = \frac{F}{A} = \frac{W}{A} = \frac{\text{weight of liquid}}{\text{area of rectangle}}$$

It does not matter what the actual area is: A larger rectangle will have a proportionally larger amount of liquid in the column above. The actual height of the column of liquid is what determines the pressure. We compute the pressure using the fact that the weight of the liquid equals the weight density D_W of the liquid times the volume V of the column.

$$F = W = D_W \times V = D_W \times l \times w \times h$$

$$A = \text{area of rectangle} = l \times w$$

Figure 4.14

In a rubber boot filled with water, the pressure of the water acts on all parts of the inner surface of the boot and forces water out of any hole punched in the boot. The speed of the water coming out of a hole depends on how far the hole is below the surface of the water.

© Vern Ostdiek / © iStockphoto.com

Figure 4.15

A tank is filled with a liquid to a depth h. Any rectangular area on the bottom supports the weight of all of the fluid directly above it. So the pressure on the bottom equals the weight of the liquid in the column divided by the area of that rectangle.

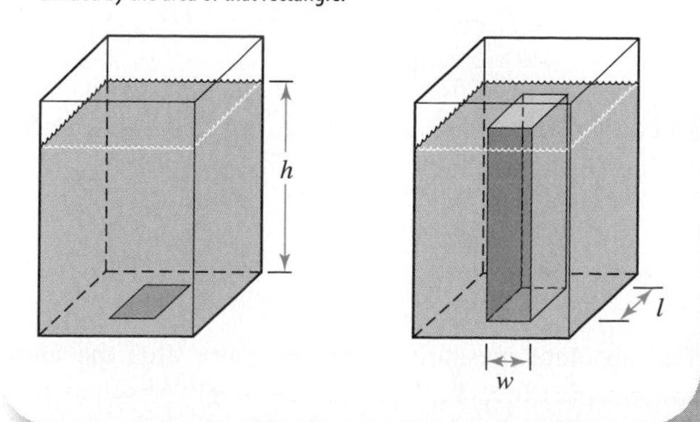

So:

$$p = \frac{W}{A} = \frac{D_W \times l \times w \times h}{l \times w}$$

$$p = D_W h \qquad \text{(gauge pressure in a liquid)}$$

Or, because $D_W = Dg$:

$$p = Dgh \qquad \text{(gauge pressure in a liquid)}$$

This gives the pressure due to the liquid above. If there is also pressure on the liquid's surface (like atmospheric pressure), this will be passed on to the bottom as well. Also, there is nothing special about the bottom: At any level above the bottom, the liquid above exerts a force and, therefore, pressure on the liquid below that level. We can summarize our result as follows.

PRINCIPLES

In a liquid, the absolute pressure at a depth h is greater than the pressure at the surface by an amount equal to the weight density of the liquid times the depth.

$$p = D_W h = Dgh \qquad \text{(gauge pressure in a liquid)}$$

EXAMPLE 4.6

Let's calculate the gauge pressure at the bottom of a typical swimming pool—one that is 10 feet (3.05 meters) deep. The gauge pressure at that depth, using Table 4.4, is

$$p = D_W h = 62.4\ \text{lb/ft}^3 \times 10\ \text{ft}$$
$$= 624\ \text{lb/ft}^2$$

To convert this to psi, we use the fact that 1 square foot equals 144 square inches.

$$p = 624\ \frac{\text{lb}}{\text{ft}^2} = 624\ \frac{\text{lb}}{144\ \text{in.}^2}$$
$$= 4.33\ \text{lb/in.}^2 = 4.33\ \text{psi (gauge pressure)}$$

The absolute pressure is this pressure plus the atmospheric pressure, 14.7 psi at sea level: Absolute pressure = 4.33 psi + 14.7 psi = 19.03 psi.

At a depth of 20 feet, the gauge pressure would be twice as large, 8.66 psi. If we do this calculation in SI units:

$$p = Dgh = 1{,}000\ \text{kg/m}^3 \times 9.8\ \text{m/s}^2 \times 3.05\ \text{m}$$
$$= 29{,}900\ \text{Pa}$$

The general result for the increase in pressure with depth in water is

$$p = 0.433\ \text{psi/ft} \times h \qquad \text{(for water, } h \text{ in ft, } p \text{ in psi)}$$

For every 10 feet of depth in water, the pressure increases 4.33 psi. Figure 4.16 shows a graph of the pressure underwater versus the depth. It is a straight line, since the pressure is proportional to the depth. In seawater, the density is slightly higher, and the pressure increases 4.47 psi for every 10 feet. Submarines and other devices that operate underwater must be designed to withstand high pressures. For example, at a depth of 300 feet in seawater, the pressure is 134 psi ($p = 0.447 \times 300$). The force on each square foot of a surface is over 9 tons.

EXAMPLE 4.7

At what depth in pure water is the gauge pressure 1 atmosphere?

$$p = 0.433\ \text{psi/ft} \times h$$
$$14.7\ \text{psi} = 0.433\ \text{psi/ft} \times h$$
$$\frac{14.7\ \text{psi}}{0.433\ \text{psi/ft}} = h$$
$$h = 33.9\ \text{ft} = 10.3\ \text{m}$$

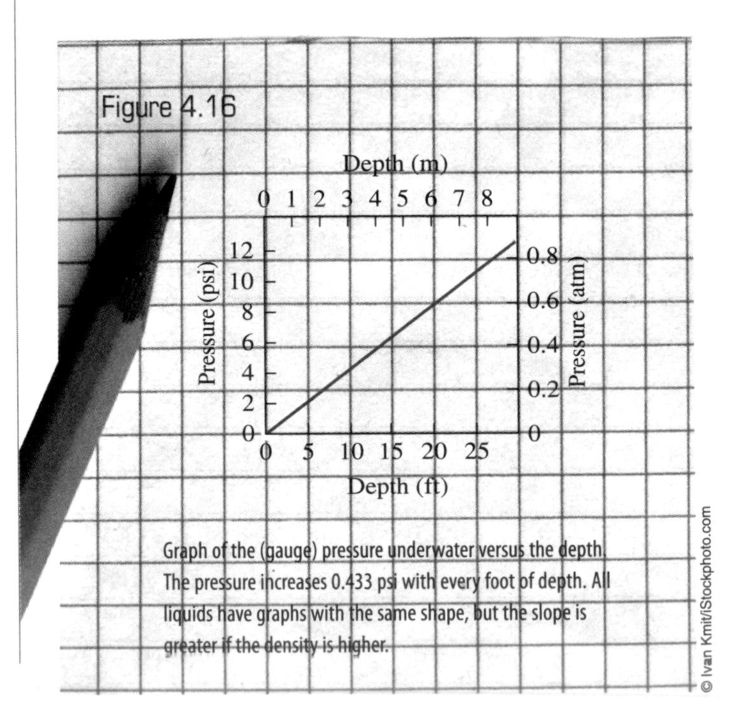

Figure 4.16

Graph of the (gauge) pressure underwater versus the depth. The pressure increases 0.433 psi with every foot of depth. All liquids have graphs with the same shape, but the slope is greater if the density is higher.

© Ivan Kmit/iStockphoto.com

Since mercury is 13.6 times as dense as water, the gauge pressure is 1 atmosphere at a depth of:

$$h = \frac{33.9\ \text{ft}}{13.6} = 2.49\ \text{ft} = 29.9\ \text{in.} = 0.76\ \text{m} \quad \text{(Figure 4.17)}$$

Fluid Pressure in the Atmosphere

The law of fluid pressure has a simple form in liquids. For gases, things are a bit more complicated. We know that the density of a gas depends on the pressure. At greater depths in a gas, the increased pressure causes increased density. The total weight of a vertical column of gas can be computed only with the aid of calculus.

The Earth's atmosphere is a relatively thin layer of a mixture of gases. The decrease in air pressure with altitude is further complicated by variations in the temperature and composition of the gas. The heating of the air by the Sun, the rotation of the Earth, and other factors cause the air pressure at a given place to vary slightly from hour to hour. Also, the pressure is not always the same at points with the same altitude.

In spite of the complexity of the situation, we can make some statements about the general variation of air pressure with altitude. Because the atmosphere is a gas, there is no upper surface or boundary. The air keeps getting progressively thinner—there are fewer molecules per unit volume—as you go higher in the atmosphere. At 9,000 meters (30,000 feet) above sea level, the density of the air is only about 35% of that at sea level. At this altitude, the average person cannot remain conscious because there is not enough oxygen in each breath. At about 160 kilometers (100 miles) above sea level, the density is down to one-billionth of the sea-level density. This is often regarded as the effective upper limit of the atmosphere. Spacecraft can remain in orbit at this altitude because the thin air causes very little air resistance.

Figure 4.18 is a graph of the air pressure versus altitude for the lower atmosphere. (When comparing this graph with Figure 4.16, remember that the depth in a liquid is measured downward from the surface, but the height in the atmosphere is measured upward from sea level.) The pressure rapidly decreases with increasing height at low altitudes where the density is still fairly high. Remember that the pressure at any elevation depends on the weight of all of the air above. At 9,000 meters, the pressure is about 0.3 atmospheres. This means that only 30% of the air is above 9,000 meters. Even though the atmosphere extends over 100 miles up, most of the air (70%) is less than 6 miles up.

Air pressure is measured with a *barometer.* The simplest type is the mercury barometer. It consists of a vertical glass tube with its lower end immersed in a bowl of mercury. All of the air is removed from the tube, so there is no air pressure acting on the surface of the mercury in the tube (see Figure 4.19). The air pressure on the surface

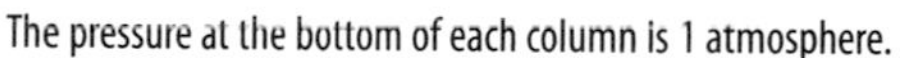
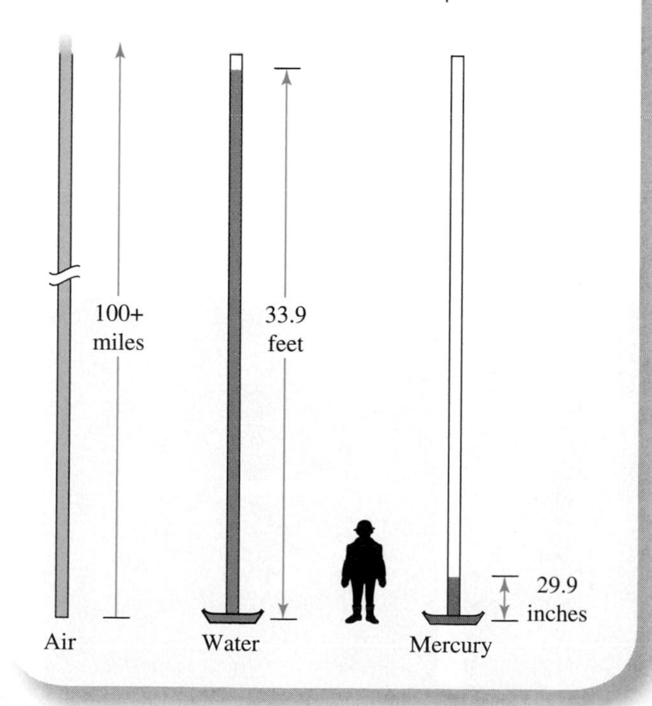

Figure 4.17
The pressure at the bottom of each column is 1 atmosphere.

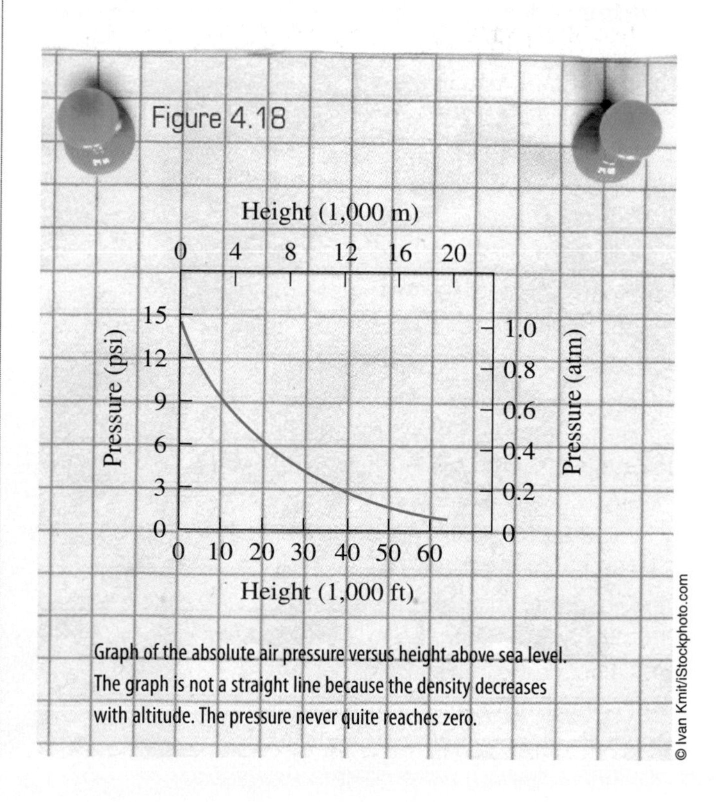

Figure 4.18
Graph of the absolute air pressure versus height above sea level. The graph is not a straight line because the density decreases with altitude. The pressure never quite reaches zero.

Figure 4.19

A mercury barometer. If there is no air in the tube, the air pressure on the mercury in the bowl forces the mercury to rise up the tube. The pressure at the bottom of the tube equals the air pressure. The higher the air pressure, the higher the column of mercury.

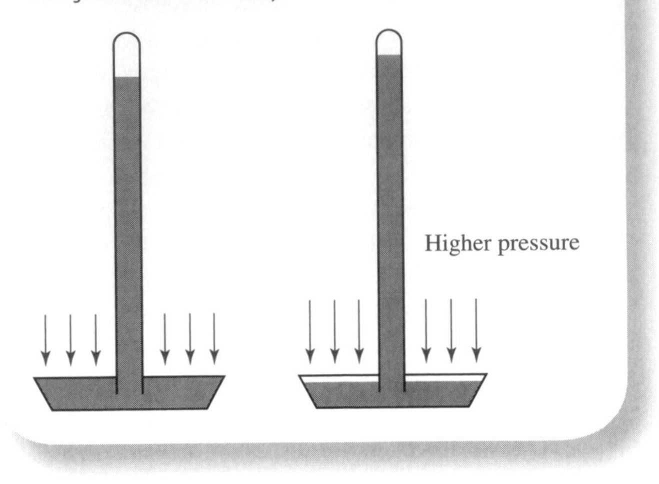

of the mercury in the bowl forces some of the liquid to rise up in the tube. (When drinking through a straw, you reduce the pressure in your mouth, and the atmospheric pressure forces the drink up the straw.) The mercury will rise in the tube until the pressure at the base of the tube equals the air pressure. Hence, the air pressure is determined by measuring the height of the column of mercury.

When the air pressure is 1 atmosphere (14.7 psi), the column of mercury is 760 millimeters (29.9 inches) long. At lower pressures, the column is shorter, so we can use the length of the column of mercury as a measure of the pressure (760 millimeters of mercury equals 1 atmosphere). Other liquids will work, but the tube has to be much longer if the liquid's density is small. For example, it would take a column of water 10.3 meters (33.9 feet) long to produce, and therefore measure, a pressure of 1 atmosphere (Figure 4.17).

This unit of measure is not limited to measuring air pressure. Blood pressure is also given in millimeters of mercury. If your blood pressure is 100 over 60, it means that the pressure drops from 100 millimeters of mercury during each heartbeat to 60 millimeters of mercury between beats.

A more portable type of barometer consists, more or less, of a very short metal can with the air removed from the inside (see Figure 4.20). The air pressure causes the ends to be squeezed inward. The higher the pressure, the greater the distortion of the ends. A pointer is attached to one end and moves along a scale as the distortion of the end changes. This is called an *aneroid barometer.*

The variation of air pressure with height is used to measure the altitude of aircraft. An *altimeter* is an aneroid barometer with a scale that registers altitude instead of pressure (left in Figure 4.21). For example, if an aneroid barometer in an airplane indicated the pressure was 0.67 atmospheres, one could infer from Figure 4.18 that the altitude was 10,000 feet. Each pressure reading on the barometer would correspond to a different altitude.

A more sophisticated instrument is used to measure the vertical speed of an aircraft—how fast it is going up or down. This device, called a *vertical airspeed indicator* in airplanes (right in Figure 4.21) and a *variometer*

Figure 4.20

Schematic of an aneroid barometer. With higher atmospheric pressure, the top of the can is pushed in more, causing the pointer to indicate higher pressure.

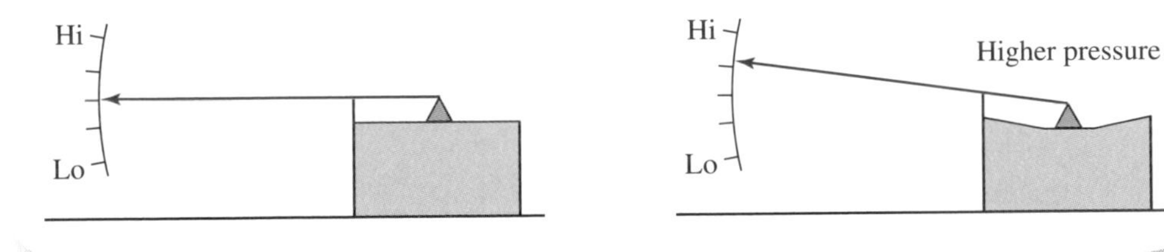

Figure 4.21

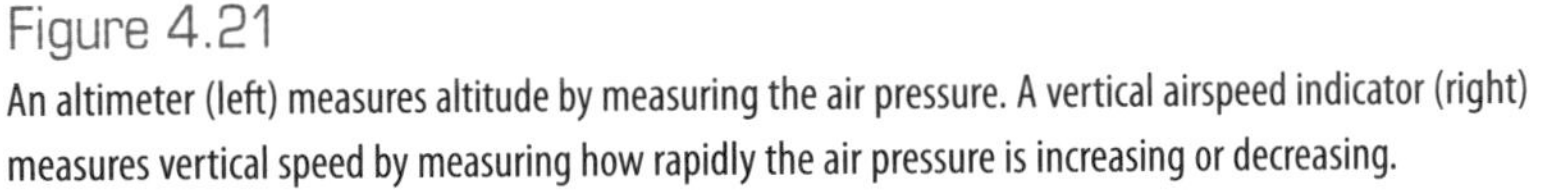
An altimeter (left) measures altitude by measuring the air pressure. A vertical airspeed indicator (right) measures vertical speed by measuring how rapidly the air pressure is increasing or decreasing.

© Pierre Landry/iStockphoto.com / © Corbis

in gliders, senses changes in the air pressure. When the aircraft is going up, the measured air pressure decreases. The instrument converts the rate of change of the air pressure into a vertical speed.

The law of fluid pressure applies to granulated solids in much the same way that it does to liquids. The walls of storage bins and grain silos are reinforced near the bottom because the pressure is higher there. Again, it is the force of gravity pulling on the material above that causes the pressure.

4.5 Archimedes' Principle

The force of gravity causes the pressure in a fluid to increase with depth. This in turn causes an interesting effect on substances partly or totally immersed in a fluid. In some cases, the substance floats—wood or oil on water, blimps or hot-air balloons in air. In other cases, the substance seems lighter—a 100-pound rock can be lifted with a force of about 60 pounds when it is underwater. Obviously, some force acts on the foreign substance to oppose the downward force of gravity (weight). This force is called the **buoyant force**.

DEFINITION

Buoyant Force *The upward force exerted by a fluid on a substance partly or completely immersed in it.*

This force acts on anything immersed in any gas or liquid. As long as there are no other forces acting on the substance, there are three possibilities (Figure 4.22). If the buoyant force is *less than* the weight of the substance, it will *sink.* If the buoyant force is *equal* to the weight, the substance will *float.* If the buoyant force is *greater than* the weight of the substance, it will *rise* upward. Examples of each case (in the same order) are a rock in water, a piece of wood floating on water, and a helium-filled balloon rising in air. Both the rock and the balloon experience a net force because the weight and the buoyant force do not cancel each other. This net force causes each to accelerate momentarily until the force of kinetic friction eventually offsets the net force and a terminal speed is reached.

At this point, we might pose two questions. First, what causes the buoyant force? Second, what determines the magnitude of the buoyant force? The answer to the first question can be arrived at rather simply in light of the previous section. Consider an object that is completely immersed (Figure 4.23). Since its bottom surface is deeper in the fluid than its top surface, the pressure on the bottom surface will be greater. Hence the upward force on its bottom surface caused by this pressure is greater than the downward force on its top surface. In short, the difference in fluid pressure acting on the surfaces of the object causes a net upward force. (The forces on the sides of the object are equal and opposite, so they cancel out.)

When something floats on the surface of a liquid, only its lower surface experiences the fluid pressure of the liquid. This causes an upward force.

As for the second question, the size of the buoyant force is determined by a law formulated in the third century BCE by the Greek scientist Archimedes.

When a piece of wood is placed in water, it displaces some of the water: Part of the wood occupies the same space or volume formerly occupied by water. The weight of this displaced water equals the buoyant force acting on the wood. Obviously, any object that is completely submerged in a fluid will displace a volume of fluid equal to its own volume.

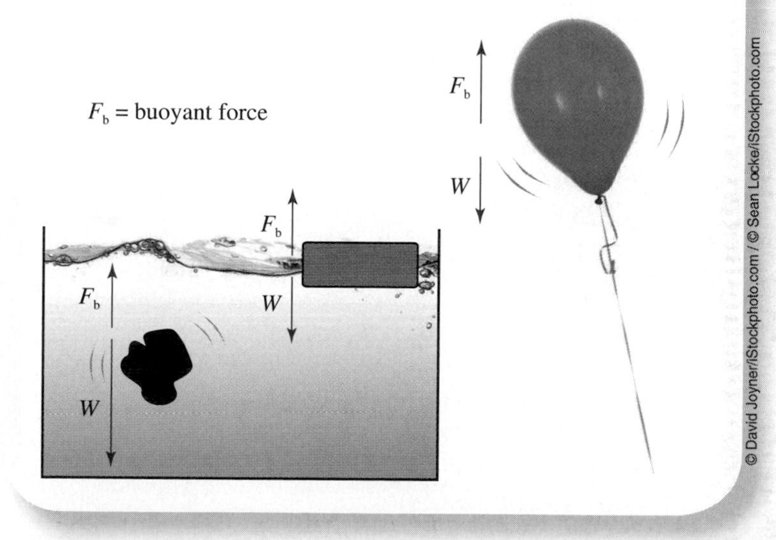

Figure 4.22
The rock sinks because the buoyant force on it is smaller than its weight. The wood floats because the buoyant force on it is equal to its weight. The helium balloon rises because the buoyant force on it is greater than its weight.

© David Joyner/iStockphoto.com / © Sean Locke/iStockphoto.com

If the buoyant force on an object is less than its weight, it sinks, but the *net* downward force is reduced. Figure 4.24 shows an object hanging from a scale. Its weight is 10 newtons. As the object is lowered into a beaker of water, it displaces some of the water. The scale reading is reduced by an amount equal to the buoyant force. When the scale shows 6 newtons, the buoyant force is 4 newtons.

$$\text{scale reading} = \text{weight} - \text{buoyant force}$$

$$6\text{ N} = 10\text{ N} - 4\text{ N}$$

The weight of the water that spills over the side of the beaker is also

Figure 4.23

The fluid pressure causes a force on each surface of an immersed object. The pressure on the lower surface is greater than the pressure on the upper surface. Consequently, the upward force on the bottom is greater than the downward force on the top. The net upward force is the buoyant force.

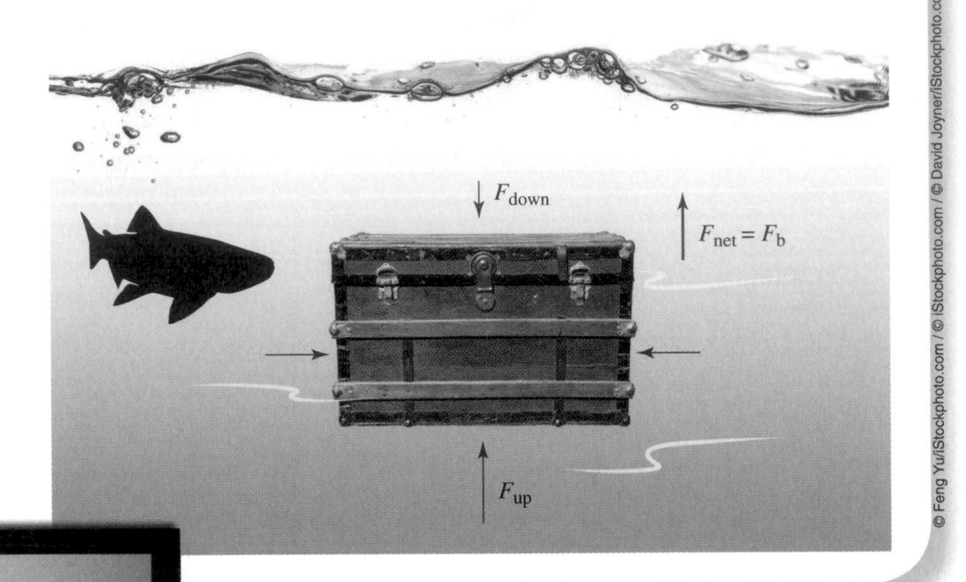

PRINCIPLES

Archimedes' Principle The buoyant force acting on a substance in a fluid at rest is equal to the weight of the fluid displaced by the substance.

F_b = weight of displaced fluid*

* For the mathematically inclined—in the simple case of a box-shaped object immersed (but level) in a liquid, we can prove Archimedes' principle. Assume the box's dimensions are l, w, and h. The downward force on the top of the box is

$$F_{top} = p_{top} \times A_{top} = p_{top} \times l \times w$$

Here p_{top} is the pressure at the top of the box. At the bottom of the box, the pressure is higher because it is deeper in the fluid. Specifically, the pressure there, p_{bottom}, is

$$p_{bottom} = p_{top} + (D_W \times h)$$

where D_W is the weight density of the liquid. Therefore, the upward force on the bottom is

$$F_{bottom} = p_{bottom} \times l \times w = [p_{top} + (D_W \times h)] \times l \times w$$
$$= (p_{top} \times l \times w) + (D_W \times h \times l \times w)$$

The buoyant force F_b is the upward force minus the downward force

$$F_b = F_{bottom} - F_{top} = (p_{top} \times l \times w) + (D_W \times h \times l \times w) - (p_{top} \times l \times w)$$
$$= D_W \times h \times l \times w = D_W \times V = W$$
$$= \text{weight of fluid displaced}$$

Archimedes (287?–212 BCE)

Figure 4.24
The weight of an object is 10 newtons. When the object is immersed in water, the scale reads a smaller force because of the upward buoyant force. The buoyant force is equal to the weight of the water that the object displaces.

4 newtons. If the object is lowered farther into the water, the buoyant force will increase, and the scale reading will decrease.

Notice that the buoyant force acting on a substance doesn't depend on what the substance is, only on how much fluid it displaces when immersed. Identical balloons filled to the same size with helium, air, and water all have the same buoyant force acting on them. Only the helium-filled balloon floats in air because its weight is less than the buoyant force. Also, the weight of a substance alone does not determine whether or not it will float. A tiny pebble will sink in water, but a 2-ton log will float.

The key is density. Let's take the simple case of a solid object submerged in a liquid. The volume of the liquid displaced by the object is, of course, equal to the object's volume. So:

$$\text{weight of object} = \text{weight density (of object)} \times \text{volume}$$

$$W = D_W(\text{object}) \times V$$

The buoyant force is

$$\text{buoyant force} = \text{weight of displaced fluid (Archimedes' principle)}$$

$$F_b = \text{weight density (of fluid)} \times \text{volume} = D_W(\text{fluid}) \times V$$

From this, we conclude that if the weight density of the object is greater than the weight density of the fluid, the object's weight is greater than the buoyant force. The object sinks. If its density is less than the fluid's density, it floats. By simply comparing the densities (or specific gravities) of the substance and the fluid, one can determine whether or not the substance will float. Any substance with a smaller density than that of a given fluid will float in (or on) that fluid.

In Table 4.4 we see that juniper wood, ice, gasoline, ethyl alcohol, and all of the gases will float on water because their densities are less than that of water. (Actually, the alcohol would simply mix in with the water. In the case of a fluid being immersed in another fluid, it is best to imagine the first being enclosed in a balloon or some other container that has negligible weight but keeps the fluids from mixing.)

The density of mercury is so high that everything in the table except gold and uranium will float on it (Figure 4.25). Hydrogen and helium both float in air, while radon and carbon dioxide sink.

Ships and blimps float, even though they are constructed out of materials with higher densities than the fluids in which they float. They are composites of substances: Most of the volume of a ship is occupied by air,

Figure 4.25
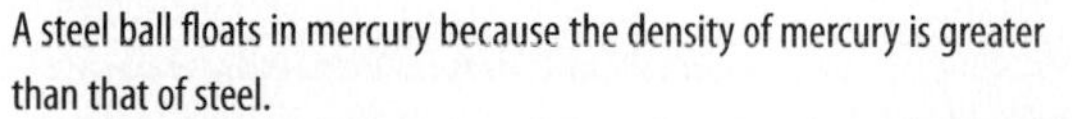
A steel ball floats in mercury because the density of mercury is greater than that of steel.

© Charles D. Winters/Photo Researchers, Inc.

and most of the volume of a blimp is occupied by helium. In both cases, the *average* density of the object is less than the density of the fluid.

As a ship is loaded with cargo, it sinks lower into the water. This makes it displace more water, thereby increasing the buoyant force and countering the added weight (see Figure 4.26).

Archimedes' principle is routinely used to measure the densities or specific gravities of solids that sink. The object is hung from a scale, and the reading is recorded. (It does not matter whether the scale gives mass or weight. Here we assume it reads weight.) Then the object is completely immersed in water, and the new scale reading is recorded (see Figure 4.24). The *difference* in these two readings equals the weight of the displaced water, by Archimedes' principle.

$$W \text{ of displaced water} = \text{scale reading out of water} - \text{scale reading in water}$$

$$W \text{ of displaced water} = \text{scale (out)} - \text{scale (in)}$$

So the weight of the object is known, and the weight of an *equal* volume of water is also known. Density is weight divided by the volume. Since the volumes are the same, the density of the object divided by the density of the water *equals* the weight of the object divided by the weight of the water.

$$\frac{\text{density of object}}{\text{density of water}} = \frac{\text{scale reading out of water}}{\text{scale (out)} - \text{scale (in)}}$$

This ratio is just the specific gravity of the object.

$$\text{specific gravity} = \frac{\text{scale (out)}}{\text{scale (out)} - \text{scale (in)}}$$

Figure 4.26
A loaded ship rides lower in the water because it must displace more water to have a larger buoyant force.

© Richard Levine/Alamy / © Gary Blakeley/iStockphoto.com / © Maria Toutoudaki/iStockphoto.com

In Section 4.3, we described how the density of the coolant in an automobile engine indicates the antifreeze content. The density is measured with a simple device that employs Archimedes' principle. Figure 4.27 shows an antifreeze tester that consists of a narrow glass tube containing five balls. The balls are specially made so that each one has a slightly higher density than the one above it. The coolant is drawn up into the glass tube with a suction bulb. Each ball will float if its density is less than the density of the coolant. The number of balls that float depends on the density of the coolant. If the antifreeze content is high, the coolant's density is high, and all of the balls float.

To check the hemoglobin content of blood, a drop of it is placed in a liquid that has the correct minimum density. If the blood sinks, its density is greater than that of the liquid, and the hemoglobin content is high enough.

The following examples use Archimedes' principle.

EXAMPLE 4.8

A contemporary Huckleberry Finn wants to construct a raft by attaching empty, plastic, 1-gallon milk jugs to the bottom of a sheet of plywood. The raft and passengers will have a total weight of 300 pounds. How many jugs are required to keep the raft afloat on water?

Figure 4.27

The number of balls that float in this antifreeze tester depends on the density of the coolant. Higher density coolant (right) has a lower freezing point.

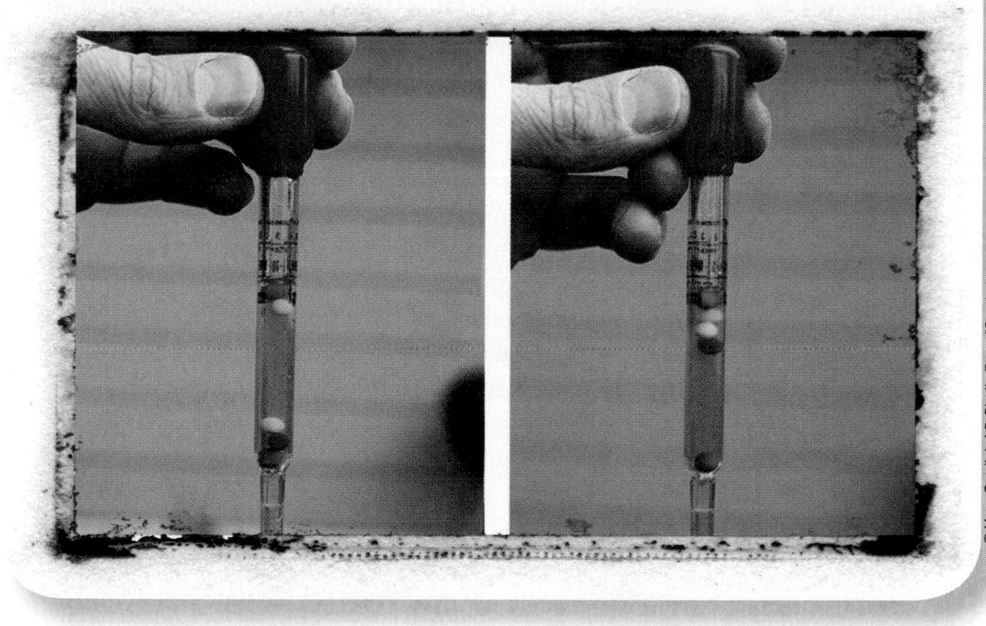

© Vern Ostdiek / © Shelly Perry/iStockphoto.com

Each cubic foot of hydrogen can lift 0.0744 pounds (see Figure 4.28). If helium is used instead, the buoyant force is still the same, but the weight of each cubic foot of helium is 0.011 pounds. So:

$$\text{net force} = 0.08 \text{ lb} - 0.011 \text{ lb}$$

$$F = 0.069 \text{ lb} \quad \text{(helium)}$$

Each cubic foot of hydrogen can lift 8% more than a cubic foot of helium. A balloon filled with hydrogen can lift 8% more than an identical balloon filled with helium. However, the big factor that tilts the scale in favor of helium is that it does not burn, as hydrogen does.

The buoyant force on the raft must be at least 300 pounds. Consequently, the raft must displace a volume of water that weighs 300 pounds.

$$F_b = \text{weight of water displaced}$$

$$= D_W \text{(water)} \times \text{volume of water displaced}$$

$$300 \text{ lb} = 62.4 \text{ lb/ft}^3 \times V$$

$$V = \frac{300 \text{ lb}}{62.4 \text{ lb/ft}^3}$$

$$= 4.8 \text{ ft}^3$$

To keep 300 pounds afloat, 4.8 cubic feet of water must be displaced. One cubic foot is equivalent to 7.48 gallons. So 4.8 × 7.48 = about 36 one-gallon jugs.

EXAMPLE 4.9

Before the spectacular and tragic destruction of the German airship* Hindenburg *on 6 May 1937, blimps, zeppelins, and balloons were filled with hydrogen. Now helium is used. Let's compare the two gases in terms of their buoyancy in air. Each cubic foot of hydrogen gas weighs 0.0056 pounds at 0°C and 1 atm (Table 4.4). When in an airship, this hydrogen will displace a cubic foot of air. So each cubic foot of hydrogen sustains a buoyant force of 0.08 pounds, the weight of 1 cubic foot of air. Therefore, the net force on each cubic foot of hydrogen gas is

$$\text{net force} = 0.08 \text{ lb} - 0.0056 \text{ lb}$$

$$F = 0.0744 \text{ lb} \quad \text{(hydrogen)}$$

The air exerts a buoyant force on everything in it, not just balloons. This force, 0.08 pounds for each cubic foot, is so small that it can be ignored except when dealing with other gases. For example, the volume of a person's body is typically around 2 or 3 cubic feet. (The density of the human body is about the same as that of

Figure 4.28

One cubic foot of hydrogen can lift 0.0744 pounds.
One cubic foot of helium can lift 0.069 pounds.

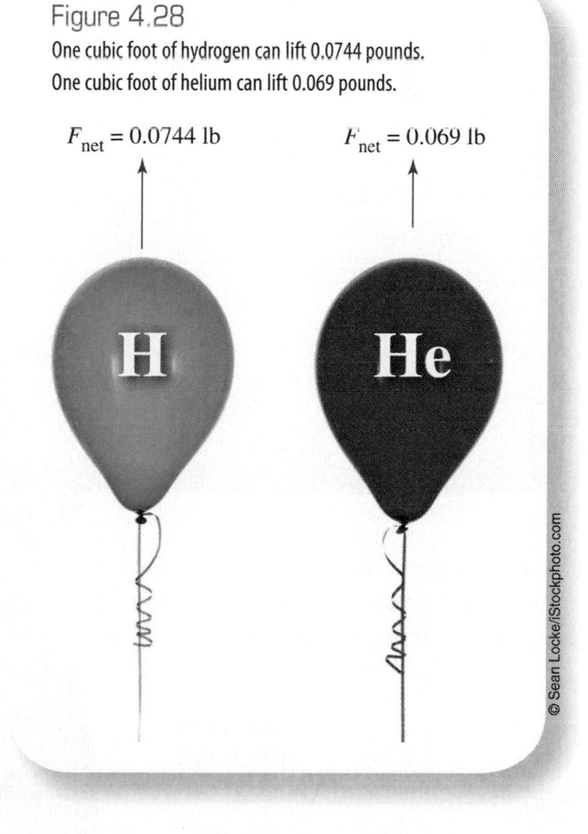

© Sean Locke/iStockphoto.com

water. That is why we just barely float in water. So the approximate volume of your body is your weight divided by the weight density of water.) This means that the buoyant force on a person due to the air is between 0.16 and 0.24 pounds.

4.6 Pascal's Principle

When a force acts on a surface of a solid, the resulting pressure is "transmitted" through the solid only in the original direction of the force. When you sit on a stool, your weight causes pressure in the stool's legs that is transmitted to the floor. This pressure doesn't act to the side of the legs, only downward.

In a fluid, any pressure caused by a force is transmitted everywhere throughout the fluid and acts in all directions. When you squeeze a tube of toothpaste, the pressure is passed on to all points in the toothpaste. An inward force on the sides of the tube can cause the toothpaste to come out of the end. This property of fluids should be familiar to you and might even be used to distinguish fluids from solids. Pascal's principle is a formal statement of this phenomenon.

PRINCIPLES

Pascal's Principle
Pressure applied to an enclosed fluid is transmitted undiminished to all parts of the fluid and to the walls of the container.

This property of fluids is exploited in widely used hydraulic systems. Hydraulic jacks and the brake systems in automobiles are common examples. The basic components of such systems are piston and cylinder combinations. Figure 4.29 shows a small piston and cylinder on the left connected via a tube to a larger piston and cylinder on the right. (For this discussion, we may ignore any effects due to gravity.) A liquid is in both cylinders and the connecting tube. When a force acts on the left piston, the resulting pressure is passed on throughout the liquid. This pressure causes a force on the right piston. Since this piston is larger than the one on the left, the force ($F = pA$) will also be larger. For example, if the area of the right piston is five times that of the left piston, the force will be five times as large. This system behaves like a lever (Figure 3.7). A small force acting at one place produces a larger force at another place. As with the lever, the smaller piston will move a correspondingly greater distance than the larger piston. The work done is the same for both pistons.

In automotive brake systems, the brake pedal is connected to a piston that slides in the "master" cylinder. This cylinder is connected via tubing to "wheel" cylinders on the brakes on the wheels (Figure 4.30). (Actually, there are two cylinders in tandem in the master cylinder. Each is connected to two wheels so that if one subsystem fails, the other two wheels will still have braking power.)

Brake fluid fills the cylinders and tubing. When the brake pedal is pushed, the piston produces pressure in the master cylinder that is transmitted to the wheel cylinders. The piston in each wheel cylinder is attached to a mechanism that applies the brakes. In disc brakes, used on the front wheels of most cars, the piston squeezes disc pads against the sides of a rotating disc attached to the wheel. This action is very similar to that of rim brakes on bicycles. The mechanism in drum brakes, used on the rear wheels of most cars, is a bit more complicated.

Using a hydraulic brake system serves two purposes. The wheel cylinders having larger diameters than those of the master cylinder yield a mechanical advantage: The forces on the wheel-cylinder pistons are larger than the force applied to the master-cylinder piston. Also, this is an efficient way to transmit a force from one location in the car to four other locations.

© The Bridgeman Art Library/Getty Images

Blaise Pascale
(1623–1662)

Figure 4.29
The pressure caused by the force on the small piston is transmitted throughout the fluid and acts on the larger piston. The resulting force on this piston is larger than the force on the smaller piston.

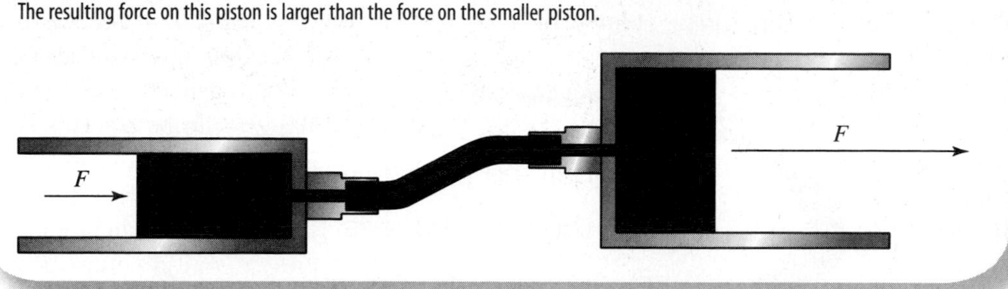

Figure 4.30
(a) Simplified diagram of an automotive hydraulic brake system. Only one wheel, equipped with disc brakes, is shown. (b) When the brake pedal is pushed, the pressure increase in the master cylinder is passed on through the fluid to the wheel cylinder. The piston is pushed outward, and the brake pads squeeze the disc.

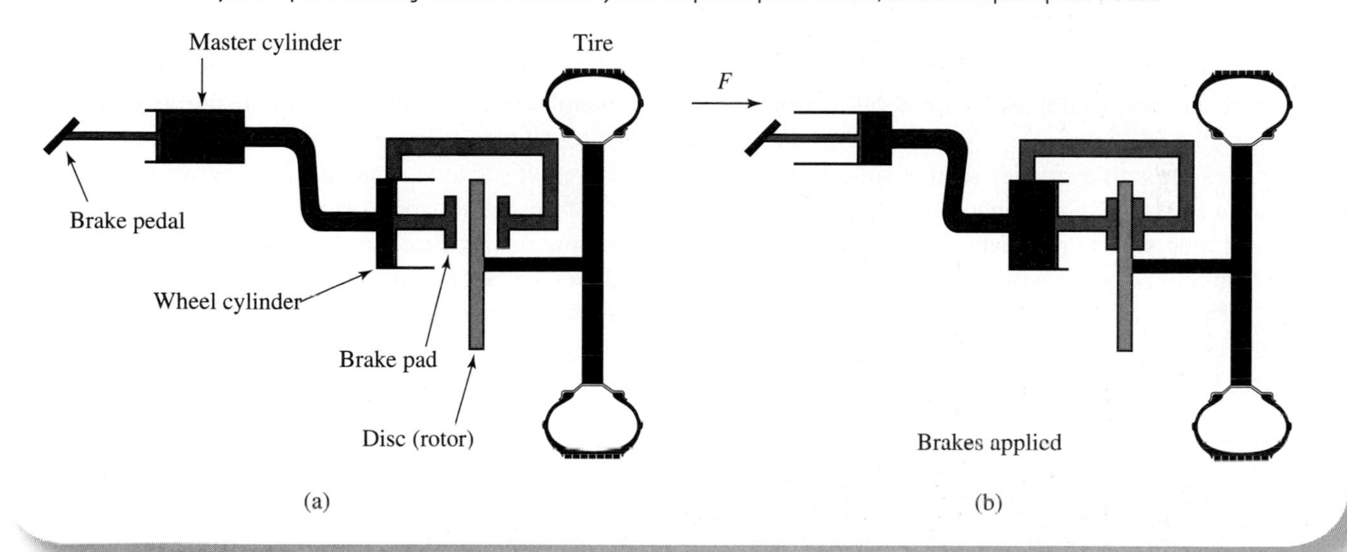

4.7 Bernoulli's Principle

In this section, we present a simple principle that applies to moving fluids. Water flowing in a stream or through pipes to a water faucet, air moving through heating ducts or as a cool breeze in the summer, blood flowing through your arteries and veins—these are all examples of fluids flowing. It is common for each of these fluids to speed up and slow down as it flows. Accompanying any change in the speed of the fluid is a change in pressure of the fluid. This is stated in the following principle, named after the Swiss physicist and mathematician Daniel Bernoulli.

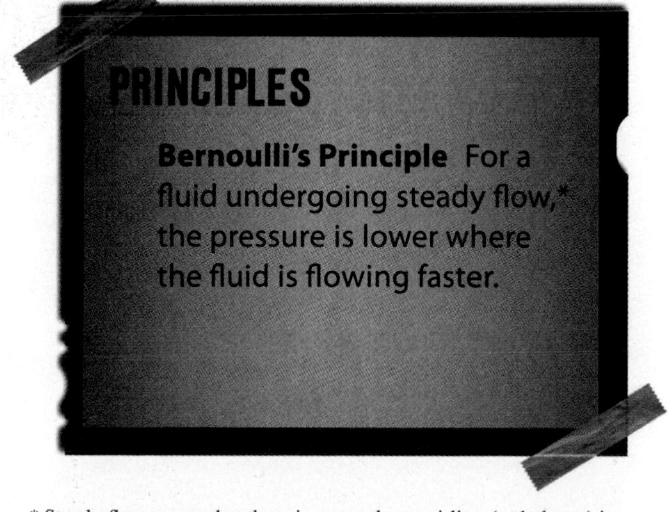

PRINCIPLES

Bernoulli's Principle For a fluid undergoing steady flow,* the pressure is lower where the fluid is flowing faster.

* Steady flow means that there is no random swirling (turbulence) in the fluid and that no outside forces act to increase or decrease the rate of flow.

This principle is based on the conservation of energy. A fluid under pressure has what can be called *pressure potential energy.* The higher the pressure, the greater the potential energy of any given volume of fluid. (When you open a faucet, water rushes out as the potential energy due to pressure is converted into kinetic energy of running water. Low water pressure makes the water come out slowly, with low kinetic energy, because it starts with low potential energy.) Moving fluids have both kinetic and potential energy. When a fluid speeds up, its kinetic energy increases. Its total energy remains constant, so its potential energy and, therefore, its pressure decrease.

The higher the pressure, the greater the potential energy of any given volume of fluid.

One of the best examples of Bernoulli's principle is that depicted in Figure 4.31. Water flowing through a pipe passes through a smaller section spliced into the pipe. The water speeds up when it enters the narrow region, and then slows down when it reenters the wide region. This occurs because the volume of fluid passing through each part of the pipe each second is the same. Where the cross-sectional area is smaller, the fluid must flow faster if the same number of cubic inches of fluid is to get through each second. You've seen this already if you've ever used your thumb to partially block water coming out of the end of a garden hose.

Bernoulli's principle tells us that the pressure in the moving water is *smaller in the narrow section* than in the wide sections upstream and downstream. Pressure gauges placed in the pipe show this to be so. This is one of those rare situations in which physical fact runs counter to one's intuition. On first thought, most people would predict that the pressure should be higher in the narrow part of the pipe because the fluid is "squeezed into a smaller stream." But that is not the case. The pressure is actually lower in the narrow part.

Atomizers on perfume bottles utilize Bernoulli's principle. When a bulb is squeezed, air moves through a horizontal tube (Figure 4.32). The air moves fast, so the air pressure is low. A small tube runs vertically downward from the horizontal tube into the perfume. Since the air pressure is reduced in the horizontal tube, the normal air pressure acting on the surface of the perfume forces the liquid to rise upward in the vertical tube and to enter the moving air. Carburetors used on lawn mowers and the automatic shutoff mechanism on gas pump nozzles also make use of Bernoulli's principle.

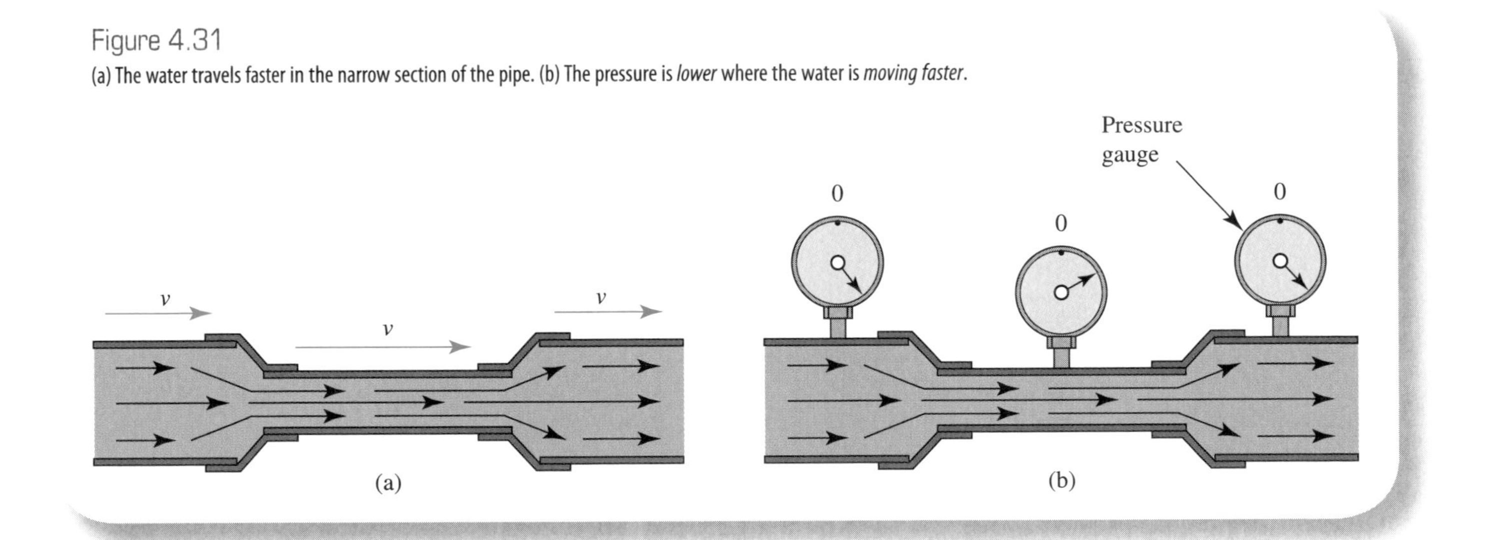

Figure 4.31

(a) The water travels faster in the narrow section of the pipe. (b) The pressure is *lower* where the water is *moving faster*.

Figure 4.32
Perfume atomizers use Bernoulli's principle. Air is forced to move through the horizontal tube when the bulb is squeezed. In this tube, the air has to move fast, so the pressure is low. Normal air pressure on the perfume in the bottle forces it to rise upward into the stream.

questions&problems

Questions

(▶ Indicates a review question, which means it requires only a basic understanding of the material to answer. The others involve integrating or extending the concepts presented thus far.)

1. ▶ Describe the four phases of matter. Compare their external, observable properties. Compare the nature of the forces between atoms or molecules (or both) in the solid, liquid, and gas phases.
2. Identify some of the elements that exist in pure form (not in compounds) around you.
3. ▶ What is the difference between a mixture of two elements and a compound formed from the two elements?
4. If you classify everything around you as an element, a compound, or a mixture, which category would have the largest number of entries?
5. ▶ Why can gases be compressed much more readily than solids or liquids?
6. Suppose you are in the International Space Station in orbit around Earth, and a fellow astronaut gives you what appears to be an inflated balloon. Describe how you could determine whether the balloon contains a gas, a liquid, or a solid.
7. ▶ Use the concept of pressure to explain why snowshoes are better than regular shoes for walking in deep snow.
8. The same bicycle tire pump is used to inflate a mountain bike tire to 40 psi and then a road bike tire to 100 psi. What difference would the user notice when using the pump on the two tires?
9. ▶ Explain the difference between gauge pressure and absolute pressure.
10. ▶ How can you use the volume of some quantity of a pure substance to calculate its mass?
11. The mass density of a mixture of ethyl alcohol and water is 950 kg/m^3. Is the mixture mostly water, mostly alcohol, or about half and half? What is your reasoning?
12. Believe it or not, canoes have been made out of concrete (and they actually float). But even though concrete has a lower density than aluminum, a concrete canoe weighs a lot more that an aluminum one of the same size. Why is that?
13. Would the weight density of water be different on the Moon than it is on Earth? What about the mass density?
14. The way pressure increases with depth in a gas is different from the way it does in a liquid. Why?
15. Workers are to install a hatch (door) near the bottom of an empty storage tank. In choosing how strong to make the hatch, does it matter how tall the tank is? How wide it is? Whether it is going to hold water or mercury? Explain.
16. If the acceleration due to gravity on the Earth suddenly increased, would this affect the atmospheric pressure?

Would it affect the pressure at the bottom of a swimming pool? Explain.

17. If the Earth's atmosphere warmed up and expanded to a larger total volume but its total mass did not change, would this affect the atmospheric pressure at sea level? Would this affect the pressure at the top of Mount Everest? Explain.

18. Is there a pressure variation (increase with depth) in a fuel tank on a spacecraft in orbit? Why or why not?

19. ▶ Explain how a barometer can be used to measure altitude.

20. Why does the buoyant force always act upward?

21. ▶ Identify some substances that would sink in gasoline but float in water?

22. It is easier for a person to float in the ocean than in a backyard swimming pool. Why?

23. ▶ A ship on a large river approaches a bridge, and the captain notices that the ship is about a foot too tall to fit under the bridge. A crew member suggests pumping water from the river into an empty tank on the ship. Would this help?

24. In "The Unparalleled Adventure of One Hans Pfaall" by Edgar Allan Poe, the hero discovers a gas whose density is "37.4 times" less than that of hydrogen. How much better at lifting would a balloon filled with the new gas be compared to one filled with hydrogen?

25. A brick is tied to a balloon filled with air and is then tossed into the ocean. As the balloon is pulled downward by the brick, the buoyant force on it decreases. Why?

26. Venus's atmosphere is much more dense than the Earth's while that of Mars is much less dense. Suppose it is decided to send a probe to each planet that, once it arrived, would be carried around in the planet's atmosphere by a helium-filled balloon. How would the size of each balloon compare to the size that would be needed on the Earth?

27. ▶ What is Pascal's principle?

28. ▶ How does a car's brake system make use of Pascal's principle?

29. ▶ What important thing happens when the speed of a moving fluid increases?

30. The pressure in the air along the upper surface of an aircraft's wing (in flight) is lower than the pressure along the lower surface. Compare the speed of the air flowing over the wing to that of the air flowing under the wing.

31. ▶ How does a perfume atomizer make use of Bernoulli's principle?

Problems

1. A grain silo is filled with 2 million pounds of wheat. The area of the silo's floor is 400 ft^2. Find the pressure on the floor in pounds per square foot and in psi.

2. A bicycle tire pump has a piston with area 0.44 in.^{2} If a person exerts a force of 30 lb on the piston while inflating a tire, what pressure does this produce on the air in the pump?

3. A large truck tire is inflated to a gauge pressure of 80 psi. The total area of one sidewall of the tire is 1,200 in.^{2} What is the outward force on the sidewall due to the air pressure?

4. The water in the plumbing in a house is at a gauge pressure of 300,000 Pa. What force does this cause on the top of the tank inside a water heater if the area of the top is 0.2 m^2?

5. A box-shaped metal can has dimensions 8 in. by 4 in. by 10 in. high. All of the air inside the can is removed with a vacuum pump. Assuming normal atmospheric pressure outside the can, find the total force on one of the 8-by-10-in. sides.

6. A viewing window on the side of a large tank at a public aquarium measures 50 in. by 60 in. (Figure 4.33). The average gauge pressure due to the water is 8 psi. What is the total outward force on the window?

Figure 4.33

© Andrea Gingerich/iStockphoto.com

7. A large chunk of metal has a mass of 393 kg, and its volume is measured to be 0.05 m^3.
   a) Find the metal's mass density and weight density in SI units.
   b) What kind of metal is it?

8. A small statue is recovered in an archaeological dig. Its weight is measured to be 96 lb and its volume 0.08 ft^3.
   a) What is the statue's weight density?
   b) What substance is it?

9. A large tanker truck can carry 20 tons (40,000 lb) of liquid.
   a) What volume of water can it carry?
   b) What volume of gasoline can it carry?

10. The total mass of the hydrogen gas in the *Hindenburg* zeppelin was 18,000 kg. Assuming standard conditions, what volume did the hydrogen occupy?

11. A large balloon used to sample the upper atmosphere is filled with 900 m^3 of helium. What is the mass of the helium?

12. A certain part of an aircraft engine has a volume of 1.3 ft^3.
    a) Find the weight of the piece when it is made of iron.
    b) If the same piece is made of aluminum, find the weight and determine how much weight is saved by using aluminum instead of iron.

13. The volume of the Drop Tower "Bremen" (a 100-meter-tall tube used to study processes during free fall) is 1,700 m^3.
    a) What is the mass of the air that must be removed from it to reduce the pressure inside to nearly zero (1 Pa compared to 100,000 Pa)?
    b) What is the weight of the air in pounds?

14. It is determined by immersing a crown in water that its volume is 26 $in.^3 = 0.015\ ft^3$.
    a) What would its weight be if it were made of pure gold?
    b) What would its weight be if half of its volume were gold and half lead?

15. Find the gauge pressure at the bottom of a swimming pool that is 12 ft deep.

16. The depth of the Pacific Ocean in the Mariana Trench is 36,198 ft. What is the gauge pressure at this depth?

17. Calculate the gauge pressure at a depth of 300 m in seawater.

18. A storage tank 30 m high is filled with gasoline.
    a) Find the gauge pressure at the bottom of the tank.
    b) Calculate the force that acts on a square access hatch at the bottom of the tank that measures 0.5 m by 0.5 m.

19. The highest point in North America is the top of Mount McKinley in Alaska, 20,320 ft above sea level. Using the graph in Section 4.4, find the approximate air pressure there.

20. The highest altitude ever reached by a glider (as of this writing) is 15,460 m. What is the approximate air pressure at that altitude?

21. An ebony log with volume 12 ft^3 is submerged in water. What is the buoyant force on it?

22. An empty storage tank has a volume of 1,500 ft^3. What is the buoyant force exerted on it by the air?

23. A blimp used for aerial camera views of sporting events holds 200,000 ft^3 of helium.
    a) How much does the helium weigh?
    b) What is the buoyant force on the blimp at sea level?
    c) How much can the blimp lift (in addition to the helium)?

24. A modern-day zeppelin holds 8,000 m^3 of helium. Compute its maximum payload at sea level.

25. A box-shaped piece of concrete measures 3 ft by 2 ft by 0.5 ft.
    a) What is its weight?
    b) Find the buoyant force that acts on it when it is submerged in water.
    c) What is the net force on the concrete piece when it is under water?

26. A juniper-wood plank measuring 0.25 ft by 1 ft by 16 ft is totally submerged in water.
    a) What is its weight?
    b) What is the buoyant force acting on it?
    c) What is the size and the direction of the net force on it?

27. The volume of an iceberg is 100,000 ft^3 (Figure 4.34).
    a) What is its weight, assuming it is pure ice?
    b) What is the volume of seawater it displaces when floating? (Hint: You know what the weight of the seawater is.)
    c) What is the volume of the part of the iceberg out of the water?

Figure 4.34

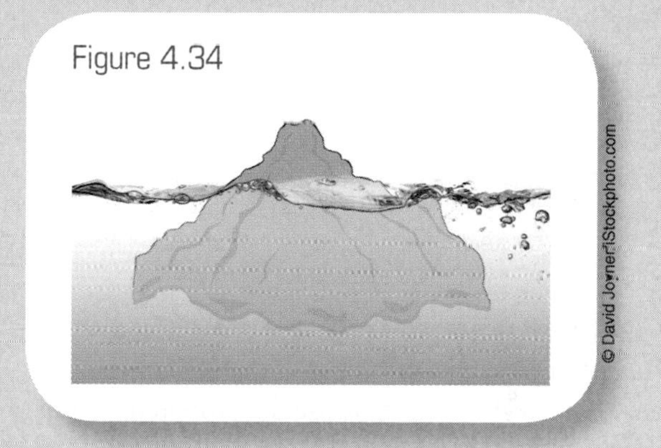

© David Joyner/iStockphoto.com

28. A boat (with a flat bottom) and its cargo weigh 5,000 N. The area of the boat's bottom is 4 m^2. How far below the surface of the water is the boat's bottom when it is floating in water?

29. A scale reads 100 N when a piece of aluminum is hanging from it. What does it read when it is lowered so that the aluminum is submerged in water?

30. A rectangular block of ice with dimensions 2 m by 3 m by 0.2 m floats on water. A person weighing 600 N wants to stand on the ice. Would the ice be able to support the individual and not sink below the surface of the water? Validate your answer.

There are more quizzes and study tools online at 4ltrpress.cengage.com/physics.

Temperature and Heat

sections

5.1 Temperature

Temperature may be the most commonly used physical quantity in our daily lives, next to time. We might loosely define temperature as a measure of hotness or coldness. But the concepts of hot and cold are themselves rather vague, subjective, and relative. In the summer, an air temperature of 70°F feels cool, while the same temperature in the winter feels warm.

The idea of temperature is distinct from that of heat or internal energy. For example, when one drop is removed from a cup of hot water, both the water in the dropper and the water in the cup have the *same* temperature, but the amount of internal energy associated with each is very different. Placing the drop in the palm of your hand would not have nearly the same effect (pain and injury) as pouring the cup of water into it.

Thermometers are devices that measure the temperature of a substance, and, for their function, they all depend on some physical property that changes with temperature. A common type exploits the fact that a liquid, usually mercury or red-colored alcohol, will expand or contract when heated or cooled, rising or falling in a glass tube as the temperature varies. Some thermometers are based on other temperature-dependent physical properties, including the volume of solids, the pressure or volume of a gas, the electrical properties of metals, the amount and frequency of radiated energy, and the speed of sound in a gas.

There are three different temperature scales in common use for calibrating thermometers: Fahrenheit, Celsius, and Kelvin. The normal freezing and boiling temperatures of water—called the *phase transition* temperatures—may be used to compare the three scales. In the Fahrenheit scale, the boiling point of water under a pressure of 1 atmosphere is 212°—designated 212°F. The freezing point of water under this pressure is 32°F. So there are 180 units, called *degrees*, that separate the two temperatures. The Celsius scale, formerly called the centigrade scale, is metric based; it uses 100 degrees between the freezing and boiling points of water. Zero degrees Celsius—designated 0°C—is the freezing temperature, and 100°C is the boiling temperature.

CHAPTER 5

physics

Most of us spend our lives subjected to temperatures within the range of −60 to 120°F (−51 to 49°C). Much higher temperatures exist in common places: the interiors of stoves and automobile engines, the filaments of light bulbs, the flame of a candle, and so on. The Sun's surface temperature is about 10,000°F (5,700°C), and its interior temperature is about 27,000,000°F (15,000,000°C). Temperatures this high have been produced on the Earth in experiments with plasmas and in nuclear explosions. There is no upper limit on temperature.

The water in the cup and the water in the dropper have the same temperature, but the water in the cup can transfer much more internal energy to its surroundings.

© Douglas Allen/iStockphoto.com / © Lise Gagne/iStockphoto.com

At the other extreme, there *is* a limit on cold

© Stock4B/Jupiterimages / © iStockphoto.com

temperatures. The coldest temperature, called **absolute zero**, is −459.67°F (−273.15°C). Because it is impossible to go below this temperature (for reasons we'll discuss later), the Kelvin scale is a convenient one because it uses absolute zero as its starting point (zero). (This scale is also referred to as the "absolute temperature scale.") The size of the unit in the Kelvin scale is the same as that in the Celsius scale, except it is called a kelvin (K) instead of a degree. Any temperature in the Kelvin scale equals the corresponding Celsius value plus 273.15. The normal boiling and freezing temperatures of water are 373.15 K and 273.15 K, respectively.

Figure 5.1 shows a comparison of the three temperature scales. Note that the Fahrenheit and Celsius scales agree at −40°. Table 5.1 lists some representative temperatures. As a physical quantity, temperature is represented by *T*.

Unfortunately, *T* is used to represent both temperature and period (Section 1.1).

PRINCIPLES

The Kelvin temperature of matter is proportional to the average kinetic energy of the constituent particles.

Kelvin scale temperature $\propto$ average *KE* of atoms and molecules

What determines the temperature of matter? In other words, what is the difference between a cup of coffee when it is hot (200°F) and the same cup of coffee when it is cold (70°F)? The atoms and molecules that compose matter have kinetic energy. In gases, they move about randomly with high speed. In liquids and solids, they vibrate much like a mass on a spring or an object oscillating in a hole (Section 3.5). At higher temperatures, the atoms and molecules in matter move faster and have higher kinetic energies.

Because of collisions between the particles, during which energy is exchanged, the particles do not all have exactly the same kinetic energy at each instant, nor does the energy of a given particle stay exactly the same from one moment to the next. But the *average* kinetic energy of all of the particles is constant as long as the temperature stays constant.

So when a cup of coffee is hot, the molecules in it have higher average kinetic energy than when it is cold. If you put your finger into hot coffee, the atoms and molecules in the coffee pass on their higher kinetic energy to the atoms and molecules in your finger by way of collisions: Your finger is warmed.

This fact—that temperature depends on the average kinetic energy of atoms and molecules—is

Figure 5.1

Comparison of the three temperature scales. The Celsius and Kelvin scales have the same sized unit. The Fahrenheit degree is five-ninths the Celsius degree.

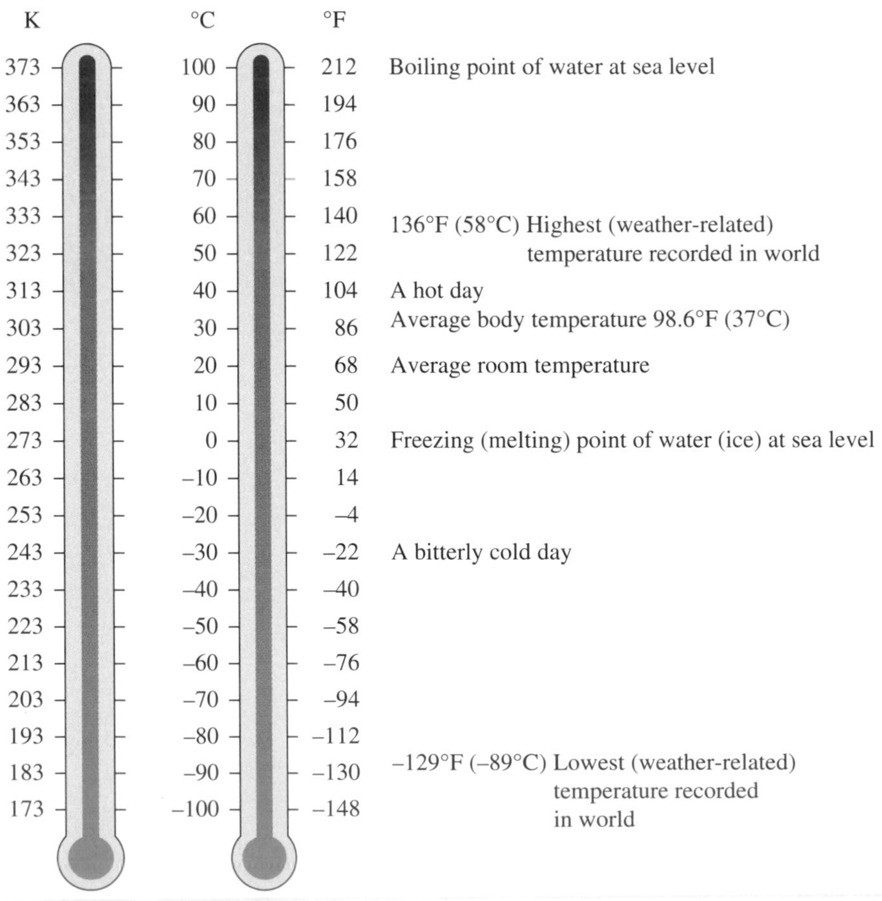

K	°C	°F	
373	100	212	Boiling point of water at sea level
363	90	194	
353	80	176	
343	70	158	
333	60	140	136°F (58°C) Highest (weather-related) temperature recorded in world
323	50	122	
313	40	104	A hot day
303	30	86	Average body temperature 98.6°F (37°C)
293	20	68	Average room temperature
283	10	50	
273	0	32	Freezing (melting) point of water (ice) at sea level
263	−10	14	
253	−20	−4	
243	−30	−22	A bitterly cold day
233	−40	−40	
223	−50	−58	
213	−60	−76	
203	−70	−94	
193	−80	−112	
183	−90	−130	−129°F (−89°C) Lowest (weather-related) temperature recorded in world
173	−100	−148	

Table 5.1 Representative Temperatures in the Three Temperature Scales

Description	°F	°C	K
Absolute zero	−459.67	−273.15	0
Helium boiling point*	−452	−268.9	4.25
Nitrogen boiling point	−320.4	−195.8	77.35
Oxygen boiling point	−297.35	−182.97	90.18
Alcohol freezing point	−175	−115	158
Mercury freezing point	−37.1	−38.4	234.75
Water freezing point	32	0	273.15
Normal body temperature	98.6	37	310.15
Water boiling point	212	100	373.15
"Red hot" (approx.)	800	430	700
Aluminum melting point	1,220	660	933
Iron melting point	2,797	1,536	1,809
Sun's surface (approx.)	10,000	5,700	6,000
Sun's interior (approx.)	27×10^6	15×10^6	15×10^6
Highest laboratory temperature	3.6×10^9	2×10^9	2×10^9

* All boiling points are for 1 atm pressure.

very important. It should help you understand many of the phenomena we will discuss in this chapter. In gases, higher kinetic energy means that the atoms and molecules move about with higher speeds. For liquids and solids, the molecules vibrate through a greater distance like a pendulum swinging through a larger arc. You may recall that when particles oscillate like this, they also have potential energy. This is the case here, too, but the potential energy of "bound" atoms and molecules in liquids and solids is not directly related to the temperature. This potential energy is important when a substance undergoes a change of phase—like freezing or boiling. More on this later.

This principle also accounts for the existence of an absolute zero. At colder temperatures, the average kinetic energy of the particles is smaller. If they stopped moving altogether, the average kinetic energy would be zero. This would be the lowest possible temperature—absolute zero (see Figure 5.2). The word "would" is used because, as it turns out, absolute zero can never be reached. The atoms and molecules in a substance cannot be completely stopped. Researchers do get very close to absolute zero—within a billionth of a degree or less—but they cannot reach it exactly.

At very low temperatures, many substances acquire unusual properties. Plastic and rubber become as brittle as glass. Below about 2 K, helium is a "superfluid" liquid; it flows without friction. Some materials become "superconductors." They conduct electricity without resistance (more on this in Chapter 7).

The nature of matter is also different at very high temperatures. Above about 6,000 K, the kinetic energies of atoms are so high that they cannot bind together; hence there can be no solids or liquids—or even molecules. Above about 20,000 K, the electrons break free from atoms, and only plasmas can exist.

Figure 5.2

At lower temperatures, atoms and molecules have lower average kinetic energy (*KE*). At absolute zero, their kinetic energies would be zero: They would be stationary.

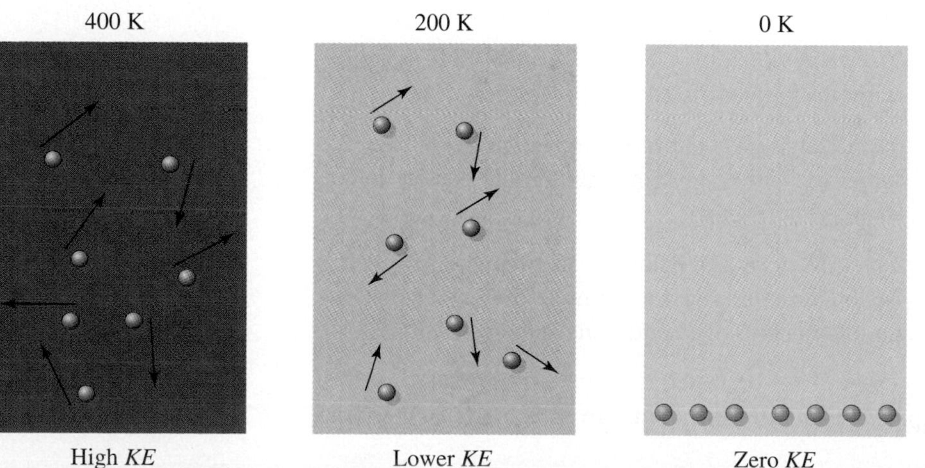

5.2 Thermal Expansion

Thermal expansion is an important phenomenon that is exploited by the common types of thermometers and by a variety of other useful devices. In almost all cases, substances that are not constrained expand when their temperatures increase. (Exceptions include water below 4°C and some compounds of tungsten.) The air in a balloon expands when heated, the mercury in a thermometer expands upward in the glass tube when placed in a hot liquid, and sections of bridges become longer in the summer. If the substance is constrained sufficiently, it will not expand, but forces and pressures will be created in response to the constraint. For example, if an empty pressure cooker is sealed and then heated, the air inside is prevented from expanding. But the pressure inside increases and causes larger and larger outward forces on the inner surfaces of the pressure cooker.

We can see qualitatively why this expansion occurs. At higher temperatures, the atoms and molecules in a solid or a liquid vibrate through a larger distance and so push each other apart slightly. In gases, they move with higher and higher speeds as the temperature rises. A balloon will expand when heated because the higher-speed air molecules will cause higher pressure and push the balloon's surface outward as they collide with it. In these cases, the expansion occurs in all three dimensions. For example, as a brick is heated, its length, width, and thickness all increase proportionally.

We can use logic and basic mathematics to predict the amount of expansion that occurs. Let us first consider the simplest case—the thermal expansion of a long, thin solid such as a metal rod. The main expansion will be an increase in its length l (see Figure 5.3). This increase, designated Δl, depends on three factors:

1. *The original length l.* The longer the rod is to begin with, the greater the change in length will be.
2. *The change in temperature, designated ΔT.* The larger the increase in temperature, the greater the increase in length.
3. *The substance.* For example, the increase in length of an aluminum rod will be more than twice that of an identical iron rod under the same conditions.

Point 3 can be tested through experimentation. The expansions of different solids are measured under similar conditions. The results are used to assign a **coefficient of linear expansion** to each material. The value of this coefficient is a fixed parameter of each substance, much like mass density or weight density. It is represented by the Greek letter alpha, α. Since aluminum expands more than iron under the same circumstances, the coefficient of linear expansion of aluminum is larger than that of iron (Figure 5.4).

The equation that gives the change in length in terms of the change in temperature and the coefficient of linear expansion is

$$\Delta l = \alpha\, l\, \Delta T$$

The change in length is proportional to the change in temperature and to the original length. The coefficient of linear expansion, α, is the constant of proportionality. Table 5.2 lists α for several different solids. In the equation, the units of l and Δl must be the same. Therefore, the units of temperature, usually °C, must cancel

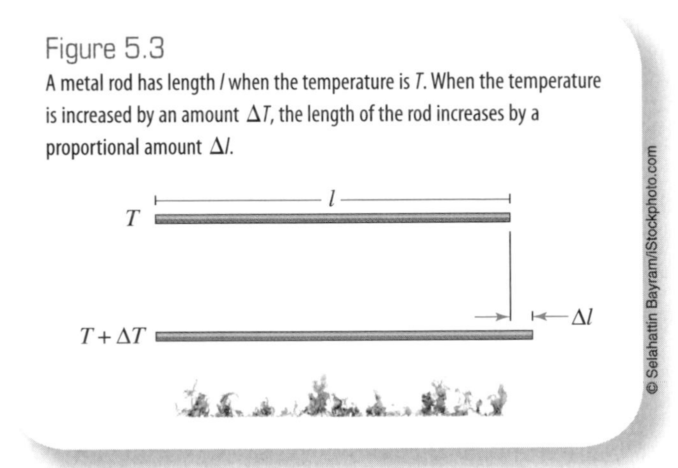

Figure 5.3
A metal rod has length l when the temperature is T. When the temperature is increased by an amount ΔT, the length of the rod increases by a proportional amount Δl.

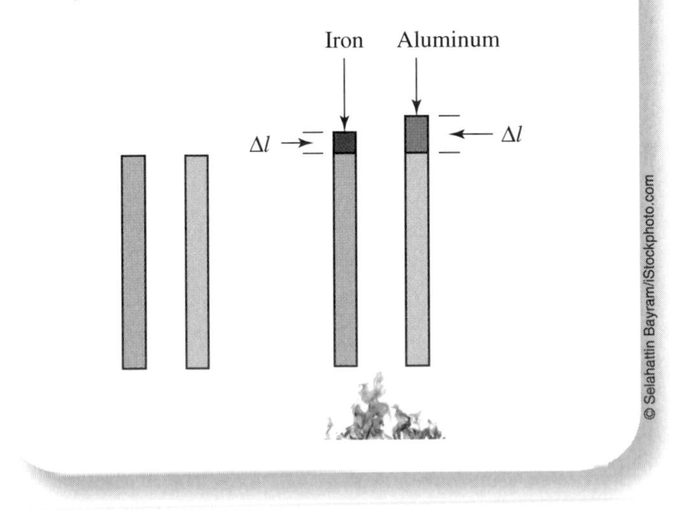

Figure 5.4
Because aluminum expands more than iron, given the same increase in temperature, its coefficient of linear expansion is larger.

those of α; that is to say, the units of α must be inverse temperature, for example, 1/°C.

EXAMPLE 5.1

The center span of a steel bridge is 1,200 meters long on a winter day when the temperature is −5°C. How much longer is the span on a summer day when the temperature is 35°C?

First, the change in temperature is the final temperature minus the initial temperature:

$$\Delta T = 35 - (-5) = 35 + 5$$
$$= 40°C$$

From Table 5.2, the coefficient of linear expansion for steel is

$$\alpha = 12 \times 10^{-6}/°C$$

So:

$$\Delta l = \alpha \, l \, \Delta T$$
$$= (12 \times 10^{-6}/°C) \times 1{,}200 \text{ m} \times 40°C$$
$$= (12 \times 10^{-6}/°C) \times 48{,}000 \text{ m-°C}$$
$$= 576{,}000 \times 10^{-6} \text{ m}$$
$$= 0.576 \text{ m}$$

Table 5.2 Some Coefficients of Linear Expansion

Solid	α (× 10^{-6}/°C)
Aluminum	25
Brass or bronze	19
Brick	9
Copper	17
Glass (plate)	9
Glass (Pyrex)	3
Ice	51
Iron or steel	12
Lead	29
Quartz (fused)	0.4
Silver	19

The change in length in Example 5.1 is considerable and must be allowed for in the bridge design. Expansion joints, which act somewhat like loosely interlocking fingers, are placed in bridges, elevated roadways, and other such structures to allow thermal expansion to safely occur (Figure 5.5).

The equation for thermal expansion also works when the temperature decreases. When this occurs, the change in temperature is negative, so the change in length is also negative: The solid becomes shorter. Another way of saying this is that thermal expansion is a reversible process. If something becomes longer when heated, it will get shorter when cooled. Most of the phenomena discussed in this chapter are reversible. If something happens when the temperature increases, the reverse will happen when the temperature decreases.

The *bimetallic strip* is an ingenious and widely used application of thermal expansion. As the name implies, a bimetallic strip consists of two strips of different metals bonded to one another (see Figure 5.6). The two metals have different coefficients of linear expansion, so they expand by different amounts when heated. The result is that the bimetallic strip bends—one way when heated and the other way when cooled. The greater the change in temperature, the greater the bending. For example, if brass and iron are used, the brass will expand and contract more than the iron will. The brass will be on the outside of the curve when the strip is hot and on the inside of the curve when it is cold.

Thermostats, thermometers, and choke-control mechanisms on automobiles often contain a bimetallic strip that is curled into a spiral. The coil will either partly unwind or wind up more tightly when the temperature changes. To make a thermometer, a pointer is attached to one end of the spiral, and the other end is held fixed. As the temperature varies, the pointer moves over a scale

Figure 5.5
Expansion joints allow for the thermal expansion of bridges and other elevated roadways. Each end of a section of the roadway is connected to a metal "comb." The teeth of the two combs fit together and can move back and forth as the lengths of the sections change. For comparison, the thermometer is 30 centimeters (1 foot) wide.

Figure 5.6

In this bimetallic strip, the metal composing the upper layer has a larger coefficient of thermal expansion than the metal composing the lower layer. When heated, the strip curves downward because the upper layer undergoes a greater change in length. When cooled, the strip curves upward.

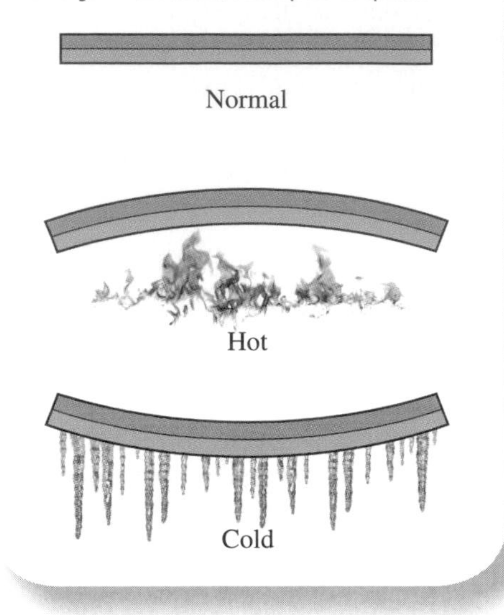

© Slavoljub Pantelic/iStockphoto.com / © Selahattin Bayram/iStockphoto.com

Figure 5.7

This thermometer uses a bimetallic strip in the shape of a spiral. Even small temperature changes cause the pointer attached to the outer end of the spiral to rotate noticeably.

© Vern Ostdiek

that indicates the temperature (Figure 5.7). In thermostats, the movement of the end of the spiral is used to turn a switch on or off. The switch might turn on a heater, an air conditioner, a fire alarm, or the cooling unit in a refrigerator.

As mentioned, thermal expansion occurs in all three dimensions. A bridge becomes longer, but also wider and thicker as the temperature increases. The area of any surface increases, as does the volume of the solid. If a solid has a hole in it, thermal expansion will make the hole bigger, contrary to most people's intuition (Figure 5.8). This is because thermal expansion causes every point in a solid to move *away* from every other point. (It is similar to what happens when a photograph is enlarged.) A point on one side of a hole moves away from any point on the other side of the hole.

Liquids

The behavior of liquids is quite similar to that of solids. Since liquids do not hold a certain shape, it is best to consider the change in volume caused by thermal expansion. In general, liquids expand considerably more than solids. This means that when a container holding a liquid is heated, the level of the liquid will usually rise because the increase in volume of the liquid typically exceeds the increase in volume of the (solid) container. When a mercury thermometer is heated, the mercury in the tube and in the glass bulb at the bottom expands more than does the confining glass envelope, so the level of mercury rises. If the glass expanded more than the mercury, the column would go down at higher temperatures instead of up.

At the beginning of this section, we implied that there are exceptions to the general rule that

Figure 5.8

The holes becomes larger when the object is heated.

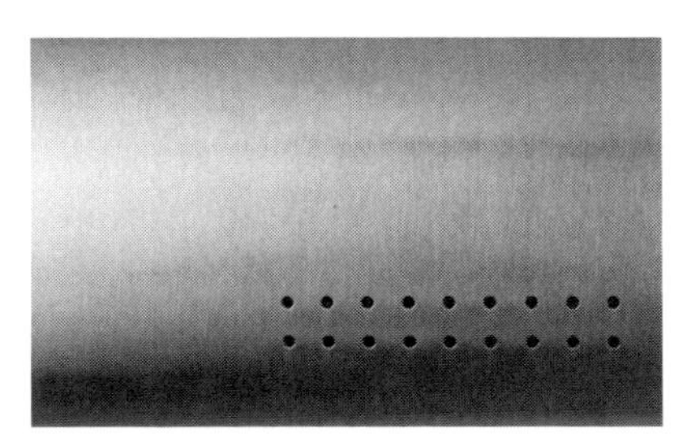

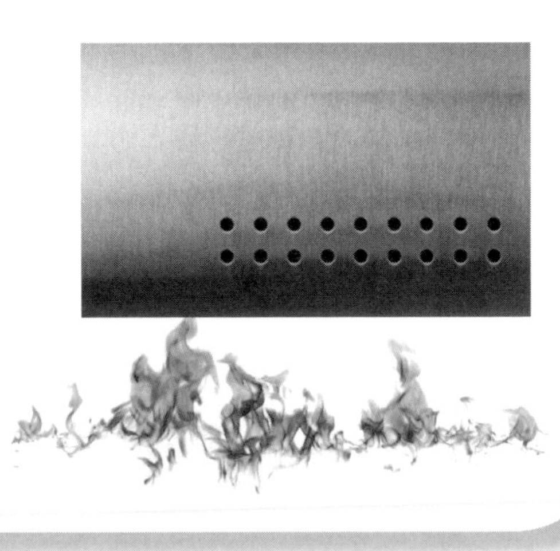

© Selahattin Bayram/iStockphoto.com / © Matjaz Boncina/iStockphoto.com

matter expands when heated. The most important example of this is water that is near its freezing temperature. Above 4°C (39°F), water expands when heated like ordinary liquids. But between 0°C and 4°C, water actually contracts when heated and expands when cooled. The volume of a given amount of water at 3°C is *less* than the volume of the same water at 1°C. This anomaly accounts for the fact that lakes, ponds, and other bodies of water freeze on top first. As long as the average water temperature is above 4°C, the warmer (less dense) water is buoyed to the surface, and the cooler water is at the bottom (see Figure 5.9). As the air gets cooler in autumn, the water at the surface is cooled. When the average temperature of the water is below 4°, the cooler water (closer to freezing) is now less dense and rises to the surface. Consequently, the surface water freezes first because it is cooler than the water below and it is in contact with the cold air.

Figure 5.9
Above 4°C, the warmer water rises to the surface. Below 4°C, the cooler water rises to the surface.

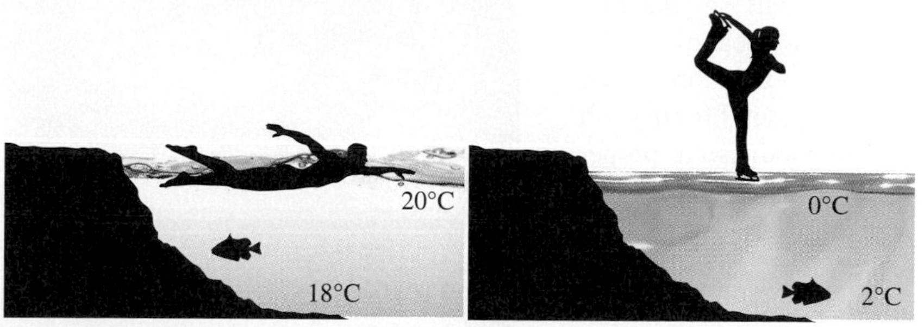

© David Joyner/iStockphoto.com / © iStockphoto.com / © iStockphoto.com / © Lepro/Shutterstock / © iStockphoto.com

Water is also unusual in that its density when in the solid phase (ice) is less than its density when in the liquid phase. Because of this, ice floats in water, whereas most solids (for example, candle wax) sink in their own liquid.

Gases

The volume expansion of gases is larger than that of solids and liquids. Also, the amount of expansion does not vary with different gases (except at very low temperatures or very high pressures). Instead of relating the expansion to a change in temperature, it is simpler to state the relationship between the volume occupied by the gas and the temperature. In particular, as long as the pressure remains constant, the volume occupied by a given amount of gas is *proportional* to its temperature (in kelvins).

$$V \propto T \quad \text{(gas at fixed pressure)}$$

The temperature must be in kelvins. When the temperature of a gas is increased by some percentage, the volume increases by the same percentage. The volume of a balloon at 303 K (86°F) is about 10% larger than the volume of the same balloon at 273 K (32°F) (see Figure 5.10). By comparison, volume change of typical solids is less than 1% over the same temperature range. In liquids, it is a maximum of 5%.

Figure 5.10
If the pressure in a gas is kept constant, the volume that the gas occupies is proportional to the Kelvin temperature. A 10% increase in temperature will cause the volume of a balloon to increase by 10%.

© iStockphoto.com / © ingmar wesemann/iStockphoto.com / © iStockphoto.com / © Maria Toutoudaki/iStockphoto.com

The *ideal gas law* expresses the interdependence of the pressure, volume and temperature of gas.

LAWS

Ideal Gas Law *In a gas whose density is low enough that interactions between its constituent particles can be ignored, the pressure, volume, and temperature of the gas are related by the following equation:*

$$pV = (\text{constant})\ T$$

The constant depends on the quantity of gas present, but not its specific type.

The utility and importance of this general relationship comes from the fact that it is applicable to most real gases under most conditions. Because the constant appearing in this equation only depends on the mass of gas present and not on what kind of gas it is, the equation can be used equally well for hydrogen, helium, oxygen, carbon dioxide, etc. Moreover, except under extreme circumstances, the ranges over which the parameters p, V, and T vary are limited enough in practice to ensure that the gas particles generally remain far enough apart so that their interactions are negligible. Together, these facts make the ideal gas law useful for describing the thermodynamic properties of diverse physical systems ranging from gases in the atmospheres of the Sun and stars to the air in a classroom and the helium in a floating blimp.

A hot-air balloon makes use of thermal expansion.

© Eliza Snow/iStockphoto.com / © Maria Toutoudaki/iStockphoto.com

A given amount of gas then can have any combination of pressure, volume and temperature so long as the three values satisfy the ideal gas law. For example, if the volume of the gas is fixed, the pressure will increase whenever the temperature increases. In other words, the pressure is proportional to the temperature (in kelvins) as long as the volume stays constant. In the earlier example of heating air in a pressure cooker, the volume of the air inside would be nearly constant. (This is because the volume expansion of the metal is quite small compared to that of a gas.) So the pressure inside would increase proportionally with the temperature.

Regardless of which phase of matter is involved, changing the temperature of a substance will not change its mass or its weight. Since thermal expansion causes the volume to increase while the mass and weight stay the same, the mass density and weight density decrease. The mass density of a hot piece of iron is slightly less than the mass density of the same piece when it is cold. The mass density of the balloon referred to earlier is about 10% less when its temperature is 303 K than when its temperature is 273 K.

The reduction in density of a gas at constant pressure because of heating is exploited in hot-air balloons. The air in the balloon, which is basically a large bag with an opening at the bottom, is heated with a burner. The pressure inside remains equal to the atmospheric pressure because of the opening. Consequently, the air inside the balloon expands and its density decreases. The balloon can float in the air because it is filled with a gas (hot air) that has a smaller density than the surrounding fluid (cooler air). As the air inside cools, the balloon will sink toward the Earth until the burner heats the air again.

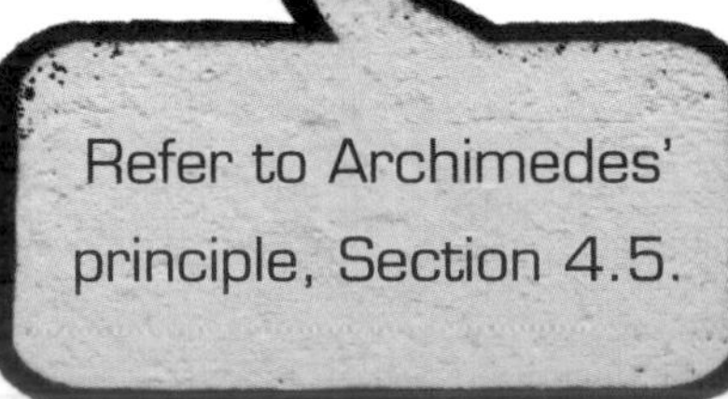

© iStockphoto.com

5.3 The First Law of Thermodynamics

We have discussed what temperature is and how changes in temperature can affect the physical properties of matter—density in particular. The next item to consider is how the temperature of matter is changed. This will lead us back to the concept of energy.

There are two general ways to increase the temperature of a substance:

1. By exposing it to something that has a higher temperature.
2. By doing work on it in certain ways.

The first way is very familiar to you. When you heat something on a stove, warm your hands over a heater, or feel the Sun warm your face, the temperature increase is caused by exposure to something that has a higher temperature: the stove, the heater, or the Sun. (More on this in Section 5.4.) As mentioned before, the atoms and molecules in the substance being warmed gain kinetic energy from those in the hotter substance. The temperature of a substance will rise only if its atoms and molecules gain kinetic energy.

Friction is an effective means of raising the temperature of a substance by the second way (doing work on it). When something is heated by kinetic friction, work is done on it. This was discussed in more detail at the end of Section 3.4.

Another example of raising the temperature of a substance by doing work on it is the compression of a gas. When a gas is squeezed (quickly) into a smaller volume, its temperature increases (see Figure 5.11). In diesel engines, air is compressed so much that its temperature is raised above the combustion temperature of diesel fuel. When the fuel is injected into the compressed hot air, it ignites.

Computation of the temperature rise of a gas when it is compressed a certain amount is beyond the scope of this text. But we can illustrate the type of heating that occurs by stating the results of one example: If air at 27°C is compressed to one-twentieth of its initial volume in a diesel engine, its temperature will increase to over 700°C (Figure 5.12).

Both processes can be reversed to cause the temperature of a substance to decrease. The cooler air in a refrigerator lowers the temperature of a pitcher of tea. Air escaping from a tire is cooled as it expands.

Temperature depends on the average kinetic energy of atoms and molecules. For matter to undergo a change in temperature, its atoms and molecules must gain energy (increase temperature) or lose energy (decrease temperature). In gases, the constituent particles have kinetic energy only. All of the energy given to the atoms and molecules acts to increase the temperature of the gas. Things are different in solids and liquids. The atoms and molecules have kinetic energy *and* potential energy because they are bound to each other and oscillate. Energy given to the particles goes to increase both their *PE* and their *KE*. The concept of **internal energy**, which we introduced in Section 3.4, incorporates both forms of energy.

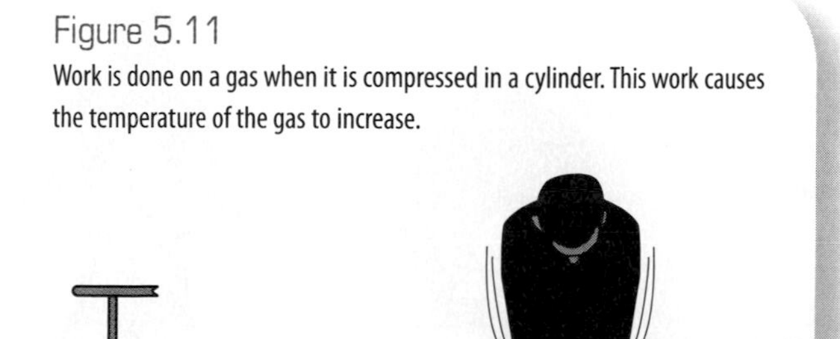

Figure 5.11
Work is done on a gas when it is compressed in a cylinder. This work causes the temperature of the gas to increase.

Figure 5.12
The air in a diesel engine has an initial temperature of 27°C. It is compressed by the piston until it occupies one-twentieth of its original volume. This raises the temperature to 721°C.

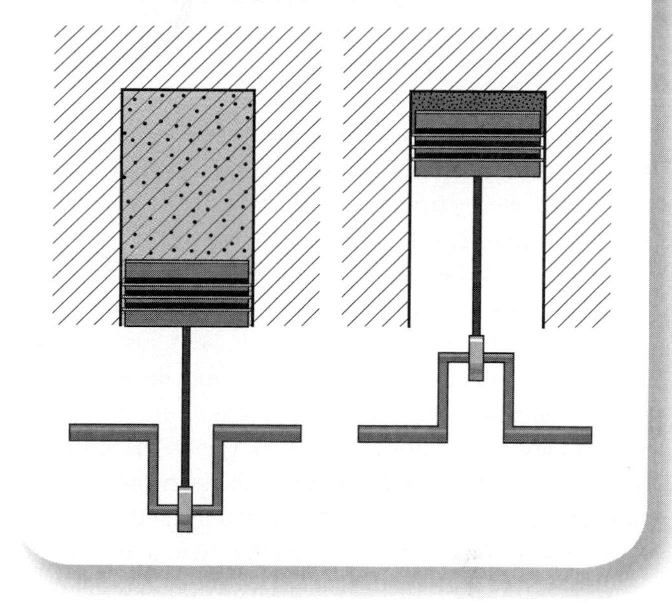

Internal energy is represented by *U*. Its units of measure are the same as those of energy and work.

In gases, the internal energy is the total of the kinetic energies only: The atoms and molecules do not have potential energy. (Gravitational potential energy is not included in internal energy.) In solids and liquids, both the kinetic energy and the potential energy of the particles contribute to the internal energy.

As the temperature of a substance rises, its internal energy increases. If this is accomplished by exposure to a hotter substance, we say that **heat** has flowed from the hotter substance into the cooler substance.

Heat is represented by *Q*. Its units are the same as those of work and energy. Traditionally, the calorie,

DEFINITION

Internal Energy *The sum of the kinetic energies and potential energies of all the atoms and molecules in a substance.**

Heat *The form of energy that is transferred between two substances because they have different temperatures.*

* Sometimes the concept of internal energy is expanded to include other forms of energy possessed by atoms and molecules. For example, batteries and gasoline have chemical energy that could be included in internal energy.

kilocalorie (also written Calorie), and British thermal unit (Btu) were used exclusively as units of heat. The joule and the foot-pound were used for work and the other forms of energy. Now the joule is becoming the standard unit for heat as well.

> Heat and work are energy in transition.

The internal energy of something can increase when heat flows into it and decrease when heat flows out of it (Figure 5.13). Heat flow is the transfer of energy from a hotter substance to a cooler substance. In this respect, heat is much like work in mechanics. As work is being done, energy is transferred from one thing to another or is transformed from one form to another. A hot pizza does not contain heat, just as a battery and a wound-up spring do not contain work. The battery and the spring contain potential energy, which means they can be used to do work (given the appropriate motor or other mechanism). In the same manner, the hot pizza contains internal energy, and it can therefore be used to heat something that is cooler. Heat and work are energy in transition, while internal energy and potential energy are stored energy.

Two substances with the same temperature are said to be in *thermal equilibrium.* No heat transfers between them. An abandoned cup of hot coffee cools off as heat is transferred from it to the air. Once the coffee and cup reach the same temperature as the surrounding air, the transfer stops and they are in thermal equilibrium.

The first law of thermodynamics is a formal summary of the preceding statements.

LAWS

First Law of Thermodynamics
The change in internal energy of a substance equals the work done on it plus the heat transferred to it.

$$\Delta U = \text{work} + Q$$

The work referred to in this law must be the type that transfers energy directly to atoms and molecules, such as compressing a gas. Work is done on an object when it is lifted, but this does not affect its internal energy—just its gravitational potential energy.

The work is positive if it is done *on* the substance and negative if the substance does work on something else. When gas in a cylinder is compressed by a piston, the work that is done is positive. If the piston is then released, the gas will expand and push the piston out. In this case, the gas does work on the piston, and the work is negative (Figure 5.14). Similarly, Q is positive when heat flows into the substance and negative when heat flows out of it. If you place a brick in a hot oven, heat flows into it, and Q is positive. Placing the brick in a freezer would result in a flow of heat out of it and a negative Q.

Figure 5.13
Heat flows into the water from the burner, increasing the water's internal energy. Heat flows out of the water and into the ice, lowering the water's internal energy.

© iStockphoto.com / © iStockphoto.com

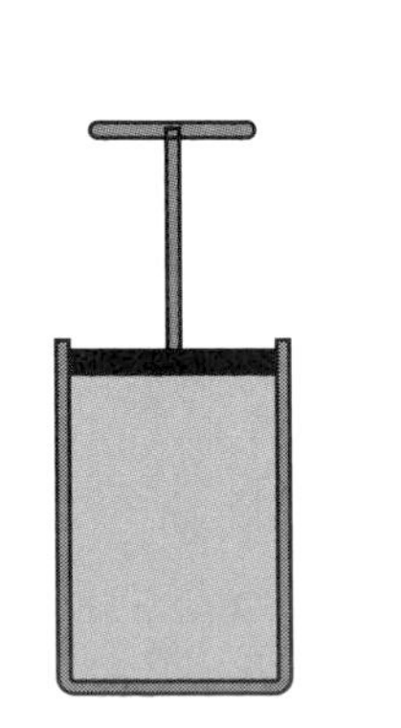

Figure 5.14
When the piston compresses the gas, the work done on the gas is positive. When the gas pushes the piston back, it does work on the piston. In this case, the work done on the gas is negative.

The first law of thermodynamics is nothing more than a restatement of the law of conservation of energy as it applies to thermodynamic systems. Work done on, or heat transferred to, a substance is "stored" in it as internal energy. In addition to its theoretical significance, the first law of thermodynamics is an important tool used in the analysis of things like internal combustion engines and air conditioners.

This section began with a consideration of how the temperature of matter is changed. How does internal energy fit in with this? Temperature depends on the kinetic energies of the atoms and molecules in a substance. Since these kinetic energies are part of the internal energy of the substance, raising the temperature of something increases its internal energy because it raises the kinetic energies of the atoms and molecules.

Perhaps you are wondering why the concept of internal energy is used at all, since temperature is determined only by the kinetic energies of particles. Internal energy is useful when considering phase transitions. When water boils on a stove, for example, heat is transferred to it, but the temperature stays the same. This means that the average kinetic energy of the water molecules also stays the same. The heat transferred during a change in phase increases the internal energy by increasing only the potential energies of the molecules. The energy is used to break the bonds between the water molecules and to free them. We will take a closer look at this in Section 5.6.

The Concept Map on your Review Card summarizes the concepts presented in this section.

5.4 Heat Transfer

Transferring heat is the more common of the two ways to change the temperature of something. Heat transfer occurs whenever there is a temperature difference between two substances or between parts of the same substance. In this section we discuss the three different mechanisms for heat transfer: **conduction**, **convection**, and **radiation**.

DEFINITION

Conduction *The transfer of heat between atoms and molecules in direct contact.*

Convection *The transfer of heat by buoyant mixing in a fluid.*

Radiation *The transfer of heat by way of electromagnetic waves.*

Conduction

Conduction occurs when a pan is placed on a hot stove, when you put your hands into cold water, and when an ice cube comes into contact with warm air. The atoms and molecules in the warmer substance transfer some of their energy directly to the particles in the cooler substance. In these examples, the conduction takes place across the boundary between the two substances, where the atoms and molecules collide with each other. Conduction also is responsible for the transfer of heat from one part of a solid to another part. (Conduction also happens in fluids but is not as important as convection.) Even though only the bottom of a pan is in contact with a hot burner, the heat flows through the metal and soon raises the temperature of all parts of the pan (Figure 5.15). As the atoms and molecules on the bottom are heated, their constant jostling passes some of their increased speed to neighboring atoms and molecules.

The ease with which heat flows within matter varies greatly. Materials through which heat moves slowly are called *thermal insulators*. Wool, Styrofoam, and bundles of fiberglass strands are all good insulators because they contain large amounts of trapped air or other gases. Conduction is poor within a gas because the atoms and molecules are not in constant contact with each other. Diamond and metals such as iron and copper are *thermal conductors:* Heat flows readily through them. Concrete, stone, wood, and glass are between the two extremes. A vacuum completely prevents conduction because no atoms or molecules are present.

Figure 5.15
Heat is conducted to the pan from the flame in contact with it. Conduction also takes place within the pan: Heat flows from the hot bottom up the sides and into the handle.

© iStockphoto.com / © Selahattin Bayram/iStockphoto.com

Metals are good conductors of heat for the same reason that they are good conductors of electricity. Some of the electrons in the atoms in metals are free to move about from one atom to the next. The motion of these "conduction electrons" constitutes an electric current (Chapter 7). These electrons can also carry internal energy from the warmer part of a metal object to the cooler parts.

The conduction of heat within matter is similar to the flow of a fluid, except that nothing material moves from one place to another. The rate at which heat flows from a hot part of an object to a cold part depends on several things—the difference in temperature between the two places, the distance between them, the cross-sectional area through which the heat flows, and how good a thermal conductor the substance is.

Heat conduction is important in our daily lives. In cold weather, we wear clothes that slow the conduction of heat from our warm bodies to the cold air. Handles on metal pans are often made of wood or plastic to reduce the conduction of heat from the burner to your hand. One reason carpets and rugs are used is that they feel warm when you step on them with bare feet. A rug is really no warmer than the bare floor next to it, but it is a poorer conductor. When you step on the rug, very little heat is conducted away from your feet, so they stay warm. When you step on bare wood or tile, materials that are better thermal conductors, your feet are cooled more because heat is conducted from them more rapidly.

Convection

Convection is the dominant mode of heat transfer within fluids. Whenever part of a fluid is heated, its thermal expansion causes its density to decrease, so it rises. (Water below 4°C is an exception.) The result is a natural mixing of the fluid. Conduction can then occur between the warmer fluid and the cooler fluid around it.

A room with a wood stove or heater is warmed by convection. The air that is heated is less dense and rises to the ceiling, and cooler, denser air near the floor moves toward the heat source to replace the rising air. The result is a natural circulation of air along the ceiling, floor, and walls of the room (Figure 5.16). The warmed air near the ceiling cools when it contacts the walls and then sinks to the floor. The same type of circulation can occur in heated aquariums and swimming pools.

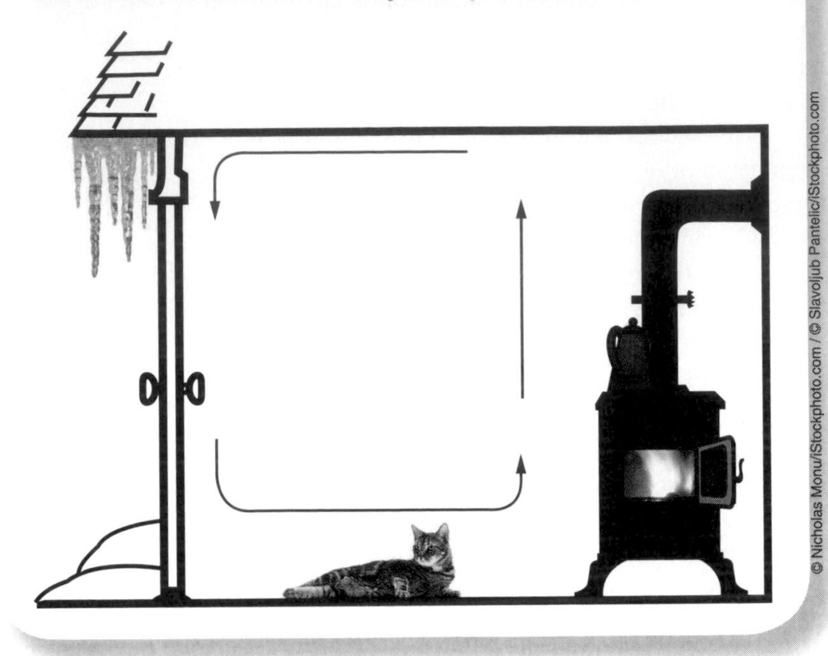

Figure 5.16

Convection causes a natural circulation of air in a room heated with a wood stove. Air heated by contact with the stove rises to the ceiling. Air cooled by contact with an outside wall sinks to the floor. This results in lateral movement of air along the ceiling and the floor.

© Nicholas Monu/iStockphoto.com / © Slavoljub Pantelic/iStockphoto.com

You may have heard the statement "heat rises." This is *not* physically correct. Heat is not something material that can rise or fall. "Heated air rises" or "heated fluids rise" are better statements for conveying the idea. Remember also that it is the denser surrounding fluid that pushes the warmer, less dense fluid upwards.

Mechanical mixing of a fluid that causes heat transfer is an example of what is called *forced* convection. Stirring cool cream into hot coffee with a spoon is forced convection—the mixing is not caused by thermal buoyancy. Another example is hot or cold air being blown around the interior of a building or vehicle, causing heat transfer by the mixing of the warmer air with the cooler air.

Convection in the Earth's atmosphere is a major cause of clouds, wind, thunderstorms, and other meteorological phenomena. White, puffy, cumulus clouds are formed when warm air rises into cooler air above.

Sea breezes—steady winds blowing into shore along coasts—are caused by convection. Sunshine warms the land more than the sea, so the air over the ground is heated and rises upward.* This reduces the air pressure over the land, so the higher pressure over the sea forces air to move inland (Figure 5.17, left). At night, the land cools

* Only the surface of the ground is heated by the Sun, because the soil is a poor conductor. Heat transferred to the sea is quickly spread deeper below the surface by currents and convective mixing. We will also see in Section 5.5 that water requires a great deal of heat transfer to raise its temperature.

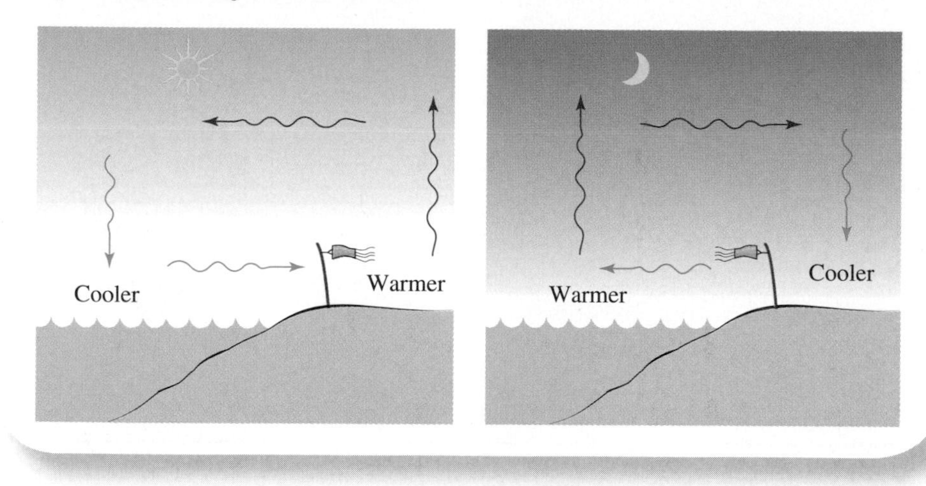

Figure 5.17
(left) Sea breeze. Warmed air rises from the heated land, causing cooler air to be drawn in from the sea. (right) The land cools at night, so the flow is reversed.

off more than the sea, so the process is reversed, and a *land breeze* is produced. Air over the cooler ground sinks and forces air to move out to sea (Figure 5.17, right).

Large-scale convection takes place within the ocean itself. Water near the equator is heated by the Sun, rises to near the ocean's surface, and flows towards the poles. There, the water cools and sinks deeper, and then flows back towards the equator.

Radiation

Radiation is the transfer of heat via electromagnetic waves. We've all felt the warmth of the Sun on our faces on a calm day and the heat radiating from a hot fire. This is heat radiation—a type of wave related to radio waves and x-rays. (We will discuss electromagnetic waves in more detail in Chapters 8 and 10.) This is the only one of the three types of heat transfer that can operate through a vacuum. The Sun's radiation passes through 150 million kilometers (93 million miles) of empty space and heats the Earth and everything on it. Without the Sun's radiation, the Earth would be a cold, lifeless rock.

Infrared heat lamps warm things by emitting radiation. From a distance, you can feel the heat from a camp fire or other heat source because the radiation from it warms your hands and face.

You might think of the radiation as a "vehicle" or "carrier" of internal energy. Internal energy of atoms and molecules is converted into electromagnetic energy—radiation. The radiation then carries the energy through space until it is absorbed by something. When absorbed, the energy in the radiation is converted into internal energy of the atoms and molecules of the absorbing substance.

© iStockphoto.com

Everything emits electromagnetic radiation: the Sun, the Earth, your body, this book. The amount and type of radiation depend on the temperature of the emitter. The hotter something is, the more electromagnetic radiation it emits. Things below about 800°F (430°C) emit mostly infrared light, which we cannot see. Hotter substances emit *more* infrared light and also visible light. That is why we can see things in the dark that are "red hot" or "white hot." Things that are even hotter, like the Sun at 10,000°F, emit more infrared and visible light but also emit ultraviolet light. We will take a closer look at heat radiation in Chapter 8.

Everything around you is emitting radiation and also absorbing the radiation emitted by other things. How does this lead to a net transfer of heat? Emission of radiation cools an object, while absorption of radiation warms it. If something absorbs radiation faster than it emits radiation, it is heated. Your face is warmed by the Sun because it absorbs more radiant energy than it emits.

Combinations

All three mechanisms of heat transfer are involved in soaring. The Sun warms the Earth via radiation. Air in contact with hot ground is heated by conduction and expands. If conditions are favorable, the hot air will form an invisible "bubble" that breaks free from the surface and rises upward into the air, causing convection. (This is similar to the formation of steam bubbles on the bottom of a pan of boiling water.) These bubbles of rising heated air are called *thermals* (see Figure 5.18). Hang-glider and sailplane pilots and soaring birds such as eagles, hawks, and vultures seek out thermals and circle around in them. The upward speed of a typical thermal is around 5 m/s (11 mph), so they provide an easy, free ride upward.

The different mechanisms of heat transfer become important in reducing heating and cooling costs in buildings.

The more heat that flows out of a building in cold weather or into a building in warm weather, the more it costs to heat or cool the building. Reducing this flow of heat saves money (Figure 5.19). Putting insulation in the walls and ceilings reduces conduction. Using thicker insulation in the ceiling counteracts one effect of convection: The warmest air in the room is at the ceiling, so that is where heat would be transferred most rapidly out of the room during the winter. Using window shades, blinds, and canopies to keep direct sunlight out during the summer reduces heat transfer by radiation.

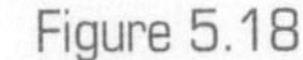
Figure 5.18

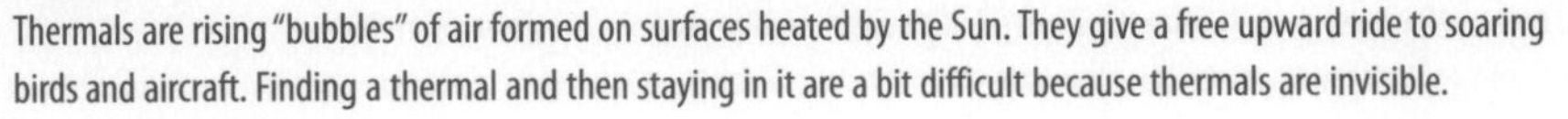
Thermals are rising "bubbles" of air formed on surfaces heated by the Sun. They give a free upward ride to soaring birds and aircraft. Finding a thermal and then staying in it are a bit difficult because thermals are invisible.

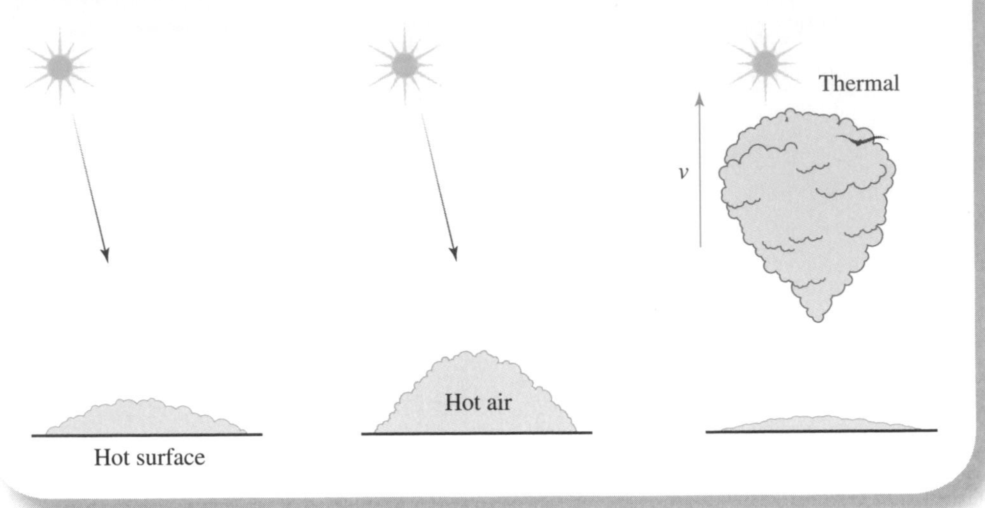

Thermos bottles are also designed to limit the flow of heat into or out of their contents. They use a near-vacuum between their inner and outer walls that almost eliminates heat flow due to conduction. The inner glass chamber is coated with silver or aluminum to reflect radiation.

The basic concepts of heat transfer are summarized in the Concept Map on your Review Card.

Figure 5.19

Heating and cooling costs for a home can be lowered by keeping in mind the mechanisms of heat transfer.

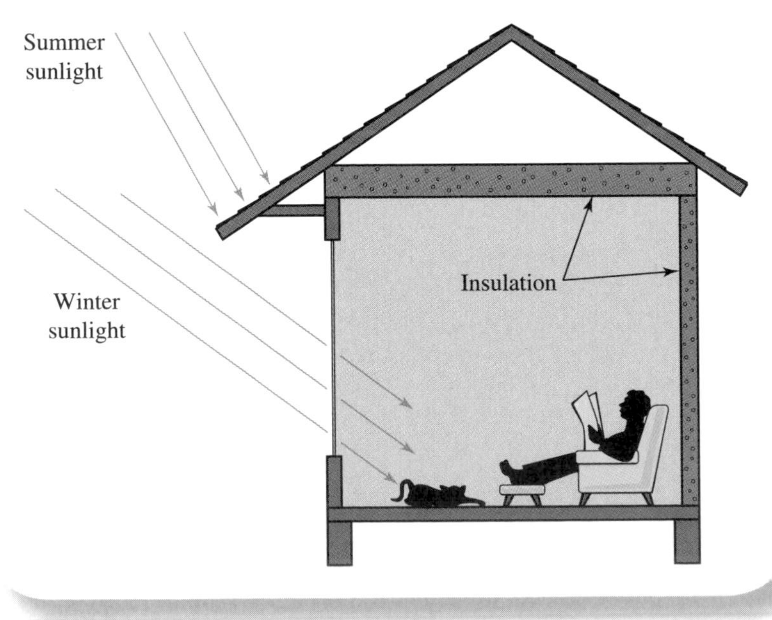

5.5 Specific Heat Capacity

Transferring heat to a substance or doing work on it increases its internal energy. In this section, we describe how the temperature of the substance is changed as a result. To simplify matters, we assume that no phase transitions take place.

We might state the topic now under consideration in the form of a question: To increase the temperature of a substance by some amount ΔT, what quantity of heat Q must be transferred to it? (We could just as well ask how much work must be done on it.)

The amount of heat needed is proportional to the *temperature increase.* It takes twice as much heat to raise the temperature 20°C as it does to raise it 10°C. So:

$$Q \propto \Delta T$$

The amount of heat needed also depends on the *quantity* (mass) of the substance to which the heat is transferred. For a given increase in temperature, 2 kilograms of water will require twice as much heat transfer as 1 kilogram. So:

$$Q \propto m$$

The required heat transfer also depends on the *substance*. It takes more heat to raise the temperature of water 1°C than it does to raise the temperature of an equal mass of iron 1°C. As with thermal expansion (Section 5.2), a characteristic value can be assigned to each substance indicating the relative amount of heat needed to raise its temperature. This number, called the *specific heat capacity C*, is determined experimentally for each substance. The larger the specific heat capacity of a substance, the greater the amount of heat transfer needed to raise its temperature by a given amount. Thus:

$$Q \propto C$$

We can combine the three proportionalities into the following equation:

$$Q = C\, m\, \Delta T$$

The amount of heat required equals the specific heat capacity of the substance times the mass of the substance times the temperature increase. The SI unit of specific heat capacity is the joule per kilogram-degree Celsius (J/kg-°C). If the specific heat capacity of a substance is 1,000 J/kg-°C, it takes 1,000 joules of energy to raise the temperature of 1 kilogram of that substance 1°C. (The kelvin is the same size unit as the °C and can be used, too.) Table 5.3 lists the specific heat capacities of some common substances.

Table 5.3 Some Specific Heat Capacities

Substance	C (J/kg-°C)
Solids	
Aluminum	890
Concrete	670
Copper	390
Ice	2,000
Iron and steel	460
Lead	130
Silver	230
Liquids	
Gasoline	2,100
Mercury	140
Seawater	3,900
Water (pure)	4,180

It takes about nine times as much energy to warm water 1° as it takes to warm the same mass of iron 1°.

EXAMPLE 5.2

Let's compute how much energy it takes to make a cup of coffee or tea. Eight ounces of water has a mass of about 0.22 kilograms. How much heat must be transferred to the water to raise its temperature from 20°C to the boiling point, 100°C? The change in temperature is

$$\Delta T = 100 - 20 = 80°C$$

The specific heat capacity C of water is 4,180 J/kg-°C (from Table 5.3). Therefore,

$$Q = C\, m\, \Delta T$$

$$= 4{,}180 \text{ J/kg-°C} \times 0.22 \text{ kg} \times 80°C$$

$$= 73{,}600 \text{ J}$$

It takes an enormous amount of energy to heat water. You may recall that in Example 3.6 we computed the kinetic energy of a small car traveling at highway speed. The answer was 364,500 joules. This much energy is only enough to bring about 5 cups of water to the boiling point from 20°C (Figure 5.20). Usually, a relatively large amount of mechanical energy does not produce a large temperature change when it is converted into heat. Example 5.3 shows this another way.

EXAMPLE 5.3

A 5-kilogram concrete block falls to the ground from a height of 10 meters. If all of its original potential energy goes to heat the block when it hits the ground, what is its change in temperature?

There are two energy conversions. The block's gravitational potential energy, $PE = mgd$, is converted into kinetic energy as it falls. When it hits the ground, the kinetic energy is converted into internal energy in the inelastic collision (see Figure 5.21). Actually, this internal energy

Figure 5.20
The energy needed to heat 5 cups of water to boiling from room temperature is about the same as the energy needed to accelerate a small car to 60 mph.

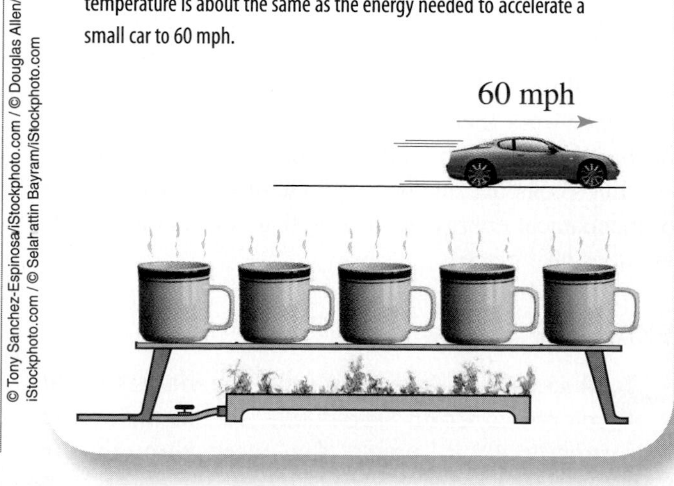

© Tony Sanchez-Espinosa/iStockphoto.com / © Douglas Allen/ iStockphoto.com / © Selahattin Bayram/iStockphoto.com

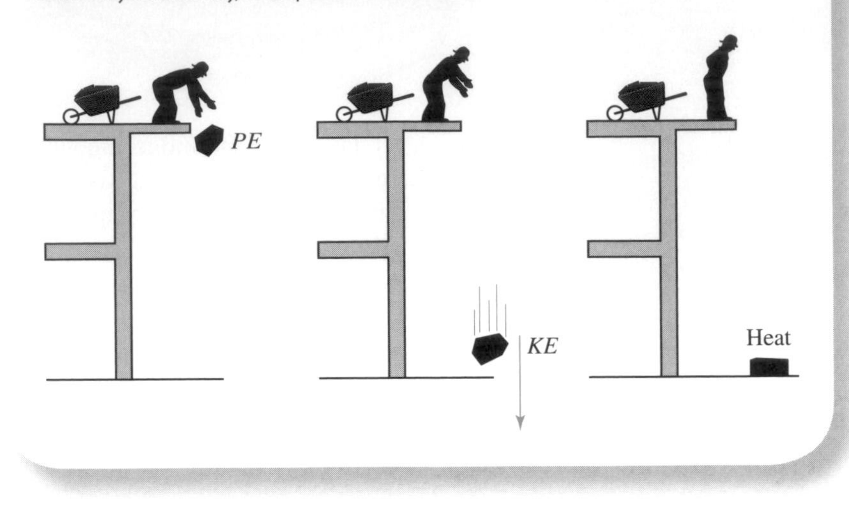

Figure 5.21

A concrete block falls 10 meters to the ground. If all of the block's energy is converted into heat that is absorbed by the block only, its temperature is raised 0.15°C.

would be shared between the block and the ground, but we assume that it all goes to the block. So the equivalent amount of heat transferred to the block equals the original potential energy.

$$Q = PE = mgd = 5 \text{ kg} \times 9.8 \text{ m/s}^2 \times 10 \text{ m} = 490 \text{ J}$$

The increase in temperature, ΔT, of the block is

$$Q = C\,m\,\Delta T$$

$$490 \text{ J} = 670 \text{ J/kg-°C} \times 5 \text{ kg} \times \Delta T$$

$$490 \text{ J} = 3{,}350 \text{ J/°C} \times \Delta T$$

$$\frac{490 \text{ J}}{3{,}350 \text{ J/°C}} = \Delta T$$

$$\Delta T = 0.15\text{°C}$$

At the end of Section 3.4, we discussed how mechanical energy is often converted into internal energy. At that time you may have wondered why you usually don't notice a temperature increase when you slide something across a floor or drop a book on a table. The quantities of energy or work that we typically deal with do not go far in changing the temperatures of the objects involved.

Let's consider one last example in which the amount of mechanical energy is so large that considerable heating does take place.

EXAMPLE 5.4

A satellite in low Earth orbit experiences slight air resistance and eventually reenters the Earth's atmosphere. As it moves downward through the increasingly dense air, the frictional force of air resistance converts the satellite's kinetic energy into internal energy. If the satellite is mostly aluminum and all of its kinetic energy is converted into internal energy, what would be its temperature increase?

Again, some of the heat is transferred to the air and some to the satellite. Just to get some idea of the potential heating, we assume that all of the heat from the friction flows into the satellite. We do not need to know the mass of the satellite, because m divides out, as we will see.

The heat transferred to the satellite equals its original kinetic energy. In Section 2.8 we computed the speed of a satellite in low Earth orbit—7,900 m/s. So:

$$Q = KE = \frac{1}{2}mv^2 = \frac{1}{2} \times m \times (7{,}900 \text{ m/s})^2$$

$$= (31{,}200{,}000 \text{ J/kg}) \times m$$

The increase in temperature caused by this much heat is

$$Q = (31{,}200{,}000 \text{ J/kg}) \times m = C\,m\,\Delta T$$

$$31{,}200{,}000 \text{ J/kg} = 890 \text{ J/kg-°C} \times \Delta T$$

$$\frac{31{,}200{,}000 \text{ J/kg}}{890 \text{ J/kg-°C}} = \Delta T$$

$$\Delta T = 35{,}000\text{°C}$$

Of course, its temperature would not actually increase this much: The satellite would start to melt. Even if 90% of the heat went to the air, the remaining 10% would still be enough to melt the satellite. (This would make ΔT = 3,500°C. The melting point of aluminum from Table 5.1 is 660 degrees Celsius.)

The purpose of Example 5.4 was to show you why unprotected satellites and meteoroids usually disintegrate when they enter the Earth's atmosphere. The friction from the air transforms their kinetic energy into enough internal energy to raise their surface temperature by several thousand degrees. The objects simply start melting on the outside. Meteors can be seen at night because of their extreme temperatures: They leave behind a trail of hot, glowing air and meteoroid fragments (see photo page 69).

Before the middle of the nineteenth century, the concepts of heat and mechanical energy were not connected to each other. It was thought that heat dealt with changes in temperature only and had nothing directly to do with mechanical energy. (The explanations of heating caused by friction were a bit strange as a result.) Water once again was used as a basis for defining a unit of measure: The calorie was defined to be the amount of heat needed to raise the temperature of 1 gram of water 1°C. Similarly, the British thermal unit (Btu) was defined to be the amount of heat needed to raise the temperature of 1 pound of water 1°F. This made the specific heat capacity of water equal to 1 cal/g-°C = 1 Btu/lb-°F.

In 1843, James Joule announced the results of experiments in which a measurable amount of mechanical energy was used to raise the temperature of water by stirring it. The amount of internal energy given to the water equaled the mechanical energy expended. The result allowed Joule to calculate what was called the "mechanical equivalent of heat"—the relationship between the unit of energy and the unit of heat:

$$1 \text{ cal} = 4.184 \text{ J}$$

It was only later that the metric unit of energy was renamed in honor of Joule.

You may have noticed that the specific heat capacity of water is quite high, nearly twice as high as that of anything else in Table 5.3. This ability to absorb (or release) large amounts of internal energy is another property of water that adds to its uniqueness—and its usefulness. Water is used as a coolant in automobile engines, power plants, and countless industrial processes partly because it is plentiful and partly because its specific heat capacity is so high. Engine parts near where the fuel burns are exposed to very high temperatures. The metal would be damaged, or even melt, if there were no way of cooling the parts. Water is circulated to these areas of the engine, where it absorbs heat from the hotter metal, thereby cooling the parts. The heated water flows to the radiator, where it transfers the heat to the surrounding air via cooling fins.

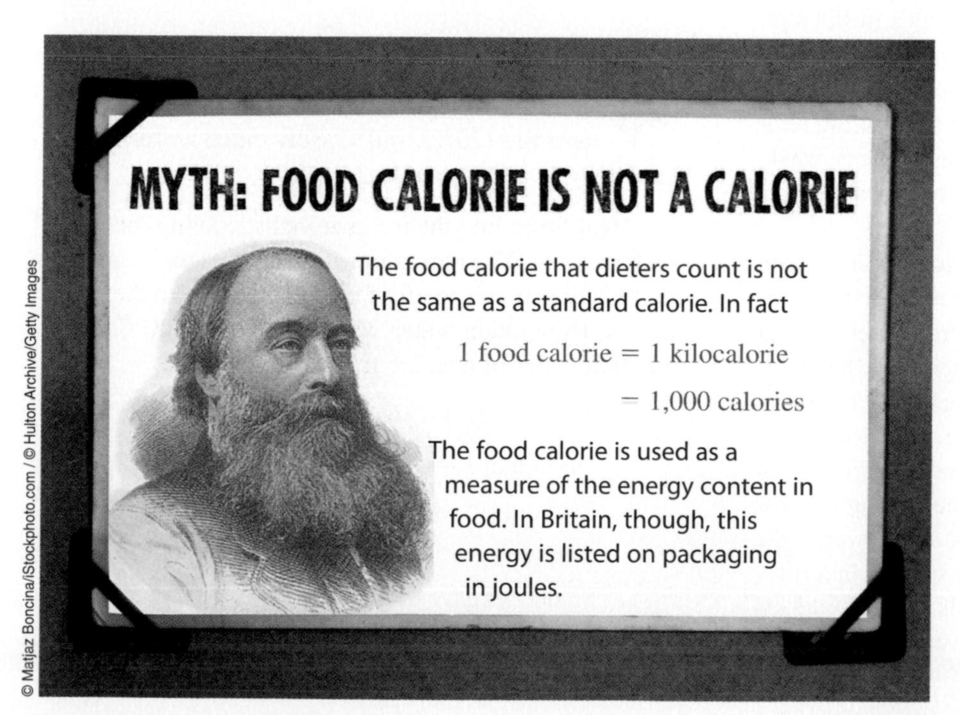

© Matjaz Boncina/iStockphoto.com / © Hulton Archive/Getty Images

5.6 Phase Transitions

A phase transition or "change of state" occurs when a substance changes from one phase of matter to another. Table 5.4 lists the common phase transitions, the phases involved, and the effect of the transition on the internal energy of the substance.*

Table 5.4 Common Phase Transitions

Name	Phases Involved	Effect
Boiling	Liquid to gas	Increases U
Melting	Solid to liquid	Increases U
Condensation	Gas to liquid	Decreases U
Freezing	Liquid to solid	Decreases U

Let's say that a pan of water is placed on a stove and that heat is transferred to the water at some rate. This causes the temperature to rise steadily until the water starts to boil. Then the temperature stays the same (100°C = 212°F at sea level) even though heat is continuing to be transferred to the water. The added energy is no longer increasing the kinetic energy of the water molecules: It is breaking the "bonds" that

* Sublimation is a fairly rare phase transition in which a solid goes directly to a gas, bypassing the liquid phase. The atoms or molecules break free from the rigid forces binding them in a crystal, and move off in the gas phase. Dry ice (solid carbon dioxide) undergoes sublimation at temperatures above −78.5°C under 1 atmosphere. This means that carbon dioxide cannot exist in the liquid phase under normal pressure. Mothballs, formerly containing naphthalene but now commonly made from dichlorobenzene crystals, also undergo sublimation at room temperature.

hold the molecules in the liquid state. The molecules are given enough energy to break free from the water's surface and become free molecules of steam (Figure 5.22).

Below the boiling temperature, the water molecules are bound to each other and so have negative potential energies. During boiling, each molecule in turn is given enough energy to break free of the bonds. The average kinetic energy of the molecules stays the same during boiling, and the temperature of the water therefore stays constant.

A similar process occurs when a solid melts. The atoms or molecules go from being rigidly bound to each other in the solid state to being rather loosely bound in the liquid state. As in boiling, the increase in internal energy that occurs during melting goes to increase *only* the potential energies of the atoms or molecules. So the temperature of ice remains at 0°C (32°F) while it is melting.

Condensation and freezing are simply the reverse processes of boiling and melting, respectively. Here the potential energies of the atoms and molecules decrease, but the kinetic energies and the temperature stay the same.

Figure 5.22
(top) The temperature of boiling water stays constant even though heat is flowing into it.
(bottom) Higher pressure inside the pressure cooker raises the boiling point of water to about 120°C.

© Peter Galbraith/iStockphoto.com / © Can Balcioglu/iStockphoto.com / © Maria Toutoudaki/iStockphoto.com

The temperature at which a particular phase transition occurs depends on the properties of the atoms or molecules in the substance—particularly the masses of the particles and the forces acting between them. When these forces are very strong, such as in table salt, the melting and boiling temperatures are quite high. When the forces are very weak, such as in helium, the phase-transition temperatures are very low—near absolute zero.

The boiling temperature of each liquid varies with the pressure of the air (or other gas) that acts on its surface. When the pressure is 1 atmosphere, water boils at 100°C. At an elevation of 3,000 meters (10,000 feet) above sea level, where the pressure is about 0.67 atmospheres, the boiling point of water is reduced to 90°C. That is why it takes longer to cook food by boiling at higher elevations. The temperature of the boiling water is lower, so conduction of heat into the food is slower. If the pressure is *increased* to 2 atmospheres such as in a pressure cooker, the boiling point of water is 120°C and food cooks faster. One type of nuclear power plant, called a *pressurized water reactor*, uses high pressure to keep water from boiling even at several hundred degrees Celsius.

This dependence of the boiling temperature on the pressure makes it possible to induce boiling or condensation simply by changing the pressure. For example, water exists as a liquid at 110°C when under 2 atmospheres of pressure. If the pressure is reduced to 1 atmosphere, the boiling temperature is lowered to 100°C, and the water begins to boil. In the same manner, steam at 110°C and 1 atmosphere pressure will start to condense if the pressure is increased to 2 atmospheres. As we will see in Section 5.7, such pressure-induced phase transitions are the basis for the operation of refrigerators and other "heat movers."

A specific amount of internal energy must be added or removed from a particular substance to complete a phase transition. For example, 334,000 joules of heat must be transferred to each kilogram of ice at 0°C to melt it. This quantity is called the **latent heat of fusion** of water. A much larger amount, 2,260,000 joules, must be transferred to each kilogram of water at 100°C to convert it completely into steam. This is the **latent heat of vaporization** of water. During the reverse processes, freezing and condensation, the same amounts of internal energy must be extracted from the water and steam, respectively.

EXAMPLE 5.5

Ice at 0°C is used to cool water from room temperature (20°C) to 0°C. How much water can be cooled by 1 kilogram of ice?

Heat flows into the ice as it melts, cooling the water in the process. The maximum amount of water it can cool to 0°C corresponds to all of the ice melting. So the question is, How much water will be cooled by 20°C when 334,000 joules of heat are transferred from it?

$$Q = C\,m\,\Delta T$$

$$-334{,}000 \text{ J} = 4{,}180 \text{ J/kg-°C} \times m \times -20\text{°C}$$

$$-334{,}000 \text{ J} = -83{,}600 \text{ J/kg} \times m$$

$$\frac{-334{,}000 \text{ J}}{-83{,}600 \text{ J/kg}} = m$$

$$m = 4.0 \text{ kg}$$

Firefighters routinely exploit water's high latent heat of vaporization.

© Phillips/Photo Researchers

Ice at 0°C can cool about four times its own mass of water from 20°C to 0°C—provided this takes place in a well-insulated container to minimize heat conduction from the outside.

These large latent heats of ice and water have common applications. The reason that ice is so good at keeping drinks cold is that it absorbs a large amount of internal energy while melting. As long as there is ice in the drink, the temperature remains near 0°C. One reason water is ideal for extinguishing fires is that it absorbs a huge amount of internal energy when it vaporizes. When poured on a hot, burning substance, the water absorbs heat as it boils, thereby cooling the substance.

Here's an example that summarizes the relationship between the temperature of water, its phase, and its internal energy. A block of ice at a temperature of −25°C is placed in a special chamber. The pressure in the chamber is kept at 1 atmosphere while heat is transferred to the ice at a fixed rate. Figure 5.23 shows a graph of the temperature of the water versus the amount of heat transferred to it.

Between points *a* and *b* on the graph, the heat transferred to the ice simply raises its temperature. At point *b*, the ice starts to melt, and the temperature stays fixed at 0°C until all of the ice is melted, point *c*. From *c* to *d*, the heat that is transferred to the water goes to increase its temperature. Between *d* and *e*, the water boils while the temperature remains at 100°C. As soon as all of the water is converted into steam, at *e*, the temperature of the steam starts to rise.

We could reverse the process by placing steam in the chamber and transferring heat from it. The result would be like moving along the graph from right to left.

Humidity

At temperatures below their boiling points, liquids can gradually go into the gas phase through a process known as **evaporation**. Water left standing will eventually "disappear" because of this. How can this phase transition occur at temperatures below the boiling point? Individual atoms or molecules in a liquid can go into the gas phase if they have enough energy. At temperatures below the boiling point, some of the atoms or molecules do have enough energy to do this. Even though the *average* energy of the particles is too low for boiling to occur, some of them have more energy than the average, and some have less. Atoms or molecules with higher-than-average energy can break free from the liquid if they are near the surface. Once in the air, the atoms or molecules can remain in the gas phase even though the temperature is below the boiling point.

Because of evaporation, water vapor is always present

Figure 5.23

Graph of the temperature of water versus the heat transferred to it. The phase transitions correspond to the two places where the graph is flat: The internal energy increases, but the temperature stays constant.

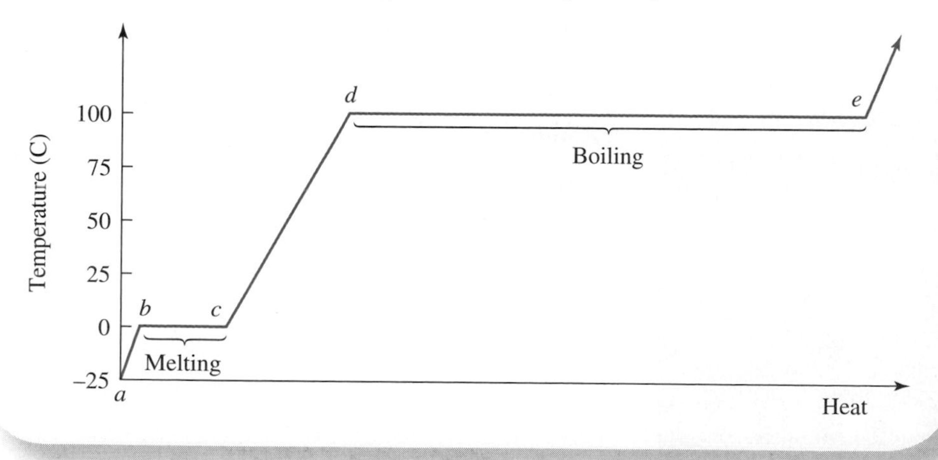

in the air. The amount varies with geographic location (proximity to large bodies of water), climate, and weather. **Humidity** is a measure of the amount of water vapor in the air.

DEFINITION

Humidity *The mass of water vapor in the air per unit volume. The density of water vapor in the air.*

Table 5.5 Saturation Density of Water Vapor in the Air

Temperature		Saturation Density	Temperature		Saturation Density
(°C)	(°F)	(kg/m^3)	(°C)	(°F)	(kg/m^3)
−15	5	0.0016	15	59	0.0128
−10	14	0.0022	20	68	0.0173
−5	23	0.0034	25	77	0.0228
0	32	0.0049	30	86	0.0304
5	41	0.0068	35	95	0.0396
10	50	0.0094	40	104	0.0511

The unit of humidity is the same as that of mass density. The humidity generally ranges from about 0.001 kg/m^3 (cold day in a dry climate) to about 0.03 kg/m^3 (hot, humid day). Note that these densities are much less than the normal density of the air, 1.29 kg/m^3. Even in humid conditions, water vapor is only a small component of the air—less than 5%.

At any given temperature, there is a maximum possible humidity, called the *saturation density.* This upper limit exists because the water molecules in the air "want" to be in the liquid phase. If there are too many molecules in the air, the probability is high that several will get close enough to each other for the attractive force between them to take over. They begin to form droplets and condense onto surfaces. This is what happens during the formation of fog and dew, or the mist in a shower. The saturation density is *higher* at higher temperatures because the water molecules are moving faster and are less likely to "stick" together when they collide. Hence, more molecules can be present in any volume of air without droplet formation. Table 5.5 lists the saturation density of water vapor in the air at several different temperatures.

The reverse of evaporation also takes place: Water molecules in the air near the surface of water can be deflected into the water and "captured." When the humidity is well below the saturation density at that temperature, evaporation occurs faster than reabsorption of water molecules. If the humidity increases to the saturation density, water molecules are reabsorbed into the water at the same rate that they evaporate. The water does not disappear. This is why damp towels dry slowly in humid environments.

Fog in the cooler air in a valley. Water droplets form when the humidity exceeds the saturation density—when there is too much water vapor in the air.

© Matt Kunz/iStockphoto.com / © Nicholas Belton/iStockphoto.com

The upshot of this is that the humidity alone doesn't determine how rapidly water evaporates. What matters is how close the humidity is to the saturation density. The **relative humidity** is a good indicator of this relation.

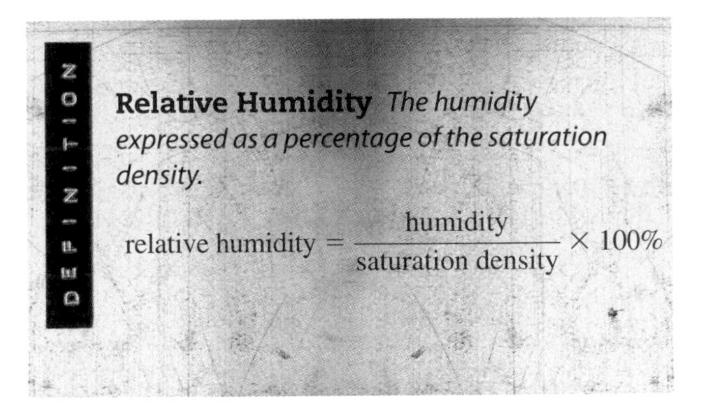

DEFINITION

Relative Humidity *The humidity expressed as a percentage of the saturation density.*

$$\text{relative humidity} = \frac{\text{humidity}}{\text{saturation density}} \times 100\%$$

When the relative humidity is 40%, this means that 40% of the maximum amount of water vapor is present.

EXAMPLE 5.6

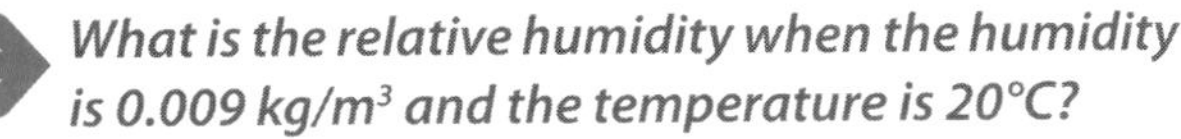

What is the relative humidity when the humidity is 0.009 kg/m^3 and the temperature is 20°C?

From Table 5.5, the saturation density at 20°C is 0.0173 kg/m^3. Therefore,

$$\text{relative humidity} = \frac{0.009\ kg/m^3}{0.0173\ kg/m^3} \times 100\%$$

$$= 52\%$$

The same humidity in air at 15°C would make the relative humidity 70%. If the air were cooled to 9°C, the relative humidity would be 100%.

When air is cooled and the water vapor content stays constant, the relative humidity *increases.* When air is heated and the humidity stays constant, the relative humidity *decreases.* This is why heated buildings often feel dry in the winter. Cold air from the outside enters the building and is heated. Unless water vapor is artificially added to this air with a humidifier, the relative humidity will be very low.

If air is cooled while the humidity stays constant, eventually condensation begins to occur. The temperature at which this happens is called the *dew point* temperature. Often on clear nights, the air temperature drops until the dew point is reached and condensation begins. The result is dew on plants and other surfaces—hence the name "dew point." The dew point is easily predicted if the humidity is known: It is the temperature at which that humidity reaches the saturation density. Table 5.5 or Figure 5.24 can be used for this purpose. For example, when the humidity is 0.0128 kg/m^3, the dew point is 15°C (59°F).

Figure 5.24b shows how the dew point can be estimated graphically. The point ○ represents the air described in Example 5.6. After the air is cooled from 20°C to 15°C, its state is marked by ●. The point moved horizontally to the lower temperature. Continued cooling would cause the air to reach the saturation density curve at a temperature of about 9°C, so that is the dew point.

Water droplets often form on the sides of cans and other containers holding cold drinks. This happens when the temperature of the container's sides is below the dew point temperature. Air in contact with the surface is cooled until the dew point is reached, and condensation begins. (This process also occurs when car windows "fog over" in cold weather.) If the air is very dry (low relative humidity), condensation doesn't occur because the dew point is below the temperature of the cold drink.

The process of evaporation cools a liquid. Atoms or molecules in the gas phase have more energy than when in the liquid phase. When water molecules evaporate, they take away some internal energy from the water, thereby cooling it. Here is another way of looking at it: Since the molecules with the higher kinetic energies are the ones that evaporate, the average kinetic energy of the water molecules that remain is lowered.

Our bodies are cooled by the evaporation of perspiration. On hot days, we feel hotter if the humidity is high because evaporation is inhibited. A breeze feels cool because it removes air near our skin that has a higher humidity due to perspiration. The drier air that replaces it allows for more rapid evaporation—and cooling.

Figure 5.24
(a) Graph of the saturation density versus temperature (data in Table 5.5). (b) Same graph, showing the situation described in Example 5.6.

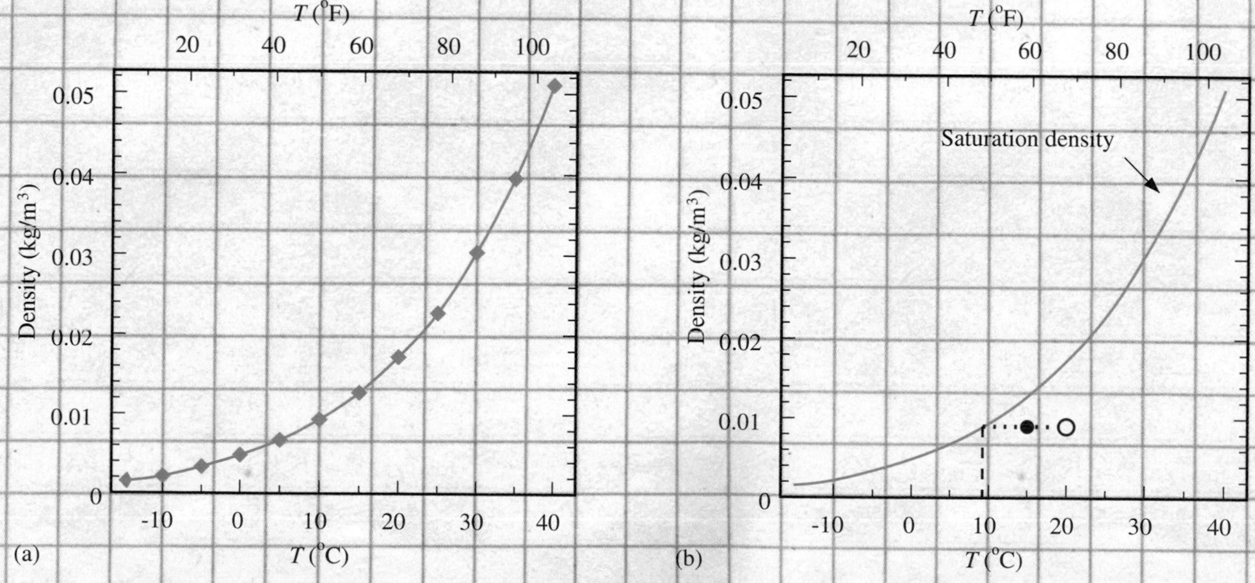

© Ivan Kmit/iStockphoto.com

The Concept Map on your Review Card summarizes the basic phase transitions and their effect on the potential energy of atoms and molecules.

5.7 Heat Engines and The Second Law of Thermodynamics

DEFINITION

Heat Engine *A device that transforms heat into mechanical energy or work. It absorbs heat from a hot source such as burning fuel, converts some of this energy into usable mechanical energy or work, and outputs the remaining energy as heat to some lower-temperature reservoir.*

Along with the efforts made in the recent past to conserve energy used to heat and cool buildings, much has been done to improve the efficiency of other ways that we use energy. In this section, we will consider some of the basic theoretical principles that are involved in energy-conversion devices that use heat and mechanical energy.

Most of the energy used in our society comes from fossil fuels—coal, oil, and natural gas. Part of these fuels is burned directly for heating, such as in gas stoves and oil furnaces. But most of these fuels are used as the energy input for devices that are classified together as **heat engines**.

Gasoline engines, diesel engines, jet engines, and steam-electric power plants are all heat engines. In gasoline and diesel engines, some of the heat from burning fuel is converted into mechanical energy. The remainder of the heat is ejected to the air from the exhaust pipe, the radiator, and the hot surfaces of the engine. Coal, nuclear, and some types of solar power plants produce electricity by using steam to turn a generator (see Figure 5.25). Heat from burning coal, fissioning nuclear fuel, or the Sun is used to boil water. The steam is piped to a turbine (basically a propeller) that is given rotational energy by the steam. A generator connected to the turbine converts rotational energy into electrical energy. After the steam leaves the turbine, it is condensed back into the liquid phase via cooling with water or the air. Cooling towers, which function somewhat like automobile radiators, are used in the latter case. Most of the heat transferred to the water to boil it is ejected from the plant as waste heat when the steam is condensed.

Even though the actual inner workings of heat engines are quite complicated, from an energy point of view we can represent them with a simple schematic diagram

Figure 5.25

Simplified sketch of an electric power plant showing the energy (heat) inputs and outputs.

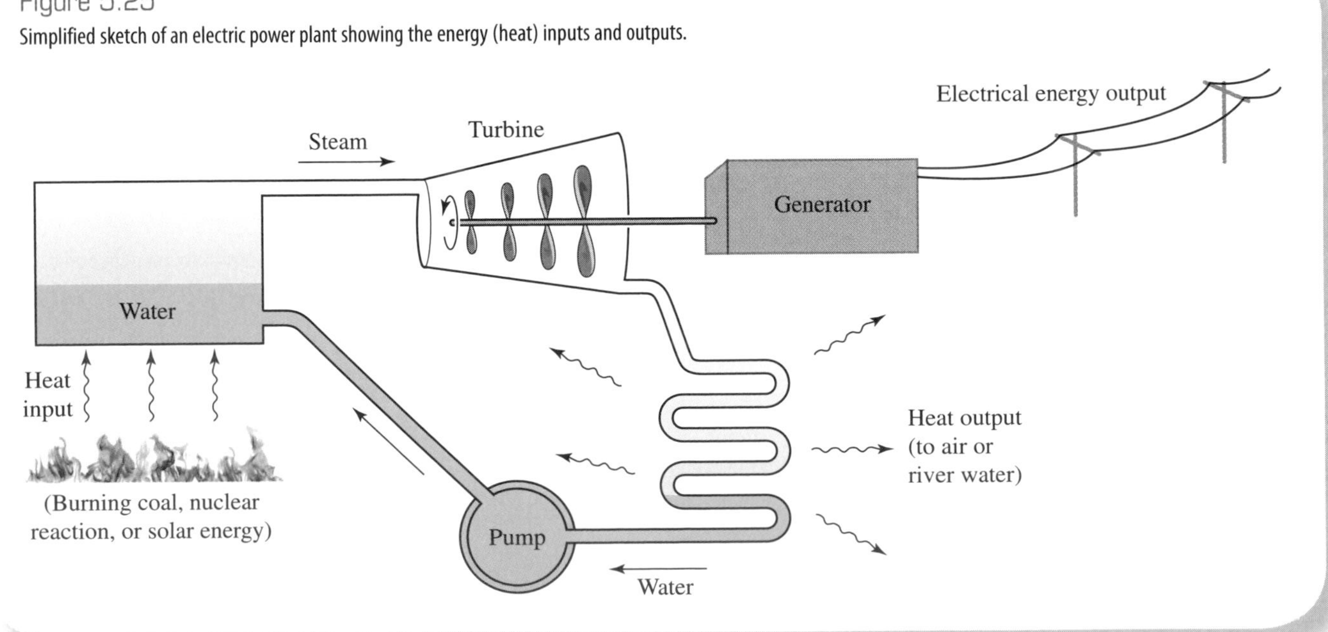

(see Figure 5.26). Energy in the form of heat from some source is input into a mechanism. The mechanism converts some of the heat into mechanical energy and rejects the remainder. In this simplified representation of a heat engine, the heat source has some fixed (high) temperature T_h and the reservoir that absorbs the rejected heat has some fixed (lower) temperature T_l.

During a given period while the engine is operating, some quantity of heat Q_h is absorbed from the heat source. Some of this energy is converted into usable work, and the remainder of the original energy input is ejected as waste heat Q_l. This wasted heat is unavoidable. The following law is a formal statement of this.

LAWS

Second Law of Thermodynamics *No device can be built that will repeatedly extract heat from a source and deliver mechanical work or energy without ejecting some heat to a lower-temperature reservoir.*

The energy efficiency of any device or process is the usable output divided by the total input, times 100%.

$$\text{efficiency} = \frac{\text{energy or work output}}{\text{energy or work input}} \times 100\%$$

Figure 5.26

Diagram of a heat engine. The mechanism absorbs energy from the heat source, uses some of it to do work, and releases the remainder to some lower-temperature reservoir.

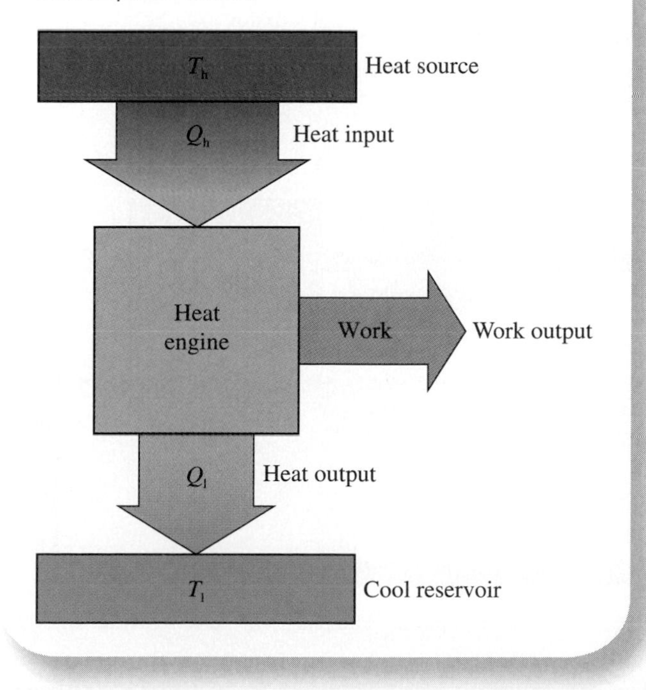

If the efficiency of a device is 25%, then one-fourth of the input energy is converted into usable form. The remaining three-fourths are "lost"—released as waste heat.

For heat engines, the input energy is Q_h, and the output is the work. Therefore,

$$\text{efficiency} = \frac{\text{work}}{Q_h} \times 100\% \qquad \text{(heat engine)}$$

As we have seen, there are many different types of heat engines. Some of them use processes that are inherently more efficient than others. However, there is a theoretical upper limit on the efficiency of a heat engine. This maximum efficiency is called the *Carnot efficiency* after the French engineer Sadi Carnot. Carnot discovered that the efficiency of a perfect heat engine is limited by the temperatures of the heat source and of the lower-temperature reservoir. In particular:

$$\text{Carnot efficiency} = \frac{T_h - T_l}{T_h} \times 100\%$$

(T_h and T_l must be in kelvins.)

Real heat engines have friction, imperfect insulation, and other factors that reduce their efficiencies. But even if these were completely eliminated, a heat engine could not have an efficiency of 100%, as this would imply that the temperature of the reservoir was zero or that the temperature of the heat source was infinitely high, neither of which is achievable in practice.

EXAMPLE 5.7

A typical coal-fired power plant uses steam at a temperature of 1,000°F (810 K). The steam leaves the turbine at a temperature of about 212°F (373 K). What is the theoretical maximum efficiency of the power plant? Here, $T_h = 810$ K *and* $T_l = 373$ K. *So:*

$$\text{Carnot efficiency} = \frac{810 - 373}{810} \times 100\%$$
$$= 54\%$$

This is the ideal efficiency. The actual efficiency of a coal power plant is usually around 35 to 40%.

Nearly two-thirds of the energy input to a power plant is lost as waste heat. The energy not used causes thermal pollution by artificially heating the environment. The actual top efficiencies of the other heat engines in common use are similar: diesel engine, 35%; jet engine, 23%; gasoline (piston) engine, 25%.

There are two general ways to improve the efficiency of a heat engine. One is to improve the process and reduce energy losses so that the efficiency gets closer to the Carnot efficiency. The other way is to increase the Carnot efficiency by raising T_h or lowering T_l. The efficiencies of steam-based heat engines have been improved a great deal through the use of higher-temperature steam.

Nearly two-thirds of the energy input to a power plant is lost as waste heat.

Heat Movers

Refrigerators, air conditioners, and heat pumps are devices that act much like heat engines in reverse. They use an input of energy to cause heat to flow from a cooler substance to a warmer substance—opposite the natural flow from hotter to colder objects. The purpose of refrigerators and air conditioners is to extract heat from an area and thereby cool it. Heat pumps are designed to heat an interior space in the winter as well as cool it in the summer. In these devices, as heat is transferred from a substance being cooled, a larger amount of heat flows into a different substance that is then warmed. A refrigerator removes heat from its interior and ejects heat into the room. We will call these devices **heat movers**, since the term describes what they do.

The mechanisms in the three devices mentioned above are much the same. A gas, called the *refrigerant*, is forced to change phases in a cyclic process. The gas must have a fairly low boiling point and be condensed easily when the pressure on it is increased. Freon (CCl_2F_2) was the most commonly used refrigerant, although it is being replaced by other compounds that do not destroy ozone after being released into the atmosphere. The gas is compressed into the liquid phase by a pump and forced to flow through a small opening called an *expansion valve* (Figure 5.27). The pressure on the other side of the valve is kept low so that the refrigerant quickly goes into the gas phase (boils). This phase transition cools the refrigerant, and in the process, it absorbs heat from the surroundings. The gas flows back to the pump, where it is recompressed into the liquid phase. This raises the temperature of the refrigerant, causing heat to flow from it to the surroundings. After the refrigerant is cooled and liquefied, it flows through the expansion valve, and the cycle repeats.

Although the net result is a flow of heat from something cooler to something warmer, the actual heat flow into, and then out of, the refrigerant is always a flow from something warmer to something cooler. Basically, the refrigerant is first cooled to a temperature below that of the substance being cooled and then heated to a temperature above that of the substance being warmed. Also note that energy must be supplied to the pump to make the process work.

Heat movers can be represented with a diagram similar to that of heat engines (Figure 5.28). Unlike heat engines, the mechanism in heat movers has an *input* of energy or work. During a given period of time, a quantity of heat Q_l is absorbed from a cool reservoir with temperature T_l. An amount of work is done on the refrigerant by the pump, and an amount of heat Q_h is released to a warm reservoir with temperature T_h. In refrigerators, the cool reservoir is the air in their interiors. The warm reservoir is the air in the room. Air conditioners absorb heat from the air that they circulate inside a building, a vehicle, and so on, and transfer heat to the warmer outside air. When in the "heating mode," heat pumps remove heat from the outside air, ground, or groundwater and release heat

Figure 5.27
Refrigerators remove heat from their interiors by exploiting the phase transition of the refrigerant. As the refrigerant vaporizes, it absorbs heat from the refrigerator's interior. As it condenses, it releases heat to the air in the room.

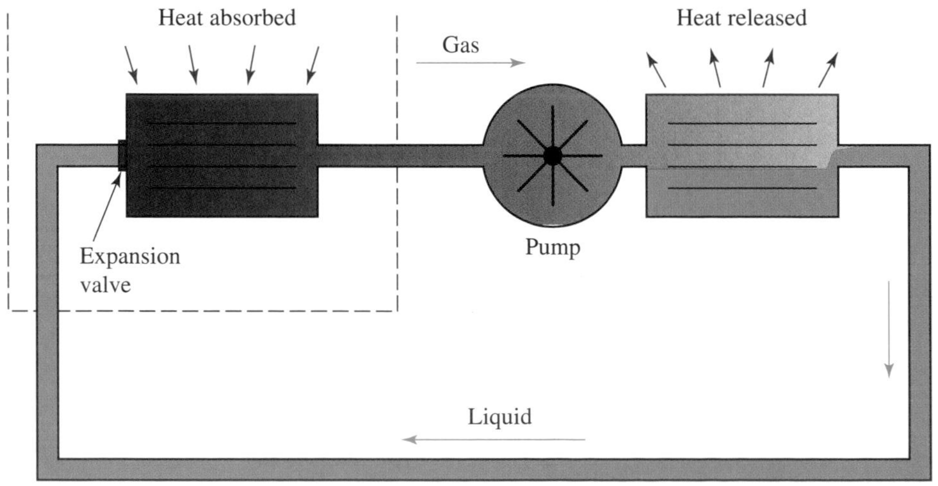

Figure 5.28

Diagram of a heat mover. A mechanism uses an energy input to extract heat from a cool source and to release heat to a higher-temperature reservoir.

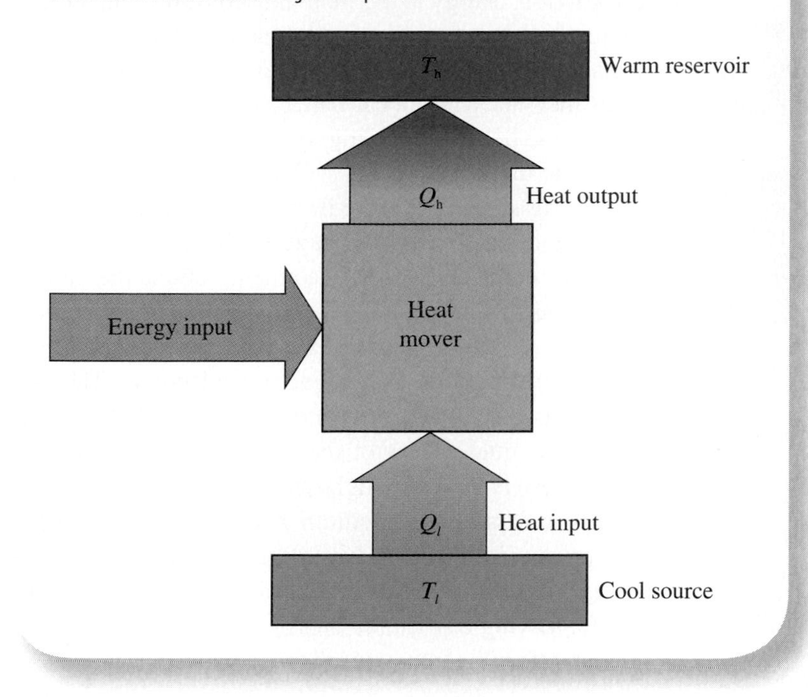

inside a building. In the "cooling mode," heat pumps reverse the heat flow: Heat is transferred from the inside of the building and ejected to the outside.

The law of conservation of energy tells us that the amount of heat released, Q_h, is equal to the amount of heat absorbed, Q_l, plus the energy input to the mechanism. The relative values of Q_h, Q_l, and the energy input depend on the efficiency of the pumping mechanism, the difference between the two temperatures, T_h and T_l, and other factors. In the average refrigerator, the amount of heat transferred from the interior is about *three times* the amount of energy input. In the heating mode, the amount of heat delivered by a good heat pump is about *three times* the amount of energy consumed by the mechanism. Energy is not being "created": The heat pump is simply extracting heat from a cooler substance and then transferring a larger amount of heat to a warmer substance. (Remember: Any system or substance with a temperature greater than absolute zero stores some internal energy, implying that it can serve as a source of heat transfer providing a suitable heat pump can be devised.)

Usable Energy

The outputs of a heat engine are heat flowing into a low-temperature reservoir, thereby increasing its internal energy, and useful work or energy. Eventually, the latter also becomes internal energy because of friction and other processes. So the overall effect of a heat engine is to convert internal energy in a higher-temperature source into internal energy in a lower-temperature reservoir. This internal energy, in turn, is no longer available to do as much useful work: Carnot tells us that lower temperatures would give us lower efficiencies. Even though energy is not truly "lost" or "destroyed" in this process, it is made unavailable for similar use again.

This is a general result of energy transformations: The energy that remains is less usable than the original energy. Electrical energy is a highly convenient, easily used, and versatile form of energy. Light bulbs, heaters, and fans convert electrical energy into other desirable forms (radiant energy by which to read a book, heat energy by which to warm a room, and kinetic energy of moving air by which to cool a room on a hot day), plus the inevitable and less beneficial internal energy associated with friction (as, for example, in the rotating parts of a fan). Certainly in the case of the internal energy, but for the other "useful" forms of energy as well, the ease with which the remaining energy can be harnessed to drive other desired processes is nowhere near that of the original electrical energy. For example, the efficiency with which the kinetic energy of the moving air from a fan can be collected and transformed into mechanical energy using a high-efficiency wind turbine system (<60%) is much smaller than the efficiency with which the original electrical energy can be converted into rotational energy via the fan motor (>90%). In this sense, we might say that the *quality* of electrical energy is higher than that associated with wind energy.

The physical quantity **entropy**, symbolized as S, can be used as a means of assessing the amount of "useless" or low-quality or *dispersed* energy in a system at a specific temperature; that is, the energy that cannot be readily harnessed to do external work. As noted above, natural processes tend to be accompanied by a dispersal of energy and hence an increase in the entropy of a system. Commonly, entropy is taken to be a measure of the disorder (or "mixedupedness" to use the terminology of 19th century American physicist J. Willard Gibbs) in a system: Increasing entropy is associated with increasing disorder. In this view, the more disordered the system, the more its energy is spread out over different possible internal states, and the greater its entropy. Considerations of this type have lead to an alternative formulation of the second law of thermodynamics.

LAWS

Second Law of Thermodynamics (alternate form) *In any thermodynamic process, the total entropy of the system and its environment either increases or remains constant: $\Delta S \geq 0$. The entropy of the universe always increases.**

Although it may not be obvious that this statement is equivalent to our earlier one for the second law (see p. 137), it can be shown to be true. Indeed, the association of the concept of entropy with the amount of "useless" energy leads to the conclusion that for any process to go forward, a quantity of energy equal to $\Delta S \times T_R$, where ΔS is the entropy change in the system and T_R is the Kelvin temperature of the external environment, **must** be transferred to the system's surroundings.** If this condition is not met, then the process (or the device designed to carry out the process) cannot be realized.

The entropy of part of a system can be decreased temporarily at the expense of the rest of the system. As living things grow, they organize the nutrients and life-sustaining chemicals they take in to become more highly ordered; their entropy decreases. But in doing so, they disperse the energy in the food they consume and thereby increase the entropy of these substances. After the organisms die, the entropy of their remains once again increases as they decay. As a car battery is charged, its entropy decreases, but the charging system is itself consuming energy to drive the process with the result that entropy is increasing somewhere else in the environment.

The term "energy shortage," which frequently appears in connection with statements about the rate at which our civilization is consuming energy resources and the risks this behavior poses for the sustainability of future generations, is something of a misnomer. There is, in fact, as much energy around now as ever before. (After all, the total energy of the universe is constant.) But our society is using up high-quality, conveniently stored energy in fossil fuels, uranium, and forests, for example, and producing lower quality, more dispersed internal energy. We are irreversibly increasing the entropy of the world, leaving less *usable* energy for our progeny. However, by tapping "renewable" energy sources such as solar, wind, geothermal, and tidal power, we can exploit continuously available (at least for the next several billion years) supplies of energy that would otherwise go to waste—and we reduce greenhouse gas emissions in the process. Solar energy eventually becomes high-entropy internal energy, whether it is absorbed by the ground or by solar cells. However, if solar energy is absorbed by high-efficiency solar cells, we get as an intermediate step to this endpoint some low-entropy, high-quality electrical energy to meet our needs.

* Entropy changes are zero only for reversible processes, but no natural, spontaneous process is truly reversible. Thus, we expect the entropy of the universe as a whole to become greater with time.

** The SI unit for entropy is the joule per Kelvin (J/K). Thus, the product of an entropy change and a temperature measured in Kelvins yields a quantity whose units are those of energy, the joule.

questions&problems

Questions

(▸ Indicates a review question, which means it requires only a basic understanding of the material to answer. The others involve integrating or extending the concepts presented thus far.)

1. ▸ What are the three common temperature scales? What are the normal boiling and freezing points of water in each scale?
2. The Fahrenheit and Celsius temperature scales agree at −40° (−40°C = −40°F). Do the Fahrenheit and Kelvin temperature scales ever agree? How about the Celsius and Kelvin scales?
3. ▸ What is the significance of absolute zero?
4. ▸ What happens to the atoms and molecules in a substance as its temperature increases?
5. ▸ Air molecules in a warm room (27°C = 300 K) typically have speeds of about 500 m/s (1100 mph). Why is it that we are unaware of these fast moving particles continuously colliding with our bodies?
6. A special glass thermometer is manufactured using a liquid that expands less than glass when the temperature increases. Assuming the thermometer does indicate the correct temperature, what is different about the scale on it?
7. ▸ Explain what a bimetallic strip is and how it functions.

8. Refer to the thermometer in Figure 5.10. Which metal in the bimetallic strip has the larger coefficient of linear expansion, the metal on the outer side of the spiral or the metal on the inner side? How can you determine this?

9. ▶ A certain engine part made of iron expands 1 mm in length as the engine warms up. What would be the approximate change in length if the part were made of aluminum instead of iron?

10. ▶ What is unusual about the behavior of water below the temperature of 4°C?

11. A company decides to make a novelty glass thermometer that uses water instead of mercury or alcohol.

    a) The thermometer would include a warning informing the user that it should not be exposed to temperatures below 0°C. Why?

    b) Suppose the thermometer is taken outside where the temperature is 1°C. Describe how the level of the water would change as it adjusts to the new temperature and how at some point it would behave very differently than a mercury- or alcohol-filled thermometer.

12. ▶ What are the two general ways to increase the internal energy of a substance? Describe an example of each.

13. Air is allowed to escape from an inflated tire. Is the temperature of the escaping air higher than, lower than, or equal to the temperature of the air inside the tire? Why?

14. Is it possible to compress air without causing its internal energy to increase? If so, how?

15. ▶ Describe the three methods of heat transfer. Which of these are occurring around you at this moment?

16. A potato will cook faster in a conventional oven if a large nail is inserted into it. Why?

17. A coin and a piece of glass are both heated to 60°C. Which will feel warmer when you touch it?

18. A submerged heater is used in an aquarium to keep the water above room temperature. Should it be placed near the surface of the water or near the bottom to be most effective?

19. On a cool night with no wind, people facing a campfire feel a breeze on their backs. Why?

20. Suppose you hang a bag of ice inside a room in which the air is at normal room temperature. If you position the palm of your hand a few inches to the side of the ice, what would you feel? Why? Would you feel anything different if you placed your palm the same distance away but below the ice?

21. When heating water on a stove, a full pan of water takes longer to reach the boiling point than a pan that is half full. Why?

22. A 1-kg piece of iron is heated to 100°C and then submerged in 1 kg of water initially at 0°C. The iron cools and the water warms until they are at the same temperature (in thermal equilibrium). Assuming there is no other transfer of heat involved, is the final temperature closer to 0°C, 50°C, or 100°C? Why?

23. In Example 5.3, is it really necessary to know the mass of the concrete block to solve this problem? Put another way, would the answer be different if it was a 10-kg block that was dropped instead of a 5-kg one?

24. A piece of aluminum and a piece of iron fall without air resistance from the top of a building and stick into the ground on impact. Will their temperatures change by the same amount? Explain.

25. The specific heat capacity of water is extremely high. If it were much lower, say, one-fifth as large, what effect would this have on processes like fighting fires and cooling automobile engines?

26. ▶ Why does the temperature of water not change while it is boiling?

27. ▶ Describe how changing the air pressure affects the temperature at which water boils.

28. One way to desalinate seawater—remove the dissolved salts so that the water is drinkable—is to distill it: Boil the seawater and condense the steam. The salts stay behind. This technique has one major disadvantage. It consumes a large amount of energy. Why is that?

29. ▶ What is saturation density? How does it change when the temperature increases?

30. ▶ What effect does heating the air in a room have on the relative humidity?

31. *Wood's metal* is an alloy of the elements bismuth, lead, tin, and cadmium that has a melting point of 70°C. Describe how it might be used in an automatic sprinkler system for fire suppression.

32. When trying to predict the lowest temperature that will be reached overnight, forecasters pay close attention to the dew point temperature. Why is the air temperature unlikely to drop much below the dew point? (The high latent heat of vaporization of water is important.)

33. ▶ Explain what a heat engine does and what a heat mover does.

34. ▶ In the winter, the amount of internal energy a heat pump delivers to a house is greater than the electrical energy it uses. Does this violate the law of conservation of energy? Explain.

35. ▶ What is entropy? In general, how does the entropy of a system change with time?

36. When a parcel of water freezes to become ice, it becomes more ordered and its entropy decreases. Does this violate the alternate form of the second law of thermodynamics? Explain.

37. Is our society truly facing an "energy crisis," assuming by this term we mean that we are running out of energy? What *is* happening to our energy resources as a result of the increasing industrialization of the world?

Problems

1. Your jet is arriving in London, and the pilot informs you that the temperature is 30°C. Should you put on your jacket? Use Figure 5.1 to determine the temperature in degrees Fahrenheit.
2. On a nice winter day at the South Pole, the temperature rises to −60°F. What is the approximate temperature in degrees Celsius?
3. An iron railroad rail is 700 ft long when the temperature is 30°C. What is its length when the temperature is −10°C?
4. A copper vat is 10 m long at room temperature (20°C). How much longer is it when it contains boiling water at 1 atm pressure?
5. A machinist wishes to insert a steel rod with a diameter of 5 mm into a hole with a diameter of 4.997 mm. By how much would the machinist have to lower the temperature of the rod to make it fit the hole?
6. An aluminum wing on a passenger jet is 30 m long when its temperature is 20°C. At what temperature would the wing be 5 cm (0.05 m) shorter?
7. A gas is compressed inside a cylinder (see Figure 5.11). An average force of 50 N acts to move the piston 0.1 m. During the compression, 2 J of heat are conducted away from the gas. What is the change in internal energy of the gas?
8. Air in a balloon does 50 J of work while absorbing 70 J of heat. What is its change in internal energy?
9. How much heat is needed to raise the temperature of 5 kg of silver from 20°C to 960°C?
10. A bottle containing 3 kg of water at a temperature of 20°C is placed in a refrigerator where the temperature is kept at 3°C. How much heat is transferred from the water to cool it to 3°C?
11. a) Compute the amount of heat needed to raise the temperature of 1 kg of water from its freezing point to its normal boiling point.
    b) How does your answer to a) compare to the amount of heat needed to convert 1 kg of water at 100°C to steam at 100°C?
12. Aluminum is melted during the recycling process.
    a) How much heat must be transferred to each kilogram of aluminum to bring it to its melting point, 660°C, from room temperature, 20°C?
    b) About how many cups of coffee could you make with this much energy (see Example 5.2)?
13. A 1,200-kg car going 25 m/s is brought to a stop using its brakes. Let's assume that a total of approximately 20 kg of iron in the brakes and wheels absorbs the heat produced by the friction.
    a) What was the car's original kinetic energy?
    b) After the car has stopped, what is the change in temperature of the brakes and wheels?
14. The 0.02-kg lead bullet in Figure 5.29 is traveling 200 m/s when it strikes an armor plate and comes to a stop. If all of the bullet's energy is converted to heat that it alone absorbs, what is its temperature change?

Figure 5.29

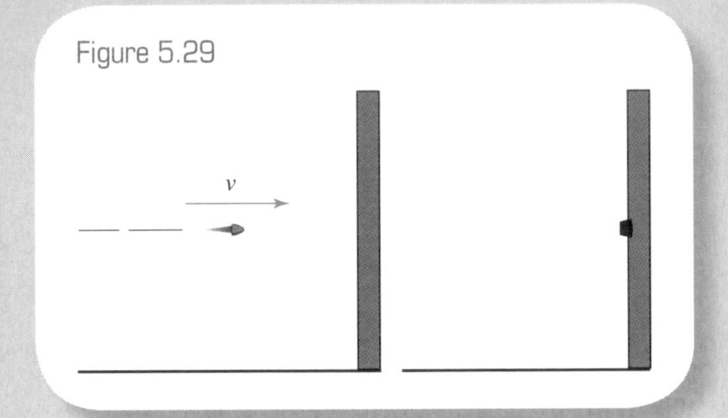

15. A 10-kg lead brick is dropped from the top of a 629-m-tall television transmitting tower in North Dakota and falls to the ground. Assuming all of its energy goes to heat it, what is its temperature increase?
16. Water flowing over the Lower Falls in Yellowstone National Park drops 94 m. If all of the water's energy goes to heat it, what is its temperature increase?
17. On a winter day, the air temperature is −15°C, and the humidity is 0.001 kg/m^3.
    a) What is the relative humidity?
    b) When this air is brought inside a building, it is heated to 20°C. If the humidity isn't changed, what is the relative humidity inside the building?
18. On a summer day in Houston, the temperature is 35°C and the relative humidity is 77%.
    a) What is the humidity?
    b) To what temperature could the air be cooled before condensation would start to take place? (i.e., What is the dew point?)
19. Inside a building, the temperature is 20°C, and the relative humidity is 40%. How much water vapor is in each cubic meter of air?
20. On a hot summer day in Washington, D.C., the temperature is 86°F, and the relative humidity is 70%. How much water vapor does each cubic meter of air contain?

21. An apartment has the dimensions 10 m by 5 m by 3 m. The temperature is 25°C, and the relative humidity is 60%. What is the total mass of water vapor in the air in the apartment?

22. The total volume of a new house is 800 m^3. Before the heat is turned on, the air temperature inside is 10°C, and the relative humidity is 50%. After the air is warmed to 20°C, how much water vapor must be added to the air to make the relative humidity 50%?

23. The temperature of the air in thermals decreases about 10°C for each 1,000 m they rise. If a thermal leaves the ground with a temperature of 30°C and a relative humidity of 31%, at what altitude will the air become saturated and the water vapor begin to condense to form a cloud? (In other words, at what altitude does the temperature equal the dew point?)

24. In cold weather, you can sometimes "see" your breath. What you are seeing is a mist of small water droplets, the same as in clouds and fog. Suppose air leaves your mouth with temperature 35°C and humidity 0.035 kg/m^3 and mixes with an equal amount of air at 5°C and humidity 0.005 kg/m^3.

    a) What is the relative humidity of the mixed air if its temperature and humidity equal the averages of those of the two original air masses?

    b) Represent what happens by plotting three points in a graph like Figure 5.24.

25. What is the Carnot efficiency of a heat engine operating between the temperatures of 300°C (573 K) and 100°C (373 K)?

26. What is the maximum efficiency that a heat engine could have when operating between the normal boiling and freezing temperatures of water?

27. As a gasoline engine is running, an amount of gasoline containing 15,000 J of chemical potential energy is burned in 1 s. During that second, the engine does 3,000 J of work.

    a) What is the engine's efficiency?

    b) The burning gasoline has a temperature of about 4,000°F (2,500 K). The waste heat from the engine flows into air at about 80°F (300 K). What is the Carnot efficiency of a heat engine operating between these two temperatures?

28. A proposed ocean thermal-energy conversion (OTEC) system is a heat engine that would operate between warm water (25°C) at the ocean's surface and cooler water (5°C) 1,000 m below the surface. What is the maximum possible efficiency of the system?

There are more quizzes and study tools online at 4ltrpress.cengage.com/physics.

Waves and Sound

sections

6.1 Waves—Types and Properties

Ripples moving over the surface of a still pond, sound from a radio speaker traveling through the air, a pulse "bouncing" back and forth on a piano string, light from the Sun illuminating and warming the Earth—these are all waves. We can feel the effects of some waves, such as earthquake tremors (called seismic waves), as they pass. Others, such as sound and light, we sense directly with our ears and eyes. Technology has given us numerous devices that produce or detect waves that we cannot sense (microwaves, ultrasound, x-rays).

What are waves? Though many and diverse, they share some basic features. They all involve *vibration or oscillation* of some kind. Floating leaves show the vibration of the water's surface as ripples move by. Our ears respond to the oscillation of air molecules and give us the perception of sound. Also, waves move and *carry energy* yet do not have mass. The sound from a loudspeaker can break a wineglass even though no matter moves from the speaker to the glass. We can define a wave as follows.

DEFINITION

Wave *A traveling disturbance consisting of coordinated vibrations that transmit energy with no net movement of matter.*

Sound, water ripples, and similar waves consist of vibrations of matter—air molecules or the water's surface, for example. The substance through which such waves travel is called the *medium* of the wave. Particles of the medium vibrate in a coordinated fashion to form the wave.

CHAPTER 6

physics

A rope stretched between two people is a handy medium for demonstrating a simple wave. A flick of the wrist sends a wave down the rope. Each short segment of the rope is pulled upward in turn by its neighboring segment. The forces between the parts of the medium are responsible for "passing along" the wave. This kind of wave is not unlike a row of dominoes knocking each other over, except that the medium of a wave does not have to be "reset" after a wave goes by.

Many waves—sound, water ripples, waves on a rope—require a material medium. They cannot exist in a vacuum. On the other hand, light, radio waves, microwaves, and x-rays can travel through a vacuum because they do not require a medium for their propagation. We will take a close look at these special waves—called *electromagnetic waves*—in Chapter 8.

Waves occur in a great variety of substances: in gases (sound), in liquids (water ripples), and in solids (seismic waves through rock). Some travel along a line (a wave on a rope), some across a surface (water ripples), and some throughout space in three dimensions (sound). Many more examples could be listed. Clearly, waves are everywhere, and they are diverse.

A wave can be short and fleeting, called a *wave pulse*, or steady and repeating, called a *continuous wave*. The sound of a bursting balloon, a tsunami (large ocean wave generated by an earthquake),

© Tova Teitelbaum/iStockphoto.com

and the light from a camera flash are examples of wave pulses. The sound from a tuning fork and the light from the Sun are continuous waves.* Figure 6.1 shows a wave pulse and a continuous wave on a long rope. You can see that a continuous wave is like a series or "train" of wave pulses, one after another.

* For now we can treat light as a simple wave. In Chapter 10, we will present the modern view of light as developed by Albert Einstein and others.

If we take a close look at many different types of waves, we find that they can be classified according to the orientation of the wave oscillations. There are two main wave types, **transverse** and **longitudinal**.

DEFINITION

Transverse Wave *A wave in which the oscillations are perpendicular (transverse) to the direction the wave travels. Examples: waves on a rope, electromagnetic waves, some seismic waves.*

Longitudinal Wave *A wave in which the oscillations are along the direction the wave travels. Examples: sound in the air, some seismic waves.*

Both types of waves can be produced on a Slinky—a short, fat spring that you may have seen "walk" down steps. If a Slinky is stretched out on a flat, smooth tabletop, a transverse wave is produced by moving one end from side to side, perpendicular to the Slinky's length (Figure 6.2a). A longitudinal wave is produced by pushing and pulling one end back and forth, first toward the other end, then back (see Figure 6.2b). For each type of wave, one can produce either a wave pulse or a continuous wave.

© iStockphoto.com

A Slinky is not the only medium that can carry both transverse and longitudinal waves. Both kinds of waves can travel in any solid. Earthquakes and underground explosions produce both longitudinal and transverse seismic waves that travel through the Earth. Simple waves that involve oscillation of atoms and molecules must be longitudinal to travel in liquids and gases because of the absence of rigid bonds between the particles.

Many waves are neither purely longitudinal nor purely transverse. Although a water ripple appears to be a simple transverse wave, individual parcels of water actually move in circles or ellipses—they oscillate forward and backward as well as up and down. Waves in plasmas and in the atmosphere are even more complicated. But the two simple types of waves described here are common and well suited for illustrating wave phenomena.

The speed of a wave is the rate of movement of the disturbance. (Do not confuse this with the speed of individual particles as they oscillate.) For a given type of wave, the speed is determined by the properties of the medium. In the waves that we have been discussing, the masses of the particles that oscillate and the forces that act between them affect the wave speed. As a longitudinal wave, for example, travels on a Slinky, each coil is

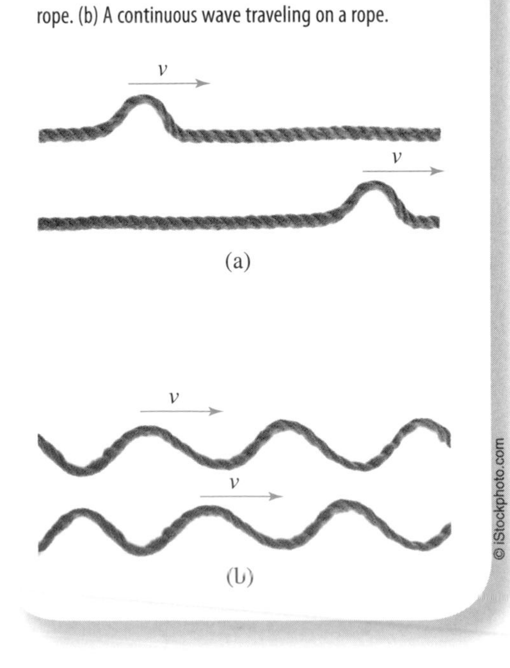

Figure 6.1
(a) Successive views of a wave pulse as it travels on a rope. (b) A continuous wave traveling on a rope.

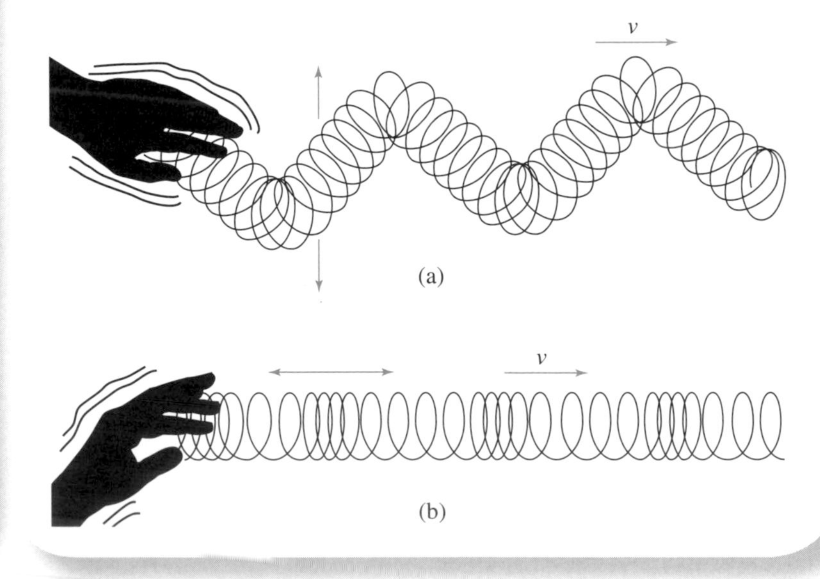

Figure 6.2
(a) A continuous transverse wave on a Slinky. Each coil oscillates up and down as the wave travels to the right. (b) A continuous longitudinal wave on a Slinky. Each coil oscillates left and right as the wave travels to the right.

accelerated back and forth by its neighbors. Basic mechanics tells us that the mass of each coil and the size of the force acting on it will determine how quickly it—and therefore the wave—moves. In general, weak forces or massive particles in a medium cause the wave speed to be low.

Often, the speed of waves in a medium can be predicted by measuring some other properties of the medium. After all, the factors that affect wave speed—particle masses and inter-particle forces—also affect other properties of a substance. For example, the speed of waves on a stretched rope or a Slinky, or on a taut wire can be computed by using the force F, which must be exerted to keep it stretched, and its **linear mass density** ρ, which equals its mass m divided by its length l. (ρ is the Greek letter rho, pronounced like *row*.) In particular:

$$v = \sqrt{\frac{F}{\rho}} \quad \left(\text{wave on a rope or spring; } \rho = \frac{m}{l}\right)$$

Increasing this force, also called the *tension*, will cause the waves to move faster. This is how stringed instruments like guitars and pianos are tuned. (More on this in Section 6.4.)

EXAMPLE 6.1

A student stretches a Slinky out on the floor to a length of 2 meters. The force needed to keep the Slinky stretched is measured and is found to be 1.2 newtons. The Slinky's mass is 0.3 kilograms. What is the speed of any wave sent down the Slinky by the student?

First, we compute the Slinky's linear mass density.

$$\rho = \frac{m}{l} = \frac{0.3 \text{ kg}}{2 \text{ m}}$$
$$= 0.15 \text{ kg/m}$$

The speed of waves on the Slinky is

$$v = \sqrt{\frac{F}{\rho}} = \sqrt{\frac{1.2 \text{ N}}{0.15 \text{ kg/m}}}$$
$$= \sqrt{8 \text{ m}^2/\text{s}^2} = 2.8 \text{ m/s}$$

The speed of sound in air or any other gas depends on the ratio of the pressure of the gas to the density of the gas. But for each gas, this ratio depends only on the temperature. In particular, the speed of sound in a gas is proportional to the square root of the Kelvin temperature. For air:

$$v = 20.1 \times \sqrt{T} \quad (\text{speed of sound in air; } v \text{ in m/s, } T \text{ in kelvins})$$

© iStockphoto.com

The speed of sound is lower at high altitudes, not because the air is thinner but because it is colder.

EXAMPLE 6.2

What is the speed of sound in air at room temperature (20°C = 68°F)? The temperature in kelvins is

$$T = 273 + 20 = 293 \text{ K}$$

Therefore:

$$v = 20.1 \times \sqrt{T}$$
$$= 20.1 \times \sqrt{293} = 20.1 \times 17.1$$
$$= 344 \text{ m/s (770 mph)}$$

The numerical factor (20.1) in the equation in Example 6.2 is determined by the properties of the molecules that comprise air and therefore applies to air only. The speed of sound in any other gas will be different, and the corresponding equation for v will have a different numerical factor. Two examples:

Speed of sound in helium:

$$v = 58.8 \times \sqrt{T} \text{ (SI units)}$$

Speed of sound in carbon dioxide (CO_2):

$$v = 15.7 \times \sqrt{T} \text{ (SI units)}$$

For the remainder of this section, we will take a look at some of the properties of a continuous wave. A convenient example is a transverse wave on a Slinky produced by moving one end smoothly side to side. Figure 6.3 shows a "snapshot" of such a wave. It shows the shape of the Slinky at some instant in time. Note that the wave has the same sinusoidal shape you've seen before (Figure 2.21).

The high points of the wave are called *peaks* or *crests*, and the low points are called *valleys* or *troughs*.

Figure 6.3

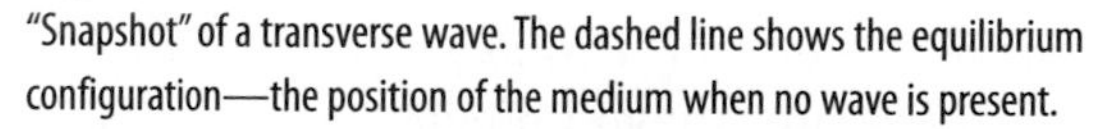

"Snapshot" of a transverse wave. The dashed line shows the equilibrium configuration—the position of the medium when no wave is present.

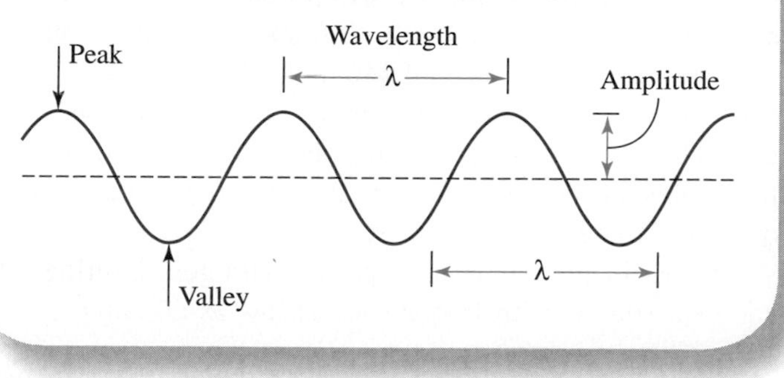

The straight line through the middle represents the equilibrium configuration of the medium—its shape when there is no wave.

In addition to wave speed, there are three other important parameters of a continuous wave that can be measured. They are **amplitude**, **wavelength**, and **frequency**. At any moment, the different particles of the medium are generally displaced from their equilibrium positions by different amounts. The maximum displacement is called the *amplitude* of the wave.

Wave with large amplitude.

© iStockphoto.com / © Shelly Perry/iStockphoto.com

The amplitude is just a distance equal to the height of a peak or the depth of a valley, which are the same. The amplitude of a particular type of wave can vary greatly. For water waves, it can be a few millimeters for ripples to tens of meters for ocean waves. When we hear a sound, its loudness depends on the amplitude of the sound wave: louder sounds have larger amplitudes.

There is also a large variation in the wavelengths of particular types of waves. The wavelengths of sound (in air) that can be heard by humans range from about 2 centimeters (very high pitch) to about 17 meters (very low pitch). Typical wavelengths for radio waves are 3 meters for FM stations and 300 meters for AM stations.

Any segment of a wave that is one wavelength long is called one *cycle* of the wave. As each cycle of a wave passes by a given point in the medium, that point makes one complete oscillation—up, down, and back to the starting position. Figure 6.3 shows three complete cycles of a wave.

Amplitude and wavelength are independent features of a wave: A short-wavelength wave can have a small or a large amplitude (see Figure 6.4).

To understand what the frequency of a wave is, we must "unfreeze" the wave and imagine it as it moves along. The rate at which the wave cycles pass a point is the frequency of the wave. Recall from Section 1.1 that the unit of measure of frequency is the hertz (Hz).

If you move the end of a Slinky back and forth three times each second, you will produce a wave with a frequency of 3 hertz. The note A above middle C on the piano has a frequency of 440 hertz. This means that 440 cycles of the sound wave reach your ear each second. The piano wires producing the sound and the air molecules in the room all vibrate with the same frequency, 440 hertz.

Under ideal conditions, a person with good hearing can hear sounds with frequencies as low as 20 hertz or

DEFINITION

Amplitude *The maximum displacement of points on a wave, measured from the equilibrium position.*

Wavelength *The distance between two successive "like" points on a wave. For example, the distance between two adjacent peaks or two adjacent valleys. Wavelength is represented by the Greek letter lambda (λ).*

Frequency *The number of cycles of a wave passing a point per unit time. The number of oscillations per second in the wave.*

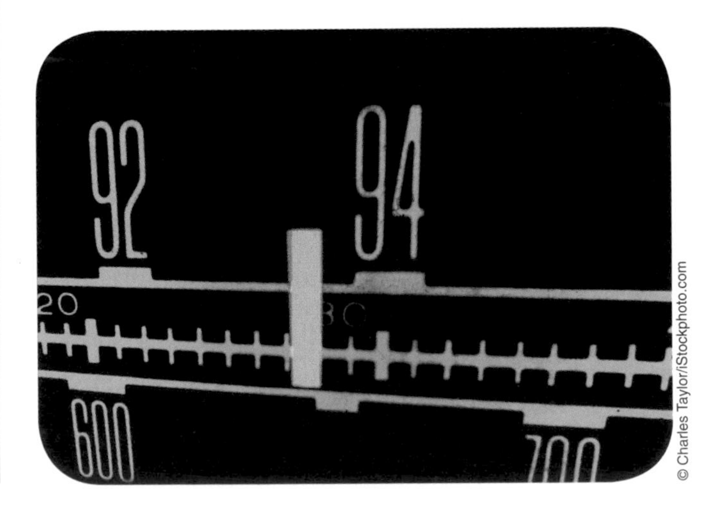

© Charles Taylor/iStockphoto.com

Figure 6.4

Transverse waves with different combinations of wavelength and amplitude.

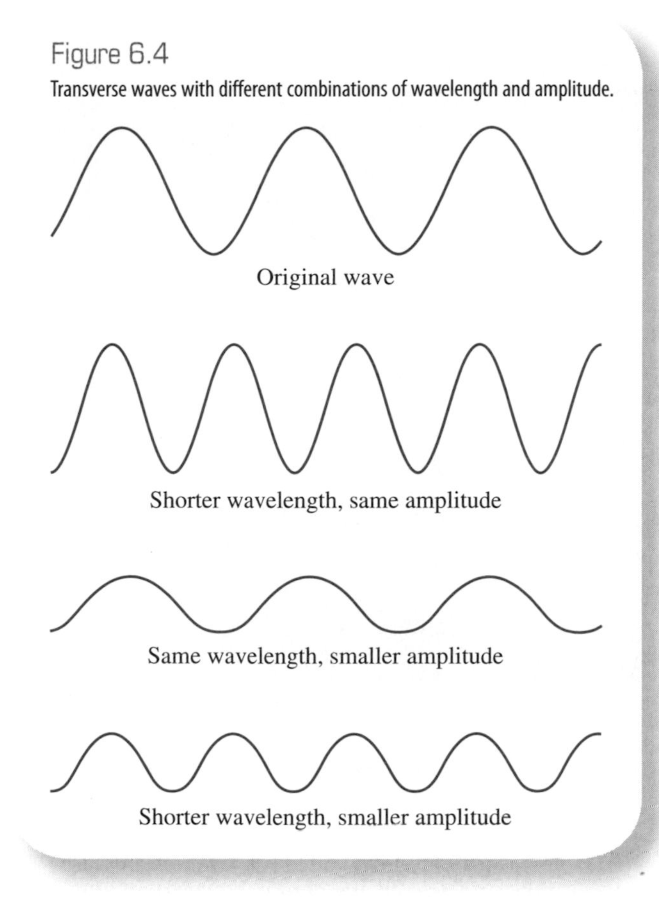

as high as 20,000 hertz. Frequency is important in other kinds of waves as well. Each radio station broadcasts a radio wave with a specific frequency, for example, 1,100 kilohertz = 1,100,000 hertz, or 92.5 megahertz = 92,500,000 hertz.

Amplitude, wavelength, and frequency can be identified for both transverse waves and longitudinal waves, although the amplitude of a longitudinal wave is a bit difficult to visualize. It is still the maximum displacement from the equilibrium position, but in this case, the displacement is along the direction the wave is traveling. Figure 6.5 shows a close-up of a Slinky with no wave and then with a longitudinal wave traveling on it. The amplitude is the farthest distance that any coil is displaced to the right or left of its equilibrium position. The regions where the coils are squeezed together are called **compressions**, and the regions where they are spread apart are called **expansions** or rarefactions. The wavelength is the distance between two adjacent compressions or two adjacent expansions.

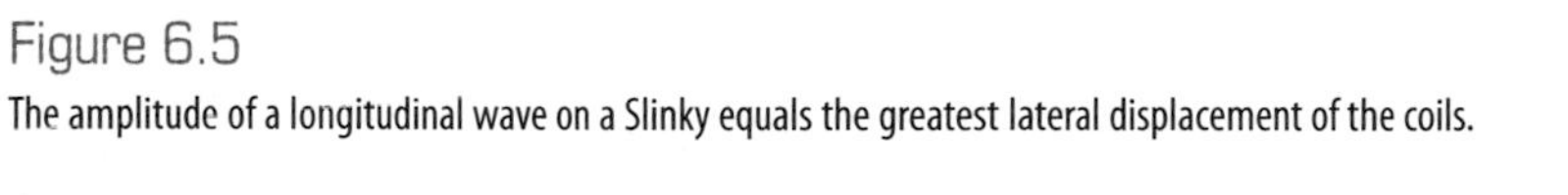

Figure 6.5

The amplitude of a longitudinal wave on a Slinky equals the greatest lateral displacement of the coils.

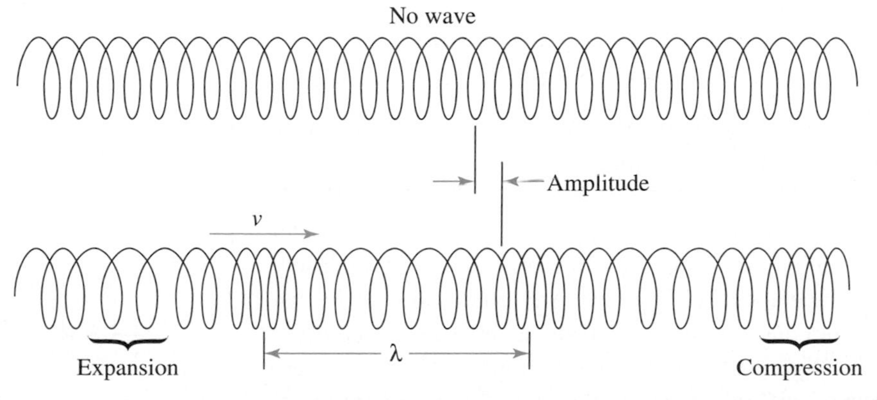

The speed of a wave (**wave speed**), its wavelength, and its frequency are related to each other in a simple way. Imagine a continuous wave passing by a point, perhaps ripples moving by a plant stem. The speed of the wave equals the number of cycles that pass by each second, multiplied by the length of each cycle. For example, if five cycles pass the stem each second and the peaks of the ripples are 0.03 meters apart, the wave speed is 0.15 m/s (see Figure 6.6). In general:

$$\text{wave speed} = \text{number of cycles per second} \times \text{length of each cycle}$$

The two quantities on the right of the equal sign are the frequency of the wave and the wavelength, respectively. Therefore:

$$v = f\lambda$$

The velocity of a continuous wave is equal to the frequency of the wave times the wavelength.

In many cases, all waves that travel in a particular medium have the same speed. Wave pulses, low-frequency continuous waves, and high-frequency continuous waves all travel with the same speed. Sound is an important example of this; sound pulses, low-frequency sounds, and high-frequency sounds travel through the air with the same speed, 344 m/s at room temperature. Similarly, light, radio waves, and microwaves travel with the same speed—3×10^8 m/s—in a vacuum. According to the equation $v = f\lambda$, when the wave speed is the same for all waves, *higher* frequency waves must have proportionally *shorter* wavelengths. A 20-hertz sound wave has a wavelength of about 17 meters, while a 20,000-hertz sound wave has a wavelength of about 1.7 centimeters.

Figure 6.6

If five cycles of a water wave pass by a plant in 1 second, and the wavelength of the wave is 0.03 meters, then the wave is traveling 0.15 m/s.

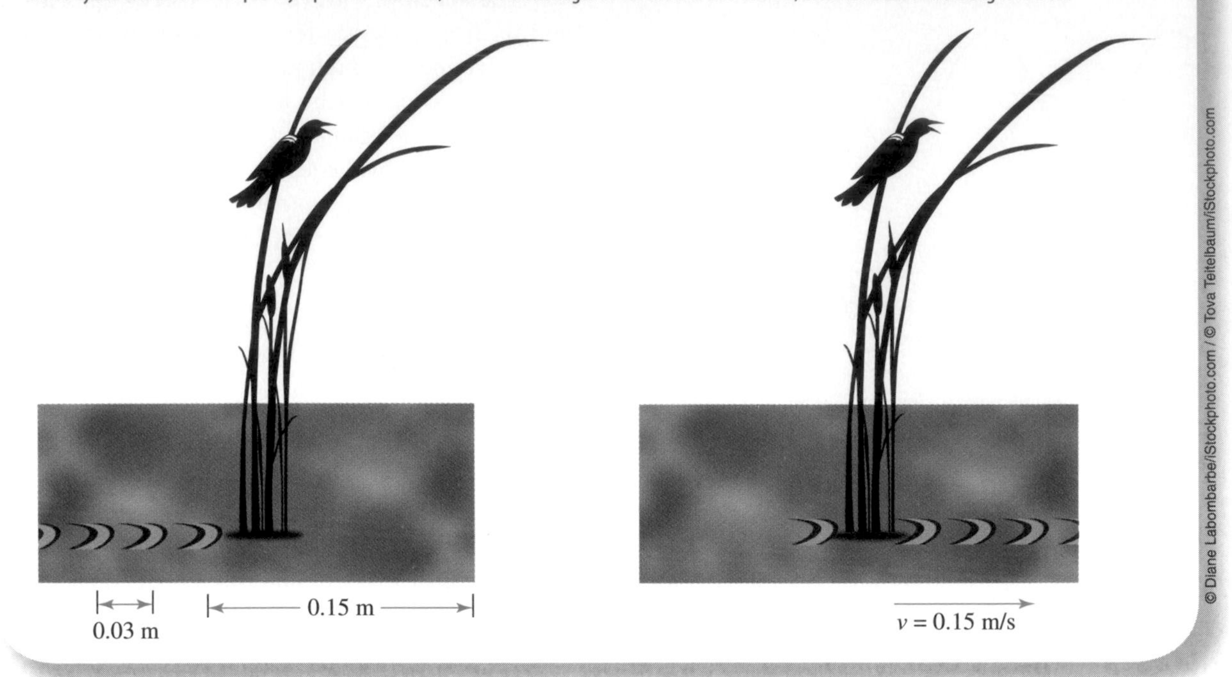

= 440 Hz

© Jeremy Voisey/iStockphoto.com

EXAMPLE 6.3

Before a concert, musicians in an orchestra tune their instruments to the note A, which has a frequency of 440 hertz. What is the wavelength of this sound in air at room temperature? The speed of sound at this temperature is 344 m/s. So:

$$v = f\lambda$$

$$344 \text{ m/s} = 440 \text{ Hz} \times \lambda$$

$$\frac{344 \text{ m/s}}{440 \text{ Hz}} = \lambda$$

$$\lambda = 0.78 \text{ m} = 2.6 \text{ ft}$$

The wavelength of sound with a frequency of 220 hertz is twice as large—1.56 meters.

Not all continuous waves have the simple sinusoidal shape shown in Figure 6.3. In fact, waves with precisely that shape are relatively rare. Any continuous wave that does not have a sinusoidal shape is called a *complex wave.* Figure 6.7 shows two examples. Note that there are three different-sized peaks in each cycle of the upper wave. The shape of a wave is called its *waveform.* The two complex waves in the figure have about the same wavelength and amplitude, but they have very different waveforms. The waveform is another feature that is needed when comparing complex waves. We will take a closer look at this in Section 6.6.

The Concept Map on your Review Card summarizes the characteristics of waves.

Figure 6.7

Two examples of complex waves. The lower wave, a "square wave," is used in many electronic devices.

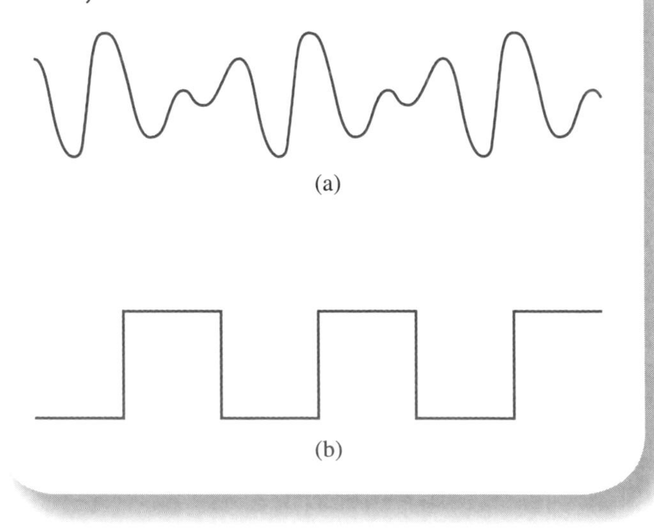

6.2 Aspects of Wave Propagation

In this section, we consider what waves do as they travel. For waves traveling along a surface or throughout space in three dimensions, it is convenient to use two different ways to represent the wave. We will call these the **wavefront model** and the **ray model**. Figure 6.8 shows how each is used to illustrate a wave pulse on water as it travels from the point where it was produced. The wavefront is a circle that shows the location of the peak of the wave pulse. A ray is a straight arrow that shows the direction a given segment of the wave is traveling. A laser beam and sunlight passing through a small hole in a window shade both approximate individual rays of light that we can see if there is dust in the air. On the other hand, the rays of water ripples are not visible, but we do see the wavefronts, as under the butterfly on the photo on page 145.

For a continuous water wave, the wavefronts are concentric circles about the point of origin (the "source" of the wave) that represent individual peaks of the wave (see Figure 6.9). The largest circle shows the position of the first peak that was produced. Each successive wavefront is smaller because it came later and has not traveled as far. The distance between adjacent wavefronts is equal to the wavelength of the wave. Again, a continuous wave is like a series of wave pulses produced one after another. The rays used to represent a continuous wave are lines radiating from the source of the wave (the blue arrows in Figure 6.9). The wavefronts arriving at a point far from the source are nearly straight lines (far right in Figure 6.9). The corresponding rays are nearly parallel.

For a wave moving in three-dimensional space, like the sound traveling outward from you in all directions as you shout or whistle, the wavefronts are spherical shells surrounding the source of the wave. The wavefront of a wave pulse, such as the sound from a hand clap, expands like a balloon that is being inflated very fast. For continuous three-dimensional waves, like a steady whistle, the wavefronts form a series of concentric spherical shells that expand like the circular wavefronts of a wave on a surface. A 440-hertz tuning fork produces 440 of these wavefronts each second. The surface of each wavefront expands outward with a speed of 344 m/s (at room temperature). As with waves on a surface, the rays used to represent a continuous wave in three dimensions are lines radiating outward from the wave source.

One inherent aspect of the propagation of waves on a surface or in three dimensions is that the amplitude of the wave necessarily decreases as the wave gets farther

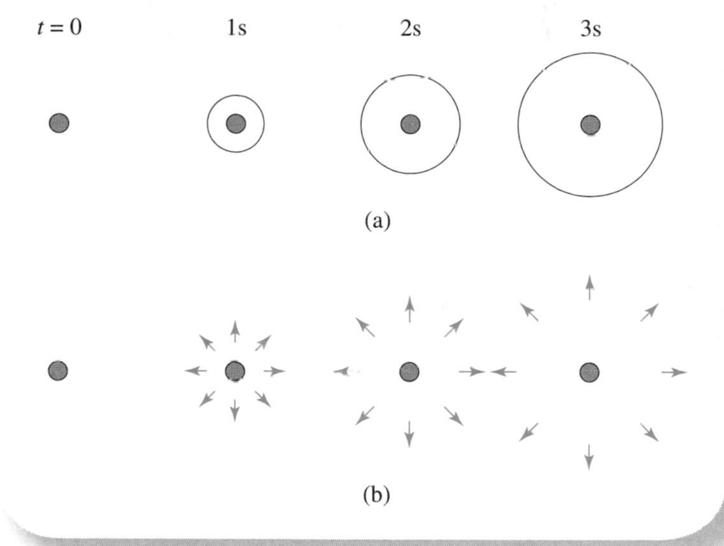

Figure 6.8
(a) A wavefront is used to show how a pulse spreads over water. (b) The same wave pulse, at the same times, represented with wave rays. The rays point in the direction the wave travels and are perpendicular to the wavefront.

Figure 6.9
Wavefronts and rays for a continuous wave on a surface. Far away from the wave source, the wavefronts are nearly straight, and the rays are nearly parallel.

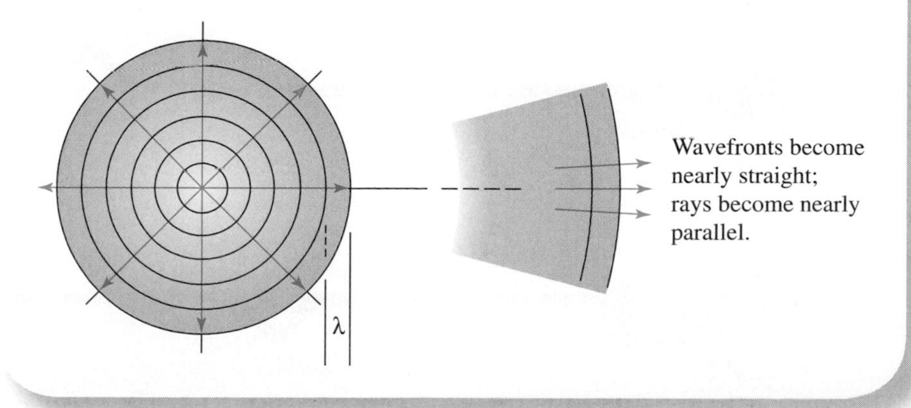

from the source. A certain amount of energy is expended to create a wave pulse or each cycle of a continuous wave. This energy is distributed over the wavefront and determines the amplitude of the wave: The greater the amount of energy given to a wavefront, the larger the amplitude. As the wavefront moves out, it gets larger, so this energy is spread out more and becomes less concentrated. This attenuation accounts for the decrease in loudness of sound as a noisy car moves away from you and for the decrease in brightness of a lightbulb as you move away from it.

One can infer when the amplitude of a wave is changing by noting changes in the wavefronts or the rays. If the wavefronts are growing larger, the amplitude is getting smaller. The same thing is indicated when the rays are diverging (slanting away from each other).

At great distances from the source of a three-dimensional wave, the wavefronts become nearly flat and are called *plane waves*. The corresponding rays are parallel, and the wave's amplitude stays constant. The light and other radiation we receive from the Sun come as plane waves because of the great distance between the Earth and the Sun.

With this background, we will look at several phenomena associated with wave propagation.

Reflection

Think about how many times you looked in a mirror today. That's a very common use we make of the reflection of waves, but it's not the only one. As we will see, the sound we hear inside rooms is affected by reflection, musical instruments like guitars make use of it when producing sound, radar and sonar systems use it for a variety of purposes (like checking how fast we are driving), and so on.

A wave is reflected whenever it reaches a boundary of its medium or encounters an abrupt change in the properties (density, temperature, and so on) of its medium. A wave pulse traveling on a rope is reflected when it reaches a fixed end (see Figure 6.10). It "bounces" off the end and travels back along the rope. Notice that the reflected pulse is inverted. When the end of the rope is attached to a very light (but strong) string instead, the reflected pulse is not inverted. (The incoming pulse causes two pulses to leave the junction, a reflected pulse and a pulse that continues into the light string.) This reflection occurs because of an abrupt change in the density of the medium from high density (for the heavy rope) to low density (for the light string).

Similarly, a wave on a surface or a wave in three dimensions is reflected when it encounters a boundary. The wave that "bounces back" is called the *reflected wave*. Rays are more commonly used to illustrate reflection because they show nicely how the direction of each part of the wave is changed. When a wave is reflected from a straight boundary (for surface waves) or a flat boundary (in three dimensions), the reflected wave appears to be expanding out from a point behind the boundary (Figure 6.11). This point is called the *image* of the original wave source. An echo is a good example: Sound that encounters a large flat surface, such as the face of a cliff, is reflected and sounds like it is coming from a point behind the cliff.

Our most common experience with reflection is that of light from a mirror. The image that you see in a mirror is a collection of reflected light rays originating from the different points on the object you see. (More on this in Section 9.2.)

Reflection from surfaces that are not flat (or straight) can cause interesting things to happen to waves.

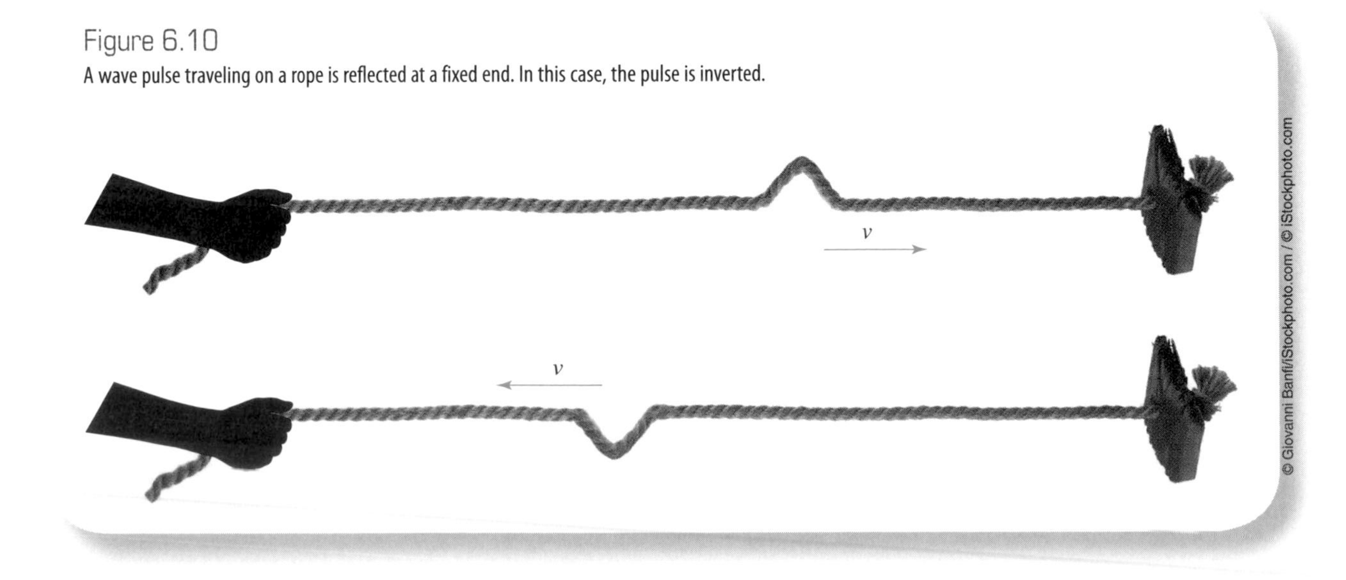

© Giovanni Banfi/iStockphoto.com / © iStockphoto.com

Figure 6.10
A wave pulse traveling on a rope is reflected at a fixed end. In this case, the pulse is inverted.

Figure 6.11
The reflection of water ripples off the side of a pool. Both models of the wave show that the reflected wave appears to diverge from a point behind the wall, called the image.

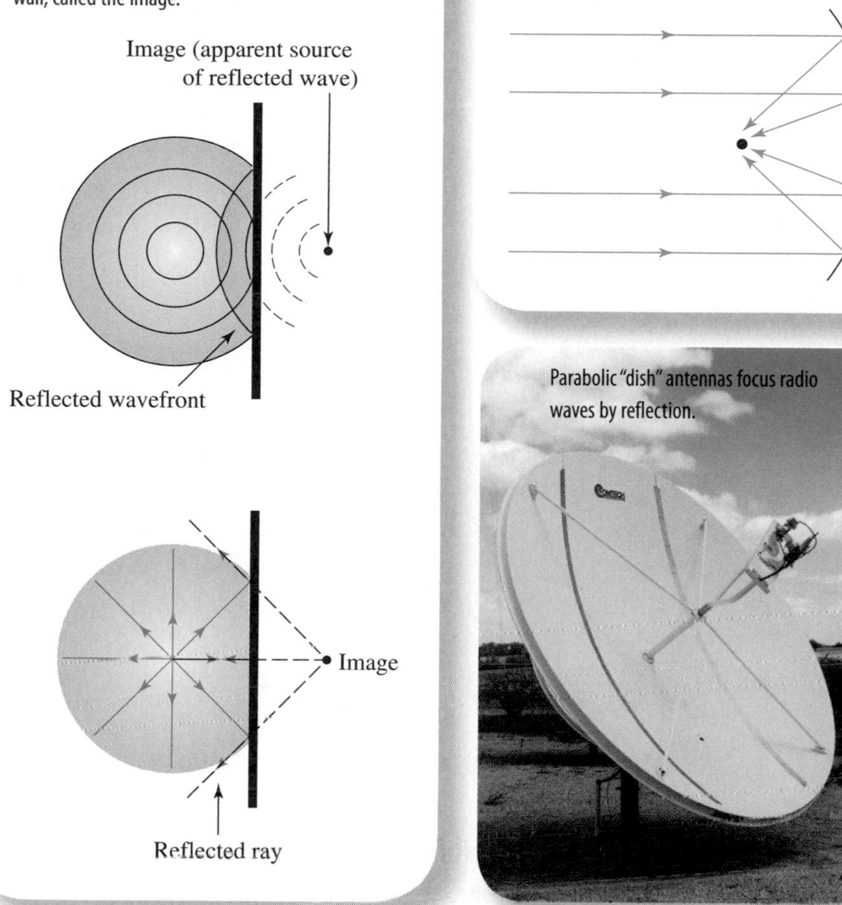

Figure 6.12
A sound wave reflecting off a concave surface. The reflected rays converge toward a point, indicating that the wave's amplitude increases.

Parabolic "dish" antennas focus radio waves by reflection.

© Vern Ostdiek

Figure 6.12 shows a wave being reflected by a curved surface. Note that the rays representing the reflected part of the wave are converging toward each other. This means that the amplitude of the wave is increasing—the wave is being "focused." Parabolic microphones seen on the sidelines of televised football games use this principle to reinforce the sounds made on the playing field. Satellite-receiving dishes do the same with radio waves.

A reflector in the shape of an ellipse has a useful property. An ellipse has two points in its interior called *foci* (plural of focus). If a wave is produced at one focus, it will converge on the other focus after reflecting off the elliptical surface. All rays originating from one focus reflect off the ellipse and pass through the other focus (Figure 6.13). A room shaped like an ellipse is called a *whispering chamber* because a person standing at one focus can hear faint sounds—even whispering—produced at the other focus.

Doppler Effect

Can you recall the last time a fast-moving emergency vehicle or train with its siren or horn blaring passed near you? If so, you may remember that the pitch or tone of its sound dropped suddenly as it went by—although you may be so used to this phenomenon that you didn't take notice of it. This is a manifestation of the Doppler effect—the apparent change in the frequency of wavefronts emitted by a moving source, perhaps a tugboat floating down a river or a train traveling along a track each blowing its whistle. As shown in Figure 6.14, each wavefront expands outward from the point where the source was when it emitted that wavefront. In contrast to what is shown in Figure 6.9 where the source is stationary, ahead of the moving source, the wavefronts are bunched together. This means that the wavelength is *shorter* than when the source is at rest and therefore the frequency of the wave is *higher*. Behind the moving source, the wavefronts are spread apart: The wavelength is *longer*, and the frequency is *lower* than when the source is motionless. In both places, the higher the speed of the wave source, the greater the change in frequency. (Note: The speed of a wave in a medium is constant and is not affected by any motion associated with the wave source. Thus, if the wavelength goes up, the frequency must go down, and vice versa, to yield a constant wave speed: $v = \lambda f$.)

Figure 6.13
The focusing property of an ellipse. Each ray of a wave produced at one focus reflects off the elliptical surface and passes through the other focus.

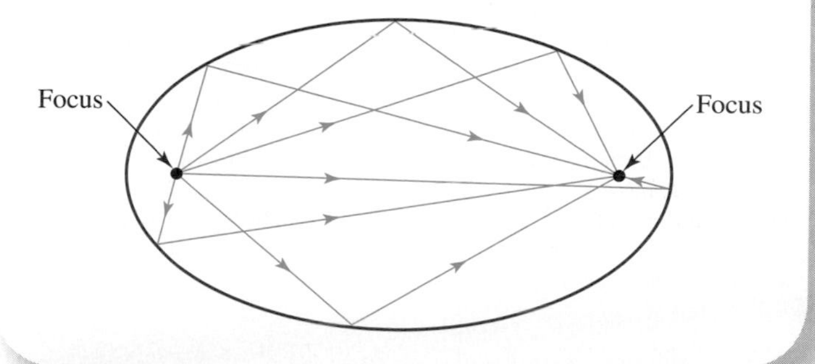

The frequency of sound that reaches a person in front of a moving train is higher than that perceived when the train is not moving. A person behind the moving train hears a lower frequency. As a train or a fast car moves by, you hear the sound shift from a higher frequency (pitch) to a lower frequency. The change in the loudness of the sound, which you also hear, is *not* part of the Doppler effect: It involves a separate process.

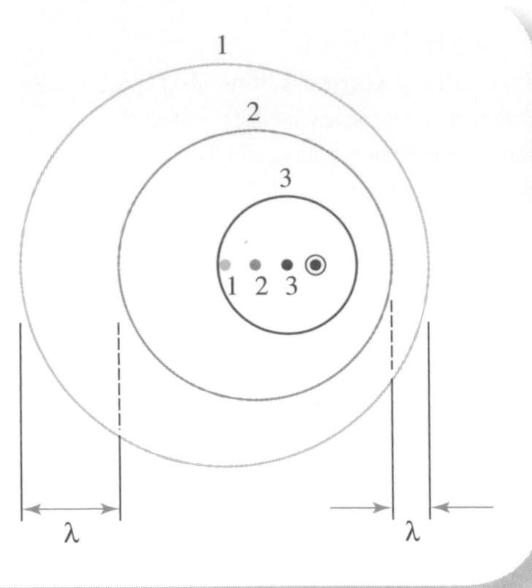

Figure 6.14
The Doppler effect with a moving source. A wave source moves to the right with constant speed. Each dot indicates the source's location when a wavefront was emitted. Wavefront 1 was emitted when the source was at position 1, and similarly for 2 and 3. Ahead of the source (to the right), the wavelength of the wave is decreased. Behind the source (to the left), the wavelength is increased.

A similar shift in frequency of sound occurs if you are moving toward a stationary sound source (see Figure 6.15). This Doppler shift happens because the speed of the wave relative to you is higher than that when you are not moving. The wavefronts approach you with a speed equal to the wave speed plus your speed. Since the wavelength is not affected, the equation $v = f\lambda$ tells us that the frequency of the wave is increased in proportion to the speed of the wave relative to you. By the same reasoning, when one is moving away from the sound source, the frequency is reduced.*

Figure 6.15
The Doppler effect with moving observers. The person in the car on the left hears a higher frequency than the pedestrian. The person in the car on the right hears a lower frequency.

The Doppler effect occurs for both sound and light, and is routinely taken into account by astronomers. The frequencies of light emitted by stars that are moving toward or away from the Earth are shifted. If the speed of the star is known, the original frequencies of the light can be computed. If the frequencies are known instead, the speed of the star can be computed from the amount of the Doppler shift. Such information is essential for determining the motions of stars in our galaxy or of entire galaxies throughout the universe.

Echolocation is the process of using the waves reflected from an object to determine its location. *Radar* and *sonar* are two examples. Basic echolocation uses reflection only: A wave is emitted from a point, reflected by an object of some kind, and detected on its return to the original point. The time between the emission of the wave and the detection of the reflected wave (the round-trip time) depends on the speed of the wave and the distance to the reflecting object. For example, if you shout at a cliff and hear the echo 1 second later, you know that the cliff is approximately 172 meters away. This is because the sound travels a total of 344 meters (172 meters each way) in 1 second (at room temperature). If it takes 2 seconds, the cliff is approximately 344 meters away, and so on (see Figure 6.16).

© José Luis Gutiérrez/iStockphoto.com / © iStockphoto.com

With sonar, a sound pulse is emitted from an underwater speaker, and any reflected sound is detected by an underwater microphone. The time between the transmission of the pulse and the reception of the reflected pulse is used to determine the distance to the reflecting object. Basic radar uses a similar process with microwaves that reflect off aircraft, raindrops, and other things.

Incorporating the Doppler effect in echolocation makes it possible to immediately determine the speed of an approaching or departing object. A moving object causes the reflected wave to be Doppler shifted. If the

* For the musically inclined: To shift the frequency of sound by a musical half-step, the speed of the source of the sound (or the listener, if the source is at rest) must be about 20 m/s (45 mph). For example, if the sound from a car's horn is the note A when the car and the listener are at rest, it will be heard as an A sharp when the car is approaching a listener at 20 m/s and as an A flat when it is moving away at 20 m/s. (The occupants of the car always hear an A.) This value applies when the temperature is around 20°C. When it is colder, the required speed would be smaller. (Why?)

Figure 6.16
Simple echolocation. You can determine the distance to the cliff by timing the echo.

frequency of the reflected wave is *higher* than that of the original wave, the object is moving *toward* the source. If the frequency is *lower*, the object is moving *away.*

Doppler radar uses this combination of echolocation and the Doppler effect. The time between transmission and reception gives the distance to the object, while the amount of frequency shift is used to determine the speed. Law-enforcement officers use Doppler radar to check the speeds of vehicles, and Doppler radar is also used in baseball, tennis, and other sports to clock the speed of a ball. Dust, raindrops, and other particles in air reflect microwaves, making it possible to detect the rapidly swirling air in a tornado with Doppler radar. Another potentially life-saving application is the detection of wind shear—drastic changes in wind speed near storms that have caused low-flying aircraft to crash. (There are a few more examples of the application of this physics at the end of Section 6.3.)

Bow Waves and Shock Waves

In the previous discussion, we have implicitly assumed that the speed of the wave source is much less than the wave speed itself. However, if you've ever heard a sonic boom, or been jostled by the wake of a passing watercraft while floating in the water, you've had experience with circumstances where the reverse is true. Figure 6.17a shows another series of wavefronts produced

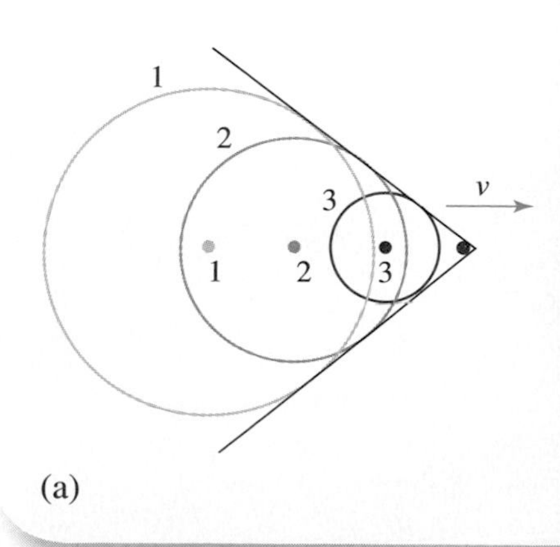

© Louise Ann Noeth/Time & Life Pictures/Getty Images

Figure 6.17
(a) A shock wave is produced when the speed of a wave source exceeds the wave speed. Parts of the wavefronts combine along the two black lines to form a V-shaped wavefront. (b) The jet-powered car traveling faster than sound. Dust can be seen being kicked up by the shock wave extending away from the car on both sides.

by a moving wave source. This time the speed of the wave source is *greater than* the wave speed. The wavefronts "pile up" in the forward direction and form a large-amplitude wave pulse called a **shock wave**. This is what causes the V-shaped bow waves produced by a swimming duck and moving boats.

Aircraft flying faster than the speed of sound produce a similar shock wave. In this case, the three-dimensional wavefronts form a conical shock wave, with the aircraft at the cone's apex. This conical wavefront moves with the aircraft and is heard as a sonic boom (a sound pulse) by persons on the ground.

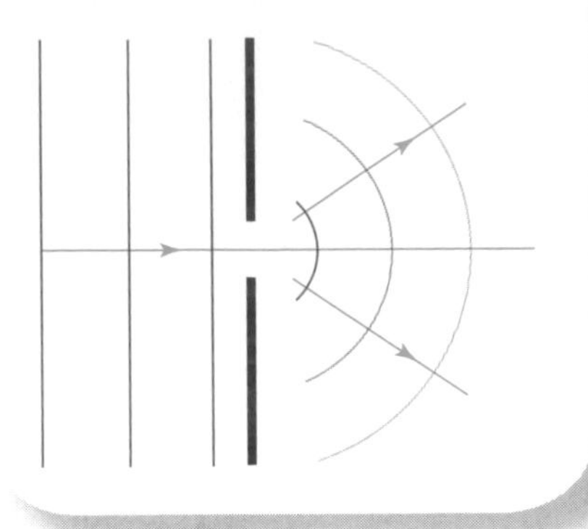

Figure 6.18
The diffraction of a wave as it passes through an opening in a barrier. The wavefronts spread out to the sides after passing through.

Diffraction

Think about walking down a street and passing by an open door or window with sound coming from inside. You can hear the sound even before you get to the opening, as well as after you've passed it. The sound doesn't just go straight out of the opening like a beam, it spreads out to the sides. This is **diffraction**. Figure 6.18 shows wavefronts as they reach a gap in a barrier. These might be sound waves passing through a door or ocean waves encountering a breakwater. The part of the wave that passes through the gap actually sends out wavefronts to the sides as well as ahead. The rays that represent this process show that the wave "bends" around the edges of the opening.

The extent to which the diffracted wave spreads out depends on the ratio of the size of the opening to the wavelength of the wave. When the opening is much larger than the wavelength, there is little diffraction: The wavefronts remain straight and do not spread out to the sides appreciably. This is what happens when light comes in through a window. The wavelength of light is less than a millionth of a meter, and consequently, there is little diffraction. When the wavelength is roughly the same size as the opening, the diffracted wave spreads out much more (see Figure 6.19). The sizes of windows and doors are well within the range of the wavelengths of sound waves, so sound diffracts a great deal after passing through them. Higher frequencies (shorter wavelengths) are not diffracted as much as the lower frequencies.

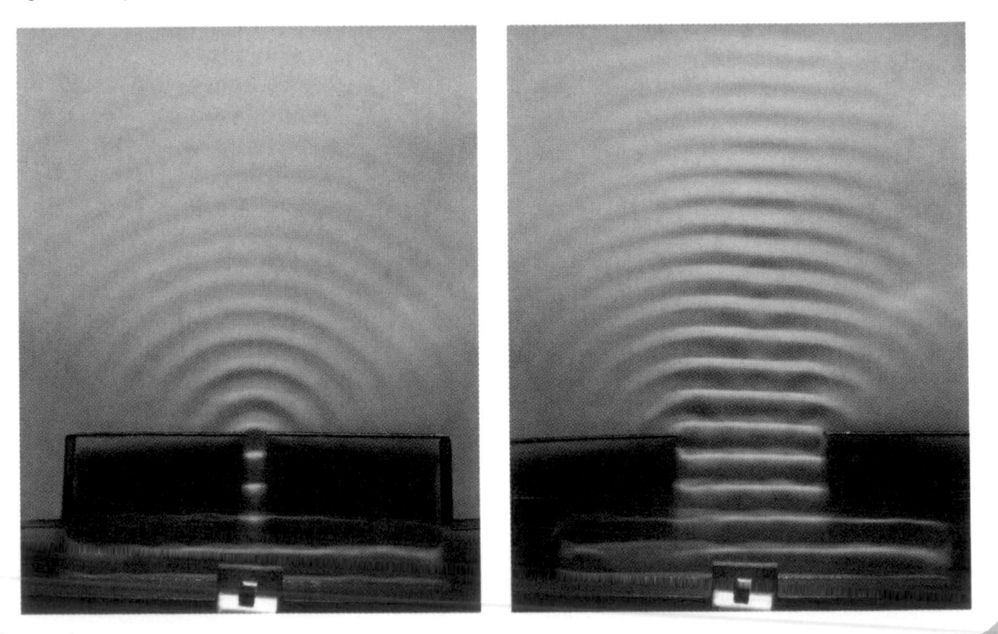

Figure 6.19
Diffraction of water waves passing through a gap in a barrier. When the size of the gap is about equal to the wavelength of the waves, the waves are appreciably diffracted (spread out) beyond the barrier (left). When the size of the opening is much larger than the wavelength, there is little diffraction (right). The waves continue to move straight ahead with little spreading to the right or left beyond the barrier.

© Andrew Lambert Photography/Photo Researchers, Inc.

Interference

Interference arises when two continuous waves, usually with the same amplitude and frequency, arrive at the same place. The sound from a stereo with the same steady tone coming from each speaker is an example of this situation. Another way to cause interference is to direct a continuous wave at a barrier with two openings in it. The two waves that emerge from the two openings will diffract (spread out), overlap each other, and undergo interference.

Consider the case of identical, continuous water waves produced by two small objects made to oscillate up and down in unison on the surface of the water. As these two waves travel outward, each point in the surrounding water moves up and down under the influence

Figure 6.20

Interference pattern of water waves from two nearby sources. The thin lines of calm water indicate destructive interference. Between these lines are regions of large-amplitude waves, caused by constructive interference.

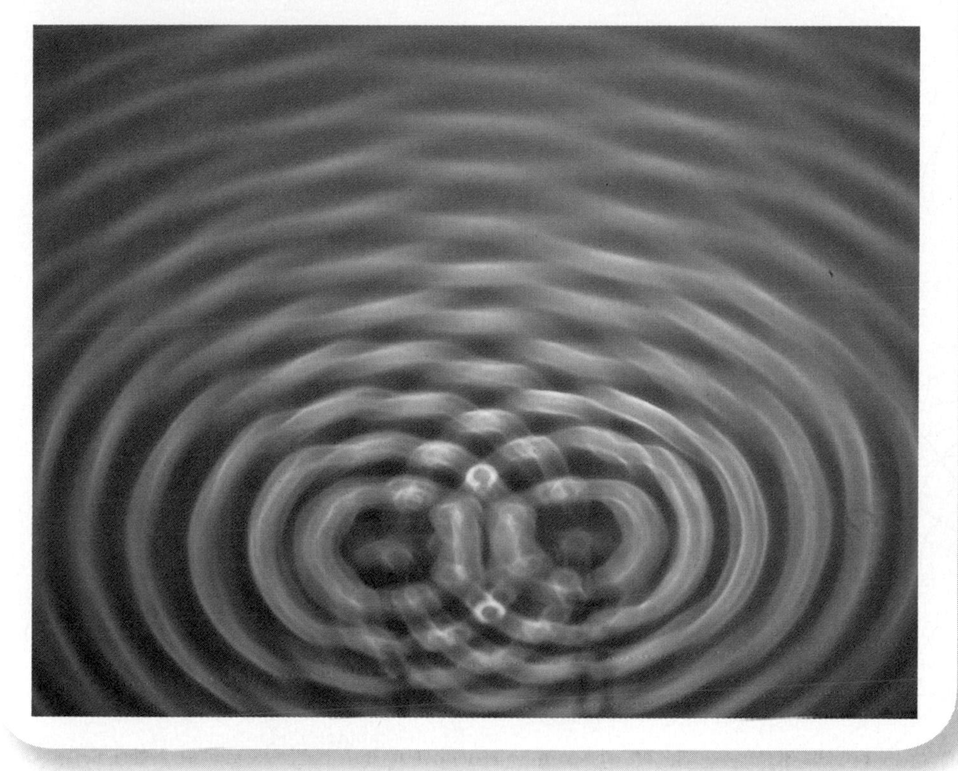

© Richard Megna/Fundamental Photographs

of both waves. If we move around in an arc about the wave sources, we find that at some places the water is moving up and down with a large amplitude. At other places, the water is actually still—it is not oscillating at all (Figure 6.20).

To see why this characteristic pattern of large-amplitude and zero-amplitude motion arises, consider Figure 6.21a—a sketch showing two waves at one moment in time. The thicker lines represent peaks of the waves, and the thinner lines represent the valleys. In Figure 6.21b, the straight lines labeled C indicate the places where the two waves are "in phase"—the peak of one wave matches the peak of the other and valley matches valley. The two waves reinforce each other and the amplitude is large. This is called **constructive interference**. On the straight lines labeled D, the waves are "out of phase"—the peak of one wave matches the valley of the other. The two waves cancel each other. (Whenever one wave has upward displacement, the other has downward displacement, and vice versa. Therefore, the net displacement is always zero.) This is called **destructive interference**. Figure 6.21c shows the same waves a short time later, after they have each traveled one-half of a wavelength. The pattern of constructive and destructive interference is not altered as the waves travel outward. If the photograph shown in Figure 6.20 had been taken earlier or later, it would look the same.

Whether the two waves are in phase or out of phase depends on the relative distances they travel. To reach any point on line C_1 in Figure 6.21b, the two waves travel the same distance and consequently arrive with peak matching peak and valley matching valley. Along the line C_2, the wave from the source on the left must travel a distance equal to one wavelength farther than the wave from the source on the right. The reverse is true along the line on the left labeled C. In general, there is *constructive interference* at all points where one wave travels one, or two, or three . . . wavelengths farther than the other wave.

On the other hand, along the line of destructive interference labeled D_1, the wave from the source on the left has to travel one-half wavelength farther than the wave from the source on the right. They arrive with peak matching valley and cancel each other. Along the far right line labeled "D," the wave from the source on the left has to travel $1\frac{1}{2}$ wavelengths farther, so the two waves again arrive out of phase. The reverse is true for the lines showing destructive interference on the left. In general, there is *destructive interference* at all points where one wave travels $\frac{1}{2}$, or $1\frac{1}{2}$, or $2\frac{1}{2}$, . . . wavelengths farther than the other wave. At places in between constructive and destructive interference, the waves are not completely in phase or out of phase so they partially reinforce or cancel each other.

Sound and other longitudinal waves can undergo interference in the same way. We can imagine Figure 6.21 representing sound waves with the peaks corresponding to compressions and the valleys corresponding to expansions. Along the lines of constructive interference, one would hear a loud, steady sound. Along the lines of destructive interference, one would hear no sound at all. In Chapter 9, we will apply a similar analysis to understand the interference of light waves.

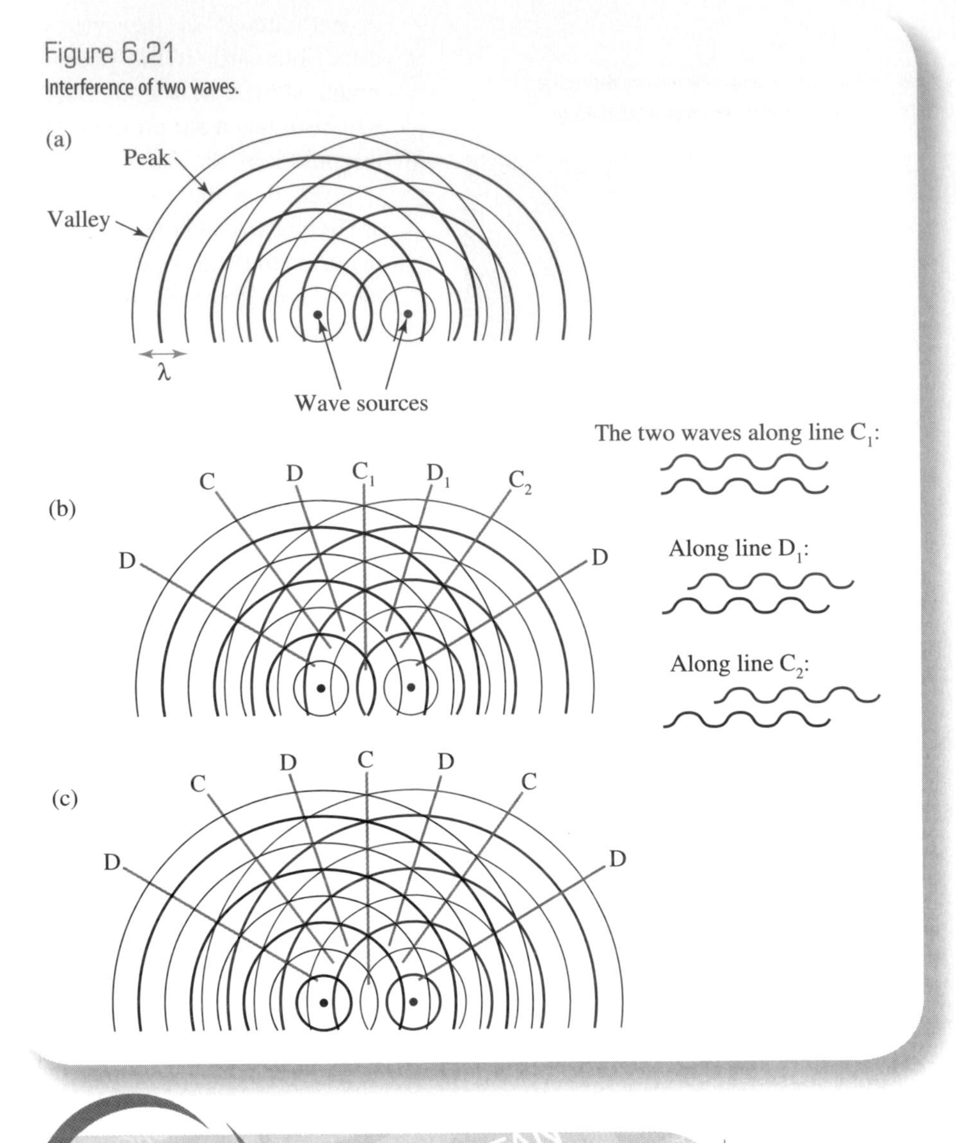

Figure 6.21
Interference of two waves.

Table 6.1 Speed of Sound in Some Common Substances

Substance	Speed* (m/s)	Speed* (mph)
Air		
At −20°C	320	715
At 20°C	344	770
At 40°C	356	795
Carbon dioxide	269	600
Helium	1,006	2,250
Water	1,440	3,220
Human tissue	1,540	3,450
Aluminum	5,100	11,400
Granite	4,000	9,000
Iron and steel	5,200	11,600
Lead	1,200	2,700

*At room temperature (20°C) except as indicated.

6.3 Sound

Our most common experience with sound is in air, but it can travel in any solid, liquid, or gas. For example, when you speak, much of what you hear is sound that travels to your ears through the bones and other tissues in your head. That is why a recording of your voice does not sound the same to you as what you hear when you are talking. Table 6.1 lists the speed of sound in some common substances.

The speed of sound in any substance depends on the masses of its constituent atoms or molecules and on the forces between them. The speed of sound is generally higher in solids than in liquids and gases because the forces between the atoms and molecules in solids are very strong. Sound in gases and liquids is a longitudinal wave, whereas in solids it can be either longitudinal or transverse. In the rest of this chapter, we will concentrate mainly on sound in air.

Sound is produced by anything that is vibrating and causing the air molecules next to it to vibrate. Figure 6.22 shows a representation of a sound wave that was emitted by a vibrating tuning fork. The shading represents the air molecules that we, of course, cannot see. The wave looks very much like a longitudinal wave on a Slinky (Figure 6.5). These compressions and expansions travel at 344 m/s (at room temperature).

The air pressure in each compression is higher than normal atmospheric pressure because the air molecules are squeezed closer together. Similarly, the pressure in each expansion is below atmospheric pressure. Beneath the sketch is a graph of the air pressure along the direction

Figure 6.22

A representation of part of the sound wave emitted by a tuning fork. The air pressure is increased in each compression and reduced in each expansion, as shown in the graph. (From *Physics in Everyday Life* by Richard Dittman and Glenn Schmeig. Used by permission of the McGraw-Hill Companies.)

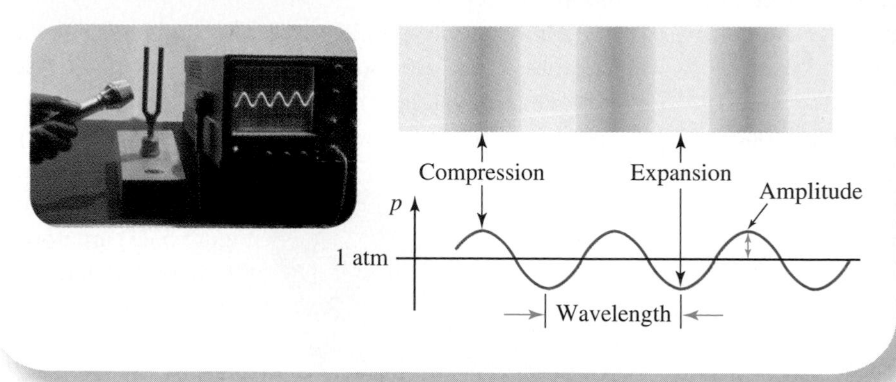

flexible membrane that responds to pressure changes. That is why your ears "pop" when you ride in a fast elevator. The oscillating pressure of a sound wave forces the eardrum to vibrate in and out. A remarkable series of organs converts this oscillation of the eardrum into an electrical signal to the brain that is perceived as sound.

The **waveform** of a sound wave is the graph of the air-pressure fluctuations caused by the sound wave. The easiest way to display the waveform of sound is to connect a microphone to an oscilloscope, an electronic device often seen displaying heartbeats in television hospital shows. (Most personal computers have this capability as well Figure 6.23a.) The oscilloscope shows a graph of the pressure variations detected by the microphone.

We can classify sounds by their waveforms (see Figure 6.23b, c, d). A **pure tone** is a sound with a sinusoidal waveform. A tuning fork produces a pure tone, as does a person carefully whistling a steady note. A **complex tone** is a complex sound wave. The waveform of a complex tone repeats itself but is not sinusoidal.

the wave is traveling. Note that it has the characteristic sinusoidal shape. So a sound wave can be represented by a series of pressure peaks and valleys—a pressure wave. It is more convenient to think of sound as regular fluctuations of air pressure than as vibrations of molecules, although it is both. The amplitude of a sound wave is the maximum pressure change. For a very loud sound, this is only about 0.00002 atmospheres.

It is these pressure variations that our ears detect and convert into the sensation of sound. The eardrum is a

Figure 6.23

Examples of the three types of waveforms. (a) Computer used to display waveforms of sounds. (b) Pure tone: sound from a tuning fork. (c) Complex tone: a spoken "ooo" sound. (d) Noise: sound of air rushing over microphone.

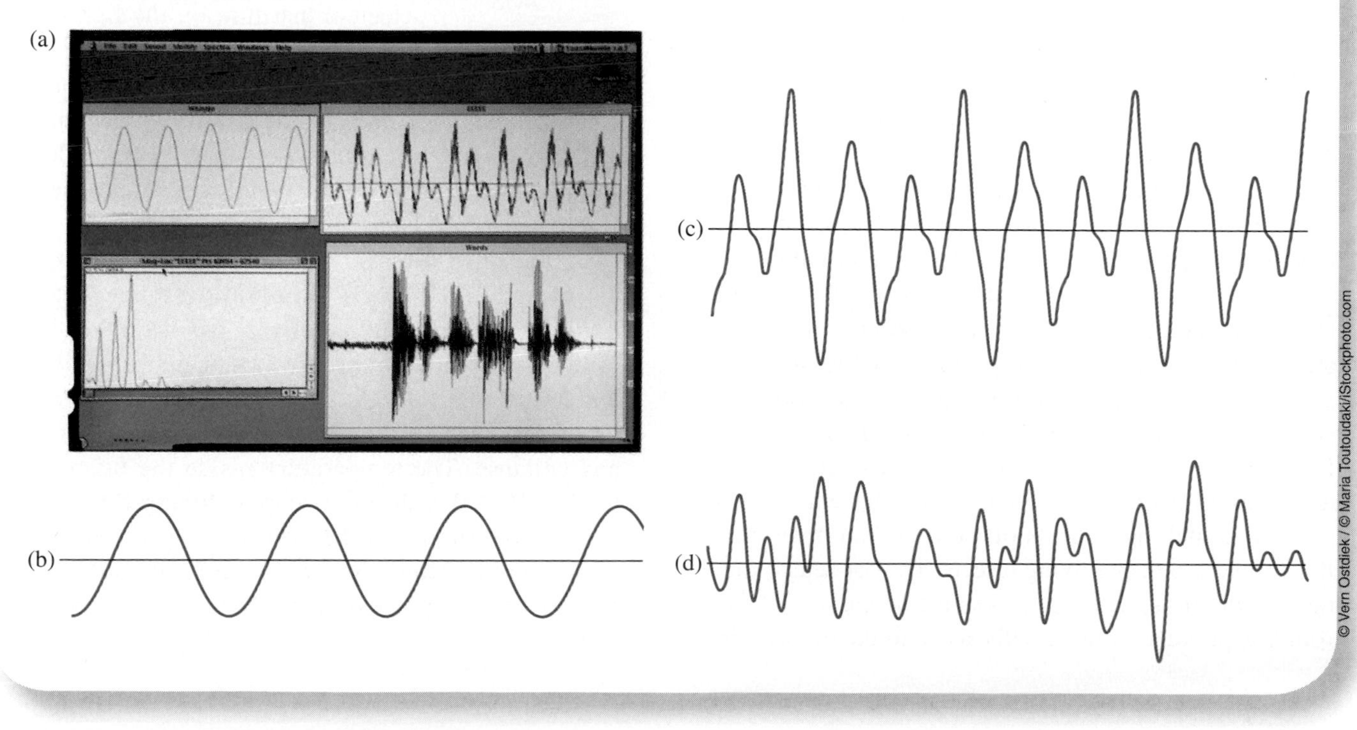

Therefore, any complex tone also has a definite wavelength and a definite frequency. Most steady musical notes are complex tones.

The third type of sound is called **noise**. Noise has a random waveform that does not repeat over and over. For this reason, noise does not have a definite wavelength or frequency. The sound of rushing air is a good example of noise. (In everyday speech, "noise" is often used to describe any unwanted sound, even if it is a pure tone or a complex tone.)

Sound with frequencies outside the range of 20 to 20,000 hertz cannot be heard by people. Inaudible sound with frequency less than 20 hertz is called **infrasound**. High-amplitude infrasound can be felt, rather than heard, as periodic pressure pulses. The hearing ranges of elephants and whales extend into the infrasound region. Sound with frequencies higher than 20,000 hertz (20 kilohertz) is called **ultrasound**. The audible ranges of dogs, cats, moths, mice, and bats extend into ultrasound frequencies; they can hear very high frequency sounds that humans cannot.

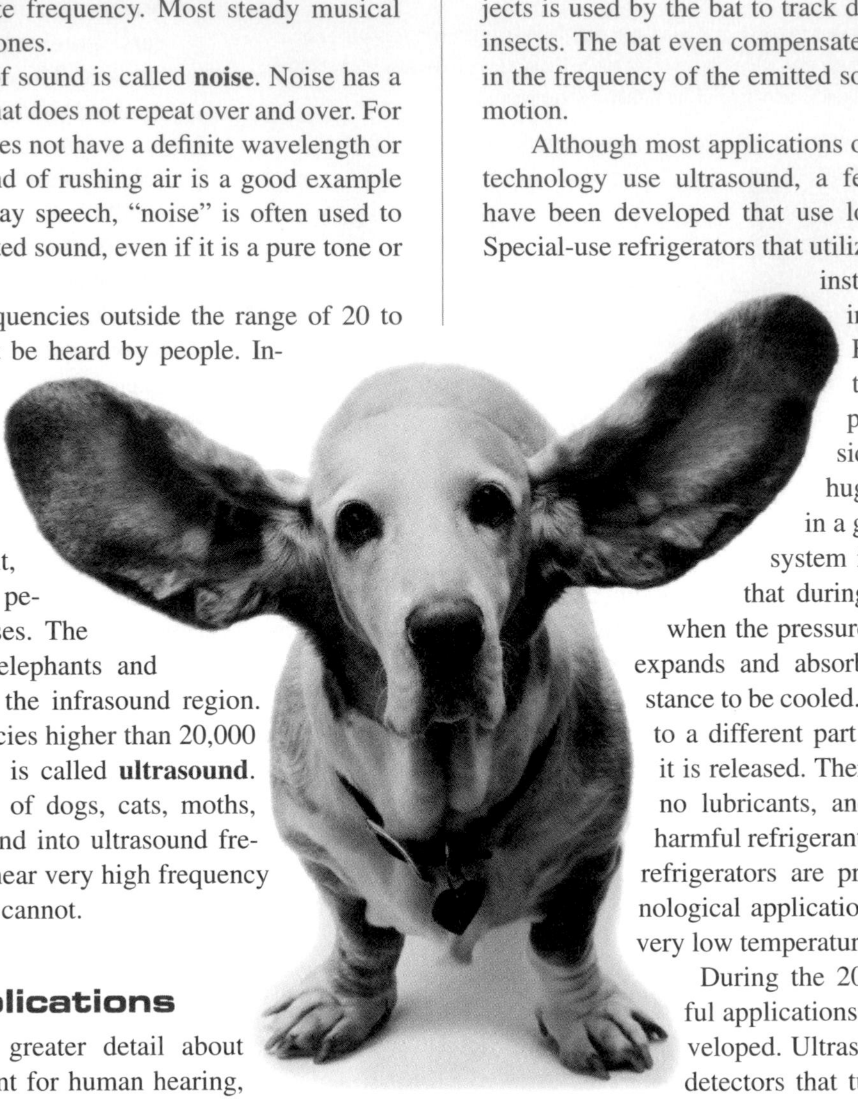

Sound Applications

Before going into greater detail about sounds mainly meant for human hearing, we consider some of the many other uses of sound.

A variety of animals use sound for echolocation—to "see" their surroundings and to find prey. Dolphins and some other marine animals emit clicking sounds that reflect off fish and other objects. By paying attention to how long it takes for reflected sound to return, the direction from which it comes, and how strong the reflected sound is, a dolphin can get a very good idea of the sizes and locations of nearby objects. Bats use very high frequency sound, usually ultrasound, in a highly sophisticated echolocation system that employs the Doppler effect. Each species uses a characteristic range of frequencies, anywhere from 16 to 150 kilohertz. The bat emits a short burst of sound that reflects off surrounding objects like the ground, trees, and flying insects. The bat detects these echoes and uses the time it takes the sound to make the round-trip to determine the distances to the objects. The shift in the frequency of sound reflected off moving objects is used by the bat to track down its dinner—flying insects. The bat even compensates for the Doppler shift in the frequency of the emitted sound caused by its own motion.

Although most applications of sound in science and technology use ultrasound, a few interesting devices have been developed that use lower-frequency sound. Special-use refrigerators that utilize sound waves in a gas instead of a pump circulating a refrigerant (refer to Figure 5.27) are now on the market. A large amplitude sound wave inside a chamber produces huge pressure oscillations in a gas such as helium. The system is tuned in such a way that during the part of the cycle when the pressure is decreasing, the gas expands and absorbs heat from the substance to be cooled. This heat is transferred to a different part of the chamber where it is released. There are no moving parts, no lubricants, and no environmentally-harmful refrigerants. Such thermoacoustic refrigerators are practical for some technological applications requiring cooling to very low temperatures.

© iStockphoto.com

During the 20th century, many useful applications of ultrasound were developed. Ultrasound is used in motion detectors that turn on the lights when a person enters a room or that set off an alarm when an intruder enters an area. It is also used to control rodents and insects; clean jewelry and intricate mechanical and electronic components; weld plastics; sterilize medical instruments; enhance certain chemical reactions; and measure the speed of the wind.

Ultrasound can also be used to produce light. First discovered in the 1930s, **sonoluminescence** (from the Latin words for sound and light) has been the subject of intense research in recent years. A bubble inside water emits flashes of light as pressure oscillations caused by low-frequency ultrasound make the bubble expand and collapse. The temperature inside the bubble rises to over 10,000 K during collapse—hotter than the surface of the Sun—and the light pulse lasts less than a billionth of a second. Recent research suggests that the light-producing process is similar to that occurring inside x-ray tubes.

Guitarists make the strings vibrate by plucking them. Most of the sound that we hear comes from the front plate of the guitar, which is made to vibrate by the strings.

© Judson Lane/iStockphoto.com

Ultrasound has several uses in medicine. It is routinely used to form images of internal organs and fetuses. High-frequency ultrasound, typically 3.5 million hertz, is sent into the body and is partially reflected as it encounters different types of tissue. These reflections are analyzed and used to form an image on a television monitor. Some sophisticated ultrasonic scanning devices also use the Doppler effect. The beating heart of a fetus and the flow of blood in arteries can be monitored by detecting the frequency shift of the reflected ultrasound.

Recently developed *acoustic surgery* uses ultrasound for tasks such as destroying tumors. Focused, high-intensity sound causes heating that destroys tissue. The precision of such an "acoustic scalpel" can exceed that of a conventional knife.

Another use of ultrasound in medicine is *ultrasonic lithotripsy*, a procedure that breaks up kidney stones that have migrated to the bladder. A large-amplitude 27,000-hertz sound wave travels through a steel tube inserted into the body and placed in contact with the stone. The ultrasound breaks the stone into small pieces, somewhat like a singer breaking a wineglass. A procedure similar to ultrasonic lithotripsy has recently been developed to break up blood clots.

6.4 Production of Sound

In the remaining sections of this chapter, we will take a brief look at the three P's of acoustics: the production, propagation, and perception of sound.

The sounds that we hear range from simple pure tones like a steady whistle to complicated and random waveforms like those found on a noisy street corner. Most of the sound we hear is a combination of many sounds from different sources. The loudness usually fluctuates, as do the frequencies of the component sounds.

Sound is produced when vibration causes pressure variations in the air. Any flat plate, bar, or membrane that vibrates produces sound. The tuning fork shown in Figure 6.22 is a nice example. A dropped garbage-can lid, a vibrating speaker cone, and a struck drumhead produce sound the same way. The tuning fork executes simple harmonic motion and produces a pure tone. The garbage-can lid and the drumhead have more complicated motions and so produce complex tones or noise.

The various musical instruments represent some of the basic types of sound producers. Drums, triangles, xylophones, and other percussion instruments produce sound by direct vibration. Each is made to vibrate by a blow from a mallet or drumstick.

Guitars, violins, and pianos use vibrating strings to produce sound. By itself, a vibrating string produces only faint sound because it is too thin to compress and expand the air around it effectively. These instruments employ "soundboards" to increase the sound production. (The electric versions of guitars and violins pick up and amplify the string vibrations electronically.) One end of the string is attached to a wooden soundboard, which is made to vibrate by the string. The vibrating soundboard, in turn, produces the sound. Figure 6.24 shows a simplified diagram of the process used in pianos. When a note is played, a hammer strikes the piano wire and produces a wave pulse. The pulse travels back and forth on the wire, being reflected each time at the ends. The soundboard receives a "kick" each time the pulse is reflected at that

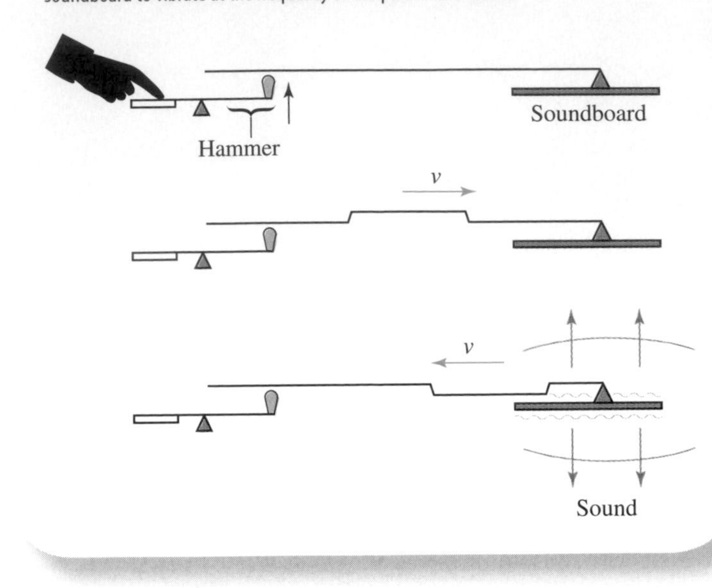

Figure 6.24
Sound production in a piano. The hammer creates a wave pulse that oscillates back and forth on the piano wire. (The amplitude is not to scale.) The pulse causes the soundboard to vibrate at the frequency of the pulse's oscillation.

end. This makes the soundboard vibrate at a frequency equal to the frequency of the pulse's back-and-forth motion. The sound dies out because the pulse loses energy to the soundboard during each cycle.

The strings on guitars are plucked instead of struck, giving the pulses a different shape. This is part of the reason why the sound of a guitar is different from that of a piano. Violin strings are bowed, resulting in even more complicated wave pulses. In all three instruments, the frequency of the pulse's motion depends on the speed of waves on the string and on the length of the string. When a string is tuned by being tightened, the wave speed is increased. The pulse moves faster on the string and makes more "round-trips" each second—the frequency of the sound is raised. Different notes are played on the same guitar or violin string by using a finger to hold down the string some distance from its fixed end. The pulse travels a shorter distance between reflections, makes more round-trips per second, and produces a higher-frequency sound.

A similar process is used in flutes, trumpets, and other wind instruments. Here it is a pressure pulse in the air that moves back and forth inside a tube (Figure 6.25). Initially, a sound pulse is produced at one end of the tube by the musician. This pulse travels down the tube and is partially reflected and partially transmitted at the other end. The transmitted part spreads out into the air, becoming the sound that we hear. The reflected part returns to the mouthpiece end, where it is reflected again and reinforced by the musician. (The sound pulse is also inverted at each end: It goes from a compression to an expansion, and vice versa.) The musician must supply pressure pulses at the same frequency that the pulse oscillates back and forth in the tube. This, of course, is the frequency of the sound that is produced.

Different notes are played by changing the length of the tube—by opening side holes in woodwinds and by using valves or slides in brasses. The speed of the pulses is determined by the temperature of the air. This is one reason why musicians "warm up" before a performance. The air inside the instrument is warmed by the musicians' breath and hands. Hence the frequencies of the notes are higher than when the air inside is cool.

The human voice uses several types of sound production and modification mechanisms. Some consonant sounds like *"sss"* and *"fff"* are technically noise: They are hissing sounds produced by air rushing over the teeth and lips. The randomly swirling air produces sounds with random, changing frequencies. The vocal cords, located inside the Adam's apple in the throat, are the primary sound producers for singing and for spoken vowel sounds. When air is blown through the vocal cords, they vibrate and produce pressure pulses (sound) much like the reed of a saxophone. This sound is modified by the shapes of the air cavities in the throat, mouth, and nasal region. Muscles in the throat are used to tighten and to loosen the vocal cords, thereby changing the pitch of the sound. Moving the tongue or jaw changes the shape of the mouth's air cavity and allows for different sounds to be produced. A sinus cold can change the sound of one's

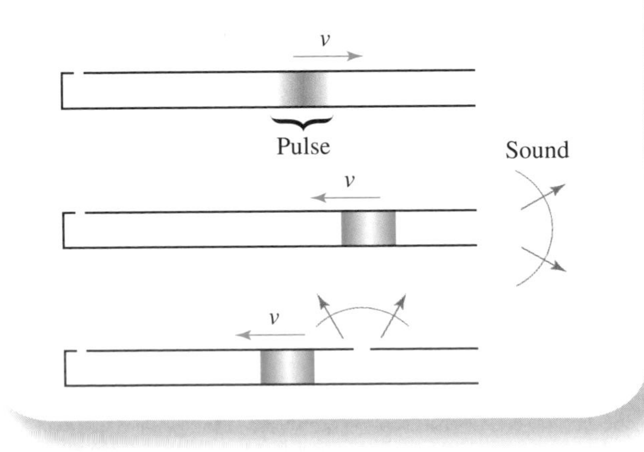

Figure 6.25
Sound production in a flute. The musician causes a sound (pressure) pulse to oscillate back and forth in the tube. Each time the pulse reaches the right end, it sends out a compression—part of the sound that we hear. Opening a side hole shortens the path of the pulse, thereby increasing the frequency.

voice because swelling changes the configuration of the nasal cavity.

Perhaps you've heard someone speak who had inhaled helium. (This is *not* a recommended exercise. It is possible to suffocate because of lack of oxygen in the lungs.) The speed of sound in helium is nearly three times that in air (refer to Table 6.1). This raises the frequencies of the sounds and gives the speaker a falsetto voice.

Sound waves carry energy, as do all waves. This means that the source of the sound must supply energy. Speaking loudly or playing an instrument for extended periods of time can tire you out for this reason. For continuous sounds, it is more relevant to consider the power of the source, since the energy must be supplied continuously. Most instruments, including the human voice, are very inefficient; typically, only a small percentage of the energy output of the performer is converted into sound energy.

6.5 Propagation of Sound

Once a sound has been produced, what factors affect the sound as it travels to our ears? The general aspects of wave propagation discussed in Section 6.2 of course apply to sound waves. Of these, reflection, diffraction, and the reduction of amplitude with distance from the sound source are most important in influencing the sound that actually reaches us.

The simplest situation is a single source of sound in an open space—such as a person talking in an empty field. The sound travels in three dimensions, and its amplitude decreases as the wavefronts expand. In particular, the amplitude is inversely proportional to the distance from the sound source.

$$\text{amplitude} \propto \frac{1}{d}$$

When you move to twice as far away from a steady sound source, the amplitude of the sound is decreased by one-half. The sound becomes quieter as you move away from the sound source.

Sound propagation is more complicated inside rooms and other enclosures. First, diffraction and reflection of sound allow you to hear sound from sources that you can't see because they are around a corner. We are so accustomed to this phenomenon that it doesn't seem mysterious. Second, even when the source is inside the room with you, most of the sound that you hear has been reflected one or more times off the walls, ceiling, floor, and any objects in the room. This has a large effect on the sound that you hear.

Figure 6.26 shows that sound emitted by a source in a room can reach your ears in countless ways. Consider a single sound pulse like a hand clap. In an open field, you would hear only a single, momentary sound as the pulse moves by you. A similar pulse produced in a room is heard repeatedly: You hear the sound that travels directly to your ears; then sound that reflects off the ceiling, floor, or a wall before reaching your ears; then sound that is reflected twice, three times, and so on. These reflected sound waves travel greater and greater distances before reaching your ears and are heard successively later than the direct sound. This process of repeated reflections of sound in an enclosure is called **reverberation**. The single hand clap is heard as a continuous sound that fades quickly.

Figure 6.27 compares the sound from a hand clap as it is heard in an open field and in a room. Each graph

Figure 6.26

Some possible pathways for the sound in a room to travel from the source to your ears; *D* represents the sound that reaches your ears directly. Rays 1 and 2 are reflected once. Ray 3 is reflected twice. In most rooms, the sound can be reflected more than a dozen times and still be heard.

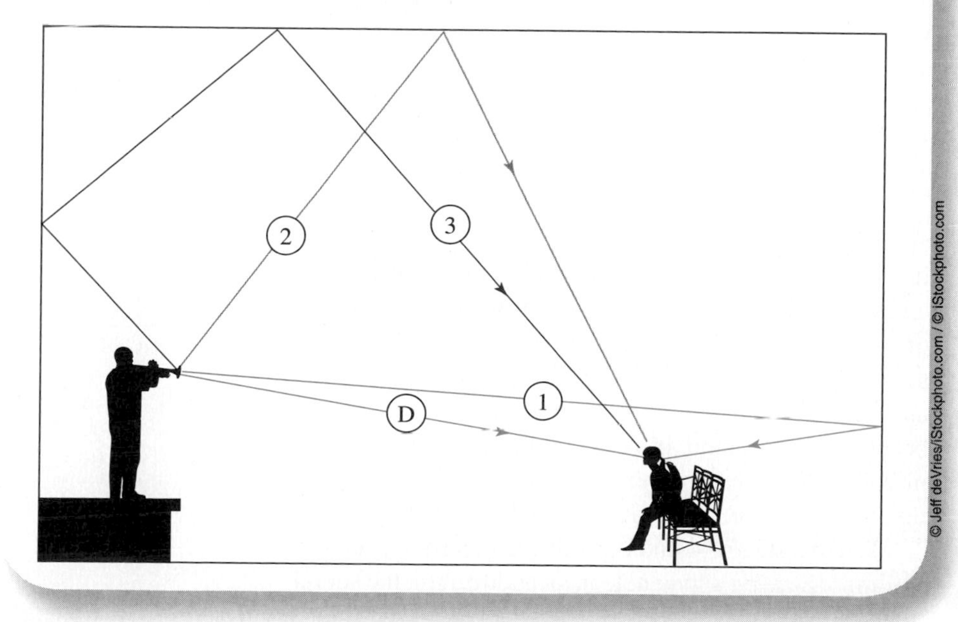

© Jeff deVries/iStockphoto.com / © iStockphoto.com

Figure 6.27

Comparison of the sound heard from a hand clap in (a) an open field and in (b) a room. A single pulse is heard in the field, whereas a continuous, fading sound is heard in the room.

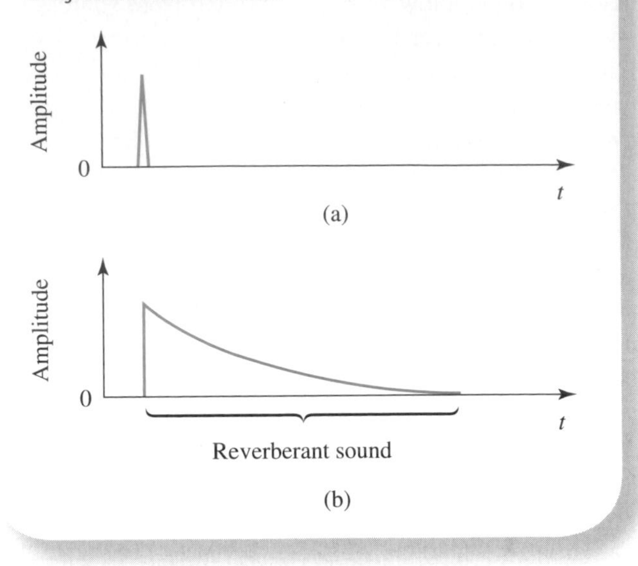

shows the amplitude of the sound that is heard versus the time after the sound pulse is produced. The reverberation causes the sound to "linger" in the room. The indirect sound that one hears after the initial direct pulse is called the *reverberant sound.* The amount of time it takes for the reverberant sound to fade out depends on the size of the room and the materials that cover the walls, ceiling, and floor. Sound is never completely reflected by a surface: Some percentage of the energy in an incoming wave is absorbed by the surface, leaving the reflected wave with a reduced amplitude. (Concrete absorbs only about 2% of the incident sound's energy, whereas carpeting and acoustical ceiling tile can absorb around 90%.) A room with a large amount of sound-absorbing materials in it will have little reverberation. After a few reflections, the sound loses most of its energy and cannot be heard.

The *reverberation time* is used to compare the amount of reverberation in different rooms. It is the time it takes for the amplitude of the reverberant sound to decrease by a factor of 1,000. It varies from a small fraction of a second for small rooms with high sound absorption to several seconds for large, brick-walled gymnasiums and similar enclosures. The Taj Mahal, a breathtakingly beautiful mausoleum in Agra, India, is made of solid marble, which absorbs very little sound. The reverberation time of its central dome is over 10 seconds.

When a steady sound is produced in a room, such as a trumpet playing a long note in an auditorium, the sound that one hears is affected by reverberation in a number of ways:

1. The sound is louder than it would be if you had heard it at the same distance in an open field.
2. The sound "surrounds" you; it comes from all directions, not just straight from the source (Figure 6.26).
3. Beyond a short distance from the sound source, the loudness does not decrease as rapidly with distance as it would in an open field. That is why one can often hear as well near the back of an auditorium as in the middle.
4. Not only does the sound fade gradually when the source stops, the sound also "builds" when the source starts.

Moderate reverberation has an overall positive effect on the sound that we hear, particularly music. However, excessive reverberation adversely affects the clarity of both speech and music. Speech and music are a series of short, steady sounds interspersed with short moments of silence. Each note, word, or syllable is followed by a brief pause. If we again graph the amplitude of sound versus time, we can see the effect of reverberation (see Figure 6.28). In an open field, one hears each syllable or note as a distinct, separate sound. In a room, the individual sounds begin to merge. As a new note is played or a new word is spoken, the reverberant sound from the preceding

Figure 6.28

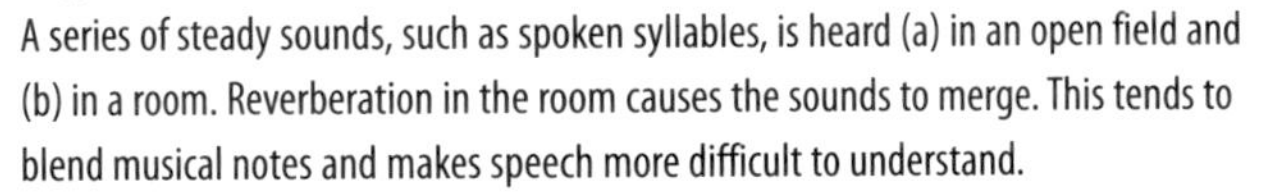

A series of steady sounds, such as spoken syllables, is heard (a) in an open field and (b) in a room. Reverberation in the room causes the sounds to merge. This tends to blend musical notes and makes speech more difficult to understand.

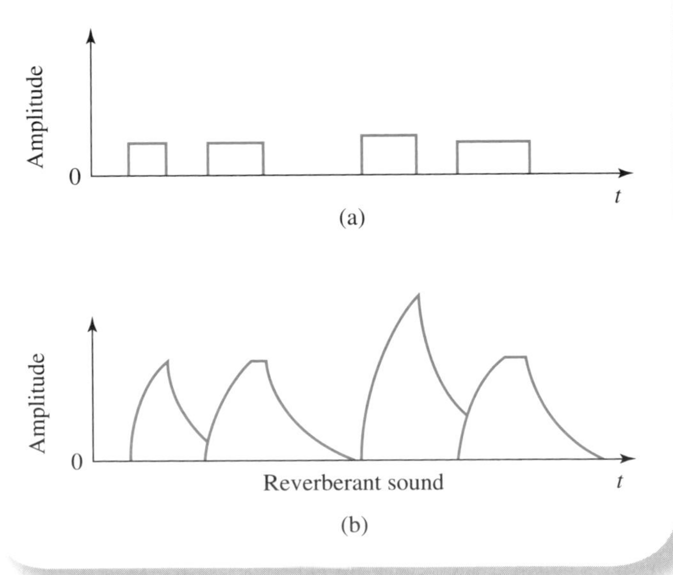

one can still be heard. The longer the reverberation time, the more the sounds overlap each other and the harder it is to understand speech. Racquetball courts have hard, smooth walls and very high reverberation times; that is why it is very difficult for players to converse unless they are close to each other. It is recommended that the reverberation time of rooms used for oral presentations and lectures should be around 0.5–1 second. For concert halls, it should be from 1 to 3 seconds, depending on the type of music being performed.

Many other factors besides reverberation time must be taken into account by architects and building designers. For example, a balcony must be high above the main floor and not extend out too far, or little reverberant sound will reach the seats below it. Also, sound is focused by concave walls and ceilings. Building an elliptical auditorium could result in the sound being concentrated in a small area of the room (Figure 6.13).

6.6 Perception of Sound

In this section, we consider some aspects of sound perception—how the physical properties of sound waves are related to the mental impressions we have when we hear sound. We will be comparing psychological sensations, which can be quite subjective, to measurable physical quantities. (A similar situation: "hot" and "cold" are subjective perceptions that are related to temperature, which is a measurable physical quantity.) To make things simple, we will limit ourselves to steady, continuous sounds. This frees us from having to include such effects as reverberation in a room—that is, how the sound builds up, how it decays, and so on.

The main categories that we use to describe sounds subjectively are **pitch**, **loudness**, and **tone quality**.

The *pitch* of a sound is the perception of highness or lowness. The sound of a soprano voice has a high pitch and that of a bass voice has a low pitch. The pitch of a sound depends primarily on the frequency of the sound wave.

The *loudness* of a sound is self-descriptive. We can distinguish among very quiet sounds (difficult to hear), very loud sounds (painful to the ears), and sounds with loudness somewhere in between. The loudness of a sound depends primarily on the amplitude of the sound wave.

The *tone quality* of a sound is used to distinguish two different sounds even though they have the same pitch and loudness. A note played on a violin does not sound quite like the same note played on a flute. The tone quality of a sound (also referred to as the *timbre* or *tone color*) depends primarily on the waveform of the sound wave.

Pitch

Pitch is perhaps the most accurately discriminated of the three categories, particularly by trained musicians. It depends almost completely on the frequency of the sound wave: The higher the frequency, the higher the pitch. Noise does not have a definite pitch, because it does not have a definite frequency.

Pitch is essential to nearly all music. There is a great deal of arithmetic in the musical scale; each note has a particular numerical frequency. Figure 6.29 shows the frequencies of the notes on the piano keyboard. There are seven octaves on the piano, each consisting of 12 different notes. Within each octave, the notes are designated A through G, plus five sharps and flats. Each note in a given octave has exactly twice the frequency of the corresponding note in the octave below. For example, the frequency of the lowest note on the piano, an A, is 27.5 hertz; for the A in the next octave, it is 55 hertz; it is 110 hertz for the third A, and so on. The frequency of middle C is 261.6 hertz.*

Certain combinations of notes are pleasing to the ear, whereas others are not. Nearly 2,500 years ago, Pythagoras indirectly discovered that two different notes are in harmony when their frequencies have a simple whole-number ratio. For example, a musical fifth is any pair of notes whose frequencies are in the ratio 3 to 2. Any E and the first A below it have this ratio of frequencies, as do any G and the first C below it.

© José Luis Gutiérrez/iStockphoto.com / © iStockphoto.com

Figure 6.29 also shows the approximate ranges of

* For the mathematically inclined: The frequencies of the notes in the most commonly used tuning scheme are based on the twelfth root of 2 ($\sqrt[12]{2} = 1.05946$). The frequency of each note equals the frequency of the note a half-step lower multiplied by $\sqrt[12]{2}$. (Put another way, the frequency of each note is about 5.9% higher than that of the next lower one.) Going up the scale, 12 half-steps yield a note whose frequency is $(\sqrt[12]{2})^{12} = 2$ times that of the initial note.

singing voices and some instruments. For normal speech, the ranges are approximately 70 to 200 hertz for men and 140 to 400 hertz for women. When whispering, you do not use your vocal cords, and you produce much higher-frequency "hissing" sounds.

Loudness

The loudness of a sound is determined mainly by the amplitude of the sound wave. The greater the amplitude of the sound wave that reaches your eardrums, the greater the perceived loudness of the sound. The actual pressure amplitudes of normal sounds are extremely small, typically around one-millionth of 1 atmosphere. This causes the eardrum to vibrate through a distance of around 100 times the diameter of a single atom. An extremely faint sound has an amplitude of less than one-billionth of 1 atmosphere, and it makes the eardrum move *less* than the diameter of an atom. The ear is an amazingly sensitive device.

There is a specially defined physical quantity that depends on the amplitude of sound but is more convenient for relating amplitude to perceived loudness. This is the sound pressure level or simply the **sound level**.

Figure 6.29

The frequencies of the notes on a piano are indicated and compared with the approximate frequency ranges of different singing voices and musical instruments. This frequency scale, and, therefore, the piano keyboard, is a logarithmic scale: The interval between 50 and 100 hertz is the same as the interval between 500 and 1,000 hertz. (From *Physics in Everyday Life* by Richard Dittman and Glenn Schmieg. Used by permission of the McGraw-Hill Companies.)

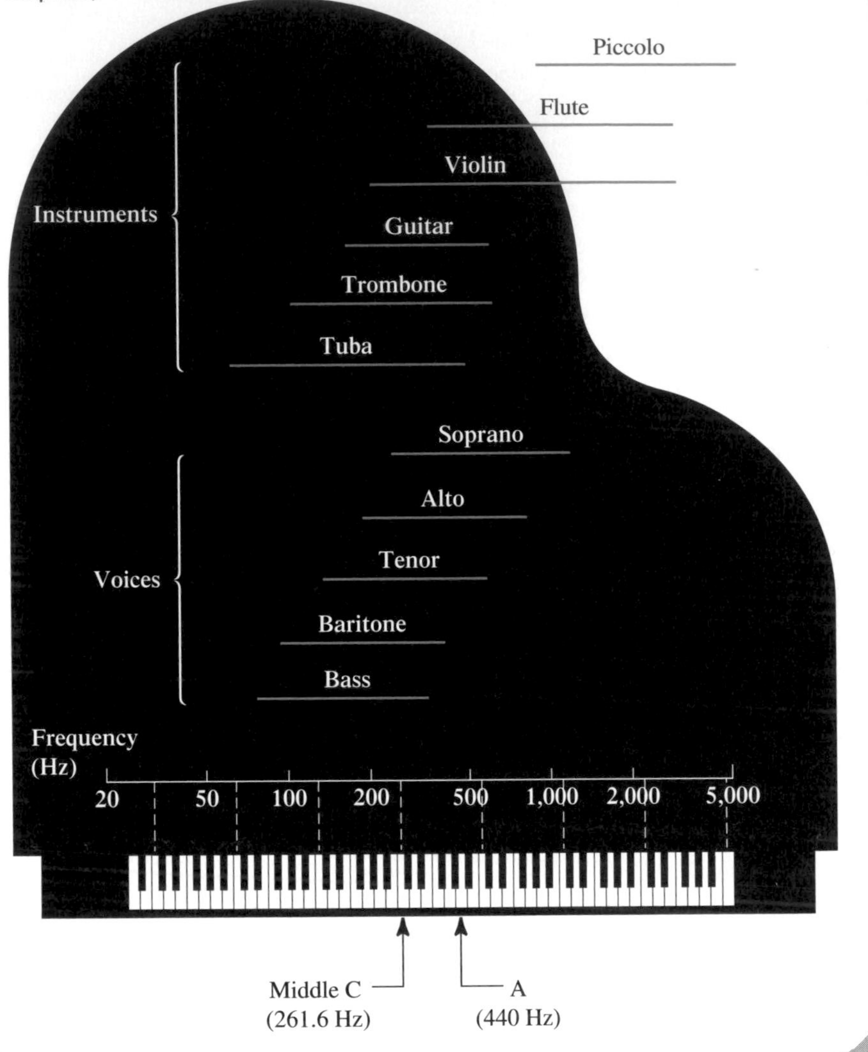

© Joachim Angeltun/ iStockphoto.com

The standard unit of sound level is the decibel (dB). The range of sounds that we are normally exposed to has sound levels from 0 decibels to about 120 decibels. Figure 6.30 shows some representative sound levels along with their relative perceived loudness. Sound level does not take into account the irritation of the sound. Your favorite music played at 100 decibels may not sound as loud as an annoying screech of fingernails on a blackboard with a sound level of 80 decibels.

© Jan Paul Schrage/iStockphoto.com

The relationship between the amplitude of a steady sound and its sound level is based on factors of 10. A sound with 10 times the amplitude of another sound has a sound level that is 20 decibels higher. A 90-decibel sound has 10 times the amplitude of a 70-decibel sound.

The following five statements describe how the perceived loudness of a sound is related to the measured sound level. These are general trends that have been identified by researchers after testing large numbers of people. For a particular person, the actual numerical values of the sound levels can vary somewhat from those listed. Also some of the values given, particularly in numbers 1 and 3 below, depend on the frequencies of the sounds.

Figure 6.30

The decibel scale with representative sounds that produce approximately each sound level. Sound levels above about 85 decibels pose a danger to hearing. Levels above 120 decibels cause ear pain and the potential for permanent hearing loss.

Jet takeoff (60 m)		120 dB	
Construction site		110 dB	*Intolerable*
Shout (1.5 m)		100 dB	
Heavy truck (15 m)		90 dB	*Very loud*
Urban street		80 dB	
Automobile interior		70 dB	*Noisy*
Normal conversation (1 m)		60 dB	
Office, classroom		50 dB	*Moderate*
Living room		40 dB	
Bedroom at night		30 dB	*Quiet*
Broadcast studio		20 dB	
Rustling leaves		10 dB	*Barely audible*
		0 dB	

1. The sound level of the quietest sound that can be heard under ideal conditions is 0 decibels. This is called the *threshold of hearing*.

2. A sound level of 120 decibels is called the *threshold of pain*. Sound levels this high cause pain in the ears and can result in immediate damage to them.

3. The minimum increase in sound level that makes a sound noticeably louder is approximately 1 decibel. For example, if a 67-decibel sound is heard and after that a 68-decibel sound, we can just perceive that the second sound is louder. If the second sound had a sound level of 67.4 decibels, we could not notice a difference in loudness.

© Cat London/iStockphoto.com

4. A sound is judged to be twice as loud as another if its sound level is about 10 decibels higher. A 44-decibel sound is about twice as loud as a 34-decibel sound. A 110-decibel sound is about twice as loud as a 100-decibel sound. This is a cumulative factor: A 110-decibel sound is about four times as loud as a 90-decibel sound, and so on.

5. If two sounds with equal sound levels are combined, the resulting sound level is about 3 decibels higher. If one lawn mower causes an 80-decibel sound level at a certain point nearby, starting up a second identical lawn mower next to the first will raise the sound level to about 83 decibels. It turns out that 10 similar sound sources are perceived to be twice as loud as a single source.

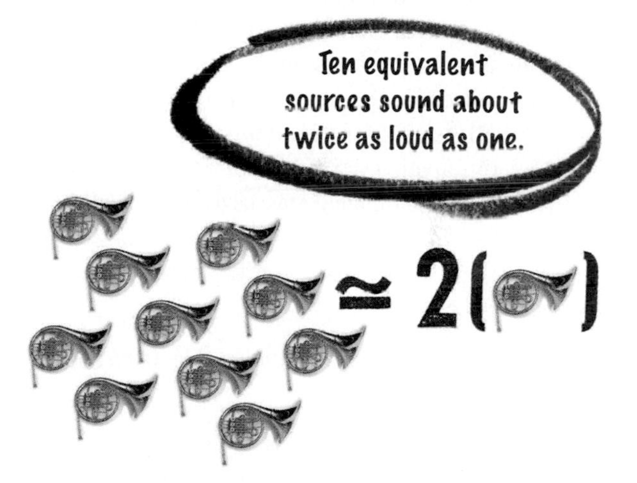

© Vladislav Lebedinski/iStockphoto.com

The loudness of pure tones and, to a lesser degree, of complex tones, also depends on the frequency. This is because the ear is inherently less sensitive to low- and high-frequency sounds. The ear is most sensitive to sounds in the frequency range of 1,000 to 5,000 hertz. For example, a 50-hertz pure tone at 78 decibels, a 1,000-hertz pure tone at 60 decibels, and a 10,000-hertz tone at 72 decibels all sound equally loud. (At very high sound levels, 80 decibels and above, the ear's sensitivity does not vary as much with frequency as it does at lower sound levels.) One reason for this variation in sensitivity is that a considerable amount of low-frequency sound is produced inside our bodies by flowing blood and flexing muscles.

The ear is less sensitive to low-frequency sounds so these internal sounds do not "drown out" the external sounds that we need to hear.

Sound levels are measured with sound-level meters. Most sound-level meters are equipped with a special weighting circuit (called the A scale) that allows them to respond to sound much as the ear does. The response to low and high frequencies is diminished. The readings on the A scale are designated dBA. When operating in the normal mode (called the C scale) a sound-level meter measures the sound level in decibels, treating all frequencies equally. When in the A scale mode, a sound-level meter responds like the human ear and therefore indicates the relative loudness of the sound.

© Vern Ostdiek / © Maria Toutoudaki/iStockphoto.com

Sound-level Meter

Loud sounds can not only damage your hearing, they can also affect the physiological and psychological balance of your body. Since the beginning of humankind, the sense of hearing has been used as a warning device: Loud sounds often indicate the possibility of danger, and the body automatically reacts by becoming tense and apprehensive. Constant exposure to loud or annoying sounds puts the body under stress for long periods of time and consequently jeopardizes the physical and mental well-being of the individual.

The Occupational Safety and Health Administration (OSHA) has established standards designed to protect workers from excessive sound levels. Workers must be supplied with sound-protection devices such as earplugs if they are exposed to sound levels of 90 dBA or higher (Table 6.2). Some communities also have noise ordinances designed to reduce the sound levels of traffic and other activities.

Table 6.2 OSHA Noise Limits

Sound Level (dBA)	Daily Exposure (hours)
90	8
92	6
95	4
97	3
100	2
102	1.5
105	1
110	0.5
115	0.25

© Alan Merrigan/iStockphoto.com

Tone Quality

The tone quality of a sound is not as easily described as loudness or pitch. Comparisons such as full versus empty, harsh versus soft, or rich versus dry are sometimes used. The tone quality of a sound is very important to our ability to identify what produced the sound. The sound of a flute is different from the sound of a clarinet, and we notice this even if they produce the same note at the same sound level. The tone quality of a person's voice helps us to identify the speaker.

The tone quality of a sound depends primarily on the waveform of the sound wave. If two sounds have different waveforms, we usually perceive different tone qualities. The simplest waveform is that of a pure tone: sinusoidal. Pure tones have a soft, pleasant tone quality (unless they are very loud or high pitched). Complex tones with waveforms that are nearly sinusoidal share the same characteristics. Unlike frequency and sound level, a waveform cannot be expressed as a single numerical factor.

What determines the waveform of a sound wave? The following mathematical principle gives us a way to comparatively analyze waveforms.

PRINCIPLES

Any complex waveform is equivalent to a combination of two or more sinusoidal waveforms with definite amplitudes. These component waveforms are called harmonics. The frequencies of the harmonics are whole-number multiples of the frequency of the complex waveform.

For our purposes, this means that any complex tone is equivalent to a combination of pure tones. These pure tones, called the **harmonics**, have frequencies that are equal to 1, 2, 3, . . . times the frequency of the complex tone. For example, Figure 6.31 shows the waveform of a complex tone whose frequency is, let's say, 100 hertz. It is equivalent to a combination of the three pure tones shown whose frequencies are 100, 200, and 300 hertz. This means that we could artificially produce this complex tone by carefully playing the individual pure tones simultaneously. This is one method that electronic music synthesizers can use to create sounds.

The tone quality of a complex tone depends on the number of harmonics that are present and on their relative amplitudes. These two factors give us a quantitative way of comparing waveforms. A spectrum analyzer is a sophisticated electronic instrument that indicates which harmonics are present in a complex tone and what their amplitudes are. In general, complex tones with a large number of harmonics have a rich tone quality. Notes played on a recorder or a flute contain only a couple of harmonics, so they sound similar to pure tones. Violins and clarinets, on the other hand, have over a dozen harmonics in their notes and consequently have richer tone qualities.

The waveform of noise is a random "scribble" that doesn't repeat. This is because noises are composed of large numbers of frequencies that are not related to each other: They are not harmonics. "White noise" contains equal amounts of all frequencies of sound. (It is so named because one way to produce "white light" is to combine equal amounts of all frequencies of light.) The sound of rushing air approximates white noise.

The existence of higher-frequency harmonics in musical notes explains why high-fidelity sound-reproduction equipment must respond to frequencies up to about 20,000 hertz. Even though the frequencies of musical notes are generally less than 4,000 hertz, the frequencies of the harmonics *do* go up to 20,000 hertz and higher. (Since our ears can't hear any harmonic with a frequency above 20,000 hertz, it isn't necessary to reproduce them.) To accurately reproduce a complex tone, each higher-frequency harmonic must be reproduced.

The study of sound perception brings together the fields of physics, biology, and psychology. The mechanical properties of sound waves interact with the physiological mechanisms in the ear to produce a psychological perception. One of the challenges is to relate subjective descriptions of different sound perceptions to measurable aspects of the sound waves. The hearing apparatus itself is one of the most remarkable in all of Nature. It responds to an amazing range of frequencies and amplitudes and still captures the beauty and subtlety of music. At the same time, it is very fragile. We need to learn more about how our sense of hearing works and how it can be protected.

The Concept Map on your Review Card summarizes the general aspects of sound perception.

Figure 6.31

The complex-tone waveform on the left is a combination of the three pure-tone waveforms on the right. The pure tones (harmonics) have frequencies that are 1, 2, and 3 times the frequency of the complex tone.

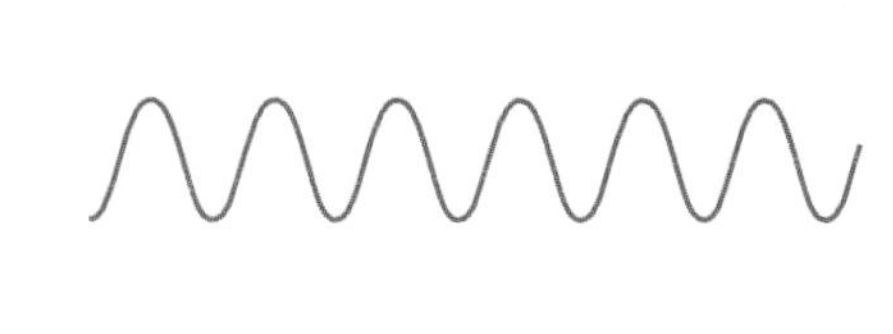

© iStockphoto.com

questions&problems

Questions

(▶ Indicates a review question, which means it requires only a basic understanding of the material to answer. The others involve integrating or extending the concepts presented thus far.)

1. Take a close look at the pulse traveling on the rope in Figure 6.32. It was produced by moving the hand quickly to the left (up in the photo), then right. Notice that the rope is blurred ahead of the peak of the pulse and behind it but that the peak itself is not. Explain why.
2. ▶ Give an example of a wave that does not need a medium in which to travel and a wave that does need a medium.
3. ▶ What is the difference between a longitudinal wave and a transverse wave? Give an example of each.
4. A popular distraction in large crowds at sporting events since the 1980s is the "wave." The people in one section quickly stand up and then sit down; the people in the neighboring sections follow suit in succession, resulting in a visible pattern in the crowd that travels around the stadium. Which of the two types of waves is this? Compared to the waves described in this chapter, how is the stadium wave different?
5. A long row of people are lined up behind one another at a service window. Joe E. Clumsy stumbles into the back of the person at the end and pushes hard enough to generate a wave in the people waiting. What type of wave is produced?
6. A person attaches a paper clip to each coil of a Slinky—about 90 in all—in such a way that waves will still travel on it. What effect, if any, will the paper clips have on the speed of the waves on the Slinky?
7. The speed of an aircraft is sometimes expressed as a Mach number: Mach 1 means that the speed is equal to the speed of sound. If you wish to determine the speed of an aircraft going Mach 2.2, for example, in meters per second or miles per hour, what additional information do you need?
8. Based on information given in Sections 6.1 and 4.3, does there seem to be a relationship between the speed of sound in a gas and its density? Explain.
9. If you were actually in a battle fought in space like the ones shown in science fiction movies, would you hear the explosions that occur? Why or why not?
10. ▶ Explain what the amplitude, frequency, and wavelength of a wave are.
11. ▶ A low-frequency sound is heard, and then a high-frequency sound is heard. Which sound has a longer wavelength?
12. The rays that could be used to represent the steady sound emitted by a warning siren on the top of a pole would look similar to the gravitational field lines (Section 2.8) representing the field around an object. Explain why they are alike, and point out the one important difference.
13. When trying to hear a faint sound from something far away, we sometimes cup a hand behind an ear. Explain why this can help.
14. ▶ What useful thing can happen to a wave when it encounters a concave reflecting surface?
15. ▶ A person riding on a bus hears the sound from a horn on a car that is stopped. What is different about the sound when the bus is approaching the car compared to when the bus is moving away from the car?
16. ▶ Explain the process of echolocation. How is the Doppler effect sometimes incorporated?
17. In the past, ships often carried small cannons that were used when approaching shore in dense fog to estimate the distance to the hidden land. Explain how this might have been done.
18. You can buy a measuring device billed as a "sonic tape measure." Describe how a device equipped with an (ultrasonic) speaker, a microphone, and a precision timer could be used to measure the distance from the device to a wall (for example).
19. While a shock wave is being generated by a moving wave source, is the Doppler effect also occurring? Explain.
20. Describe some of the things that would happen if the speed of sound in air suddenly decreased to, say, 20 m/s. What would it be like living next to a freeway?
21. ▶ If a boat is producing a bow wave as it moves over the water, what must be true about its speed?
22. When a wave passes through two nearby gaps in a barrier, interference will occur, provided that there is also diffraction. Why must there be diffraction?

Figure 6.32

A wave traveling on a ropelike spring.

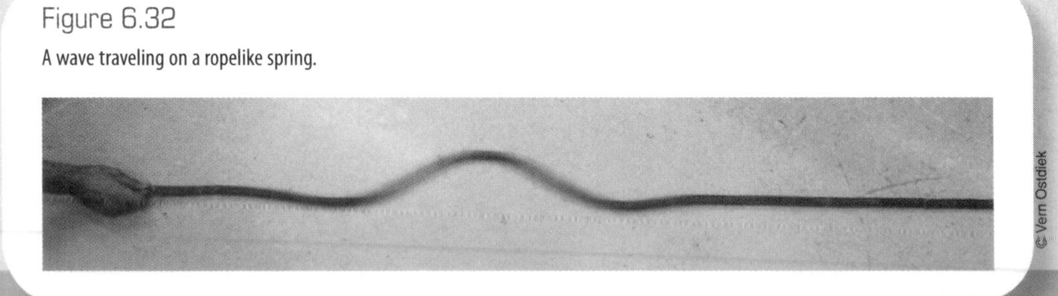

© Vern Ostdiek

23. A recording of a high-frequency pure tone is played through both speakers of a portable stereo placed in an open field. A person a few meters in front of the stereo walks slowly along an arc around it. How does the sound that is heard change as the person moves?

24. As a loud, low-frequency sound wave travels past a small balloon, the balloon's size is affected. Explain what happens. (The effect is too small to be observed under ordinary circumstances.)

25. ▶ Describe the waveforms of pure tones, complex tones, and noise.

26. ▶ What is ultrasound, and what is it used for?

27. ▶ Describe how sound is produced in string instruments. Why does tightening a string change the frequency of the sound it makes?

28. A conditioning drill consists of repeatedly running from one end of a basketball court to the other, turning around and running back. Sometimes the drill is changed and the runner turns around at half court, or perhaps at three-fourths of the length of the court. Describe how the number of round-trips a runner can do each minute changes when the distance is changed and how this is related to a guitarist changing the note generated by a string by pressing a finger on it at some point.

29. A special room contains a mixture of oxygen and helium that is breathable. Two musicians play a guitar and a flute in the room. Does each instrument sound different from when it is played in normal air? Why or why not?

30. ▶ What is reverberation? How does reverberation affect how we hear sounds?

31. ▶ What are the three categories used to describe our mental perception of a sound? Upon what physical properties of sound waves does each depend?

32. A 100-Hz pure tone at a 70-dB sound level and a 1,000-Hz pure tone at the same sound level are heard separately. Do they sound equally loud? If not, which is louder, and why?

33. A sound is produced by combining three pure tones with frequencies of 200 Hz, 400 Hz, and 600 Hz. A second sound is produced using 200 Hz, 413 Hz, and 600 Hz pure tones. What important difference is there between the two sounds?

34. ▶ The highest musical note on the piano has a frequency of 4,186 Hz. Why would a tape of piano music sound terrible if played on a tape player that reproduces frequencies only up to 5,000 Hz?

35. Normal telephones do not transmit pure tones with frequencies below about 300 Hz. But a person whose speaking voice has a frequency of 100 Hz can be heard and understood over the phone. Why is that?

Problems

1. Two children stretch a jump rope between them and send wave pulses back and forth on it. The rope is 3 m long, its mass is 0.5 kg, and the force exerted on it by the children is 40 N.
   a) What is the linear mass density of the rope?
   b) What is the speed of the waves on the rope?

2. The force stretching the D string on a certain guitar is 150 N. The string's linear mass density is 0.005 kg/m. What is the speed of waves on the string?

3. What is the speed of sound in air at the normal boiling temperature of water?

4. The coldest and hottest temperatures ever recorded in the United States are −83°F (210 K) and 135°F (330 K), respectively. What is the speed of sound in air at each temperature?

5. A 4-Hz continuous wave travels on a Slinky. If the wavelength is 0.5 m, what is the speed of waves on the Slinky?

6. A 500-Hz sound travels through pure oxygen. The wavelength of the sound is measured to be 0.65 m. What is the speed of sound in oxygen?

7. A wave traveling 80 m/s has a wavelength of 3.2 m. What is the frequency of the wave?

8. What frequency of sound traveling in air at 20°C has a wavelength equal to 1.7 m, the average height of a person?

9. Verify the figures given in Section 6.1 for the wavelengths (in air) of the lowest and highest frequencies of sound that people can hear (20 and 20,000 Hz, respectively).

10. What is the wavelength of 3.5 million Hz ultrasound as it travels through human tissue?

11. The frequency of middle C on the piano is 261.6 Hz.
    a) What is the wavelength of sound with this frequency as it travels in air at room temperature?
    b) What is the wavelength of sound with this frequency in water?

12. A steel cable with total length 30 m and mass 100 kg is connected to two poles. The tension in the cable is 3,000 N, and the wind makes the cable vibrate with a frequency of 2 Hz. Calculate the wavelength of the resulting wave on the cable.

13. In a student laboratory exercise, the wavelength of a 40,000-Hz ultrasound wave is measured to be 0.868 cm. Find the air temperature.

14. A 1,720-Hz pure tone is played on a stereo in an open field. A person stands at a point that is 4 m from one of the speakers and 4.4 m from the other. Does the person hear the tone? Explain.

15. A person stands directly in front of two speakers that are emitting the same pure tone. The person then moves to one side until no sound is heard. At that point, the person is 7 m from one of the speakers and 7.2 m from the other. What is the frequency of the tone being emitted?

16. Ultrasound probes can resolve structural details with sizes approximately equal to the wavelength of the ultrasound waves themselves. What is the size of the smallest feature observable in human tissue when examined with 20 MHz ultrasound. The speed of sound in human tissue is 1,540 m/s.

17. A sonic depth gauge is placed 5 m above the ground. An ultrasound pulse sent downward reflects off snow and reaches the device 0.03 seconds after it was emitted. The air temperature is −20°C.

    a) How far is the surface of the snow from the device?

    b) How deep is the snow?

18. Ultrasound can be used to measure the thickness of protective coatings applied to surfaces as part of industrial processes because it reflects well from boundaries separating regions with different densities. How thick is a layer of plastic coating on a metal substrate that produces ultrasound echoes from the upper and lower surfaces having round-trip times that differ by 0.75 μs. Assume the speed of sound in the plastic is 900 m/s. (Remember: In making a round-trip, the wave reflected off the lower surface travels a distance equal to twice the thickness of the film.)

19. The huge volcanic eruption on the island of Krakatoa, Indonesia, in 1883 was heard on Rodrigues Island, 4,782 km (2,970 miles) away. How long did it take the sound to travel to Rodrigues?

20. A baseball fan sitting in the "cheap seats" is 150 m from home plate. How much time elapses between the instant the fan sees a batter hit the ball and the moment the fan hears the sound (see Figure 6.33)?

21. A geologist is camped 8,000 m (5 miles) from a volcano as it erupts.

    a) How much time elapses before the geologist hears the sound from the eruption?

Figure 6.34

© David Joyner/iStockphoto.com / © iStockphoto.com

    b) How much time does it take the seismic waves produced by the eruption to reach the geologist's camp, assuming the waves travel through granite as sound waves do?

22. A person stands at a point 300 m in front of the face of a sheer cliff. If the person shouts, how much time will elapse before an echo is heard?

23. A sound pulse emitted underwater reflects off a school of fish and is detected at the same place 0.01 s later. How far away are the fish (see Figure 6.34)?

24. The sound level measured in a room by a person watching a movie on a home theater system varies from 65 dB during a quiet part to 95 dB during a loud part. Approximately how many times louder is the latter sound?

25. Approximately how many times louder is a 100-dB sound than a 60-dB sound?

26. What are the frequencies of the first four harmonics of middle C (261.6 Hz)?

27. The frequency of the highest note on the piano is 4,186 Hz.

    a) How many harmonics of that note can we hear?

    b) How many harmonics of the note one octave below it can we hear?

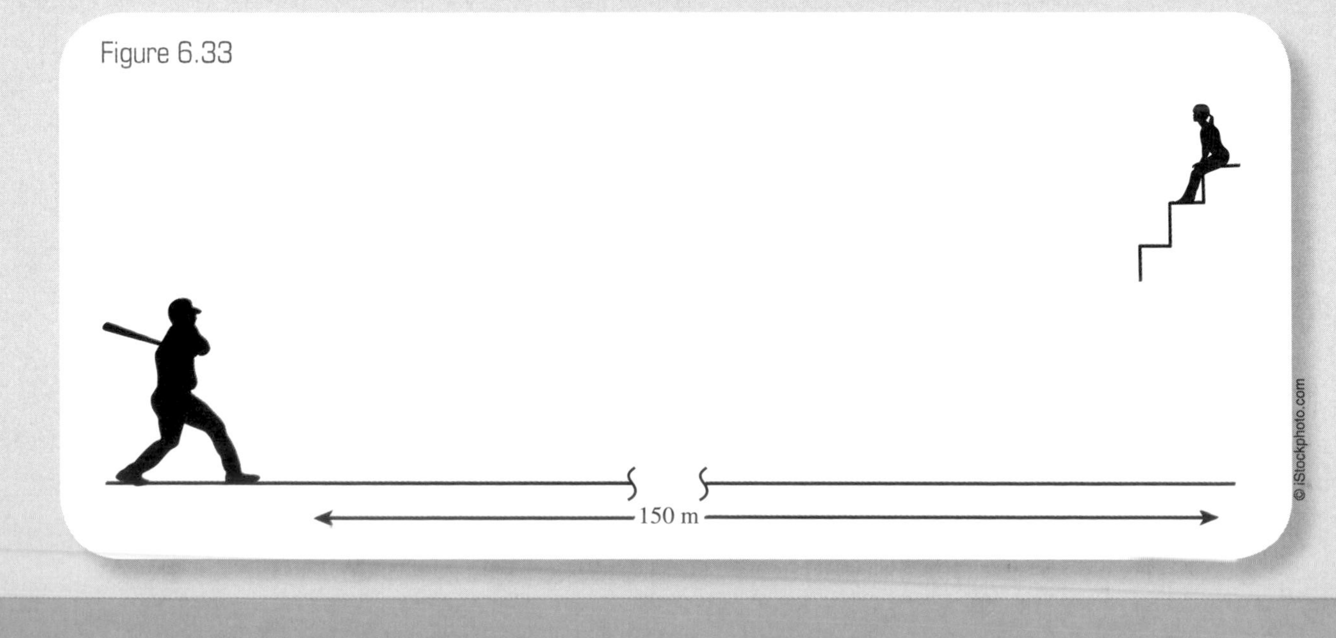

Figure 6.33

© iStockphoto.com

Through conversations, focus groups, surveys, and interviews, we collected data that drove the creation of PHYSICS. Here are just a few quotes from real students like you:

"With the cost of education rising, it's such a relief to use an affordable textbook!"

"The magazine style is eye-catching and interesting to read."

"Interactive quizzing is so helpful!"

LISTEN UP!

SHE DID

PHYSICS was designed for students just like you – busy people who want choices, flexibility, and multiple learning options.

PHYSICS delivers concise, focused information in a fresh and contemporary format. And…

PHYSICS gives you a variety of online learning materials designed with you in mind.

At **4ltrpress.cengage.com/physics,** you'll find electronic resources such as **simulations, interactive quizzes,** and **flashcards** for each chapter.

These resources will help supplement your understanding of core physics concepts in a format that fits your busy lifestyle. Visit **4ltrpress.cengage.com/physics** to learn more about the multiple resources available to help you succeed!

Electricity

sections

7.1 Electric Charge

Since its introduction in 2001, the iPod has revolutionized how people purchase and consume music, videos, and images. iPods and iPhones now support hundreds of "apps" that you can use to locate the nearest coffee shop, monitor your blood pressure, get the latest "tweets" from your favorite celebrities, manage your bank accounts, find the most recent results from the world of sports, and much, much more. Such devices represent a triumph of the applications of electricity, magnetism, and optics.

Several key components of these "iProducts" function by exploiting electric fields. For example, the click wheel, the primary interface for controlling most iPod models, detects the presence and motion of a user's finger by sensing changes the finger causes in the electric field in the device. The LCD display, like those on cell phones and laptop computers, creates images by using electric fields to selectively activate individual pixels. Digital information stored on the iPod is then converted into audio or video signals. This huge computational task is accomplished by millions of transistors embedded in integrated circuit chips. These transistors perform their jobs by using electric fields to control the flow of electric charges through them.

Electricity is not reserved for electronic devices, however. Do you realize that as you read these words, the information sent to your brain and processed there also relies upon electric charges and electric signals? When you turn this page, your brain will communicate to the muscles in your hand in the same way. Electricity governs your life in ways you probably rarely think about. Even the properties of all matter that surrounds you—the air you breathe, the water you drink, the chair you sit in—are largely determined by electrical forces acting in and between atoms. From your newest electronic gizmo to the "glue" that holds matter together, electricity is inextricably woven into your life. So let's take a closer look at what electricity is.

The word *electricity* comes from *electron*, which is based on the Greek word for amber. Amber is a fossil resin that attracts bits of thread, paper, hair, and other things after it has been rubbed with fur. You may have noticed that a comb can do the same after you run it through your hair. This phenomenon, known as the "amber effect," was documented by the ancient Greeks, but its cause remained a mystery for more than two millennia. The results of numerous experiments, some conducted by the

CHAPTER 7

> **Electric Charge** *An inherent physical property of certain subatomic particles that is responsible for electrical and magnetic phenomena. Charge is represented by* q*, and the SI unit of measure is the* coulomb *(C).*

American scientist and statesman Benjamin Franklin, indicated that matter possessed a "new" property not connected to mass or gravity. This property was eventually traced to the atom and is called **electric charge**.

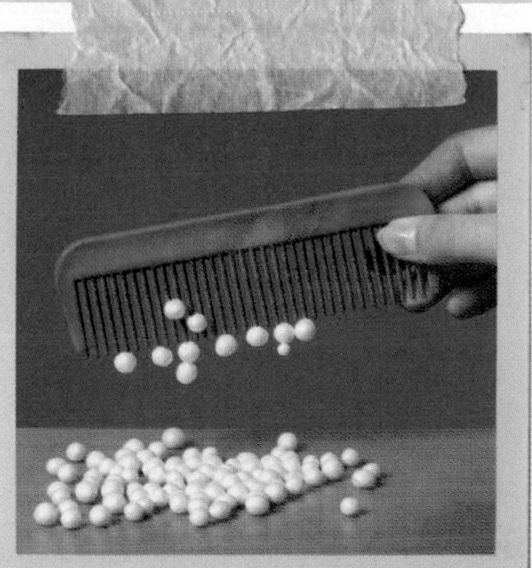

The amber effect. The comb attracts the beads because it was charged by rubbing.

© Martin F. Chillmaid/SPL/Photo Researchers / © iStockphoto.com

Early on, we stated that there are three fundamental things that physicists can quantify or measure: space, time, and properties of matter (Section 1.1). Until now, mass has been the only fundamental property of matter that we have used. Electric charge is another basic property of matter, but it is intrinsically possessed only by electrons, protons, and certain other subatomic or "elementary" particles (more on this in Chapter 12). Unlike mass, there are two different (and opposite) types of electric charge, appropriately named *positive* and *negative* charge. One coulomb of positive charge will "cancel" 1 coulomb of negative charge. In other words, the *net* electric charge would be zero ($q = -1\text{ C} + 1\text{ C} = 0\text{ C}$).

Recall that every atom is composed of a nucleus surrounded by one or more electrons (Section 4.1). The nucleus itself is composed of two types of particles, protons and neutrons (Figure 7.1). Every electron has a charge of -1.6×10^{-19} C, and every proton has a charge of $+1.6 \times 10^{-19}$ C. (To have a total charge of -1 C, over 6 billion billion electrons are needed.) Neutrons are so named because they are neutral; they have no net electric charge. Normally, an atom will have the same number of electrons as it has protons, which means the atom as a whole is neutral. The number of protons in the nucleus is the element's *atomic number*, which determines the atom's identity. For example, an atom of the element helium has two protons in the nucleus and two electrons in orbit around the nucleus. The positive charge possessed by the two protons is exactly balanced by the negative charge possessed by the two electrons, so the net charge is zero. Most of the substances that we normally encounter are electrically neutral simply because the total number of electrons in all of the atoms is equal to the total number of protons.

A variety of physical and chemical interactions can cause an atom to gain one or more electrons or to lose one or more of its electrons. In these cases, the atom is said to be *ionized.* For example, if a helium atom gains one electron, it has three negative particles (electrons) and two positive particles (protons). This atom is said to be a **negative ion**, since it has a net negative charge (Figure 7.2a). The value of its net charge is just the charge on the "extra" electron, $q = -1.6 \times 10^{-19}$ C. Similarly, if a neutral helium atom loses one electron, it becomes a **positive ion**, since it has two positive particles and only one negative particle. Its net charge is $+1.6 \times 10^{-19}$ C (Figure 7.2b).

In many situations, ions are formed on the surface of a substance by the action of friction. When a piece of amber, plastic, or hard rubber is rubbed with fur, negative ions are formed on its surface. The contact between the fur and the material causes some of the electrons in the atoms of the fur to be transferred to some of the atoms on the surface of the solid. The fur acquires a net positive

Figure 7.1

Simplified model of an atom of the element helium. The nucleus contains two protons; that is why helium's atomic number is 2. Also in the nucleus are uncharged neutrons, two in this case. Orbiting the nucleus are electrons. As long as the number of electrons equals the number of protons, the atom has no net charge.

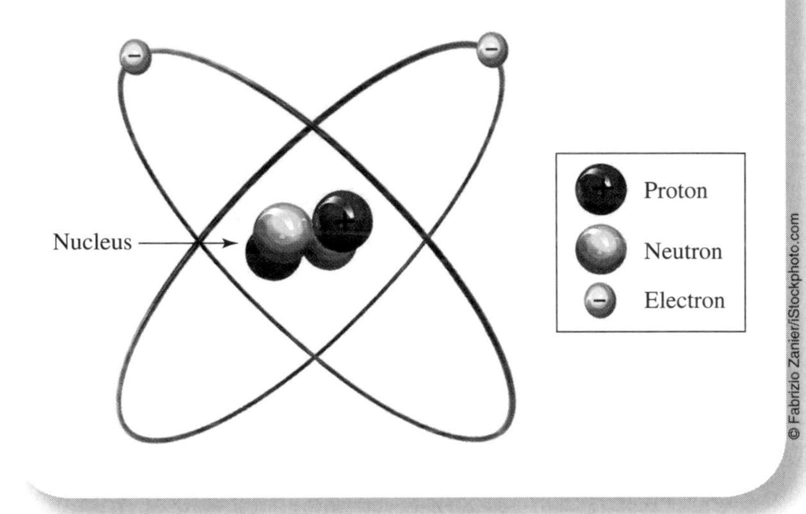

Figure 7.2

An atom is ionized when the number of electrons does not equal the number of protons. (a) A negative helium ion. (b) A positive helium ion.

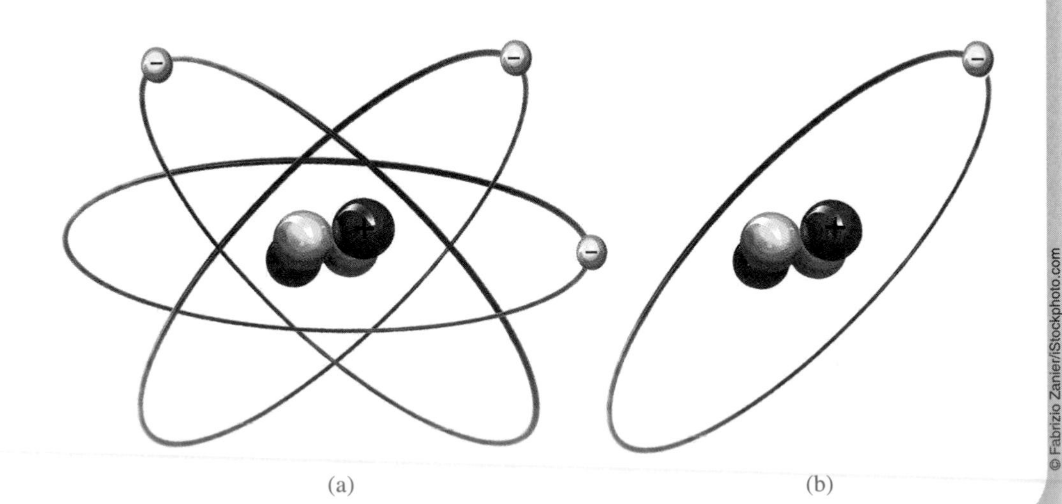

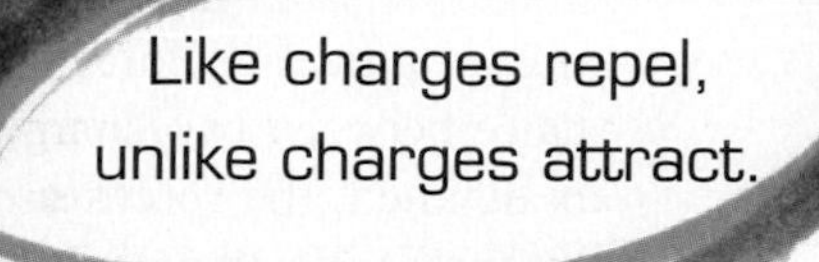

charge, because it has fewer electrons than protons. Similarly, the amber, plastic, or hard rubber acquires a net negative charge since it has an excess of electrons (Figure 7.3). Combing your hair can charge the comb in the same way. Rubbing glass with silk causes the glass to acquire a net positive charge. Some of the electrons in the surface atoms of the glass are transferred to the silk, which becomes negatively charged.

Ion formation by friction is a complicated phenomenon that is not completely understood. It is affected by many factors such as which materials are used and what the relative humidity is.

7.2 Electric Force and Coulomb's Law

The original amber effect illustrates that electric charges can exert forces. You may have noticed hair being pulled toward a charged comb or "static cling" between items of clothes removed from a dryer. These are the most common situations—two objects with opposite charges attracting each other (Figure 7.4). The negatively charged comb exerts an attractive force on the positively charged hair. In addition, two objects with the same kind of charge (both positive or both negative) repel each other. When two similarly charged combs are suspended from threads, they push each other apart. Just remember this simple rule: *Like charges repel, unlike charges attract.*

This force between charged objects is extremely important in the physical world, particularly at the atomic level. It holds atoms together and makes it possible for them to exist. In each atom, the positively charged protons in the nucleus exert attractive forces on the negatively charged

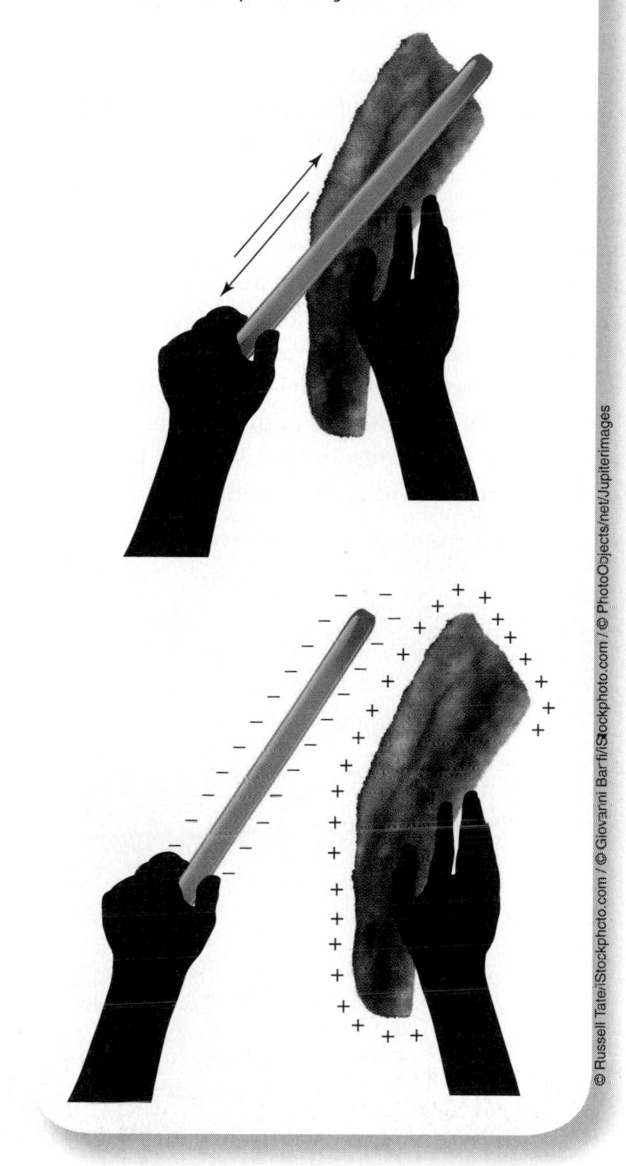
© Russell Tate/iStockphoto.com / © Giovanni Barfi/iStockphoto.com / © PhotoObjects.net/Jupiterimages

Figure 7.3
The contact between fur and plastic when they are rubbed together causes some of the electrons in the fur to be transferred to the plastic. This leaves the plastic with a net negative charge and the fur with a net positive charge.

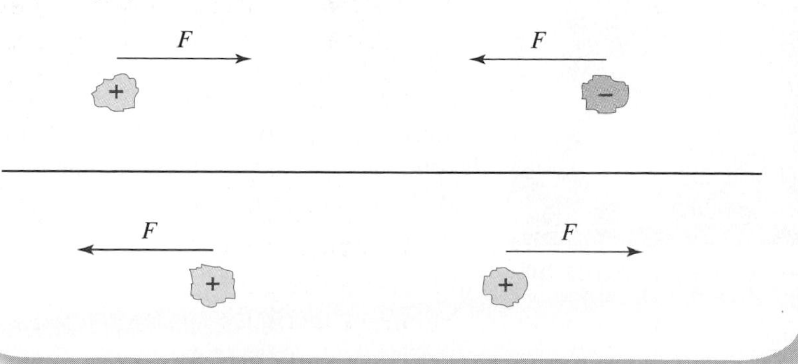

Figure 7.4
Objects with opposite electric charges exert attractive forces on one another. Objects with the same kind of charge repel each other.

electrons. The electric force on each electron keeps it in its orbit, much as the gravitational force exerted by the Sun keeps the Earth in its orbit. The forces between the atoms in many compounds arise because opposite charges attract. For example, when salt is formed from the elements sodium and chlorine, each sodium atom gives up an electron to a chlorine atom. The resulting ions exert attractive forces on one another because they are oppositely charged (Figure 7.5). Matter as we know and experience it would not exist without the electrical force.

Figure 7.5
Ordinary table salt consists of positive sodium ions and negative chlorine ions. The strong attractive forces between the oppositely charged ions hold them in a rigid crystalline array.

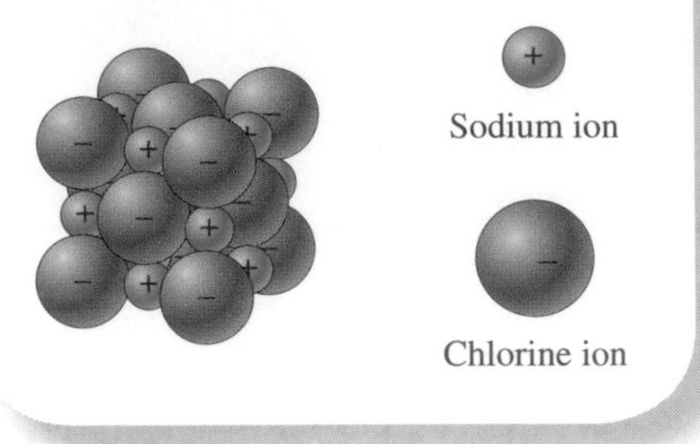

Recall Newton's law of universal gravitation in Section 2.8, which gives the size of the force acting between two masses. The corresponding law for the electrical force, **Coulomb's law**, is very similar.

LAWS

Coulomb's Law *The force acting on each of two charged objects is directly proportional to the net charges on the objects and inversely proportional to the square of the distance between them.*

$$F \propto \frac{q_1 q_2}{d^2}$$

The constant of proportionality in SI units is 9×10^9 N-m²/C². Therefore:

$$F = \frac{(9 \times 10^9) q_1 q_2}{d^2}$$

SI units—F in newtons, q_1 and q_2 in coulombs, and d in meters

Charles-Augustin de Coulomb (1736–1806)

© Emilio Segre Visual Archives/American Institute of Physics/SPL/Photo Researchers Inc.

The force on q_1 is equal and opposite to the force on q_2, by Newton's third law of motion. Note that if both objects have the same kind of charge (both positive or both negative), the force F is positive. This indicates a repulsive force. If one charge is negative and the other positive, the force F is negative, indicating an attractive force. If the distance between two charged objects is doubled, the forces are reduced to one-fourth their original values (Figure 7.6).

Perhaps it is not a surprise that Coulomb's law has the same form as Newton's law of universal gravitation. After all, mass and charge are both fundamental properties of the particles that comprise matter. We must remember, however, that the (gravitational) force between two bodies due to their masses is *always* an attractive force, while the (electrostatic) force between two bodies due to their electric charges can be attractive or repulsive, depending on whether they have opposite charges or not. Also, all matter has mass and so experiences and exerts gravitational forces, whereas the electrostatic force normally acts between objects only when there is a net charge on one or both of them. Generally, when two objects have electric charges, the electrostatic force between them is much stronger than the gravitational force. For example, the electrostatic force between an electron and a proton is about 10^{39} times as large as the gravitational force between them.

It is possible for a charged object to exert a force of attraction on a second object that has no net charge. This is what happens when a charged comb is used to pick up bits of paper or thread. Here, the negatively charged comb attracts the nuclei of the atoms and repels the electrons. The orbits of the electrons are distorted so that the electrons are, on the average, farther away from the charged comb than the nuclei (Figure 7.7). This results in a net attractive force because the repulsive force on the slightly more distant, negatively charged electrons is smaller than the attractive force on the closer, positively charged protons. The process of inducing a small charge separation (or displacement) between the nucleus of an atom and its electrons is called *polarization.*

Some molecules are naturally polarized; that is, they have a net negative charge displaced to one side of the net positive nuclear charge. They are called *polar molecules.* Water molecules have this property. If a polar molecule is free to rotate—as in a liquid—it will be attracted to a charged object. Its side with the charge opposite that on the object will turn toward the object, and the attractive force on that side will be stronger than the repulsive force on the other side, as with the atoms in Figure 7.7.

Figure 7.6
Coulomb's law. Doubling the charge on one object, (b), doubles the force on both. Doubling the distance between the charges, (c), reduces the forces to one-fourth their original values.

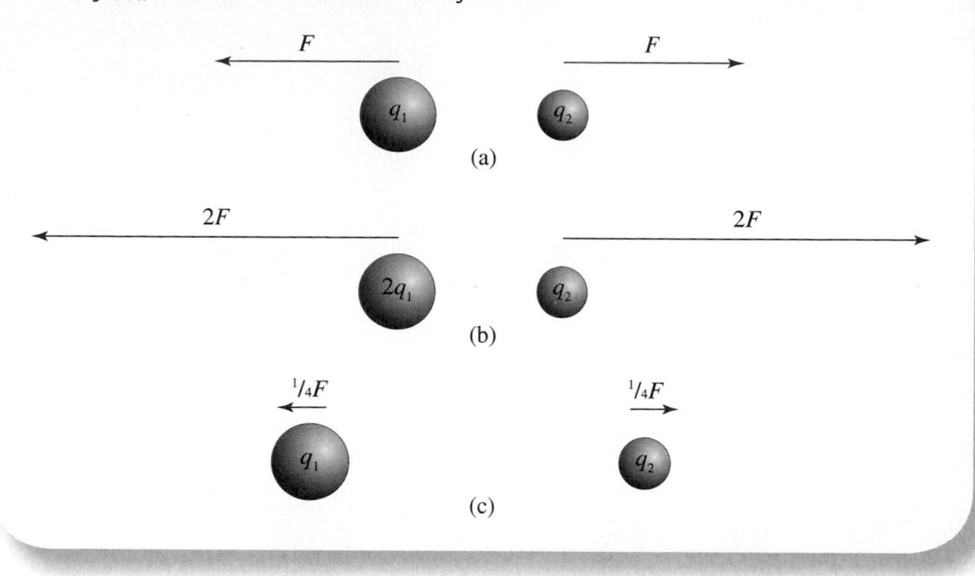

The electrostatic force is another example of "action at a distance." As with gravitation, the concept of a field is useful. In the space around any charged object, there is an *electric field.* This field is the "agent" of the electrostatic force: It will cause any charged object to experience a force. The electric field around a charged particle is represented by lines that indicate the direction of the force that the field would exert on a positive charge. Thus the *electric field lines* around a positively charged particle point radially outward, and the field lines around a negatively charged particle point radially inward (Figure 7.8).

The strength of an electric field at a point in space is equal to the size of the force that it would cause on a given charged object placed at that point, divided by the size of the charge on the object.

$$\text{electric field strength} = \frac{\text{force on a charged object}}{\text{charge on the object}}$$

Where the field is strong, a charged object will experience a large force. The strength of the electric field is indicated by the spacing of the field lines: Where the lines are close together, the field is strong. The electric field around a charged particle is clearly weaker at greater distances from it.

Any time a positive charge is in an electric field, it experiences a force in the *same* direction as the field lines. A negative charge in an electric field feels a force in the *opposite* direction of the field lines (Figure 7.9). Remember

Figure 7.7
A charged comb exerts a net attractive force on a neutral piece of paper because the electrons are displaced slightly away from the negatively charged comb. (The distorted electron orbits are exaggerated in this drawing.) The attractive force on the closer nuclei is stronger than the repulsive force on the electrons.

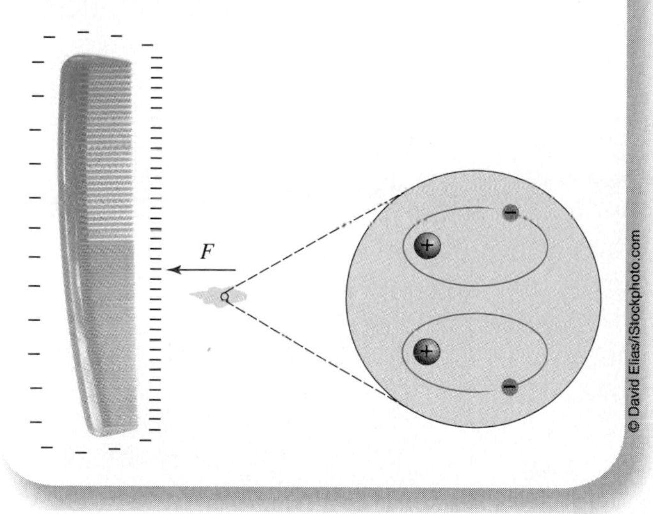

Figure 7.8
Electric field lines in the space around a positive charge (a) and around a negative charge (b). The arrows show the direction of the force that would act on a positive charge placed in either field.

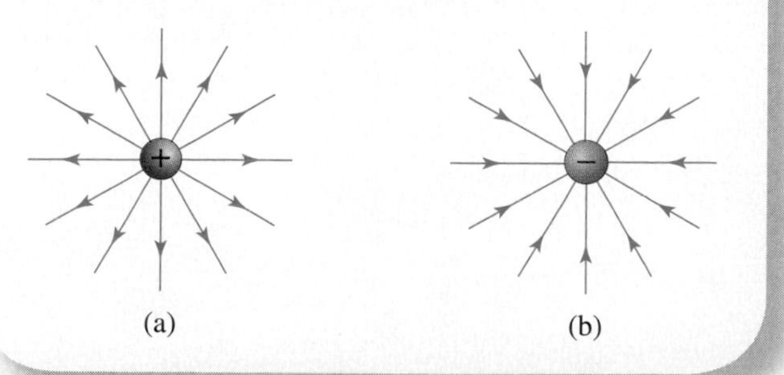

Figure 7.9
In an electric field, the force on a positively charged body is parallel to the field, while the force on a negatively charged body is opposite to the direction of the field.

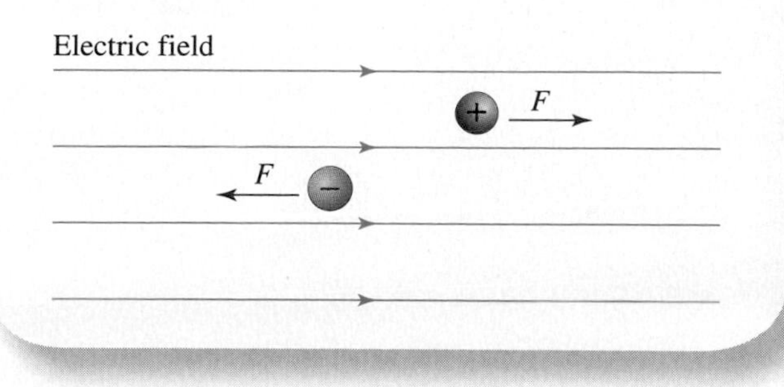

At any given time, there are roughly 2,000 thunderstorms in progress around the world, producing some 30 to 100 cloud-to-ground lightning flashes each second, totaling about 5 million lightning strikes a day. The physics of electricity can explain this natural wonder.

Lightning is akin to the short sparks we experience when we touch a metal doorknob after walking across a wool rug in a dry environment in the winter. Cumulonimbus clouds, the most common thunderstorm cloud, and the Earth effectively acquire opposite electrical charges (like your finger and the doorknob), the air between serving as an insulating material. When the separated charge grows sufficiently large, the strong electric field between the cloud and the Earth, typically some 300,000–400,000 volts/meter, causes an electrical breakdown of the insulating air creating an ionized path between the cloud and the ground along which charge can flow, that is, a lightning discharge. Usually during this discharge, negative charge is transferred from the lower part of the cloud to neutralize the induced positive charge on the Earth below. The average maximum current in such lightning flashes is about 30,000 amperes, lasting for only about 30 microseconds and delivering about a coulomb of charge. It is the large currents carried by lightning discharges that make them so dangerous to human beings. For more on how lightning is formed go to 4ltrpress.cengage.com/physics.

© Clayton Hansen/iStockphoto.com / © ryan burke/iStockphoto.com / © Dale Robins/iStockphoto.com / © Charles Uibel/iStockphoto.com

© Sheila Terry/Photo Researchers, Inc. / © Maria Toutoudaki/iStockphoto.com

Figure 7.10

A large spark created in a lab. The spark is produced because there is a very strong electric field in the space between the two shiny, spherical objects.

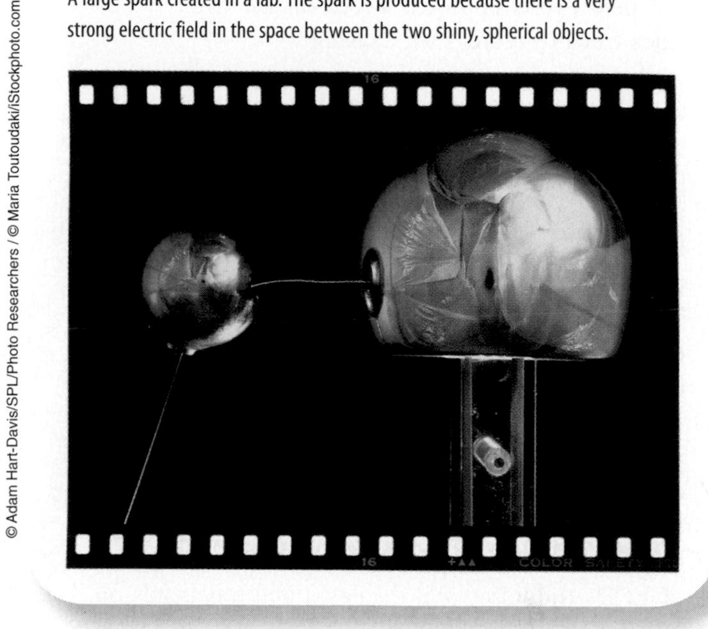

© Adam Hart-Davis/SPL/Photo Researchers / © Maria Toutoudaki/iStockphoto.com

that the electric field exerts the force on a charged object; in Chapter 8, we will show that electric fields can be produced without directly using electric charges.

Perhaps you have had the experience of walking across a carpeted floor and receiving a shock when you touched a metal doorknob. This is more likely to happen in winter than in summer because the relative humidity is usually lower, and electrostatic charging takes place more readily. The shock is due to charges flowing between you and the doorknob, and it may be accompanied by a visible spark. Air normally does not allow charges to flow through it. A spark occurs when there is an electric field strong enough to ionize atoms in the air. Freed electrons accelerate in a direction opposite to the direction of the electric field, and positive ions accelerate in the same direction as the field. The electrons and ions pick up speed and collide with other atoms and molecules, ionizing them or causing them to emit light (see Figure 7.10). Lightning is produced in this same way on a much larger scale as Benjamin Franklin demonstrated using kites, keys, and metal rods in the middle of the 18th century. In Chapter 10, we will discuss how atomic collisions cause the emission of light.

Electronic signs that behave like electronically erasable paper make use of electric fields to form letters and other images. One type, called SmartPaper™, consists of millions of tiny beads between two thin plastic sheets (see Figure 7.11). One side of each bead is

Figure 7.11

Simplified edge view of a segment of SmartPaper™ electronic paper. One half of each tiny bead has a negative charge and is some color (red in this example), and the other half has a positive charge and is a contrasting color. An electric field exerts opposite forces on the two sides of each bead, so the bead rotates and aligns with the field. Letters are formed by using electric fields to turn the red sides up in parts of the display and the blue sides up elsewhere.

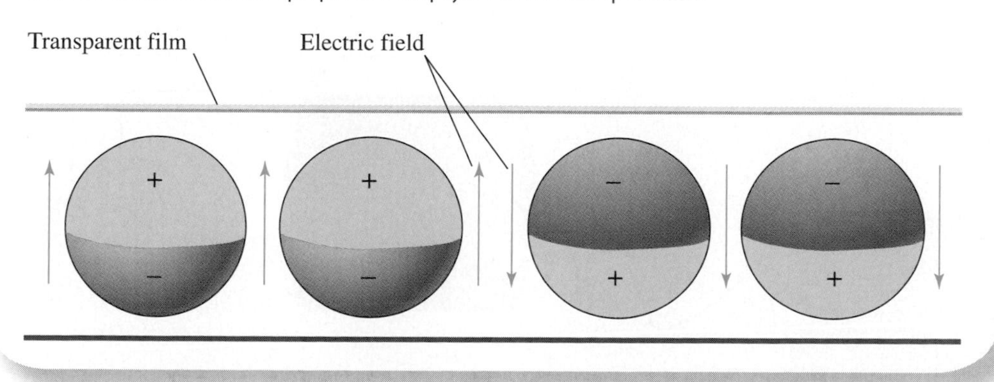

some color and negatively charged, and the other side is a contrasting color and positively charged. An electric field exerts opposite forces on the two sides (recall Figure 7.9), causing the beads to rotate until they are aligned with the field. (As we will see in Chapter 8, this is just like what a compass needle does in a magnetic field.) Letters are formed on the electronic paper by selectively applying upward and downward electric fields at different places, so parts of the display are one color and the rest are the other color.

The type of transistor used most widely in computers and similar devices is the field-effect transistor (FET). In FETs, an electric field controls the flow of electricity through the transistor. Electric fields play crucial roles in the operation of LCDs (liquid crystal displays) and touchpads on laptop computers as well.

The Concept Map on your Review Card summarizes the ideas presented in these first two sections of Chapter 7.

© Alan Merrigan/iStockphoto.com

7.3 Electric Currents—Superconductivity

An **electric current** is a flow of charged particles. The cord on an electrical appliance encloses two separate metal wires covered with insulation. When the appliance is plugged in and operating, electrons inside each wire move back and forth. Inside a television picture tube, free electrons are accelerated from the back of the tube to the screen at the front. There is a near vacuum inside the picture tube, so the electrons can travel without colliding with gas molecules (Figure 7.12). When salt is dissolved in water, the sodium and chlorine ions separate and can move about just like the water molecules. If an electric field is applied to the water, the positive sodium ions will flow one way (in

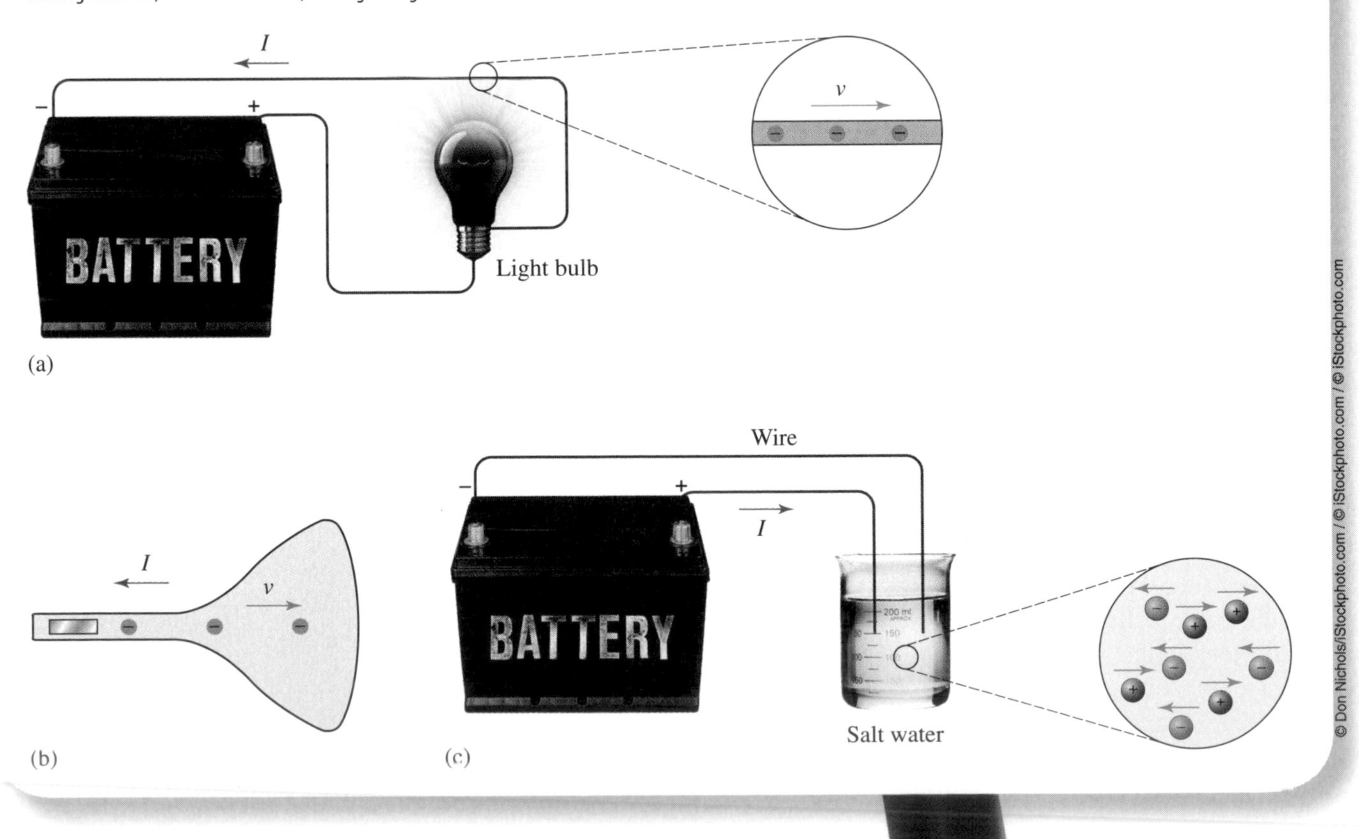

Figure 7.12

Examples of electric current. (a) Electrons flowing inside a metal wire. (b) Electrons moving through the near-vacuum inside a television picture tube. (c) Positive ions and negative ions, from dissolved salt, flowing through water.

© Don Nichols/iStockphoto.com / © iStockphoto.com / © iStockphoto.com

the direction of the field), and the negative chlorine ions will flow the other way.

Regardless of the nature of the moving charges, the quantitative definition of electric current is as follows.

DEFINITION

Current *The rate of flow of electric charge. The amount of charge that flows by per second.*

$$\text{current} = \frac{\text{charge}}{\text{time}} \qquad I = \frac{q}{t}$$

The SI unit of current is the ampere *(A or amp), which equals 1 coulomb per second. Current is measured with a device called an* ammeter.

A current of 5 amperes in a wire means that 5 coulombs of charge flow through the wire each second. (Table 7.1 lists some representative currents.) Either positive charges or negative charges can comprise a current. The effect of a positive charge moving in one direction is the same as that of an equal negative charge moving in the opposite direction. Formally, an electric current is represented as a flow of positive charge. This is because it was originally believed that current consists of positive charges that move through metals. Even after it was discovered that it is the negatively charged electrons that flow in a wire to comprise the current, the convention of defining the direction of current flow as that which would be associated with positive charges was retained. If positive ions are flowing to the right in a liquid, then the current is to the right. If negative charges (like electrons) are flowing to the right, the direction of the current is to the left. In Figure 7.12, the current is to the left in (a) and (b), and to the right in (c).

Table 7.1 Typical Currents in Common Devices

Device	Current (A)
Calculator	0.0001
Spark plug	0.001
Clock radio	0.03
60-watt light bulb	0.5
Color television	2
1,800-watt hair dryer	15

The ease with which charges move through different substances varies greatly. Any material that does not readily allow the flow of charges through it is called an electrical *insulator.* Substances like plastic, wood, rubber, air, and pure water are insulators because the electrons are tightly bound in the atoms, and electric fields are usually not strong enough to rip them free so they can move. Our lives depend on insulators: The electricity powering the devices in our homes could kill us if insulators, like the covering on power cords, didn't keep it from entering our bodies.

An electrical *conductor* is any substance that readily allows charges to flow through it. Metals are very good conductors because some of the electrons are only loosely bound to atoms and so are free to "skip along" from one atom to the next when an electric field is present. In general, solids that are good conductors of heat are also good conductors of electricity. As mentioned before, liquids such as water are conductors when they contain dissolved ions. Most drinking water has some natural minerals and salts dissolved in it and so conducts electricity. Solid insulators can become conductors when wet because of ions in the moisture. The danger of being electrocuted by electrical devices increases dramatically when they are wet.

Semiconductors are substances that fall in between the two extremes. The elements silicon and germanium, both semiconductors, are poor conductors of electricity in their pure states, but they can be modified chemically ("doped") to have very useful electrical properties. Transistors, solar cells, and numerous other electronic components are made out of such semiconductors (Figure 7.13). The electronic revolution in the second half of the twentieth century, including the development of inexpensive calculators, computers, sound-reproduction systems, and other devices, came about because of semiconductor technology.

Resistance

What makes a 100-watt light bulb brighter than a 60-watt bulb? The size of the current flowing through the filament determines the brightness. That, in turn, depends on the filament's **resistance**.

DEFINITION

Resistance *A measure of the opposition to current flow. Resistance is represented by* R, *and the SI unit of measure is the* ohm *(Ω).*

Figure 7.13

Image of semiconductor integrated circuit chip and a wafer.

In general, a conductor will have low resistance and an insulator will have high resistance. The actual resistance of a particular piece of conducting material—a metal wire, for example—depends on four factors:

1. *Composition.* The particular metal making up the wire affects the resistance. For example, an iron wire will have a higher resistance than an identical copper wire.
2. *Length.* The longer the wire is, the higher its resistance.
3. *Diameter.* The thinner the wire is, the higher its resistance.
4. *Temperature.* The higher the temperature of the wire, the higher its resistance.

The filament of a 100-watt bulb is thicker than that of a 60-watt bulb, so its resistance is lower. As we will see in later sections, this means a larger current normally flows through the 100-watt bulb, and consequently, it is brighter.

Resistance can be compared to friction. Resistance inhibits the flow of electric charge, and friction inhibits relative motion between two substances. In metals, electrons in a current move among the atoms and in the process collide with them and give them energy. This impedes the movement of the electrons and causes the metal to gain internal energy. The consequence of resistance is the same as that of kinetic friction—heating. The larger the current through a particular device, the greater the heating.

Superconductivity

In 1911, the Dutch physicist Heike Kamerlingh Onnes made an important discovery while measuring the resistance of mercury at extremely low temperatures. He found that the resistance decreased steadily as the temperature was lowered, until at 4.2 K (−452.1°F) it suddenly dropped to zero (Figure 7.14). Electric current flowed through the mercury with *no* resistance. Onnes named this phenomenon *superconductivity* for good reason: Mercury is a perfect conductor of electric current below what is called its *critical temperature* (referred to as T_c) of 4.2 K. Subsequent research showed that hundreds of elements, compounds, and metal alloys become superconductors, but only at very low temperatures. Until 1985, the highest known T_c was 23 K for a mixture of the elements niobium and germanium.

Superconductivity seems too good to be true: electricity flowing through wires with *no* loss of energy to heating. Once a current is made to flow in a loop of superconducting wire, it can flow *for years* with no battery or other source of energy because there is no energy loss due to resistance. A great deal of the electrical energy that is wasted as heat in wires could be saved if conventional conductors could be replaced with superconductors. But the superconducting state for a given material has limitations. Resistance returns if the temperature is raised above the superconductor's T_c, if the current through the substance becomes too large, or if it is placed in a magnetic field that is too strong.

Practical superconductors were developed in the 1960s and are now widely used in science and medicine.

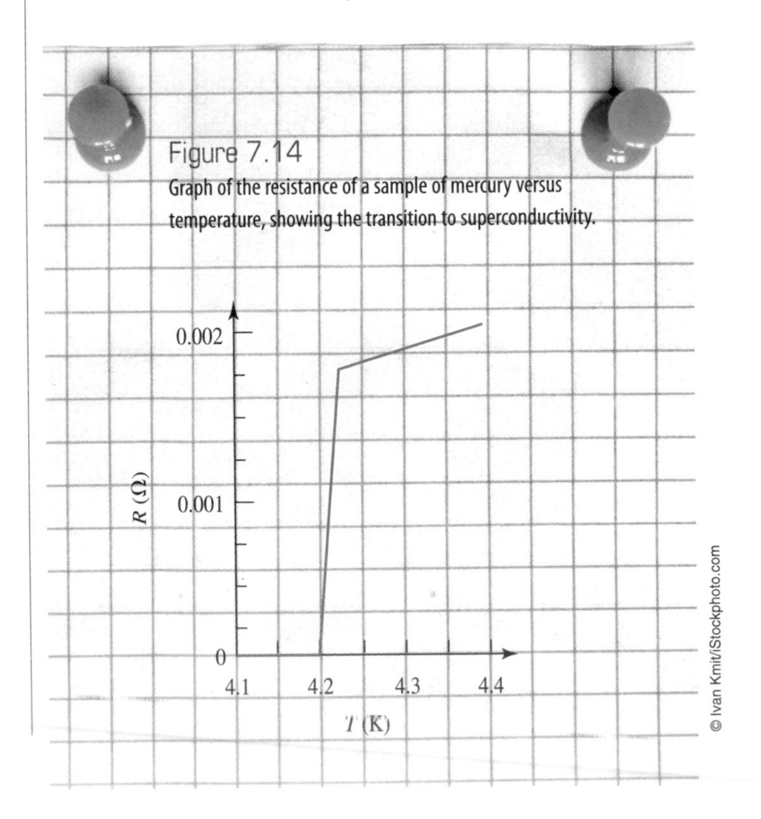

Figure 7.14

Graph of the resistance of a sample of mercury versus temperature, showing the transition to superconductivity.

Most of them are compounds of the element niobium. *Superconducting electromagnets*, the strongest magnets known, are used to study the effects of magnetic fields on matter and to direct high-speed charged particles. The world's highest-energy particle accelerator, located near the city of Geneva, Switzerland, uses superconducting electromagnets to guide subatomic particles as they are accelerated to nearly the speed of light. An entire experimental passenger train was built that levitated by superconducting electromagnets. *Magnetic resonance imaging* (MRI) uses superconducting electromagnets to form incredibly detailed images of the body's interior. (Magnetism and electromagnets are discussed in Chapter 8.)

Widespread practical use of these superconductors is severely limited because they must be kept cold using liquefied helium. Helium is very expensive and requires sophisticated refrigeration equipment to cool and to liquefy. Once a superconducting device is cooled to the temperature of liquid helium, bulky insulation equipment is needed to limit the flow of heat into the helium and the superconductor. These factors combine to make the so-called low-T_c superconductors unwieldy or uneconomical except in certain special applications when there are no alternatives.

But hope for wider use of superconductivity blossomed in 1987 when a new family of "high-T_c" superconductors was developed with critical temperatures as high as 90 K, since then pushed upward to about 140 K. This was an astounding breakthrough because these materials can be made superconducting through the use of liquid nitrogen (boiling point 77 K). Liquid nitrogen is widely available, inexpensive to produce compared to liquid helium, and can be used with much less sophisticated insulation. However, the new high-T_c superconductors are handicapped by a couple of unfortunate properties: They are brittle and consequently are not easily formed into wires, and they aren't very tolerant of strong magnetic fields or large electric currents. If these problems can be overcome, a new revolution in superconducting technology will likely occur.

7.4 Electric Circuits and Ohm's Law

An electric current will flow in a light bulb, a radio, or other such device only if an electric field is present to exert a force on the charges. A flashlight works because the batteries produce an electric field that forces electrons to flow through the light bulb. An *electric circuit* is any such system consisting of a battery or other electrical *power supply*, some electrical device like a light bulb, and wires or other conductors to carry the current to and from the device (see Figure 7.15). The power supply acts like a "charge pump": It forces charges to flow out of one terminal, go through the rest of the circuit, and flow into the other terminal. Electrons typically move through a circuit quite slowly, about 1 millimeter per second. In this respect, an electric circuit is much like the cooling system in a car in which the water pump forces coolant to flow through the engine, radiator, and the hoses connecting them.

The concepts of energy and work are used to quantify the effect of a power supply in a circuit. In a flashlight, for instance, the batteries cause electrons to flow through the bulb's filament. Since a force acts on the electrons and causes them to move through a distance, *work is done* on the electrons *by the batteries*. In other words, the batteries give the electrons energy. This energy is converted into internal energy and light as the electrons go through the light bulb. This leads to the concept of electric **voltage**.

DEFINITION

Voltage *The work that a charged particle can do divided by the size of the charge. The energy per unit charge given to charged particles by a power supply.*

$$V = \frac{\text{work}}{q} \qquad V = \frac{E}{q}$$

The SI unit of voltage is the volt (V), *which is equal to* 1 joule per coulomb. *Voltage is measured with a device called a* voltmeter.

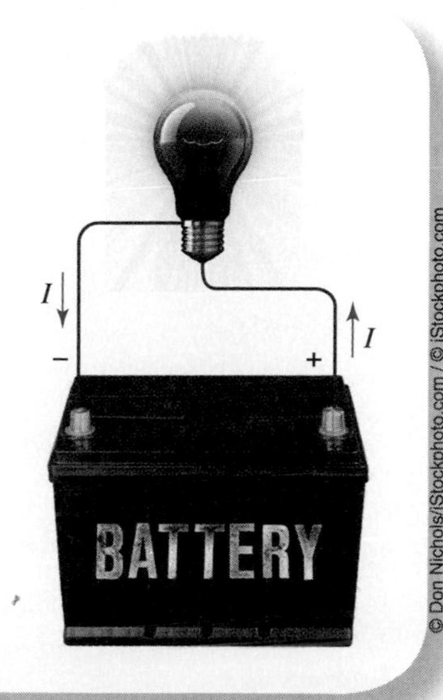

© Don Nichols/iStockphoto.com / © iStockphoto.com

Figure 7.15
A simple electric circuit. The battery supplies the energy or "pressure" needed to move the charges through the circuit. Charge does not build up anywhere; for each coulomb of charge that leaves the positive terminal, 1 coulomb enters the negative terminal.

A 12-volt battery gives 12 joules of energy to each coulomb of electric charge that it moves through a circuit. Each coulomb does 12 joules of work as it flows through the circuit. (Table 7.2 lists some typical voltages.)

If we return to the analogy of a battery as a charge pump, the voltage plays the role of pressure. A high voltage causing charges to flow in a circuit is similar to a high pressure causing a fluid to flow (Figure 7.16). Even when the circuit is disconnected from the power supply and there is no charge flow, the power supply still has a voltage. In this case, the electric charges have potential energy. (Voltage is also referred to as electric *potential.*)

The size of the current that flows through a conductor depends on its resistance and on the voltage causing the current. **Ohm's law**, named after its discoverer, Georg Simon Ohm, expresses the exact relationship.

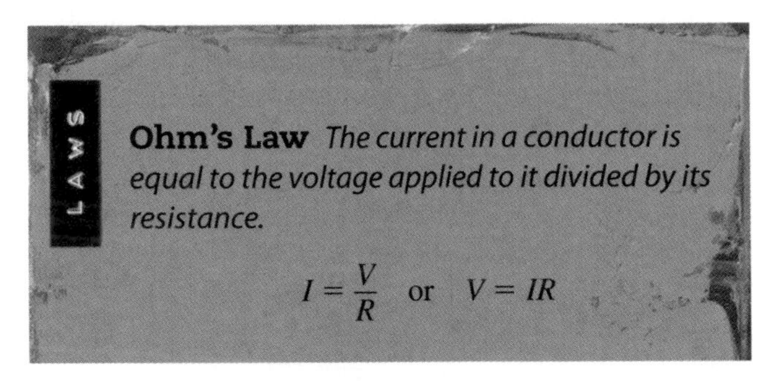

The units of measure are consistent in the two equations: If I is in amperes and R is in ohms, V will be in volts.

By Ohm's law, the higher the voltage for a given resistance, the larger the current. The larger the resistance for a given voltage, the smaller the current. By applying different-sized voltages to a given conductor, one can produce different-sized currents. A graph of the voltage versus the current will be a straight line whose slope is equal to the conductor's resistance (Figure 7.17). Reversing the polarity of the voltage (switching the "+" and "−" terminals) will cause the current to flow in the opposite direction.

Table 7.2 Examples of Common Voltages

Description	Voltage (V)
Nerve impulse	0.1
D-cell battery	1.5
Car battery	12
Wall socket (varies)	120
TV or computer monitor picture tube (typical)	20,000
Power plant generator (typical)	24,000
High-voltage transmission line (typical)	345,000

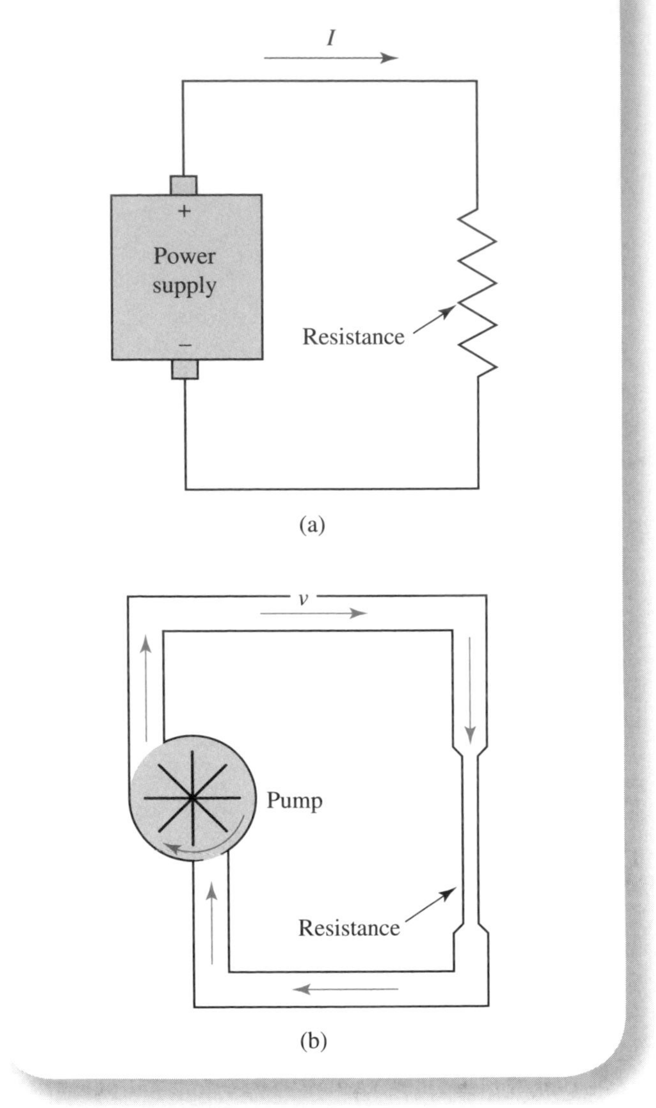

Figure 7.16
(a) The flow of charge in an electric circuit is much like (b) the flow of water through a closed pipe. The power supply corresponds to the water pump, and the resistance corresponds to the narrow segment of pipe. The pressure on the output side of the pump is much like the voltage on the "+" terminal of the power supply. The electric current corresponds to the rate of flow of the water.

EXAMPLE 7.1

A light bulb used in a 3-volt flashlight has a resistance equal to 6 ohms. What is the current in the bulb when it is switched on? By Ohm's law:

$$I = \frac{V}{R} = \frac{3\ \text{V}}{6\ \Omega}$$
$$= 0.5\ \text{A}$$

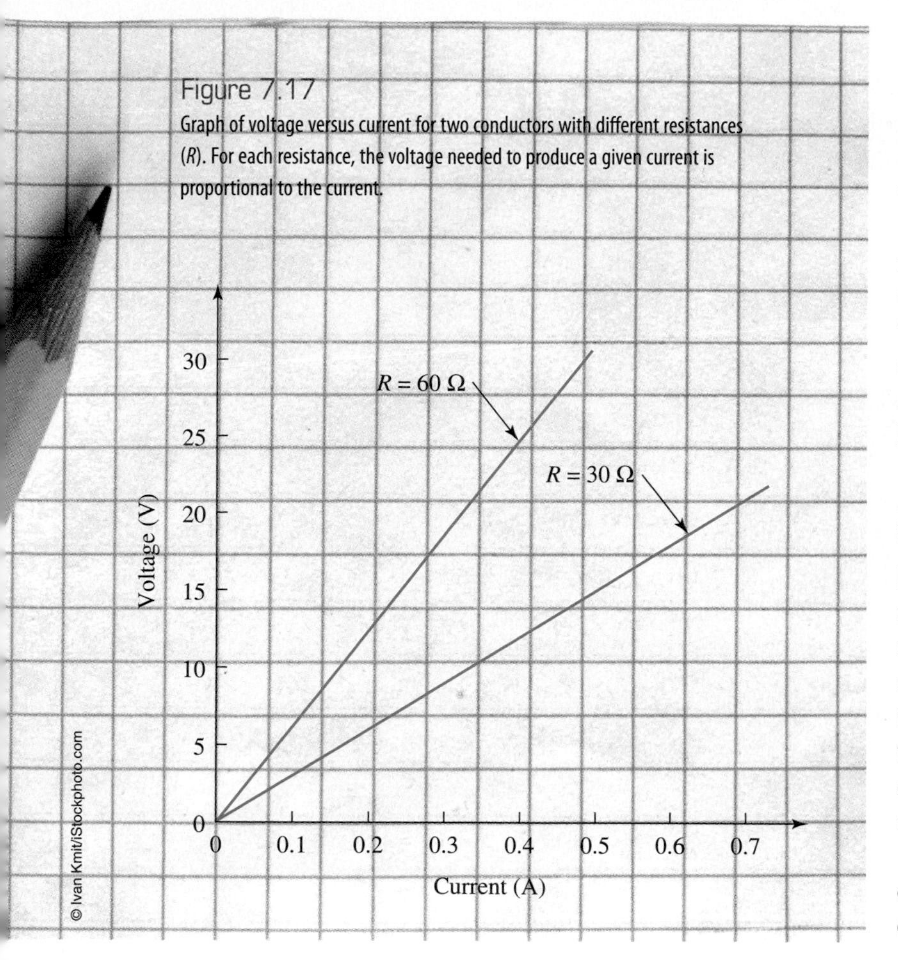

Figure 7.17
Graph of voltage versus current for two conductors with different resistances (R). For each resistance, the voltage needed to produce a given current is proportional to the current.

it generally has lower resistance when higher voltages are applied to it: Doubling the voltage will more than double the current. A graph of V versus I for ordinary tap water is less steep at higher voltages.

Many electrical devices are controlled by changing a resistance. The volume control on a radio or a television simply varies the resistance in a circuit. Turning up the volume reduces the resistance, so more current flows in the circuit, resulting in louder sound. A dimmer control used to change the brightness of the lights in a room works the same way.

Series and Parallel Circuits

In many situations, several electrical devices are connected to the same electrical power supply. A house may have a hundred different lights and appliances all connected to one cable entering the house. An automobile has dozens of devices connected to its battery. There are two basic ways in which more than one device can be connected to a single electrical power supply—by a series circuit and by a parallel circuit.

In a *series circuit*, there is only one path for the charges to follow, so the same current flows in each device (see Figure 7.19). In such a circuit, the voltage

EXAMPLE 7.2

A small electric heater has a resistance of 15 ohms when the current in it is 2 amperes. What voltage is required to produce this current?

$$V = IR = 2\text{ A} \times 15\ \Omega$$
$$= 30\text{ V}$$

Not all devices remain "ohmic"—that is, obey Ohm's law—as the voltage applied to them changes. Often, instead of remaining constant, the resistance of a conductor changes when the voltage changes. At higher voltages, a larger current flows through the filament of a light bulb, so its temperature is also higher. The resistance of the hotter filament is consequently greater (Figure 7.18). Some semiconductor devices, called *diodes*, are designed to have very low resistance when current flows through them in one direction but very high resistance when a voltage tries to produce a current in the other direction. Water with salt dissolved in

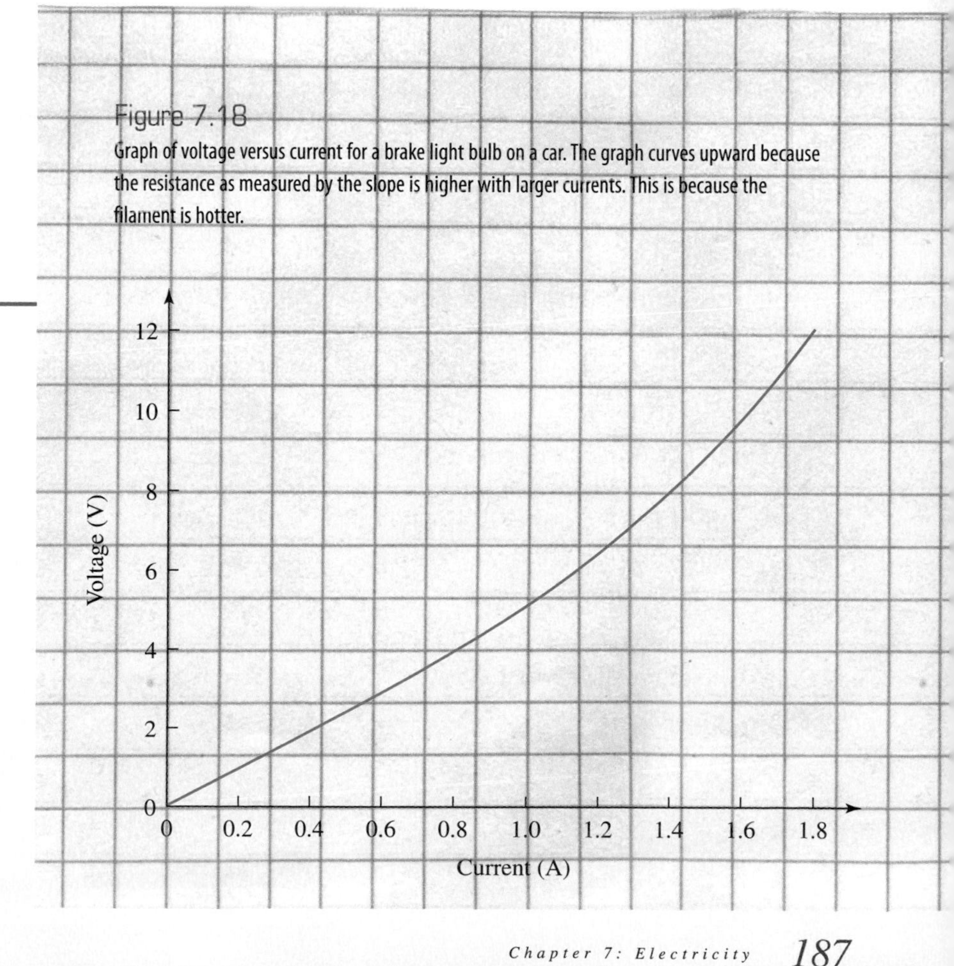

Figure 7.18
Graph of voltage versus current for a brake light bulb on a car. The graph curves upward because the resistance as measured by the slope is higher with larger currents. This is because the filament is hotter.

is divided among the devices: The voltage on the first device plus the voltage on the second device, and so on, equals the voltage of the power supply. For example, if three light bulbs with the same resistance are connected in series to a 12-volt battery, the voltage on each bulb is 4 volts. If the bulbs had different resistances, each one's "share" of the voltage would be proportional to its resistance.

Notice that the current in a series circuit is stopped if any of the devices breaks the circuit (Figure 7.20). A series circuit is not normally used with, say, a number of light bulbs because if one of them burns out, the current stops and all of the bulbs go out. A string of Christmas lights that flash at the same time uses a series circuit so that all the bulbs go on and off together.

In a *parallel circuit*, the current through the power supply is "shared" among the devices while each has the same voltage (Figure 7.21). The current flowing in the first device plus the current in the second device, and so on, equals the current output by the power supply. There is more than one path for the charges to follow—in this case, three. If one of the devices burns out or is removed, the others still function. The light bulbs in multiple-bulb light fixtures are in parallel so that if one bulb burns out, the others remain lit. Often, the two types of circuits are combined: One switch may be in series with several light bulbs that are in parallel.

EXAMPLE 7.3

Three light bulbs are connected in a parallel circuit with a 12-volt battery. The resistance of each bulb is 24 ohms. What is the current produced by the battery?

The voltage on each bulb is 12 volts. Therefore, the current in each bulb is

$$I = \frac{V}{R} = \frac{12\text{ V}}{24\ \Omega}$$
$$= 0.5\text{ A}$$

Figure 7.19

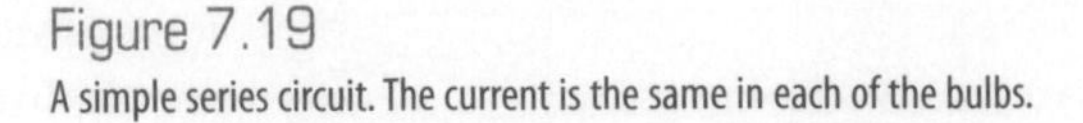
A simple series circuit. The current is the same in each of the bulbs.

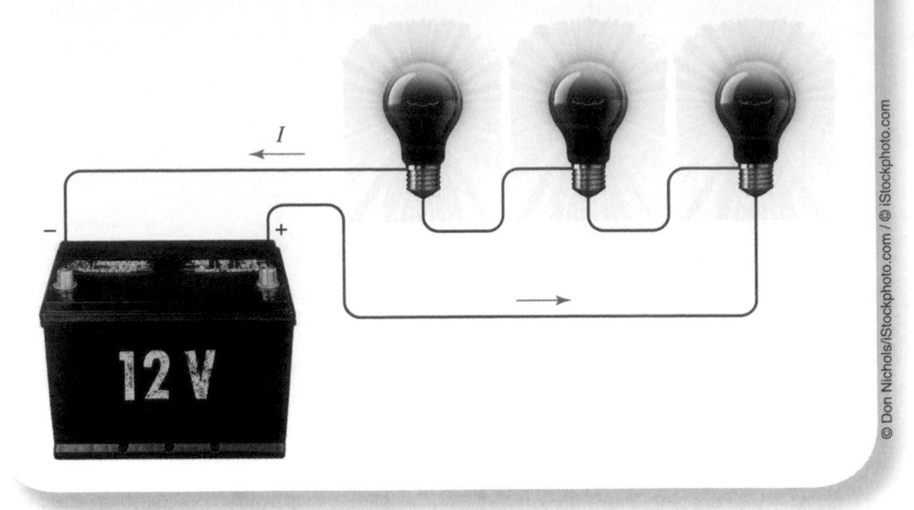

Figure 7.20

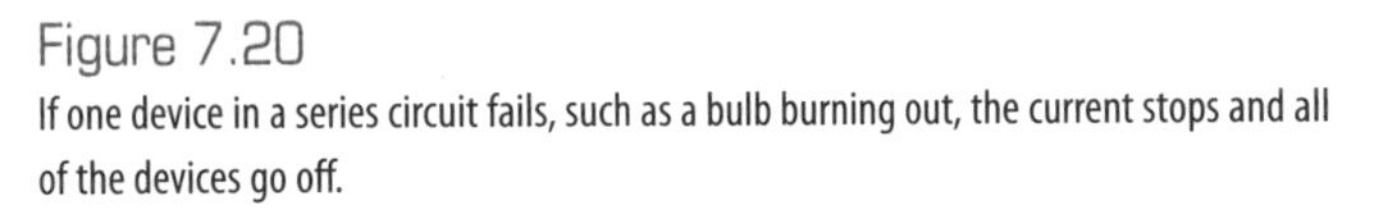
If one device in a series circuit fails, such as a bulb burning out, the current stops and all of the devices go off.

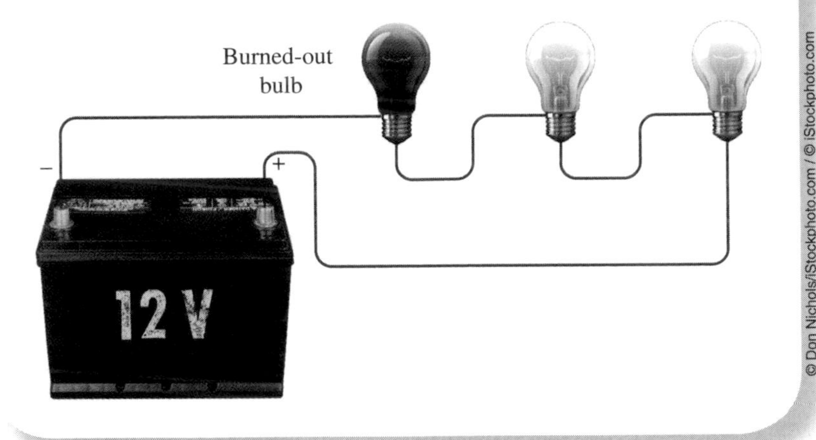

Figure 7.21

A simple parallel circuit. Each bulb has the same voltage (the voltage of the power supply). Because there are three separate pathways for the current, should one of the bulbs burn out (not shown), the others will remain lit.

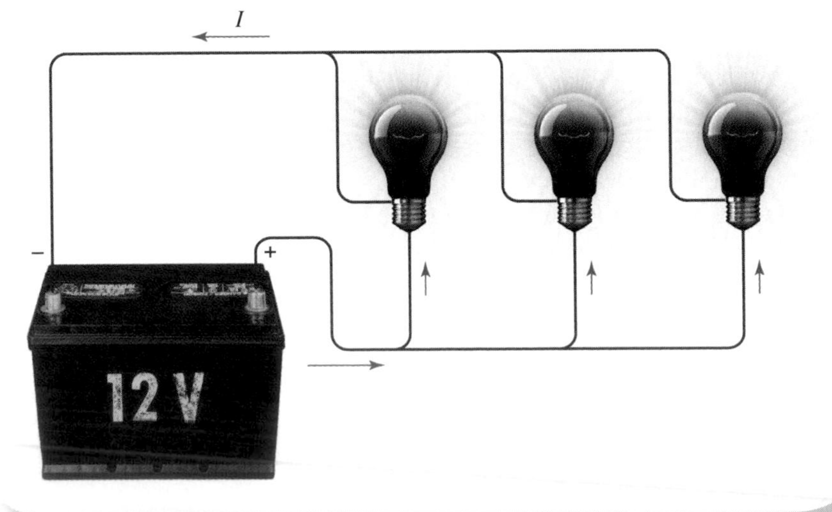

The total current supplied by the battery equals the sum of the currents in the three bulbs.

$$I = 0.5\ \text{A} + 0.5\ \text{A} + 0.5\ \text{A}$$
$$= 1.5\ \text{A}$$

The concept of voltage is quite general and is not restricted to electrical power supplies and electric circuits. Whenever there is an electric field in a region of space, a voltage exists because the field has the potential to do work on electric charges. The strength of an electric field can be expressed in terms of the voltage change per unit distance along the electric field lines.* For example, air conducts electricity when the electric field is strong enough to ionize atoms in the air. The minimum electric field strength required for this to happen is between 10,000 and 30,000 volts per centimeter, depending on the conditions. This means that if there is a spark one-fourth of an inch long between your finger and a doorknob, the voltage that causes the spark is at least 7,500 volts.

As transistors and other components on integrated circuit chips (ICs) are made smaller, even the low voltages that are used to make them operate (typically around 1 volt) produce very strong electric fields. Inside modern ICs, electric field strengths can reach 400,000 V/cm. Designers of ICs must keep this in mind because electric fields only about 25% stronger than this can disrupt circuit processes.

7.5 Power and Energy in Electric Currents

Since a battery or other electrical power supply must continually put out energy to cause a current to flow, it is important to consider the **power output**—the rate at which energy is delivered to the circuit. The power is determined by the voltage of the power supply and the current that is flowing. Think of it this way: The power output is the amount of energy expended per unit amount of time. The power supply gives a certain amount of energy to each coulomb of charge that flows through the circuit. Consequently, the energy output per unit time equals the *energy given to each coulomb of charge* multiplied by the *number of coulombs that flow through the circuit per unit time:*

$$\frac{\text{energy}}{\text{unit time}} = \frac{\text{energy}}{\text{coulomb}} \times \frac{\text{number of coulombs}}{\text{unit time}}$$

These three quantities are just the power, voltage, and current, respectively. Consequently, the power output of an electrical power supply is

$$\text{power} = \text{voltage} \times \text{current}$$
$$P = VI$$

The units work out correctly in this equation also: joules per coulomb (volts) multiplied by coulombs per second (amperes) equals joules per second (watts).

The power output of a battery is proportional to the current that it is supplying: the larger the current, the higher the power output.

EXAMPLE 7.4

In Example 7.1, we computed the current that flows in a flashlight bulb. What is the power output of the batteries?

Recall that the batteries produce 3 volts and that the current in the light bulb is 0.5 amperes. The power output is

$$P = VI = 3\ \text{V} \times 0.5\ \text{A}$$
$$= 1.5\ \text{W}$$

The batteries supply 1.5 joules of energy each second.

What happens to the energy delivered by an electrical power supply? In a light bulb, less than 5% is converted into visible light, and the rest becomes internal energy. Even the visible light emitted by a light bulb is absorbed eventually by the surrounding matter and transformed into internal energy. (Interior lighting is actually used to heat some buildings.) Electric motors in hair dryers, vacuum cleaners, and the like convert about 60% of their energy input into mechanical work or energy while the remainder goes to internal energy. The mechanical energy is generally dissipated as internal energy through friction. In a similar way, we can trace the energy conversions in other electrical devices and the outcome is the same: Most electrical energy eventually becomes internal energy.

* As defined in Section 7.2, the SI unit for the electric field is the newton per coulomb. Here we see an alternative, equivalent unit for the electric field, the volt/m. 1 N/C = 1 V/m.

AC is a kind of "wave."

Ordinary metal wire converts electrical energy into internal energy whenever there is a current flowing. You may have noticed when using a hair dryer that its cord becomes warm. This heating, called **ohmic heating**, occurs in any conductor that has resistance, even when the resistance is quite small. The huge cables used to conduct electricity from power plants to cities are heated by this effect. This heating represents a loss of usable energy.

The temperature that a current-carrying wire reaches due to ohmic heating depends on the size of the current and on the wire's resistance. Increasing the current in a given wire will raise its temperature. Many devices utilize this effect. The resistances of heating elements in toasters and electric heaters are chosen so that the normal operating current is large enough to heat them until they glow red hot and can toast bread or heat a room. The filament in an incandescent light bulb is made so thin that ohmic heating causes it to glow white hot and emit enough light to illuminate a room.

Ohmic heating is a major consideration in the design of sophisticated integrated circuit chips. Even though the currents flowing through the tiny transistors are extremely small, there are so many circuits in such a small space that special steps must be taken to make sure the heat that is produced is conducted away. Since a superconductor has zero resistance, there is no ohmic heating. The overall efficiencies of most electrical devices could be improved if regular wires could be replaced by superconductors. Superconducting transmission lines would allow electricity to be carried from a power plant to a city with no loss of energy. The limitations of currently known superconductors make such uses impracticable.

A sufficiently large current in any wire can cause it to become very hot—hot enough to melt any insulation around it or to ignite combustible materials nearby. Fuses and circuit breakers are put into electric circuits as safety devices to prevent dangerous overheating of wires. If something goes wrong or if too many devices are plugged into the circuit and the current exceeds the recommended safe limit for the size of wire used, the fuse or circuit breaker will automatically "break" the circuit and the current will stop. Designers of electric circuits in cars, houses, and other buildings must choose wiring that is large enough to carry the currents needed without overheating. They must also include fuses or circuit breakers that will disconnect a circuit if it is overloaded.

© Ruslan Olinchuk/iStockphoto.com

Most electrical devices are rated by the power that they consume in watts. The equation $P = VI$ can be used to determine how much current flows through the device when it is operating.

EXAMPLE 7.5

An electric hair dryer is rated at 1,875 watts when operating on 120 volts. What is the current flowing through it?

$$P = VI$$

$$1{,}875 \text{ W} = 120 \text{ V} \times I$$

$$\frac{1{,}875 \text{ W}}{120 \text{ V}} = I$$

$$I = 15.6 \text{ A}$$

The wires in the electric cord must be large enough to allow 15.6 amperes to flow through them without becoming dangerously hot.

The highest current that can flow in a particular wire without causing excessive heating depends on the size of the wire. This is one reason why electric utilities use high voltages in their electrical power supply systems. The electricity delivered to a city, subdivision, or individual house must be transmitted with wires. Since $P = VI$, using a large voltage makes it possible to transmit the same power with a smaller current. If low voltages

© Shane Cummins/iStockphoto.com

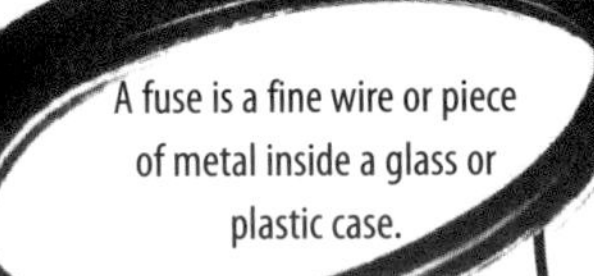

If current exceeds fuse's design limit, say 20 amperes, the metal melts away and the fuse "burns" out.

Burnt-out fuse = broken circuit

were used (say, 100 volts instead of the more typical 345,000 volts) much larger cables would have to be used to handle the larger currents.

Customers pay for the electricity supplied to them by electric companies based on the amount of energy they use. An electric meter keeps track of the total energy used by monitoring the power (rate of energy use) and the amount of time each power level is maintained. Recall the equation used to define power:

$$P = \frac{E}{t}$$

Therefore:

$$E = Pt$$

The amount of energy used is equal to the power times the time elapsed. If P is in watts and t is in seconds, E will be in joules.

EXAMPLE 7.6

If the hair dryer discussed in Example 7.5 is used for 3 minutes, how much energy does it use?

The power is 1,875 watts. To get E in joules, we must convert the 3 minutes into seconds.

$$t = 3 \text{ min} = 3 \times 60 \text{ s} = 180 \text{ s}$$

So:

$$E = Pt = 1{,}875 \text{ W} \times 180 \text{ s}$$

$$= 340{,}000 \text{ J}$$

This is a large quantity of energy—about the same as the kinetic energy of a small car going 60 mph (Example 3.6). Another comparison: A 150-pound person would have to climb 1,700 feet (about 170 floors) to gain 340,000 joules of potential energy.

A typical household can consume more than a billion joules of electrical energy each month. For this reason, a more appropriately sized unit of measure is used for electrical energy—the kilowatt-hour (kWh). Energy in kilowatt-hours is computed by expressing power in kilowatts (kW) and time in hours. The conversion factor between joules and kilowatt-hours is

$$1 \text{ kWh} = 1 \text{ kW} \times 1 \text{ h}$$

$$= 1{,}000 \text{ W} \times 3{,}600 \text{ s}$$

$$= 3{,}600{,}000 \text{ J}$$

In Example 7.6, the energy used by the hair dryer is about 0.1 kilowatt-hours. The cost of electricity varies from region to region, but it is typically around 10 cents per kilowatt-hour. This means that it costs about 1 cent to run the hair dryer 3 minutes. (Would you climb 1,700 feet for 1 cent?)

Perhaps you have wondered why a common 1.5-volt D-cell battery is larger than a 9-volt battery used in smoke alarms and other common electrical devices. The voltage of a battery really has nothing to do with its physical size. Different 1.5-volt batteries range from the size of a button (for wristwatches) to larger than a beer can. The size is more an indication of the amount of electrical energy stored in the battery. A large battery can supply the same current (and the same power) for a longer time than a small battery with the same voltage. In applications with small power requirements, like hearing aids and calculators, even a small battery can provide enough stored electrical energy to operate the device for up to a year or more.

7.6 AC and DC

The electric current supplied by a battery is different from the current supplied by a normal household wall socket. Batteries supply **direct current (DC)**, and household outlets supply **alternating current (AC)**. A DC power supply, such as a battery, causes a current to flow in a fixed direction in a circuit (Figure 7.22). The current flows out of the positive (+) terminal of the power supply, moves through the circuit, and flows into the negative (−) terminal of the power supply. If the total resistance in the circuit doesn't change, the size of the current remains constant (as long as the battery doesn't run down). A graph of the current I versus time t is simply a horizontal line.

Radio-frequency electric meters, like this one, allow electric companies to remotely collect information on a household's energy usage.

Courtesy of Chapel House Photography

In an AC power supply, the polarity of the two output terminals switches back and forth—the voltage alternates. This causes the current in any circuit connected to the power supply to alternate as well. It flows counterclockwise, then clockwise, then back to counterclockwise, and so on. All the while the size of the current is increasing, then decreasing, and so forth. A graph of the current in an AC circuit shows this variation in the size and direction of the current. (When I goes below zero, it means that the direction has reversed; see Figure 7.23).

We have seen this kind of oscillation before in Section 2.6 and throughout Chapter 6. One can even think of

Figure 7.22

Direct current. The current flows in one direction and doesn't increase or decrease, as shown in the graph of current versus time.

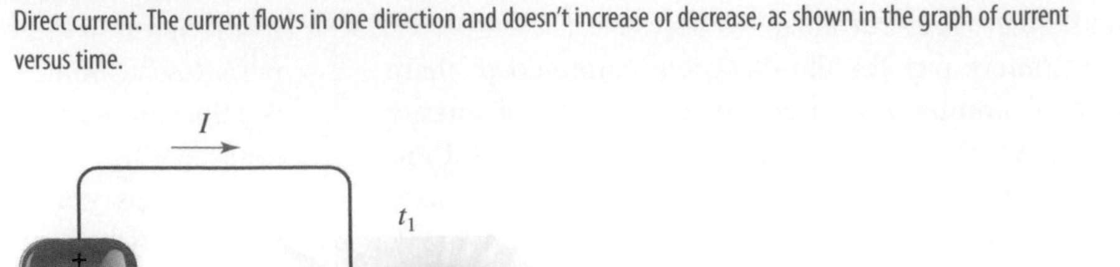

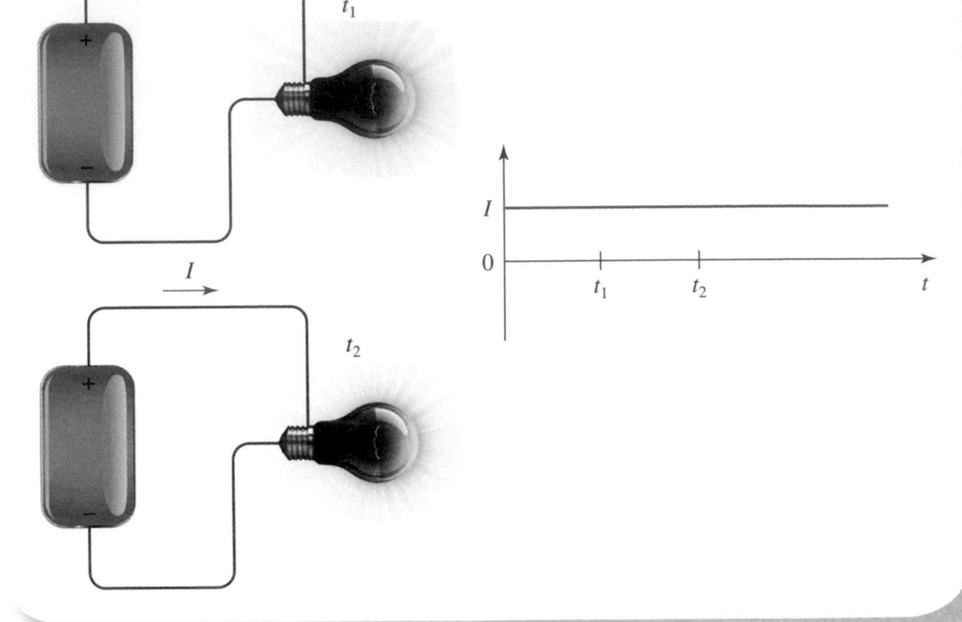

© iStockphoto.com / © Russell Tate/iStockphoto.com

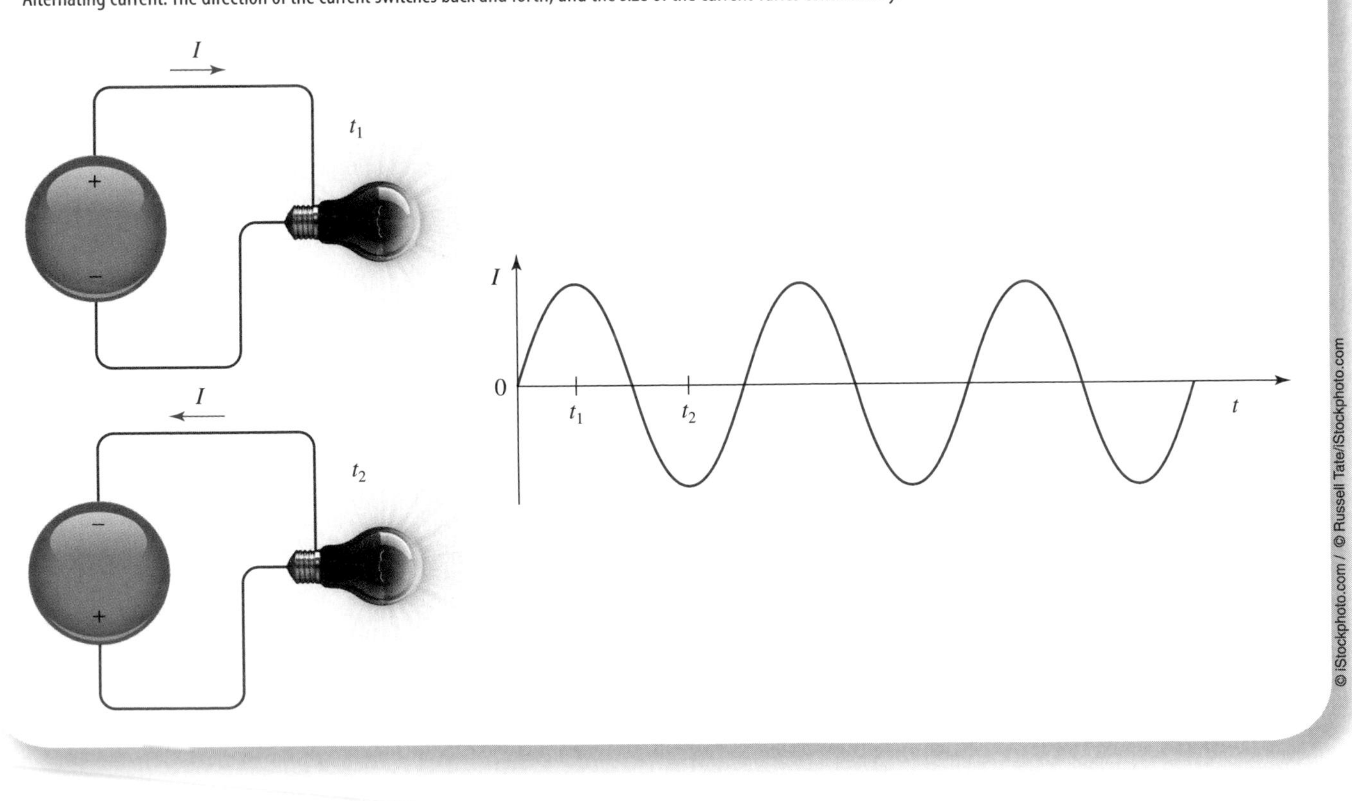

Figure 7.23

Alternating current. The direction of the current switches back and forth, and the size of the current varies continuously.

© iStockphoto.com / © Russell Tate/iStockphoto.com

AC as a kind of "wave" causing the charges in a conductor to oscillate back and forth. Almost all public electric utilities in the United States supply 60-hertz AC. The voltage between the two slots in a wall outlet oscillates back and forth 60 times per second. (In Europe, the standard frequency of AC is 50 hertz.)

Some electronic devices (such as light bulbs) can operate on AC or DC, while others require one or the other. Electric motors and generators must be designed to operate on or to produce either AC or DC. There are devices that can convert an AC voltage to a DC voltage, and vice versa. Batteries can produce direct current only. For this reason, automobiles have DC electrical systems. (The alternator in an automobile generates AC, which is then converted into DC to be compatible with the battery.)

Alternating current has one distinct advantage over DC: Simple, highly efficient devices called *transformers* can "step up" or "step down" AC voltages. This makes it possible to generate AC at a power plant at some intermediate voltage, step it up to a very high voltage (typically over 300,000 volts) for economical transmission, and then step it down again to lower voltages for use in homes and industries. There is no counterpart of the transformer for DC. Another important use of AC is in electronic sound equipment. For example, if a 440-hertz tone is recorded on tape and then played back, the "signal" going to the speaker will be an alternating current with a frequency of 440 hertz. (We will discuss transformers and sound reproduction in Chapter 8.)

questions&problems

Questions

(▶ Indicates a review question, which means it requires only a basic understanding of the material to answer. The others involve integrating or extending the concepts presented thus far.)

1. ▶ All matter contains both positively and negatively charged particles. Why do most things have no net charge?
2. A particular solid is electrically charged after it is rubbed, but it is not known whether its charge is positive or negative. How could you determine which charge it has by using a piece of plastic and fur?
3. ▶ What is a positive ion? A negative ion?
4. What remains after a hydrogen atom is positively ionized?
5. ▶ Describe the similarities and the differences between the gravitational force between two objects and the electrostatic force between two charged objects.
6. In otherwise empty space, what would happen if the size of the electrostatic force acting between two positively charged objects was exactly the same as that of the gravitational force acting between them? What would happen if they were moved closer together or farther apart?
7. a) A negatively charged iron ball (on the end of a plastic rod) exerts a strong attractive force on a penny even though the penny is neutral. How is that possible?
   b) The penny accelerates toward the ball, hits it, and then is immediately repelled. What do you think causes the sudden change from an attractive to a repulsive force on contact between the penny and the ball?
8. ▶ What is an electric field? Sketch the shape of the electric field around a single proton.
9. ▶ At one moment during a storm, the electric field between two clouds is directed toward the east. What is the direction of the force on any electron in this region? What is the direction of the force on a positive ion in this region?
10. ▶ Explain what an electrostatic precipitator is and how it works.
11. Salt water contains an equal number of positive and negative ions. When salt water is flowing through a pipe, does it constitute an electric current?
12. ▶ Materials can be classified into four categories based on the ease with which charges can flow through them. Give the names of these categories and describe each one.
13. A solid metal cylinder has a certain resistance. It is then heated and carefully stretched to form a longer, thinner cylinder. After it cools, will its resistance be the same as, greater than, or less than what it was before?
14. A student using a sensitive meter that measures resistance finds that the resistance of a thin wire is changed slightly when it is picked up with a bare hand. What causes the change in the resistance, and does it increase or decrease?
15. If a new material is found that is a superconductor at all temperatures, what parts of some common electric devices would definitely *not* be made out of it?
16. ▶ Explain what current, resistance, and voltage are.
17. ▶ Make a sketch of a simple electric circuit.

18. ▶Describe Ohm's law.

19. A power supply is connected to two bare wires that are inserted into a glass of salt water. The resistance of the water decreases as the voltage is increased. Sketch a graph of the voltage versus the current in the water showing this type of behavior.

20. ▶There are two basic schemes for connecting more than one electrical device in a circuit. Name, describe, and give the advantages of each.

21. Make a sketch of an electric circuit that contains a switch and two light bulbs connected in such a way that if either bulb burns out the other still functions, but if the switch is turned off, both bulbs go out.

22. Two 1.5-volt batteries are connected in series in an electric circuit. Use the concept of energy to explain why this combination is equivalent to a single 3-volt battery. When connected in parallel, what are two 1.5-volt batteries equivalent to?

23. An electrical supply company sells two models of 100-watt power supplies (the maximum power output is 100 W), one with an output of 12 V and the other 6 V. What can you conclude about the maximum current that the two power supplies can produce?

24. A simple electric circuit consists of a constant-voltage power supply and a variable resistor. What effect does reducing the resistance have on the current in the circuit and on the power output of the power supply?

25. ▶What is the purpose of having fuses or circuit breakers in electric circuits? How should they be connected in circuits so they will be effective?

26. A 20-A fuse in a household electric circuit burns out. What catastrophe could occur if it is replaced by a 30-A fuse?

27. ▶Why is it economical to use extremely high voltages for the transmission of electrical power?

28. ▶Explain what AC and DC are. Why is AC used by electric utilities? Why is DC used in flashlights?

29. If AC in a circuit can be thought of as a wave, which kind is it, longitudinal or transverse?

30. If the electric utility company where you live suddenly changed the frequency of the AC to 20 Hz, what problems might this cause?

Problems

1. During 30 seconds of use, 250 C of charge flow through a microwave oven. Compute the size of the electric current.

2. A defibrillator used to send an electrical shock to the heart of a patient suffering a heart attack delivers 0.16 C of charge to the body in a time of 10 ms. What is the current associated with this potentially life-saving action? (Note: Only a small portion of this current actually flows through the heart in a typical defibrillating procedure, insofar as much of the charge is diffused throughout the patient's skin and thoracic cavity.)

3. A current of 0.7 A goes through an electric motor for 1 min. How many coulombs of charge flow through it during that time?

4. A calculator draws a current of 0.0001 A for 5 min. How much charge flows through it?

5. A current of 12 A flows through an electric heater operating on 120 V. What is the heater's resistance?

6. A 120-V circuit in a house is equipped with a 20-A fuse that will "blow" if the current exceeds 20 A. What is the smallest resistance that can be plugged into the circuit without causing the fuse to blow?

7. The resistance of each brake light bulb on an automobile is 6.6 Ω. Use the fact that cars have 12-V electrical systems to compute the current that flows in each bulb.

8. The light bulb used in a computer projector has a resistance of 80 Ω. What is the current through the bulb when it is operating on 120 V?

9. The resistance of the skin on a person's finger is typically about 20,000 Ω. How much voltage would be needed to cause a current of 0.001 A to flow into the finger?

10. A 150-Ω resistor is connected to a variable-voltage power supply.
    a) What voltage is necessary to cause a current of 0.3 A in the resistor?
    b) What current flows in the resistor when the voltage is 18 V?

11. Compute the power consumption of the electric heater in Problem 5.

12. An electric eel can generate a 400-V, 0.5-A shock for stunning its prey. What is the eel's power output?

13. An electric train operates on 750 V. What is its power consumption when the current flowing through the train's motor is 2,000 A?

14. All of the electrical outlets in a room are connected in a single parallel circuit (Figure 7.24). The circuit is equipped with a 20-A fuse, and the voltage is 120 V.
    a) What is the maximum power that can be supplied by the outlets without blowing the fuse?
    b) How many 1,200-W appliances can be plugged into the sockets without blowing the fuse?

Figure 7.24

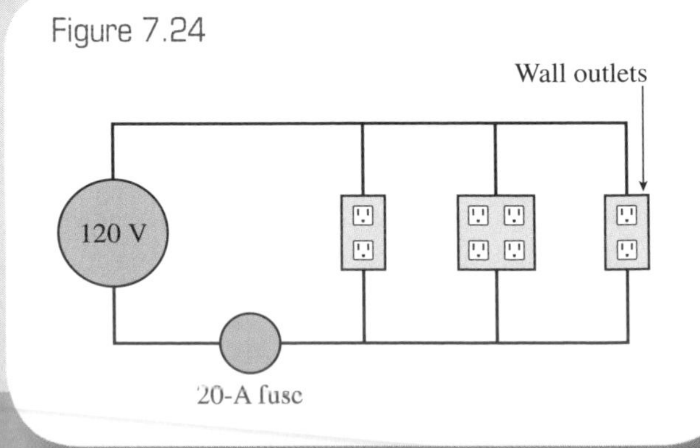

15. A car's headlight consumes 40 W when on low beam and 50 W when on high beam.
    a) Find the current that flows in each case ($V = 12$ V).
    b) Find the resistance in each case.
16. Find the current that flows in a 40-W bulb used in a house ($V = 120$ V). Compare this with the answer to the first part of Problem 15.
17. An electric clothes dryer is rated at 4,000 W. How much energy does it use in 40 min?
18. A clock consumes 2 W of electrical power. How much energy does it use each day?
19. Which costs more, running a 1,200-W hair dryer for 5 min or leaving a 60-W lamp on overnight (10 h)?
20. A representative lightning strike is caused by a voltage of 200,000,000 V and consists of a current of 1,000 A that flows for a fraction of a second. Calculate the power.
21. A toaster operating on 120 V uses a current of 9 A.
    a) What is the toaster's power consumption?
    b) How much energy does it use in 1 min?
22. A certain electric motor draws a current of 10 A when connected to 120 V.
    a) What is the motor's power consumption?
    b) How much energy does it use during 4 h of operation? Express the answer in joules and in kilowatt-hours.
23. The generator at a large power plant has an output of 1,000,000,000 W at 24,000 V.
    a) If it were a DC generator, what would be the current in it?
    b) What is its energy output each day—in joules and in kilowatt-hours?
    c) If this energy is sold at a price of 10 cents per kilowatt-hour, how much revenue does the power plant generate each day?
24. A light bulb is rated at 60 W when connected to 120 V.
    a) What current flows through the bulb in this case?
    b) What is the bulb's resistance?
    c) What would be the current in the bulb if it were connected to 60 V, assuming the resistance stays the same?
    d) What would be its power consumption in this case?
25. An electric car is being designed to have an average power output of 4,000 W for 2 h before needing to be recharged. (Assume there is no wasted energy.)
    a) How much energy would be stored in the charged-up batteries?
    b) The batteries operate on 30 V. What would the current be when they are operating at 4,000 W?
    c) To be able to recharge the batteries in 1 h, how much power would have to be supplied to them?
26. The resistance of an electric heater is 10 Ω when connected to 120 V. How much energy does it use during 30 min of operation?
27. Your cell phone typically consumes about 400 mW of power while used to text a friend. If the phone is operated using a lithium-ion battery with a voltage of 3.6 V, what is the current flowing through the cell phone circuitry under these circumstances? Compare your result with the data given in Table 7.1.
28. A common three-way bulb has two filaments and is wired with three contacts in its base. When used in a three-way lamp, the switch successively sends current first through a high-resistance filament to produce a dim bulb, then through a lower resistance filament to yield a brighter bulb, and finally through both filaments to give the highest brightness bulb.
    a) If a particular three-way bulb is rated at 40-, 60-, and 100-watts power and is operated as part of a circuit delivering 120 V, how much current is drawn by the light fixture when used on each of the three possible power/brightness settings?
    b) If the bulb is left on for 90 minutes in the second switch position, how much energy, measured in both joules and kilowatt-hours, does it consume? If the local cost of electrical energy is $0.08 per kWh, how much does it cost the homeowner to run the lamp for this period of time?

***remember**

There are more quizzes and study tools online at 4ltrpress.cengage.com/physics.

Electromagnetism and EM Waves

sections

8.1 Magnetism

Magnetism was first observed in a naturally occurring ore called *lodestone*. Lodestones were fairly common around Magnesia, an ancient city in Asia Minor. Small pieces of iron, nickel, and certain other metals are attracted by lodestones, much as pieces of paper are attracted by charged plastic. The Chinese were probably the first to discover that a piece of lodestone will orient itself north and south if suspended by a thread or floated in water on a piece of wood. The *compass* revolutionized navigation because it allowed mariners to determine the direction of north even at sea in cloudy weather. It was also one of the few useful applications of magnetism up to the nineteenth century.

Now magnets are made into a variety of sizes and shapes out of special alloys that exhibit much stronger magnetism than lodestone. For example, they are the first line of defense against people trying to smuggle weapons onto passenger planes or into schools, government buildings, and many other places. Metal detectors probe your clothing and body without physically touching you, looking for metal that could be part of a gun, a knife, or other dangerous object. Although metal detectors operate on electricity, it is magnetism that probes you. Brief magnetic pulses are sent through you and around you, typically about 100 times a second. The device carefully monitors how swiftly each magnetic pulse dies out. Any metal object encountered by a pulse is induced to produce its own magnetic pulse, which affects how rapidly the total pulse dies out. Sophisticated electronics in the metal detectors sense this change and signal that metal is present. They detect iron and other metals that ordinary magnets attract as well as metals like aluminum and gold that do not respond to magnets.

All simple magnets exhibit the same compass effect—one end or part of it is attracted to the north, and the opposite end or part is attracted to the south. The north-seeking part of a magnet is called its **north pole**, and the south-seeking part is its **south pole**. All magnets have both poles. If a magnet is broken into pieces,

© Visuals Unlimited/ Getty Images

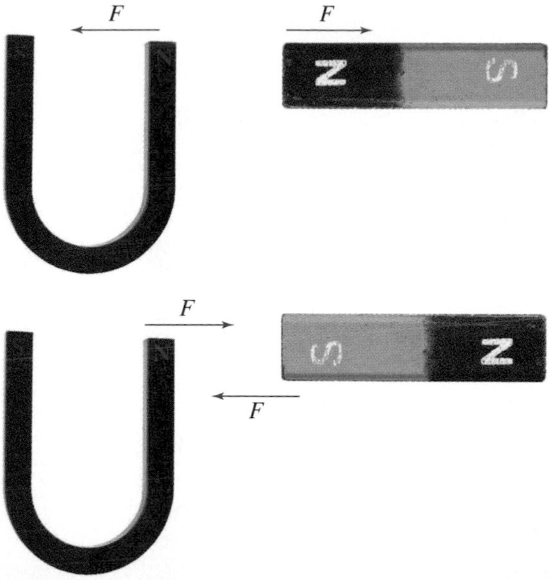

Figure 8.1

The poles of two magnets exert forces on each other. Like poles repel each other, and unlike poles attract each other.

© iStockphoto.com / © iStockphoto.com

each part will have its own north and south poles. The south pole of one magnet exerts a mutually attractive force on the north pole of a second magnet. The south poles of two magnets repel each other, as do the north poles (Figure 8.1). Simply: *Like poles repel, unlike poles attract* (just as with electric charges).

Metals that are strongly attracted by magnets are said to be *ferromagnetic*. Such materials have magnetism induced in them when they are near a magnet. If a piece of iron is brought near the south pole of a magnet, the part of the iron nearest the magnet has a north pole induced in it, and the part farthest away has a south pole induced in it (see Figure 8.2). Once the iron is removed from the vicinity of the magnet, it loses most of the induced magnetism. Some

© iStockphoto.com

Figure 8.2
When a piece of ferromagnetic material (like iron) is brought near a magnet, it has magnetism induced in it. That is why it is attracted by the magnet.

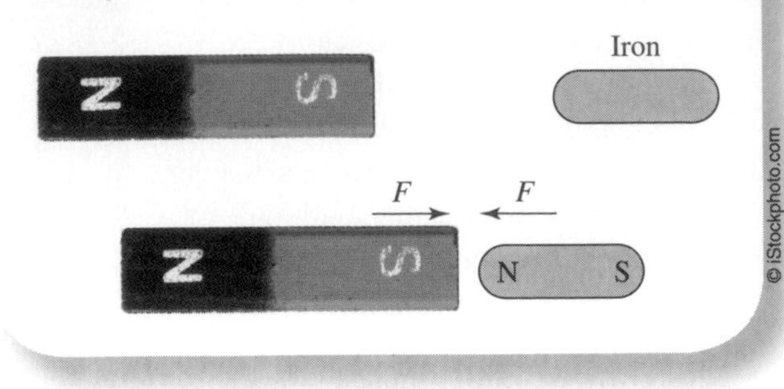

Figure 8.3
The forces on the poles of a compass placed in a magnetic field are in opposite directions. This causes the needle to turn until it is aligned with the field.

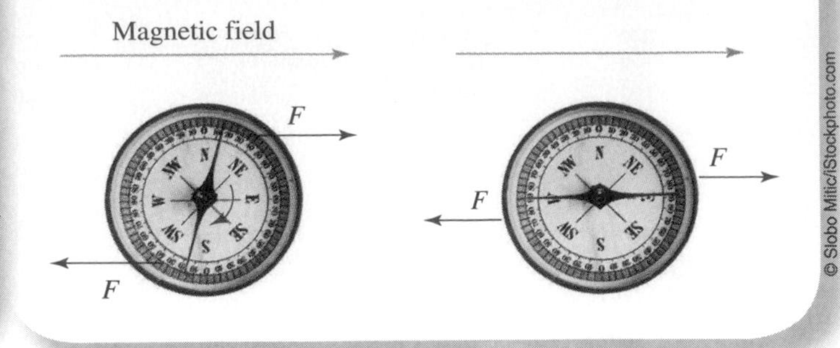

ferromagnetic metals actually retain the magnetism induced in them—they become *permanent magnets.* Common household magnets and compass needles are made of such metals. Ferromagnetism is also the basis of magnetic data recording, but more on this later.

As with gravitation and electrostatics, it is useful to employ the concept of a field to represent the effect of a magnet on the space around it. A *magnetic field* is produced by a magnet and acts as the agent of the magnetic force. The poles of a second magnet experience forces when in the magnetic field: Its north pole has a force in the same direction as the magnetic field, but its south pole has a force in the opposite direction. A compass can be thought of as a "magnetic field detector" because its needle will always try to align itself with a magnetic field (Figure 8.3). The shape of the magnetic field produced by a magnet can be "mapped" by noting the orientation of a compass at various places nearby. Magnetic *field lines* can be drawn to show the shape of the field, just as electric field lines are used to show the shape of an electric field (Figure 8.4). The direction of a field line at a particular place is the direction that the north pole of a compass needle at that location points.

Because magnets respond to magnetic fields, the fact that compass needles point north indicates that the Earth itself has a magnetic field. The shape of the Earth's field has been mapped carefully over the course of many centuries because of the importance of compasses in navigation. The Earth's magnetic field has the same general shape as the field around a bar magnet with its poles tilted about 11° with respect to the axis of rotation (Figure 8.5). The direction of "true north" shown on maps is determined by the orientation of the Earth's axis of rotation. (The axis is aligned closely with Polaris, the North Star.)

Because of the tilt of the Earth's "magnetic axis," at most places on Earth, compasses do not point to true north. For example, in the western two-thirds of the

Figure 8.4
(a) Sketch of the magnetic field in the space around a bar magnet. Note that the field lines point toward the south pole and away from the north pole. (b) Photograph of iron filings around a magnet. Each tiny piece of iron becomes magnetized and aligns itself with the magnetic field.

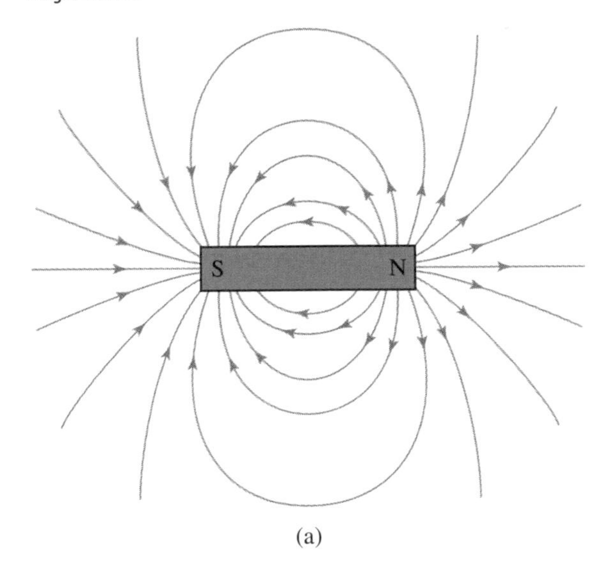

(a)

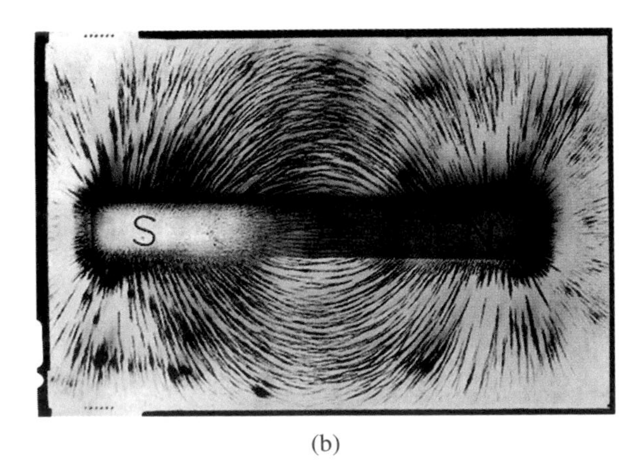

(b)

United States, compasses point to the right (east) of true north, while in New England compasses point to the left (west) of true north. The difference, in degrees, between the direction of a compass and the direction of true north varies from place to place and is referred to as the *magnetic declination.* (In parts of Alaska, the magnetic declination is as high as 25° east.) This must be taken into account when navigating with a compass.*

Incidentally, the Earth's field is responsible for the magnetism in lodestone. This naturally occurring ferromagnetic ore is weakly magnetized by the Earth's magnetic field. Another thing to note about the Earth's magnetic field: The Earth's *north* magnetic pole is at (near) its *south* geographic pole, and vice versa. Why?

1. The north pole of a magnet is attracted to the south pole of a second magnet.
2. The north pole of a compass needle points to the north.

Therefore, a compass's north pole points at the Earth's south magnetic pole. This is not a physical contradiction: It is a result of naming the poles of a magnet after directions instead of, say, + and −, or A and B.

Some organisms use the Earth's magnetic field to aid navigation. Although the biological mechanisms that they employ have not yet been fully identified, certain species of fish, frogs, turtles, birds, newts, and whales are able to sense the strength of the Earth's field or its direction (or both). The former allows the animal to determine its approximate latitude (how far north or south it is) because the Earth's magnetic field is stronger near the magnetic poles (Figure 8.5). Some migratory species travel thousands of miles before returning home, guided—at least in part—by sensing the Earth's magnetic field.

Superconductors, so named because of their ability to carry electric current with zero resistance, react to magnetic fields in a rather startling fashion. When in the superconducting state, the material will expel any magnetic field from its interior. This phenomenon, known as the *Meissner effect,* is why strong magnets are levitated when placed over a superconductor. When trying to determine whether a material is in the superconducting state, it is easier to test for the presence of the Meissner effect than it is to see if the resistance is exactly zero.

You have probably noticed that magnetism and electrostatics are very similar: There are two kinds of poles and two kinds of charges. Like poles repel as do like charges. There are magnetic fields and electric fields. However, there are some important differences. Each kind of charge can exist separately, while magnetic poles always come in pairs. (Modern theory indicates the possible existence of a particular type of subatomic "elementary particle" that does have a single magnetic pole. As of this writing, such a "magnetic monopole" has not been found.) Furthermore, all conventional matter contains positive and negative charges (protons and electrons) and can exhibit electrostatic effects by being "charged." But, with the exception of

Figure 8.5

The Earth's magnetic field. It is shaped as if there were a huge bar magnet deep inside the Earth, tilted 11° relative to the Earth's axis of rotation.

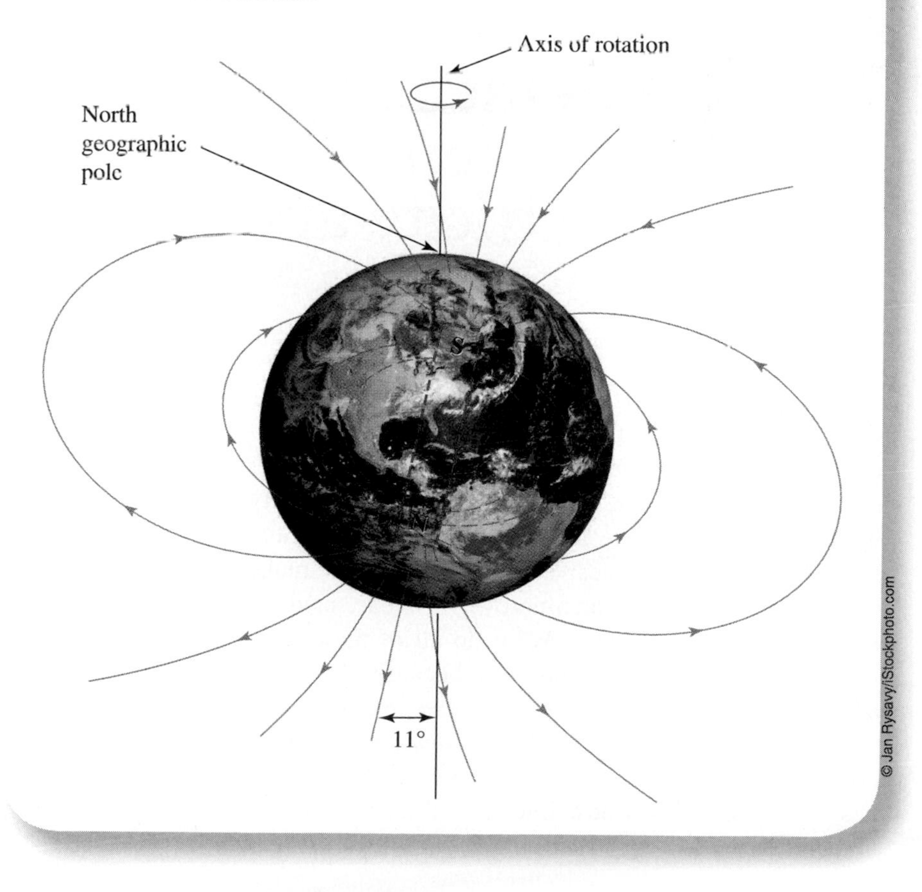

© Jan Rysavy/iStockphoto.com

* The Earth's magnetic poles actually move around: The south magnetic pole is not exactly where it was 20 years ago. At most places on Earth, a compass does not point exactly in the same direction it did 20 years ago. Detailed maps used for navigation are corrected periodically to show any change in magnetic declination.

The Meissner effect. A magnet levitating above a high-T_c superconductor (left). One example of the Meissner effect in everyday life is the levitating, high-speed train, like this one (right) in Germany.

© Nigel Treblin/AFP/Getty Images / © Bill Pierce/Time & Life Pictures/Getty Images / © Nicholas Belton/iStockphoto.com

ferromagnetic materials, most matter shows very little response to magnetic fields.

We should also point out that the electrostatic and magnetic effects described so far are completely independent. Magnets have no effect on pieces of charged plastic, for instance, and vice versa. This is the case as long as there is no motion of the objects or changes in the strengths of the electric and magnetic fields. As we shall see in the following sections, a number of fascinating and useful interactions between electricity and magnetism take place when motion or change in field strength occurs.

The Concept Map on your Review Card summarizes the similarities and differences between electrostatics and magnetism.

8.2 Interactions between Electricity and Magnetism

Consider the following items that we usually take for granted: electric motors in hair dryers, vacuum cleaners, computer disk drives, elevators, and countless other devices; generators that produce most of the electricity we use; speakers, audio and videotape recorders, and high-fidelity microphones; and the waves that make radios, cell phones, radar, microwave ovens, medical x-rays, and our eyes work. What do all of these have in common? They all are possible because electricity and magnetism interact with each other in basic—and very useful—ways. The word *electromagnetic*, which appears dozens of times in this chapter, is perhaps the best indication of just how intertwined these two phenomena are.

© iStockphoto.com

Before we delve into these interactions, let's summarize and review the key aspects of electrostatics and magnetism presented in Sections 7.1, 7.2, and 8.1:

- Electric charges produce electric fields in the space around them (see Figure 7.8).
- An electric field, regardless of its origin, causes a force on any charged object placed in it (see Figure 7.9).
- Magnets produce magnetic fields in the space around them (see Figure 8.4).
- A magnetic field, regardless of its origin, causes forces on the poles of any magnet placed in it (see Figure 8.3).

These statements have been worded in a particular way because, as we shall see, it is the electric and magnetic *fields* that are involved in the interplay between electricity and magnetism. In this section, we describe three basic observations of these interactions and discuss some useful applications of them. In Section 8.3, we summarize the underlying concepts in the form of two principles like the four statements above. We emphasize the ways in which electricity and magnetism interact and how these help us understand such things as how many electrical devices work and what light and other electromagnetic waves are. Fortunately, we can do this without having to go into the complex underlying causes of these interactions.

The first of the three observations is the basis of electromagnets.

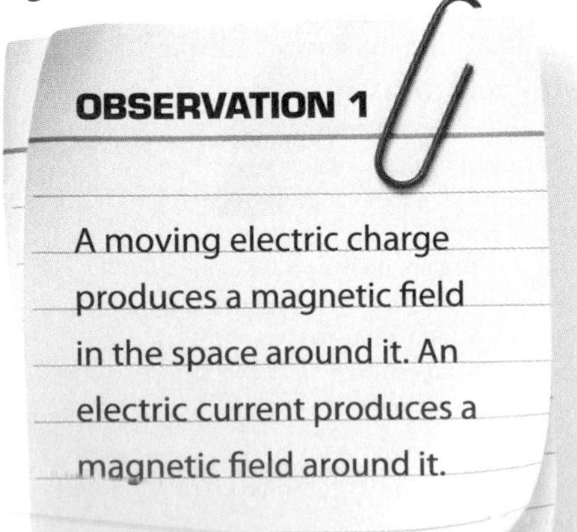

OBSERVATION 1

A moving electric charge produces a magnetic field in the space around it. An electric current produces a magnetic field around it.

Figure 8.6
Magnetic field produced by (a) a moving charge and by (b) a wire carrying DC. The field lines are circles concentric with the path of the charges. (The power source for the current is not shown.)

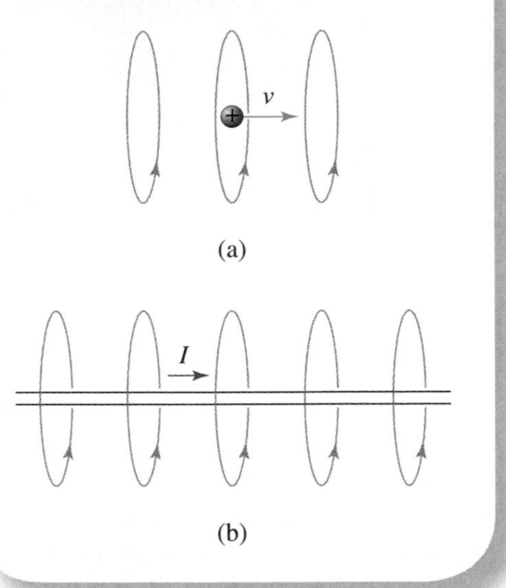

A single charged particle creates a magnetic field only when it is moving. The magnetic field produced is in the shape of circles around the path of the charge (see Figure 8.6). For a steady (DC) current, which is basically a succession of moving charges, in a wire, the field is steady and its strength is proportional to the size of the current and inversely proportional to the distance from the wire. (The field is quite weak unless the current is large. A current of 10 amperes or more will produce a field strong enough to be detected with a compass; see Figure 8.7.)

Figure 8.8
(a) The magnetic field produced by a current in a coil of wire. The field has the same shape as that produced by a bar magnet. When the direction of the current is reversed (b), the polarity of the magnetic field is also reversed.

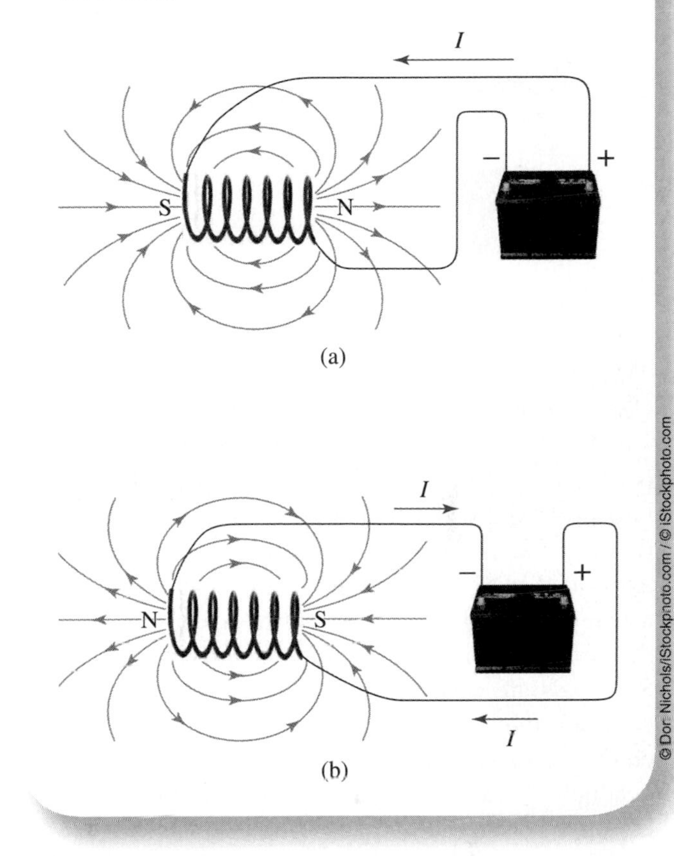

Reversing the direction of the current in the wire will reverse the directions of the magnetic field lines.

Most applications of this phenomenon use coils—long wires wrapped in the shape of a cylinder, often around an iron core. The magnetism induced in the iron greatly enhances the magnetic field of the coil. The magnetic field of such a coil (when carrying a direct current) has the same shape as the field around a bar magnet. (See Figure 8.8 and compare it to Figure 8.4.) This device is an *electromagnet.* It behaves just like a permanent magnet as long as there is a current flowing. One end of the coil is a north pole, and the other is a south pole. Electromagnets have an advantage over permanent magnets in that the magnetism can be

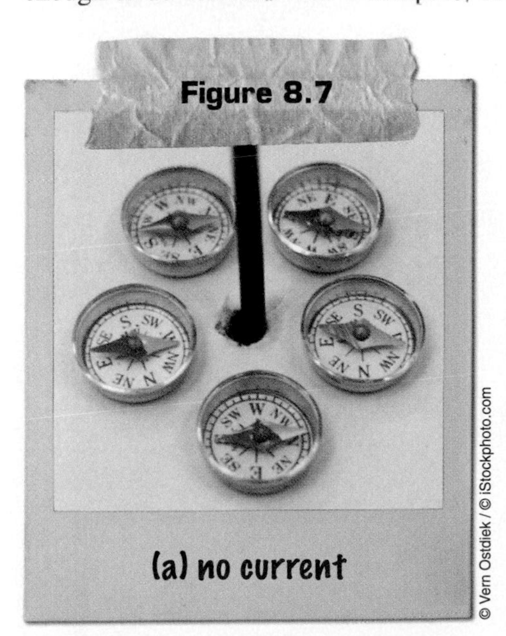

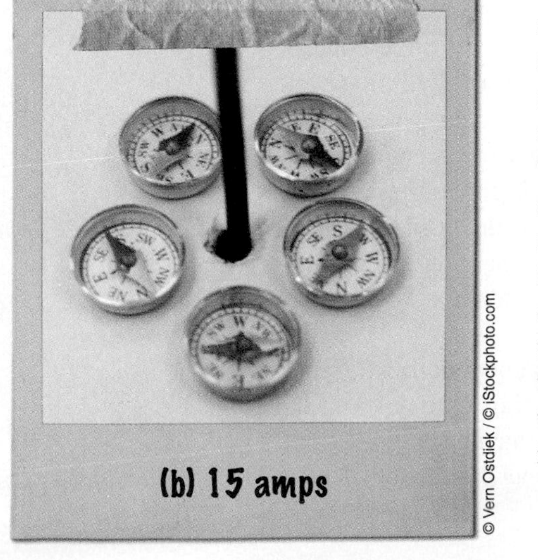

(a) The compass needles align with the Earth's magnetic field when no current is in the wire. (b) The compass needles show the circular shape of the magnetic field produced by a large current (15 amperes) flowing in the wire.

Figure 8.9

(a) The iron rod is pulled into the coil (solenoid) when the current flows. (b) The solenoid (red) in this doorbell chime pulls in the iron rod when a current flows in it. The rod strikes the black bar at the left, producing sound. When the current is shut off, the spring retracts the rod, and it returns to strike the bar at the right. If the bars are tuned to different frequencies, together they can produce the familiar "ding-dong" sound.

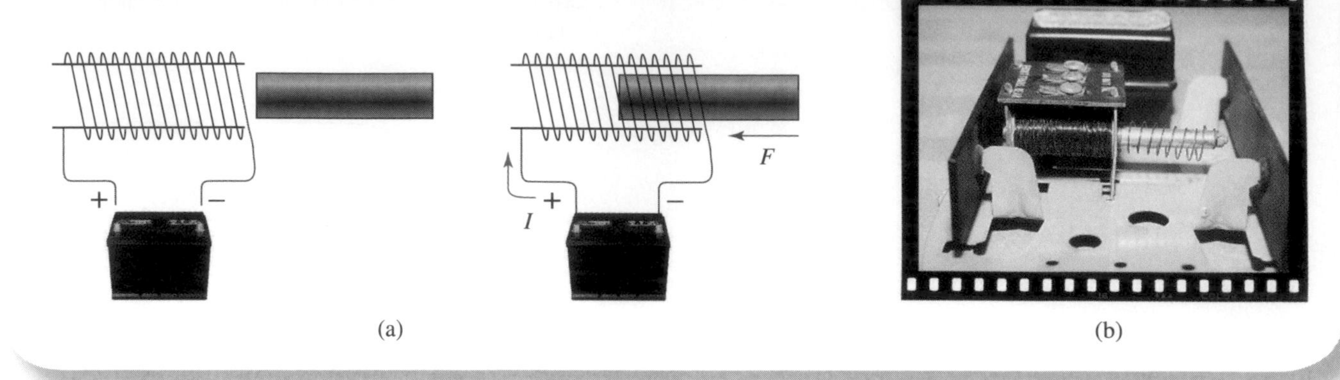

© Vern Ostdiek / © Don Nichols/iStockphoto.com / © Maria Toutoudaki/iStockphoto.com

"turned off" simply by switching off the current. (Large electromagnets are routinely used to pick up scrap iron.)

A coil whose length is much greater than its diameter is called a *solenoid.* If an iron rod is partially inserted into a solenoid with a hollow core, the rod will be pulled in when the current is switched on: the magnetic pole associated with the coil's field nearest the rod induces a magnetic field with the opposite polarity in the rod, thus exerting an attractive force on it (see Figure 8.9). Solenoids are used in common devices for striking doorbell chimes, opening valves to allow water to enter and to leave washing machines, withdrawing deadbolts in electric door locks, and engaging starter motors on car and truck engines.

Electromagnets are used to produce the strongest magnetic fields on Earth. Two factors contribute to stronger fields: wrapping more coils around the cylinder and using a larger electric current. The former suggests the use of thinner wire so that more coils can fit into the same amount of space. But smaller wire requires smaller electric current so the wire does not overheat and melt. This limitation is overcome in *superconducting electromagnets* (Figure 8.10). When the wire used in an electromagnet is a superconductor, it can carry huge electric currents with no ohmic heating because there is no resistance. Very small superconducting electromagnets can generate very strong magnetic fields while using much less electrical energy than a conventional electromagnet. (We described some uses of superconducting electromagnets in Section 7.3 and will introduce others later in this section.)

Superconducting electromagnets do have limitations, though. The superconducting state is lost if the temperature, electric current, or magnetic field strength exceeds certain values. Most superconducting electromagnets now found in laboratories throughout the world use a compound of niobium and tin that must be kept cold with liquid helium ($T = 4$ K). The added cost of the liquid helium system is offset by the high magnetic fields achieved and the great reduction in use of electric energy compared to conventional electromagnets.

The polarity of an electromagnet is reversed if the direction of the current is reversed (see Figure 8.8b). An alternating current in a coil will produce a magnetic field

Figure 8.10

The pattern of magnetic field lines surrounding a pair of powerful superconducting electromagnets is made visible by scattering iron nails on a piece of white plywood.

© Bettmann/CORBIS / © Maria Toutoudaki/iStockphoto.com

Figure 8.11

The arrows represent the magnetic fields of electrons in individual atoms. (a) The fields remain randomly oriented in nonferromagnetic materials. (b) Inside ferromagnetic material that is magnetized, the individual magnetic fields are aligned.

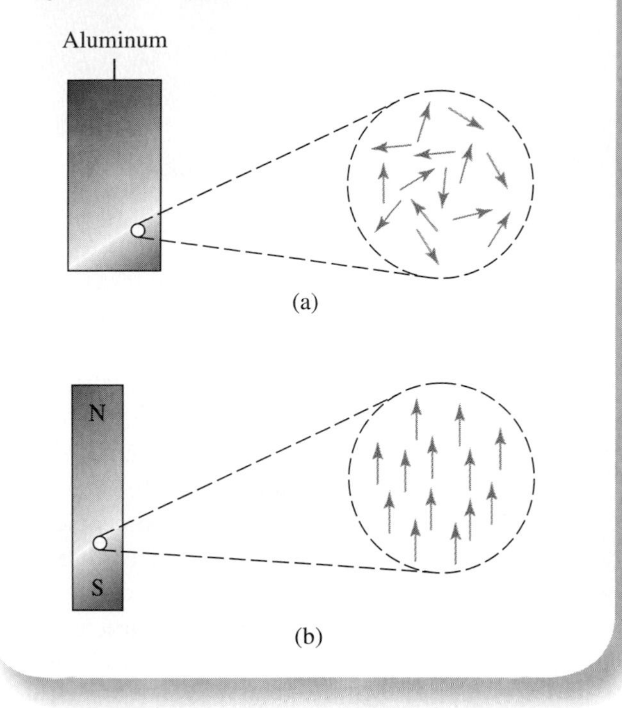

that oscillates: It increases, decreases, and switches polarity with the same frequency as the current. Such an oscillating magnetic field will cause a nearby piece of iron to vibrate. The oscillating magnetic field of a coil with AC in it is used in many common devices, as we shall see in the following sections.

Not only does this first interaction explain how electromagnets work, it gives us new insight into permanent magnets as well. Since electrons in atoms are charged particles in motion about the nucleus, they produce magnetic fields. Also, the electrons have their own magnetic fields associated with their spin (more on this in Chapter 12). In any unmagnetized material, the individual magnetic fields of the electrons are randomly oriented and cancel each other out (Figure 8.11). In ferromagnetic materials, these fields can be aligned with one another by an external magnetic field; the material then produces a net magnetic field. So we can conclude that moving electric charges are the causes of magnetic fields even in ordinary bar and horseshoe magnets.

This brings us back to a statement made at the beginning of Chapter 7: Electric charges are the cause of both electrical and magnetic effects. We might regard electricity and magnetism as two different manifestations of the same thing—charge.

© iStockphoto.com / © iStockphoto.com

OBSERVATION 2

A magnetic field exerts a force on a moving electric charge. Therefore, a magnetic field exerts a force on a current-carrying wire.*

* EXCEPTION: If a charge's velocity or the direction of a current is parallel to the magnetic field or in the opposite direction, the magnetic field does not exert this force.

The second observation helps us understand how things like electric motors and speakers work.

A stationary electric charge is not affected by a magnetic field, but a moving charge usually is. Note that this second observation is a logical consequence of the first: Anything that produces a magnetic field will itself be affected by other magnetic fields.

A curious characteristic of electromagnetic phenomena is that the effects are often perpendicular to the causes. The direction of the magnetic field from a current-carrying wire is perpendicular to the direction the current is flowing (Figure 8.6). Similarly, the force that a magnetic field exerts on a moving charge or on a current-carrying wire is perpendicular to both the direction of the magnetic field and the direction the charge is flowing. For example, if a horizontal magnetic field is directed away from you and a wire is carrying a current to your right, the force on the wire is *upward* (see Figure 8.12). If the direction of the current is reversed, the direction of the force is reversed (downward). An alternating current would cause the wire to experience a force that alternates up and down.

Electric motors—like those in hair dryers and elevators—exploit this electromagnetic interaction. The simplest type of electric motor consists of a coil of wire mounted so that it can rotate in the magnetic field of a horseshoe-shaped magnet (see Figure 8.13). A direct current flows through the coil, the magnetic field causes forces on the sides of the coil, and it rotates. Once the coil has completed half of a rotation, a simple mechanism reverses the direction of the current. This reverses the force on the coil, causing it to rotate another half-turn. This process is repeated, and the coil spins continuously. Motors designed to run on AC can exploit the fact that the direction of the current is automatically reversed 120 times each second (sixty-cycles-per-second AC, with two reversals each cycle).

Liquid metals, such as the molten sodium used in certain nuclear reactors, can be moved through pipes

Figure 8.12

(a) The force on a current-carrying wire in a magnetic field. (b) When the direction of the current is reversed, the direction of the force is also reversed. (c) A current-carrying wire levitating in a magnetic field.

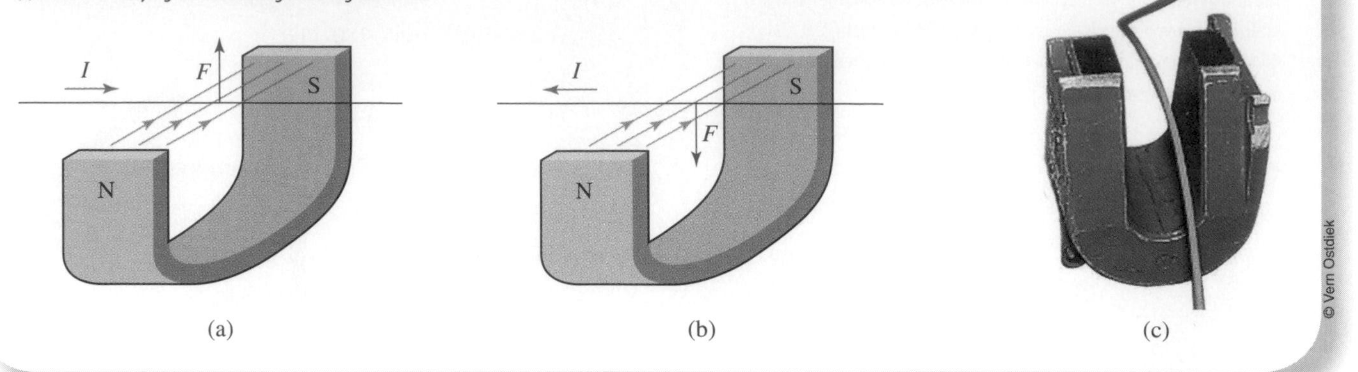

Figure 8.13

Simplified sketch of an electric motor. The loop of wire rotates because of the forces on its sides. Each time the loop becomes horizontal, the direction of the current is reversed, and the rotation continues.

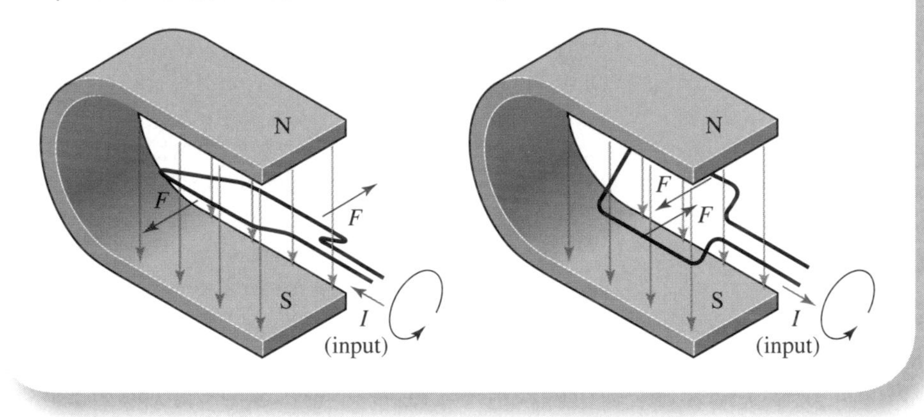

In the absence of other forces, a charged particle moving perpendicularly to a magnetic field will travel in a circle: The force on the particle is always perpendicular to its velocity and is therefore a centripetal force. An electron, proton, or other charged particle can be forced to move in a circle by a magnetic field and then gradually accelerated during each revolution. Particle accelerators used for experiments in atomic, nuclear, and elementary particle physics, as well as for producing radiation for cancer treatments at some large hospitals, operate on this principle.

using an electromagnetic pump that has no moving parts. If the metal has to be moved in a pipe that is oriented north-south, for example, a large electric current can be sent across the pipe—east to west perhaps. Then, if a strong magnetic field is directed downward through the same section of pipe, the current-carrying metal will be forced to move southward.

Several large-scale devices used in experimental physics make use of the effect of magnetic fields on moving charged particles. High-temperature plasmas cannot be kept in any conventional metal or glass container because the container would melt. Since plasmas are composed of charged particles, magnetic fields can be used to contain them in what is known as a "magnetic bottle." This is one approach being employed in the attempt to harness nuclear fusion as an energy source (Section 11.7).

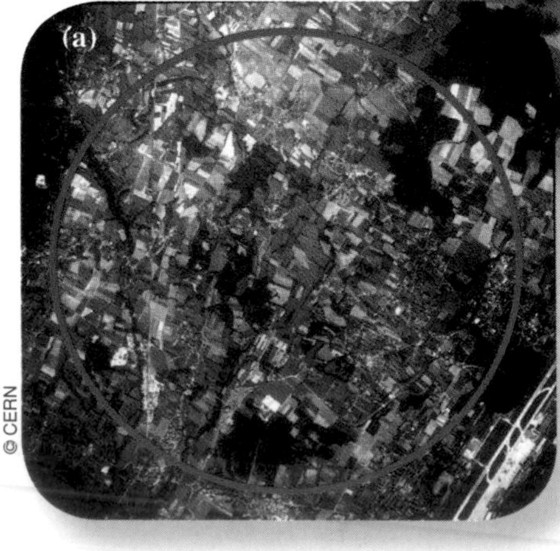

(a) The Large Hadron Collider, near Geneva, Switzerland. Inside the subterranean tunnel, charged particles are accelerated to nearly the speed of light. Magnetic fields are used to keep the charges moving in a circle. (b) Surface buildings at Echenevex, France, point 4 along the LHC track, which house instrumentation and support services for experiments involving high energy particle collisions.

Figure 8.14

(a) The electric field produced by a moving magnet is circular. (The magnetic field of the magnet is not shown.) (b) A magnet will induce a current in a coil as it passes through. (c) The ammeter shows there is no current in the coil of wire (yellow) when the horseshoe magnet (blue) is not moving. (d) A current flows when the magnet is moving.

Theory

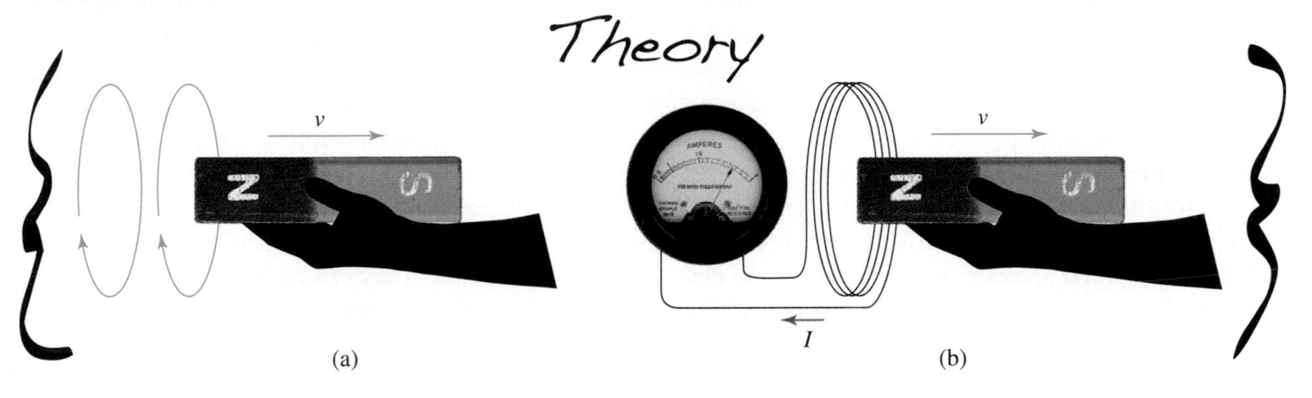

Practice

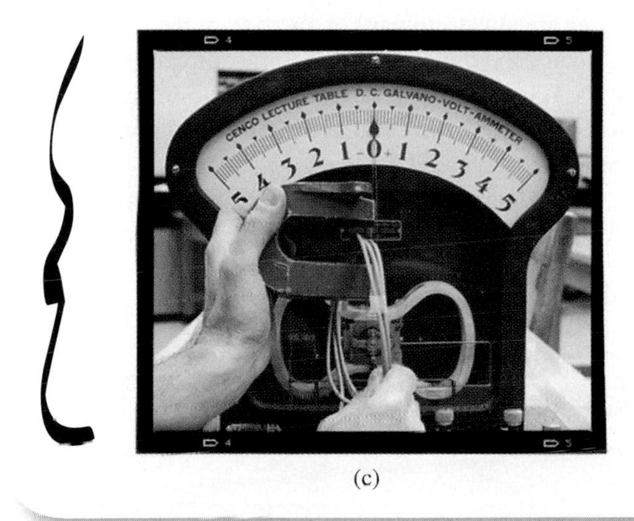

(c)

(d)

© Vern Ostdiek / © Vern Ostdiek / © Mara Toutoudaki/iStockphoto.com / © Giovanni Banfi/iStockphoto.com / © Dale Robins/iStockphoto.com

The world's highest-energy particle accelerator is the Large Hadron Collider (LHC) located along the border between France and Switzerland near the city of Geneva. This device comprises a circular tunnel with a diameter of 8.6 kilometers (5.3 miles) buried 50 to 175 meters beneath the Earth's surface in which two counter-rotating beams of charged particles travel in a vacuum, guided by superconducting magnets. The head-on collisions between the particles in these oppositely moving beams yield information about the fundamental forces and interactions in Nature. More will be said on this topic in Chapter 12.

The third observed interaction between electricity and magnetism is used by electric generators. Recall that the first observation tells us that moving charges create magnetic fields. The third one is a similar statement about moving magnets.

© iStockphoto.com

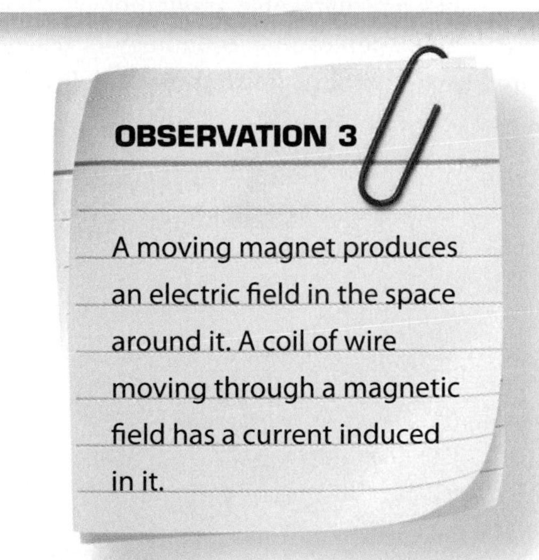

OBSERVATION 3

A moving magnet produces an electric field in the space around it. A coil of wire moving through a magnetic field has a current induced in it.

The electric field around a moving magnet is in the shape of circles around the path of the magnet. This circular electric field will force charges in a coil of wire to move in the same direction—as a current (Figure 8.14).

The process of inducing an electric current with a magnetic field is known as **electromagnetic induction**. All that is required is that the magnet and coil move relative to each other. If the coil moves and the magnet remains stationary, a current is induced. If the motion is steady in either case, the induced current is in one direction. If either the coil or the magnet oscillates back and forth, the current alternates with the same frequency—it is AC.

Electromagnetic induction is used in the most important device for the production of electricity, the generator. The simplest generator is basically an electric motor. When the coil is forced to rotate, it moves relative to the magnet, so a current is induced in it (see Figure 8.15). We might call this device a "two-way energy converter." When electrical energy is supplied to it, it is a motor. It converts this electrical energy into mechanical energy of rotation. When it is mechanically turned (by hand cranking, by a fan belt on a car engine, or by a turbine in a power plant), it is a generator. It converts mechanical energy into electrical energy.

This motor-generator duality is used in dozens of *pumped-storage* hydroelectric power stations. During the night when there is a surplus of electrical energy available from other power stations, the motor mode is used to pump water from one reservoir to another that is at a higher elevation. Most of the electrical energy is converted into "stored" gravitational potential energy. During the peak time of electric use the next day, water flows in the opposite direction, and the generator mode is used as the moving water turns the pumps (now acting as turbines) that turn the motors (now acting as generators), thereby producing electricity. You might say that the system functions like a rechargeable gravitational battery.

Another application of this technology is *regenerative braking*, used in electric and hybrid vehicles. While accelerating and cruising, electric motors turn the wheels using electricity from batteries. During braking, the motors function as generators: The wheels turn them, and the electricity that is generated can partially recharge the batteries. Instead of all of the vehicle's kinetic energy being converted into wasted heat—the case with conventional friction brakes—some of it is saved for reuse.

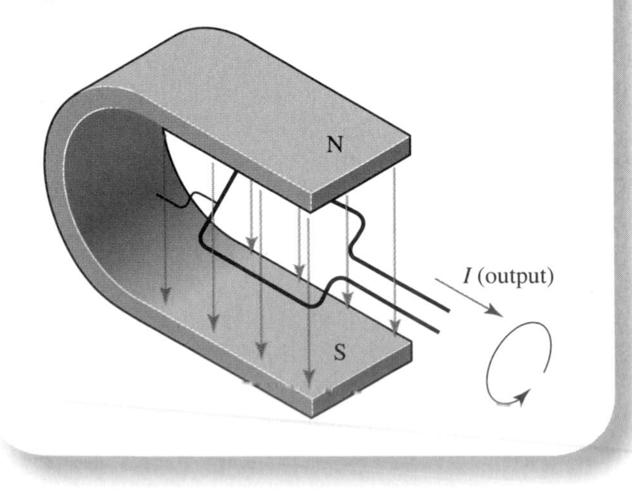

Figure 8.15
Simplified sketch of a generator. As the loop of wire rotates relative to the magnetic field, a current is induced in it.

In summary, when electric charges or magnets are in motion, electricity and magnetism are no longer independent phenomena. The three observations given here are statements of experimental facts that illustrate this interdependence. They can be demonstrated easily using a battery, wires, a compass, a large magnet, and a sensitive ammeter. The fact that electricity and magnetism interact only when there is motion (and then the effects are perpendicular to the causes) is somewhat startling when compared to, say, gravitation and electrostatics. As we saw in Chapters 2 and 7, gravitational and electrostatic forces are always toward or away from the objects causing them, and they act whether or not anything is moving or changing. These basic yet surprising interactions between electricity and magnetism are crucial to our modern electrified society.

The Concept Map on your Review Card summarizes the interactions between electricity and magnetism.

The interactions between electricity and magnetism described in the previous section, along with other similar observations, suggest the following two general statements. We might call these the **principles of electromagnetism**:

PRINCIPLES

1. An electric current or a changing electric field induces a magnetic field.
2. A changing magnetic field induces an electric field.

These two statements summarize the previous observations and also emphasize the symmetry that exists. In both cases, a "changing" field means that the strength or the direction of the field is changing. The first principle can be used to explain the first observation: As a charge moves by a point in space, the strength of the electric field increases and then decreases. All the time the direction of the field is changing as well (Figure 8.16). The effect of this is to cause a magnetic field to be produced. Similarly, the second principle explains electromagnetic induction.

As mentioned in Section 7.6, a transformer is a device used to step up or step down AC voltages. It represents one of the most elegant applications of electromagnetism. In essence, a transformer consists of two separate coils of wire in close proximity. An AC voltage is applied to one of the coils, called the "input" or "primary" coil, and an AC voltage appears at the other coil, called the "output" or "secondary" coil (Figure 8.17). The AC in the primary coil produces an oscillating magnetic field through both coils. Most transformers have both coils wrapped around a single ferromagnetic core to intensify the magnetic field and guide it from one coil to the other. This oscillating (and therefore changing) magnetic field induces an AC current in the output coil. Note that a DC input would produce a steady magnetic field that would not induce a current in the output coil. Transformers do not work with DC.

Figure 8.16
The electric field at the point *P* changes as the charged particle moves by. The upper blue arrows indicate the magnitude and direction of the electric field for three different locations of the particle. The magnetic field that is induced at point *P* is directed straight out of the paper.

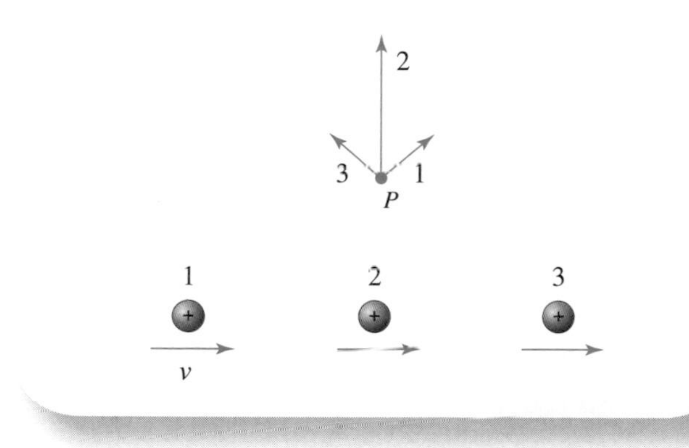

Figure 8.17
Simplified diagram of a transformer. The alternating magnetic field produced by the AC in the input coil induces an alternating current in the output coil. In this case, the output voltage would be higher than the input voltage.

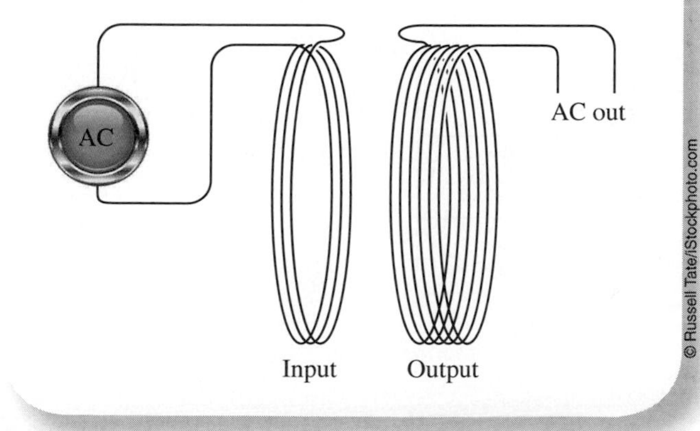

© Russell Tate/iStockphoto.com

Now, how can the voltage of the output be different from the voltage of the input? Each "loop" or "turn" of the output coil has the same voltage induced in it. The voltages in all of the turns add together so that the more turns there are in the output coil, the higher the total voltage. The ratio of the number of turns in the two coils determines the ratio of the input and output voltages. In particular:

$$\frac{\text{voltage of output}}{\text{voltage of input}} = \frac{\text{number of turns in output coil}}{\text{number of turns in input coil}}$$

$$\frac{V_o}{V_i} = \frac{N_o}{N_i}$$

If there are twice as many turns in the output coil as in the input coil, the output voltage will be twice the input voltage. If there are one-third as many turns in the output coil, the output voltage will be one-third the input voltage. Thus, the AC voltage can be stepped up or stepped down by any desired amount by adjusting the ratio of the number of turns in the two coils.

EXAMPLE 8.1

A transformer is being designed to have a 600-volt output with a 120-volt input. If there are to be 800 turns of wire in the input coil, how many turns must there be in the output coil?

$$\frac{V_o}{V_i} = \frac{N_o}{N_i}$$

$$\frac{600\ V}{120\ V} = \frac{N_o}{800}$$

$$800 \times 5 = N_o$$

$$N_o = 4{,}000 \text{ turns}$$

In addition to being used to change voltages in electrical distribution systems, transformers are used in a wide variety of electrical appliances. Most electrical components used in radios, calculators, and the like require voltages much smaller than 120 volts. Appliances designed to operate on household AC must include transformers to reduce the voltage accordingly. High-intensity desk lamps also use transformers; that is what makes their bases so heavy. The spark used to ignite gasoline in automobile engines is generated using a type of transformer called a "coil." The number of turns in the output coil is many times the number of turns in the input coil. A spark is produced by first sending a brief current into the input. A magnetic field is produced that quickly disappears. This induces a very high voltage (around 25,000 volts) in the output, which is conducted to the spark plugs to ignite the fuel.

Understanding electromagnetism allows us to better appreciate how the metal detectors introduced at the start of the chapter work. The magnetic pulses are produced by sending an electric current through a coil of wire for a short period of time. When the current stops, the magnetic field that was created dies out quickly, and this decreasing field induces an electric current in the coil. This current is used to monitor how swiftly the magnetic pulse dies out.

Metals are detected because the rapidly changing magnetic field of each pulse induces electrons in the metal to move—as in the secondary coil in a transformer—and this current produces an opposite magnetic pulse. This change in the total magnetic field affects the current induced in the coil. The electronics are designed to detect any such change and signal an alarm.

8.4 Applications to Sound Reproduction

A hundred years or so ago, the only people who listened to music performed by world-class musicians were those few who could attend live performances. Today, people in the most remote corners of the world can hear concert-quality sound from large home entertainment systems, pocket-sized MP3 players, and many devices in between. The first Edison phonographs were strictly mechanical and did a fair job of reproducing sound. It was the invention of electronic recording and playback machines that brought true high fidelity to sound reproduction. The sequence that begins with sound in a recording studio and ends with the reproduced sound coming from a speaker in your home, headphones, or car includes components that use electromagnetism.

The key to electronic sound recording and playback is first to translate the sound into an alternating current and then later re-translate the AC back into sound. The first step requires a microphone, and the second step requires a speaker. Although there are several different types of microphones, we will take a look at what is called a dynamic microphone. It consists of a magnet surrounded by a coil of wire attached to a diaphragm (see Figure 8.18). The coil and diaphragm are free to oscillate relative to the stationary magnet. When sound waves reach the microphone, the pressure variations in the wave push the diaphragm back and forth, making it and the coil oscillate. Since the coil is moving relative to the magnet, an oscillating current is induced in it. The frequency of the AC in the coil is the same as the frequency of the diaphragm's oscillation, which is the same as the frequency of the original sound. That is all it takes. This type of dynamic microphone is referred to as a "moving coil" microphone. The alternative is to attach a small magnet to the diaphragm and keep the coil stationary—a "moving magnet" microphone.

Let's skip ahead now to when the sound is played back. The output of the CD player, radio, or other audio component is an alternating current that has to be converted back into sound by a speaker. The basic speaker is quite similar to a dynamic microphone. In this case, the coil (called the "voice coil") is connected to a stiff paper cone instead of to a diaphragm (Figure 8.19). Recall from Section 8.2 that an alternating current in the voice coil in the presence of the magnet will cause the coil to

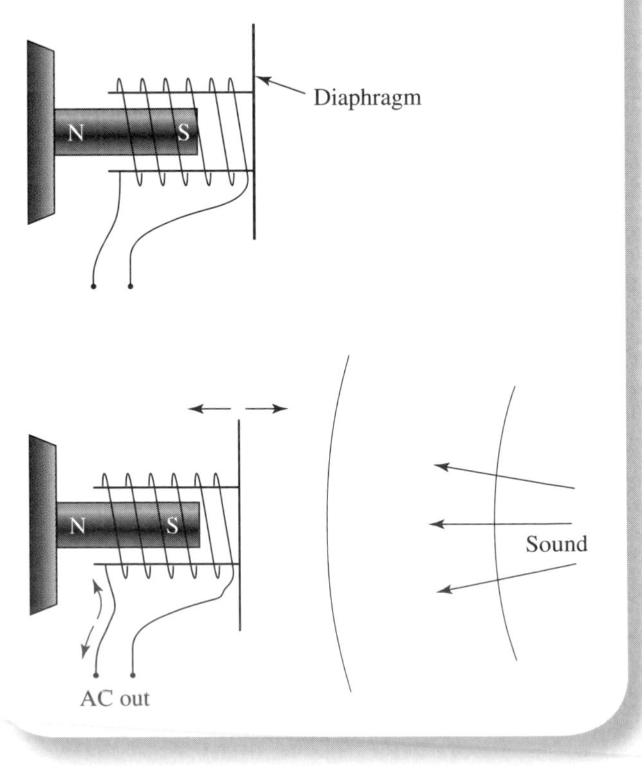

Figure 8.18
Simplified diagram of one type of microphone. Sound waves cause the diaphragm and coil to oscillate relative to the magnet. This induces AC in the coil.

Figure 8.19

Simplified sketch of a speaker. AC in the voice coil causes the cone to be forced in and out, thereby producing sound.

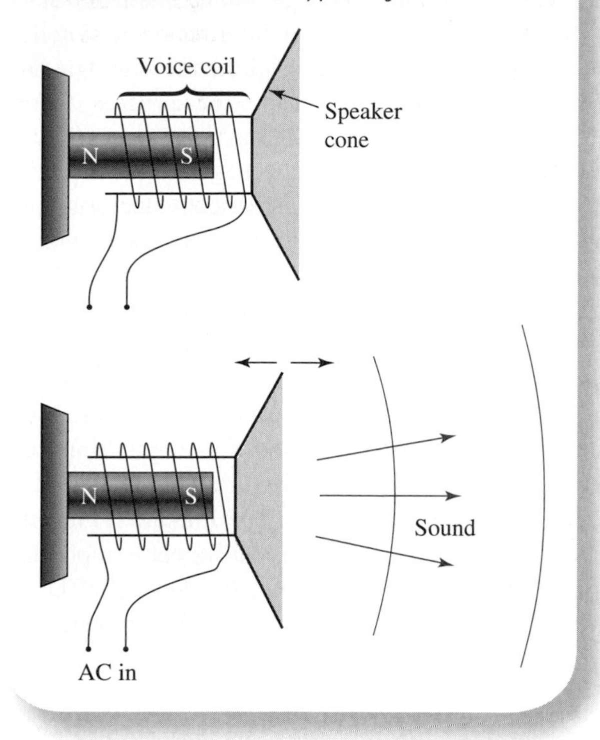

experience an alternating force. The voice coil and the speaker cone oscillate with the same frequency as the AC input. The oscillating paper cone produces a longitudinal wave in the air—sound.

Microphones and speakers are classified as *transducers:* They convert mechanical oscillation due to sound into AC (microphone), or they convert AC into mechanical oscillation and sound (speaker). They are almost identical. In fact, a microphone can be used as a speaker, and a speaker can be used as a microphone. But, as with motors and generators, each is best at doing what it is designed to do.

VCR Interior

© iStockphoto.com

Magnetic recording is not limited to sound reproduction. VCRs use this technology, too.

Most sound recording, from simple cassette recorders to sophisticated studio tape machines, is done on magnetic tape. The tape is a plastic film coated with a thin layer of fine ferromagnetic particles that retain magnetism. Sound is recorded on the tape using a recording head, a ring-shaped electromagnet with a very narrow gap (Figure 8.20). During recording, an AC signal (from a microphone, for example) produces an alternating magnetic field in the gap of the recording head. As the tape is pulled past the gap, the particles in each part of the tape are magnetized according to the polarity of the head's magnetic field at the instant they are in the gap. The polarity of the particles changes from north-south to south-north, and so on, along the length of the tape.

To play back the recording, the tape is pulled past a playback head, often the same head used for recording. The magnetic field of the particles in the tape oscillates back and forth and induces an oscillating magnetic field in the tape head (Figure 8.21). This oscillating magnetic

Figure 8.20

(a) Simplified sketch of a tape head. During recording, (b) and (c), AC in the coil induces alternating magnetism in the tape.

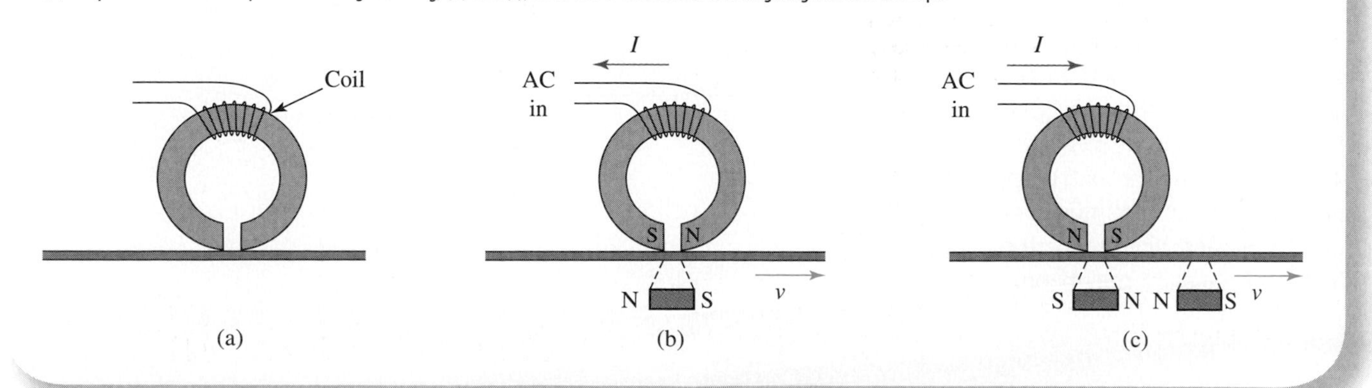

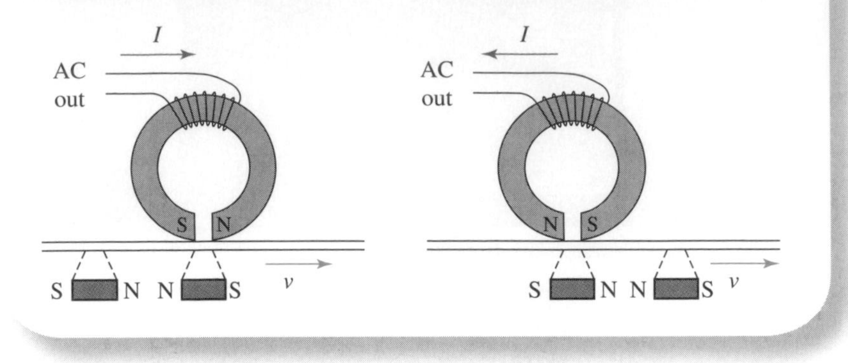

Figure 8.21
During playback, the alternating magnetism in the tape induces AC in the coil on the tape head.

field induces an oscillating current (AC) in the coil—electromagnetic induction again.

Magnetic recording is not limited to sound reproduction. Television video recorders (VCRs) record both sound and visual images on magnetic tape. Computers store information magnetically on tape and hard disks/drives.

The AC signals produced by microphones, CD players, and tape playback heads are quite weak. Amplifiers are used to increase the power of these signals before they are sent to speakers. Amplifiers also allow the listener to modify the sound by adjusting its loudness with the volume control and its tone quality with the bass and treble controls.

Digital Sound

A revolution in sound reproduction occurred in the 1980s with the advent of digital sound reproduction, the method used in *compact discs* (CDs) and various computer sound file formats, including MP3. In a process known as analog-to-digital conversion, the sound wave to be recorded is measured and stored as numbers. For CDs, the actual voltage of the AC signal from a microphone is measured 44,100 times *each second* (Figure 8.22). Note that this frequency is more than twice the highest frequency that people can hear. The waveform of the sound is "chopped up" into tiny segments and then recorded as numerical values. These numbers are stored as binary numbers using 0s and 1s, just as information is stored in computers. To play back the sound, a digital-to-analog conversion process reconstructs the sound wave by generating an AC signal whose voltage at each instant in time equals the numerical value originally recorded. After being "smoothed" with an electronic filter, the waveform is an almost perfect copy of the original.

A huge amount of data is associated with digital sound reproduction—millions of numbers for each minute of music. CDs (and DVDs) store these data in the form of microscopic pits in a spiral line several miles long (see Figure 8.23). A tiny laser focused on the pits reads them as 0s and 1s. The amount of information stored on a 70-minute CD is equivalent to more than a dozen full-length encyclopedias. A standard DVD can store about seven times as much data. Little wonder that

Figure 8.22
Digital sound reproduction in CDs. (a) To record the sound, the voltage of the waveform is measured 44,100 times each second. The resulting numbers are stored for later use. (b) During playback, the voltage of the output at each time is set equal to the numerical value that was stored originally. The reconstructed waveform is then "smoothed" using an electronic filter. The resulting waveform is an almost perfect reproduction of the original.

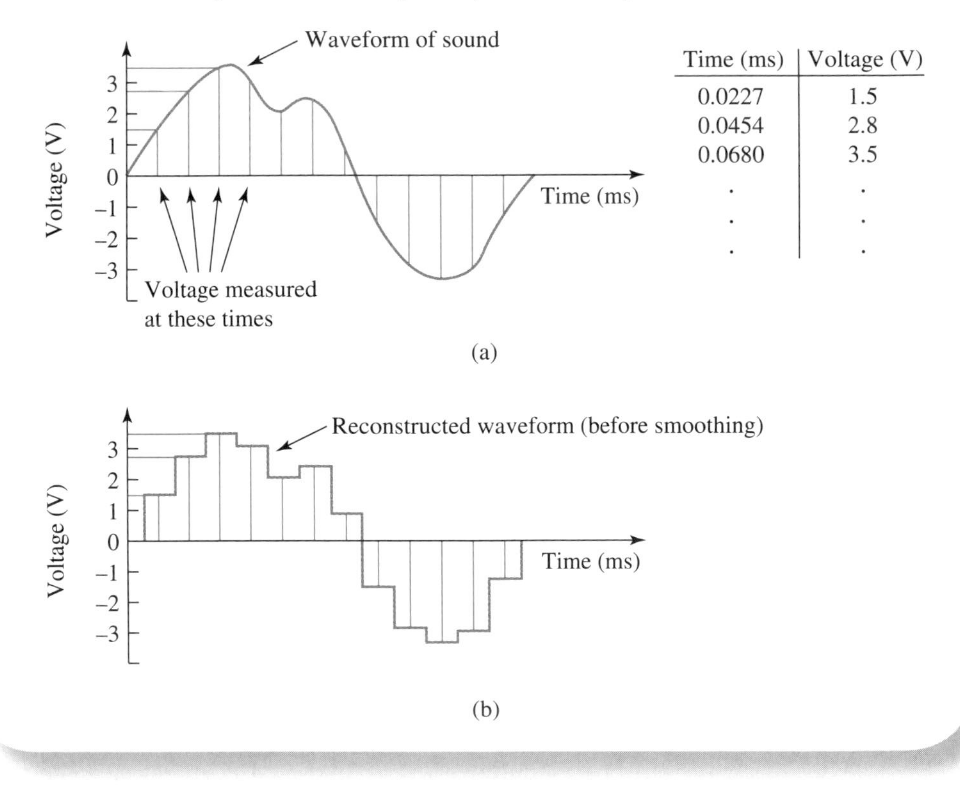

Figure 8.23
Information is stored on CDs and DVDs in digital form, as microscopic pits used to represent 0s and 1s. It is read from the spinning disc by the beam from a tiny laser. (From *College Physics,* 2e, by Urone. Used by permission.)

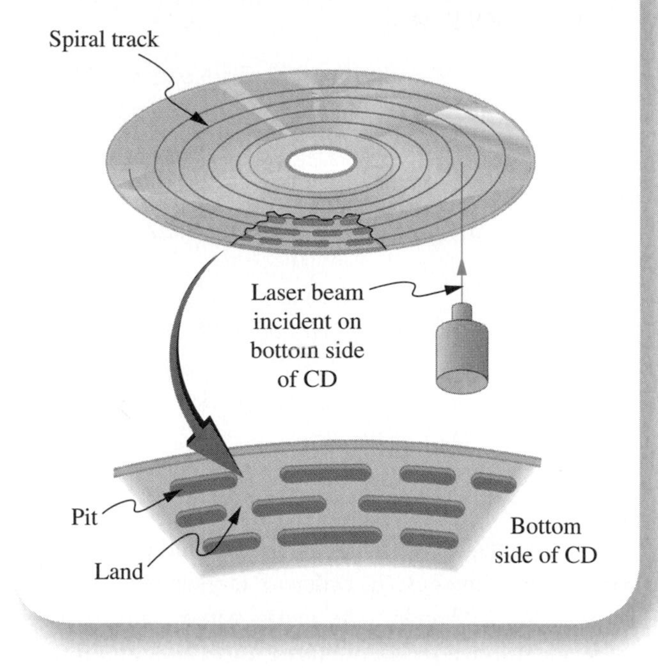

CDs and DVDs have also been embraced by the personal-computer industry as a way to store huge amounts of information in durable, portable form.

The superior quality of digital sound comes about because the playback device looks only for numbers. It can ignore such things as imperfections in the disk or tape, the weak random magnetization in a tape that becomes tape hiss on cassettes, and the mechanical vibration of motors that we hear as a rumble on phonographs. A sophisticated error-correction system can even compensate for missing or garbled numbers. Because the pickup device in a CD player does not touch the disc, each CD can be played over and over without the slow deterioration in quality that results from a needle moving in a phonograph groove or from the constant unwinding and rewinding of a cassette tape over the recorder heads. This combination of high fidelity and disk durability made the CD system an immediate hit with consumers.

This is just a glimpse of some of the factors in state-of-the-art high-fidelity sound reproduction. Perhaps we are all so accustomed to it that we cannot appreciate how much of a technological miracle it really is. The next time you listen to high-quality recorded music, remember that it is all possible because of the basic interactions between electricity and magnetism described in Section 8.2.

8.5 Electromagnetic Waves

Eyes, radios, televisions, radar, x-ray machines, microwave ovens, heat lamps, . . . What do all of these things have in common? They all use **electromagnetic (EM) waves**. EM waves occupy prominent places both in our daily lives and in our technology. These waves are also involved in many natural processes and are essential to life itself. In the rest of this chapter, we will discuss the nature and properties of electromagnetic waves and will look at some of their important roles in today's world.

As the name implies, EM waves involve both electricity and magnetism. The existence of these waves was first suggested by the nineteenth-century physicist James Clerk Maxwell while he was analyzing the interactions between electricity and magnetism. Consider the two principles of electromagnetism stated in Section 8.3. Let's say that an oscillating electric field is produced at some place. The electric field switches back and forth in direction while its strength varies accordingly. This oscillating electric field will induce an oscillating magnetic field in the space around it. But the oscillating magnetic field will then induce an oscillating electric field. This will then induce an oscillating magnetic field and so on in an endless "loop": The principles of electromagnetism tell us that a continuous succession of oscillating magnetic and electric fields will be produced. These fields travel as a wave, an EM wave.

DEFINITION

Electromagnetic Wave (EM Wave)
A transverse wave consisting of a combination of oscillating electric and magnetic fields.

Electromagnetic waves are transverse waves because the oscillation of both of the fields is perpendicular to the direction the wave travels. (Some EM waves in plasmas can be longitudinal.) Figure 8.24 shows a "snapshot" of an EM wave traveling to the right. (The three axes are perpendicular to each other.) In this particular case, the

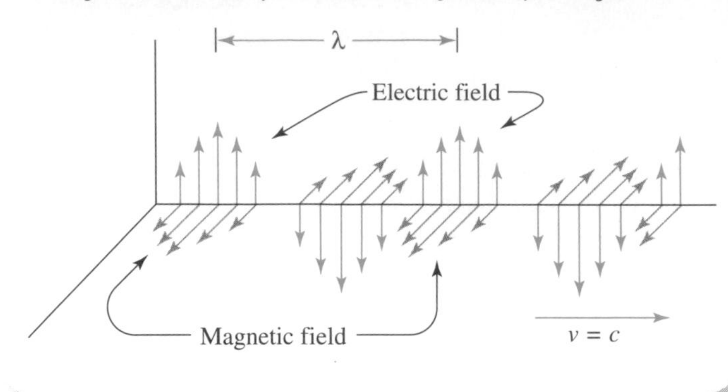

Figure 8.24

Representation of a portion of an EM wave. The electric field is always perpendicular to the magnetic field. The entire pattern moves to the right at the speed of light.

electric field is vertical. As the wave travels by a given point in space, the electric field oscillates up and down, the way a floating petal oscillates on a water wave. The magnetic field at the point oscillates horizontally.

Figure 8.24 should remind you of the transverse waves we described in Chapter 6 (Figure 6.3, for example). Electromagnetic waves do differ from mechanical waves in two important ways. First, they are a combination of two waves in one: an electric field wave and a magnetic field wave. These cannot exist separately. Second, EM waves do not require a medium in which to travel. They can travel through a vacuum: The light from the Sun does this. They can also travel through matter. Light through air and glass, and x-rays through your body, are common examples.

Electromagnetic waves travel at an extremely high speed. Their speed in a vacuum, called the "speed of light" because it was first measured using light, is represented by the letter c. Its value is

$$c = 299{,}792{,}458 \text{ m/s} \qquad \text{(speed of light)}$$

or

$$c = 3 \times 10^8 \text{ m/s} \qquad \text{(approximately)}$$

$$= 300{,}000{,}000 \text{ m/s}$$

$$= 186{,}000 \text{ miles/s} \qquad \text{(approximately)}$$

All of the parameters introduced for waves in Chapter 6 apply to EM waves. The wavelength can be readily identified in Figure 8.24. The amplitude is the maximum value of the electric field strength. The equation $v = f\lambda$ holds with v replaced by c. There is an extremely wide range of wavelengths of EM waves, from the size of a single proton, about 10^{-15} meters, to almost 4,000 kilometers for one type of radio wave. The corresponding frequencies of these extremes are about 10^{23} hertz and 76 hertz, respectively. Most EM waves used in practical applications have extremely high frequencies compared to sound.

EXAMPLE 8.2

An FM radio station broadcasts an EM wave with a frequency of 100 megahertz. What is the wavelength of the wave?

The prefix "mega" stands for 1 million. Therefore, the frequency is 100 million hertz.

$$c = f\lambda$$

$$300{,}000{,}000 \text{ m/s} = 100{,}000{,}000 \text{ Hz} \times \lambda$$

$$\frac{300{,}000{,}000 \text{ m/s}}{100{,}000{,}000 \text{ Hz}} = \lambda$$

$$\lambda = 3 \text{ m}$$

Electromagnetic waves are named and classified according to frequency. In order of increasing frequency, the groups, or "bands," are **radio waves**, **microwaves**, **infrared radiation**, **visible light**, **ultraviolet radiation**, **x-rays**, and **gamma rays (γ rays)**. (Use of the word "radiation" instead of "waves" is not significant here.) Figure 8.25 shows these groups along with frequency and wavelength scales. This is called the **electromagnetic spectrum**. Notice that the groups overlap. For example, a 10^{17}-hertz EM wave could be ultraviolet radiation or an x-ray. In cases of overlap, the name applied to an EM wave depends on how it is produced.

We will briefly discuss the properties of each group of waves in the electromagnetic spectrum—how they are produced, what their uses are, and how they can affect us. The great diversity of uses of EM waves arises from the variety of ways in which they can interact with different kinds of matter. All matter around us contains charged particles (electrons and protons), so it seems logical that EM waves can affect and be affected by matter. The oscillating electric field can cause AC currents in conductors; it can stimulate vibration of molecules, atoms, or individual electrons; or it can interact with the nuclei of atoms. Which sort of interaction occurs, if any, depends on the frequency (and wavelength) of the EM wave and on the properties of the matter through which it is traveling—its density, molecular and atomic structure, and so on.

In principle, an electromagnetic wave of any frequency could be produced by forcing one or more charged particles to oscillate at that frequency. The oscillating field of the charges would initiate the EM wave. The

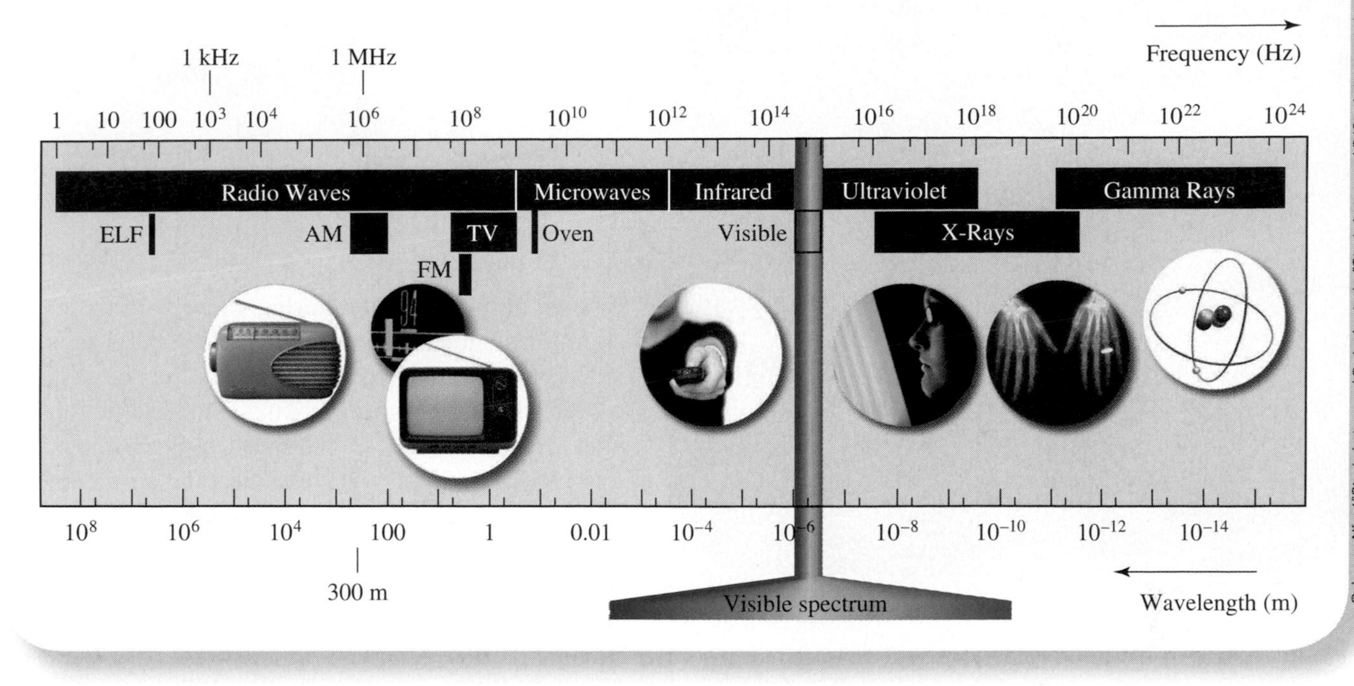

Figure 8.25
The electromagnetic (EM) spectrum.

© James Alfred/iStockphoto.com / © charles taylor/iStockphoto.com / © iStockphoto.com / © iStockphoto.com / © Brand X Pictures/Jupiterimages / © Fabrizio Zanier/iStockphoto.com

"lower-frequency" EM waves (radio waves and microwaves) are produced this way: A transmitter generates an AC signal and sends it to an antenna. At higher frequencies, this process becomes increasingly difficult. Electromagnetic waves above the microwave band are produced by a variety of processes involving molecules, atoms, and nuclei. Note that charged particles are present in all of these processes.

Electromagnetic waves are a form of energy.

There is one other factor to keep in mind: Electromagnetic waves are a form of energy. Energy is needed to produce EM waves, and energy is gained by anything that absorbs EM waves. The transfer of heat by way of radiation is one example.

Radio Waves

Radio waves, the lowest frequency EM waves, extend from less than 100 hertz to about 10^9 Hz (1 billion hertz or 1,000 megahertz). Within this range are a number of frequency bands that have been given separate names—for example, ELF (extremely low frequency), VHF (very high frequency), and UHF (ultra high frequency). Most frequencies are given in kilohertz (kHz) or megahertz (MHz). Sometimes radio waves are classified by wavelength: long wave, medium wave, or short wave.

As mentioned earlier, radio waves are produced using AC with the appropriate frequency. Radio waves propagate well through the atmosphere, which makes them practical for communication. Lower-frequency radio waves cannot penetrate the upper atmosphere, so higher frequencies are used for space and satellite communication. Only the very lowest frequencies can penetrate ocean water.

By far the main application of radio waves is in communication. The process involves broadcasting a certain frequency of radio wave with sound, video, or other information "encoded" in the wave. The radio wave is then picked up by a receiver, which recovers the information. Sometimes, this is a one-way process (commercial AM and FM radio and television), but in most other applications, it is two-way: Each party can broadcast as well as receive. Narrow frequency bands are assigned for specific purposes. For example, frequencies from 88 to 108 megahertz (88 million hertz to 108 million hertz) are reserved for commercial FM radio. There are dozens of bands assigned to government and private communication.

Microwaves

The next band of EM waves, with frequencies higher than those of radio waves, is the microwave band. The frequencies extend from the upper limit of radio waves to

Three-dimensional view of the surface of Saturn's moon, Titan, generated using radar data accumulated since 2005 from flybys of the *Cassini* spacecraft. This topographical image, about 450 kilometers (280 miles) across, shows a portion of the equatorial 'sand sea" called Belet.

Cooking with microwaves is fast because energy is given directly to all of the molecules.

the lower end of the infrared band, about 10^9–10^{12} hertz. The wavelengths range from about 0.3 m to 0.3 mm.

One use of microwaves is in communication. Bluetooth and Wi-Fi signals that interconnect computers, cell phones, and other devices are microwaves. Early experiments with microwave communication led to the most important use of microwaves, *radar* (*ra*dio *d*etection *a*nd *r*anging), after the discovery that microwaves are reflected by the metal in ships and aircraft. As we discussed in Section 6.2, radar is echolocation using microwaves. The time it takes microwaves to make a round-trip from the transmitter to the reflecting object and back is used to determine the distance to the object. Radar systems are quite sophisticated: Doppler radar can determine the speed of an object moving toward or away from the transmitter by measuring the frequency shift of the reflected wave. Such radars are essential tools for air traffic control and monitoring severe weather. Since 2005, the *Cassini* spacecraft has used imaging radar to penetrate the dense, perpetual smog that envelopes Titan, Saturn's largest moon, and to map its surface topology. Similar radar equipment placed in orbit around the Earth is used to form images of its surface, for such purposes as monitoring changes in the global environment and searching for archaeological sites.

Microwaves have gained wide acceptance as a way to cook food. The goal of cooking is to heat the food, in other words, to increase the energies of the molecules in the food. Conventional ovens heat the air around the food and rely on conduction (in solids) or convection (in liquids) to transfer the energy throughout the food. Microwave ovens send microwaves (typically with f = 2,450 megahertz and λ = 0.122 meters) into the food. The microwaves penetrate the food and raise the energies of the molecules directly. Recall from Section 7.2 that water consists of polar molecules—they have a net positive charge on one side and a net negative charge on the other side (Figure 8.26). The electric field of a microwave exerts forces on the two sides of the water molecules in food. These forces are in opposite directions and twist the molecule. Since the electric field is oscillating, the molecules are alternately twisted one way and then the other. This process increases the kinetic energy of the molecules and thereby raises the temperature of the food. Cooking with microwaves is fast because energy is given directly to all of the molecules. It does not rely

Figure 8.26

(a) Simplified sketch of a water molecule showing the net charges on its sides. (b, c) The oscillating electric field of a microwave twists the molecule back and forth, giving it energy.

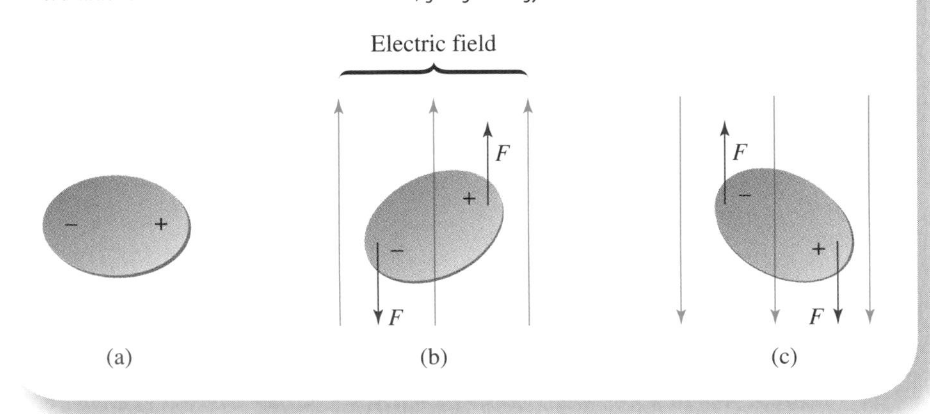

completely on the conduction of heat from the outside to the inside of the food—a much slower process.

Infrared

Infrared radiation (IR; also called *infrared light*) occupies the region between microwaves and visible light in the electromagnetic spectrum. The frequencies are from about 10^{12} hertz to about 4×10^{14} hertz (400,000,000 megahertz). The wavelengths of IR range from approximately 0.3 to 0.00075 millimeters.

Infrared radiation is ordinarily the main component of heat radiation (introduced in Section 5.4). Everything around you is both absorbing and emitting infrared radiation, just as you are. The warmth you feel from a fire or heat lamp is due to your skin absorbing the IR. Infrared radiation is constantly emitted by atoms and molecules because of their thermal vibration. Absorption of IR by a cooler substance increases the vibration of the atoms and molecules, thus raising the temperature. We will take a closer look at heat radiation and its uses in Section 8.6.

Infrared radiation is commonly used in wireless remote-control units for televisions, and for short-distance wireless data transfer between such devices as personal digital assistants (PDAs) and laptop computers. These units emit coded IR that is detected by other devices. In this capacity, IR is used much like radio waves. Another use of IR is in lasers; some of the most powerful ones in use emit infrared light (see Section 10.8).

Visible Light

Visible light is a very narrow band of frequencies of EM waves that happens to be detectable by human beings. Certain specialized cells in the eye, called *rods* and *cones*, are sensitive to EM waves in this band. They respond to visible light by transmitting electrical signals to the brain, where a mental image is formed. (The visible ranges of some animals such as hummingbirds and bees extend into the ultraviolet band. Some flowers that seem plain to humans are quite attractive to these nectar eaters.)

This remote-control unit uses infrared radiation to transmit information.

© iStockphoto.com

Table 8.1 Approximate Frequencies and Wavelengths of Different Colors

Color	Frequency Range ($\times 10^{14}$ Hz)	Wavelength Range ($\times 10^{-7}$ m)
Red	4.0–4.8	7.5–6.3
Orange	4.8–5.1	6.3–5.9
Yellow	5.1–5.4	5.9–5.6
Green	5.4–6.1	5.6–4.9
Blue	6.1–6.7	4.9–4.5
Violet	6.7–7.5	4.5–4.0

Visible light is a component of the heat radiation emitted by very hot objects. About 44% of the Sun's radiation is visible light: It glows white hot. Incandescent light bulbs produce visible light in the same way. Fluorescent and neon lights use excited atoms that emit visible light. In Chapter 10, we will discuss this process and describe how infrared and ultraviolet light and even x-rays are emitted by excited atoms.

Within the narrow band of visible light, the different frequencies are perceived by people as different colors. The lowest frequencies of visible light, next to the infrared band, are perceived as the color red. The highest frequencies are perceived as violet. Table 8.1 shows the approximate frequencies and wavelengths of the six main colors in the rainbow.

Note how narrow the frequency band is: The highest frequency of light we can see is less than twice the lowest. By comparison, the range of frequencies of sound that can be heard is huge: The highest is 1,000 times the lowest.

Most colors that you see are combinations of many different frequencies. White represents the extreme: One way to produce white light is to combine equal amounts of all frequencies (colors) of light. Rainbow formation involves reversing the process: White light is separated into its component colors. When no visible light reaches the eye, we perceive black.

In our daily lives, visible light is the most important of all electromagnetic waves. The entire next chapter is dedicated to optics, the study of light and its interaction with matter.

Tanning requires a source of ultraviolet light.

© iStockphoto.com

Ultraviolet Radiation

Ultraviolet (UV) radiation, also called *ultraviolet light*, is a band of EM waves that begins just above the frequency of violet light and extends to the x-ray band. The frequency range is from about 7.5×10^{14} hertz to 10^{18} hertz.

Ultraviolet light is also part of the heat radiation emitted by very hot objects. About 7% of the radiation from the Sun is UV. This part of sunlight is responsible for suntans and sunburns. Ultraviolet radiation does not warm the skin as much as IR, but it does trigger a chemical process in the skin that results in tanning. Overexposure leads to sunburn as a short-term effect, and repeated overexposure during a person's lifetime increases the chances of developing skin cancer. In Section 8.7, we describe how the ozone layer protects us from excessive UV in sunlight.

Some substances undergo "fluorescence" when irradiated with UV: They emit visible light. The inner surfaces of fluorescent lights are coated with such a substance. The UV emitted by excited atoms in the tube strikes the fluorescent coating, and visible light is produced. The same process is used in plasma TVs. Some fluorescent materials appear to be colorless under normal light and can be used as a kind of invisible ink. They can be seen under a UV lamp but are invisible otherwise.

UV radiation has many practical applications. For example, it is used as an investigative tool at crime scenes to help identify bodily fluids like blood and bile. Ultraviolet lights are used by entomologists to attract and collect nocturnal insects for cataloguing and study. Ultraviolet lamps are used to sterilize workspaces and tools used in biology laboratories and medical facilities. And, increasingly, ultraviolet lasers (see Chapter 9) are finding use in many fields from metallurgy (engraving) to medicine (dermatology and optical keratectomy) to computing (optical data storage).

X-Rays

The next higher frequency electromagnetic waves are x-rays. They extend from about 10^{16} to 10^{20} hertz. An important feature of x-rays is that their range of wavelengths (about 10^{-8} to 10^{-11} meters) includes the size of the spacing between atoms in solids. X-rays are partially reflected by the regular array of atoms in a crystal and so can be used to determine the arrangement of the atoms. X-rays also travel much greater distances through most types of matter compared to UV, visible light, and other lower-frequency EM waves.

X-rays are produced by smashing high-speed electrons into a "target" made of tungsten or some other metal (Figure 8.27). The electrons spontaneously emit x-rays as they are rapidly decelerated on entering the metal. X-rays are also emitted by some of the atoms excited by the high-speed electrons.

Medical and dental "x-ray" photographs are made by sending x-rays through the body. Typically, x-rays with frequencies between 3.6×10^{18} hertz and 12×10^{18} hertz are used. As x-rays pass through the body, the degree to which they are absorbed depends on the material through which they pass. Tissue containing elements with relatively large atomic numbers (Z), like calcium ($Z = 20$), tend to absorb x-rays more effectively than those that contain predominantly light elements like carbon ($Z = 6$), oxygen ($Z = 8$), or hydrogen ($Z = 1$). Lead, with atomic number 82, is a particularly good shield for blocking x radiation.

Figure 8.27
Simplified sketch of an x-ray tube. Electrons are accelerated to a very high speed by the high voltage. X-rays are emitted as the electrons enter the metal target.

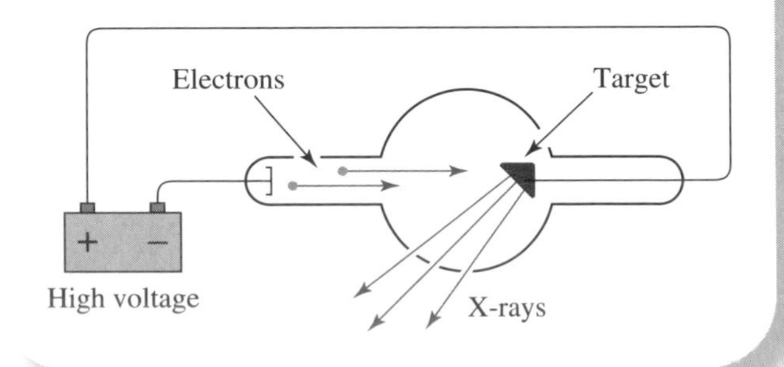

Bones, which are rich in calcium, absorb x-rays better than soft tissue such as muscle or fat, and hence show up more clearly on x-rays (Figure 8.28).

X-rays (and gamma rays) can be harmful because they are **ionizing radiation**—radiation that produces ions as it passes through matter. Such radiation can "kick" electrons out of atoms, leaving a trail of freed electrons and positive ions. This process can break chemical bonds between atoms in molecules, thereby altering or breaking up the molecule. Living cells rely on very large, sophisticated molecules for their normal functioning and reproduction. Disruption of such molecules by ionizing radiation can kill the cell or cause it to mutate, perhaps into a cancer cell. The human body can (and does) routinely replace dead cells, but massive doses of x-rays or other ionizing radiation can overwhelm this process and cause illness, cancer, or death. Because medical x-rays are the largest source of artificially produced radiation in the United States, comprising about 10% of the total annual radiation dose for the average resident, it is little wonder that protecting the public from unnecessary exposure to damaging radiation in diagnostic radiology is one of the greatest challenges to health and radiological physicists. We will discuss this more in Section 11.3.

Figure 8.28

X-ray photograph of a human hand. In this image, areas that appear white are those that strongly absorbed the incident x radiation. Bones are much more efficient at absorbing x-rays because of their calcium content. Elements like gold and silver found in most jewelry are even better absorbers of x-rays because of their higher atomic numbers.

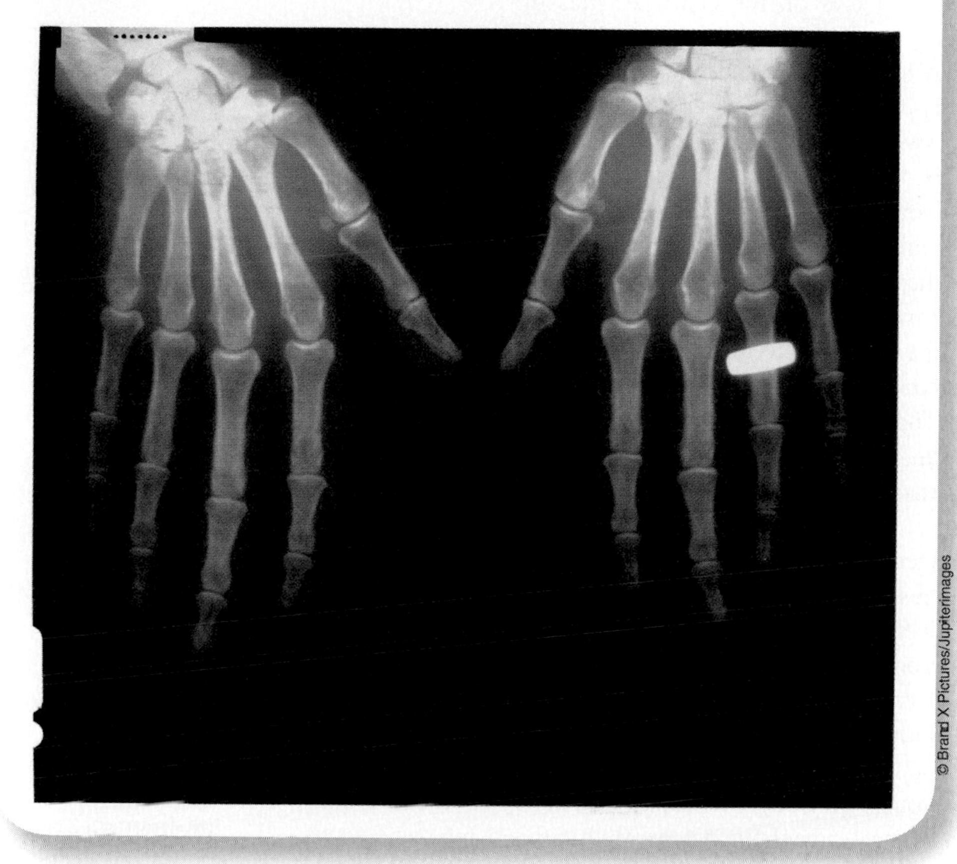

© Brand X Pictures/Jupiterimages

Gamma Rays

The highest-frequency EM waves are gamma rays (γ rays). The frequency range is from about 3×10^{19} hertz to beyond 10^{23} hertz. The wavelength of higher-frequency gamma rays is about the same distance as the diameter of individual nuclei. Gamma rays are emitted in a number of nuclear processes: radioactive decay, nuclear fission, and nuclear fusion, to name a few. We will study these processes in detail in Chapter 11.

This concludes our brief look at the electromagnetic spectrum. Even though the various types of waves are produced in different ways and have diverse uses, the only real difference in the waves themselves is their frequency and, therefore, their wavelength.

8.6 Blackbody Radiation

Every object emits electromagnetic radiation because of the thermal motion of its atoms and molecules. We have already discussed how this radiation offers one method of transferring heat (Section 5.4). Without radiation from the Sun, the Earth would be a frozen rock. In this section, we take a closer look at heat radiation and consider some of its uses.

The nature of the radiation emitted by a given object—the range of frequencies or wavelengths of EM waves present and their intensities—depends on the temperature of the object and on the characteristics of its surface (for example, its color). A hypothetical object that is perfectly black—one that absorbs all EM waves that strike it—would actually be the best at emitting heat radiation. Referred to as a *blackbody*, it would emit radiant energy at a higher rate than any other object at the same temperature, and the intensities of all of the wavelengths of EM waves emitted could be predicted quite accurately. The heat radiation emitted by such an object is referred to as **blackbody radiation (BBR)**.

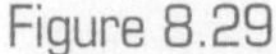
Figure 8.29

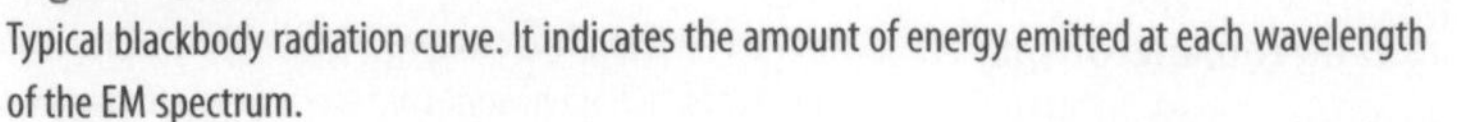
Typical blackbody radiation curve. It indicates the amount of energy emitted at each wavelength of the EM spectrum.

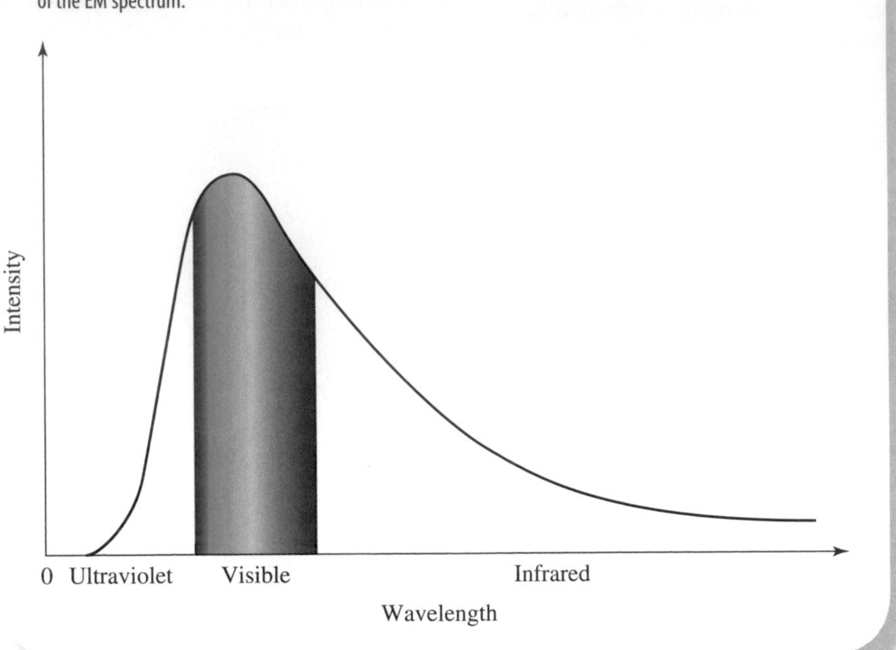

Blackbody radiation, an idealized representation of heat radiation, has been analyzed thoroughly and is well understood. The actual radiation emitted by real objects usually is not too much different from BBR, so we can use it as a model of heat radiation.

The heat radiation emitted by any object (such as your own body, the Sun, a blackbody) is a broad band of electromagnetic waves. Within this band, some wavelengths are emitted more strongly than others: The intensity of the different wavelengths, the amount of energy released per square meter of emitting surface per second, varies with wavelength. For example, the heat radiation from the Sun contains more energy in each wavelength of visible light than in each wavelength of IR radiation. The intensity of the visible wavelengths is higher than that of the IR wavelengths.

A graph showing the intensity of each wavelength of radiation emitted by a blackbody is called a *blackbody radiation curve* (Figure 8.29). The size and shape of the graph change with the object's temperature. The graph for a real object (not a blackbody) would be similar.

The total amount of radiation emitted per second by an object obviously depends on how large it is. A 100-watt light bulb is brighter (it emits more light) than a 10-watt light bulb because its filament is larger. Aside from this factor, it is the object's *temperature* that has the greatest influence on the amount and types of radiation emitted. Three aspects of heat radiation are affected by the object's temperature.

1. The amount of each type of radiation (such as microwave and IR) emitted increases with temperature.

An incandescent light bulb fitted with a dimmer control illustrates this well. Dimming the light causes the filament temperature to decrease. This reduces the amount of visible light emitted: The bulb's brightness is decreased. The amount of infrared is also reduced.

2. The total amount of radiant energy emitted per unit area per unit time increases rapidly with any increase in temperature. For a particular blackbody, the total radiant energy emitted per second (power) is proportional to the temperature (in kelvins) raised to the fourth power.

$$P \propto T^4 \qquad (T \text{ in kelvins})$$

Doubling the Kelvin temperature of an object will cause it to emit 16 times as much radiant energy each second. The human body at 310 K (98.6°F) radiates about 25% more power than it would at room temperature, 293 K (68°F). If you could see infrared, humans would appear to glow more brightly than their cooler surroundings.

3. At higher temperatures, more of the power is emitted at successively shorter wavelengths (higher frequencies) of electromagnetic radiation. For a

blackbody, the wavelength that is given the maximum power (the peak of the blackbody radiation curve) is inversely proportional to its temperature.

$$\lambda_{max} = \frac{0.0029}{T} \qquad (\lambda_{max} \text{ in meters, } T \text{ in kelvins})$$

Objects cooler than about 700 K (about 800°F) emit mostly IR with smaller amounts of microwaves, radio waves, and visible light. There is not enough visible light to be detected by the human eye. Above this temperature, objects emit enough visible light to glow. They appear red hot because more red (longer wavelength) radiation is emitted than any other visible wavelength. The peak wavelength is still in the infrared region. (Several factors influence the minimum temperature needed to cause an object to glow. These include the size and color of the object, the brightness of the background light, and the acuity of the observer's eyesight.)

The filament of an incandescent light bulb can be as hot as 3,000 K. The peak wavelength of its heat radiation curve is in the infrared region, not far from the visible band. The visible light emitted is a bit stronger in the longer wavelengths, so it has a slightly reddish tint. At 6,000 K, the Sun's surface emits heat radiation that peaks in the visible band (the wavelength of the peak is one-half that of the light bulb's). It appears to be white hot (Figure 8.30). When comparing the curves, keep in mind that the Sun's is higher *not* because the Sun is bigger than a light bulb but because it is hotter.

EXAMPLE 8.3

Assuming that the Sun is a blackbody with a temperature of 6,000 K, at what wavelength does it radiate the most energy?

$$\lambda_{max} = \frac{0.0029}{T}$$

$$= \frac{0.0029}{6{,}000 \text{ K}}$$

$$= 4.8 \times 10^{-7} \text{ m}$$

Table 8.1 shows this to be in the blue-green part of the visible band.

Some stars are hot enough to appear bluish. Sirius and Vega, two of the brightest stars in the northern sky, are good examples. The peaks of their radiation curves are in the UV, so they emit more of the shorter wavelength visible light (blue) than the longer wavelengths.

The temperature dependence of blackbody radiation is responsible for a number of interesting phenomena, and it is used in some ingenious ways. The following are examples.

Figure 8.30
Blackbody radiation curves for the Sun and a light bulb.

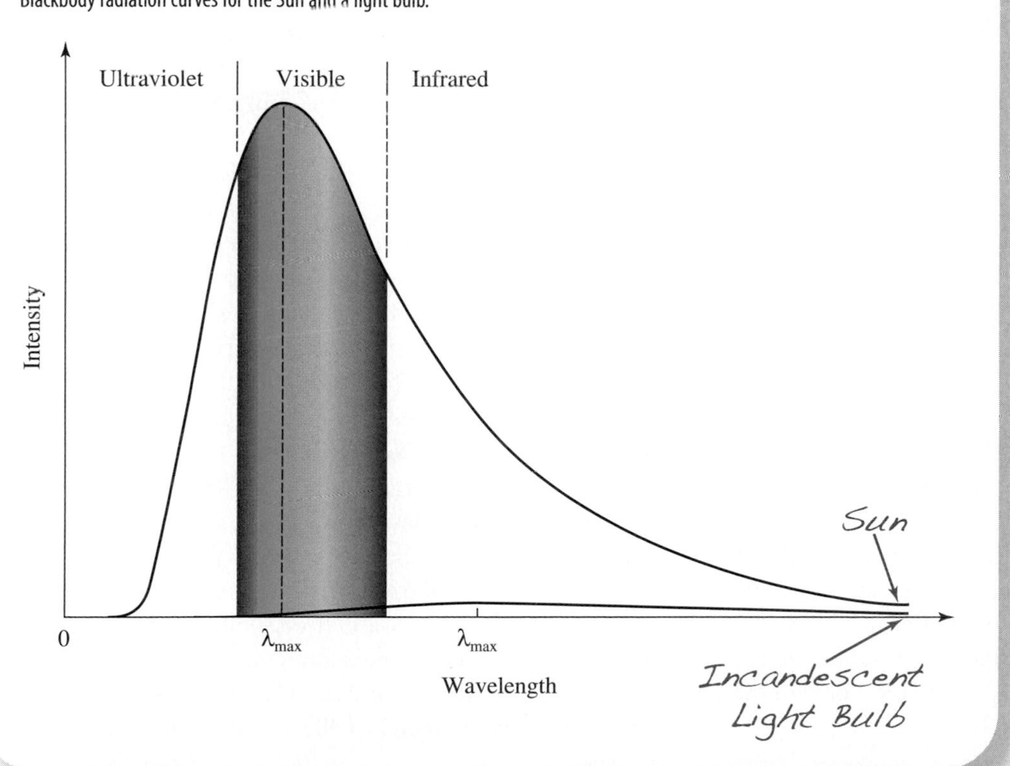

Temperature Measurement

The temperature of an object can be determined by examining the radiation that it emits. This is particularly useful when very high temperatures are involved, as in a furnace, because nothing has to come into contact with hot matter. Special devices called *pyrometers* measure the amount and types of radiation emitted and use the rules mentioned above to determine the temperature. A similar process is used to measure the temperature of the Sun and other stars.

The electronic ear thermometer works in a similar way: It determines the patient's body temperature by measuring the intensity of infrared radiation emitted by an eardrum.

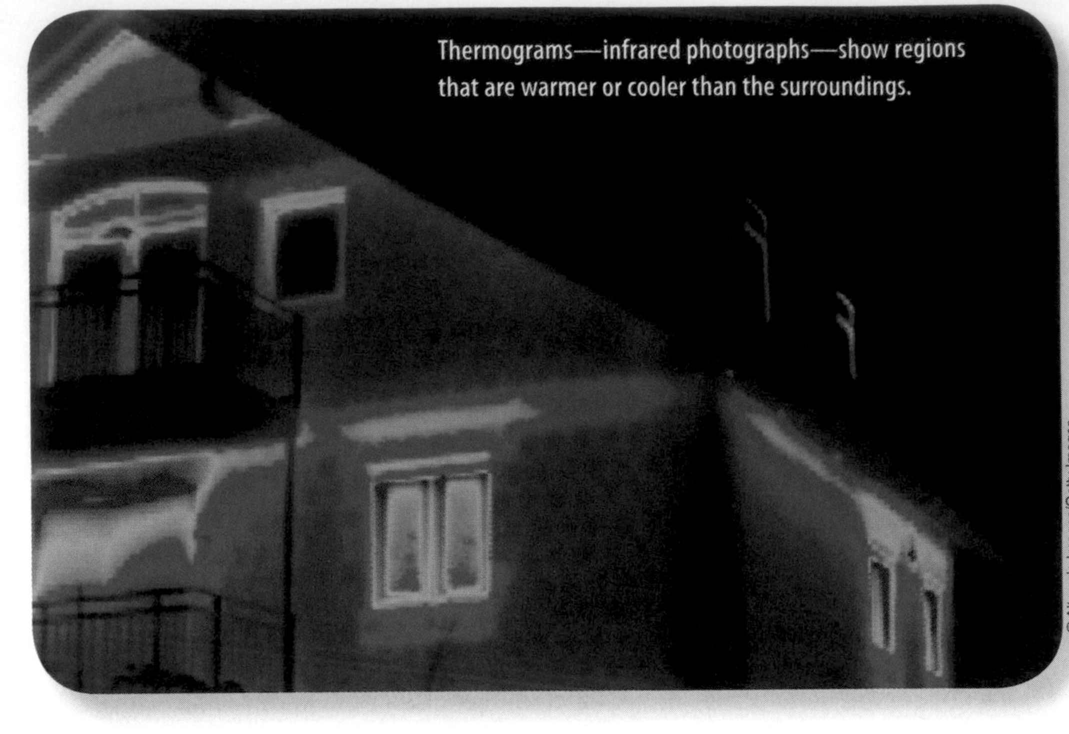
Thermograms—infrared photographs—show regions that are warmer or cooler than the surroundings.

© Altrendo Images/Getty Images

Detection of Warm Objects

Most things on Earth have temperatures that cause them mainly to emit infrared light. Anything that can detect IR can use this fact to locate warmer-than-average objects, since they will emit more IR. Rattlesnakes, for example, use IR to hunt mice and other warm-blooded animals at night. These snakes have sensitive organs that detect the higher-intensity infrared emitted by objects warmer than their surroundings.

Infrared-sensitive photographic film, video cameras, and other detection devices have many practical uses. For instance, IR photographs, called *thermograms*, can show where heat is escaping from a poorly insulated house and can detect ohmic heating caused by a short circuit in an electrical substation. They can locate warmer areas on the human body that might be caused by tumors. The military uses IR detectors to locate soldiers at night. Heat-seeking missiles automatically steer themselves toward the hot exhaust of aircraft.

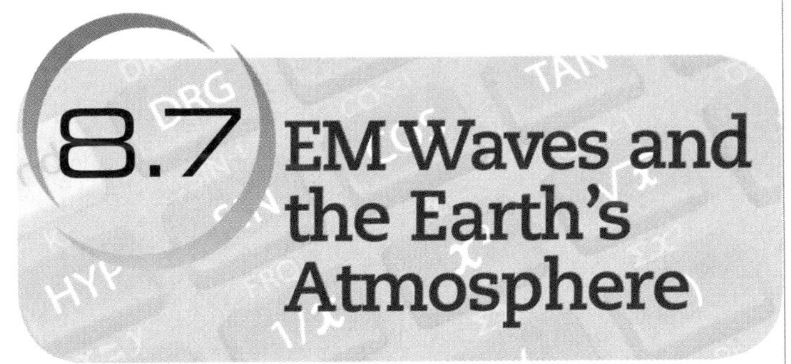

8.7 EM Waves and the Earth's Atmosphere

Many substances in the atmosphere surrounding the Earth interact with EM waves in important ways. Some of these interactions are crucial to the existence of life on this planet, another helps us communicate, and some add to the beauty that characterizes life on Earth. The visible phenomena, rainbows, for example, are described in the chapter dedicated to visible light, Chapter 9. Some of the others are discussed here.

Ozone Layer

The sunlight that keeps the Earth warm and provides the energy for plants to grow also contains ultraviolet radiation that is harmful to living things. But life has evolved on this planet because the atmosphere has protected it from this UV. In the region between about 20 and 40 kilometers above the Earth, known as the *ozone layer*, there is a comparatively high concentration of *ozone* (O_3), the form of oxygen with three atoms in each molecule. This ozone absorbs most of the harmful UV in sunlight. The ozone layer has been a shield protecting living things on Earth.

In 1974, it was reported that *chlorofluorocarbons* (CFCs), chemical compounds such as Freon used in refrigerators, air conditioners, and as an aerosol propellant in spray cans, could be depleting the ozone layer. The CFCs that are released into the air drift upward to the ozone layer and chemically break up the ozone molecules with alarming efficiency. The 1995 Nobel Prize in chemistry was awarded to the discoverers of this effect. Because of this, CFCs were banned for use as aerosol propellants in the United States. In 1985 it was discovered that a "hole" developed in the ozone layer over Antarctica and grew during the later part of each year. The concentration of ozone in a huge section of the atmosphere was reduced by about one-half (Figure 8.31). A review of old satellite measurements revealed that this hole had developed in previous years as well and that the measurement taken in 1982 was twice the area of the

Figure 8.31

False-color image of the ozone hole over the southern hemisphere, based on satellite measurements taken in 5 October 2006. The region of greatly reduced ozone, shown in blue, extends well beyond Antarctica. The total loss of ozone in 2006 was the largest ever recorded.

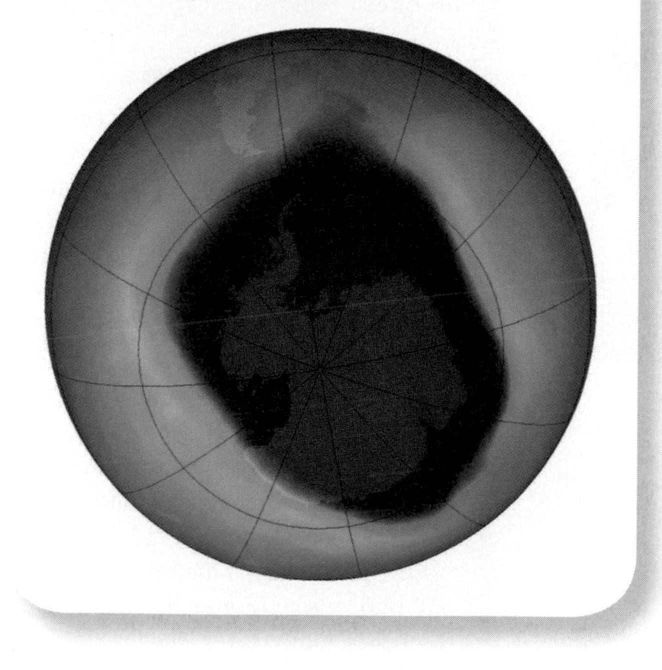

United States. Scientists now believe that the ozone hole is produced by a complex set of processes involving chlorine that originates in CFCs. During the southern winter, chlorine molecules (Cl_2) are released by chemical reactions that take place in extremely high clouds, called *polar stratospheric clouds* (PSCs), over the sunless South Pole. The return of sunlight in spring then triggers the reactions that cause chlorine to break up ozone molecules.

Until recently it was thought that a similar ozone hole would not form over the North Pole. But in March 1997 the ozone level over the Arctic dropped to a record low for that time of year. The depletion was not as severe as in the southern ozone hole, but it suggests that the damage to the ozone layer is increasing.

Global monitoring of the ozone layer revealed an overall decline during the latter part of the twentieth century, not just over the poles. Continued reduction in ozone levels could have tragic consequences. Rates of occurrence of skin cancer could rise. Crop yields could decrease, since increased UV adversely affects many plant species. But unprecedented international cooperation led to the banning of CFCs in developed countries. Levels of CFCs in the atmosphere seem to be declining, but the extremely high chemical stability of these compounds means they will continue to do damage for many years to come.

Greenhouse Effect

The **greenhouse effect** is so named because it is partly responsible for keeping greenhouses warm in cold weather. Glass and certain other materials allow visible light to pass through them while they absorb or reflect the longer-wavelength infrared radiation. A glass wall or roof on a building allows visible light to enter and to warm the interior. As the temperature of things inside increases, they emit more IR. Without the glass, this IR would escape from the enclosure and carry away the added energy. But the glass blocks the IR, so the added internal energy is trapped and the enclosure is warmed. (The glass not only reduces heat loss by radiation, it also eliminates convection: The heated air that rises cannot leave and take internal energy with it.) A car parked in the sun with the windows up is much warmer than the outside air because of this heating. Windows on the sunny side of a building help to keep it warm in the same way.

The greenhouse effect occurs naturally in the Earth's atmosphere. Water vapor, carbon dioxide (CO_2), and other gases in the air act somewhat like glass in that they allow visible light from the Sun to pass through to the Earth's surface while they absorb part of the infrared radiation emitted by the warmed surface (see Figure 8.32a). The atmosphere is heated by the IR that is absorbed; it is about 35°C (65°F) warmer than it would be without this effect.

From 1958 to 2008, the carbon-dioxide content of the atmosphere rose 21% (see Figure 8.32b). Evidence points to human activity as the likely cause. During the last century, huge quantities of fossil fuels—coal, oil, and natural gas—have been burned. This has released vast amounts of carbon dioxide into the air. At the same time, forests have been cut down for building materials and cleared for farming and human occupation. This contributes to the problem because trees and other plants take carbon dioxide out of the air. The concentration of methane is also increasing. Methane gas, released into the air by some animals and by rice as it grows, and the ozone-threatening chlorofluorocarbons are also "greenhouse gases" like water vapor and CO_2. The result of this is the possibility of global warming—the entire atmosphere heating up.

Anyone who has lived through a harsh winter might think that the warming of the atmosphere is not such a bad idea. The main concern is that we might be starting a runaway greenhouse effect that could lead to global disaster. The temperature of the Earth's atmosphere could rise enough to trigger changes in the weather patterns. Once-verdant areas could become deserts, and vice versa. The polar ice caps could start melting, thereby raising

Figure 8.32

(a) The Earth's atmosphere produces a greenhouse effect. Increased carbon dioxide content in the air could cause too much heating. (b) Graph of CO_2 content in the air versus time, measured at the observatory atop Mauna Loa in Hawaii. The concentration varies throughout each year because plants in the northern hemisphere take in more CO_2 in the summer when they are most active.

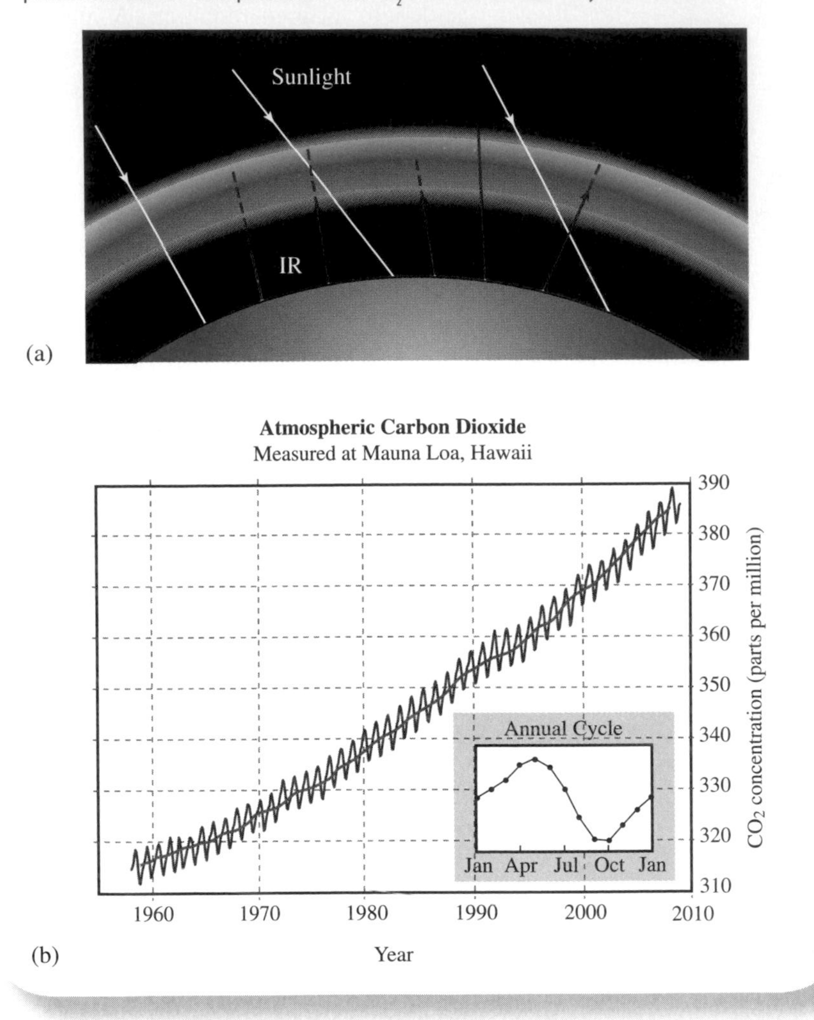

the level of the oceans and flooding heavily populated coastal areas around the world. A runaway greenhouse effect did occur naturally on the planet Venus, where the surface temperature is now 460°C. Conditions on Venus prevented excessive atmospheric carbon dioxide from being trapped in carbonate rocks or dissolved in oceans, as is the case on Earth.

The atmospheric greenhouse effect is such a complex phenomenon that it is nearly impossible to predict exactly what will happen. Different scenarios have been proposed: As the Earth heats up, huge amounts of CO_2 dissolved in the oceans could be released and could exacerbate the global warming. Increased evaporation of water would raise the water-vapor content of the atmosphere and possibly cause more heating. Or the higher humidity could lead to more cloud cover, which might cool the Earth as less sunlight reaches its surface. About all we can be sure of is that we are altering the life-sustaining blanket of air in which we live even though we cannot predict exactly what the effects will be.

The Ionosphere

Did you ever wonder why you can pick up AM radio stations that are several hundred kilometers away but usually cannot receive FM radio or television signals more than about 80 kilometers away? About 50–90 kilometers above the Earth's surface, in a region of the atmosphere known as the *ionosphere*, there is a relatively high density of ions and free electrons. Radio waves from transmitters travel upward into the ionosphere. Higher-frequency radio waves, such as those used in FM radio and television, pass through the ionosphere and out into space (Figure 8.33b). Lower-frequency radio waves, such as the 500–1,500 kilohertz waves used in AM radio, reflect off the ionosphere and return to Earth (Figure 8.33a). The range for high-frequency radio waves is limited to "line-of-sight" reception. The curvature of the Earth eventually blocks the signal from the transmitting tower. Low-frequency radio waves can skip off the ionosphere and travel farther around the planet. The radio waves used to communicate with spacecraft must have high frequency to pass through the ionosphere.

The ionosphere is also the home of the *auroras*—the northern and southern lights. Charged particles from the Sun excite atoms and molecules in the ionosphere, causing them to emit light. (More on this in Chapter 10.)

Astronomy

Stars, galaxies, and other objects in space emit all types of electromagnetic radiation. Astronomers were originally limited to studying only visible light through optical telescopes. Now, they examine the entire spectrum of EM radiation to gain a more complete understanding about the universe. The atmosphere is a hindrance to some of these investigations. Ultraviolet light from stars and galaxies is absorbed by ozone and other gases. Infrared light is also absorbed, mainly by water vapor. Lower-frequency radio waves are absorbed in the ionosphere. Even visible light is affected by the random swirling of the air, which causes stars to twinkle and degrades the images formed in large telescopes. Microwaves and higher-frequency radio waves are about the only EM waves from space that are not affected by the atmosphere.

Since the early 1960s, dozens of telescopes and other astronomical instruments have been placed in orbit to overcome the deleterious effects of Earth's atmosphere. Among the most important and sophisticated of recent space missions are the four that comprise NASA's Great Observatories Program. Each was designed to examine different parts of the EM spectrum, from infrared light for studying cool stars and interstellar dust, to gamma radiation emitted in high-energy processes associated with supernova explosions and compact star collisions. The most famous of these, the *Hubble Space Telescope* (HST), was launched in 1990 and is equipped with instruments that analyze visible, ultraviolet, and shorter-wavelength infrared light (see Figure 8.34). After a shaky start, HST has achieved enormous success: Research astronomers, as well as millions of people worldwide, have been captivated by the stunning images it has returned. (There is more on HST in Section 9.2.) The other spacecraft in the program, along with the year in which each was launched, are the *Compton Gamma-Ray Observatory* (1991), the *Chandra X-Ray Observatory* (1999), and the *Spitzer Space Telescope* (infrared, 2003). Astronomy textbooks for years to come are likely to contain illustrations and findings supplied by these observatories. Space exploration continues to give astronomers access to the entire electromagnetic spectrum.

Figure 8.33

(a) Low-frequency radio waves are bent back to the Earth's surface by the ionosphere. This allows commercial AM radio to be transmitted hundreds of kilometers. (b) High-frequency radio waves pass through the ionosphere. Because of this, transmission of commercial FM radio and television is limited to about 100 km.

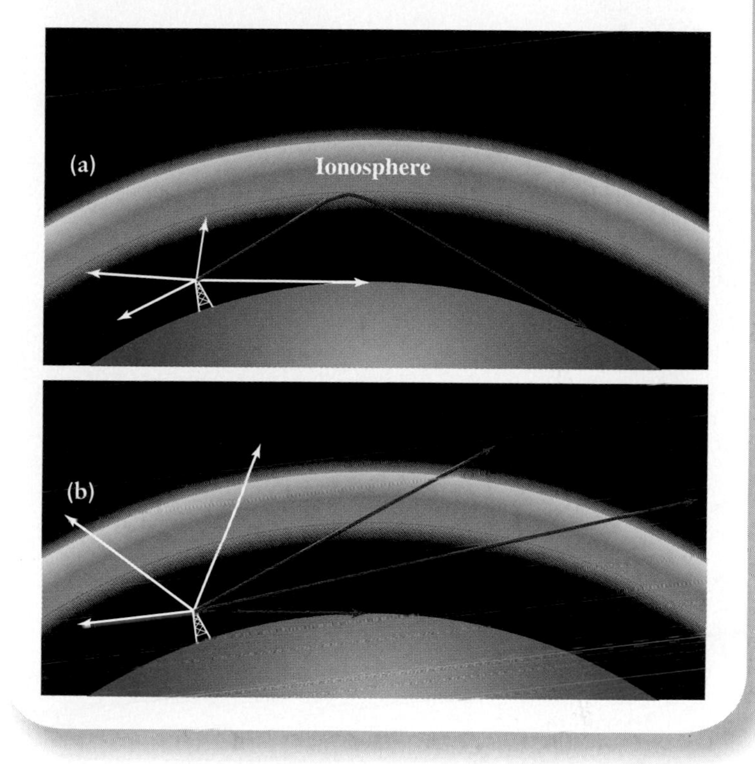

Figure 8.34

(Left) The *Hubble Space Telescope* during deployment from the space shuttle *Discovery*. (Right) *Hubble Space Telescope* image of the Cat's Eye Nebula showing gas and dust ejected from a dying star.

© NASA/CXC/SAO/STScI

questions&problems

Questions

(▶ Indicates a review question, which means it requires only a basic understanding of the material to answer. The others involve integrating or extending the concepts presented thus far.)

1. Three bar magnets are placed near each other along a line, end to end, on a table. The net magnetic force on the middle one is zero, and its north pole is to the left. Make two sketches showing the possible arrangements of the poles of the other magnets.
2. ▶ Sketch the shape of the magnetic field around a bar magnet.
3. ▶ What happens to a ferromagnetic material when it is placed in a magnetic field?
4. What causes magnetic declination? Is there a place where the magnetic declination is 180° (a compass points south)? If so, approximately where?
5. ▶ Describe the three basic interactions between electricity and magnetism.
6. ▶ Explain what *superconducting electromagnets* are. What advantages do they have over conventional electromagnets? What disadvantages do they have?
7. ▶ Name five different *basic* devices that use at least one of the electromagnetic interactions.
8. ▶ In many cases, the effect of an electromagnetic interaction is perpendicular to its cause. Describe two different examples that illustrate this.
9. To test whether a material is a superconductor, a scientist decides to make a ring out of the material and then to see whether a current will flow around in the ring with no steady energy input.
   a) Explain how a magnet could be used to initiate the current.
   b) At some later time, how could the scientist check to see whether the current is still flowing in the ring but without touching the ring?
10. ▶ A coil of wire has a large alternating current flowing in it. A piece of aluminum or copper placed near the coil becomes warm even if it does not touch the coil. Explain why.
11. ▶ What is the "motor-generator duality"? Explain how it is used.
12. Explain why two wires, each with a current flowing in it, exert forces on each other even when they are not touching each other.
13. ▶ Explain why a transformer doesn't work with DC.
14. What would an audio speaker like that shown in Figure 8.19 on page 209 do at the instant a low-voltage battery is connected to it?
15. The type of microphone described in Section 8.4 can be thought of as a speaker "operating in reverse." Explain.
16. ▶ What is *analog-to-digital conversion,* and how is it used in sound reproduction?
17. ▶ Explain how electromagnetic waves are a natural outcome of the principles of electromagnetism.
18. An electromagnetic wave travels in a region of space occupied only by a free electron. Describe the resulting motion of the electron.
19. ▶ List the main types of electromagnetic waves in order of increasing frequency. Give at least one useful application for each type of wave.
20. ▶ Alternating current with a frequency of 1 million Hz flows in a wire. What in particular could be detected traveling outward from the wire?
21. ▶ What are the main uses of microwaves? Explain how each process works.
22. A liquid compound is not heated by microwaves the way water is. What can you conclude about the nature of the compound's molecules?
23. Pilots of aircraft equipped with powerful radar units are forbidden from turning the radar units on when on the ground near people. Explain why this is so.
24. ▶ Which type of EM wave does your body emit most strongly?
25. ▶ What is different about our perceptions of the different frequencies within the visible light band of the EM spectrum?
26. A heat lamp is designed to keep food and other things warm. Would it also make a good tanning lamp? Why or why not?
27. ▶ How are x-rays produced?
28. ▶ Why are x-rays more strongly absorbed by bones than by muscles and other tissues?
29. ▶ What is *blackbody radiation?* How does the radiation emitted by a blackbody change as its temperature increases?
30. A light bulb manufacturer makes bulbs with different "color temperatures," meaning that the spectrum of light they emit is similar to a blackbody with that temperature. What would be the difference in appearance of the light from bulbs with color temperatures of 2,700 K and 4,000 K?
31. Explain how infrared light can be used to detect some types of animals in the dark. Can you think of a situation in which this would not work?
32. ▶ Stars emit radiation whose spectrum is very similar to that of a blackbody. Imagine two stars identical in size, each of which is at the same distance from us. One of the stars appears reddish in color, while the other one looks distinctly bluish. Based on this information, what can you say about the relative temperatures of the two stars? Which is the hotter? How do you know?

33. ▶ What effect does the ozone layer have on the EM waves from the Sun? What is currently threatening the ozone layer?

34. ▶ Describe the greenhouse effect that is occurring in the Earth's atmosphere.

35. ▶ How does the ionosphere affect the range of radio communications?

Problems

1. The charger cord used to recharge a cellphone contains a transformer that reduces 120 V AC to 5 V AC. For each 1,000 turns in the input coil, how many turns are there in the output coil?

2. The generator at a power plant produces AC at 24,000 V. A transformer steps this up to 345,000 V for transmission over power lines. If there are 2,000 turns of wire in the input coil of the transformer, how many turns must there be in the output coil?

3. Compute the wavelength of the carrier wave of your favorite radio station.

4. What is the wavelength of the 60,000-Hz radio wave used by "radio-controlled" clocks and wristwatches?

5. Compute the frequency of an EM wave with a wavelength of 1 in. (0.0254 m).

6. The wavelength of an electromagnetic wave is measured to be 600 m.
   a) What is the frequency of the wave?
   b) What type of EM wave is it?

7. Determine the range of wavelengths in the UV radiation band.

8. A piece of iron is heated with a torch to a temperature of 900 K. How much more energy does it emit as blackbody radiation at 900 K than it does at room temperature, 300 K?

9. The filament of a light bulb goes from a temperature of about 300 K up to about 3,000 K when it is turned on. How many times more radiant energy does it emit when it is on than when it is off?

10. What is the wavelength of the peak of the blackbody radiation curve for the human body ($T = 310$ K)? What type of EM wave is this?

11. What wavelength EM wave would be emitted most strongly by matter at the temperature of the core of a nuclear explosion, about 10,000,000 K? What type of wave is this?

12. What is the frequency of the EM wave emitted most strongly by a glowing element on a stove with temperature 1,500 K?

13. The blackbody radiation emitted from a furnace peaks at a wavelength of 1.2×10^{-6} m (0.0000012 m). What is the temperature inside the furnace?

14. What is the lowest temperature that will cause a blackbody to emit radiation that peaks in the infrared?

There are more quizzes and study tools online at 4ltrpress.cengage.com/physics.

Optics

sections

© Nic Taylor/iStockphoto.com /
© Douglas Palmer/iStockphoto.com

9.1 Light Waves

The basis for many practical devices is found in optics—from cameras, telescopes, and liquid-crystal displays, to striking natural phenomena like rainbows, soap-bubble iridescence, and sundogs. The field of *optics* involves the study of light and its interaction with matter. Light generally refers to the narrow band of electromagnetic (EM) waves that can be seen by human beings. These are transverse waves with frequencies from about 4×10^{14} hertz to 7.5×10^{14} hertz (see Table 8.1 on p. 215). The corresponding wavelengths are so small that we will find it useful to express them in *nanometers* (nm). One nanometer is one-billionth of a meter.

$$1 \text{ nanometer} = 10^{-9} \text{ meter} = 0.000000001 \text{ meter}$$

$$1 \text{ nm} = 10^{-9} \text{ m}$$

The wavelengths of visible light (in a vacuum or in air) range from about 750 nanometers for low-frequency red light to about 400 nanometers for high-frequency violet light. Two important points from Section 8.5 that you should keep in mind are (1) different frequencies of light are perceived as different colors, and (2) white light is typically a combination of all frequencies in the visible spectrum.

The various properties of waves described in the first part of Chapter 6 also apply to light. (You may find it useful to review Sections 6.1 and 6.2 at this time.) As with sound and water ripples, we will use both wavefronts and rays to represent light waves. Recall that a wavefront shows the location in space of one particular phase (peak or valley, for example) of the wave. For a light bulb, the wavefronts are spherical shells (not unlike balloons) expanding outward at the speed of light (see Figure 9.1). A *light ray* is a line drawn in space representing a "pencil" of light that is part of a larger beam. Rays are represented as arrows and indicate the direction the light is traveling. A laser beam can often be thought of as a single light ray. The light from a light bulb can be represented by light rays radiating outward in all directions. (Be careful not to confuse these light rays with the electric and magnetic field lines discussed in previous chapters.)

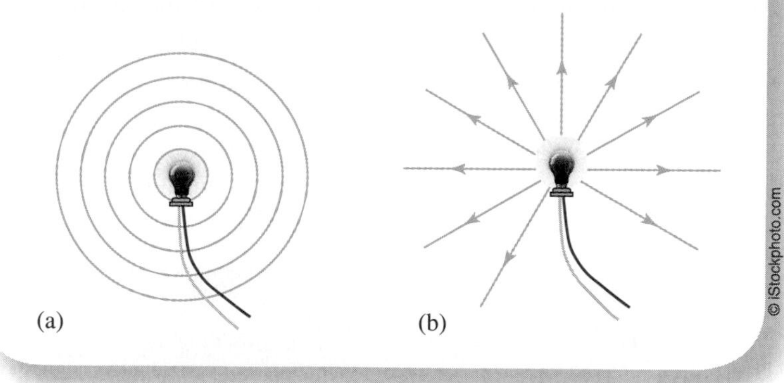

Figure 9.1
The light from a light bulb represented (a) by wavefronts and (b) by light rays.

Some of the general characteristics of wave propagation, like reflection, are readily observed with light waves. But other phenomena are more rare in everyday experience because of two factors:

1. The speed of light is extremely high (3×10^8 m/s in a vacuum).
2. The wavelengths of light are extremely short.

For example, we must turn to distant galaxies moving away from us at high speeds to easily observe the Doppler

effect with light. As we saw in Section 6.2, the Doppler effect with sound is, on the other hand, quite common because the speed of sound is only about 350 m/s and the wavelengths of sound (in air) are in the centimeter to meter range.

In these first two sections, we will describe some of the phenomena that can occur when light encounters matter. The remaining sections of the chapter deal with important things that occur after light has traveled inside transparent material.

Reflection

Reflection of light waves is extremely common: Except for light sources like the Sun, light bulbs, fires, etc., everything we see is reflecting light to our eyes. There are two types of reflection: specular and diffuse.

Specular reflection is the familiar type that we see in a mirror or in the surface of a calm pool of water. A mirror is a very smooth, shiny surface, usually made by coating glass with a thin layer of aluminum or silver. Specular reflection occurs when the direction the light wave is traveling changes (Figure 9.2). By changing the angle of the incident (incoming) light ray and observing the reflected ray, we see that the light behaves somewhat like a billiard ball bouncing off a cushion on a pool table.

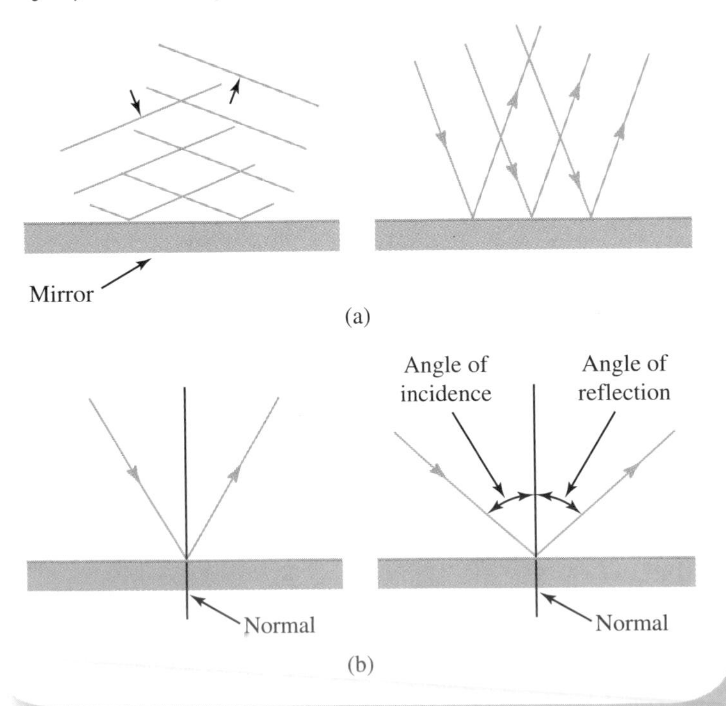

Figure 9.2
(a) Specular reflection using wavefronts and light rays. (b) Specular reflection of a single light ray with different angles of incidence.

Figure 9.2b shows an imaginary line drawn perpendicular to the mirror and touching it at the point where the incident ray strikes it. This line is called the **normal**. The angle between the incident ray and the normal is called the **angle of incidence**, and the angle between the reflected ray and the normal is called the **angle of reflection**. Observations indicate that these angles are always equal. The following law, first described in a book entitled *Catoptrics*, thought to have been written by Euclid in the third century BCE, states this formally.

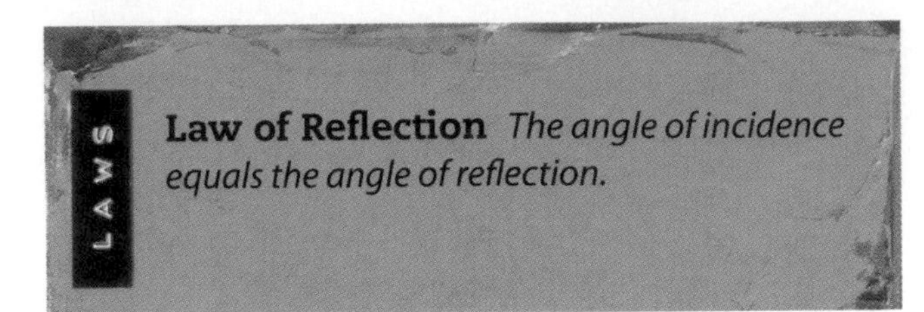

Law of Reflection *The angle of incidence equals the angle of reflection.*

So specular reflection of light is much like sound echoing off a cliff, as described in Chapter 6.

The other type of reflection, **diffuse reflection**, occurs when light strikes a surface that is not smooth and polished but uneven, like the bottom of an aluminum pan or the surface of this paper. The light rays reflect off the random bumps and nicks in the surface and scatter in all directions (see Figure 9.3). The law of reflection still applies, but the rays encounter segments of the irregular surface oriented at different angles and therefore leave the surface in different directions. That is why you can shine a flashlight on an aluminum pan and see the reflected light from different angles around the pan. With specular reflection off a mirror, you could see the reflected light from only one direction.

Except for light sources and smooth, shiny surfaces like mirrors, every object we see is reflecting light diffusely. This diffuse reflection causes light

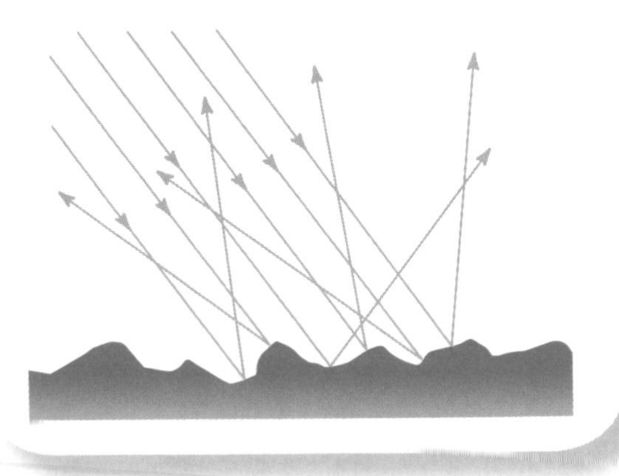
Figure 9.3
Diffuse reflection of light from a rough surface.

to radiate outward from each point on a surface. You can see every point on your hand as you turn it in front of your face because each point on your skin is reflecting light in all directions.

Things can have color because light actually penetrates into the material and is partially reflected and partially absorbed along its way into and out of the material. The reflected light that leaves the surface will have color if pigments in the material absorb some frequencies (colors) more efficiently than others. A white surface, like this paper, reflects all frequencies of light nearly uniformly. If you shine just red light on it, it will appear red. With just blue light, it will appear blue. A colored surface, like that of a red fire extinguisher, "removes" some frequencies of the light. A red surface reflects the lower-frequency light (red) most effectively and absorbs much of the rest (see Figure 9.4). If you shine red light on it, it will appear red. With blue or any other single color, it will appear black: Very little of the light will be reflected.

Figure 9.4

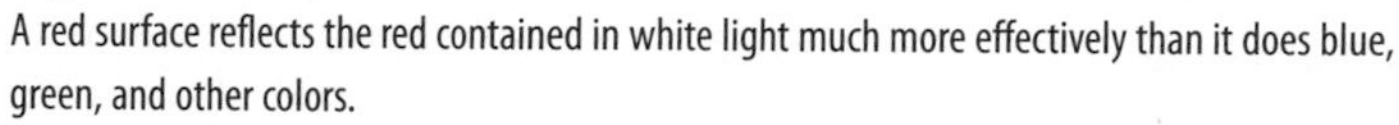

A red surface reflects the red contained in white light much more effectively than it does blue, green, and other colors.

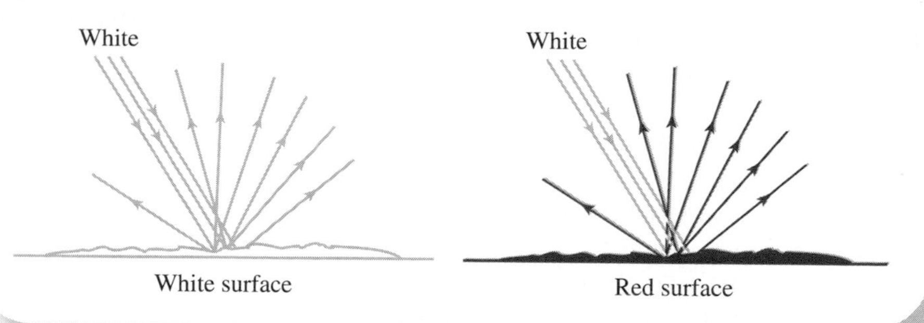

Diffraction

As with all waves, diffraction of light as it passes through a hole or slit is observable only when the width of the opening is not too much larger than the wavelength of the light. This means that light doesn't spread out after passing through a window nearly as much as sound does, but diffraction is observed when a very narrow slit (about the width of a human hair) is used (Figure 9.5). The narrower the slit, the more the light spreads out.

Figure 9.5

Diffraction of light. (a) Light passing through a narrow slit spreads out, as shown on the screen. (b) Photograph of laser light projected onto a screen after passing through a slit 0.008 cm wide. The screen was 10 m from the slit and shows the laser beam spreading out to a size of about 12 cm.

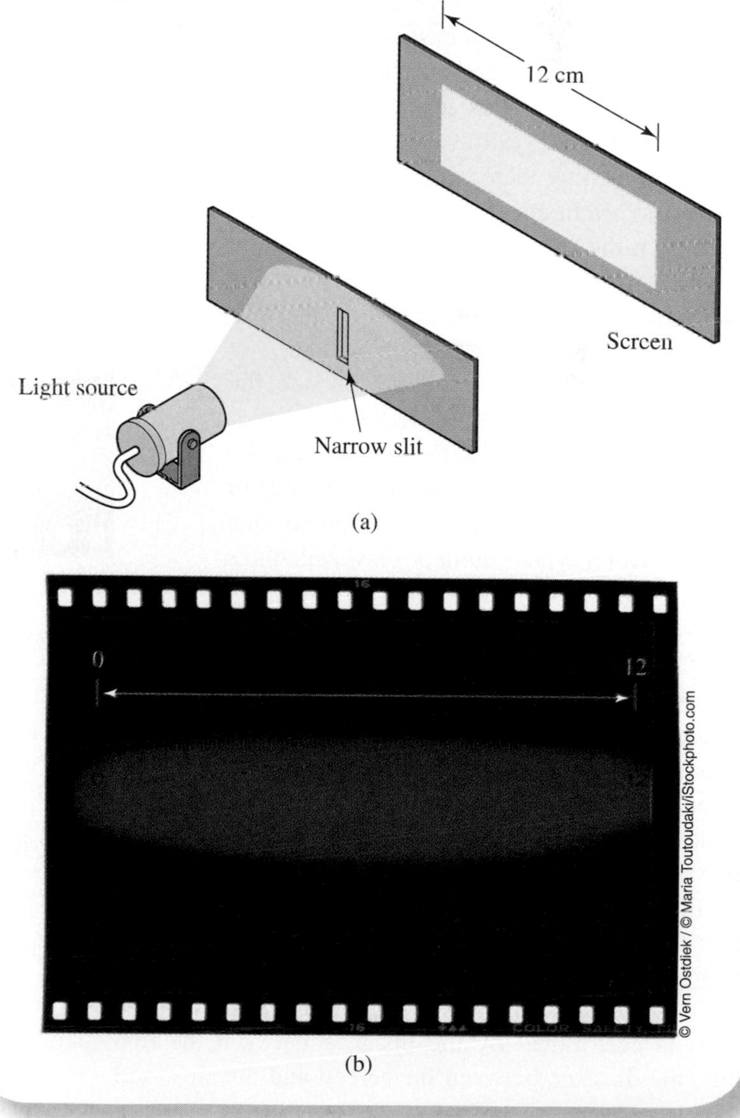

Interference

In Section 6.2, we described how waves can undergo interference. Recall that when two identical waves arrive at the same place, they add together. If the two waves are "in phase"—peak matches peak—the resulting amplitude is doubled. This is called **constructive interference**. At any point where the two waves are "out of phase"—peak matches valley—they cancel each other. This is **destructive interference**.

Interference of light waves is an important phenomenon for two reasons. First, in experiments conducted around 1800, the British physician Thomas Young used interference to prove that light is indeed a wave. Second, interference is routinely used to measure the wavelength of light. We will consider two types of interference, *two-slit* interference and *thin-film* interference.

Figure 9.6

Two-slit interference. (a) Light passing through two narrow slits forms an interference pattern on the screen. The bright areas occur at places where the light waves from the two slits arrive in phase. (b) Photograph of an interference pattern formed by laser light passing through two narrow slits 0.025 centimeter apart. The screen was 10 meters from the slits.

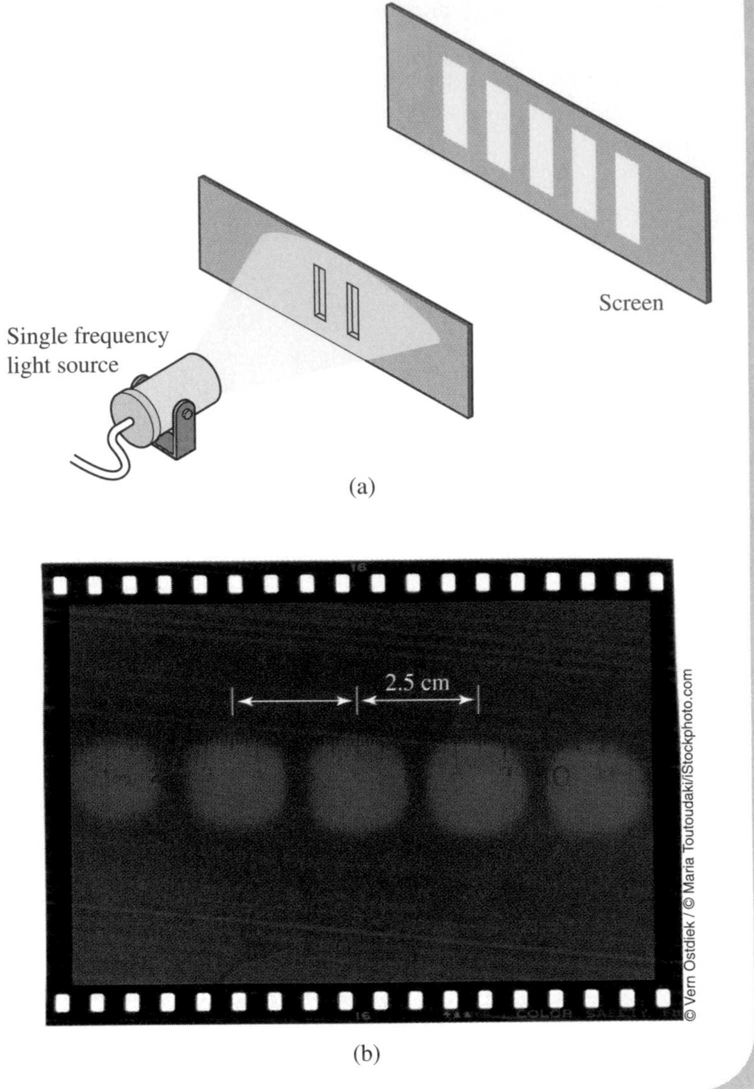

When a light wave passes through two narrow slits that are close together, the two waves emerging from the slits diffract outward and overlap. If the light consists of a single frequency (color), a screen placed behind the slits where the two light waves overlap will show a pattern of bright areas alternating with dark areas (Figure 9.6). At each bright area, the two waves from the slits are completely in phase and undergo constructive interference. Conversely, at each dark area the two waves are completely out of phase and undergo destructive interference—they cancel each other. There is a bright area at the center of this *interference pattern* because the two waves travel exactly the same distance in getting there, so they are in phase. At the first bright area to the left of center, the wave from the slit on the right has to travel a distance exactly equal to one wavelength farther than the wave from the slit on the left. This puts them in phase as well. Similarly, at each successive bright area on the left side, the wave from the right slit has to travel 2, 3, 4, . . . wavelengths farther than the wave from the left slit. At each bright area on the right side of the pattern, it is the wave from the left slit that has to travel a whole number of wavelengths farther.

At the first dark area to the left of the center of the pattern, the wave from the right slit travels one-half wavelength farther than the wave from the left slit. The two waves are out of phase and interfere destructively. At the next dark area on the left, the additional distance is $1\frac{1}{2}$ wavelengths, then $2\frac{1}{2}$ wavelengths at the next, and so on. True constructive and destructive interference actually occurs only at the centers of the bright and dark areas. At points in between, the waves are neither exactly in phase nor exactly out of phase, so they partially reinforce or partially cancel each other.

The distance between two adjacent bright or dark areas is determined by the distance between the two slits, the distance between the screen and the slits, and the wavelength of the light. Since the first two can be measured easily, their values can be used to compute the wavelength of the light.*

* For the mathematically inclined: The separation, δx, between two adjacent bright areas in the interference pattern is determined by the distance, a, between the slits, the distance, S, between the plane of the slits and the screen on which the pattern is projected, and the wavelength, λ, of light used in the experiment. Specifically, the relationship among these variables is given by:

$$\delta x = \left(\frac{S}{a}\right)\lambda.$$

For the experimental setup used to produce Figure 9.6b, $a = 0.025$ cm, $S = 10$ m, and $\lambda = 633$ nm, the wavelength of red light from a He-Ne laser. Under these circumstances, $\delta x = 2.5$ cm, as observed.

The swirling colors you see in oil or gasoline spills floating on wet pavement are caused by *thin-film interference.* Part of the light striking a thin film of oil is reflected from it, and part passes through to be reflected off the water (Figure 9.7). The light wave that passes through the film before being reflected travels a greater distance than the wave that reflects off the upper surface of oil. If the two waves emerge in step, there is constructive interference. If they emerge out of step, there is destructive interference.

Figure 9.7

Interference of light striking a thin film of oil. The light reflecting from the upper surface of the film undergoes interference with the light that passes through to, and is reflected by, the lower surface. Part of the light continues on through the lower surface.

© Garrett Hamilton/iStockphoto.com

The wavelength of the light, the thickness of the film, and the angle at which the light strikes the film combine to determine whether the interference is constructive, destructive, or in between. With single-color (one wavelength) light, one would see bright areas and dark areas at various places on the film. With white light, one sees different colors at different places on the film. At some places, the film thickness and angle of incidence will cause constructive interference for the wavelength of red light, at other places for the wavelength of green light, and so on.

Interference in thin films in hummingbird and peacock feathers is the cause of their iridescent colors. Soap bubbles are also colored by interference of light reflecting off the front and back surfaces of the soap film.

© Tom England/iStockphoto.com / © iStockphoto.com / © iStockphoto.com

Polarization

The fact that light could undergo diffraction and interference convinced Young and other scientists of his time that light can behave like a wave. The other model of light, elaborated by Newton, held that light is a stream of tiny particles, but this approach could not account for these distinctively wavelike phenomena. **Polarization** reveals that light is a *transverse* wave rather than a longitudinal wave like sound.

A rope secured at one end can be used to demonstrate polarization. If you pull the free end tight and move it up and down, a wave travels on the rope that is *vertically polarized.* Each part of the rope oscillates in a vertical plane (Figure 9.8). In a similar manner, moving the free end horizontally produces a *horizontally polarized* wave on the rope. Moving the free end at any other angle with the vertical will also produce a polarized wave. Polarization is possible only with transverse waves.

The fact that light can be polarized reveals its transverse nature. A *Polaroid filter*, like the lenses of Polaroid sunglasses, absorbs light passing through it unless the light is polarized in a particular direction. This direction is coincident with the *transmission axis* of the filter. Light polarized in this direction passes through the Polaroid largely unaffected, light polarized perpendicular to this direction is blocked (absorbed), and light polarized in some direction in between is partially absorbed (Figure 9.9). (We will assume for simplicity that our Polaroid filters are 100% efficient in absorbing light polarized in a direction perpendicular to the transmission axis.)

Interference in thin films in peacock feathers is the cause of their iridescent colors. Soap bubbles are also colored by interference of light reflecting off the front and back surfaces of the soap film.

The light that we get directly from the Sun and from ordinary light fixtures is a mixture of light waves polarized in all different directions. The light is said to be "natural" or "unpolarized" because it has no preferred plane of vibration. When natural light encounters a Polaroid filter, it emerges polarized along the transmission axis (Figure 9.10). The filter allows only that portion of the incident light that oscillates along this direction to pass through; the rest of the radiation is absorbed.

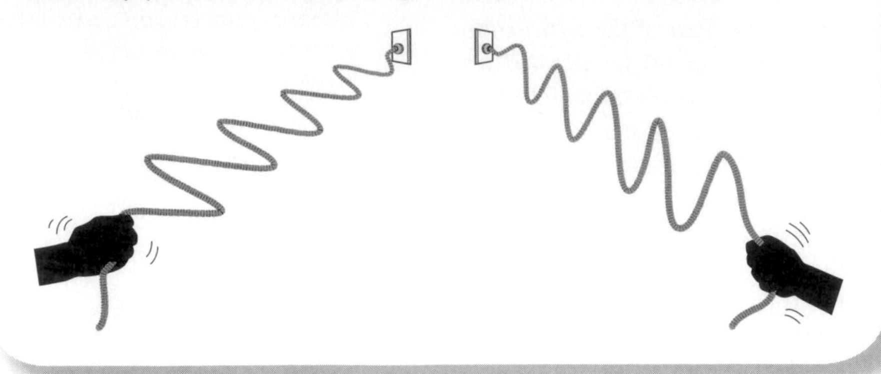

Figure 9.8
Waves on a rope polarized (left) horizontally and (right) vertically.

Now, if the emergent light encounters a second Polaroid filter, the amount of light that passes through will depend on the orientation of the transmission axis of the second filter. If the axis of the second filter is aligned with that of the first, the light will continue on unimpeded. If the axis of the second is perpendicular to that of the first, all of the light will be blocked by the second filter. This is referred to as "crossed Polaroids" (Figure 9.11). When the angle between the transmission axes of the two filters is other than zero or 90°, some of the light will pass through, with the intensity becoming progressively less for angles closer to 90°.

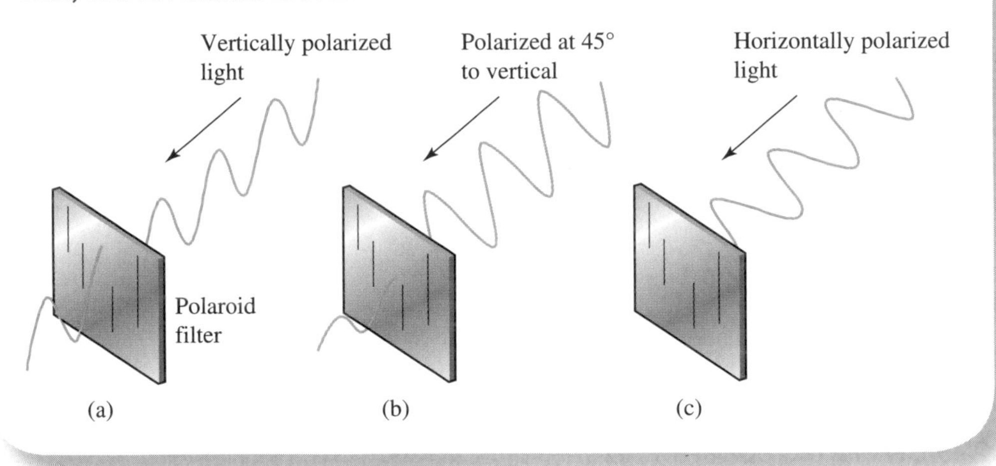

Figure 9.9
Light polarized in different directions encountering a Polaroid filter with its transmission axis vertical. The lines are not actually visible in a Polaroid filter but are drawn here to show the direction of the filter's transmission axis.

Polaroid sunglasses are very useful because light is polarized to some extent when it reflects off a smooth surface like that of water, asphalt, or the paint on the hood of a car. In particular, the reflected sunlight is partially polarized horizontally. This reflected light, called *glare*, is usually bright and annoying. Sunglasses using Polaroid lenses with their transmission axes vertical will block most of this reflected light, which makes it easier to see the surface itself.

Liquid-crystal displays (LCDs) used in calculators, digital watches, laptop computers, some flat-screen TVs, and video games also use polarization. The liquid crystal is sandwiched between crossed Polaroids, and this assembly is placed in front of a mirror. Without the liquid crystal present, the display would be dark: Light passing through the first Polaroid is polarized vertically and would be totally absorbed when it reached the sec-

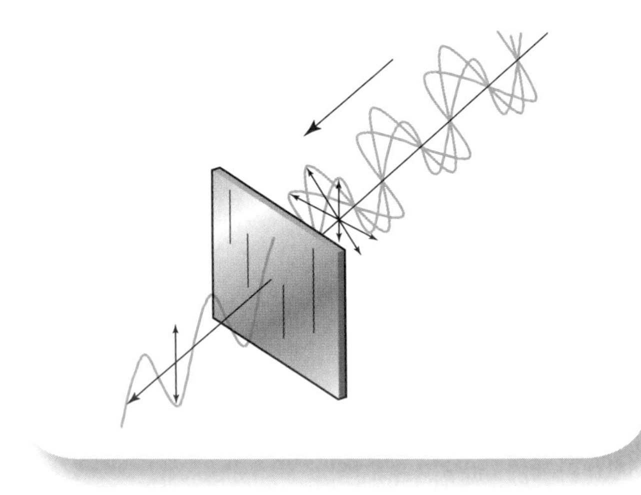

Figure 9.10
Unpolarized light being vertically polarized by a Polaroid filter. The star-shaped cluster of arrows represents a combination of light waves with all different directions of polarization. The filter blocks all but the vertically polarized components in the light.

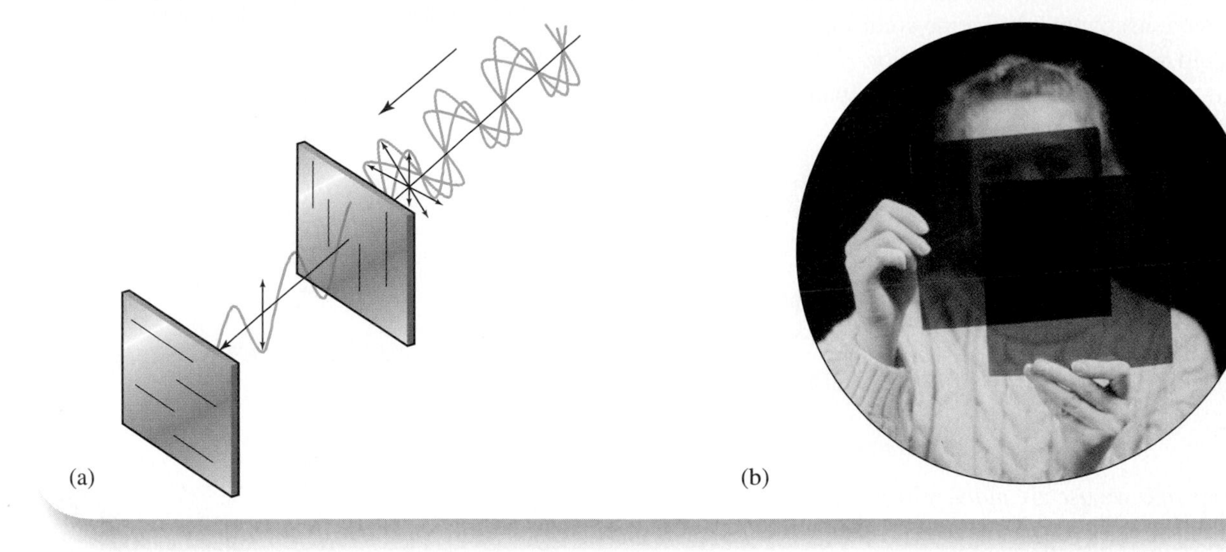

Figure 9.11

Crossed Polaroids. (a) Light emerging from the first Polaroid is polarized vertically. The second Polaroid, with its transmission axis horizontal, blocks the light. (b) A photograph of crossed Polaroid sheets demonstrating the effect visually.

ond Polaroid. No light would reach the mirror, so none would be reflected back from the display. The specially chosen liquid crystal material between the Polaroids is arranged so that in its normal state its molecules change the polarization of the light passing through it. The liquid crystal twists the polarized light 90°. The polarization is changed from vertical to horizontal for light passing through from the front and from horizontal to vertical for light passing through from the rear. The light that was vertically polarized by the first Polaroid is made horizontally polarized by the liquid crystal, passes through the second Polaroid, is reflected back through the display, and emerges polarized vertically (see Figure 9.12a).

To make images on the display, segments of it are darkened. Here is where the liquid crystal property is used. An electric field is switched on in the parts of the display to be darkened. This causes the molecules of the liquid crystal to rotate and to become aligned in one direction. As a result, they no longer change the polarization of light passing through. Light going through this segment is absorbed by the second Polaroid, and that part of the display is dark (see Figure 9.12b). By the way, the transmission axis of the first Polaroid is

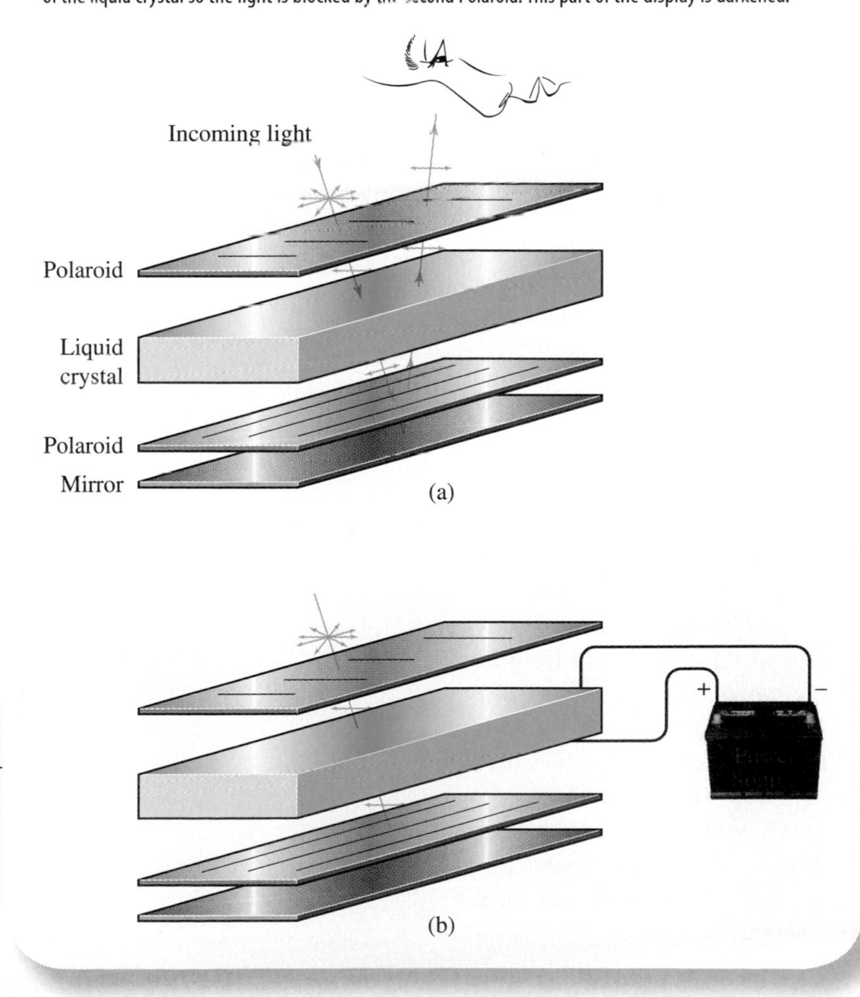

Figure 9.12

Expanded view of part of a liquid-crystal display. (a) The liquid crystal changes the polarization of the light so that it can make a round trip through the crossed Polaroids. (b) The voltage neutralizes this part of the liquid crystal so the light is blocked by the second Polaroid. This part of the display is darkened.

oriented vertically so that you can see the display while wearing Polaroid sunglasses. If you rotate an LCD, say like that on a digital watch, and look at it while wearing Polaroid sunglasses, at one point all of it will be dark.

It has been discovered that some color-blind animals such as squid can sense the polarization of light. This helps squid "see" plankton, which are transparent.

9.2 Mirrors: Plane and Not So Simple

Most mirrors that we use are *plane* mirrors: They are flat, smooth, almost perfect reflectors of light. When we use a mirror to "see ourselves," light that is diffusely reflected off our clothes and face strikes the mirror and undergoes specular reflection. Some of the rays leaving the mirror are going in the proper direction to enter our eyes and to give us an image of ourselves. The image appears to be on the other side of the mirror. Figure 9.13 uses a mannequin to show what happens when a person views his image in a plane mirror. Instead of showing every light ray traveling outward from every point on the person, we show selected rays that happen to enter the person's eyes. The dashed lines from the image show the apparent paths taken by the rays when traced back to the image. Applying the law of reflection and a little mathematics, it can be demonstrated that the image of an object in a plane mirror is as far behind the mirror's surface as the object is in front of the mirror. (If you want to test this conclusion, point your index finger in the direction of a mirrored surface, and slowly move your hand toward the mirror. The image of your finger will approach the mirror's

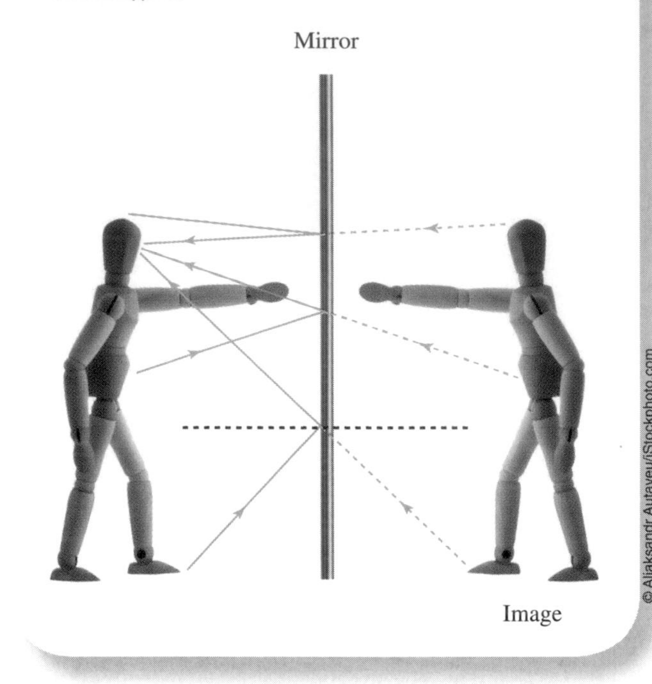

Figure 9.13
Image formed by a plane mirror. The rays reaching the viewer's eyes form an image that appears to be on the other side of the mirror. (The normal for the lower ray is shown as a dashed line to illustrate that the law of reflection applies.)

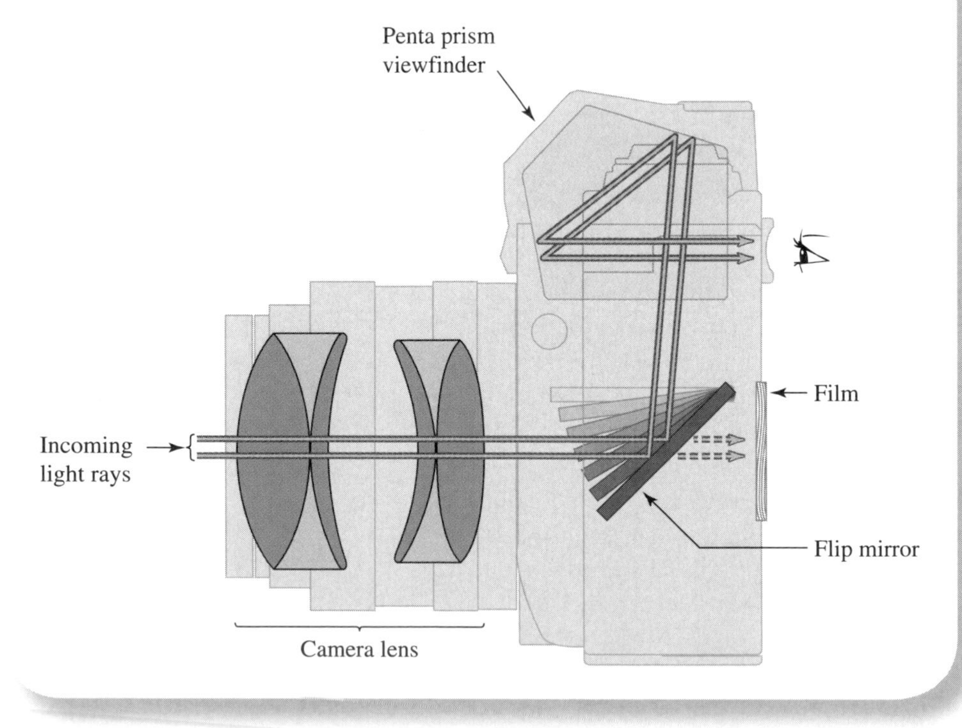

Figure 9.14
A typical reflex camera design. Light enters through the lens, strikes the mirror, travels to the prism and then to the photographer's eye. When the shutter is released, the mirror pops up, allowing light to pass directly to the film. The mirror then drops down.

surface at the same rate that your real finger does, and it will arrive at the surface just as your finger touches it.)

A number of important, practical devices employ plane mirrors. They are used as "optical levers" to amplify small rotations in specialized laboratory instruments. For example, as the mirror rotates through an angle θ, the reflected beam will be turned through an angle 2θ. A plane mirror is used in many reflex cameras (Figure 9.14) to redirect light from the lens to the viewfinder. When the shutter is pressed, the mirror tilts up allowing the light to reach the film. More recently, *micromirrors*, small enough to fit through the eye of a needle (0.5 mm or less), have become indispensable parts of modern telecommunications networks, where they are used in high-speed optical switches to control the flow of information through optical fibers (see Section 9.3).

"One-Way Mirror"

A "one-way mirror" is made by partially coating glass so that it reflects some of the light and allows the rest to pass through. This is called a *half-silvered mirror* (Figure 9.15). When used as a window or wall between two rooms, it will function as a one-way mirror if one of the rooms is brightly lit and the other is dim. It will appear to be an ordinary mirror to anyone in the bright room, but it will appear to be a window to anyone in the dim room. This is because, in the bright room, the light reflected off the half-silvered mirror is much more intense than the light that passes through from the other room. In the dim room, the transmitted light from the bright room dominates. A person in the dim room can see what is happening in the bright room without being seen by anyone in the bright room.

This device is often used in interview and interrogation rooms and as a means of observing customers in stores and gambling casinos. Note that if a bright light is turned on in the dimmer room, the one-way effect is destroyed. Ordinary window glass is a crude one-way mirror because it does reflect some of the light that strikes it. At night, one can see into a brightly lit room through a window, but anyone in the room has difficulty seeing out due to room light reflected by the window pane.

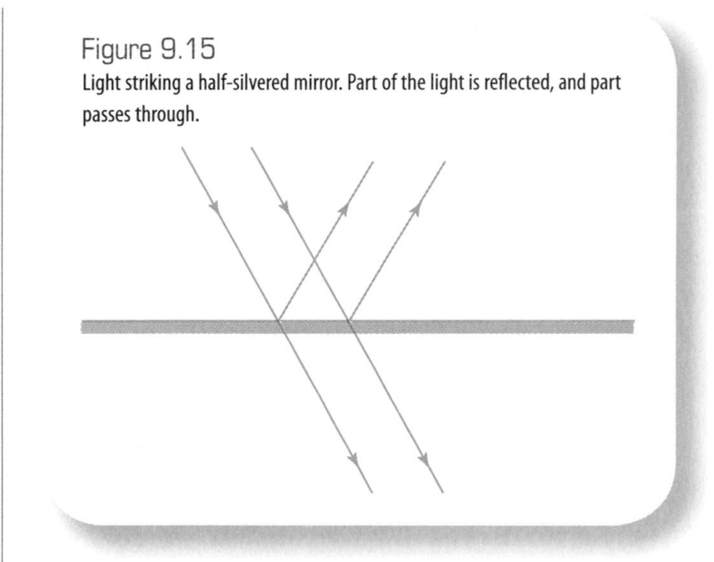

Figure 9.15
Light striking a half-silvered mirror. Part of the light is reflected, and part passes through.

Curved Mirrors

As we saw in Section 6.2, reflectors—mirrors in this case—that are curved have useful properties. Parallel light rays that reflect off a properly shaped *concave mirror*—a mirror that is curved inward—are focused at a point called the **focal point** (Figure 9.16a). The energy in the light is concentrated at that point. Sunlight focused by a concave mirror can heat things to very high temperatures (Figure 9.16b). Even when a mirror's surface is curved, the law of reflection still holds at each point that a ray strikes the mirror. If a normal line is drawn at each point (as was done in Figure 9.2b), the angle of incidence equals the angle of reflection. One such normal line is shown in Figure 9.16a.

A concave mirror can be used to form images that are enlarged (magnified). Magnifying makeup and shaving

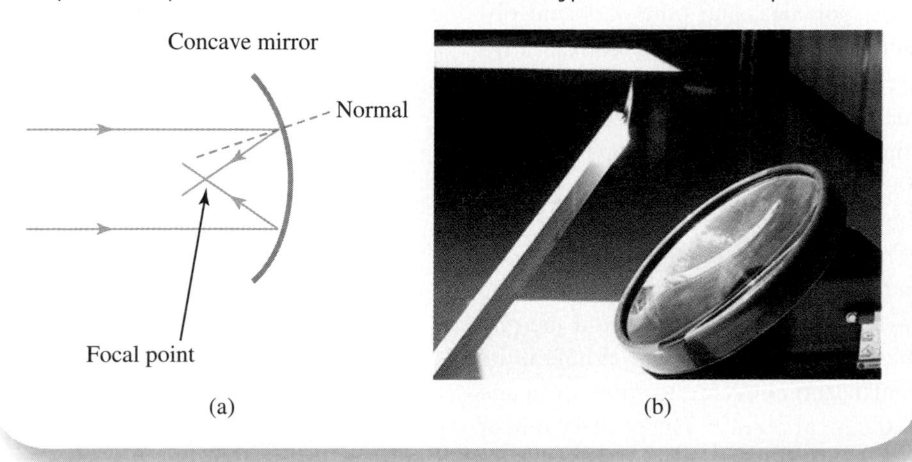

Figure 9.16
(a) Parallel rays reflecting off a concave mirror converge on a point called the *focal point*. (The normal for the upper ray shows that the law of reflection applies.) (b) Sunlight focused by a concave mirror can generate high temperatures. This piece of wood was in flames a few seconds after being placed at the mirror's focal point.

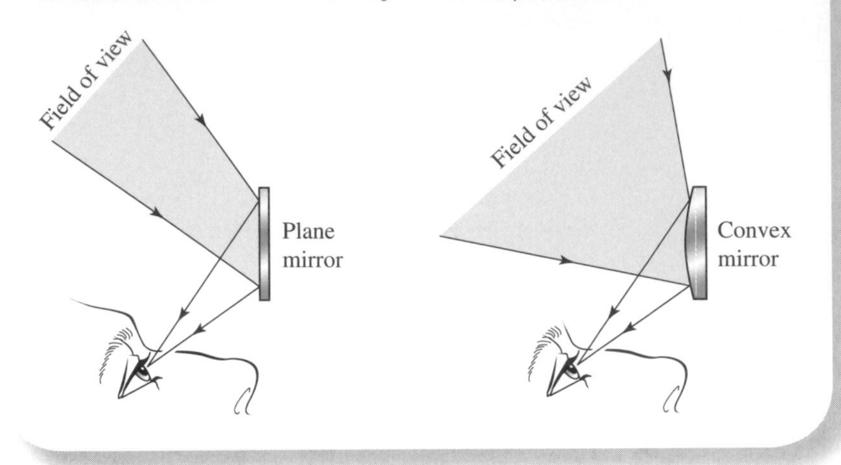

Figure 9.17
The field of view of a convex mirror is much larger than that of a plane mirror.

mirrors are concave mirrors, as are the large mirrors used in astronomical telescopes.

A *convex mirror* is one that is curved outward. The image formed by a convex mirror is *reduced*—it is smaller than the image formed by a plane mirror. The advantage of a convex mirror is that it has a wide field of view—images of things spread over a wide area can be viewed in it. Figure 9.17 shows the fields of view for a convex mirror and a plane mirror of the same size. One glance at a well-placed convex mirror on a bike path allows quick surveillance of a large area. Passenger-side rearview mirrors on cars and auxiliary "wide-angle" rearview mirrors on trucks and other vehicles are convex so the driver can view a large region to the rear. Care must be taken when using such a mirror because the reduced image makes any object appear to be farther away than it actually is.

Astronomical Telescope Mirrors

The largest telescopes used by astronomers to examine stars, galaxies, and other celestial objects make use of curved mirrors. Figure 9.18 shows a common design for such telescopes. Light from the distant source enters the telescope and reflects off a large concave mirror called the *primary mirror.* The reflected rays converge onto a much smaller convex mirror called the *secondary mirror.* The rays are reflected back toward the primary mirror, pass through a hole in its center, and converge to form an image at the focal point *F.* The primary mirror is the key component of the telescope.

Figure 9.18
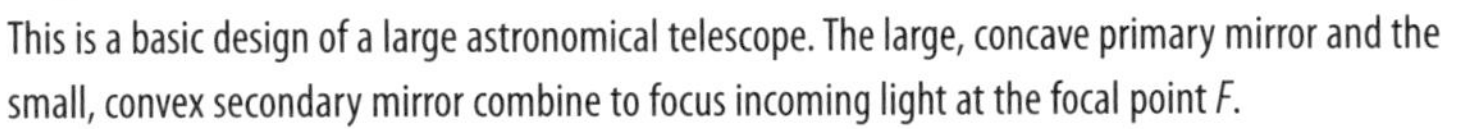
This is a basic design of a large astronomical telescope. The large, concave primary mirror and the small, convex secondary mirror combine to focus incoming light at the focal point *F*.

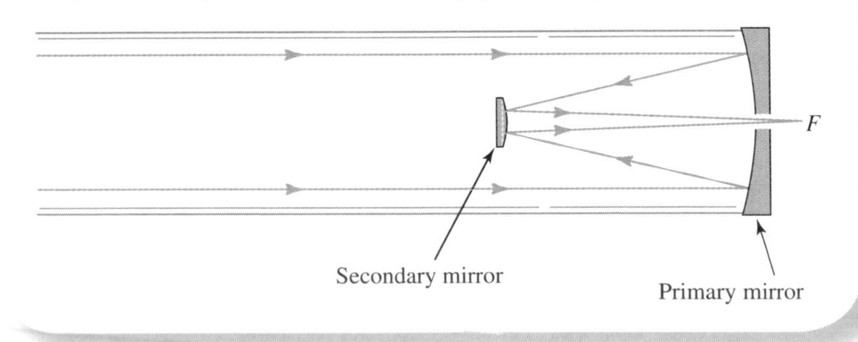

Telescope mirrors have as their basic functions the gathering of light and the concentration of that light to a point. The ability of a mirror to collect light increases with its surface area. To acquire enough radiation to study faint objects adequately, astronomers have sought to build instruments with larger and larger apertures (openings) and, hence, larger light-collecting areas.

The quality of the images produced by telescopes is greatly affected by the shapes of the mirrors. The easiest curved mirror to make is one whose surface has the shape of part of a sphere. But such a *spherical mirror* is not perfect for the task of focusing light rays. Figure 9.19a shows that parallel light rays reflecting off a spherical mirror are not all focused at the same point. An image formed using such a mirror will be somewhat blurred. This phenomenon is called **spherical aberration**. We will see in Section 9.4 that the same thing happens with lenses.

As the name implies, spherical aberration is a defect associated with spherical surfaces. A concave mirror in the shape of a parabola does not have this aberration. (You may recall that we saw the parabola in Section 2.6.) A *parabolic mirror* will concentrate all the rays coming from a distant source at the same point (Figure 9.19b). Thus, the ideal surface for a telescope mirror (or for that matter, reflectors in auto headlamps and household flashlights) is one shaped like a parabola. Fabricating very large mirrors, some as big as 10 meters (33 feet) in diameter, with the precise parabolic shape is an enormous technical challenge. One technique called *spin casting* exploits the fact that the surface of a liquid rotating at a steady rate has the required parabolic shape.

Since the 1980s, several telescopes have been constructed that employ rotating liquid mirrors, the most

Figure 9.19

(a) A spherical mirror does not reflect all incoming rays from a distant source to the same point. Rays located well off the symmetry axis of the mirror are brought to a focus closer to the mirror than those rays nearer the axis. This effect is called *spherical aberration.* (b) A parabolic mirror focuses all the light from a distant source to a single point. It is the ideal shape for a telescope mirror.

Light from a distant source

Circle

(a)

Parabola

Circle

(b)

ambitious being the 6-meter Large Zenith Telescope (LZT) in British Columbia. In the simplest designs, a large bowl of mercury is spun at the proper rate to produce a surface with the desired parabolic shape. Although such liquid mirror telescopes can't be tilted and can only examine the sky nearly directly above them, they are relatively cheap to build: The 6-meter LZT cost only about $1 million, 100 times less than a comparable conventional glass mirror telescope.

Shortly after the Hubble Space Telescope (HST; see Figure 8.34) was placed in orbit on 25 April 1990, scientists discovered that its primary mirror was afflicted with a type of spherical aberration. At the edge of the 2.4-meter-diameter mirror, its surface is misshapened by 0.002 millimeters from what it is supposed to be. This seemingly minuscule error drastically reduced the telescope's ability to form sharp images. In December 1993, space shuttle astronauts installed corrective optics on the instrument platform of the Space Telescope to correct this problem and allow the observatory to perform as designed. Figure 9.20 shows the dramatic improvement in the ability of the Space Telescope to resolve fine detail as a result of the repairs. Further improvements in the performance of HST have occurred with the installation of the Advanced Camera for Surveys (ACS) in the spring of 2002. The ACS replaced an earlier camera and, with its higher resolution, larger field of view, and greater light collection efficiency, provides about a factor-of-ten improvement in the capability of the Space Telescope.

Figure 9.20

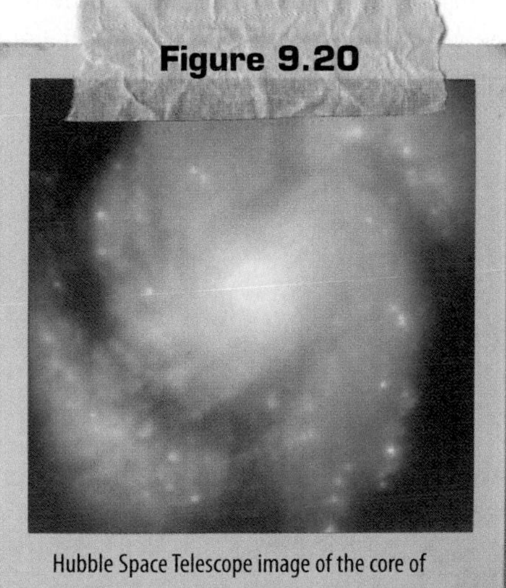

Hubble Space Telescope image of the core of the galaxy M100 taken before the installation of corrective optics.

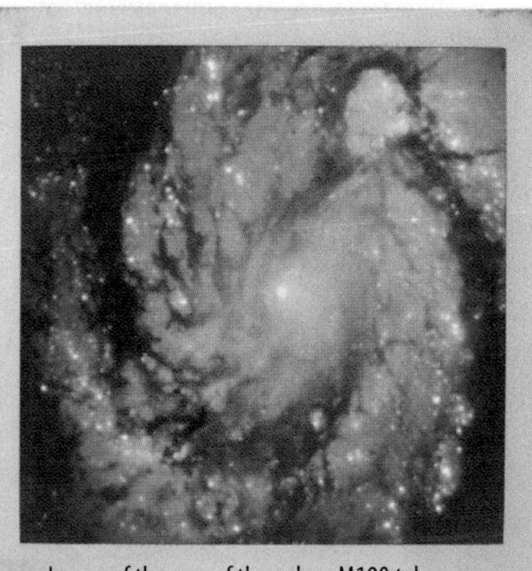

Image of the core of the galaxy M100 taken after the installation of corrective optics in December 1993.

NASA / NASA / © iStockphoto.com

From the ground, recent efforts to combat the blurring effects of the Earth's atmosphere (see the discussion on p. 222) and thereby to increase the resolution of optical instruments, have involved the use of "adaptive optics" (AO). Here, elaborate wavefront sensors, fast computers, and deformable mirrors are used to produce images so sharp that it's as though the atmosphere had disappeared entirely. Figure 9.21a shows a typical design of an AO system, similar

Figure 9.21

(a) An adaptive optics (AO) system. The distorted wavefronts Σ_1 are analyzed and reshaped by the use of a deformable "rubber" mirror. The corrected wavefronts Σ_2 are transmitted to the scientific instruments and to a wavefront sensor that provides feedback for controlling the deformable mirror. (b) Laser used in conjunction with a telescope equipped with adaptive optics to compensate for atmospheric turbulence.

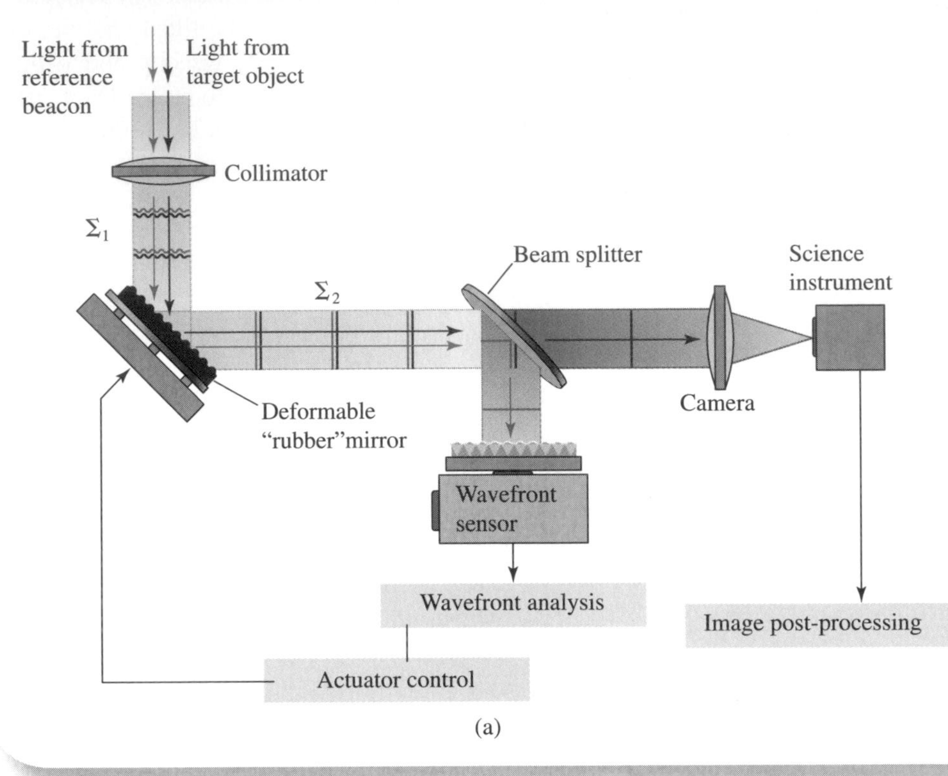

(a)

© Ray Nelson/Phototake

(b)

to those in use with the two 10-m Keck Telescopes in Hawaii, the 8.2-m Gemini North (Hawaii) and South (Chile) instruments, and the Very Large Telescope array (Chile). The key optical element is the "rubber mirror"—a thin glass mirror whose surface can be slightly deformed by up to a hundred tiny actuators attached to the back. If the tiny deformations in the rubber mirror can be made to counteract the deformations in the wavefronts from distant sources caused by turbulence in the Earth's atmosphere, then the original wavefronts can be restored and the blurring in the image removed. Fast computers are required to monitor continuously the rapidly changing conditions in the atmosphere, compute the necessary corrections, and signal the accuators to move appropriately.

This technique normally uses a bright star in the field of view of the telescope as the reference beacon for the wavefront assessment. In the absence of such a star, an artificial one can be created using a high-power, highly focused laser beam projected up through the telescope (Figure 9.21b). Called "laser guidestars," the light from the beam that is scattered downward either from air molecules at altitudes of 10 to 40 km or from sodium atoms residing as high as 90 km produces a faint but measurable target for the wavefront sensors. As indispensable as AO systems are for optimizing the light-gathering and resolving power of large, 10-m class telescopes, still greater advances will be needed in this field to harness the full capabilities of the *really* large telescopes with 20-m apertures planned for the future. To meet the challenges, new initiatives described as "atmospheric tomography"—similar to "medical tomography" where a 3-D view of a patient is produced—will be required. The Concept Map on your Review Card summarizes the types and uses of mirrors.

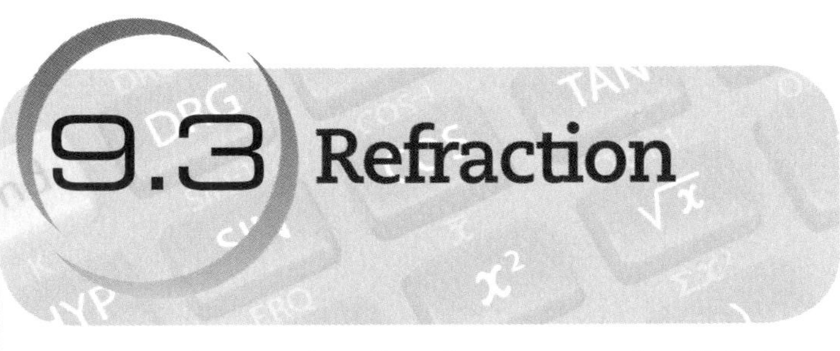

9.3 Refraction

In this section, we describe what happens when light interacts with glass and other transparent material. Imagine a light ray from some source in air arriving at the surface of a second transparent substance like glass. The boundary between the air and glass is called the *interface*. As mentioned earlier, some of the incident light is reflected back into the air, but the rest of it is transmitted across the interface into the glass (see Figure 9.22). The law of reflection gives us the direction of the reflected light ray, but what about the transmitted light ray?

The light that passes into the glass is *refracted*—the transmitted ray is bent into a different direction than the incident ray. (Only if the incident ray is perpendicular to the interface is there no bending.) This bending is caused by the fact that light travels slower in glass than in air. We again draw in the normal, a line perpendicular to the interface. The angle between the transmitted ray and the normal, called the **angle of refraction**, is smaller than the angle of incidence. We can also have the reverse process: A light ray traveling in glass reaches the interface and is transmitted into air (Figure 9.23). In this case, the angle of refraction is larger than the angle of incidence. The following law summarizes these observations.

LAWS

Law of Refraction *A light ray is bent toward the normal when it enters a transparent medium in which light travels more slowly. It is bent away from the normal when it enters a medium in which light travels faster.*

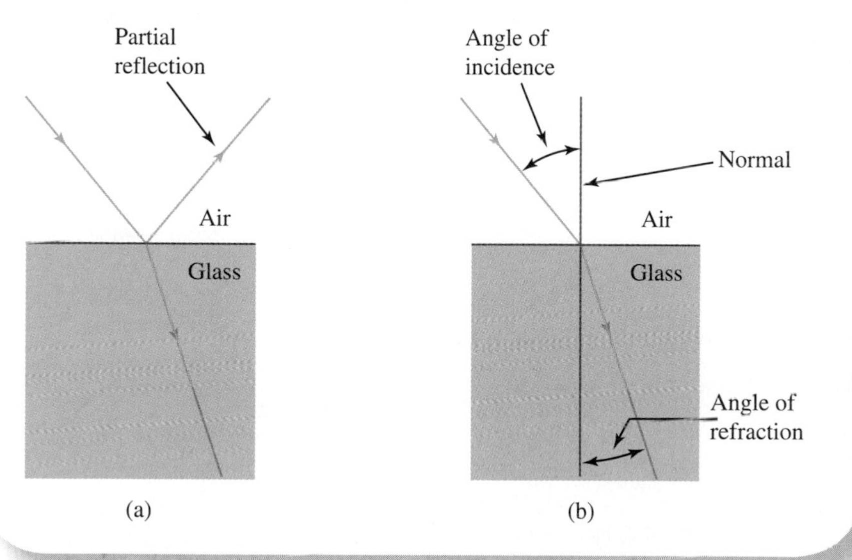

Figure 9.22
(a) Refraction of light as it enters glass, with partial reflection shown. (b) The normal and angles of incidence and refraction. Note that the ray is bent toward the normal.

© Lepro/Shutterstock

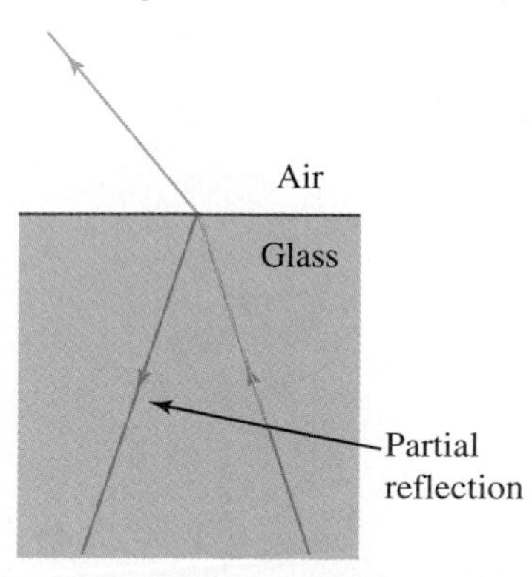

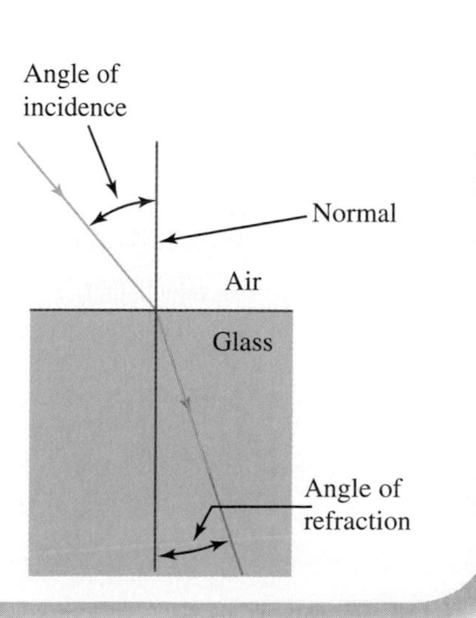

Figure 9.23
Refraction of light as it goes from glass into air. The path of the refracted ray is the reverse of the path shown in Figure 9.22.

© Lepro/Shutterstock

Notice the symmetry in Figures 9.22 and 9.23. This is one example of the *principle of reversibility*: The path of a light ray through a refracting surface is reversible. The path that a ray takes when going from air into glass is the same path it would take if it turned completely around and went from glass into air. In all of the figures in this and the following sections, the arrows on the light rays could be reversed and the paths would be the same.

Refraction of light affects how you see things that are in a different medium, such as under water. Figure 9.24 shows a pencil partially submerged in a glass of

Figure 9.24
A pencil under water appears closer and larger to an observer.

© Rob Walls/Alamy

water. Notice first the apparent discontinuity of the pencil's length that occurs at the interface between the water's surface and the air above. This effect is due to the refraction of light from the pencil as it emerges into the air. Moreover, the portion of the pencil that is submerged in water appears larger than the portion in air because the image formed in the water is closer to the observer than the image formed in air. Figure 9.25 displays a simplified ray diagram for this circumstance revealing how the bending of the rays of light from the edges of the pencil as they enter the air from the water causes the pencil to appear to be closer to the surface of the containing vessel than it really is. The magnification effect in Figure 9.24 is also aided by the fact that the surface of the glass container is curved creating what amounts to a liquid cylindrical lens (see Section 9.4 for more on lenses), but even if the walls of the vessel were planar, the same basic phenomena would be observed.

The speed of light in any transparent material is less than the speed of light c in a vacuum. For example: $c = 3 \times 10^8$ m/s, while the speed of light in water is 2.25×10^8 m/s, and the speed of light in diamond is only 1.24×10^8 m/s. Table 9.1 gives the speed of yellow light in selected transparent media.* (The critical angle and the variation of speed with color will be discussed later.)

At this point, you may have two questions in mind. Why is the speed of light lower in transparent media such as water and glass, and why does this cause refraction? To answer the first question, we must remember that light is an electromagnetic wave—a traveling combination of oscillating electric and magnetic fields. As light enters glass, the oscillating electric field in the wave causes electrons in the atoms of the glass to oscillate at the same frequency. These oscillating electrons emit secondary EM waves (light) that travel outward to neighboring electrons. However, the wave emitted by each electron is not exactly "in step" with the primary incident wave that is causing the

Figure 9.25

Ray diagram showing why the portion of the object in water appears magnified to the observer when seen through water. The object looks larger because its image lies closer to the observer.

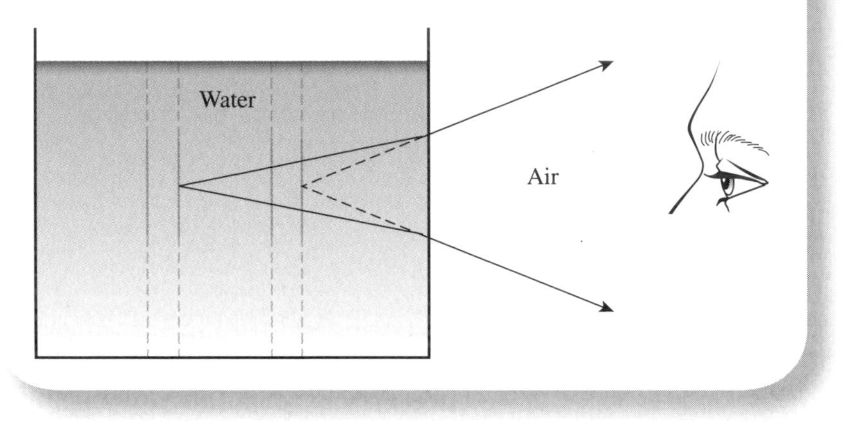

Table 9.1 Speed of (Yellow) Light and Critical Angles for Selected Materials

Phase	Substance	Speed	Critical Angle†
Gases at 0°C and 1 atm	Carbon dioxide	2.9966×10^8 m/s	
	Air	2.9970	
	Hydrogen	2.9975	
	Helium	2.9978	
Liquids at 20°C	Benzene	1.997×10^8 m/s	41.8°
	Ethyl alcohol	2.203	47.3°
	Water	2.249	48.6°
Solids at 20°C	Diamond	1.239×10^8 m/s	24.4°
	Glass (dense flint)	1.81	37.0°
	Glass (light flint)	1.90	39.3°
	Glass (crown)	1.97	41.1°
	Table salt	2.00	41.8°
	Fused silica	2.05	43.2°
	Ice	2.29	49.8°

†Critical angles are for air as the second medium.

* For the mathematically inclined: Each transparent material has an *index of refraction*, n, given by the equation

$$n = \frac{c}{v}$$

(v = speed of light in the material)

Using the information in Table 9.1, the index of refraction of water is 1.33 and that of diamond is 2.42. The exact mathematical relationship between the angles of incidence and refraction is:

$$n_1 \sin A_1 = n_2 \sin A_2$$

(Snell's law)

where n_1 and A_1 are the index of refraction and the angle between the ray and the normal, respectively, in medium 1, and n_2 and A_2 are the corresponding quantities for medium 2. Looking ahead to Figure 9.29b, for example, $n_1 = 1.00$ (air), $A_1 = 14.7°$ (angle of refraction), $n_2 = 1.46$ (glass), and $A_2 = 10°$ (angle of incidence).

electron to oscillate: The secondary emitted wave lags the incident wave slightly. This means, for example, that a secondary wave crest arrives at a given point in the medium a little bit later in time than the corresponding primary wave crest. As the process repeats from electron to electron, these phase lags accumulate, yielding a net EM disturbance formed from the addition of the primary and secondary waves whose speed of propagation through the medium is reduced.

As to the second question, we can see why a change in the speed of light causes bending by carefully following wavefronts as they cross an interface (see Figure 9.26). A good analogy to this is a marching band, lined up in rows, crossing (at an angle) the boundary between dry ground and a muddy area. Each person in the band is slowed as his or her feet slip and sink in the mud. To remain aligned and properly spaced, each person has to turn a bit, and the whole band ends up marching in a slightly different direction. A similar change in direction occurs for light entering glass. Adding together the slower moving wavefronts emitted by the electron oscillators located at different distances from the interface in a way that respects their relative phases produces a net (total) wavefront that propagates in a different direction compared to that of the incident wave. Note that the reduction in speed also makes the wavelength *shorter*. This has to be the case because the frequency of the wave remains the same (why?) and we must have $v = f\lambda$.

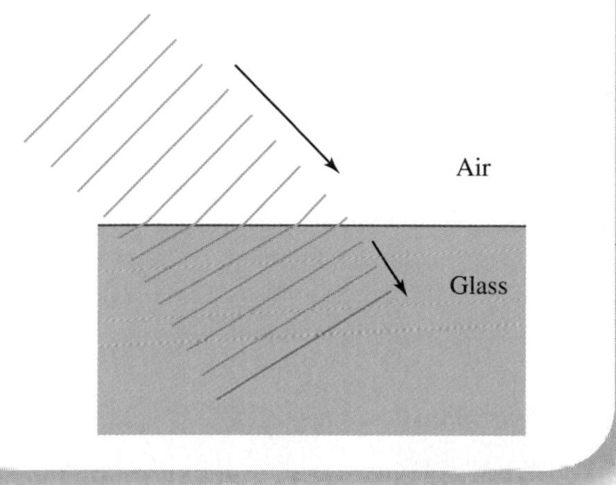

Figure 9.26
Wavefront representation of light refracting as it enters glass. The change in direction occurs because the light travels slower in the glass.

The law of refraction was discovered in 1621 by the Dutch physicist Willebrord Snell and later expressed in its common form by the French mathematician René Descartes (see the footnote on the previous page). It gives the exact value of the angle of refraction when the angle of incidence and the speeds of light in the two media are known.

The graph in Figure 9.27 shows the relationship between the two angles for an air-glass boundary. (A similar graph could be constructed for different pairs of media.) The angle between the normal and the light ray in air is

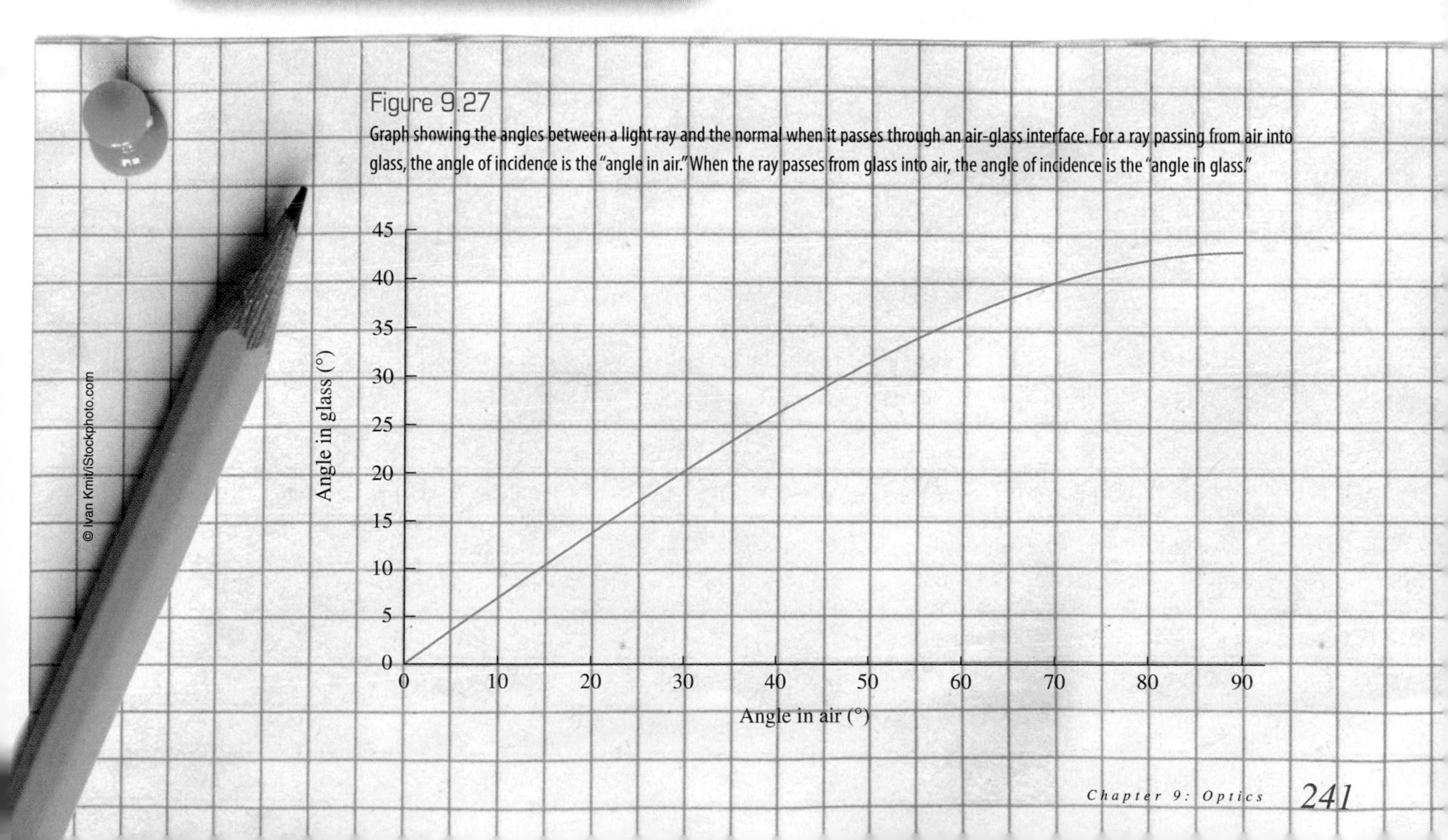

Figure 9.27
Graph showing the angles between a light ray and the normal when it passes through an air-glass interface. For a ray passing from air into glass, the angle of incidence is the "angle in air." When the ray passes from glass into air, the angle of incidence is the "angle in glass."

plotted on the horizontal axis, and the corresponding angle for the ray in glass is plotted on the vertical axis. Notice that the angle in glass is smaller than the angle in air. For a ray going from air into the glass, this means that the angle of refraction is smaller than the angle of incidence, as shown in Figure 9.22. For the largest possible angle of incidence, 90°, the angle of refraction is about 43°.

Figure 9.28
The light ray is refracted twice as it passes through the block of glass. Its final path is parallel to its initial path if the two surfaces of the glass are parallel. (Note the partial reflection at the upper surface where the ray initially enters the glass.)

EXAMPLE 9.1

Figure 9.28 depicts a light ray going from air into glass with an angle of incidence of 60°. Find the angle of refraction.

We use Figure 9.27 with the angle in air equal to 60°. Locate 60° on the horizontal axis, follow the grid line up to the curve, and then move horizontally to the left. The point on the vertical scale indicates the angle in glass is about 36°. Therefore, the angle of refraction is 36°.

But what happens to light that passes completely through a sheet of glass like a windowpane? Each light ray is bent toward the normal when it enters the glass and then is bent away from the normal when it reenters the air on the other side. No matter what the original angle of incidence is, the two bends exactly offset each other, and the ray emerges from the glass traveling parallel to its original path (see Figure 9.28).

Total Internal Reflection

Consider now the following experiment shown schematically in Figure 9.29. Rays of light originating in glass strike the boundary (separating it from the surrounding air) at ever-increasing angles. In this case, since the speed of light in glass is smaller than the speed of light in air, we see that the ray is bent *away* from the normal. In other words, the angle of refraction is *larger* than the angle of incidence, as shown in Figure 9.29b. Moreover, the angle of refraction increases as the angle of incidence does.

Because of reversibility we can use the same graph, Figure 9.27, to find the angle of refraction when the angle of incidence is known. In this situation, the angle of incidence is the "angle in glass," while the angle of refraction is the "angle in air." For example, when the angle of incidence is 20°, we locate 20° on the vertical scale, move horizontally to the curve, and then drop down to the horizontal axis where we see that the angle of refraction would be about 30°. This is shown in Figure 9.29c.

What distinguishes this case from the previous one, in which the incident ray was in air, is the existence of a **critical angle** of incidence for which the angle of refraction

Figure 9.29
A light ray traveling in glass strikes a glass-air interface with different angles of incidence. The angle of incidence in (e) is the *critical angle,* for which the angle of refraction is 90°. Total internal reflection (e and f) occurs when the angle of incidence is equal to, or greater than, the critical angle. No light enters the air.

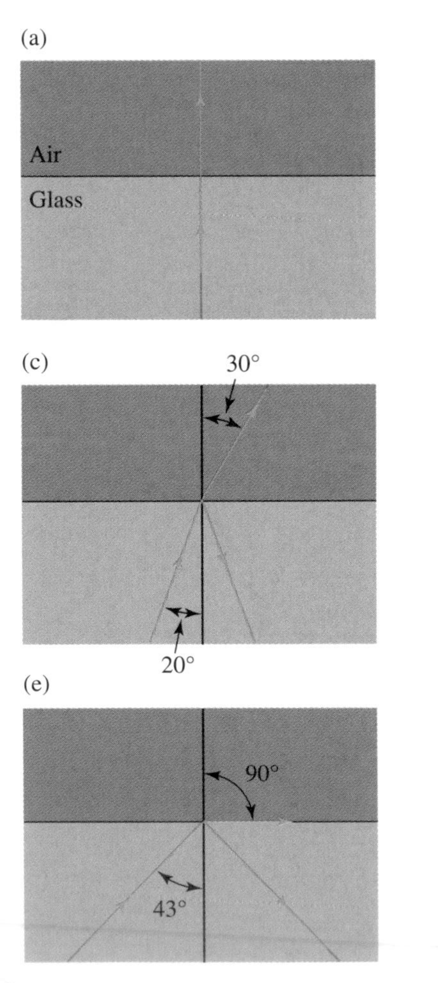

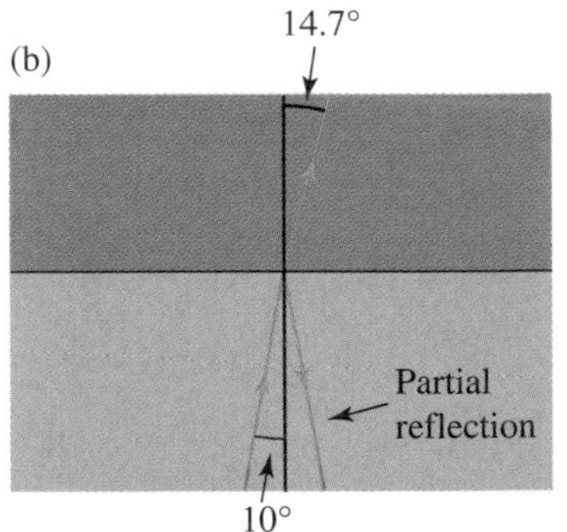

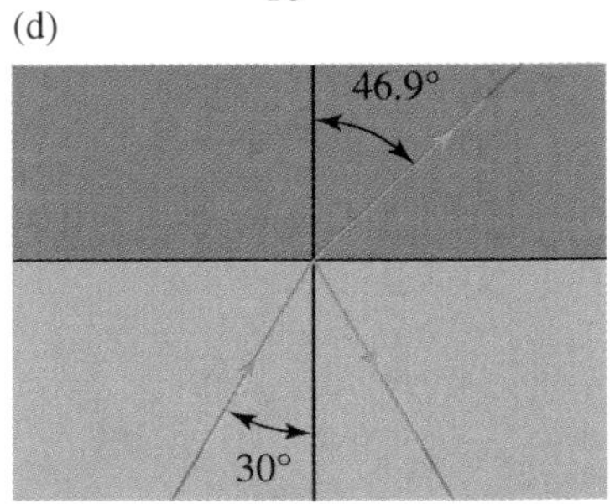

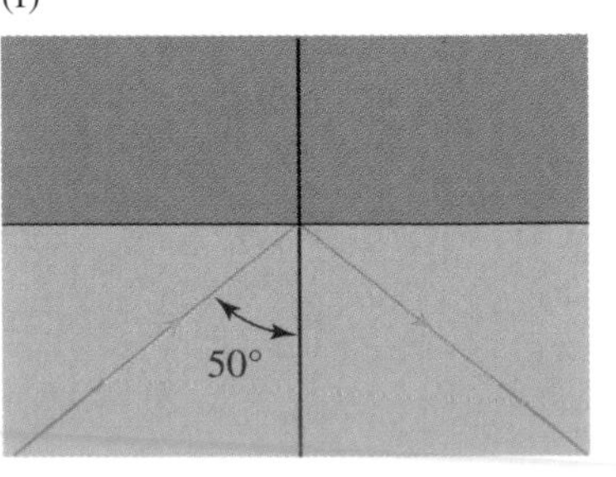

© David Joyner/iStockphoto.com / © Valerie Loiseleux/iStockphoto.com / © iStockphoto.com

reaches 90°. When the angle of incidence equals this critical angle, the transmitted ray travels out *along* the interface between the two media (Figure 9.29e). For angles of incidence greater than the critical angle, the formerly transmitted ray is bent back into the incident medium and does not travel appreciably into the second medium. When this happens, we have a condition called **total internal reflection**. This is shown in Figure 9.29e and f.

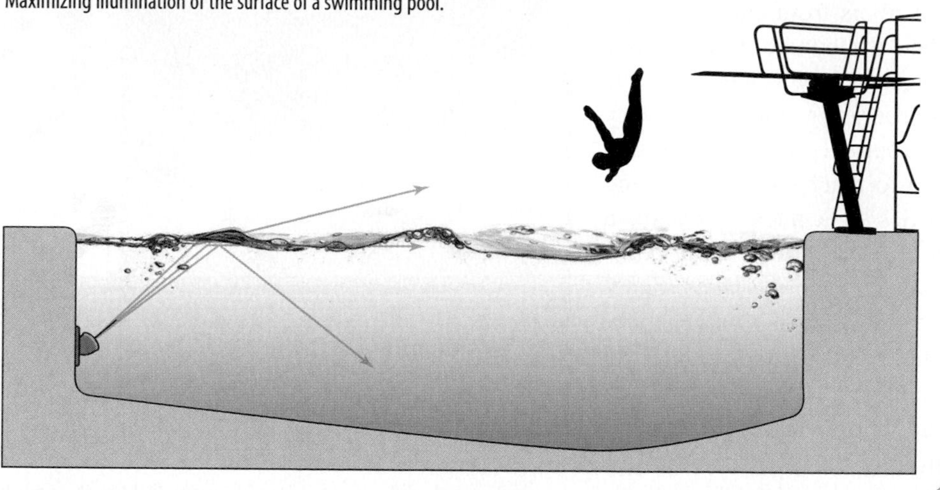

Figure 9.30
Maximizing illumination of the surface of a swimming pool.

The graph in Figure 9.27 indicates that the critical angle for a glass-air interface, the angle in glass that makes the angle in air equal to 90°, is about 43°. In general, different media have different critical angles. Table 9.1 includes the values of the critical angle for light rays entering air from different transparent media. Notice that the critical angle is smaller for materials in which the speed of light is lower. Diamond, with the lowest speed of light in the list, has a critical angle of less than 25°—little more than half that of glass.

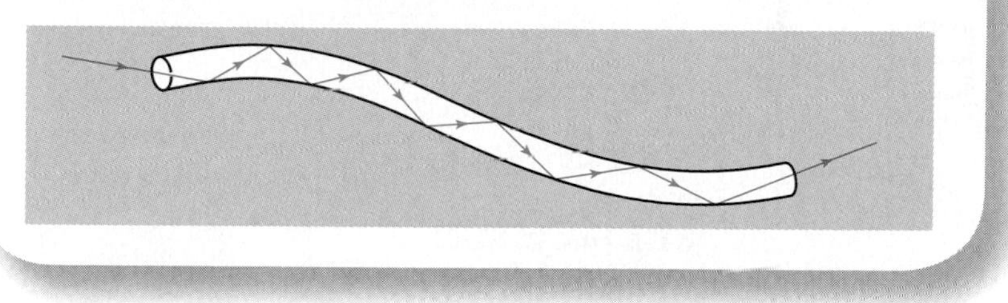

Figure 9.31
Multiple internal reflections within an optical fiber.

EXAMPLE 9.2

A homeowner wishes to mount a floodlight on a wall of a swimming pool under water so as to provide the maximum illumination of the surface of the pool for use at night (see Figure 9.30). At what angle with respect to the wall should the light be pointed?

To illuminate the surface of the water, the refracted ray at the water-air interface should just skim the water's surface. This means that the angle of refraction must be 90°, so the incident angle must be the critical angle.

Table 9.1 shows that the critical angle for water is about 49°. The homeowner should direct the floodlight upward so that the beam makes an angle of roughly 49° with the vertical side wall of the pool.

Optical fibers are flexible, coated strands of glass that utilize total internal reflection to channel light. In a sense, an optical fiber does for light what a garden hose does for water. Figure 9.31 shows the path of a particular light ray as it enters one end of an optical fiber. When this ray strikes the wall of the fiber, it does so with an angle of incidence that is greater than the critical angle, so the ray undergoes total internal reflection. This is repeated each time the ray encounters the fiber's wall, so the light is trapped inside until it emerges from the other end. You may have seen decorative lamps with light bursting from the ends of a "bouquet" of optical fibers. Fiber-optic cables consisting of dozens or even hundreds of individual optical fibers are now in common use to transmit information. Telephone conversations, audio and video signals, and computer information are encoded ("digitized") and then sent through fiber-optic cables as pulses of light produced by tiny lasers. A typical fiber-optic cable can transmit thousands of times more information than a conventional wire cable that is much larger in diameter. For example, the first fiber-optic transatlantic cable, which went into service in 1988, was designed to carry 40,000 simultaneous conversations over just two pairs of glass fibers. By contrast, the last of the large copper bundles installed for overseas communications (1983) could handle only about 8000 calls.

As noted above, one significant advantage of using optical fibers to conduct light from one place to another is their flexibility. If bundles of free fibers are bound together in a way that preserves the relative positions of adjacent fibers from one end to the other, the bundle may transmit images as well as light. Devices incorporating this technology are routinely used to examine hard-to-reach places like the insides of a nuclear reactor, a jet engine, or the human body. When used in the latter capacity, the instrument is generally called an *endoscope.* Specific examples of endoscopes include bronchoscopes (for examining lung tissue), gastroenteroscopes (for checking the stomach and digestive tract), and colonoscopes (for surveying the bowel).

© Spencer Grant/Photo Researchers / © Maria Toutoudaki/iStockphoto.com

A fiber-optic light-guide cable (below) can typically carry many more telephone conversations than a much larger wire cable (above).

9.4 Lenses and Images

In Section 9.2, we described how curved mirrors are used in astronomical telescopes and other devices to redirect light rays in useful ways. Microscopes, binoculars, cameras, and many other optical instruments use specially shaped pieces of glass called *lenses* to alter the paths of light rays. As was the case with mirrors, the key to making glass or other transparent substances redirect light is the use of curved surfaces (interfaces) rather than flat (planar) ones.

Suppose we grind a block of glass so that one end takes the shape of a segment of a sphere as shown in cross section in Figure 9.32. Let parallel rays strike the convex spherical surface at various points above and below the line of symmetry (called the *optical axis*) of the system. If one applies the law of refraction at each point to determine the angle of refraction of each ray, the results shown in Figure 9.32 are found. In particular, rays traveling along the optical axis emerge from the interface undeviated. Rays entering the glass at points successively above or below the optical axis are deviated ever more strongly toward the optical axis. The result is to cause the initially parallel bundle of rays to gradually *converge* together—to become focused—into a small region behind the interface. This point is called the *focal point* and is labeled F in the figure. (Compare this to Figure 9.16a in which a concave mirror causes light rays to converge.)

Figure 9.33 shows the behavior of parallel rays refracted across a spherical interface that, instead of

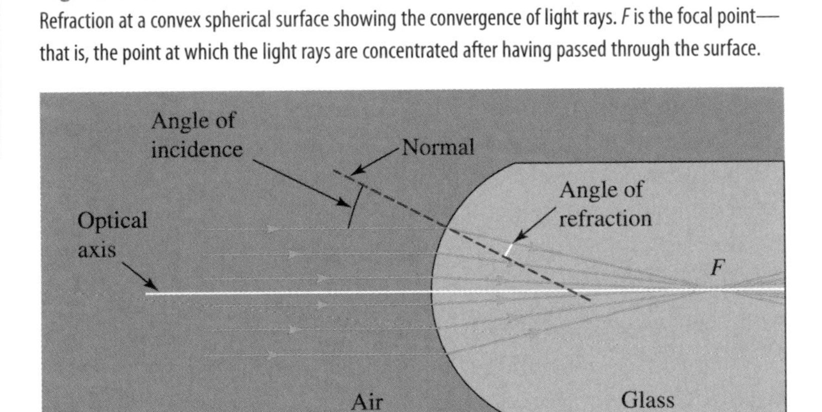

Figure 9.32
Refraction at a convex spherical surface showing the convergence of light rays. F is the focal point—that is, the point at which the light rays are concentrated after having passed through the surface.

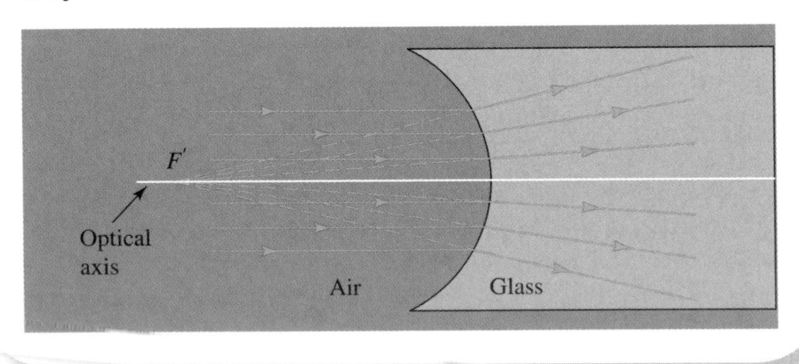

Figure 9.33
Refraction at a concave spherical surface showing the divergence of a beam of parallel light. F' is the focal point—that is, the point from which the rays appear to diverge after having passed through the surface.

bowing outward, curves inward. In this case, the emergent rays *diverge* outward as though they had originated from a point F' to the left of the interface. The ability to either bring together or spread apart light rays is the basic characteristic of lenses, be they camera lenses, telescope lenses, or the lenses in human eyes.

Although the situation shown in Figure 9.32 does occur in the eye, in most devices the light rays must enter and then leave the optical element (lens) that redirects them. Common lenses have two refracting surfaces instead of one, with one surface typically in the shape of a segment of a sphere and the second either spherical as well or flat (planar). The effect on parallel light rays passing through both surfaces is similar to that in the previous examples with one refracting surface. A *converging lens* causes parallel light rays to converge to a point, called the *focal point* of the lens (Figure 9.34). The distance from the lens to the focal point is called the **focal length** of the lens. A more sharply curved lens has a shorter focal length. Conversely, if a tiny source of light is placed at the focal point, the rays that pass through the converging lens will emerge parallel to each other. This is the principle of reversibility again.

A *diverging lens* causes parallel light rays to diverge after passing through it. These emergent rays appear to be radiating from a point on the other side of the lens. This point is the focal point of the diverging lens (Figure 9.35). The distance from the lens to the focal point is again called the focal length, but for a diverging lens it is given as a negative number, −15 centimeters, for example. If we reverse the process and send rays converging toward the focal point into the lens, they emerge parallel.

For both types of lenses there are two focal points, one on each side. Clearly, if parallel light rays enter a converging lens from the right side in Figure 9.34, they will converge to the focal point to the left of the lens. Whether a lens is diverging or converging can be determined quite easily: If it is *thicker* at the center than at the edges, it is a converging lens; if it is *thinner* at the center, it is a diverging lens (Figure 9.36).

Figure 9.35
Parallel light rays diverging after passing through a diverging lens. The rays appear to radiate from the focal point, to the left of the lens.

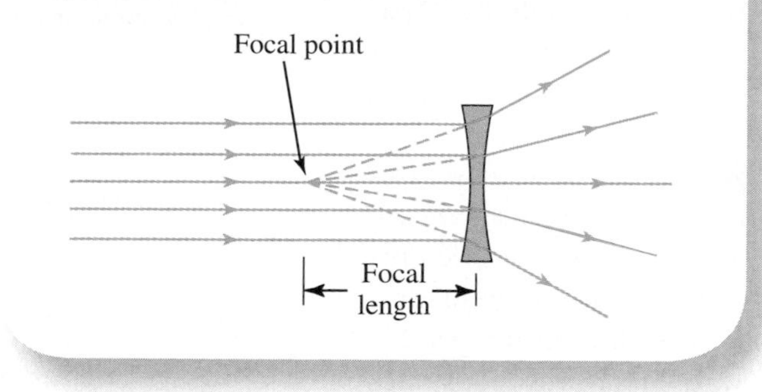

Figure 9.34
Converging lenses focusing parallel light rays at their focal points. A more sharply curved lens has a shorter focal length.

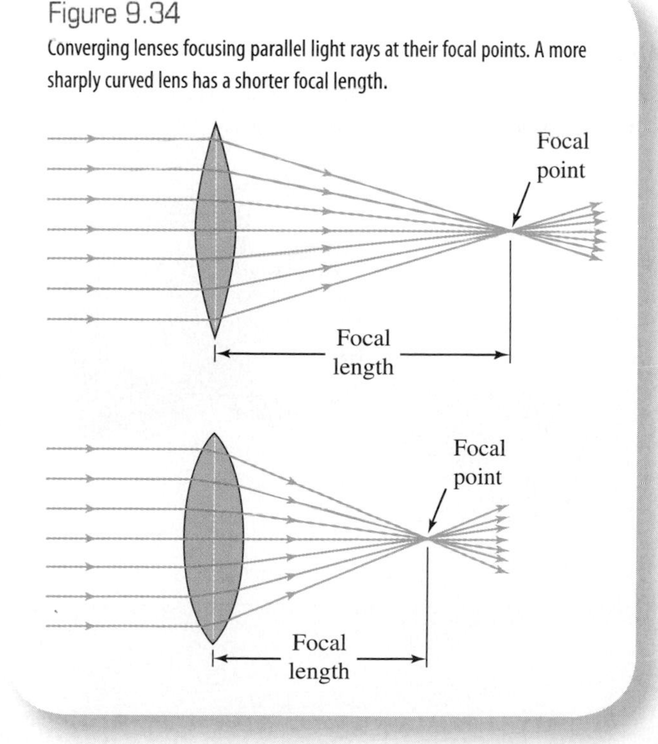

Figure 9.36
Examples of different types of lenses: (a) bi-convex (left), bi-concave (right); (b) plano-convex (left), plano-concave (right); (c) meniscus-convex (left), meniscus-concave (right).

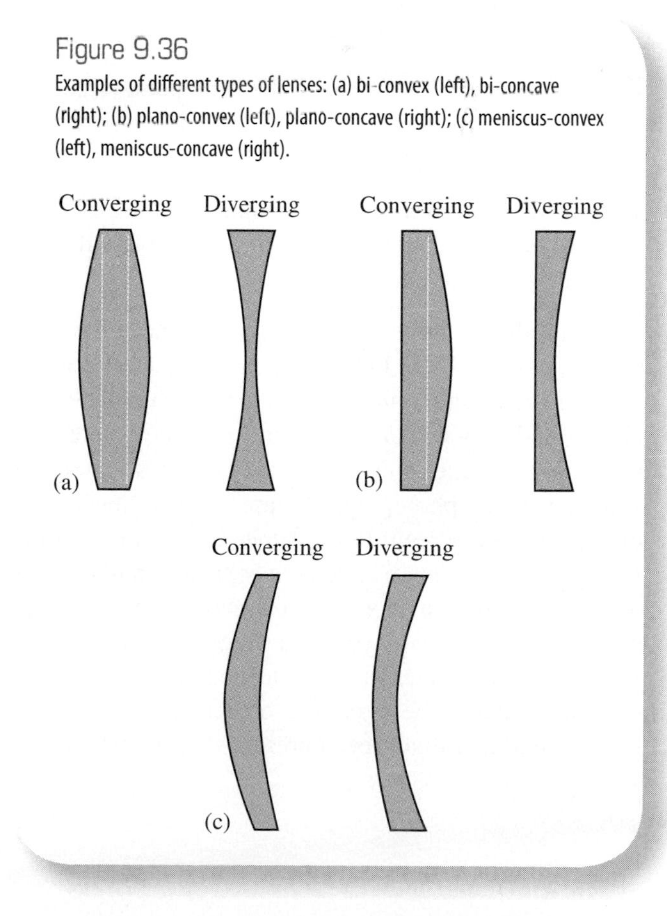

Image Formation

The main use of lenses is to form images of things. First, let's consider the basics of image formation when a symmetric converging lens is used. Our eyes, most cameras (both still and video), slide projectors, movie projectors, and overhead projectors all form images this way. Figure 9.37 illustrates how light radiating from an arrow, called the *object*, forms an *image* on the other side of the lens. One way of setting up this example would be to point a flashlight at the arrow so that light would reflect off the arrow and pass through the lens. The image could be projected onto a piece of white paper placed at the proper location to the right of the lens.

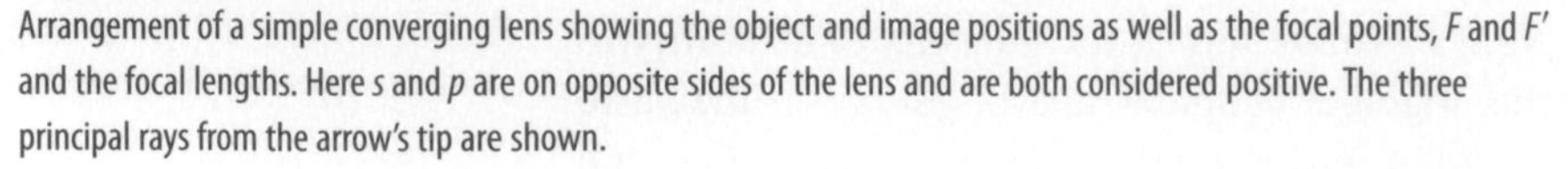

Figure 9.37
Arrangement of a simple converging lens showing the object and image positions as well as the focal points, F and F' and the focal lengths. Here s and p are on opposite sides of the lens and are both considered positive. The three principal rays from the arrow's tip are shown.

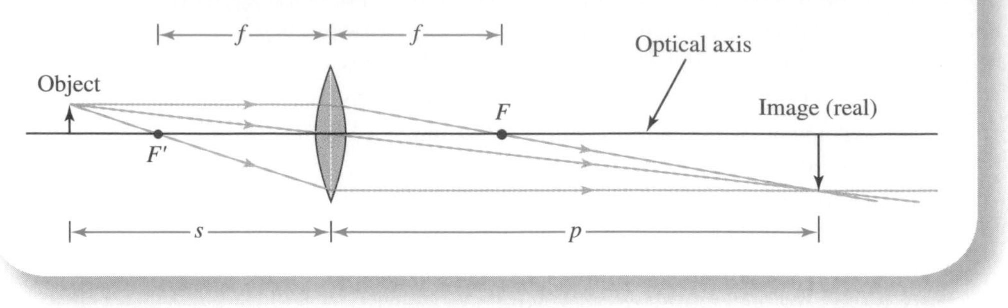

Although each point on the object has countless light rays spreading out from it in all directions, it is easier first to consider only three particular rays from a single point—the arrow's tip. These rays, shown in Figure 9.37, are called the **principal rays**.

1. The ray that is initially parallel to the optical axis passes through the focal point (F) on the other side of the lens.
2. The ray that passes through the focal point (F') on the same side of the lens as the object emerges parallel to the optical axis.
3. The ray that goes exactly through the center of the lens is undeviated because the two interfaces it encounters are parallel.

Note that the image is *not* at the focal point of the lens. Only parallel incident light rays converge to this point.

We could draw principal rays from each point on the object, and they would converge to the corresponding point on the image. This kind of image formation occurs when you take a photograph (Figure 9.38) or view a slide on a projection screen. In the latter case, light radiating from each point on the slide converges to a point on the image on the screen. Note that the image is inverted (upside down). That's why you load slides in upside down if you want their images to be right side up.

The distance between the object and the lens is called the **object distance**, represented by s, and the distance between the image and the lens is called the **image distance**, represented by p. By convention (with the light traveling from left to right), s is positive when the object is to the left of the lens, and p is positive when the image is to the right of the lens. If we place the object at a different point on the optical axis, the image would also be formed at a different point. In other words, if s is changed, p changes. Using a lens with a different focal length for fixed s would also cause p to change. For example, the image would be closer to the lens if the focal length were shorter.

The following equation, known as the *lens formula*, relates the image distance p to the focal length f and the object distance s.

$$p = \frac{sf}{s - f} \qquad \text{(lens formula)}$$

Figure 9.38
Image formation in a camera. The central principal ray from a point at the top and from a point at the bottom of the object are shown.

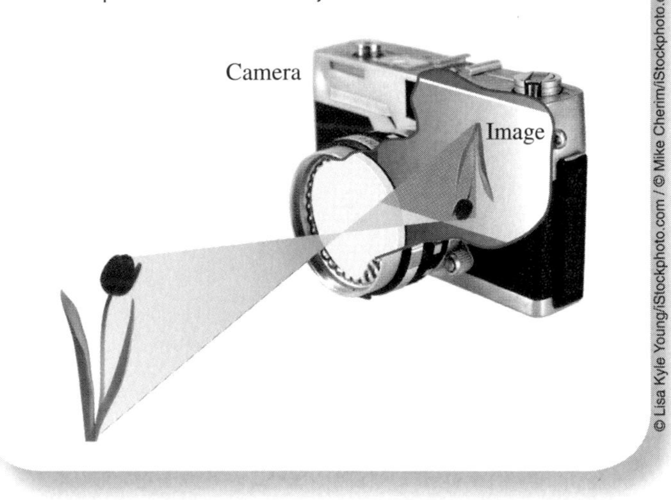

© Lisa Kyle Young/iStockphoto.com / © Mike Cherim/iStockphoto.com

The following example shows how the lens formula can be used.

EXAMPLE 9.3

In a slide projector, a slide is positioned 0.102 meter from a converging lens that has a focal length of 0.1 meter. At what distance from the lens must the screen be placed so that the image of the slide will be in focus?

The screen needs to be placed a distance p from the lens, where p is the image distance for the given focal length and object distance. So:

$$p = \frac{sf}{s - f} = \frac{0.102 \text{ m} \times 0.1 \text{ m}}{0.102 \text{ m} - 0.1 \text{ m}}$$

$$= \frac{0.0102 \text{ m}^2}{0.002 \text{ m}}$$

$$= 5.1 \text{ m}$$

If the slide-to-lens distance is increased to 0.105 meter, the distance to the screen (p) would have to be reduced to 2.1 meters.

If the lens is replaced by one that has a shorter focal length, the distance to the screen would have to be reduced as well.

The images formed in the manner just described are called **real images**. Such images can be projected onto a screen. We see the image on the screen because the light striking the screen is diffusely reflected to our eyes. A simple magnifying glass is a converging lens, but the image that it forms under normal use is *not* a real image—it can't be projected onto a screen. We see the image by looking *into* the lens, just as we see a mirror image by looking *into* the mirror. This type of image is called a **virtual image**. Figure 9.39 shows how an image is formed in a magnifying glass. In this case, the object is between the focal point F' and the lens, so the object distance s is *less than* the focal length f of the lens.

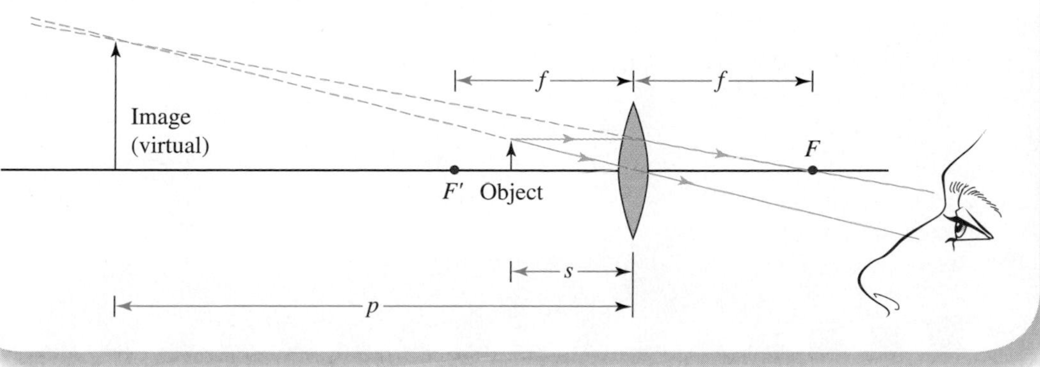

Figure 9.39
Image formation when the object is between the focal point and the lens. A virtual image of the arrow is formed, seen by looking through the lens toward the object. (Only two of the principal rays are shown here.)

Note that the image is enlarged and that it is upright. It is also on the same side of the lens as the object, which means that p is negative. This situation is very much like the image formation with a concave mirror (think of a makeup mirror).

EXAMPLE 9.4

A converging lens with focal length 10 centimeters is used as a magnifying glass. When the object is a page of fine print 8 centimeters from the lens, where is the image?

$$p = \frac{sf}{s - f} = \frac{8 \text{ cm} \times 10 \text{ cm}}{8 \text{ cm} - 10 \text{ cm}}$$

$$= \frac{80 \text{ cm}^2}{-2 \text{ cm}}$$

$$= -40 \text{ cm}$$

The negative value for p indicates that the image is on the same side of the lens as the object. Therefore, it is a virtual image and must be viewed by looking through the lens.

A virtual image is also formed when you look at an object through a diverging lens (Figure 9.40). In this case, the image is smaller than the object, as it is with a convex mirror.

Magnification

In Figures 9.37 through 9.40, the size of the image is not the same as the size of the object. This is one of the most useful properties of lenses: They can be used to produce images that are enlarged (larger than the original object) or reduced (smaller than the original object). Figure 9.41 shows this.

Figure 9.40
Image formation with a diverging lens. The image is virtual, so it must be viewed by looking through the lens. (Again, only two of the principal rays are shown.)

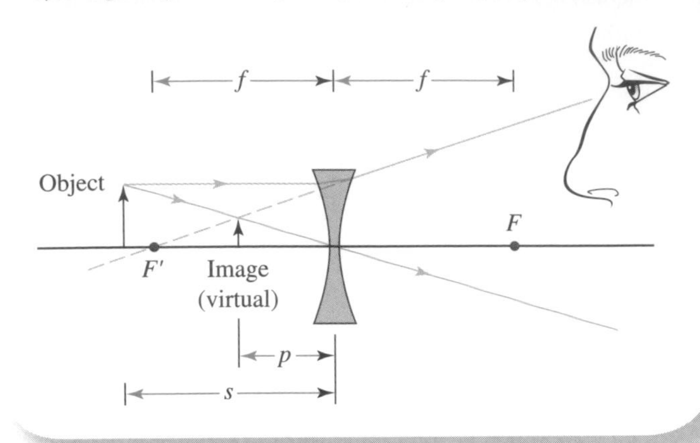

DEFINITION

Magnification (M) *The magnification of a particular configuration is the height of the image divided by the height of the object.*

$$M = \frac{\text{image height}}{\text{object height}}$$

$$M = \frac{-p}{s}$$

Thus, if the image is twice the height of the object, the **magnification** is 2. If the image is upright, the magnification is positive. If the image is inverted, the magnification is negative (because the image height is negative).

The magnification that one gets with a particular lens changes if the object distance is changed. Simple geometry permits the magnification to be written in an alternative but equivalent way in terms of the object and corresponding image distances.

From this we can conclude that:

1. If p is positive (image is to the right of the lens and real), M is negative; the image is inverted (Figures 9.37 and 9.38).
2. If p is negative (image is to the left of the lens and virtual), M is positive; the image is upright (see Figures 9.39 and 9.40).

EXAMPLE 9.5

Compute the magnification for the slide projector in Example 9.3 and for the magnifying glass in Example 9.4.

In the first case in Example 9.3, $s = 0.102$ meters and $p = 5.1$ meters. Therefore:

$$M = \frac{-p}{s} = \frac{-5.1 \text{ m}}{0.102 \text{ m}}$$
$$= -50$$

The image is 50 times as tall as the object, but it is inverted (since M is negative). A slide that is 35 millimeters tall has an image on the screen that is 1,750 millimeters (1.75 meters) tall. When s is 0.105 meters, p is 2.1 meters, and the magnification is -20.

In Example 9.4, $s = 8$ centimeters and $p = -40$ centimeters. Consequently:

$$M = \frac{-p}{s} = \frac{-(-40 \text{ cm})}{8 \text{ cm}}$$
$$= +5$$

The image of the print seen in the magnifying glass is five times as large as the original, and it is upright (since M is positive).

Figure 9.41
The same sheet music viewed through (a) a converging lens and (b) a diverging lens. Compare the images to those shown in the ray diagrams in Figure 9.39 and 9.40, respectively. In both cases, the images are upright and virtual, but in (a) the notes are larger in size (that is, magnified), while in (b) they are smaller in size (reduced) compared to those adjacent to the lens positions.

© Andrew Lambert Photography/SPL/Photo Researchers, Inc. / © Nicholas Belton/iStockphoto.com

Telescopes and microscopes can be constructed by using two or more lenses together. Figure 9.42 shows a simple telescope consisting of two converging lenses. The real image formed by lens 1 becomes the object for lens 2. The light that could be projected onto a screen to form the image for lens 1 simply passes on into lens 2. In essence, lens 2 acts as a magnifying glass and forms a virtual image of the object. In this telescope, the image is magnified but inverted. Replacing lens 2 in Figure 9.42 with a diverging lens yields a telescope that produces an upright image.

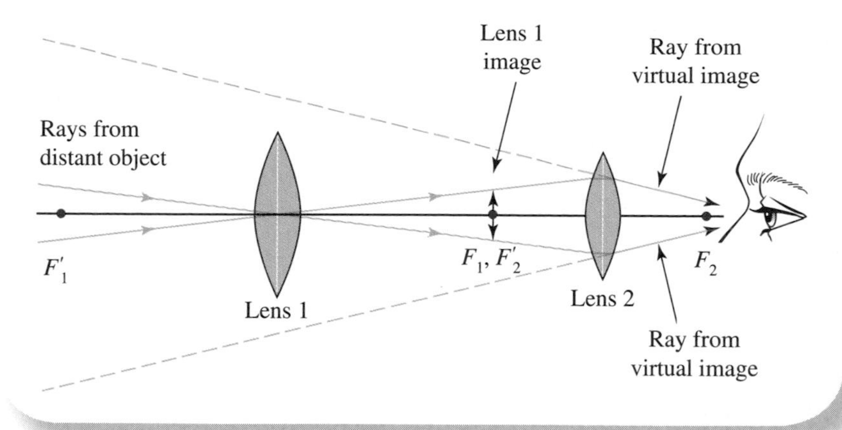

Figure 9.42
Simple telescope. Only the third principal rays from the top and from the bottom of the distant object are shown. The dashed lines show that the virtual image subtends a larger angle than the incoming rays from the object do—the image is magnified.

Aberrations

In real life, lenses do not form perfect images. Suppose we carefully apply the law of refraction to a number of light rays, all initially parallel to the optical axis, as they pass through a real lens that has a surface shaped like a segment of a sphere (Figure 9.43). We would find that the lens exhibits the same flaw we saw in Section 9.2 with spherically shaped curved mirrors: *spherical aberration.* Figure 9.43 shows that rays striking the lens at different points do not cross the optical axis at the same place (compare Figure 9.19a). In other words, there is no single focal point. This causes images formed by such lenses to be somewhat blurred. Lens aberrations of this type can be corrected, but this process is complicated and often necessitates the use of several simple lenses in combination.

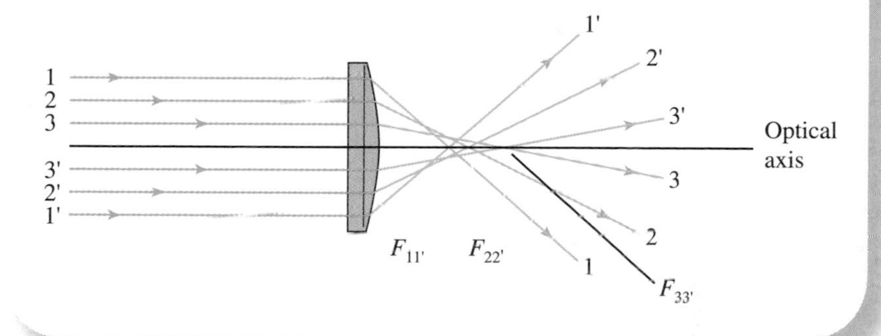

Figure 9.43
Spherical aberration for a convex lens illuminated by a beam of parallel light. Rays 1 and 1′ are brought to a focus at $F_{11'}$, while rays 2 and 2′, and 3 and 3′ are focused at $F_{22'}$, and $F_{33'}$, respectively.

Another type of aberration shared by all simple lenses even when used under ideal conditions is *chromatic aberration.* A lens affected by chromatic aberration, when illuminated with white light, produces a sequence of more or less overlapping images, varying in size and color. If the lens is focused in the yellow-green portion of the EM spectrum where the eye is most sensitive, then all the other colored images are superimposed and out of focus, giving rise to a whitish blur or fuzzy overlay (see Figure 9.44). For a converging lens, the blue images would form closer to the lens than the yellow-green images, while the reddish images would be brought to a focus farther from the lens than the yellow-green ones.

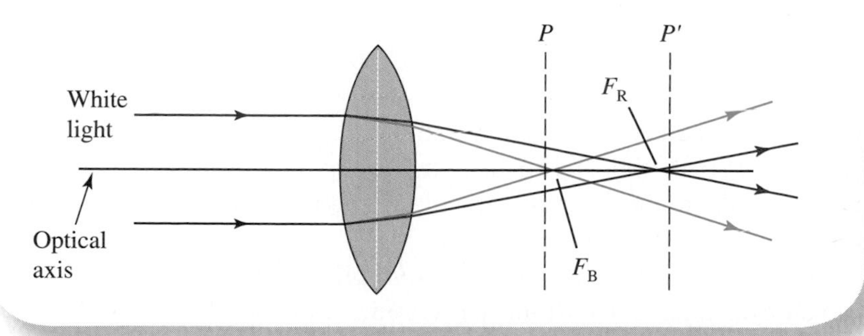

Figure 9.44
Diagram showing the positions of the focal points for two colors for a simple lens. The focal point for blue light is closer to the lens than that for red light. A screen placed at point *P* will show an image whose outside edges are tinged red-orange, while a screen placed at point *P*′ will reveal an image whose outside edges are bluish.

The cause of chromatic aberration has its roots in the phenomenon of *dispersion,* to be discussed in

Figure 9.45
A Fraunhofer cemented achromatic doublet lens.

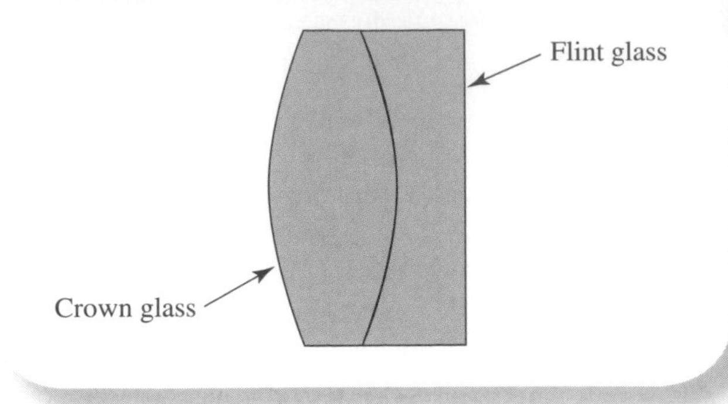

Section 9.6. The remedy for this problem, originally thought to be insoluble by none other than Newton himself, was discovered around 1733 by C. M. Hall and later (in 1758) developed and patented by John Dolland, a London optician. It involves using two different types of glass mounted in close proximity. Figure 9.45 shows a common configuration called a *Fraunhofer cemented achromat* (meaning "not colored"). The first lens is made of crown glass, the second of dense flint glass (see Table 9.1). These materials are chosen because they have nearly the same dispersion. To the extent to which this is true, the excess convergence exhibited by the first lens at bluish wavelengths is compensated for by the excess divergence produced by the second lens at these same wavelengths. Similar effects occur at the other wavelengths in the visible spectrum, permitting cemented doublets of this type to correct more than 90% of the chromatic aberration found in simple lenses.

9.5 The Human Eye

The eye may be the most sophisticated of our sense organs. But up to the point when the light rays are absorbed and the signal to the brain is formed, the eye is a fairly simple optical instrument.

Figure 9.46 shows a simplified cross section of the human eye. Light enters from the left and is projected onto the retina at the rear of the eyeball. The iris, the part of the eye that is colored, controls the amount of light that enters the eye. Its circular opening, the pupil, is large in dim light and small in bright light. As light enters the cornea, it converges because of the cornea's convex shape (Figure 9.32 shows this effect). The eye's *auxiliary lens*, simply called the *lens*, makes the light converge even more. This lens is used by the eye to ensure that the image is in focus on the retina. The effect of the cornea and the lens is the same as if the eye were equipped with a single converging lens, so the image formation process shown in Figure 9.37 applies.

For an object to be seen clearly, the image must be in focus on the retina. Within the eye, the image distance is always the same—the diameter of the eyeball. We are able to focus on objects near and far (with small and large object distances) because the focal length of the lens can be varied by changing its shape. When the eye is focused on a distant object (farther than, say, 5 meters away) the lens is thin and has a long focal length (Figure 9.47). For a near object, special muscles make the lens thicker. This shortens the focal length of the lens so that the image of the near object is focused on the retina. Unlike many cameras and other optical devices, the eye has a constant value for p, the image distance, but it accommodates different values for s, the object distance, by changing its focal length f.

The two most common types of poor eyesight, nearsightedness and farsightedness, are the result of improper focusing. *Nearsightedness*, or *myopia*, occurs when close objects are in focus but distant objects are not. The light rays from a distant object are brought into focus before they reach the retina. When the rays do reach the retina, they are out of focus (Figure 9.48). The problem is remedied by placing a properly chosen *diverging lens* in front of the eye. The corrective lens is either held in place by a frame (glasses) or placed in direct contact with the cornea (contact lenses). This lens makes the light rays diverge slightly before they enter the eye. This moves the point

Figure 9.46
The human eye. Light passes through the opening in the iris, the pupil, and forms an image on the retina. The cornea and the lens act as a single converging lens.

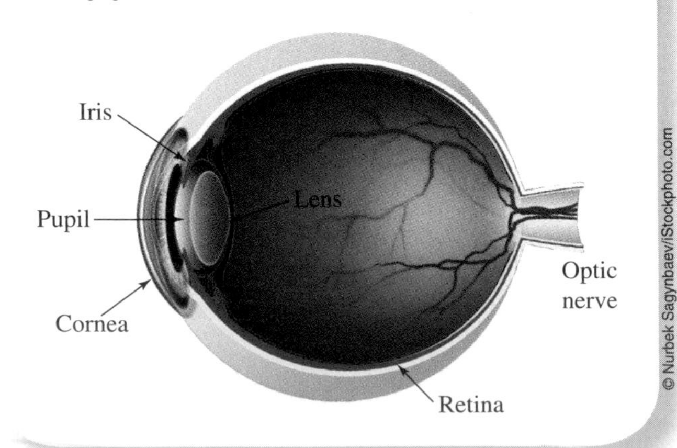

where the rays meet back to the retina.

Farsightedness, or *hyperopia*, is the opposite: Distant objects are in focus, but near objects are not. The cornea and the lens do not make the rays from a near object converge enough. The rays reach the retina before they meet, and the image is out of focus (Figure 9.49). The remedy for this condition is a properly chosen *converging lens* placed in front of the eye. This lens makes the light rays converge slightly before they enter the eye, thereby bringing the light rays into focus on the retina.

Another common problem is *astigmatism*, which occurs when the cornea is not symmetric. For example, the cornea's focal length might be shorter for two parallel rays that enter the eye one above the other compared to the focal length for two parallel rays that enter side by side. Objects appear distorted. This condition can often be corrected by using a specially shaped lens that has the opposite asymmetry.

© Johanna Goodyear/iStockphoto.com

Humans, like most mammals, accommodate by changing the shape of the eye lens. Fish, however, move the lens itself toward or away from the retina, thus changing the value of p.

Figure 9.47

The eye can form images on the retina of either distant or near objects by changing the thickness (and therefore the focal length) of the lens.

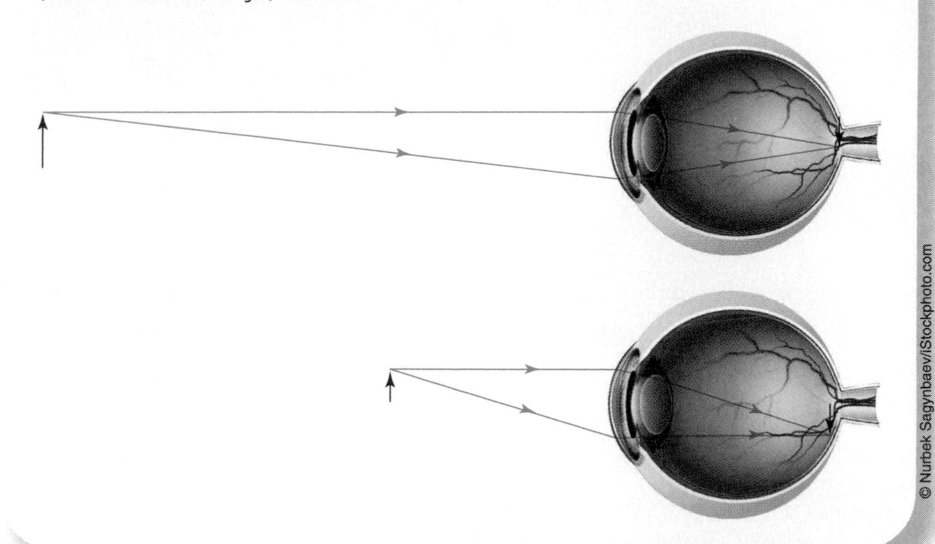

© Nurbek Sagynbaev/iStockphoto.com

Figure 9.48

(a) A nearsighted eye causes rays from a distant object to converge too quickly. (b) A diverging lens corrects the problem.

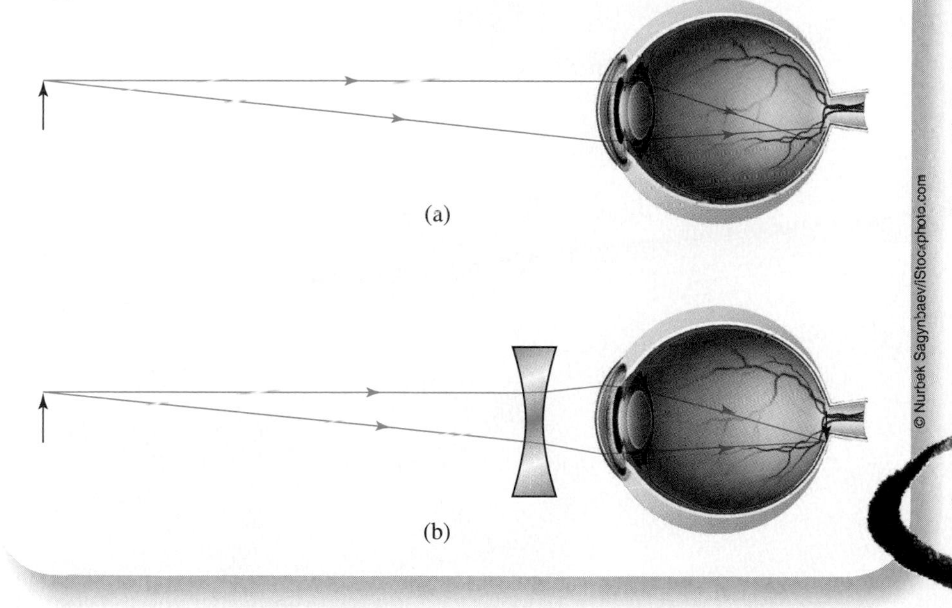

© Nurbek Sagynbaev/iStockphoto.com

Figure 9.49

(a) A farsighted eye does not cause light rays from a near object to converge enough. (b) A converging lens corrects the problem.

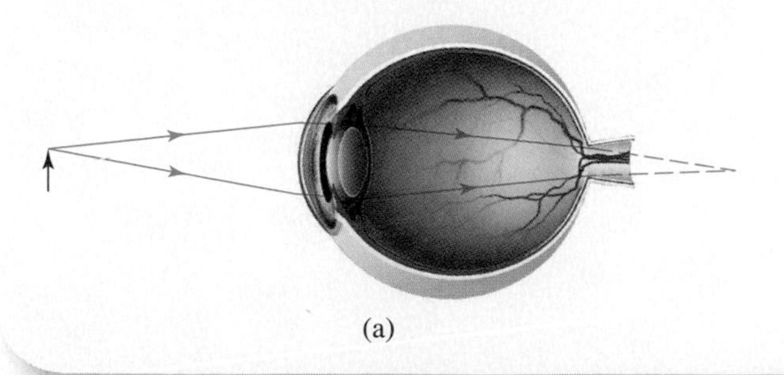

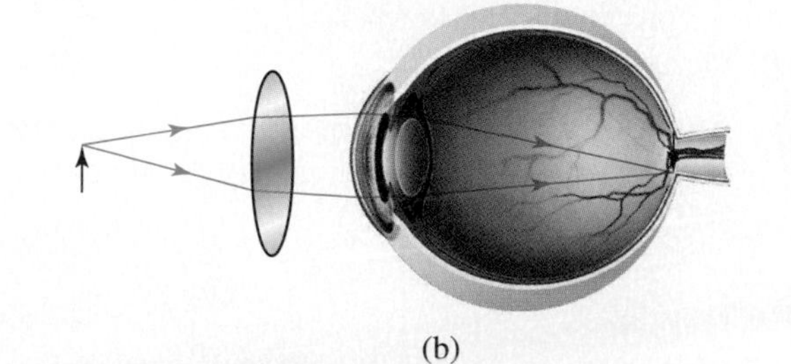

corrective eye surgery

When the cornea's shape is imperfect, light travelling through it does not focus properly. That is, the resulting image is out of focus on the retina. During the later part of the twentieth century, several types of corrective eye surgery became commonplace. In the 1970s, radial keratotomy (RK) was developed to correct myopia. During RK, the surgeon creates radial incisions in the cornea which regularizes its curvature as it heals. Today, most vision correction surgeries use lasers to reshape the cornea. (More about lasers in Chapter 10.) Here are some of the different types of surgeries available.

LASIK (*laser-assisted in situ keratomileusis*)—In LASIK surgery, a flap is made in the outer layer of the cornea, and a laser is used to reshape the underlying tissue.

Uses: nearsightedness, farsightedness, and astigmatism.

PRK (*photorefractive keratectomy*)—A cool pulsing beam of ultraviolet light is applied to the surface of the cornea, requiring no cutting as in LASIK.

Uses: mild to moderate nearsightedness, farsightedness, and astigmatism.

LASEK (*laser epithelial keratomileusis*)—A procedure similar to LASIK and PRK but without cutting. Instead, a 20% alcohol solution is used to create an epithelial flap.

Uses: nearsightedness, farsightedness, and astigmatism.

ALK (*automated lamellar keratoplasty*)—A flap is created in the cornea (like in LASIK) and a second incision (without a laser) is created in the underlying tissue to reshape the cornea.

Uses: severe nearsightedness and slight farsightedness.

LTK (*laser thermokeratoplasty*)—A laser heats the cornea, shrinking and reshaping it. No cutting or tissue removal is involved.

Uses: farsightedness and astigmatism.

AK (*astigmatic keratotomy*)—Not a laser surgery. An irregularly shaped cornea is reshaped by making one or two incisions at its steepest part, causing the cornea to relax and become rounder. This procedure can be used with PRK, LASIK, and RK.

Uses: astigmatism.

Source: "Eye Health: Overview of Refractive and Laser Eye Surgery," ed. Tracy C. Shuman, MD, *WebMD Eye Health Center*, http://www.webmd.com/eye-health/overview-refractive-laser-eye-surgery (accessed July 23, 2009).

© iStockphoto.com / © Charles O'Rear/Corbis

9.6 Dispersion and Color

Most of you at one time or another have seen decorative glass pendulums hanging in windows through which sunlight is streaming. If so, you probably noticed patches of bright, rainbow-hued light playing about the room as the pendulums slowly turned in response to air currents. Did you ever wonder how such beauty was produced? Sir Isaac Newton did, and he performed several experiments in an attempt to answer this question. He concluded that sunlight—white light—was a mixture of all the colors of the rainbow and that upon being refracted through transparent substances like glass, it could be *dispersed*, or separated, into its constituent wavelengths (colors). The process of refraction was seen to be color dependent!

This color dependence of refraction comes about because the speed of light in any medium is slightly different for each color. In glass, diamond, ice, and most other common transparent materials, *shorter* wavelengths of light travel slightly *slower* than longer wavelengths. Violet travels a little slower than blue, blue travels slower than green, and so on. In the case of common glass, the speed of violet light is about 1.95×10^8 m/s, while the speed of red light is about 1.97×10^8 m/s. The speeds of the other colors lie between these two values. This slight difference in the speeds of different colors causes dispersion. In diamond, the difference in speeds of violet and red light is comparatively larger—about 2%, compared to about 1% for glass—so the dispersion is greater. That is why one sees such brilliant colors in diamonds. The difference in speeds of violet and red light in water is comparatively smaller than it is in glass.

Up until this time, we have ignored the fact that the speed of light in a medium depends on wavelength. (You can think of our previous study of refraction as having been done using light of only a single color or wavelength, something known as *monochromatic* light.) What effect does this now have on how violet light rays are refracted at an air-glass interface relative to how red rays are refracted?

Consider Figure 9.50, showing an incoming ray of light that we will assume is a mixture of only blue and red wavelengths. Because the speeds of both blue light *and* red light are lower in glass than they are in air, we expect that both rays will be bent toward the normal based on our analysis in Section 9.3. But we know that the speed of blue light in glass is lower than the speed of red light in the same medium, so that the blue light will be bent slightly *more* toward the normal than the red light. In Figure 9.50, this results in the angle of refraction for blue light being a bit smaller than the angle of refraction for red light. Thus, although both rays are bent toward the normal upon passing into the glass, the blue ray is refracted more strongly and emerges from the interface along a different path than the red ray. The colors have been dispersed, or separated, as a result of the refraction process because of the wavelength dependence of the speed of light.

If the incoming beam is now allowed to contain the remaining colors between red and blue, the emergent rays for each will fall between the limits set by the red and blue rays. What is produced is a *spectrum*—the different colors spread over a range of angles. It is the process of dispersion, acting to sort out the different colors, that causes the chromatic aberration in simple lenses described at the end of Section 9.4.

The difference between the angles of refraction for the red and blue rays above amounts to less than 0.5°. This may not sound like much, but it is some 30 times the minimum angular separation between rays that the human eye can detect under bright conditions and, therefore, would certainly be noticeable. If additional air-glass surfaces are introduced, more refractions may occur, and the angular spread in the emerging rays may be increased. The light is said to be more highly dispersed in this case.

A *prism* is a common device used to disperse light and form a spectrum (Figure 9.51). Prisms were well known and highly prized by the Chinese from the early 1600s for their ability to generate color. Today, they are highly valued by scientists for much the same reason. For example, one can analyze the radiation emitted by a source of light

Figure 9.50

Dispersion at an air-glass interface produces a separation of red and blue rays. The angle of refraction of the blue ray is smaller than the angle of refraction of the red ray.

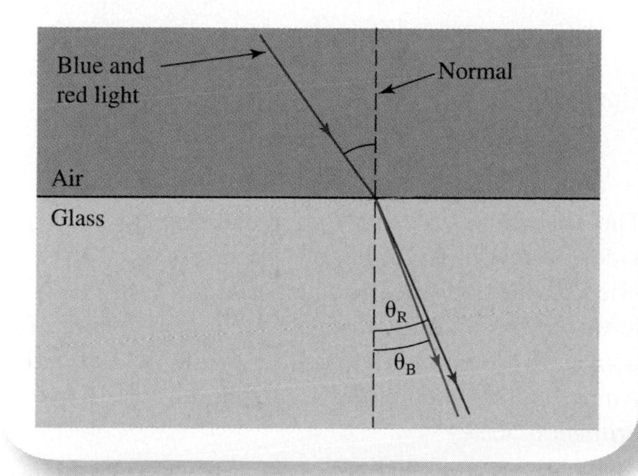

Figure 9.51

Dispersion of white light into a spectrum of colors as it passes through an equilateral prism. Each color is bent a different amount because the speed is slightly different for each wavelength.

by dispersing its light into a spectrum and measuring the intensity (amount) of radiation coming off in the various wavelengths (colors). If the source radiates like a blackbody (see Section 8.6), this information might be used to determine the temperature of the source. As we shall see in Chapter 10, it is also possible to determine the chemical composition of a source by examining its spectrum.

9.7 Atmospheric Optics: Rainbows, Halos, and Blue Skies

The applications of optics principles that we have provided thus far have largely been practical, device-oriented ones. Anyone who has spent any time outdoors, however, has undoubtedly witnessed many natural phenomena whose roots run deeply into the field of optics. Among the most common are rainbows, halos, and the brilliant blue sky.

Rainbows

How do the elements of water and sunlight combine to produce the spectacle of a rainbow?

Before going into detail, we need to point out some rainbow basics. First, rainbows consist of arcs of colored light (spectra) stretching across the sky, with the red part of the spectrum lying on the outside of the bow and the blue-violet part lying on the inside. Second, rainbows are always seen against a background of water droplets with the Sun typically at our backs. These two basic characteristics of rainbows are what we seek to understand.

Imagine a beam of light from the Sun striking a raindrop. For simplicity, we will assume that raindrops are spherical, although real falling raindrops are shaped more like the squashed circular pillows that decorate sofas. If we apply the law of refraction, concentrating only on those rays that are internally reflected at the back of the drop and return in the general direction of the Sun (to be consistent with the second rainbow basic above), we find the result shown in Figure 9.52. The ray striking the drop at its center (ray 1) returns directly back along its incident direction and defines the *axis* of the drop. Rays entering above the axis exit below the axis, and vice versa. The farther above the axis the ray enters, the greater its emergent angle up to a point defined by ray 7. This ray is called the *Descartes ray*, after René Descartes, who first suggested this explanation for rainbows in 1637.

For rays entering above the Descartes ray, the exit angles are *less* than that of the Descartes ray. Thus rays entering the drop on either side of the Descartes ray emerge at about the same angle as the Descartes ray itself, leading to a concentration of rays leaving the droplet at a maximum angle corresponding to that of the Descartes ray. This angle is about 41° for rays 6 through 10 in Figure 9.52.

This concentration of reflected and refracted sunlight at exit angles near 41° produces rainbows. The Descartes model predicts that rainbows should consist of circles of light of angular radii equal to 41°, centered on a point opposite the Sun in the sky—the *antisolar point.* Thus, to see a rainbow, we need to look for these concentrated rays in a direction about 41° from the "straight back" direction with the Sun behind us (Figure 9.53). Notice, if the Sun is above the horizon, the antisolar point will be below the horizon along the direction of your shadow. In this case, the rainbow circle intersects the horizon, and we see only an arc of the circle. For earthbound observers, the best rainbow apparitions occur when the Sun is on the horizon, for then we see half of the rainbow circle. If the Sun is higher in the sky than about 41°

above the horizon, then no rainbow can be seen from the ground because the antisolar point lies 41° or more below the horizon, and the rainbow circle never reaches above the horizon. This is why observers throughout most of the continental United States rarely see rainbows at noon. When viewed from an aircraft, a rainbow can form a complete circle.

So far, we have addressed several aspects of the shape and location of rainbows but not their colors. To do so, we must include the phenomenon of dispersion. Recall from Section 9.6 that blue light is more strongly deviated in passing through transparent media than is red light. This means that the maximum emergent angle from the raindrop for blue light will be smaller than the maximum emergent angle for red light (Figure 9.54). Therefore, the blue light is concentrated at slightly smaller angles than is the red light. Calculations show that blue-violet light is concentrated in a circle of angular radius of about 40°, while red light is concentrated at an angle of about 42°. The other colors of the rainbow fall in between. A more detailed model, including dispersion, thus predicts that real rainbows should consist of bands of color in the sky a total of some 2° or so wide, with blue-violet colors on the inside and red-orange colors on the outside. And this is precisely what is seen.

The application of the laws of reflection and refraction (including dispersion) to falling raindrops gives us an explanation for what is called the *primary* rainbow. The primary results from *one* internal reflection of the rays in the drop at the rear surface. Higher-order rainbows may be produced by rays executing two or more internal reflections before leaving the droplet. Some of you no doubt have seen *secondary* rainbows lying outside the primaries along

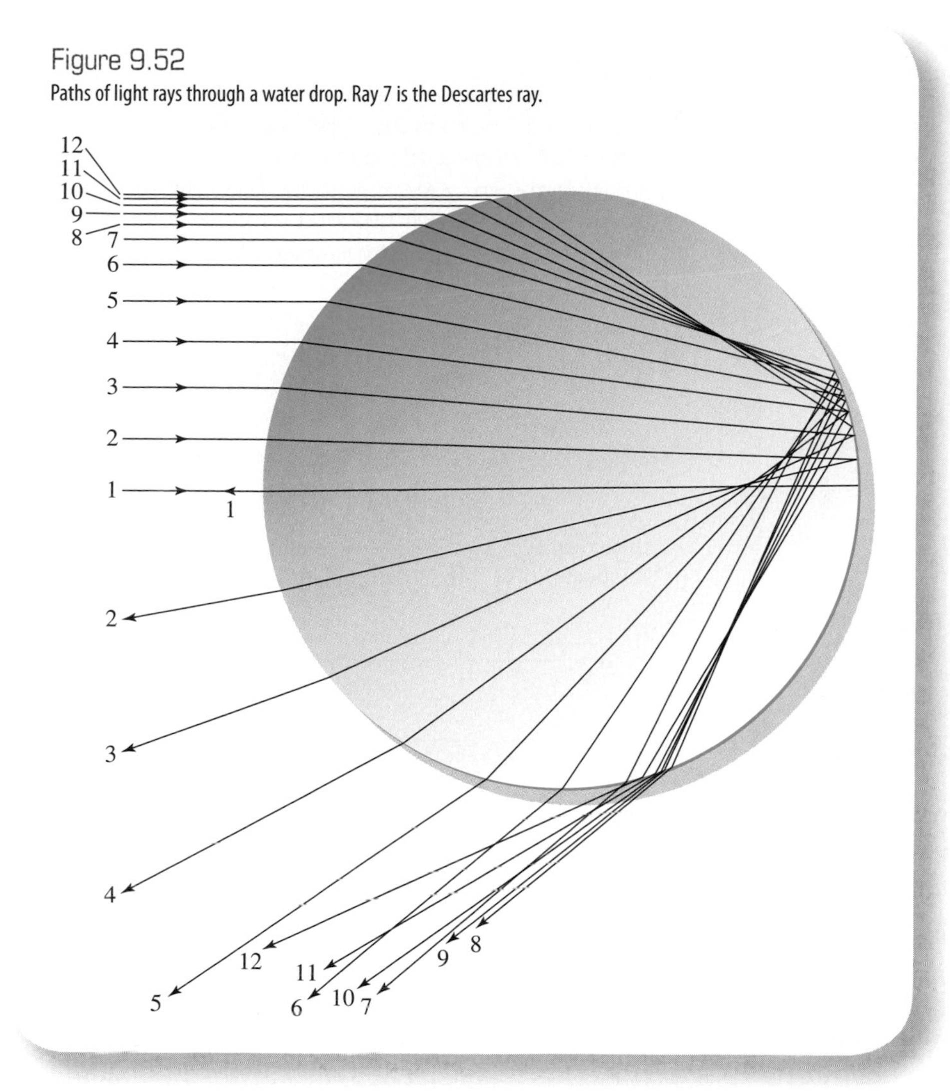

Figure 9.52
Paths of light rays through a water drop. Ray 7 is the Descartes ray.

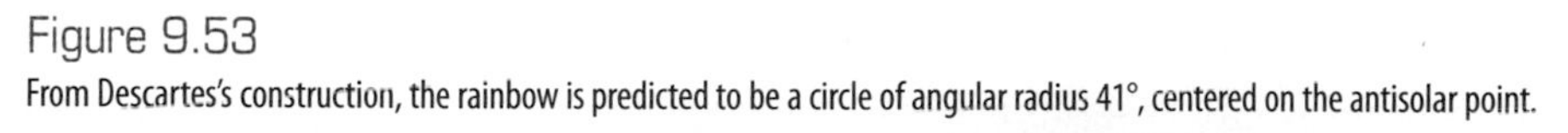

Figure 9.53
From Descartes's construction, the rainbow is predicted to be a circle of angular radius 41°, centered on the antisolar point.

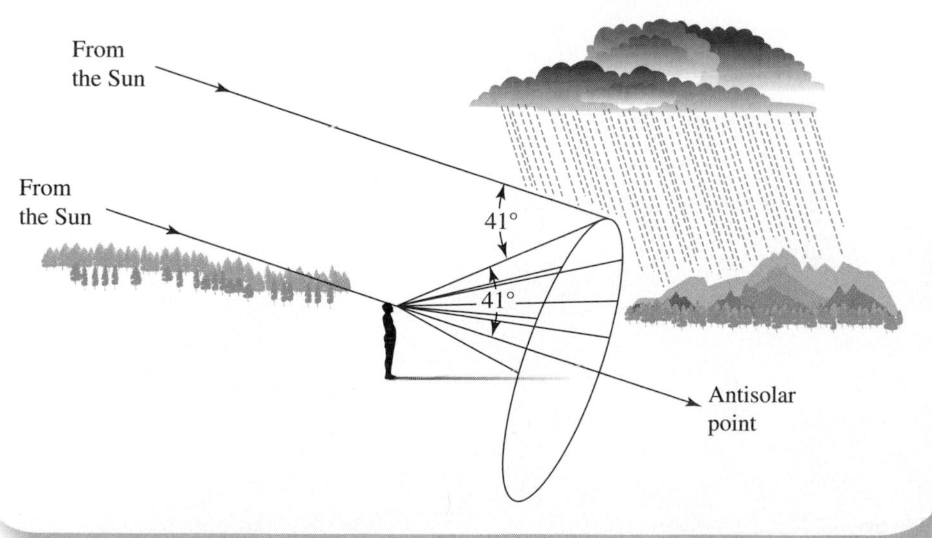

Figure 9.54
Dispersion of sunlight by a spherical raindrop.

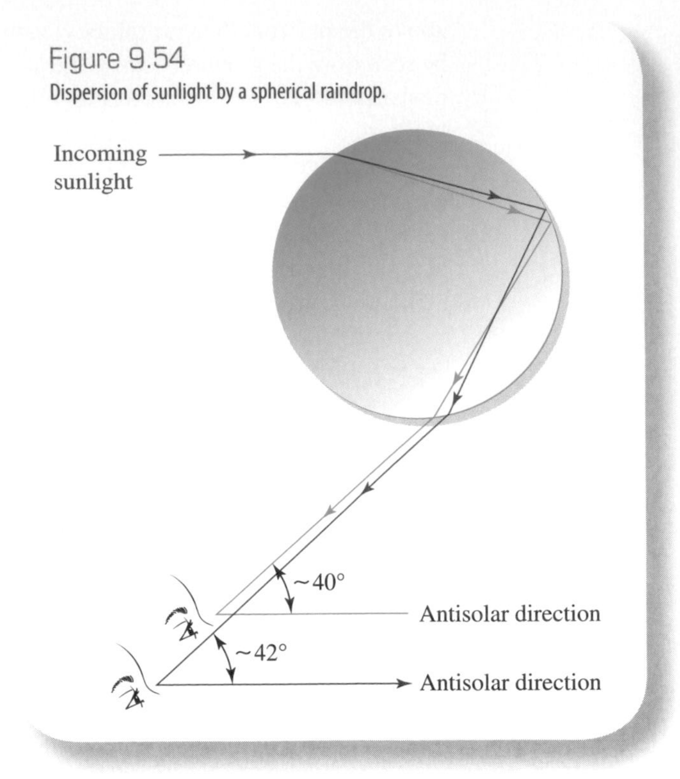

Figure 9.55
Schematic diagram showing the production of a secondary rainbow from two internal reflections in a raindrop. The path shown is for a typical ray of yellow light. The ray emerges at an angle of about 51° with respect to the antisolar direction. Red rays emerge with slightly smaller angles, while blue rays emerge with somewhat larger angles, thus accounting for the reversal of the color-ordering in the secondary bows.

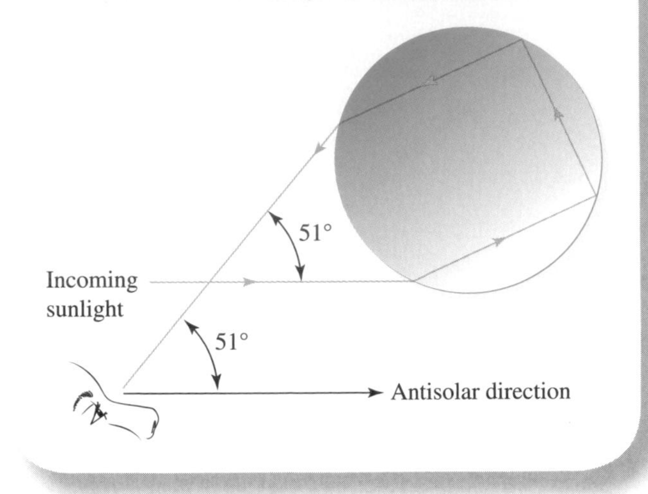

arcs of circles having angular radii of approximately 51°. The ordering of the colors of these rainbows is reversed from that of the primaries. All of these properties are explicable in terms of the laws of optics with which we have become familiar (Figure 9.55).

Halos

Halos, circular arcs of light, often with reddish inner edges, surrounding the Sun or full Moon might be considered winter's answer to rainbows. When the temperature in the upper atmosphere drops below freezing, ice crystals form. At high elevations in the temperate regions of the Earth, one common shape exhibited by such crystals is that of a hexagon, similar to a short, unsharpened pencil (Figure 9.56a). When seen in cross section, such crystals may be considered as pieces of equilateral (60°) prisms, and they deviate light like them (see Figure 9.56b).

As in the case of rays entering raindrops, if one traces the paths of rays entering such a crystal at various incident angles, one finds that there is a concentration of exiting rays with deviation angles near 22°. Thus, when light from the Sun or the Moon enters a cloud of such ice crystals having all possible orientations, the emergent

Primary and secondary rainbows.

NOAA Photo Library, NOAA Central Library; OAR/ERL/National Severe Storms Laboratory (NSSL) / © Shelly Perry/iStockphoto.com

Actual 22° halo around the Sun (itself partially obscured by the intervening tree).

© Catherine dée Auvil/iStockphoto.com / © Stefan Klein/iStockphoto.com

© Jerry Schad/SPL/Photo Researchers

© Norbert Rosing/National Geographic/Getty Images

Like 22° halos, sundogs arise by refraction and dispersion in ice crystals, but not randomly oriented pencil-like ones. Instead, plate-like crystals (see Fig. 9.56a lower) with their large flat sides aligned parallel to the ground are responsible for concentrating (and coloring) the light at angular positions along the horizon 22° ahead of and/or behind the Sun.

rays tend to be clustered into circular arcs having angular radii of 22° centered on the source of illumination.

To see a ray of light forming part of a halo, we should look in a direction 22° away from the Sun or Moon (see Figure 9.57). When doing so, you may notice that the inner edge of the halo circle is tinted red. This is again the result of dispersion. At each refraction, the blue component of sunlight (or moonlight, which is merely reflected sunlight) is more strongly refracted than is the red component. Consequently, the angle of concentration for the blue light is somewhat greater than it is for red light, and the latter piles up preferentially at the inner edge of the halo, as indicated in Figure 9.58.

What we have described is the well-known 22° halo. There are also 46° halos, which result from light entering one face of the pencil crystal and leaving through one end. These halos are much fainter than the 22° halos and are much harder to see—partly because they occupy such a large portion of the sky, having angular diameters of more than 90°! And these are but two of many, many phenomena

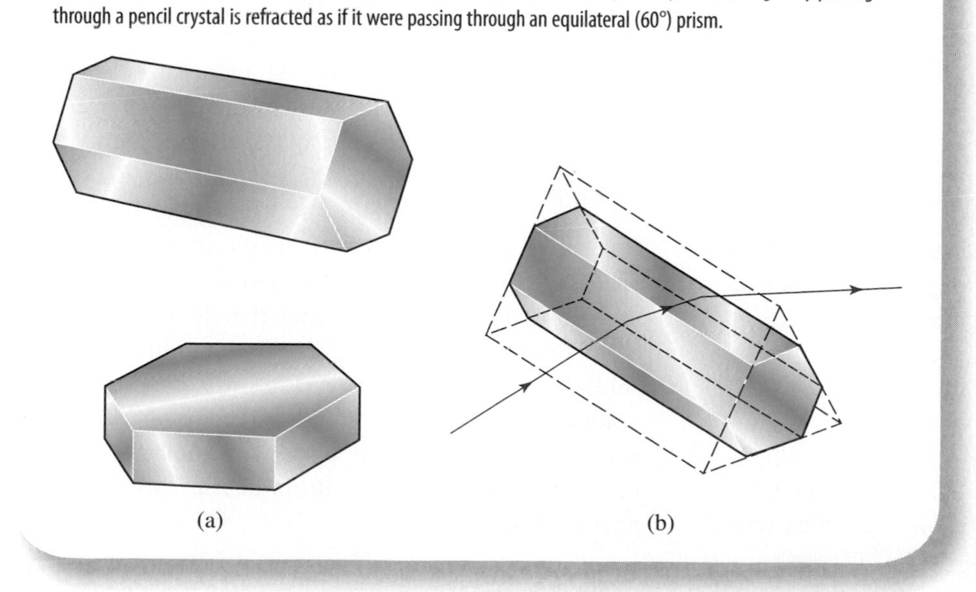

Figure 9.56

(a) Two simple ice crystal forms: top, a columnar or pencil crystal; bottom, a plate crystal. (b) A light ray passing through a pencil crystal is refracted as if it were passing through an equilateral (60°) prism.

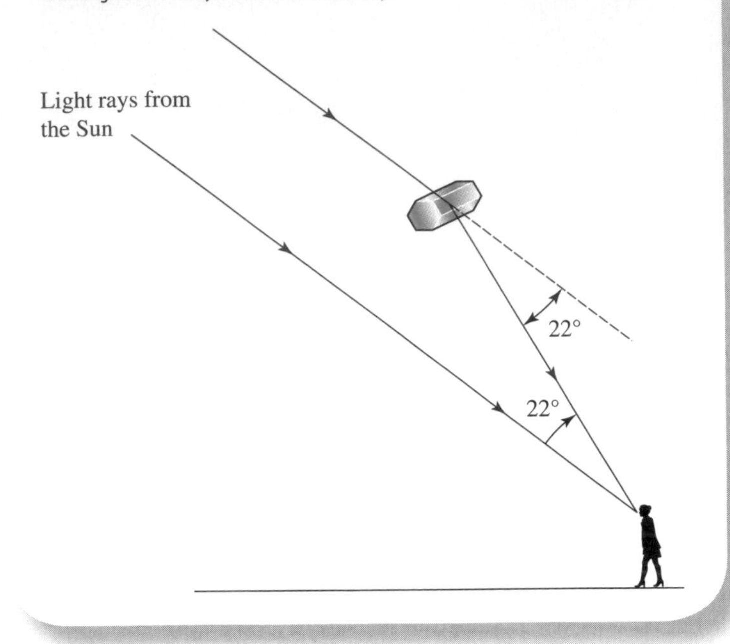

Figure 9.57
Schematic representation of how a 22° halo is produced, showing an enlarged image of the ice crystals typically responsible for this phenomenon. To see sunlight that is refracted through an angle of 22° by the ice crystals to form part of the halo, one looks to an angle of 22° away from the Sun in the sky.

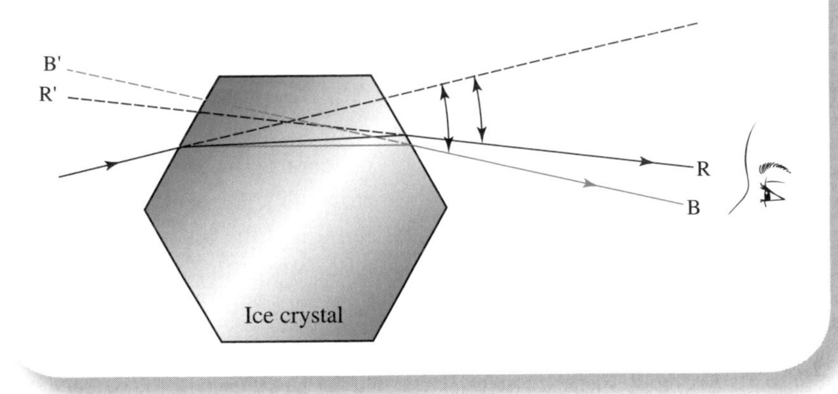

Figure 9.58
Dispersion of sunlight by ice crystals produces halos with reddish inner edges. Because the deviation of blue rays is larger than that of red rays, the blue rays appear to originate along the line *B′B*, farther away from the Sun than the red rays, which appear to come from the direction *R′R*. Thus, the inner edge of the halo—that is, the part nearest the Sun—is tinged reddish.

associated with ice crystal reflection and refraction. Such magnificence surrounds us daily if only we allow our eyes to be open to it. A knowledge of physics can help us to appreciate these natural wonders more deeply.

Blue Skies

The most common of all atmospheric optical phenomena is the blue sky. It is caused by air molecules *scattering* sunlight in all directions. As a light wave travels through the atmosphere, the wave's oscillating electric field causes the electrons in air molecules to oscillate with the same frequency. From the discussion in Chapter 8, we know that oscillating electric charges (electrons, in this case) emit electromagnetic radiation (light, in this case). This emitted light, which travels outward in all directions (hence the use of the term *scattered*), is what we see.

But why is it blue instead of white like the incident sunlight? It turns out that the electrons in air molecules are much more efficient at absorbing and radiating higher frequencies of light. When blue light makes an electron in an air molecule oscillate, it absorbs and scatters much more of the incident radiant energy than when red light makes it oscillate. In particular, the amount of radiant energy scattered per second by air molecules is inversely proportional to the wavelength of the light raised to the fourth power.

$$\frac{E_{\text{scattered}}}{t} \propto \frac{1}{\lambda^4}$$

The wavelength of blue light is about 0.7 times that of red light (from Table 8.1). Consequently, there is about 4.2 ($= 1/0.7^4$) times as much blue in scattered sunlight as there is red. The sky is pale blue rather than "pure" blue since all frequencies of light still remain present to some extent in the scattered sky light.

Not only does molecular scattering give us beautiful blue skies, it is also responsible for the stunning orange and red colors often seen in clouds shortly after sunset and shortly before sunrise (Figure 9.59a). The sunlight that reaches such clouds must travel hundreds of kilometers through the atmosphere. Along the way, more and more of the radiant energy is removed from the sunlight and transferred to scattered light. The higher frequencies (like blue) are much more strongly attenuated by this process than are the lower frequencies (like red). Consequently, the light that reaches the cloud is reddish in color since comparatively less of the red in the original sunlight has been removed (Figure 9.59b). The same scattering process that produces reddish-hued clouds near sunrise and sunset is also responsible for the fiery red-orange appearance of the Sun itself when located near the horizon.

Figure 9.59

(a) Photo of clouds shortly after sunset. (b) Sunlight on its way to a cloud at sunset has most of the blue removed by scattering.

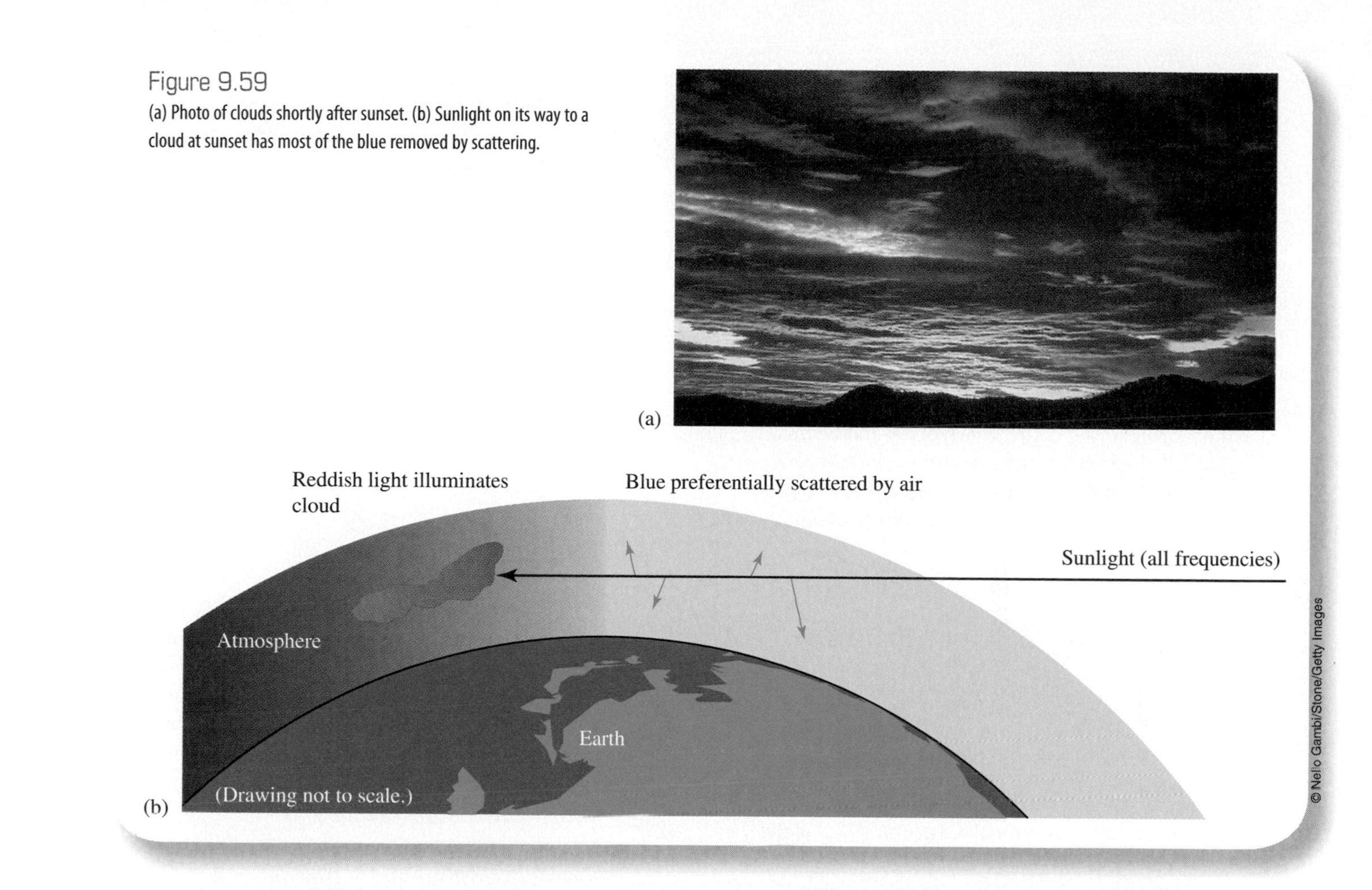

questions&problems

Questions

(▶ Indicates a review question, which means it requires only a basic understanding of the material to answer. The others involve integrating or extending the concepts presented thus far.)

1. ▶ Why are the Doppler effect and diffraction not as commonly observed with light as they are with sound?
2. ▶ Distinguish between specular reflection and diffuse reflection.
3. ▶ The law of reflection establishes a definite relationship between the angle of incidence of a light ray striking the boundary between two media and its angle of reflection. Describe this relationship.
4. The cover of a book appears blue when illuminated with white light. What color will it appear in blue light? Red light? Explain.
5. A ballet dancer is on stage dressed in a green body suit. How could you light the stage so that the dancer's costume looked black to the audience?
6. ▶ Describe how light passing through two narrow slits can produce an interference pattern.
7. A person looking straight down on a thin film of oil on water sees the color red. How thick could the film be to cause this? How thick could it be at another place where violet is seen?
8. If the wavelength of visible light were around 10 cm instead of 500 nm and we could still see it, what effect would this have on our ability to see diffraction and interference effects?
9. An interference pattern is formed by sending red light through a pair of narrow slits. If blue light is then used, the spacing of the bright areas (where constructive interference takes place) won't be the same. How will it be different? Why?
10. ▶ What is polarized light? How do Polaroid sunglasses exploit polarization?
11. Describe how you could use two large, circular Polaroid filters in front of a circular window as a kind of window shade.

12. Just before the Sun sets, a driver encounters sunlight reflecting off the side of a building. Will Polaroid sunglasses stop this glare? Why?

13. You get a new pair of sunglasses as a birthday gift from a friend. When wrapping the present, your friend has removed the price tag and all the other labels from the sunglasses. How can you tell whether the pair of shades has polarizing lenses in them or not?

14. Compared to a person's height, what is the minimum length (top to bottom) of a mirror that will allow the person to see a complete image from head to toe?

15. An observer *O* stands in front of a plane mirror as shown in the accompanying figure. Which of the numbered locations 1 through 5 best represents the location of the image of the source *S* seen by the observer? Justify your choice by appealing to the appropriate law(s) of optics.

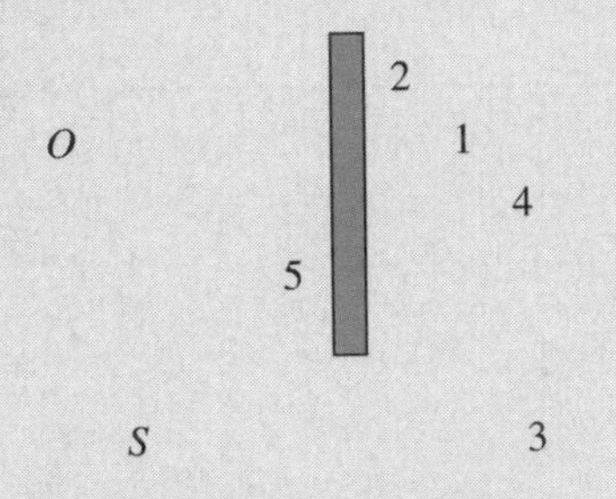

16. Light traveling along direction *SO* in the accompanying figure strikes the surface of a plane mirror. Which of the paths *OP, OQ, OR*, or *OT* best describes the path of the light reflected from the mirror? Defend your choice by appealing to the appropriate law(s) of optics.

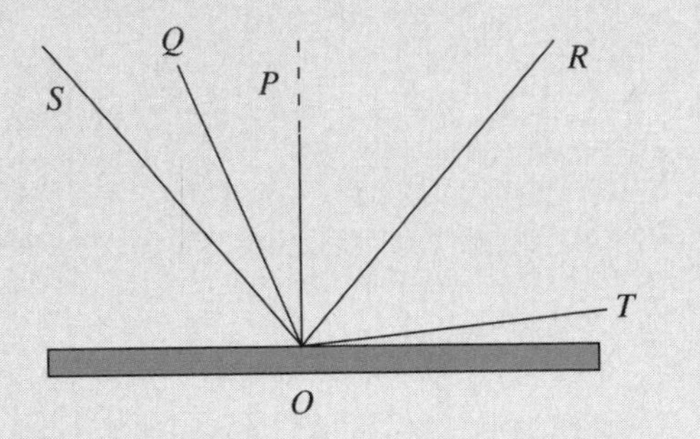

17. Consider the figure below. At which of the lettered positions would an observer be able to see the image of the dot in the mirror?

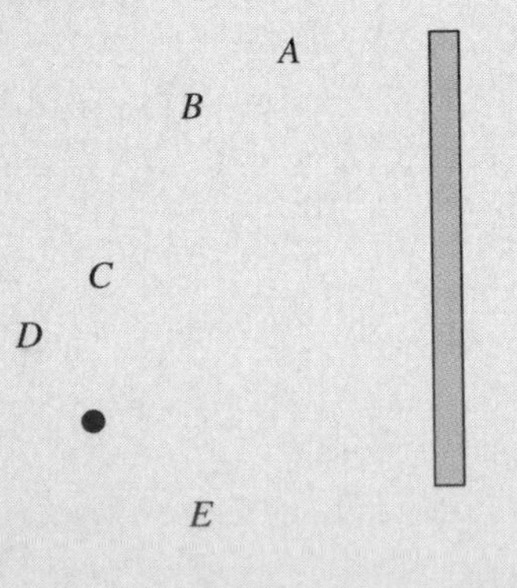

18. Two plane mirrors are hinged along one edge and set at right angles to one another as shown in the figure below. A light ray enters the system, striking mirror 1 with an angle of incidence of 45°. Draw a diagram showing the direction in which the ray will exit the system. This device is called a *corner reflector* and is the basis for the design of bicycle reflectors and some highway signs.

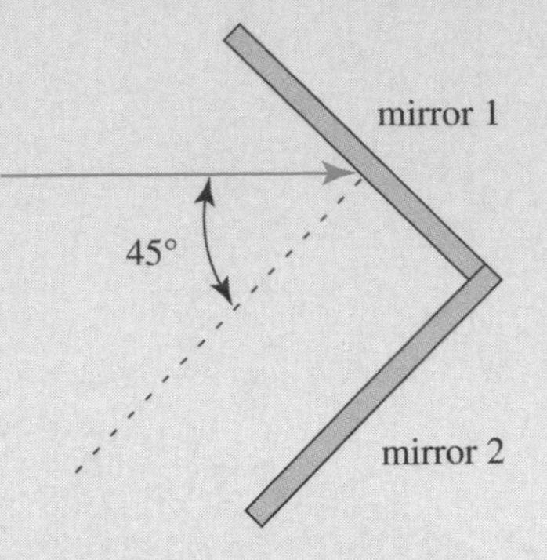

19. If you hold up your right hand in front of a plane mirror, what do you see? (Be cautious now. Describe the image carefully.) If you're having trouble, place a tube of toothpaste or other object with writing on it in front of the mirror. Now what can you say? (The process you're witnessing is called *reversion* and is one of the common characteristics of plane mirrors.)

20. ▶ What is different about an image (of a nearby object) formed with a convex mirror compared to an image formed with a concave mirror? What are the advantages of each type of mirror?

21. ▶ What is the ideal shape of concave mirrors used in telescopes?

22. If you were lost in the forest and wanted to start a small fire to keep warm or cook a meal using sunlight and a small lens, what type of lens—converging or diverging—would you use and why?

23. What kind of mirror (concave, convex, or plane) do you think dental care workers use to examine patients' teeth and gums for disease? Explain your choice.

24. ▶ Describe how the path of a ray is deviated as it passes (at an angle) from one medium into a second medium in which the speed of light is lower. Contrast this with the case when the speed of light in the second medium is higher.

25. How would Figure 9.22 be different if the glass were replaced by water or by diamond?

26. The speed of light in a certain kind of glass is exactly the same as the speed of light in benzene—a liquid. Describe what happens when light passes from benzene into this glass, and vice versa.

27. A piece of glass is immersed in water. If a light ray enters the glass from the water with an angle of incidence greater than zero, in which direction is the ray bent?

28. A light ray in air enters a block of clear plastic as shown in the figure below. Which of the numbered paths represents the correct one for the ray in the plastic? Justify your choice by appealing to the appropriate law(s) of optics.

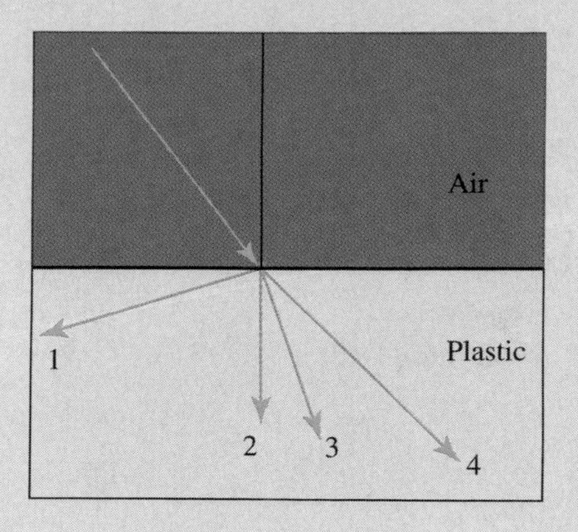

29. After hitting the ball into a water trap, a golfer looks into the pond and spies the ball within apparent easy reach. Reaching in to retrieve the ball, the golfer is surprised to find that it cannot be grasped even with a fully extended arm. Explain why the golfer was deceived into thinking that the location of the ball was close at hand.

30. One sometimes hears the expression, "It was like shooting fish in a barrel!" This usually is taken to mean that the task, whatever it was, was easy to complete. But is it really easy to shoot fish in a barrel? Only if you know some optics! Suppose you're in a boat and spy a large fish a few meters away. If you want to shoot the fish, how should you aim? Above the image of the fish? Below it? Directly at the image? Explain your choice. (You may assume that the path of the projectile you fire will not be deviated from a straight line upon entering the water, unlike light.)

31. Rank (from smallest to largest) the angle of refraction for a light ray in air entering each of the following substances with an angle of incidence equal to 30°: (i) water; (ii) benzene; (iii) dense flint glass; (iv) diamond.

32. ▶ What is total internal reflection, and how is it related to the critical angle?

33. Would the critical angle for a glass-water interface be less than, equal to, or greater than the critical angle for a glass-air interface? Explain your choice.

34. Explain why images seen through flat, smooth, uniform, plate-glass windows are undistorted.

35. ▶ For transparent solids, distinguish between effects of surfaces that curve inward and those that curve outward on the paths of a parallel bundle of light rays incident on each.

36. ▶ Distinguish converging lenses from diverging lenses, and give examples of each type.

37. Of the three converging lenses shown in Figure 9.36, which would you expect to have the shortest focal length? Why?

38. ▶ Describe the three principal rays used to locate an image.

39. ▶ Contrast real images with virtual images in as many ways as you can.

40. Indicate whether each of the following is a real image or a virtual image.

    a) Image on the retina in a person's eye.

    b) Image one sees in a rearview mirror.

    c) Image one sees on a movie screen.

    d) Image one sees through a magnifying lens.

41. A convex lens forms a clear, focused image of some small, fixed object on a screen. If the screen is moved closer to the lens, will the lens have to be (i) moved closer to the object, (ii) moved farther from the object, or (iii) left at the same location to produce a clear image on the screen at its new location? Explain your answer.

42. Sketch the image of the letter **F** formed on the film in a simple camera.

43. ▶ How is the magnification of a lens related to the object distance? How is it related to the image distance? What is the significance of the sign of the magnification? What is the significance of its magnitude (size)?

44. Estimate the values of the magnification in Figures 9.37, 9.39, and 9.40.

45. ▶ What is chromatic aberration? How can it be remediated?

46. ▶ How is the eye able to form focused images of objects that are different distances away?

47. ▶ When a person is nearsighted, what happens in the eye when the person is looking at something far away? How is this condition commonly corrected?

48. ▶ Describe the phenomenon of dispersion, and explain how it leads to the production of a spectrum.

49. Two light waves that have wavelengths of 700 and 400 nm enter a block of glass (from air) with the same angle of incidence. Which has the larger angle of refraction? Why? Would the answer be different if the light waves were going from glass into air?

50. Suppose a 20-m long tube is filled with benzene and the ends sealed off with thin disks of glass. If pulses of red and blue light are admitted simultaneously at one end of the tube, will they emerge from the opposite end together, that is, at the same time? If not, which pulse will arrive at the far end of the tube first? Why?

51. The difference in speed between red light and violet light in glass is smaller than the difference in speed between the same two colors in a certain type of plastic. For which material, glass or plastic, would the angular

spread of the two colored rays after entering the material obliquely from air be the largest? Why?

52. ▶ What is a prism? Why are such devices useful to scientists?

53. Would a prism made of diamond be better at dispersing light than one made of glass? Why or why not?

54. ▶ Describe the accepted model of rainbows. Specifically, discuss how the model accounts for the size, shape, location, and color ordering of primary rainbows.

55. Suppose an explosion at a glass factory caused it to "rain" tiny spheres made of glass. Would the resulting rainbow be different from the normal one? If so, how might it be different and why?

56. ▶ Compare the primary and secondary rainbows in regard to their angular size, color ordering, and number of internal reflections that occur in the rain droplets.

57. Suppose you are told by close friends that they had witnessed a glorious rainbow in the west just as the Sun was setting. Would you believe them? Why or why not?

58. ▶ How is a 22° halo formed?

59. Water droplets in clouds scatter sunlight similar to the way air molecules do. Since clouds are white (during the day), what can you conclude about the frequency dependence of scattering by cloud particles?

Problems

1. A light ray traveling in air strikes the surface of a slab of glass at an angle of incidence of 50°. Part of the light is reflected and part is refracted. Find the angles the reflected and refracted rays make with respect to the normal to the air-glass interface.

2. A ray of yellow light crosses the boundary between glass and air, going from the glass into air. If the angle of incidence is 20°, what is the angle of refraction?

3. Using Figure 9.27, find the angles of refraction for a light ray passing from air into glass with the following angles of incidence: 5°, 10°, and 20°. Do you notice a trend in the resulting values? If so, describe it. Based on your observations, what would you predict the angle of refraction to be for an angle of incidence of 40°? How does your value compare with that inferred from Figure 9.27?

4. Using the definition of the index of refraction contained in the footnote on p. 240 and the data in Table 9.1, compute the index of refraction, n, for the following substances: (i) air; (ii) benzene; and (iii) glass (any type).

5. A fish looks up toward the surface of a pond and sees the entire panorama of clouds, sky, birds, and so on, contained in a narrow cone of light, beyond which there is darkness. What is going on here to produce this vision, and how large is the opening angle of the cone of light received by the fish?

amera is equipped with a lens with a focal length m. When an object 2 m (200 cm) away is being photographed, how far from the film should the lens be placed?

7. A 2.0-cm-tall object stands in front of a converging lens. It is desired that a virtual image 2.5 times larger than the object be formed by the lens. How far from the lens must the object be placed to accomplish this task, if the final image is located 15 cm from the lens?

8. When viewed through a magnifying glass, a stamp that is 2 cm wide appears upright and 6 cm wide. What is the magnification?

9. A person looks at a statue that is 2 m tall. The image on the person's retina is inverted and 0.005 m high. What is the magnification?

10. What is the magnification in Problem 6?

11. A small object is placed to the left of a convex lens and on its optical axis. The object is 30 cm from the lens, whose focal length is 10 cm. Determine the location of the image formed by the lens. Describe the image.

12. If the object in Problem 11 is moved toward the lens to a position 8 cm away, what will the image position be? Describe the nature of this new image.

13. a) In a camera equipped with a 50-mm focal-length lens, the maximum distance that the lens can be from the film is 60 mm. What is the closest an object can be to the camera if its image on the film is to be in focus? What is the magnification?

    b) An extension tube is added between the lens and the camera body so that the lens can be positioned 100 mm from film. How close can the object be now? What is the magnification?

14. The focal length of a diverging lens is negative. If $f = -20$ cm for a particular diverging lens, where will the image be formed of an object located 50 cm to the left of the lens on the optical axis? What is the magnification of the image?

15. The equation connecting s, p, and f for a simple lens can be employed for spherical mirrors, too. A concave mirror with a focal length of 8 cm forms an image of a small object placed 10 cm in front of the mirror. Where will this image be located?

16. If the mirror described in the previous problem is used to form an image of the same object now located 16 cm in front of the mirror, what would the new image position be? Assuming that the magnification equations developed for lenses also apply to mirrors, describe the image (size and orientation) thus formed.

17. If the wavelength of light is doubled, by what fraction does the amount of the light scattered by the Earth's atmosphere change? Is the amount of light scattered at the new wavelength larger or smaller than the amount of light scattered at the old wavelength?

18. Compute the approximate ratio of the amount of blue light in scattered sky light to the amount of orange.

"It's easy to read, it outlines important topics, and it's relevant. Thanks for the good stuff on the website, I think it will **really help with tests**."

– Thomas Scholtes, Student at University of Maryland, College Park

REVIEW

HE DID

PHYSICS puts a multitude of study aids at your fingertips. After reading the chapters, check out these resources for further help:

- **Chapter Review Cards**, found in the back of your book, include definitions, laws, principles, equations, topic summaries, and concept maps for each chapter.
- **Online printable flash cards** give you three additional ways to check your comprehension of key physics concepts.

Other great ways to help you study include **simulations, interactive quizzes,** and **flashcards**.

You can find it all at **4ltrpress.cengage.com/physics**.

Atomic Physics

sections

10.1 The Quantum Hypothesis

Fluorescent light and plasma TV. One is a familiar, trusted, often taken-for-granted device that's been providing efficient lighting for decades. The other became one of the first status symbols of the twenty-first century. One is typically hidden in fixtures and noticed only if it doesn't work correctly (it flickers or buzzes). The other can cost thousands of dollars and is often the center of attention—even becoming part of the studio set for TV news programs and military press briefings. Its inventors shared an Emmy Award in 2002. One is mundane, the other glamorous. And yet they both produce light using the same physical process. In fact, a plasma display is basically millions of tiny fluorescent lamps emitting light in a coordinated fashion. These products and many, many more are examples of devices that take advantage of modern atomic physics and the quantum nature of light.

By the end of the nineteenth century, physicists were quite satisfied with the great progress that had been made in the study of physics. Many questions that had been puzzling scientists for thousands of years had been answered. Advances made possible by Newton's great treatise on mechanics and the clarification of electromagnetism by Maxwell gave physicists cause to celebrate the deep understanding of the physical world that they had acquired. In fact, some physicists worried that the field might be dying; they feared that *all* of their questions might soon be answered.*

But some problems defied solution, and advancements in experimental equipment and techniques led to new discoveries. The implications arising from attempts to explain these phenomena were so revolutionary that by the early 1900s many of these same physicists were bewildered and wondered just how much they really did know.

In the first five sections of this chapter, we take an historical approach to three of these ...blems. We describe how investigations into them led to a reinterpretation of the very nature

...vent of *grand unified theories* and *theories of everything* (TOEs; see Section 12.5) within the last three decades or so, similar ...been raised again by some physicists. While the views expressed by these scientists may not be universally shared, the issues ...oncerning our progress toward viable TOEs are important.

of light and the way it interacts with matter. The phenomena are (1) **blackbody radiation**, (2) the **photoelectric effect**, and (3) **atomic spectra**.

Physicists knew a great deal about these phenomena, but the fundamental understanding of their *causes* had eluded them. It was much like the period before Newton: Astronomers knew a lot about the shapes of the orbits of the Moon and the planets, but they didn't know what caused these particular shapes. Newton's mechanics and his law of universal gravitation provided the answer. Similarly, scientists some two centuries later were seeking the theoretical basis for these three phenomena.

Blackbody Radiation

Blackbody radiation, the characteristics of which we have already discussed at some length in Section 8.6, had been studied carefully by scientists in the nineteenth century, and its properties were well-known but not understood. For example, it had been determined that a perfectly "black" body, one that would absorb all light and other EM radiation incident upon it, would also be a perfect emitter of EM radiation. Such an ideal *blackbody* was said to emit *blackbody radiation* (BBR).

© Michal Skowronski/iStockphoto.com

The well-known characteristics of the BBR emitted by a given blackbody at a particular temperature can be conveniently illustrated with a graph. Imagine examining each wavelength of radiation in turn, from very short wavelength ultraviolet to much longer wavelength microwaves, and measuring the energy that is emitted each second (the power) from the blackbody. The graph of the resulting data, plotted as the intensity of the radiation versus wavelength, is called a *blackbody radiation curve.* Figure 10.1 shows two BBR curves for blackbodies at two different temperatures. These graphs illustrate two important ways that BBR changes when the temperature of the body is increased:

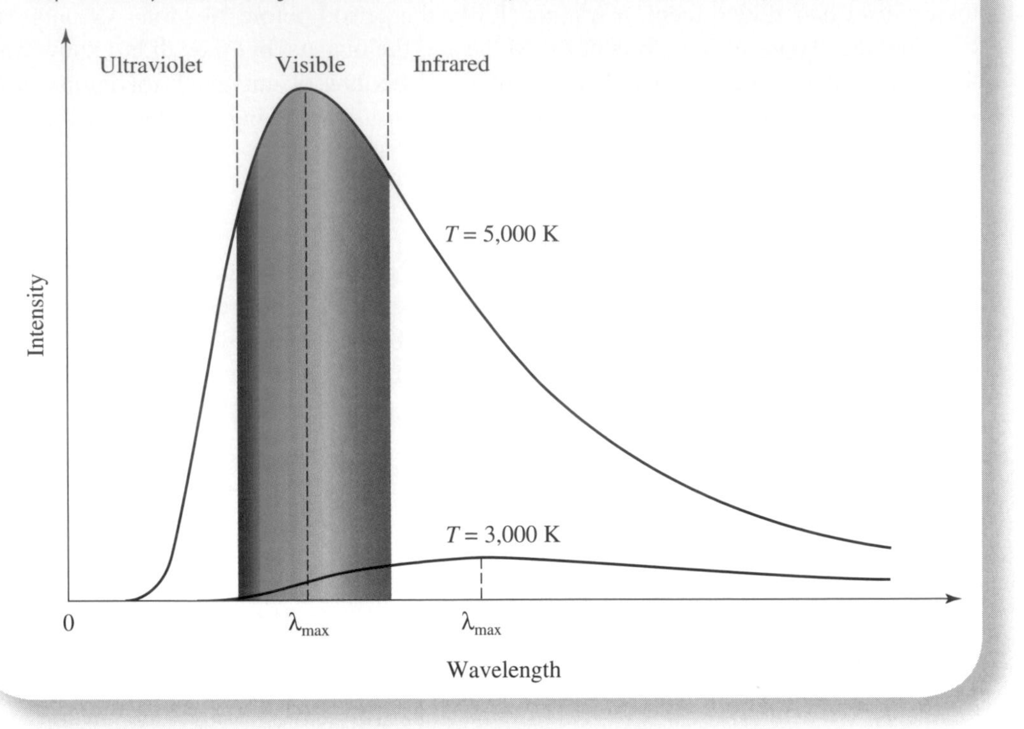

Figure 10.1
Graph of intensity versus wavelength for the EM waves emitted by blackbodies at two different temperatures (cf. Figure 8.30).

1. More *energy* is emitted per second *at each wavelength* of EM radiation.
2. The *wavelength* at which the *most energy* is emitted per second (in other words, the *peak* of the BBR curve) *shifts to smaller values.* That is why toaster elements glow red hot, while extremely hot stars glow blue hot.

Why does a blackbody emit different amounts of EM radiation at different wavelengths in precisely this way? Using the principles of electromagnetism, we can easily see why EM waves are emitted. Atoms and molecules are continually oscillating, and they contain charged particles—electrons and protons. We saw in Chapter 8 that this kind of system will produce EM waves. But the exact mechanism involved, one that would account for the two features above, was a mystery to scientists at the end of the nineteenth century.

The clue to solving the puzzle was discovered by the German physicist Max Planck in the year 1900. First, by trial and error, Planck derived a mathematical equation that fit the shape of the BBR curves. (This is a common first step in theoretical physics. The mathematical shapes of the planetary orbits—ellipses—were known about a century before Newton explained *why* they were ellipses.) The equation did little to increase the understanding of the fundamental process, but Planck also developed a model that would account for his equation.

Planck proposed that an oscillating atom in a blackbody can have only certain fixed values of energy. It can have zero energy or a particular energy E, or an amount equal to $2E$, or $3E$, or $4E$, and so on. In other words, the energy of each atomic oscillator is **quantized**. The energy E is called the fundamental quantum of energy for the oscillator. The allowed values of energy for the atom are integral multiples of this quantum:

$$\text{allowed energy} = nE, \text{ where } n = 0, 1, 2, 3, \ldots$$

Max Planck (1858–1947)

© CORBIS

This is quite different from an ordinary oscillator, such as a mass hanging from a spring or a child on a swing, which can have a continuous range of energies, not just certain values.

We can illustrate the difference between quantized

and continuous energy values by comparing a stairway and a ramp. A cat lying on a stairway has quantized potential energy: It can be on only one of the steps, and each step corresponds to a particular *PE*. A cat lying on a ramp does not have quantized potential energy. It can be anywhere on the ramp. Its height above the ground can have any value within a certain range, and therefore, its *PE* is one of a continuous range of values (Figure 10.2).

The concept of energy quantization for the oscillating atoms was revolutionary. There seemed to be no logical reason for it. But it worked. Planck's quantized atomic oscillators could emit light only in bursts as they went from a higher energy level to a lower one. He showed that light emitted in this fashion by a blackbody resulted in the correct BBR curves. Moreover, he determined that the basic quantum of energy was proportional to the oscillator's frequency $E \propto f$. In particular:

$$E = (6.63 \times 10^{-34}) f = hf$$

The constant h is called *Planck's constant* and has SI units of joule-seconds (J-s). The allowed energies can now be written as:

$$\text{allowed energy} = nhf, \text{ with } n = 0, 1, 2, 3, \ldots$$

Planck was unsure of the implication of his model. He regarded it mainly as a helpful construct that gave him the correct result. But it turned out to be the first of many scientific discoveries showing the existence of quantized effects at the atomic level.

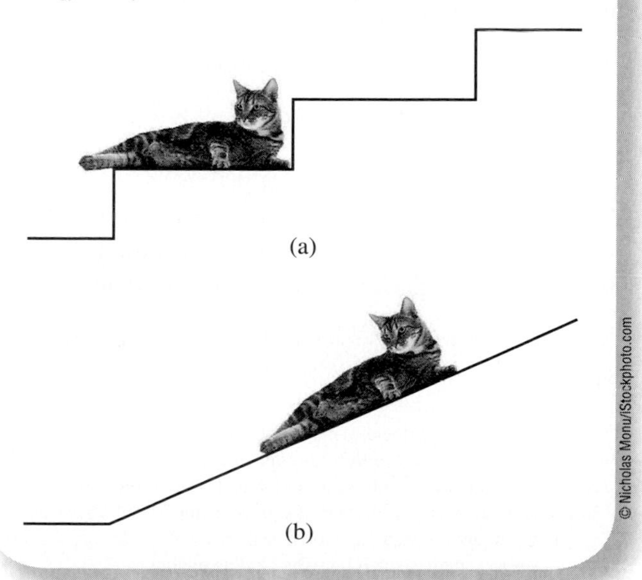

Figure 10.2
(a) A "quantized" cat. Its potential energy is restricted to certain values, one for each step. (b) The cat can be anywhere on the ramp, so its potential energy is not quantized.

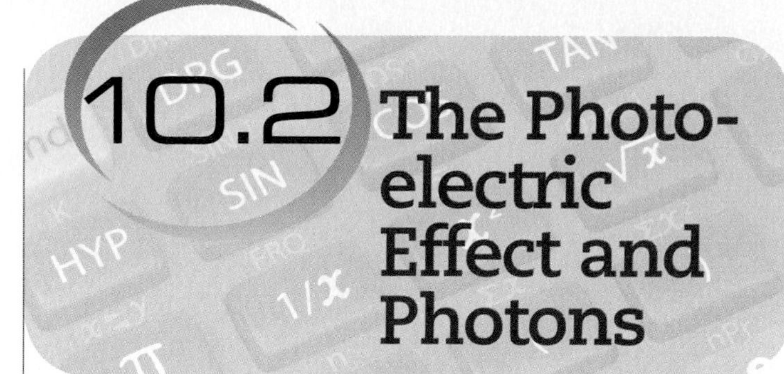

10.2 The Photoelectric Effect and Photons

The second phenomenon that puzzled physicists at the beginning of the twentieth century was the photoelectric effect. It seemed to be unrelated to blackbody radiation, other than that it also involved light, but the concept of quantization turned out to be the key to explaining it, too. The photoelectric effect was accidentally discovered by Heinrich Hertz during his experiments with electromagnetic waves. Hertz was producing EM wave pulses by generating a spark between two conductors. The EM wave would travel out in all directions and induce a spark between two metal knobs used to detect the wave. Hertz noticed that when these knobs were illuminated with ultraviolet light, the sparks were much stronger. The UV was somehow increasing the current in the spark. This is one example of the photoelectric effect.

The photoelectric effect is exhibited by metals when exposed to x-rays, UV light, or (for some metals including sodium and potassium) high frequency visible light. By some means, the EM waves give energy to electrons in the metal, and the electrons are ejected from the surface (Figure 10.3). (It was these freed electrons that enhanced the spark in Hertz's apparatus.) Because of the nature of EM waves, we shouldn't be too surprised that this sort of thing can happen. But again, some of the characteristics of the photoelectric effect that scientists saw

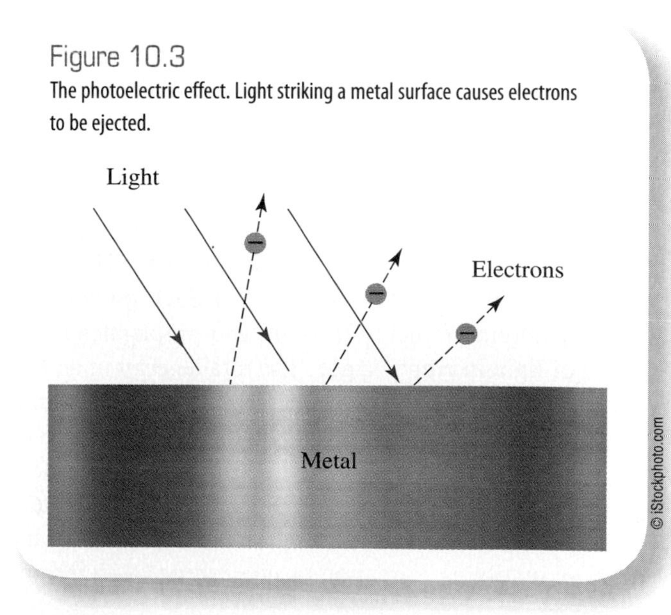

Figure 10.3
The photoelectric effect. Light striking a metal surface causes electrons to be ejected.

showed them that the fundamental process was not entirely understood. The biggest puzzle was the relationship between the speed or energy of the electrons and the incident light. It might seem reasonable to suppose that brighter light would cause the electrons to gain more energy and be ejected from the surface with higher speeds. Yet while it was found that brighter light caused *more* electrons to be ejected each second, it did not increase their energies. Even more of a surprise was the finding that only the frequency (color) of the light affected the electron energies. Higher-frequency light ejected electrons with higher energy, and even extremely dim light with high enough frequency would immediately cause electrons to be ejected.

The explanation of the photoelectric effect was supplied in 1905 by Albert Einstein. Planck had suggested that light is emitted in discrete "bursts" or "bundles." Einstein took this idea one step further and proposed that the light itself remains in bundles or "packets" and is absorbed in this form. The electrons in metal can only take in light energy by absorbing one of these discrete quanta of radiation. The amount of energy in each quantum of light depends on the frequency. In particular:

$$E = hf \qquad \text{(energy of a quantum of EM radiation)}$$

This is the same equation for the energy of Planck's quantized atomic oscillators. Einstein suggested that light and other electromagnetic waves are quantized, just like the energy of oscillating atoms.

This idea, that the energy of an EM wave is quantized, allows us to picture the wave as being made up of individual particles now called **photons**. (The word was coined by G. N. Lewis in 1926.) Each photon carries a quantum of energy, $E = hf$, and propagates at the speed of light in empty space. The total energy in an EM wave is just the sum of the energies of all the photons in the wave. Notice how different this picture is from the wave picture of light developed in Chapter 8. Both pictures are correct and present mutually complementary aspects of EM waves. Under certain circumstances, such as those involving refraction or interference of light, the wave nature of light is manifested. In other circumstances, including those involving the emission or absorption of light, the particle aspects of light are demonstrated. These two different sides of light are like the front and back sides of a person. Sometimes, we see only the front of an individual. We recognize the individual by facial structure, eye color, hair color, and so on. At other times, we see only the back of the person. Here again we may recognize the person, but now by virtue of body structure and gait. In each case, we see different aspects of the same person, but we are still able to recognize the person, although for the most part the clues leading to recognition are not the same in the two instances. The situation with light is analogous. Different experiments reveal different aspects of what we recognize to be the same type of EM radiation.

With Einstein's proposal, the observed aspects of the photoelectric effect came together. Higher-frequency light ejects electrons with more energy because each photon has more energy to give. Brighter light simply means that more photons strike the metal each second. This results in more electrons being ejected each second, but it does not increase the energy of each electron.* Einstein's explanation of the photoelectric effect earned him the 1921 Nobel Prize in physics. The new understanding of the nature of light and the way it interacts with matter profoundly altered the course of twentieth-century physics.

Just how much energy does a typical photon have? Not very much. The energy of a photon of visible light is only about 3×10^{-19} joules. On this minute scale, it is convenient to use a much smaller unit of energy, called the *electronvolt* (eV). One electronvolt is the potential energy of each electron in a 1-volt battery. Since voltage is energy per charge, the charge of an electron multiplied by voltage equals energy:

$$1 \text{ electronvolt} = 1.6 \times 10^{-19} \text{ coulomb} \times 1 \text{ volt}$$

$$= 1.6 \times 10^{-19} \text{ coulomb} \times 1 \text{ joule/coulomb}$$

$$1 \text{ eV} = 1.6 \times 10^{-19} \text{ J}$$

The energy of visible photons is on the order of 2 electronvolts, a much easier number to deal with. In terms of the electronvolt, the value of Planck's constant is

$$h = 6.63 \times 10^{-34} \text{ J-s} = 4.14 \times 10^{-15} \text{ eV/Hz}$$

* For each metal, there does exist a threshold (or cutoff) frequency below which no photoelectrons will be emitted, regardless of how intense the light is. Photons with frequencies less than this threshold possess too little energy to eject any electrons from the metal. The electrons are too tightly bound to the metal atoms, and the available energy from the photons is too small to overcome the binding energy

Figure 10.4

Photon energies in the EM spectrum. Visible light photons range from approximately 1.7 to 3.1 electronvolts (red light to violet light).

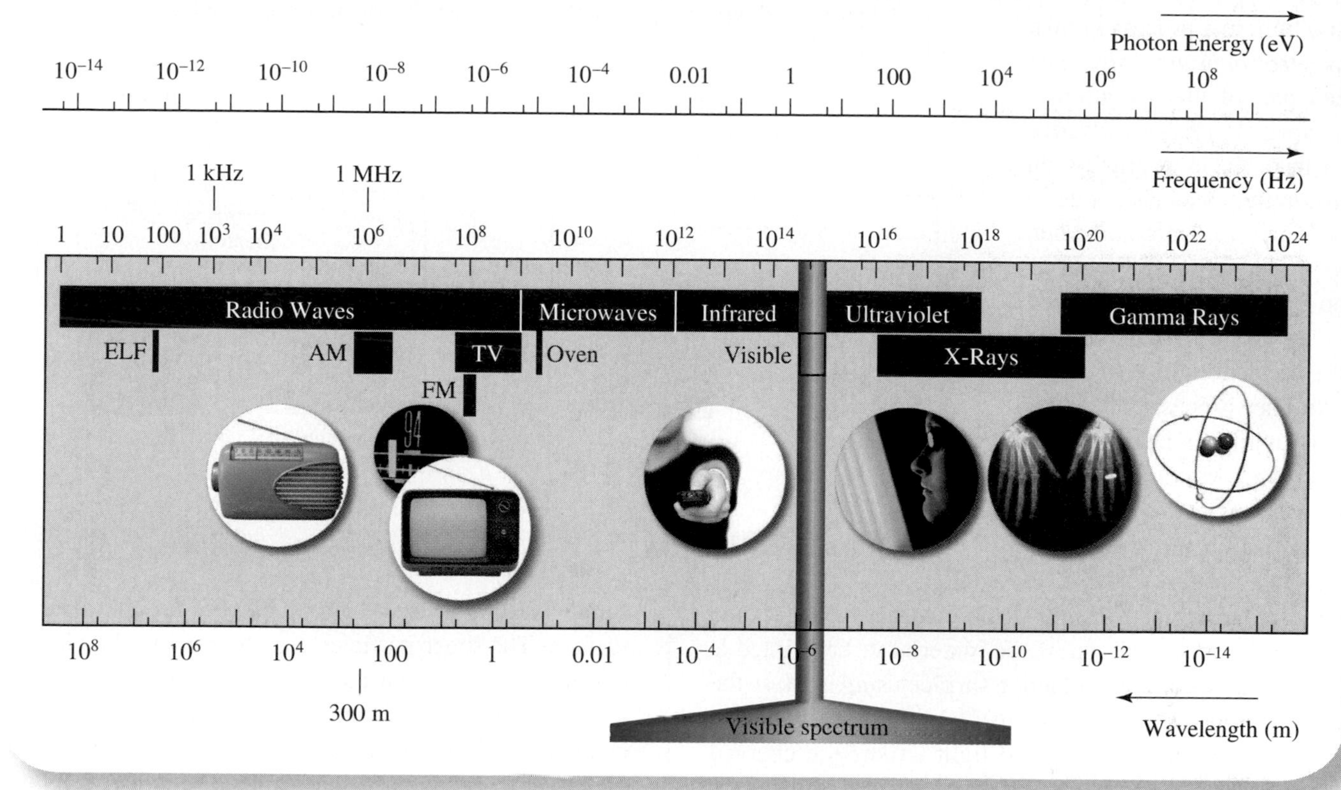

© James Alfred/iStockphoto.com / © charles taylor/iStockphoto.com / © iStockphoto.com / © iStockphoto.com / © Brand X Pictures/Jupiterimages / © Fabrizio Zanier/iStockphoto.com

Figure 10.4 shows the photon energies corresponding to different types of EM waves.

EXAMPLE 10.1

Compare the energies associated with a quantum of each of the following types of EM radiation.

$$\text{red light: } f = 4.3 \times 10^{14} \text{ Hz}$$

$$\text{blue light: } f = 6.3 \times 10^{14} \text{ Hz}$$

$$\text{x-ray: } f = 5 \times 10^{18} \text{ Hz}$$

For each one, we use Planck's original equation for a quantum of energy.

$$E = hf$$

$$= 4.136 \times 10^{-15} \text{ eV/Hz} \times 4.3 \times 10^{14} \text{ Hz}$$

$$= 1.78 \text{ eV} \qquad \text{(red light)}$$

Using the same equation for blue light and x-rays, we get:

$$E = 2.61 \text{ eV} \qquad \text{(blue light)}$$

$$E = 20{,}700 \text{ eV} \qquad \text{(x-ray)}$$

Notice how much greater the energy of a quantum of x radiation is compared to that of either red or blue light. Little wonder then that high doses of x-rays can be harmful to the human body.

Before we move on to the third puzzle that faced scientists in the early 1900s, let's look at some applications of the photoelectric effect. This phenomenon is the key to "interfacing" light with electricity. Just as the principles of electromagnetism make it possible to convert motion into electrical energy and vice versa, the ability of electrons to absorb the energy in photons makes it possible to detect, measure, and extract energy from light. Figure 10.5 shows a schematic of a device that can detect light. When no light strikes the metal, no current flows in the circuit because there is nothing to carry the charge through the tube. When light strikes the metal, electrons are ejected and attracted to the positive terminal of the tube. The result is a current flowing in the circuit. This sort of detector could be used to automatically count people entering a building. The detector is placed on one side of the door, and a light source is placed on the other side, pointed toward the detector. When something blocks the light going to the detector, the current stops. A counter connected to an ammeter in the circuit then automatically tallies one count. A similar setup could be used to automatically open and close a door.

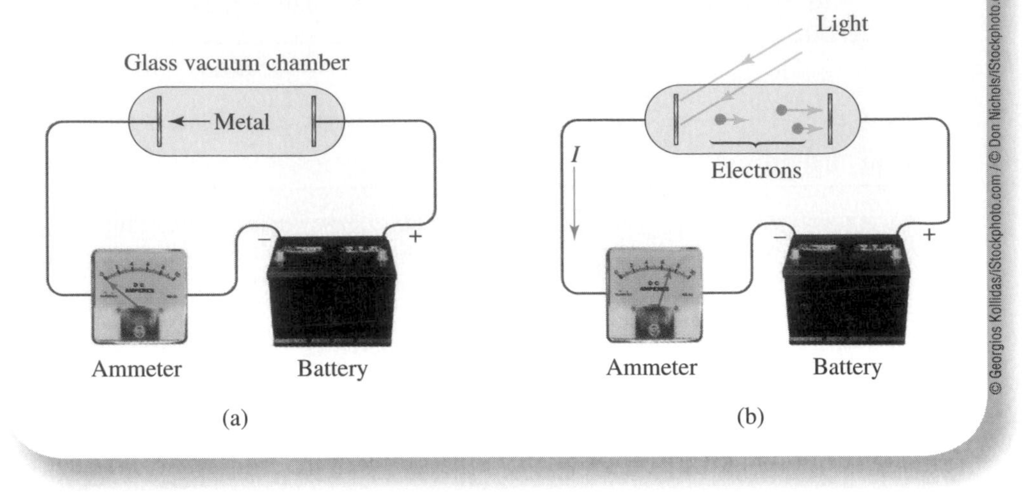

Figure 10.5
Schematic of a light detector. (a) No current flows in the circuit because there is nothing to carry charge through the tube. (b) Light releases electrons from the metal, allowing a current to flow. The size of the current indicates the brightness of the light.

© Georgios Kollidas/iStockphoto.com / © Don Nichols/iStockphoto.com

Photocopying machines and laser printers use the interplay between electricity and light in a process known as *electrophotography*. The key part of the operation is a special *photoconductive* surface. It is normally an insulating material, but it becomes a conductor when exposed to light. Electrons bound to atoms are freed when they absorb photons in the incident light.

The process of forming an image begins when the photoconductive surface is charged electrostatically. The surface retains the charge until light strikes it, and the freed electrons allow the charge to flow off the surface. A mirror image of the material to be printed is formed on the photoconductive surface using light. In the case of a photocopying machine, a bright light shines on the original, and the reflected light strikes the charged surface (Figure 10.6). White areas on the original reflect most of this light onto the corresponding areas of the photoconductive surface, which are consequently discharged by the large number of incident photons. Dark parts of the original—printed letters, for example—reflect very little light. Consequently, the corresponding regions on the photoconductive surface retain their electrical charge. Then fine particles of toner, somewhat like a solid form of ink, are brought near the surface. The toner particles are attracted to the charged regions and collect on them, while the discharged areas remain clear. A blank piece of paper, also electrically charged, is placed in contact with it. The toner particles are attracted to the paper and collect on it. The final image is "fused" on the paper by melting the toner particles into the paper.

In a laser printer, a laser under computer control illuminates and discharges those parts of the photoconductive surface that will not be dark in the image to be printed. The rest of the process parallels the steps in photocopiers.

Solar cells use the photoelectric effect to convert light into electricity.

Other specialized "photosensitive" materials, many of them semiconductors, have been devised for diverse uses. Light meters in cameras, the light-sensing elements in video and digital cameras that convert optical images into electrical signals, scanning elements in fax machines, sensitive photodetectors used by astronomers to measure the faint light from distant stars and galaxies, and solar cells used to convert energy in sunlight into electricity are just some of the devices that rely on extracting the energy in photons.

© Lena Andersson/iStockphoto.com

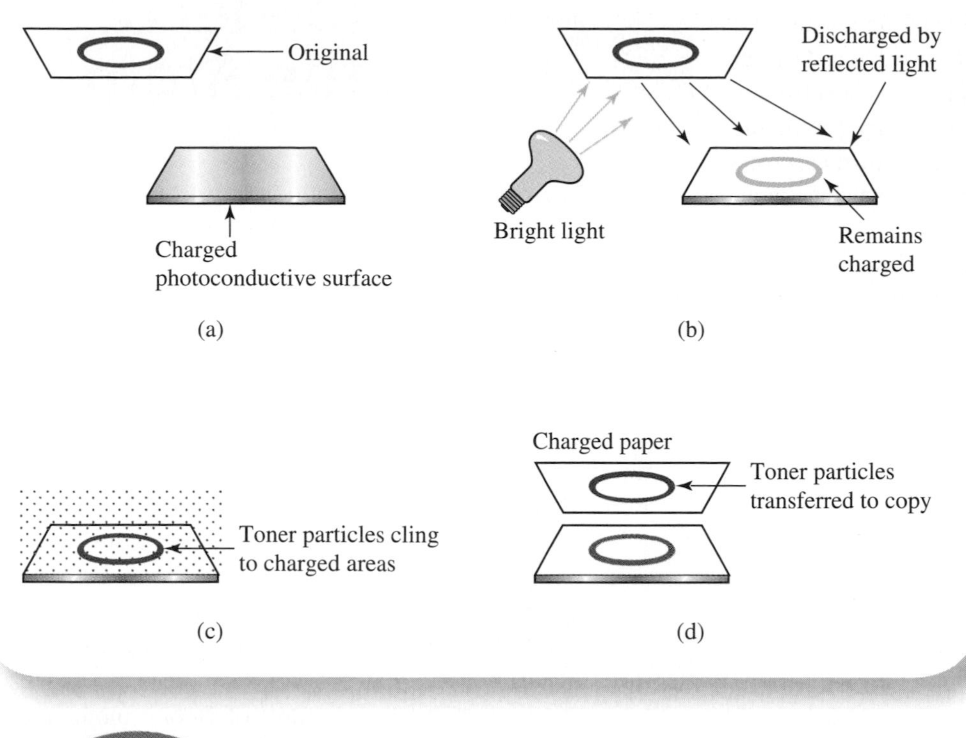

Figure 10.6
The main steps in the photocopying process.

10.3 Atomic Spectra

The third problem that defied explanation at the beginning of the twentieth century involved the spectra produced by the various chemical elements. Suppose we use a prism to examine the light produced by the heated filament of an ordinary incandescent bulb. The light from the bulb will be dispersed upon passing through the prism, as described in Chapter 9, and a spectrum will be produced (see Figure 9.51). The spectrum will appear as a continuous band of the colors of the rainbow, one color smoothly blending into the next. Such a spectrum is called a **continuous spectrum**, and it is characteristic of the radiation emitted by a hot, luminous solid.

Imagine now a sample of some gas confined in a glass tube and induced to emit light, as by heating (Figure 10.7). If we examine the light from the luminous gas with a prism, we do not see a continuous spectrum. Instead we see an **emission-line spectrum** consisting of a few, isolated, discrete lines of color. In this case, the source is not emitting radiation at all wavelengths (colors) but only at certain selected wavelengths. If a different type of gas is investigated, one finds that the resultant line spectrum is different. Each type of gas has its own unique set of spectral lines (Figure 10.8).

The fact that luminous, vaporized samples of material produce line spectra when their light is dispersed (as by a prism) was discovered in the 1850s. Once chemists recognized that each element has its own special spectrum, it became possible for them to determine the compositions of substances in the laboratory by examining their spectra. Robert Bunsen (of Bunsen burner fame) and Gustav Kirchhoff, two German scientists, were pioneers in this new field of *spectroscopy*, the study of spectra, during the last half of the nineteenth century. The problem that remained was to understand why

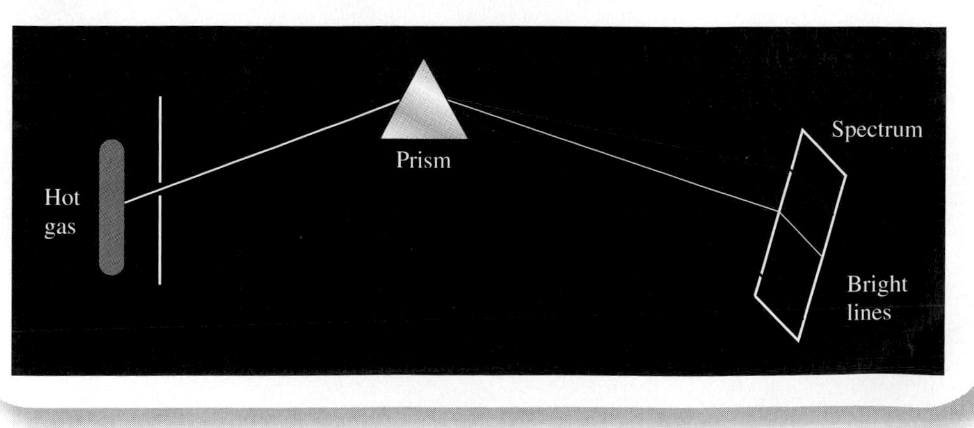

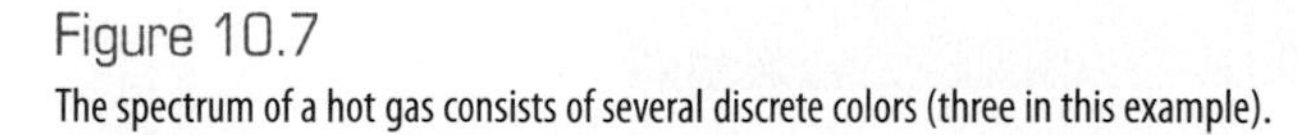
Figure 10.7
The spectrum of a hot gas consists of several discrete colors (three in this example).

luminous gases produce line spectra and not continuous spectra, and how it is that each element has its own unique spectral "fingerprint" by which it can be identified.

Spectroscopy has grown to be one of the most useful tools for chemical analysis in fields as diverse as law enforcement and astronomy. Suspected poisons or substances found at a crime scene can be identified by comparing their line spectra to those in a catalog of known elements and compounds. Astronomers can determine what chemicals exist in the atmospheres of stars by examining spectra of the starlight with the aid of telescopes.

The explanation of atomic spectra inspired one of the most crucial periods of advancement in physics ever. In the remainder of this chapter, we describe how a picture of the structure of the atom was created and refined to account for atomic spectra. This effort spawned a "new" physics for dealing with matter on the scale of atoms—quantum mechanics.

Figure 10.8
Emission spectra of selected elements.

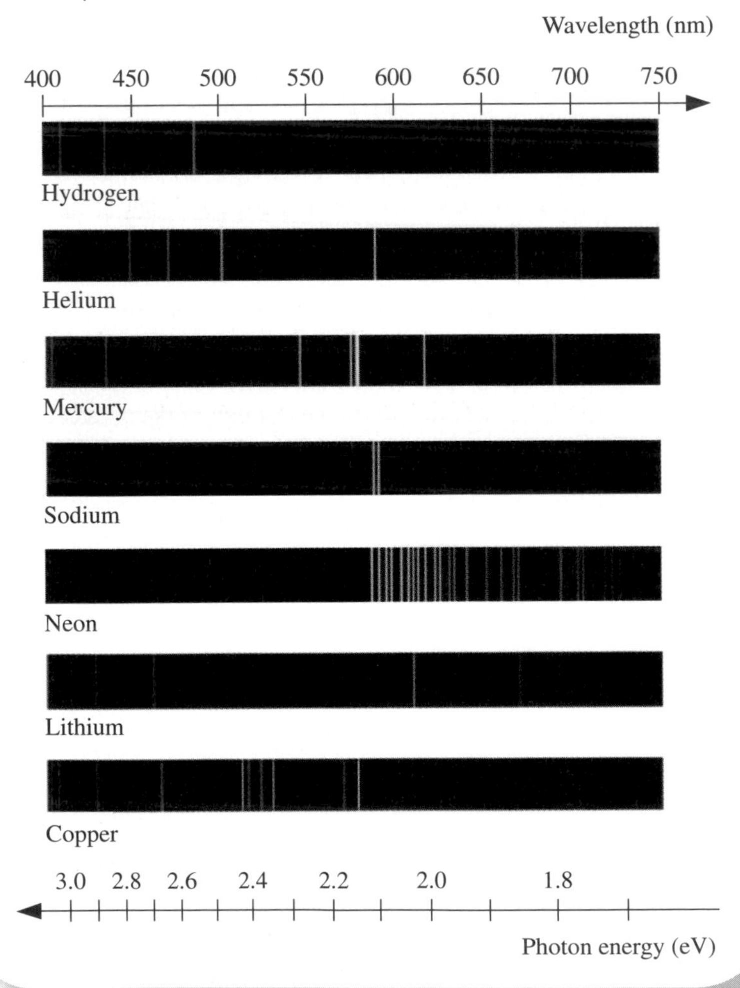

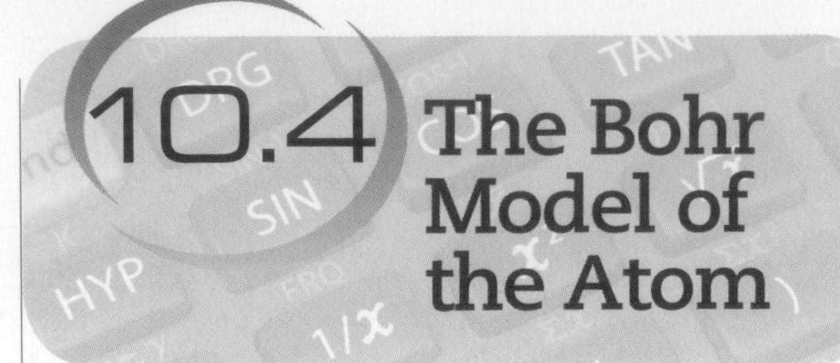

10.4 The Bohr Model of the Atom

At the beginning of the twentieth century, little was known about the atom. In fact, many doubted that atoms existed at all. This made the origin of atomic spectra a mystery, in spite of the fact that spectroscopy was a booming field. In 1911, an important experiment performed by Ernest Rutherford in England revealed that the positive charge in an atom is concentrated in a tiny core—the nucleus. This result was quickly followed in 1913 by a model of the atom put forth by the great Danish physicist Niels Bohr. **Bohr's model** of the atom, later modified, as we shall see in the next section, successfully explained the nature of atomic spectra. But like Planck's explanation of blackbody radiation, Bohr's model of the atom was based on assumptions that did not seem sensible at the time. The basic features of Bohr's model are the following:

1. The atom forms a miniature "solar system," with the nucleus at the center and the electrons moving about the nucleus in well-defined orbits. The nucleus plays the role of the Sun and the electrons are like planets.

© Science Source/Photo Researchers, Inc. / © iStockphoto.com

Figure 10.4

Photon energies in the EM spectrum. Visible light photons range from approximately 1.7 to 3.1 electronvolts (red light to violet light).

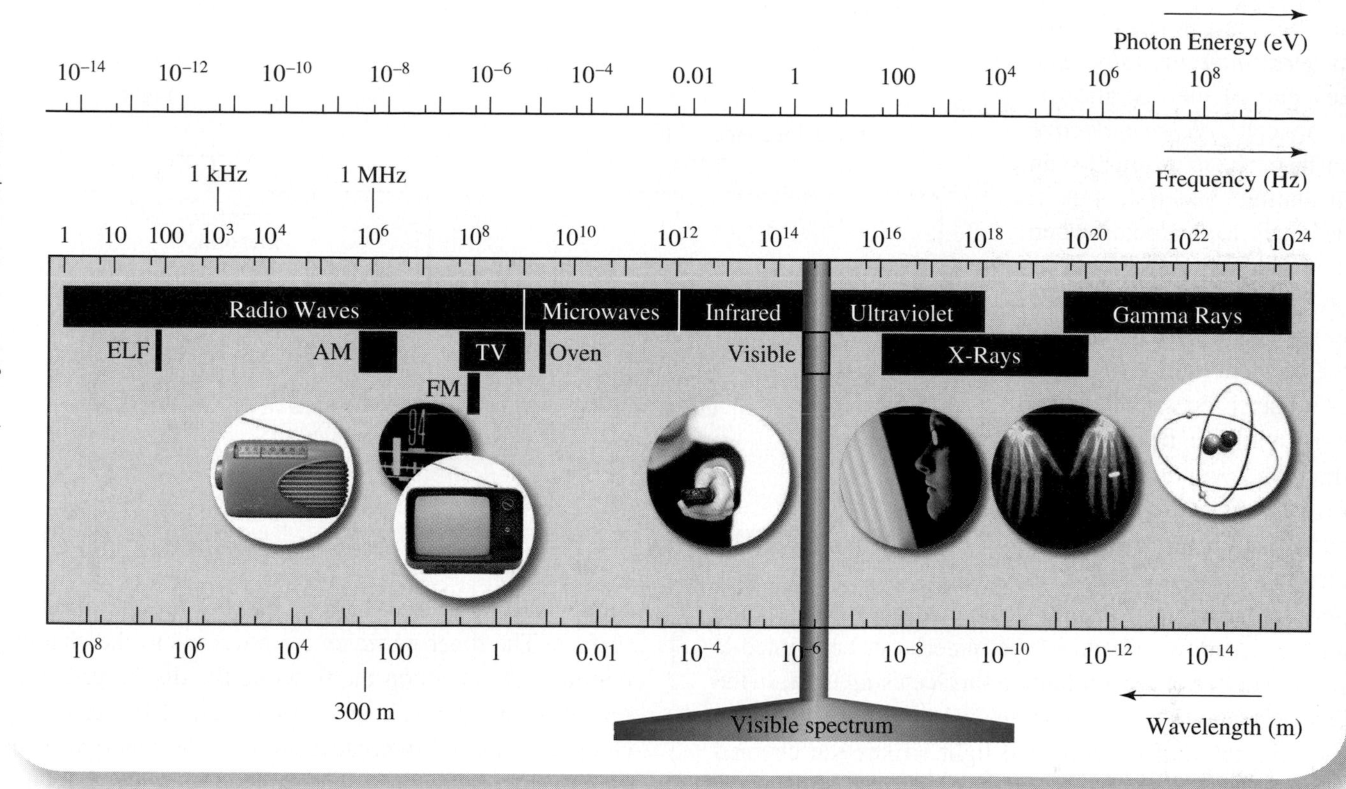

© James Alfred/iStockphoto.com / © charles taylor/iStockphoto.com / © iStockphoto.com / © iStockphoto.com / © Brand X Pictures/Jupiterimages / © Fabrizio Zanier/iStockphoto.com

Figure 10.4 shows the photon energies corresponding to different types of EM waves.

EXAMPLE 10.1

Compare the energies associated with a quantum of each of the following types of EM radiation.

$$\text{red light: } f = 4.3 \times 10^{14}\ \text{Hz}$$

$$\text{blue light: } f = 6.3 \times 10^{14}\ \text{Hz}$$

$$\text{x-ray: } f = 5 \times 10^{18}\ \text{Hz}$$

For each one, we use Planck's original equation for a quantum of energy.

$$E = hf$$

$$= 4.136 \times 10^{-15}\ \text{eV/Hz} \times 4.3 \times 10^{14}\ \text{Hz}$$

$$= 1.78\ \text{eV} \qquad \text{(red light)}$$

Using the same equation for blue light and x-rays, we get:

$$E = 2.61\ \text{eV} \qquad \text{(blue light)}$$

$$E = 20{,}700\ \text{eV} \qquad \text{(x-ray)}$$

Notice how much greater the energy of a quantum of x radiation is compared to that of either red or blue light. Little wonder then that high doses of x-rays can be harmful to the human body.

Before we move on to the third puzzle that faced scientists in the early 1900s, let's look at some applications of the photoelectric effect. This phenomenon is the key to "interfacing" light with electricity. Just as the principles of electromagnetism make it possible to convert motion into electrical energy and vice versa, the ability of electrons to absorb the energy in photons makes it possible to detect, measure, and extract energy from light. Figure 10.5 shows a schematic of a device that can detect light. When no light strikes the metal, no current flows in the circuit because there is nothing to carry the charge through the tube. When light strikes the metal, electrons are ejected and attracted to the positive terminal of the tube. The result is a current flowing in the circuit. This sort of detector could be used to automatically count people entering a building. The detector is placed on one side of the door, and a light source is placed on the other side, pointed toward the detector. When something blocks the light going to the detector, the current stops. A counter connected to an ammeter in the circuit then automatically tallies one count. A similar setup could be used to automatically open and close a door.

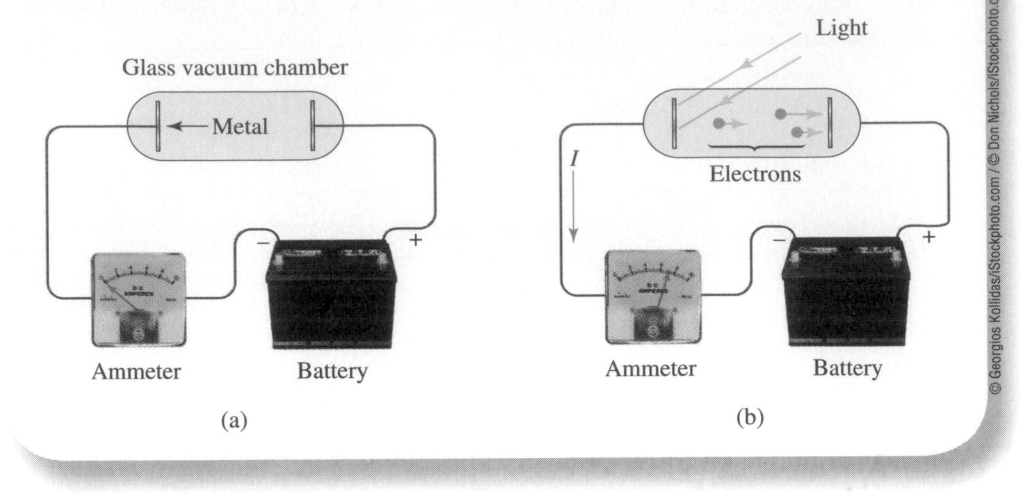

Figure 10.5

Schematic of a light detector. (a) No current flows in the circuit because there is nothing to carry charge through the tube. (b) Light releases electrons from the metal, allowing a current to flow. The size of the current indicates the brightness of the light.

© Georgios Kollidas/iStockphoto.com / © Don Nichols/iStockphoto.com

Photocopying machines and laser printers use the interplay between electricity and light in a process known as *electrophotography.* The key part of the operation is a special *photoconductive* surface. It is normally an insulating material, but it becomes a conductor when exposed to light. Electrons bound to atoms are freed when they absorb photons in the incident light.

The process of forming an image begins when the photoconductive surface is charged electrostatically. The surface retains the charge until light strikes it, and the freed electrons allow the charge to flow off the surface. A mirror image of the material to be printed is formed on the photoconductive surface using light. In the case of a photocopying machine, a bright light shines on the original, and the reflected light strikes the charged surface (Figure 10.6). White areas on the original reflect most of this light onto the corresponding areas of the photoconductive surface, which are consequently discharged by the large number of incident photons. Dark parts of the original—printed letters, for example—reflect very little light. Consequently, the corresponding regions on the photoconductive surface retain their electrical charge. Then fine particles of toner, somewhat like a solid form of ink, are brought near the surface. The toner particles are attracted to the charged regions and collect on them, while the discharged areas remain clear. A blank piece of paper, also electrically charged, is placed in contact with it. The toner particles are attracted to the paper and collect on it. The final image is "fused" on the paper by melting the toner particles into the paper.

Solar cells use the photoelectric effect to convert light into electricity.

© Lena Andersson/iStockphoto.com

In a laser printer, a laser under computer control illuminates and discharges those parts of the photoconductive surface that will not be dark in the image to be printed. The rest of the process parallels the steps in photocopiers.

Other specialized "photosensitive" materials, many of them semiconductors, have been devised for diverse uses. Light meters in cameras, the light-sensing elements in video and digital cameras that convert optical images into electrical signals, scanning elements in fax machines, sensitive photodetectors used by astronomers to measure the faint light from distant stars and galaxies, and solar cells used to convert energy in sunlight into electricity are just some of the devices that rely on extracting the energy in photons.

2. The electron orbits are quantized, that is, electrons can only be in certain orbits about a given atomic nucleus. Each allowed orbit has a particular energy associated with it, and the larger the orbit, the greater the energy. Electrons do not radiate energy (emit light) while in one of these stable orbits.

3. Electrons may "jump" from one allowed orbit to another. In going from a low-energy orbit to one of higher energy, the electron must *gain* an amount of energy equal to the difference in energy it has in the two orbits. When passing from a high-energy orbit to a lower-energy one, the electron must *lose* the corresponding amount of energy.

Figure 10.9 The Bohr model of the hydrogen atom. The electron can have only certain orbits. Four of these are shown. The figure is not drawn to scale: The fourth orbit is actually 16 times the size of the first orbit.

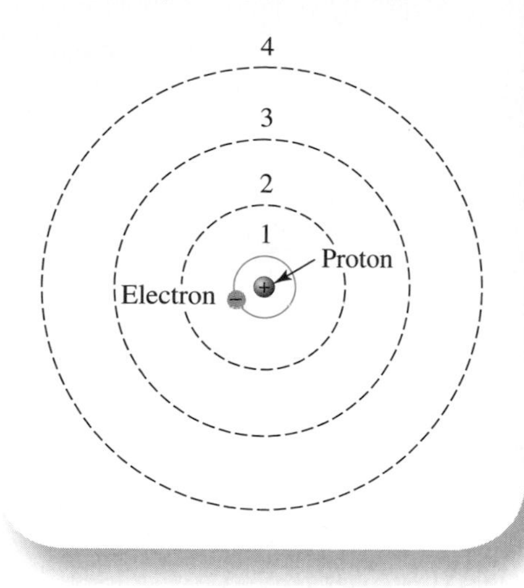

Figure 10.9 shows the Bohr model for the simplest atom, that of the element hydrogen (atomic number = 1). A lone electron orbits the nucleus, in this case a single proton. The electron can be in any one of a large number of orbits (four are shown). In each orbit, the electrical force of attraction between the oppositely charged electron and proton supplies the centripetal force needed to keep the electron in orbit.

When in orbit 1, the electron has the lowest possible energy. The electron has more energy in each successively larger orbit (Figure 10.10). The electron's energy while in any of the orbits is negative because it is bound to the nucleus (see the end of Section 3.5). To go to a larger orbit, the electron must gain energy. The maximum energy the electron can have and still remain bound to the proton is called the **ionization energy**. If the electron acquires more than this energy, it breaks free from the nucleus, and the atom is ionized. The resulting positive ion in this case is a bare proton.

How does the Bohr model account for the characteristic spectra emitted by luminous hydrogen gas? When an electron in a larger (higher-energy) orbit "jumps" to a smaller orbit, it loses energy. One of the ways it can lose this energy is by emitting light. This process is the origin of atomic spectra.

Imagine that the electron in our hydrogen atom is in the sixth allowed orbit, with its energy represented by E_6. Suppose the electron makes a transition (a jump) to an inner orbit, say, the second one, where its energy is E_2. To do so, the electron must lose an amount of energy, ΔE, equal to:

$$\Delta E = E_6 - E_2$$

In what is called a *radiative transition*, the electron loses this energy through the emission of a photon whose energy is just equal to ΔE (Figure 10.11). The photon, spontaneously created during the transition of the electron from orbit 6 to orbit 2, carries off the excess energy into space in the form of EM radiation. This is much like Planck's atomic oscillators emitting photons. Because the energy of a photon equals h times its frequency, we have:

$$\text{photon energy} = \Delta E = E_6 - E_2$$

And:

$$\text{photon energy} = hf$$

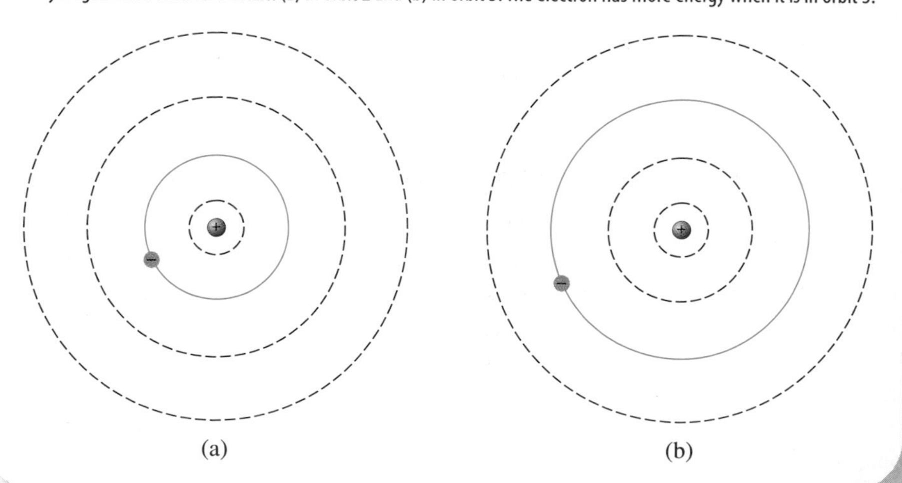

Figure 10.10 A hydrogen atom with its electron (a) in orbit 2 and (b) in orbit 3. The electron has more energy when it is in orbit 3.

Therefore:

$$hf = E_6 - E_2$$

The frequency of the emitted light is directly proportional to the difference in the energy of the orbits between which the electron jumped. The larger the energy difference, the higher the frequency of the light given off in the process. A downward transition from orbit 3 to orbit 2 will produce a photon with lower energy and lower frequency, because the energy difference between orbit 3 and orbit 2 is smaller.

$\Delta E = E_3 - E_2$ is smaller than

$\Delta E = E_6 - E_2$

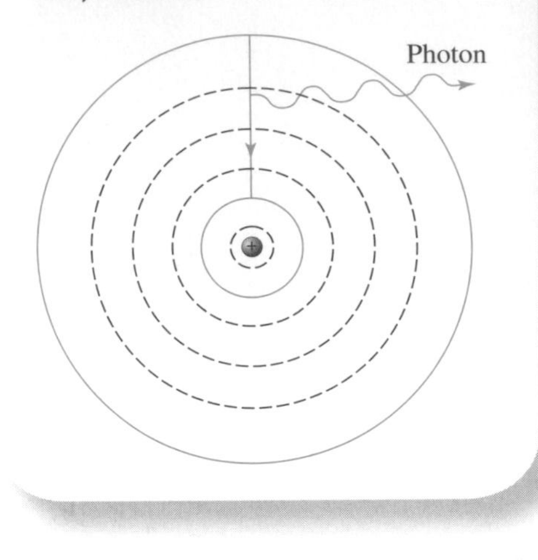

Figure 10.11
Photon emission. The electron makes a transition from orbit 6 to orbit 2 and emits a photon. The photon's energy equals the energy lost by the electron.

For the 6-to-2 transition in hydrogen, a violet-light photon is emitted. For the 3-to-2 transition, it is a red-light photon.

Each possible downward transition from an outer orbit to an inner orbit results in the emission of a photon with a particular frequency. An appropriately heated sample of hydrogen gas will emit light with these different frequencies but no other radiation. This is the line spectrum of hydrogen. (We will discuss this more in Section 10.6.)

A downward electron transition can also occur without the emission of light through what is called a *collisional transition.* In this case, a collision between a hydrogen atom and another particle (perhaps another hydrogen atom) can induce the electron in the outer orbit to spontaneously jump to an inner orbit. The energy that the electron loses can be transferred to the other particle, or it can be converted into increased kinetic energy of both colliding particles. (This is much like the collision described in Figure 3.19.) Collision-induced transitions are generally important in dense gases where the numbers of atoms or molecules per unit volume of gas are large and the likelihood of two or more gas particles colliding is therefore quite high.

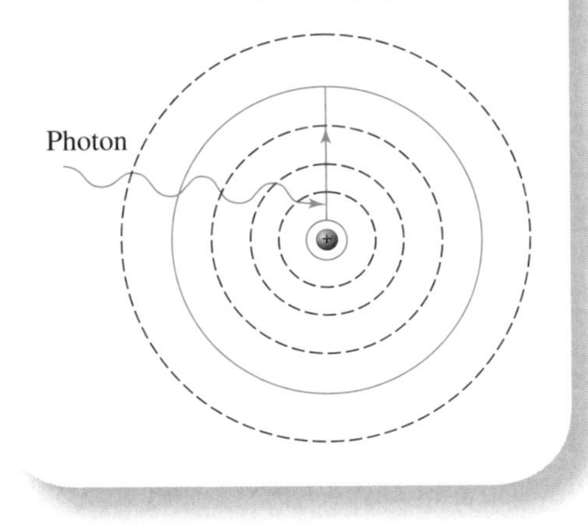

Figure 10.12
Photon absorption. The electron makes a transition from orbit 1 to orbit 5 by absorbing a photon.

So far, we have focused on how an electron in an outer orbit may lose energy by jumping to an inner orbit. But the reverse of this process also occurs: An electron in an inner orbit can gain just the right amount of energy and jump to an outer orbit. For example, a hydrogen atom in the lowest energy state might gain the amount of energy needed for its electron to jump from orbit 1 to, say, orbit 5. To do this, the electron would have to acquire energy:

$$\Delta E = E_5 - E_1$$

One way to do this would be through a collision: the hydrogen atom could collide with another atom with more energy.

Another way an electron can jump to an outer orbit is by absorbing a photon with the proper energy (Figure 10.12). For example, an electron in orbit 1 can jump to orbit 5 if it absorbs a photon with energy:

$$\text{photon energy} = \Delta E = E_5 - E_1$$

If a sample of hydrogen gas is irradiated with a broad band of EM waves (like blackbody radiation), many such transitions to outer orbits will occur. Some of the photons in the incident radiation will have just the right energies to induce transitions from inner orbits to outer orbits. In the process, the number of photons with these particular energies will be reduced, and the intensity of the EM radiation at the corresponding frequencies will decrease. This reduction of intensity of light at only certain frequencies after it passes through a gas results in an **absorption spectrum** (Figure 10.13). For example, if white light is passed through hydrogen gas and then dispersed with a prism, dark bands appear at certain frequencies. They are exactly the same frequencies that are in the emission spectrum of luminous hydrogen.

This is how the element helium was discovered. In 1868, some of the absorption lines in the spectrum of sunlight were found not to correspond with those of any elements known at that time. The existence of a new element in the Sun's atmosphere was suggested to account for these lines. This new element was named after the Greek word for Sun (*helios*).

The Bohr model was successful at explaining the origin of atomic spectra. But Bohr's model of electron orbits rested on two unexplained assumptions. First, there are only certain allowed orbits. While one can place a satellite in orbit about the Earth with any radius, the electron orbits were restricted to specific radii. These radii were determined by a seemingly arbitrary but nonetheless effective condition: that the angular momentum of the electron in its orbit is quantized. Much like the energy of Planck's quantized atomic oscillators, the orbital angular momentum of Bohr's electrons could only be integer multiples of Planck's constant divided by 2π. That is:

Refer to the discussion on angular momentum in Section 3.8.

© Вячеслав Крисанов/iStockphoto.com

$$\text{allowed angular momentum} = n\left(\frac{h}{2\pi}\right),$$
$$\text{with } n = 1, 2, 3, \ldots$$

So quantization was popping up everywhere: in the energy of oscillating atoms, the energy of EM waves, and now in the orbital angular momentum of atomic electrons.

The second assumption that physicists of the time found objectionable was that as long as an electron remained in one of its allowed orbits, it did not emit EM radiation. Maxwell's work had indicated that whenever a charged object undergoes acceleration, including centripetal acceleration, it will radiate. Put another way, an electron in a periodic orbit is much like a charge oscillating back and forth, and the latter results in the production of an EM wave. An orbiting electron should be continually radiating, and in the process it should lose energy and spiral into the nucleus. According to the physical laws known at the time, atoms could not remain stable and should collapse in a fraction of a second!

Even though Bohr had a model that worked, clearly the physics behind it was not fully elucidated. What was needed was a revolution in our understanding of how Nature works at the atomic level. The Concept Map on your Review Card summarizes the Bohr Model of the atom.

10.5 Quantum Mechanics

The success of Bohr's model of the atom, even though it was at odds with accepted principles of physics at the time, indicated that perhaps a new physics was needed to describe what goes on at the atomic level. The first step in this direction came during the summer of 1923. While working on his doctorate in physics, a French aristocrat named Louis Victor de Broglie (rhymes with "Troy") proposed that electrons and other particles possess wavelike properties. Einstein had shown that light has both wavelike and particlelike properties, so why not electrons too? The wave associated with any moving particle has a specific wavelength (called the **de Broglie wavelength**) that depends on the particle's momentum, mv:

$$\lambda = \frac{h}{mv} \qquad \text{(de Broglie wavelength)}$$

The higher the momentum of a particle, the shorter its wavelength. High-speed electrons have shorter wavelengths than low-speed electrons.

Figure 10.13

Absorption spectrum of a gas. The gas absorbs photons in the light passing through it, so at those frequencies (colors) the spectrum is fainter.

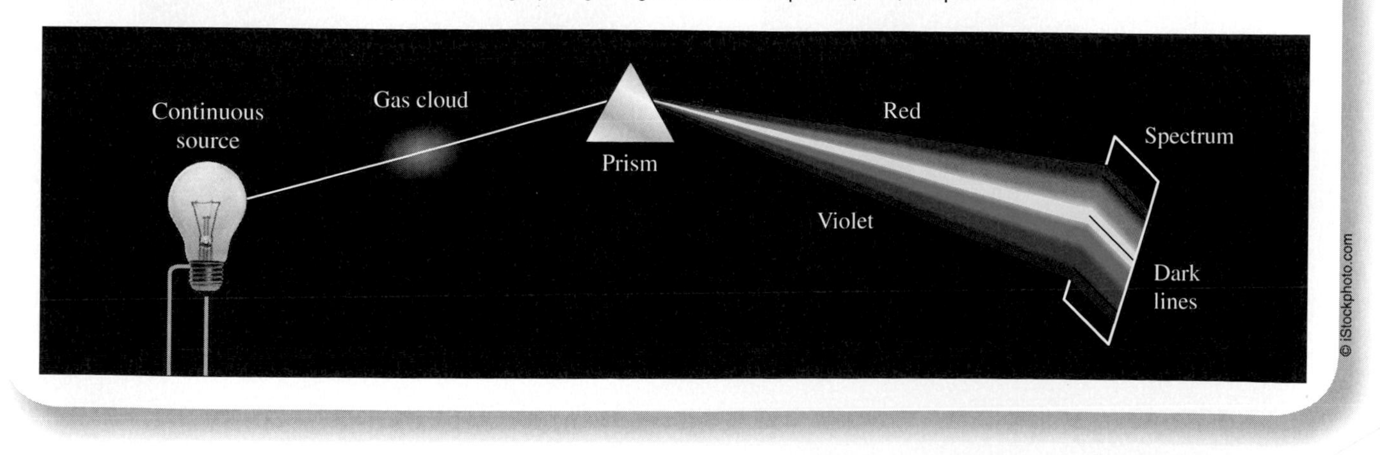

© iStockphoto.com

Once again, Planck's constant shows up. Since h is such a tiny number, de Broglie wavelengths are extremely small. This means that the wave properties of particles are only manifested at the atomic and subatomic level.

EXAMPLE 10.2

➔ ***What is the de Broglie wavelength of an electron with speed 2.19×10^6 m/s? (This is the approximate speed of an electron in the smallest orbit in hydrogen.)***

Using the electron mass given on the periodic table tear-out card, the electron's momentum is

$$mv = 9.11 \times 10^{-31} \text{ kg} \times 2.19 \times 10^6 \text{ m/s}$$
$$= 1.99 \times 10^{-24} \text{ kg-m/s}$$

Using the value of h in SI units (Section 10.1):

$$\lambda = \frac{h}{mv} = \frac{6.63 \times 10^{-34} \text{ J-s}}{1.99 \times 10^{-24} \text{ kg-m/s}}$$
$$= 3.32 \times 10^{-10} \text{ m} = 0.332 \text{ nm}$$

This distance is in the same range as the diameters of atoms.

Another radical theory had entered the arena of physics. Although submitted by a newcomer, de Broglie's hypothesis was not completely rejected by the physics community, because Einstein himself found it plausible. Then, in 1925, the puzzling results of some experiments were interpreted as proof of the existence of de Broglie's matter waves. In a series of experiments, American physicist Clinton Davisson (with various collaborators) showed that a beam of high-speed electrons underwent diffraction when sent into a nickel target. The electrons were behaving just like waves. In fact, one gets the same kind of scattering pattern using x-rays or electrons (see Figure 10.14).

Other experiments have verified the wave properties of electrons, protons, and other particles. Young's classic two-slit experiment, by which he proved that light is a wave, can be used to make particles undergo interference (Figure 10.15). The *electron microscope* exploits the wave properties of electrons. Instead of using ordinary light like a conventional microscope, an electron microscope uses a beam of electrons that acts like a beam of electron waves. The magnification can be much higher with an electron microscope because the de Broglie wavelength of the electrons is much shorter than the wavelengths of visible light (Figure 10.16). Diffraction and interference of waves passing around and between small objects such as cells strongly affect image clarity. Shorter-wavelength waves (electrons) diffract and interfere much less so than do longer-wavelength waves (light).

A more recently developed microscope uses the wavelike nature of electrons to form tantalizing images of individual atoms on the surfaces of solids. The *scanning tunneling microscope* (STM) uses an extremely fine

Figure 10.14

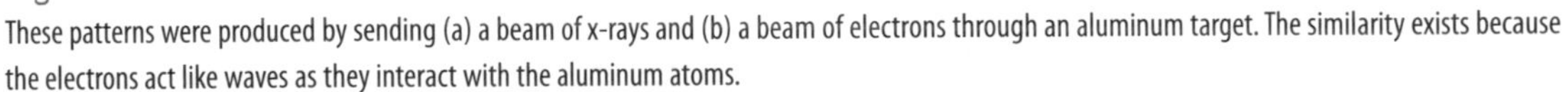

These patterns were produced by sending (a) a beam of x-rays and (b) a beam of electrons through an aluminum target. The similarity exists because the electrons act like waves as they interact with the aluminum atoms.

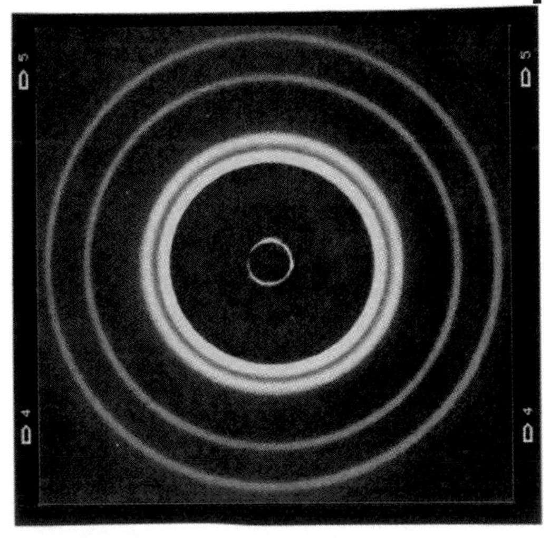

(a)

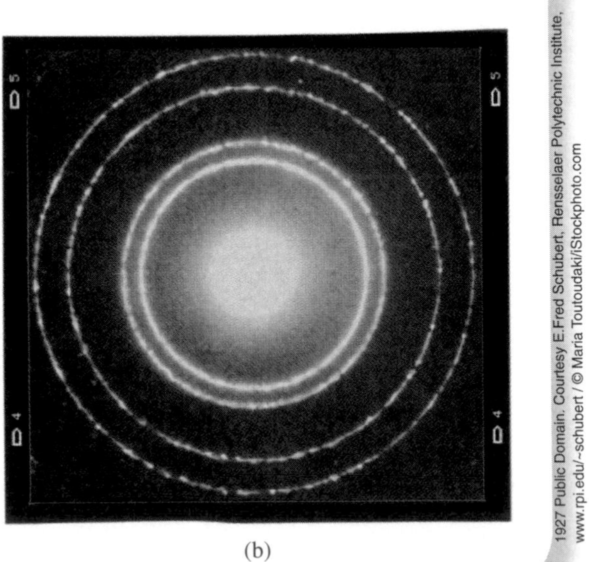

(b)

1927 Public Domain. Courtesy E.Fred Schubert, Rensselaer Polytechnic Institute, www.rpi.edu/~schubert / © Maria Toutoudaki/iStockphoto.com

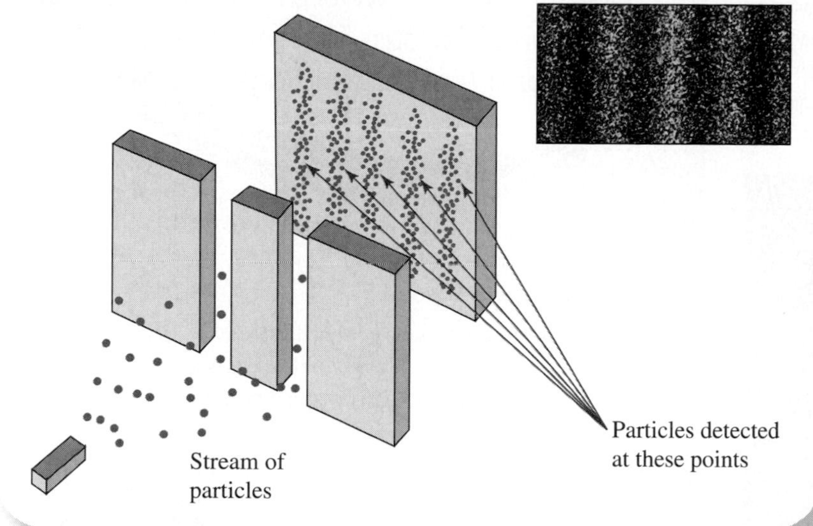

Figure 10.15
Particles, like electrons, undergo interference when passed through narrow slits. They exhibit a property of waves (compare to Figure 9.6).

© 2009, National Academy of Sciences, U.S.A.

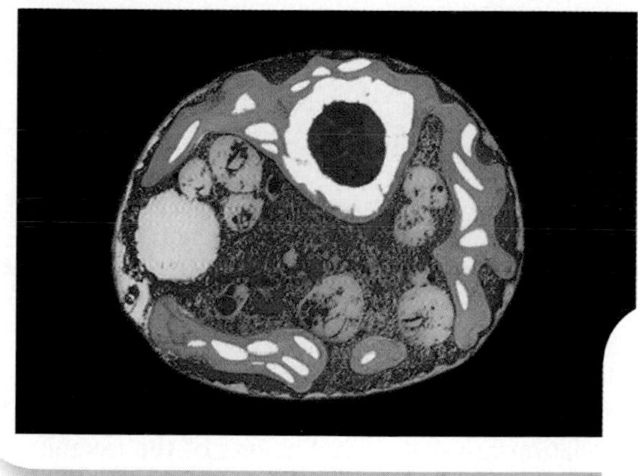

Figure 10.16
A false-color image of an alga cell taken with an electron microscope.

© Jeremy Burgess/Photo Researchers

pointed needle that scans back and forth over a surface. A positive voltage maintained on the needle attracts the electrons at the surface of the sample. If electrons were simply particles, they could not traverse the gap—less than 1 nanometer wide—between the surface and the needle. But their wave nature allows the electrons to penetrate or to "tunnel" through the gap and to reach the needle (Figure 10.17). The tiny current of electrons that flows decreases rapidly if the gap widens. As the needle scans back and forth over the surface (like someone mowing a lawn), a feedback mechanism moves the needle up and down to keep this tunneling current constant. The varying height of the needle is recorded and used to draw a contour map of the surface that clearly shows individual molecules and atoms (Figure 10.18 and Figure 4.5c). The 1986 Nobel Prize in physics was shared by the inventor of the electron microscope and the codevelopers of the STM.

Another major triumph for de Broglie's wave hypothesis was its ability to explain Bohr's quantized orbits. De Broglie reasoned that since an orbiting electron acts like a wave, its wavelength has to affect the circumference of its orbit. In particular, the electron's wave wraps around on itself, and in doing so it must interfere with itself. Only if the wave interferes constructively (peak matches peak) can the electron's orbit remain stable (Figure 10.19). This means that the circumference of its orbit must equal exactly one de Broglie wavelength, or two, or three, and so on.

$$\text{circumference of orbit} = n\lambda, \text{ with } n = 1, 2, 3, \ldots$$

If r is the radius of a circular orbit, then the circumference is $2\pi r$. Thus, the radius is

$$r = n\left(\frac{\lambda}{2\pi}\right), \text{ with } n = 1, 2, 3, \ldots$$

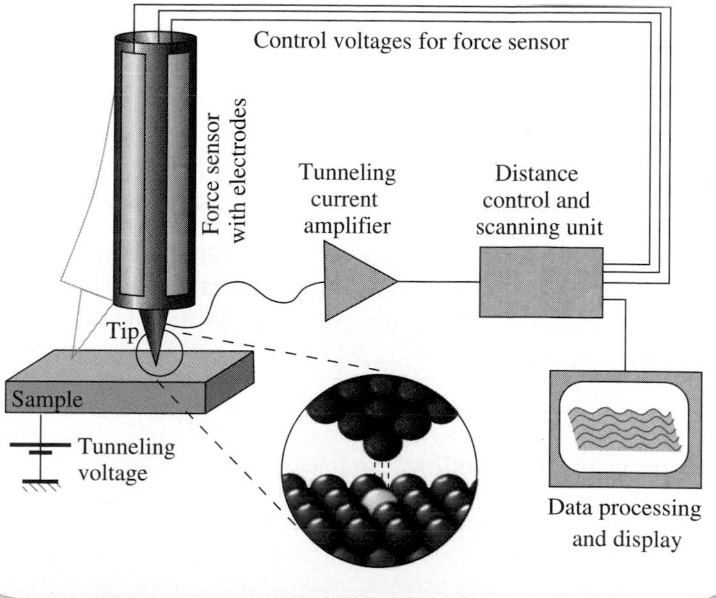

Figure 10.17
Simplified sketch of a scanning tunneling microscope (STM). The wavelike nature of the electrons allows them to cross the gap from the surface to the needle.

Figure 10.18
False-color scanning tunneling micrograph (STM) of DNA. A sample of uncoated, double-stranded DNA was dissolved in a salt solution and deposited on graphite prior to being imaged in air by the STM. An STM image is formed by scanning a fine point just above the specimen surface and electronically recording the height of the point as it moves. This image is dominated by the row of orange and yellow peaks that represent the ridges of the double helix in the short section of this right-handed DNA molecule.

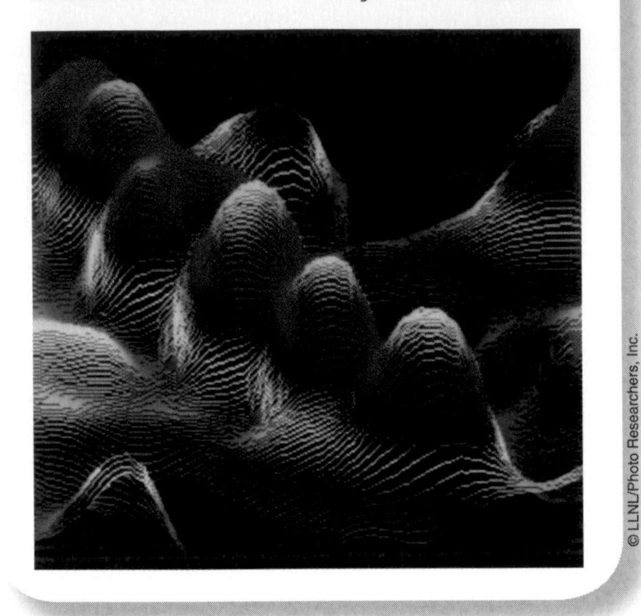

© LLNL/Photo Researchers, Inc.

EXAMPLE 10.3

Using the results of Example 10.2, find the radius of the hydrogen atom.

In de Broglie's model, the circumference of the smallest orbit ($n = 1$) must equal the de Broglie wavelength of the electron. We calculated this wavelength to be 0.332 nanometers for the smallest orbit. So:

$$r = \frac{\lambda}{2\pi}$$

$$r = \frac{0.332 \text{ nm}}{2\pi} = \frac{0.332 \text{ nm}}{2 \times 3.14} = \frac{0.332 \text{ nm}}{6.28}$$

$$= 0.0529 \text{ nm}$$

The allowed values for the circumference (or radius) of an electron's orbit leads to Bohr's allowed values for the electron's angular momentum. Since the de Broglie wavelength is given by

$$\lambda = \frac{h}{mv}$$

the radius can have the following values.

$$r = \frac{n}{2\pi}\left(\frac{h}{mv}\right), \text{ with } n = 1, 2, 3, \ldots$$

Want to see more STM images? Go to http://www.almaden.ibm.com/vis/stm/gallery.html for dozens of other images.

Figure 10.19
Simplified wave representation of an electron orbiting a nucleus. The orbit in (a) is not possible because the wave interferes destructively with itself. The orbit in (b) is allowed. This corresponds to orbit 5, since it has 5 wavelengths fitted into the orbit.

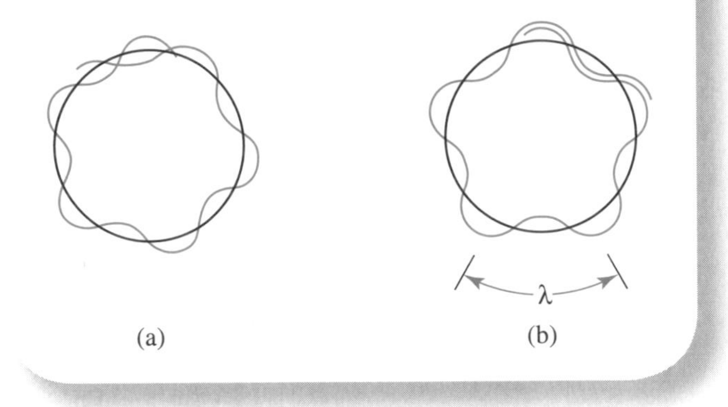

When we multiply both sides by mv, we get:

$$mvr = n\left(\frac{h}{2\pi}\right), \text{ with } n = 1, 2, 3, \ldots$$

But, the quantity mvr is the angular momentum of the electron (see Section 3.8), and we have recovered the same relationship that served as one of the assumptions in Bohr's theory of the atom at the end of Section 10.4. De Broglie's condition on the size of the electron's orbit turns out to be identical to Bohr's condition on the angular momentum of the electron.

The success of the concepts of quantization and wave-particle duality at the atomic level could not be overlooked. In the later half of the 1920s, a flurry of activity resulted in a formal mathematical model that incorporated these ideas—**quantum mechanics**. The two principal founders were Werner Heisenberg and Erwin Schrödinger.

One of the main contributions of Heisenberg was the **uncertainty principle**. Because electrons and other particles on the atomic scale have wavelike properties, we can no longer think of them as being like very tiny marbles or ball bearings. They are a bit spread out in space, more like tiny, fuzzy cotton balls. In the old particle model of the electron, it was possible, at least in theory, to state exactly where an electron is and exactly what its momentum is at any instant in time. Heisenberg stated that the wave nature of particles makes this impossible. One cannot specify both the position *and* the momentum of an electron to arbitrarily high precision. The more precisely you know the position of the electron, the less precisely you can determine its momentum and vice versa. In other words, you are constrained by the uncertainty principle.

PRINCIPLES

If we let Δx represent the uncertainty in the position of a particle, and Δmv represent the uncertainty in the momentum of the particle, then:

$$\Delta x \, \Delta mv \gtrsim h$$

(uncertainty principle)

The product of the two uncertainties is greater than or equal to Planck's constant. No matter how good the experimental apparatus, the best precision we can achieve in the measurement of a particle's position and momentum is limited by this principle. On the atomic scale, particles cannot be localized in the same way that they can be on a large scale.

Schrödinger established a mathematical model for the waves associated with particles. In his model, a quantum system like a hydrogen atom can be described by a *wave function.* The wave function contains all the information needed to predict the characteristics and future evolution of the system once the nature of the interactions in which the system is involved are known. For example, knowledge of the wave function allows one to calculate the most probable position of the electron with respect to the nucleus, as well as the electron's average energy. Although often difficult to determine in practice, the wave function of a system is the key ingredient to understanding its behavior on an atomic scale.

During the early decades of the twentieth century, the view of what matter is like at the submicroscopic level changed dramatically. Electromagnetic waves have particlelike properties, and electrons and other particles have wavelike properties. Clearly, Nature is very different at this level compared to what we commonly experience on a macroscopic scale.

10.6 Atomic Structure

The findings of de Broglie, Heisenberg, and Schrödinger force us to revise the simple Bohr model of the atom with its planetary structure. We can no longer represent electrons as little particles with well-defined boundaries moving in nice circular orbits. In reality, each electron is described by a wave that determines the probability of finding it at different locations. The atom is pictured as a tiny nucleus surrounded by an "electron cloud" (Figure 10.20). The density of the cloud at each point in space indicates the likelihood of finding the electron there. The different allowed orbits of the electrons appear as clouds with different sizes and shapes. So the simple drawings of atoms that we used earlier in this text (such as Figure 7.1) are not strictly correct. They did, however, serve the purpose of presenting the basic structure of the atom without complicating the picture with the wave nature of the electrons.

Because we can't say exactly where the bound electron is with respect to the nucleus at any given moment, it is more useful to concentrate on the electron's energy which can be precisely predicted. After all, when determining the frequency of radiation emitted or absorbed by an atom, the important quantity is the difference between the electron's initial and final energy. In the new model, we describe the electrons as being in certain allowed energy states or **energy levels**. For the hydrogen atom, the lowest energy level (corresponding to the innermost orbit in the Bohr theory) is called the *ground state*. The higher energy states are referred to as *excited states*. We now represent the structure of the atom schematically using an *energy-level diagram*, like the one for hydrogen shown in Figure 10.21. Each energy level is labeled with a **quantum number**, n, beginning with the ground state having $n = 1$ and continuing on up. As the quantum number increases, so does the energy associated with the state. Moreover, as n gets larger, the difference in energy between the adjacent states becomes smaller. The difference in energy between the $n = 4$ and the $n = 3$ states is smaller

Figure 10.20

Computer-generated plots of the "electron clouds" for four atomic orbits. (a) $n = 1$ (ground state) probability density. (b)–(d) Three of the four probability distributions for $n = 2$ (first excited state; see the discussion of the Pauli exclusion principle on page 284). The probability of finding the electron at a given location is specified by the color of the display: Red represents the highest probability, followed by yellow, green, light blue, and finally dark blue. The cloud sizes show that the electron has a greater chance of being found farther from the nucleus (the center of the distributions) in the $n = 2$ orbits than in the $n = 1$ orbit. This is consistent with the Bohr model of the atom. (The scale of these images has been magnified about 50 million times.)

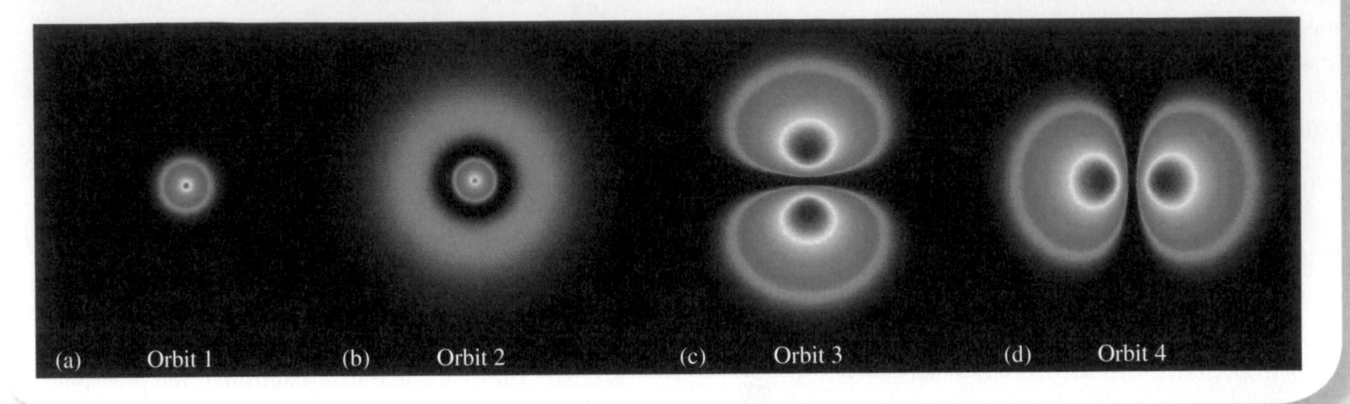

than the difference in energy between the $n = 3$ and the $n = 2$ states. Finally, we again note the existence of a maximum allowed energy above which the electron is no longer bound to the nucleus. The state is designated $n = \infty$, and its energy is the ionization energy.

The numbers on the left in the energy-level diagram are the electron energies for each state. (The negative values indicate that the electron is bound to the nucleus, as we mentioned in Section 10.4.) The transition of an electron from one orbit to another corresponds to the atom going from one energy level to another. The change in energy of the electron as a result of the "energy-level transition" is found by comparing the energies of the two states.

Using the energy-level diagram, we can give a more complete picture of the emission spectrum of hydrogen. Suppose an atom in the $n = 2$ state undergoes a transition to the $n = 1$ state by emitting a photon. (Typically, an atom will remain in an excited state for only about a billionth of a second.) The transition is represented by an arrow drawn from the initial level to the final level (Figure 10.22). The photon that is emitted has an energy equal to the difference in energy between the two levels. Since electron transitions may occur only from one allowed energy state to another, the arrows representing such transitions *must* begin and end on allowed levels within the energy-level diagram. In this case:

$$\begin{aligned}\text{photon energy} &= \Delta E = E_2 - E_1 \\ &= -3.4\text{ eV} - (-13.6\text{ eV}) \\ &= 10.2\text{ eV}\end{aligned}$$

This is a photon of UV light (refer to Figure 10.4).

Similarly, we can represent all possible downward transitions from higher energy levels to lower energy levels. Figure 10.23 is an enlarged energy-level diagram for hydrogen showing various possible energy-level transitions from higher levels to lower ones. The number with each arrow is the wavelength (in nanometers) of the photon that is emitted.

Figure 10.21

Energy-level diagram for hydrogen. Each level corresponds to one of the allowed electron orbits.

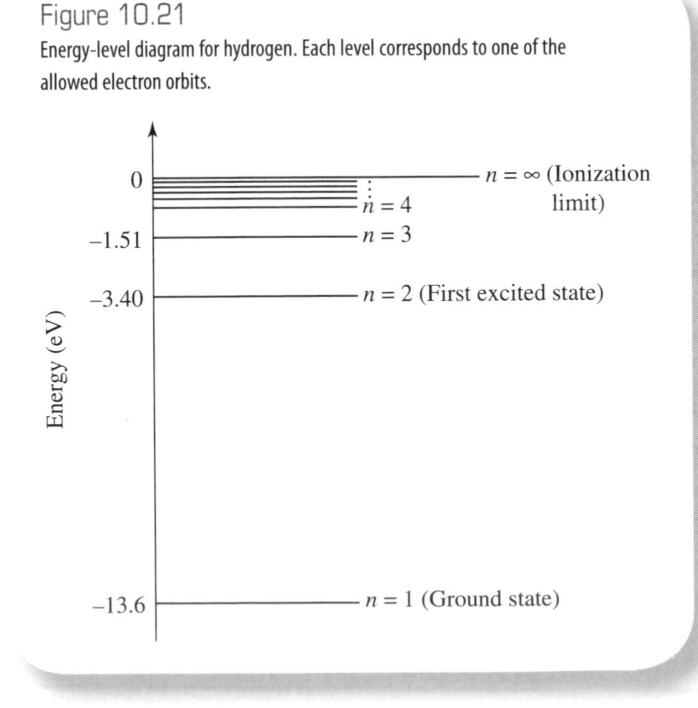

EXAMPLE 10.4

Find the frequency and wavelength of the photon emitted when a hydrogen atom goes from the n = 3 state to the n = 2 state.

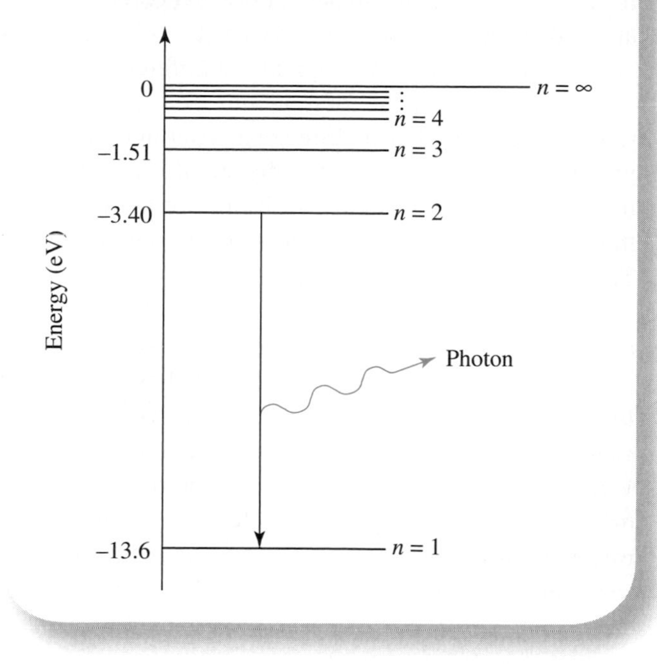

Figure 10.22

Downward electronic transition from level $n = 2$ to level $n = 1$ in hydrogen. The transition is accompanied by the emission of a photon with energy 10.2 electronvolts.

We first find the energy of the photon, then use that to determine its frequency. The wavelength we get from the equation $c = f\lambda$.

$$\text{photon energy} = hf = E_3 - E_2$$
$$= -1.51 \text{ eV} - (-3.4 \text{ eV})$$
$$= 1.89 \text{ eV}$$

Therefore:

$$f = \frac{1.89 \text{ eV}}{h} = \frac{1.89 \text{ eV}}{4.136 \times 10^{-15} \text{ eV/Hz}}$$
$$= 4.57 \times 10^{14} \text{ Hz}$$

For the wavelength:

$$\lambda = \frac{c}{f} = \frac{3 \times 10^{8} \text{ m/s}}{4.57 \times 10^{14} \text{Hz}}$$
$$= 6.56 \times 10^{-7} \text{ m} = 656 \text{ nm}$$

From Figure 10.4, we see that this is a photon of visible light. (To be more precise, Table 8.1 shows that it is red light.)

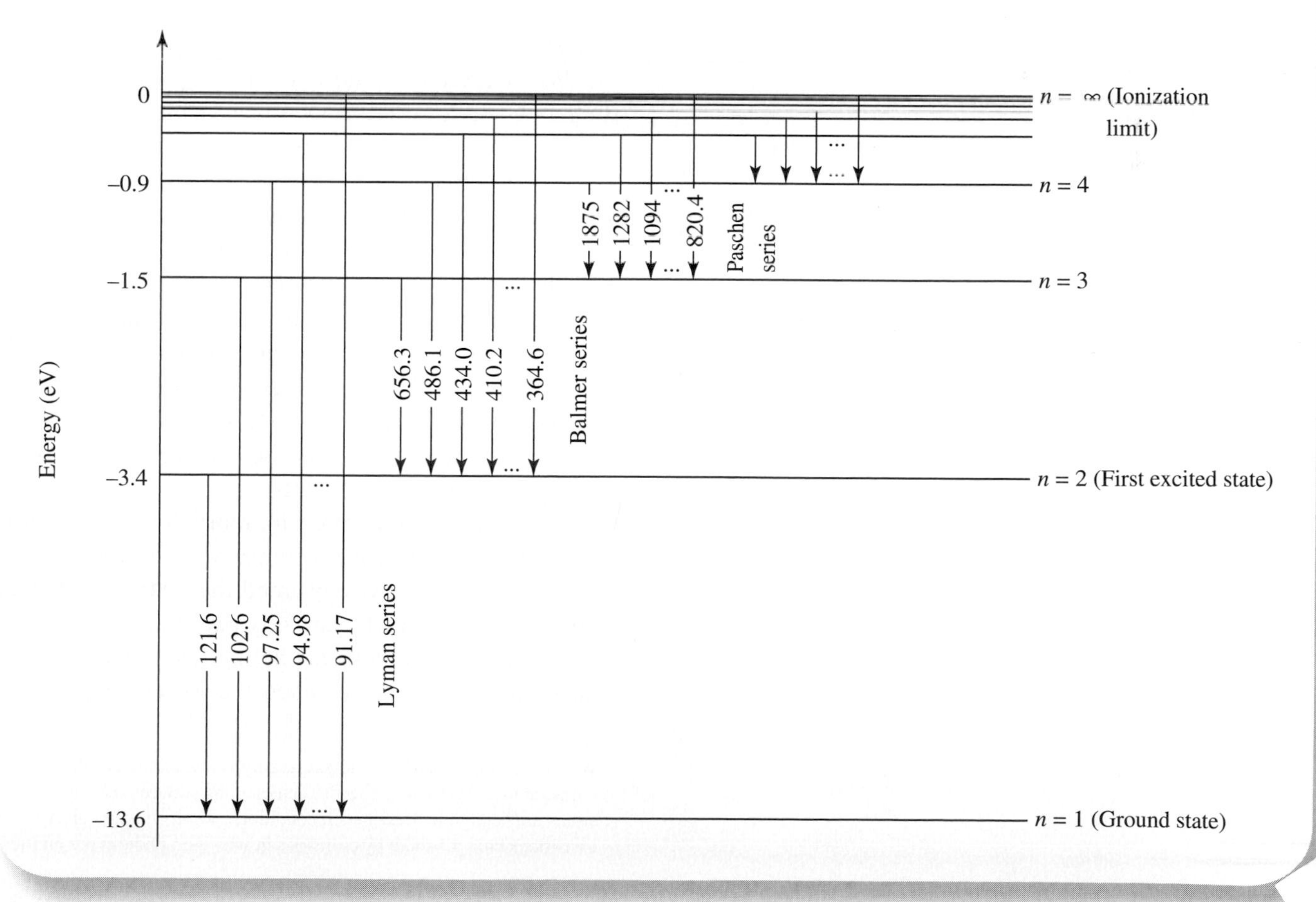

Figure 10.23

Energy-level diagram for hydrogen showing different possible energy-level transitions. The number with each arrow is the wavelength (in nanometers) of the photon that is emitted.

By doing similar calculations for the other transitions, we can draw the following conclusions about the light that can be emitted by excited hydrogen atoms.

1. Transitions from higher energy levels to the *ground* state ($n = 1$) result in the emission of *ultraviolet* photons. This series of emission lines is referred to as the Lyman series.
2. Transitions from higher energy levels to the $n = 2$ state result in the emission of *visible* photons. This series of emission lines is referred to as the Balmer series (see Figure 10.8 and Example 10.4).
3. Transitions from higher energy levels to the $n = 3$ state result in the emission of infrared photons. This series of emission lines is referred to as the Paschen series.
4. Downward transitions to other states with n greater than 3 result in the emission of other, lower-energy photons.

An atom in the $n = 3$ state or higher can make transitions to intermediate energy levels instead of jumping directly to the ground state. For example, an atom in, say, the $n = 4$ state may jump to the $n = 1$ state, or it may go from $n = 4$ to $n = 2$ and then from $n = 2$ to $n = 1$. In the latter case, two different photons would be emitted (Figure 10.24). Atoms in the higher energy levels can undergo different "cascades" in returning to the ground state.

Figure 10.24

Two different ways for a hydrogen atom in the $n = 4$ state to return to the ground state. The left arrow represents a direct transition with the emission of one photon. The two arrows on the right show a transition to the $n = 2$ level, followed by a transition to the ground state. Two lower energy (longer wavelength) photons are emitted.

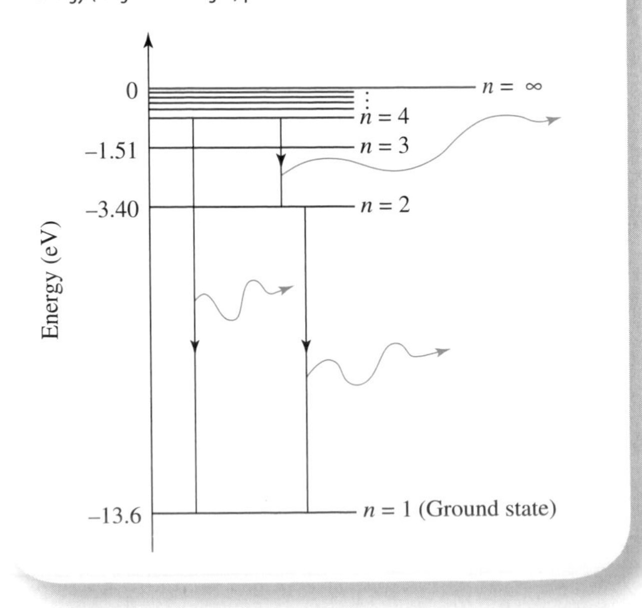

We can envision what happens when hydrogen gas in a tube is heated to a high temperature, as in Figure 10.7, or is excited by passing an electric current through it. The billions and billions of atoms will be excited, some to each of the possible higher energy levels. The excited atoms then undergo spontaneous transitions to lower levels, with different atoms "stopping" at different levels according to a complex set of quantum-mechanical transition probabilities. The hydrogen gas continuously emits photons corresponding to all of the possible energy-level transitions, although some transitions are more favored than others depending in part upon the gas temperature.* This is hydrogen's emission spectrum.

Upward energy-level transitions occur when the atom gains energy, by absorption of a photon with the proper energy, or perhaps by collision. For example, a hydrogen atom in the ground state can jump to the $n = 2$ state by absorbing a photon with energy 10.2 electronvolts. When white light—containing photons with many different energies—passes through hydrogen, those photons with just the right energy can be absorbed, leading to the observed absorption spectrum (see Figure 10.13).

An atom is ionized if it absorbs enough energy to make the electron energy greater than zero. For example, a hydrogen atom in the ground state is ionized if it absorbs a photon with energy greater than 13.6 electronvolts. This process, referred to as "photoionization," is essentially the same as the photoelectric effect. Any excess energy the electron has appears as kinetic energy.

We've presented a fairly complete picture of the hydrogen atom, but what of the other elements? The presence of more than one electron complicates things, but the general principles of atomic structure are much the same. Each element has its own atomic energy-level diagram, with correspondingly different energy values for each level. The various downward transitions between these levels produce the element's characteristic emission spectrum. Atoms with larger atomic numbers have more protons in the nucleus, so the force on the inner electrons is stronger. This means that the electrons are more tightly bound and that their energies have larger magnitudes but are still negative.

The structure of atoms with more than one electron is governed by a principle formulated by Wolfgang Pauli in 1925. Pauli, who was awarded the 1945 Nobel Prize in physics for his work, carefully analyzed the emission spectra of different elements noting that some expected transitions did not occur when the lower energy level

* We have assumed here that the gas density is low enough or the temperature is high enough so that interactions between the gas atoms are negligible. Under such conditions, we may ignore electron transitions brought about by collisions in which no photons are emitted.

EMISSION SPECTRA

in the Real World

One very common device that uses the emission spectra of elements is the neon sign—so named because neon is one of the most commonly used elements in them. These signs are made by placing low-pressure gas in a sealed glass tube. A high-voltage alternating current power supply connected across the ends of the tube causes electrons to move back and forth through the tube. The electrons excite the atoms of the gas by collision, and the atoms emit photons as they return to their ground states. Since different elements have different emission spectra, signs can be made to emit different colors by being filled with different gases. Neon-filled signs are red because there are several bright red lines in neon's emission spectrum (see Figure 10.8).

Fluorescent lights and plasma displays (TVs) make use of ultraviolet emission lines. As in neon lights, the atoms of a gas (mercury in the former and xenon or neon in the latter) are collisionally excited by a current. The photons that are emitted are mostly ultraviolet, so a second substance, called a phosphor, is used to convert these to the visible-light photons that we see. Atoms in the phosphor, which is coated onto the inside surface of the chamber, absorb the UV photons and jump to higher energy levels. They return to the ground state via two steps: First, they jump to intermediate energy levels, which gives energy to neighboring atoms, then they return to the ground state as they emit visible-light photons. Different phosphors produce different frequencies of light. Fluorescent lights use phosphors that emit several different colors, which combine to approximate white light. Each pixel in a plasma display consists of three separate subpixel cells, one with a red phosphor, another with green, and the third with blue. Different colors are produced by varying the brightness of the red, green, and blue light. This is done by varying the current in each cell.

Nature provides us with a spectacular example of emission spectra, the *aurora borealis* (northern lights) and the *aurora australis* (southern lights). These light displays, seen at night mainly in the far north and the far south, arise from a fascinating combination of physical processes. It begins at the Sun, where ionized atoms and free electrons are flung out into space as part of the solar wind. After a journey of several days they reach the Earth, where the charged particles are guided by the magnetic field surrounding our planet (recall Figure 8.5). The lightweight electrons spiral around the magnetic field lines and follow them toward the north and south poles. If the conditions are right, these energetic electrons enter the upper atmosphere (down to about 100 kilometers above the Earth) and collide with atoms, molecules, and ions, thereby exciting them to emit spectra. The most common emission is a whitish-green glow from oxygen atoms. Energy-level transitions in excited nitrogen molecules are responsible for a pink that is often seen.* Much of the emitted radiation is in the ultraviolet and infrared bands, which we can't see.

* It should be noted that the mechanism that creates the aurorae is much like that used in conventional televisions and computer monitors. In their picture tubes, fast-moving electrons are guided along their paths to the screen by magnetic fields. The image we see on the screen is comprised of light emitted by atoms that are given energy when these electrons collide with them.

© iStockphoto.com / © Alex Slobodkin/iStockphoto.com / © Roman Krochuk/iStockphoto.com / © Catherine dee Auvil/iStockphoto.com / © iStockphoto.com

was already occupied by a certain number of electrons. From this and other evidence, Pauli concluded that only a limited, fixed number of electrons could populate each energy level in an atom. (By analogy, we can't sit comfortably on a sofa if it is already full of people.) Pauli stated his conclusion in the **exclusion principle**:

PRINCIPLES

Two electrons cannot occupy precisely the same quantum state at the same time.*

(Pauli exclusion principle)

For each energy level, there exists a set number of quantum states available to the electrons. Once all of the quantum states corresponding to a given energy are filled, any remaining electrons in the atom must occupy other energy levels that have vacancies. The number of quantum states associated with each energy level is related to the electron's orbital angular momentum and spin (see Section 12.2). In the $n = 1$ level, the number of distinct quantum states is 2, so that the maximum number of electrons that can occupy this level is 2. For the $n = 2$ energy level, the maximum number of allowed states and, hence, electrons is 8; for the $n = 3$ level it is 18. The general rule is, for level n, the occupation limit of electrons is $2n^2$. The ground state of a multi-electron atom is one in which all electrons are in the lowest energy levels consistent with the Pauli exclusion principle. If any one electron is in a higher energy level, the atom is in an excited state. Table 10.1 shows the ground-state energy-level populations for several atoms.

The properties of each element are determined to a great extent by the ground-state configuration of its atoms, particularly the number of electrons in the highest energy level that is occupied. Table 10.1 shows that helium in the ground state has the $n = 1$ energy level filled and that neon has the $n = 2$ energy level filled. Because of this, helium and neon have similar properties: They are both gases at normal room temperature and pressure and are very stable. They don't burn or react chemically in other ways except under special circumstances. Hydrogen, lithium, and sodium have similar properties because they all have one electron in the highest occupied energy level (when in their respective ground states).

* The exclusion principle applies to an entire class of elementary particles called *fermions*, which includes electrons. More on this in Chapter 12.

Table 10.1 **Ground-State Configurations of Several Atoms**

Element	Atomic Number	Number of Electrons in Level		
		$n = 1$	$n = 2$	$n = 3$
Hydrogen	1	1	0	0
Helium	2	2	0	0
Lithium	3	2	1	0
Carbon	6	2	4	0
Oxygen	8	2	6	0
Neon	10	2	8	0
Sodium	11	2	8	1

The periodic table of the elements (tear-out card) was developed by Dmitri Mendeleev in 1869. He arranged the elements known at the time according to their properties: Elements with similar properties were placed in the same column. After Pauli's discovery, it was determined that the elements in each column have similar ground-state configurations, and that is why they are alike. (For example, hydrogen, lithium, and sodium are in the first column, called Group 1A, because each has one electron in its highest occupied energy level.) This is one indication of how understanding quantum mechanics can play a crucial role in other scientific disciplines like chemistry. The Concept Map on your Review Card summarizes the types of transitions atomic electrons can undergo.

10.7 X-Ray Spectra

We have seen that the Bohr model of the atom (as refined by quantum mechanics) had great success in explaining the spectrum produced by hydrogen. Another early triumph of the Bohr atom came in connection with the study of x-rays. Soon after Bohr published his model, a young English physicist named H. G. J. Moseley used it to explain the characteristic spectra of x-rays.

In Section 8.5, we described how x-rays are produced. Electrons are accelerated to a high speed and then directed into a metal target (Figure 8.27). A band of x-rays is emitted, with different wavelengths having

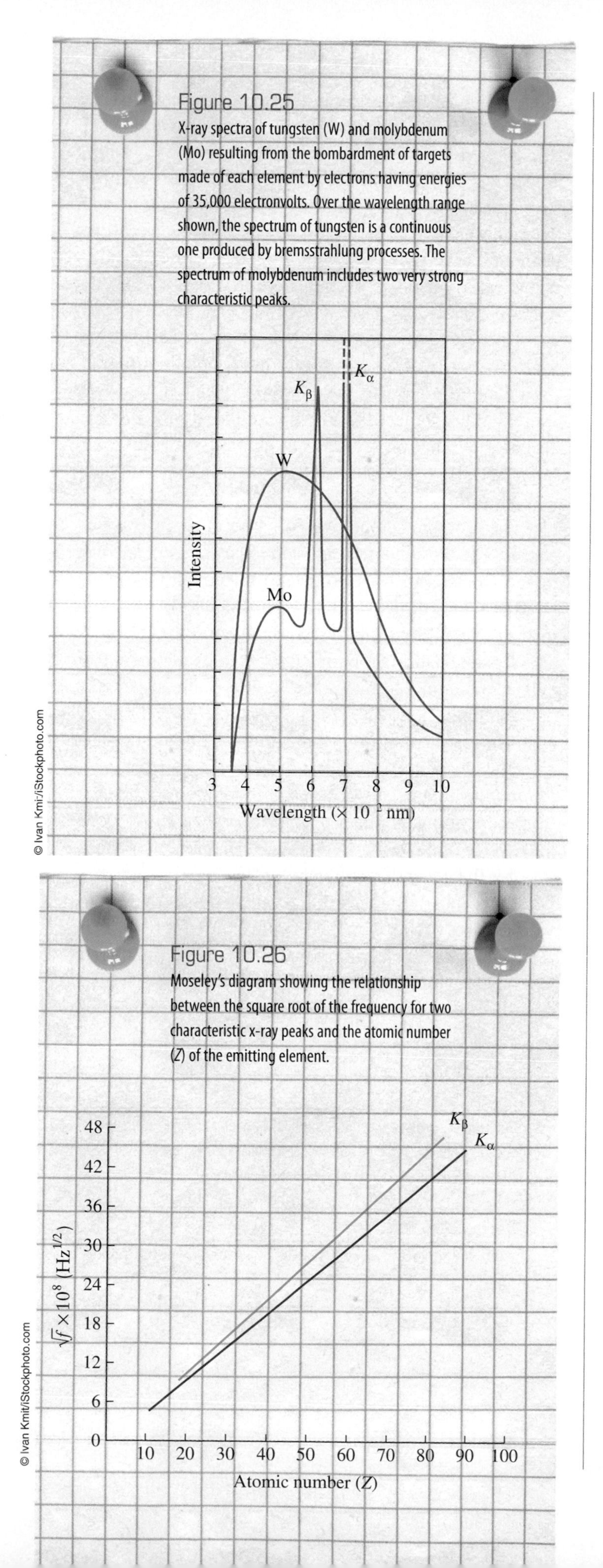

Figure 10.25
X-ray spectra of tungsten (W) and molybdenum (Mo) resulting from the bombardment of targets made of each element by electrons having energies of 35,000 electronvolts. Over the wavelength range shown, the spectrum of tungsten is a continuous one produced by bremsstrahlung processes. The spectrum of molybdenum includes two very strong characteristic peaks.

Figure 10.26
Moseley's diagram showing the relationship between the square root of the frequency for two characteristic x-ray peaks and the atomic number (Z) of the emitting element.

© Ivan Kmit/iStockphoto.com
© Ivan Kmit/iStockphoto.com

different intensities. Figure 10.25 shows the x-ray spectra when the elements tungsten (W) and molybdenum (Mo) are used as targets. These graphs are like blackbody radiation curves in that they are plots of intensity versus wavelength. Most of the x-rays are produced as the electrons are rapidly decelerated upon entering the target. (This is one example of Maxwell's finding that accelerated charges emit EM waves.) This *bremsstrahlung* (German for "braking radiation") appears as the smooth part of the spectra covering the full range of wavelengths. Bremsstrahlung spectra are much the same for different elements. But the two sharp peaks for the molybdenum target are unique to this element and constitute its characteristic spectrum. Other elements exhibit characteristic spectra but at different wavelengths.

Moseley compared the characteristic x-ray spectra of different elements and found a simple relationship between the frequencies of the peaks and the atomic number of the element. For each peak, a graph of the square root of the frequency versus the atomic number of the element was a straight line (Figure 10.26). This was a very practical discovery because it allowed him to determine the atomic number, and therefore, the identity, of unknown elements. All he had to do was determine the wavelengths of the characteristic x-ray peaks and use his graph. This method was a key factor in the discovery of several new elements, including promethium (atomic number 61) and hafnium (atomic number 72).

After Bohr developed his model of the atom, Moseley quickly realized that the characteristic x-rays are much like the emission lines of hydrogen. He showed that they are produced when one of the innermost electrons jumps from one orbit (energy level) to another. Different elements have x-ray peaks with different wavelengths because the electron energy levels have different energy values. In particular, atoms with higher atomic numbers have more protons in the nucleus and bind the inner electrons more tightly; their energies are greater.

Moseley proposed the following scenario to account for the x-ray spectra of the elements. If a bombarding electron collides with, say, a molybdenum atom and knocks out an electron from the $n = 1$ level, electrons in the upper levels will cascade down to fill the vacancy left by the ejected electron, and photons will be emitted. The lowest frequency (longest wavelength) x-ray photon corresponds to the transition from $n = 2$ to $n = 1$. Because the lowest energy level was called the "K shell" by x-ray experimenters, this is called the K_α (K-alpha) peak. Electrons jumping from $n = 3$ to $n = 1$ emit photons that form the K_β (K-beta) peak. Notice that the K_α and K_β peaks correspond to the two lowest-frequency lines in hydrogen's Lyman series.

With this knowledge, we can see why high-speed electrons are needed to produce the characteristic x-rays of heavy elements. The inner electrons are so tightly bound that it takes very high energy electrons to knock them out.

10.8 Lasers

The word *laser* is an acronym derived from the phrase "light amplification by stimulated emission of radiation." By examining this phrase one piece at a time, we will describe how the laser operates. In doing so we will find yet another example of how our theory of atomic structure provides the basis for understanding a process whose practical application has touched nearly every aspect of modern life.

Imagine that an electron in an atom has been excited to some higher energy level by either a collision or the absorption of a photon with the proper energy, E. Generally, such an electron will remain in this excited state for only a short time (around a billionth of a second) before returning to a lower energy level. If this decay to a lower state occurs spontaneously by radiation (which is usually the case in low-density gases where collisions are rare), a photon will be emitted in some random direction.

It turns out, however, that an electron can be *stimulated* to return to its original energy level through the intercession of a second photon with the same energy, E (Figure 10.27). If a group of atoms all having their electrons in this same excited state is "bathed" in light consisting of photons with energy E, the atoms will be stimulated to decay by emitting additional photons with the same energy E. By this process, the intensity of a beam of light is increased or amplified as a result of stimulated emission by atoms in the region through which the radiation passes. We have a laser.

© Karin Lau/iStockphoto.com

To achieve this amplification, we must first arrange for the majority of the atoms through which the stimulating radiation travels to have their electrons in the same excited state. Otherwise, the unexcited atoms will just absorb the radiation. This is not an easy situation to arrange because the excited atoms normally don't stay that way for long.

The problem can be overcome because many atoms possess excited states that are said to be *metastable*. Once in such a state, the electrons tend to remain there for a relatively long time (perhaps a thousandth of a second instead of a billionth) before spontaneously decaying to a lower state. During the time it takes to excite more than half of the atoms to that state, the ones already excited don't jump back down. This process, called "pumping," can result in the majority of the atoms being in the same metastable, excited state. This condition is called a *population inversion* because there are more atoms with electrons populating an upper energy level than a lower one, the reverse of the usual situation. This condition is necessary for the generation of laser light.

If we now irradiate the population-inverted atoms with photons of the correct energy, a chain reaction can be established that greatly amplifies the incident light beam. One photon stimulates the emission of another identical one, these two stimulate the emission of two more, these

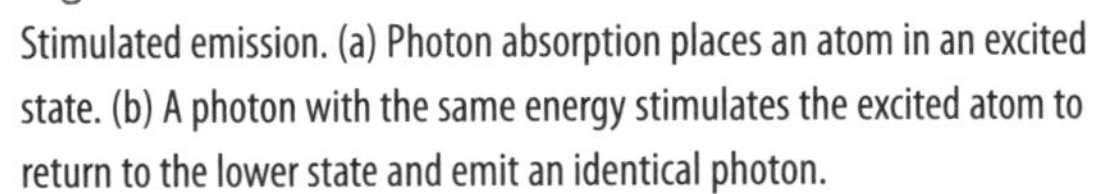

Figure 10.27
Stimulated emission. (a) Photon absorption places an atom in an excited state. (b) A photon with the same energy stimulates the excited atom to return to the lower state and emit an identical photon.

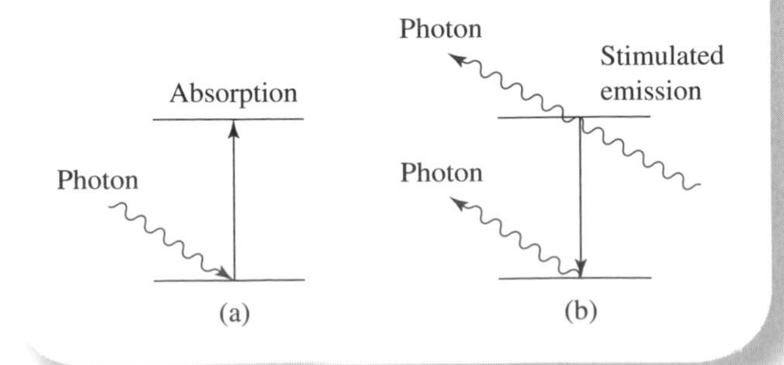

four stimulate the emission of four more, and so on. In the end, an avalanche of photons is produced, all having the same frequency (color). The beam is said to be *monochromatic.* The light amplification can produce a very high-intensity beam of light of a single wavelength.

Besides being intense and monochromatic, laser light has an additional property that makes it extremely useful: *coherence.* This means that the stimulating radiation and the additional emitted laser radiation are in phase: The crests and troughs of the EM waves at a given point all match up (Figure 10.28b). An ordinary light source such as an incandescent bulb emits incoherent light (different parts of the beam are not in phase with one another) because the excited atoms giving off the radiation do so independently of each other. The emitted photons in this case may be considered to be individual, short "wave trains" bearing no constant phase relation to one another (Figure 10.28a). By contrast, stimulated emission of radiation by excited atoms produces photons that are in phase with the stimulating radiation. The coherence of laser light contributes to its high intensity. Coherent sources like lasers are important for producing interference patterns and *holograms* (see p. 288).

Although not in widespread use except for some specialized scientific applications, the ruby laser is simple in concept and illustrates all the important design characteristics of most other types of lasers. This device is shown schematically in Figure 10.29. It consists of a ruby rod whose ends are polished and silvered to become mirrors, one of which is partially (1–2%) transparent. The ruby rod is composed of aluminum oxide (Al_2O_3) in which some of the aluminum atoms have been replaced by chromium atoms. The chromium atoms produce the lasing effects. The rod is surrounded by a flash tube capable of producing a rapid sequence of short, intense bursts containing green light with a wavelength of 550 nanometers. The chromium atoms are excited by these flashes from their lower state E_0 to state E in a process referred to as *optical pumping.* (The high-intensity flash tube "pumps" energy into the chromium atoms.)

Figure 10.28

(a) Excited atoms in incoherent sources like ordinary light bulbs emit photons independently. The directions of propagation and the relative phases of emitted light are random. (b) Excited atoms in a laser emit coherent light. Photons produced by stimulated emission all travel in the same direction in phase with one another.

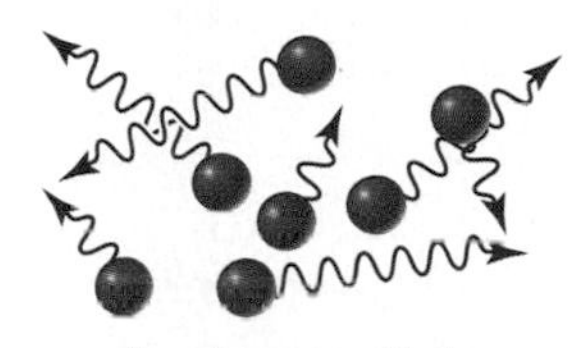

(a) Incoherent radiation from excited atoms

(b) Coherent radiation from excited atoms

Figure 10.29

Schematic of a ruby laser system. The pumping radiation is provided by a high-intensity flash tube. The stimulating photons are reflected back and forth between the parallel end mirrors to build up a beam of high intensity. The laser beam consists of photons that escape through the partially transmitting end reflector at the right. Lasers of this type must be pulsed to avoid overheating the ruby rod and possibly cracking it.

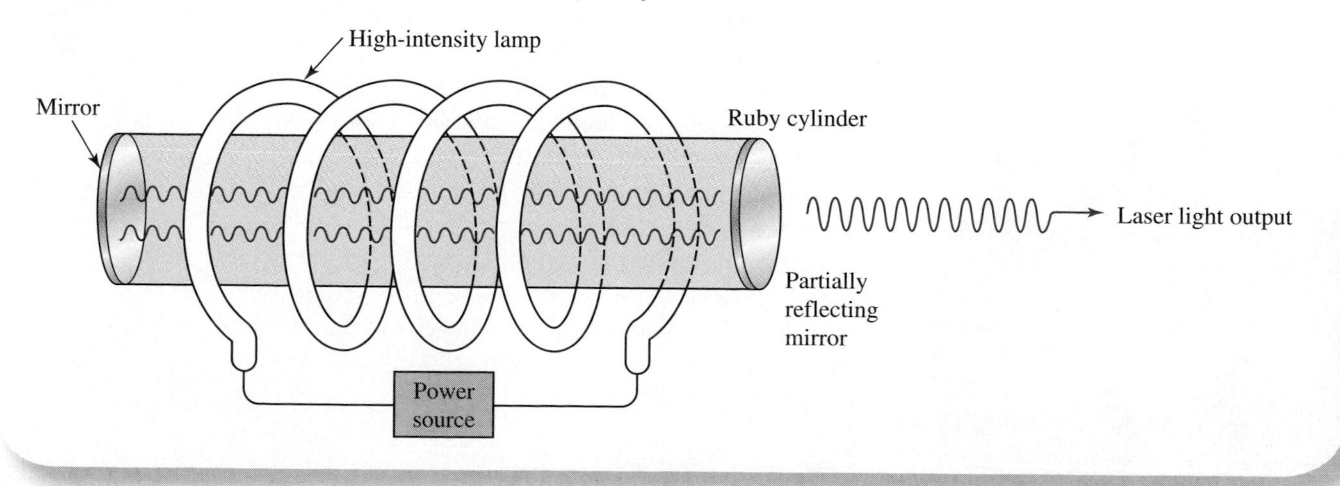

HOLOGRAMS

3-D Images, No Glasses Required

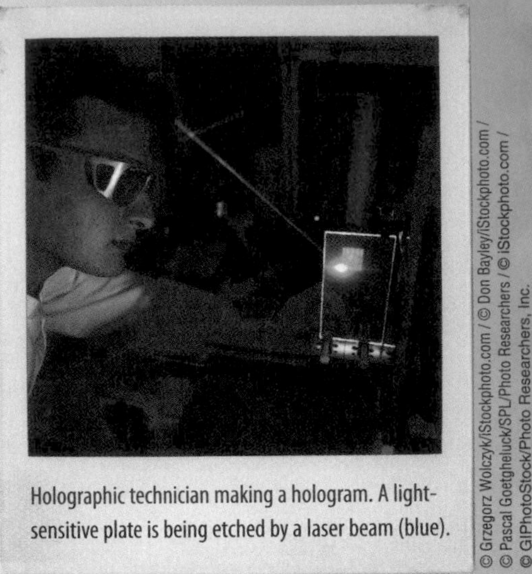

Holographic technician making a hologram. A light-sensitive plate is being etched by a laser beam (blue).

© Grzegorz Wolczyk/iStockphoto.com / © Don Bayley/iStockphoto.com / © Pascal Goetgheluck/SPL/Photo Researchers / © iStockphoto.com / © GIPhotoStock/Photo Researchers, Inc.

Increasingly, 3-D imagery is becoming the norm in video gaming and web-based displays for everything from architectural renderings to Martian landforms to zoological nanostructures.

The production of true 3-D images that faithfully reproduce all aspects of the original objects was first accomplished in 1947 by Dennis Gabor. Gabor developed a way to preserve information about the relative phases of light beams emitted by or reflected from different points on an object in a complex interference pattern. A simple version of the experimental setup is shown in Figure 10.31. A beam of coherent radiation, in modern applications usually from a laser, is split in two, one beam traveling directly to a piece of photographic film, and the second reaching the film after being reflected off some object. At the position of the film, the two beams recombine, interfering with one another to produce a complex pattern of bright and dark areas (see "Interference" in Section 9.1). This pattern is recorded on the film and contains information not only about the relative intensities of the object and reference beam, but also about their phase differences.

To see the image (to reconstruct the object), light of the same type as that used originally is shone on the film. Depending on the details of the original exposure process, the transmitted or reflected light is viewed obliquely to reveal the image. The resulting hologram (from the Greek word *holos*, meaning "whole") is a true 3-D image of the original object. Its aspect and/or color changes as you move your head; some parts of the object become visible while others disappear. The apparent shape, size, and brightness of the image change just as they would if you were examining the genuine article. Gabor's research earned him the Nobel Prize in physics in 1971.

When first perfected during the 1960s, holography—the science of producing holograms—was viewed as little more than a curiosity, a clever way to generate interesting 3-D images and to illustrate simultaneously some basic physical concepts. In recent years, however, the ability of holograms to store huge quantities of information has been recognized by those in the computer industry involved in data storage and retrieval. Other applications involve component testing and analysis in which two holograms of an object taken at different times are superimposed. Any changes in the object, like deformations produced by, say, forces exerted during industrial stress tests or by tumor growth beneath the surface of the object, show up immediately as interference patterns. More commonly, banks and credit unions regularly place embossed holograms on their credit cards to reduce fraud. The presence of these highly detailed holographic images makes card duplication nearly impossible and attempts at counterfeiting readily apparent.

The Visa Dove hologram

Figure 10.30

Energy-level transitions used by a ruby laser. The chromium atoms are excited by green pumping radiation. A transition to the metastable state follows. Stimulated emission from this level produces the red laser light.

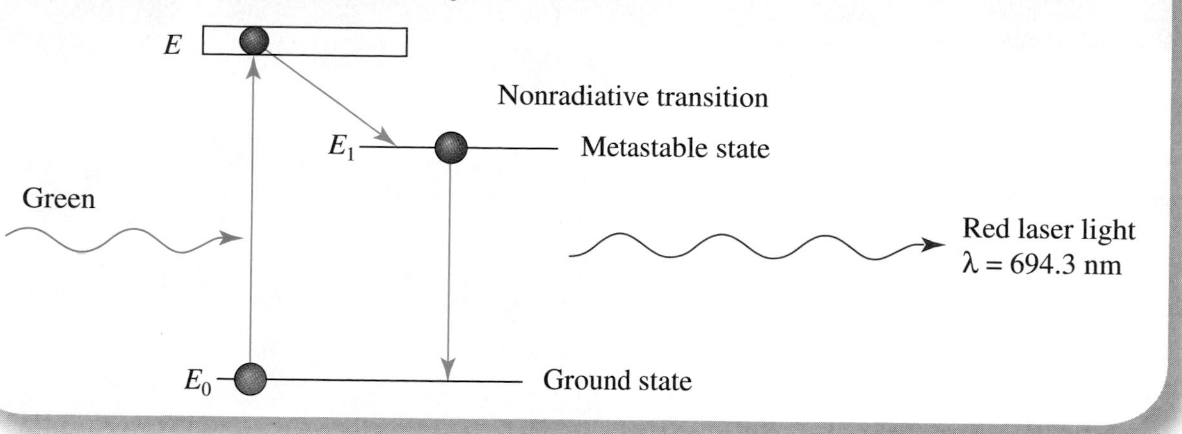

The chromium atoms quickly and spontaneously decay back to level E_0, or, in some cases, to the metastable level E_1 (Figure 10.30). With very strong pumping, a majority of the atoms can be forced into state E_1, and a population inversion is produced. Eventually, a few of the chromium atoms in state E_1 decay to the state E_0, thus emitting photons that stimulate other excited chromium atoms to execute the same transition. When these photons strike the end mirrors, most of them are reflected back into the tube. As they move back in the opposite direction, they cause more stimulated emission and increased amplification. A small fraction of the photons oscillating back and forth through the rod is transmitted through the partially silvered end and makes up the narrow, intense, coherent laser beam. The beam produced by a ruby laser has a wavelength of 694.3 nanometers, a deep red color.

The ruby laser is an example of a pulsed solid-state laser: It produces a single, short pulse of laser light each time the flash tube fires, and the lasing material (the chromium atoms) is distributed in a solid matrix. Neodymium:yttrium-aluminum garnet (Nd:Yag) lasers are also of this type. Another common type of laser is the gas laser, of which the helium-neon (He-Ne) laser is a good example. Here, the lasing material is a mixture

Figure 10.31

One method of hologram production. Laser beams reflected from the mirror and from the object combine at the photographic film to produce an interference pattern. The developed film is the hologram.

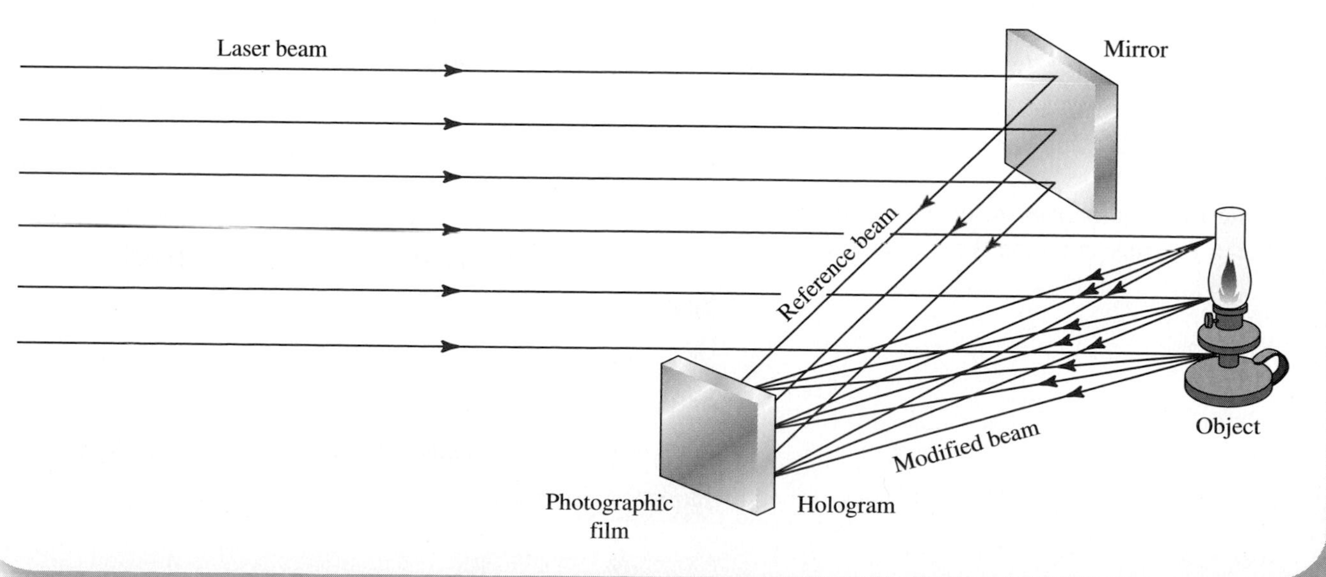

Lasers have assumed increasingly large and important roles in a variety of areas since their invention in the late 1950s. Today, lasers are used to:

- perform various surgeries
- precisely cut metals and other materials
- induce nuclear fusion reactions
- accurately measure distances separating objects (in space and on Earth)
- transmit telephonic information along optical fibers
- replay recorded music, movies, and data (CD, CD-ROM, and DVD players)
- determine the price of goods at the supermarket checkout

All lasers—even in inexpensive laser levels like the one shown—operate according to the same basic physics.

© Veronika Bakos/iStockphoto.com / © Tom McNemar/iStockphoto.com / © Greg Nicholas/iStockphoto.com

of about 15% helium gas and 85% neon gas. In most applications, the neon gas produces coherent radiation of wavelength 632.8 nanometers (red) as a result of stimulated emission from a metastable level to which it has been excited by collisions with the helium atoms. Helium-neon lasers can also be designed to yield green light with a wavelength of 543.5 nanometers. The He-Ne laser is a continuous laser in that it produces a steady beam of laser light.

Table 10.2 summarizes the characteristics of some of the most common types of lasers.

Table 10.2 Characteristics of Some Common Lasers

Active Medium	Wavelength (nm)	Type
Argon fluoride	193	Pulsed
Nitrogen	337.1	Pulsed
Helium-cadmium	441.6	Continuous
Argon	476.5, 488.0, 514.5	Continuous
Krypton	476.2, 520.8, 568.2, 647.1	Continuous
Helium-neon	543.5, 632.8	Continuous
Rhodamine 6G dye	570–650	Pulsed
Ruby	694.3	Pulsed
Gallium arsenide	780–904* (IR)	Continuous
Neodymium:Yag	1060 (IR)	Pulsed
Carbon dioxide	10,600 (IR)	Continuous

*Depends on temperature.

questions&problems

Questions

(▶ Indicates a review question, which means it requires only a basic understanding of the material to answer. The others involve integrating or extending the concepts presented thus far.)

1. ▶ What does it mean when we say that the energy of something is "quantized"?
2. Of the things that a car owner has to purchase routinely—gasoline, oil, antifreeze, tires, sparkplugs, and so on—which are normally sold in quantized units, and which are not?
3. How might you explain the concept of quantization to a younger child (brother, sister, niece, etc.), using money as the quantized entity?
4. ▶ What assumption allowed Planck to account for the observed features of blackbody radiation?
5. Two common controls in a car are the steering wheel and the station selector on the radio. Indicate whether each is controlling something in a quantized or a continuous fashion.
6. ▶ What is a photon? How is its energy related to its frequency? To its wavelength?
7. ▶ Describe the photoelectric effect. Name some devices that make use of this process.
8. If Nature suddenly changed and Planck's constant became a much larger number, what effect would this have on things like solar cells, atomic emission and absorption spectra, lasers, and so on?
9. Sodium undergoes the photoelectric effect, with one electron absorbing a photon of violet light and another absorbing a photon of ultraviolet light. What is different about the two electrons afterward?
10. Can you think of a reason why metals exhibit the photoelectric effect most easily? [Hint: Do you see any connection between this phenomenon and the properties of a good electrical or thermal conductor?]
11. What aspects of the photoelectric effect can be explained without recourse to the concept of the photon? What aspects of this phenomenon require the existence of photons for their explanation?
12. Based on what you learned about image formation in Chapter 9, describe how you might design a photocopying machine that could make a copy that is enlarged or reduced compared to the size of the original.
13. If sunlight can be conceived of as a beam of photons, each of which carries a certain amount of energy and momentum, why don't we experience (or feel) any recoil as these particles collide with our bodies when, say, we're at the beach?
14. ▶ What is the difference between a continuous spectrum and an emission-line spectrum?
15. A mixture of hydrogen and neon is heated until it is luminous. Describe what is seen when this light passes through a prism and is projected onto a screen.
16. ▶ What are the basic assumptions of the Bohr model? Describe how the Bohr model accounts for the production of emission-line spectra from elements like hydrogen.
17. ▶ Discuss what is meant by the term *ionization*. Give two ways by which an atom might acquire enough energy to become ionized.
18. ▶ Compare the emission spectra of the elements hydrogen and helium (see Figure 10.8). Which element emits photons of red light that have the higher energy?
19. If an astronomer examines the emission spectrum from luminous hydrogen gas that is moving away from the Earth at a high speed and compares it to a spectrum of hydrogen seen in a laboratory on Earth, what would be different about the frequencies of spectral lines from the two sources? (See Section 6.2 if you need a little help in answering this question.)
20. The spectrum of light from a star that is observed by the Hubble Space Telescope is not exactly the same as that star's spectrum observed by a telescope on Earth. Explain why this is so.
21. A high-energy photon can collide with a free electron and give it some energy. (This is called the Compton effect.) How are the photon's energy, frequency, and wavelength affected by the collision?
22. ▶ What is the de Broglie wavelength? What happens to the de Broglie wavelength of an electron when its speed is increased?
23. ▶ An electron and a proton are moving with the same speed. Which has the longer de Broglie wavelength? (You may want to look ahead at some useful information in the table on your periodic card in the back of the book.)
24. ▶ In an electron microscope, electrons play the role that light does in optical microscopes. What makes this possible?
25. ▶ What is the uncertainty principle?
26. Explain why the Bohr model of the atom is incompatible with the Heisenberg uncertainty principle.
27. ▶ What does it mean when a hydrogen atom is in its "ground state"?
28. ▶ Explain why a hydrogen atom with its electron in the ground state cannot absorb a photon of just any energy when making a transition to the second excited state ($n = 3$).
29. Describe the spectrum produced by ionized hydrogen, that is, a sample of hydrogen atoms all of which have lost one electron.

30. ▶ Will the energy of a photon that ionizes a hydrogen atom from the ground state be larger than, smaller than, or equal to, the energy of a photon that ionizes another hydrogen atom from the first excited state ($n = 2$)? Explain.

31. What would an energy-level diagram look like for the quantized cat in Figure 10.2?

32. ▶ In what part of the EM spectrum does the Lyman series of emission lines from hydrogen lie? The Balmer series? The Paschen series? Describe how each of these series is produced. In what final state do the electron transitions end in each case?

33. Radioactive strontium (Sr) tends to concentrate in the bones of people who ingest it. Why might one expect that strontium would behave like calcium (Ca) chemically and thus be preferentially bound in bone material, which is predominantly calcium in composition?

34. ▶ The x-ray spectrum of a typical heavy element consists of two parts. What are they? How is each produced?

35. ▶ Will the frequency of the K_α peak in the x-ray spectrum of copper (Cu) be higher or lower than the frequency of the K_α peak of tungsten (W)? Explain how you arrived at your answer.

36. ▶ Describe how the Bohr model may be used to account for characteristic x-ray spectra in heavy atoms.

37. ▶ What is the origin of the word *laser*?

38. ▶ Distinguish laser light from the light emitted by an ordinary light bulb in as many ways as you can.

39. ▶ Distinguish between a metastable state and a normally allowed energy state within an atom. Discuss the role of metastable states in the operation of laser systems.

40. ▶ Define what is meant by the term *population inversion*. Why must this condition be achieved before a system can successfully function as a laser?

41. ▶ Describe the operation of a pulsed ruby laser.

42. A friend tells you that physicists have just invented a new pulsed laser that is pumped with yellow light and produces laser light in the ultraviolet portion of the EM spectrum. Why might you be a little skeptical of this claim? Explain.

Problems

1. Find the energy of a photon whose frequency is 1×10^{16} Hz.

2. Things around you are emitting infrared radiation that includes the wavelength 9.9×10^{-6} m. What is the energy of the photons?

3. In what part of the EM spectrum would a photon of energy 9.5×10^{-25} J be found? What is its energy in electronvolts?

4. Gamma rays (γ-rays) are high-energy photons. In a certain nuclear reaction, a γ-ray of energy 0.511 MeV (million electronvolts) is produced. Compute the frequency of such a photon.

5. Electrons striking the back of a TV screen travel at a speed of about 8×10^7 m/s. What is their de Broglie wavelength?

6. In a typical electron microscope, the momentum of each electron is about 1.6×10^{-22} kg-m/s. What is the de Broglie wavelength of the electrons?

7. During a certain experiment, the de Broglie wavelength of an electron is 670 nm $= 6.7 \times 10^{-7}$ m, which is the same as the wavelength of red light. How fast is the electron moving?

8. If a proton were traveling the same speed as electrons in a TV picture tube (see Problem 5), what would its de Broglie wavelength be? The mass of a proton is 1.67×10^{-27} kg.

9. a) A hydrogen atom has its electron in the $n = 4$ level. The radius of the electron's orbit in the Bohr model is 0.847 nm. Find the de Broglie wavelength of the electron under these circumstances.
   b) What is the momentum, mv, of the electron in its orbit?

10. a) A small ball with a mass of 0.06 kg moves along a circular orbit with a radius of 0.5 m at a speed of 3.0 m/s. What is the angular momentum of the ball?
    b) If the angular momentum of this ball were quantized in the same manner that the angular momentum of electrons in the Bohr model of the atom is, what would be the approximate value of the quantum number n in such a case?

11. A hydrogen atom has its electron in the $n = 2$ state.
    a) How much energy would have to be absorbed by the atom for it to become ionized from this level?
    b) What is the frequency of the photon that could produce this result?

12. A hydrogen atom initially in the $n = 3$ level emits a photon and ends up in the ground state.
    a) What is the energy of the emitted photon?
    b) If this atom then absorbs a second photon and returns to the $n = 3$ state, what must the energy of this photon be?

13. The following is the energy-level diagram for a particularly simple, fictitious element, Kansasium (Ks). Indicate by the use of arrows all allowed transitions leading to the emission of photons from this atom and order the frequencies of these photons from highest (largest) to lowest (smallest).

14. A neutral calcium atom (Z = 20) is in its ground state electronic configuration. How many of its electrons are in the $n = 3$ level? Explain how you arrived at your figure.

15 An atom of neutral zinc possesses 30 electrons. In its ground configuration, how many fundamental energy levels are required to accommodate this number of electrons? That is, what is the smallest value of n needed so that all 30 of zinc's electrons occupy the lowest possible quantum energy states consistent with the Pauli exclusion principle?

16. Referring to Figure 10.25, notice that the bremsstrahlung x-ray spectrum of both Mo and W cut off (have zero intensity) at about 0.035 nm. X-rays with this wavelength are produced by the target element when bombarding electrons are promptly stopped in a single collision and give up all their energy in the form of EM waves. Confirm that electrons having energies of 35,000 eV will produce x-ray photons with wavelengths near 0.035 nm by this process.

17. The characteristic K_α and K_β lines for copper have wavelengths of 0.154 nm and 0.139 nm, respectively. What is the ratio of the energy difference between the levels in copper involved in the production of these two lines?

18. Referring to Figure 10.26, we see that the atomic number Z is proportional to $f^{1/2}$ or that Z^2 is proportional to f. Since the frequency of the characteristic x-ray lines is itself proportional to the energy of the associated x-ray photon, we are led to conclude that ΔE and hence the energies of the atomic levels also scale as Z^2. Based on this analysis (and ignoring any differences due to the masses of their nuclei), how much greater is the energy associated with the ground state of helium ($Z = 2$) than that of hydrogen ($Z = 1$)? Make a similar comparison between the ground state energies for sodium ($Z = 11$) and hydrogen.

19. Characteristic x-rays emitted by molybdenum have a wavelength of 0.072 nm. What is the energy of one of these x-ray photons?

20. In a helium-neon laser, find the energy difference between the two levels involved in the production of red light of wavelength 632.8 nm by this system.

21. The carbon dioxide laser is one of the most powerful lasers developed. The energy difference between the two laser levels is 0.117 eV.

    a) What is the frequency of the radiation emitted by this laser?

    b) In what part of the EM spectrum is such radiation found?

22. If you bombard hydrogen atoms in the ground state with a beam of particles, the collisions will sometimes excite the atoms into one of their upper states. What is the *minimum* kinetic energy the incoming particles must have if they are to produce such an excitation?

23. a) Which of the following elements emits a K-shell x-ray photon with the highest frequency?

    b) Which emits a K-shell photon with the lowest frequency?

    i) Silver (Ag) iii) Iridium (Ir)

    ii) Calcium (Ca) iv) Tin (Sn)

There are more quizzes and study tools online at 4ltrpress.cengage.com/physics.

Nuclear Physics

sections

11.1 The Nucleus

Unimaginably tiny, yet incredibly dense, the nucleus occupies the very center of the atom. More than 99.9% of the atom's mass is compressed into roughly *one-trillionth* of its total volume to yield a density of approximately 2×10^{17} kg/m^3. This is about 2×10^{14} times the density of water. If an atom could be enlarged until it were 2,000 feet (0.38 mi) across, its nucleus would only be about the size of a pea. The nucleus is impervious to the chemical and thermal processes that affect its electrons. But it is seething with energy—energy, for example, that makes the Sun and the other stars shine.

The nucleus contains two kinds of particles, protons and neutrons. These particles have nearly the same mass, about 1,840 times that of an electron. The masses are extremely small, so it becomes convenient to introduce an appropriate unit of mass, the *atomic mass unit*, u.

$$1\ \text{u} = 1.66 \times 10^{-27}\ \text{kg}$$

Using this measure, the masses of a proton and a neutron are each about 1 u. The proton carries a positive charge, and the neutron is uncharged (Table 11.1). The number of protons in the nucleus is the **atomic number** Z of the atom. There is one proton in the nucleus of each hydrogen atom ($Z = 1$), two in that of helium ($Z = 2$), eight in oxygen ($Z = 8$), and so on. This number determines the identity of the atom.

The neutral neutrons have much less influence on the properties of the atom: their main effect is on the atom's mass. In fact, the number of neutrons in the nuclei of a particular element can vary. Most helium atoms

Table 11.1 Properties of the Particles in the Atom

Particle	Mass	Charge
Electron	9.109×10^{-31} kg = 0.00055 u	-1.602×10^{-19} C
Proton	1.67262×10^{-27} kg = 1.00730 u	1.602×10^{-19} C
Neutron	1.67493×10^{-27} kg = 1.00869 u	0

have two neutrons in the nucleus, but some have one, three, four, or even six (Figure 11.1). Each is still a helium atom; the only difference is the mass. This allows us to associate another number with each nucleus, the **neutron number**, which is the number of neutrons contained in a nucleus.

The symbol N is used to represent the neutron number just as Z is used to represent the atomic number. For all helium atoms, $Z = 2$, but N can be 1, 2, 3, 4, or 6.

DEFINITION

Atomic Number (Z)
The number of protons contained in a nucleus.

Neutron Number (N)
The number of neutrons contained in a nucleus.

Mass Number (A)
The total number of protons and neutrons in a nucleus.

The mass of an atom is determined primarily by how many protons and neutrons there are in the nucleus. (The electrons are so light that they contribute only a negligible amount to the total mass.) Therefore, it is useful to define yet a third number for each nucleus: the **mass number**.

The mass number, represented by the letter A, is the sum of the atomic number and neutron number. So,

$$A = Z + N$$

© Marek Uliasz/iStockphoto.com

The mass number indicates what the mass of a nucleus is, just as the atomic number indicates the amount of electric charge in the nucleus. Protons and neutrons are collectively referred to as **nucleons**. (The terms *nucleon number* and *atomic mass number* are sometimes used instead of mass number.) The mass number is just the total number of nucleons in the nucleus. The possible mass numbers for helium are $A = 3, 4, 5, 6$, or 8. Each different possible "type" of helium is called an **isotope**.

Figure 11.1
The different possible nuclei of helium atoms. All have two protons, but the number of neutrons varies. The electrons in orbit around the nuclei are not shown. With this scale, the radius of such an orbit would be about 1,000 feet.

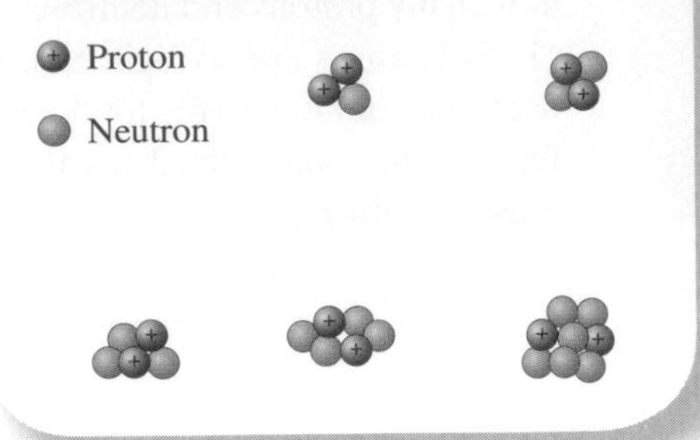

DEFINITION

Isotopes *Isotopes of a given element have the same number of protons in the nucleus but different numbers of neutrons.*

The different isotopes of an element have essentially the same atomic properties: These are determined by the atomic number Z. For example, most of the carbon atoms in your body have six neutrons in their nuclei, but small numbers of the isotopes with seven and eight are also present. The three different carbon isotopes are indistinguishable as far as chemical processes (such as burning) are concerned. The different isotopes of an element often *do* have vastly different nuclear properties. For example, most nuclear power plants rely on the splitting of uranium-235 nuclei, that is, uranium atoms with mass number $A = 235$. Uranium-238 will not work.

Most of the 117 different elements have several isotopes. Some have only a few (hydrogen has 3), and others have more than 20 (iodine, silver, and mercury, to name a few). More than 3,100 different isotopes have been identified and studied. Of these, only about 340 occur naturally. The rest are produced in certain nuclear processes, such as nuclear explosions. Of the known isotopes, only 256 are long-lived. The remaining 2800 or so are unstable; the nuclei eventually transform into different nuclei through a process known as *radioactive decay.* (More on this in Section 11.2.)

Each different isotope of an element is designated by its mass number. The most common isotope of helium has $A = 4$ ($Z = 2$ and $N = 2$) and is called helium-4. The other helium isotopes are helium-3, helium-5, helium-6, and helium-8. The three isotopes of carbon in your body are carbon-12, carbon-13, and carbon-14. Two of the isotopes of hydrogen have been given special names: hydrogen-2 is called deuterium and hydrogen-3 is called tritium. (The prefixes come from the Greek words for *second* and *third.*)

Freezing, boiling, burning, crushing, and other chemical and physical processes do not affect the nuclei of atoms. These processes are influenced by the forces between different atoms, forces that involve only the outer electrons. But nuclei are not indestructible: A number of nuclear processes do affect them. Nuclei can lose or gain neutrons and protons, absorb or emit gamma rays, split into smaller nuclei, or combine with other nuclei to form larger ones. These processes are called *nuclear reactions.* Some occur around us naturally, but most are produced artificially in laboratories. Most of the remainder of this chapter deals with the nature of nuclear reactions and their applications.

To diagram nuclear reactions, we use a special notation to represent each isotope. It consists of the atom's chemical symbol with a subscript and a superscript on the left side. The *subscript* is the atom's atomic number Z, and the *superscript* is the atom's mass number A. Some examples:

Helium-4	${}^{4}_{2}He$	Carbon-14	${}^{14}_{6}C$
Carbon-12	${}^{12}_{6}C$	Uranium-235	${}^{235}_{92}U$

The two numbers indicate the relative mass and charge that the nucleus possesses. The neutron number N can be found by subtracting the lower number (Z) from the upper number (A). For example, each uranium-235 nucleus contains 143 neutrons ($235 - 92 = 143$). Note that both the chemical symbol and the subscript indicate what the element is, so they must always agree.

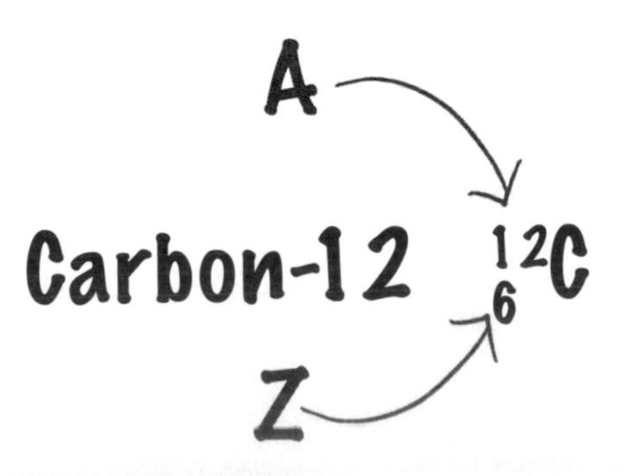

Regardless of what the superscript is, if the subscript is 6, the element is carbon and the symbol must be C.*

This notation can be extended to represent individual particles as well. The designations for neutrons, protons, and electrons are

Neutron	$^{1}_{0}\text{n}$
Proton	$^{1}_{1}\text{p}$
Electron	$^{0}_{-1}\text{e}$

No known stable isotope has an atomic number larger than 83.

For the electron, 0 is used for the mass number because its mass is so small. The −1 indicates that it has the same size charge as a single proton, except it is negative.

Perhaps you've wondered how protons can be bound together in a nucleus. After all, like charges do repel each other. There is indeed a large electrostatic force acting to push the protons apart, but another force many times stronger acts to hold them together inside a nucleus. This is called the *strong nuclear force*, one of the four fundamental forces in Nature. (The *weak nuclear force* is involved in certain nuclear processes, but it is not important in holding the nucleus together. More on this in Chapter 12.) The strong nuclear force is an attractive force that acts between nucleons: Every proton and neutron in a nucleus exerts an attractive force on every other proton and neutron. Compared to the gravitational and electromagnetic forces, the strong nuclear force is rather strange. It is much stronger than the others, but it has an extremely short range: The attractive nuclear force between particles in the nucleus begins to weaken appreciably when the particles are more than about 3×10^{-15} meters apart, and it effectively disappears if the particles become separated by more than 10^{-14} meters. This puts an upper limit on the size that a nucleus can have and still be stable. In a large nucleus, protons on opposite sides are far enough apart that they are near the limit of the effective range of the strong attractive nuclear force; in these instances, the repulsive electric force between the protons becomes important. No known stable isotope has an atomic number larger than 83.

The effectiveness of the nuclear force at holding a nucleus together also depends on the relative numbers of neutrons and protons. If there are too many or too few neutrons compared to the number of protons, the nucleus will not be stable. For example, carbon-12 and carbon-13 are stable, but carbon-11 and carbon-14 are not. The ratio of N to Z for stable nuclei is about 1 for small atomic numbers and increases to about 1.5 for large nuclei (Figure 11.2). The stable isotope lead-208 has 126 neutrons and 82 protons in each nucleus.

* There is a certain redundancy in this notation, insofar as the atomic number and the proper chemical symbol really convey the same information about the isotope. For example, we could eliminate the use of the atomic number and designate carbon-14 as simply ^{14}C. This simplification is adopted in many books, but we will continue to include the atomic number in our notation throughout this chapter because it is helpful in verifying charge conservation in the course of nuclear reactions (see Section 11.2).

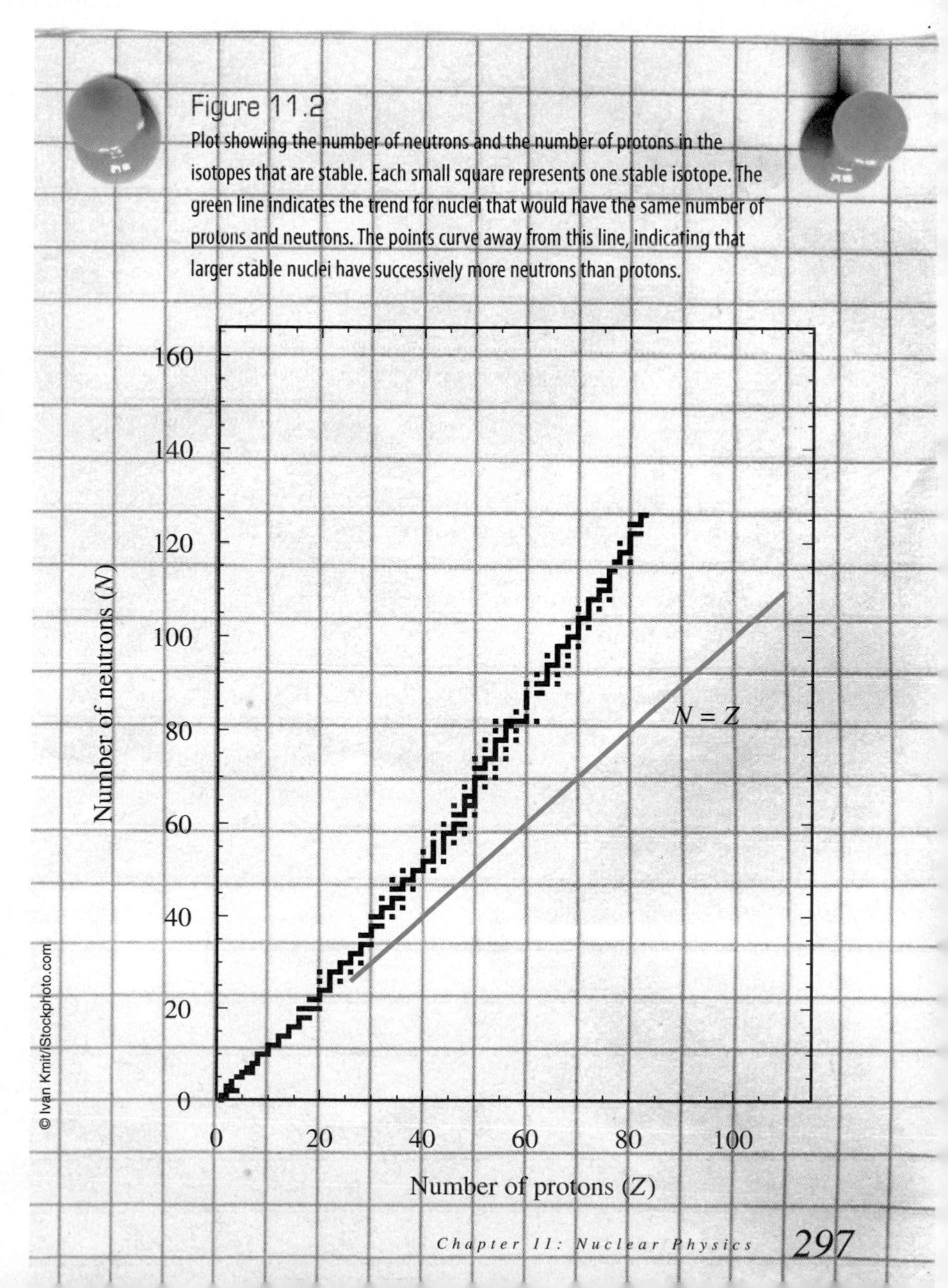

© Ivan Kmit/Stockphoto.com

Figure 11.2 Plot showing the number of neutrons and the number of protons in the isotopes that are stable. Each small square represents one stable isotope. The green line indicates the trend for nuclei that would have the same number of protons and neutrons. The points curve away from this line, indicating that larger stable nuclei have successively more neutrons than protons.

recap

The nucleus is a collection of neutrons and protons held together by the strong nuclear force. This force is limited in its ability to hold nucleons together. Nuclei that are too large or that do not have the proper ratio of neutrons to protons are unstable. They eject particles and release energy.

© Catherine dée Auvil/iStockphoto.com

Eject — There are several different mechanisms used by various unstable nuclei to eject particles and release energy. The main ones are discussed next. — Forward

© Russell Tate/iStockphoto.com

11.2 Radioactivity

Radioactivity, also called *radioactive decay*, is a process wherein an unstable nucleus emits particles and/or EM radiation. Isotopes with unstable nuclei are called **radioisotopes**. The majority of all isotopes are radioactive. For the moment, we will put aside the question of where radioisotopes come from and concentrate on the manner in which radioactive decay occurs.

When it was first investigated, nuclear radiation was found to be similar to x-rays. For example, it exposes photographic film. Soon it was determined that there were actually three different types of nuclear radiation, named **alpha** (α), **beta** (β), and **gamma** (γ) **radiation**. Alpha rays and beta rays are actually high-speed charged particles. They can be deflected with magnetic and electric fields (Figure 11.3). Gamma rays are extremely high-frequency electromagnetic waves (high-energy photons). It is now known that there are many different *decay channels* that unstable nuclei may use when undergoing radioactive decay. For simplicity, we will focus here on only a few of the most common mechanisms of radioactivity: alpha, beta, and gamma decay.

The emission of each type of radiation has a different effect on the nucleus. Both alpha decay and beta decay alter the identity of the nucleus (the atomic number is changed) as well as release energy. Gamma-ray emission does not in itself change the nucleus: It simply carries away excess energy in much the same way that photon emission carries away excess energy from excited atoms. Gamma-ray emission often accompanies alpha decay

Figure 11.3

The three common types of nuclear radiation are affected differently by a magnetic field. Alpha and beta rays are deflected as they go through the field because they are charged particles. Gamma rays are not deflected.

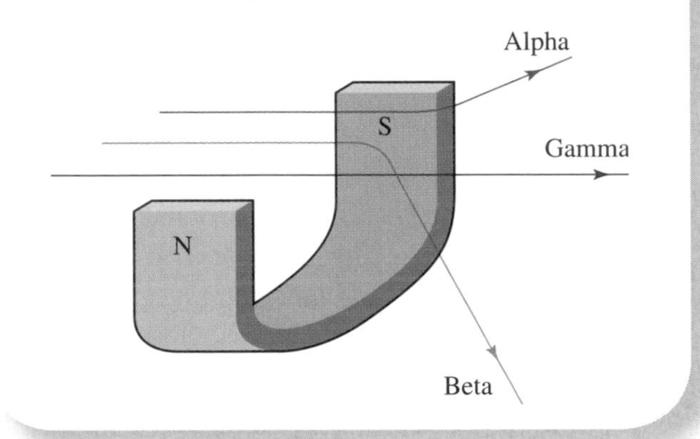

Figure 11.4

(Left) Simplified sketch of a Geiger counter. Nuclear radiation ionizes the gas in the cylinder. The freed electrons are accelerated to the wire and produce a current pulse. (Right) Geiger counter detecting beta rays emitted by an isotope in a lead cup (gray). The count rate is enhanced by the magnet (blue), since beta rays are deflected by the magnetic field into the detector (marked G).

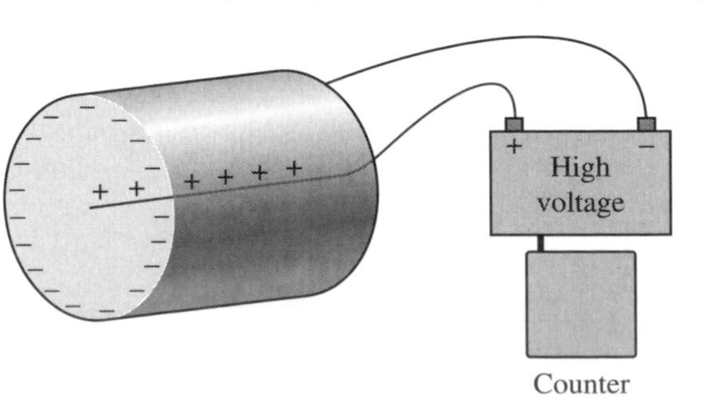

© Vern Ostdiek / © Maria Toutoudaki/iStockphoto.com

and beta decay, as these two processes often leave the nucleus in an excited state with excess energy that is quickly lost through the emission of high-energy photons. Table 11.2 lists several radioisotopes and their modes of decay.

Table 11.2 **Properties of Selected Isotopes**

Element	Isotope	Decay Mode(s)	Relative Abundance (%)
Hydrogen	$^{1}_{1}\text{H}$	. . . (stable)	99.985
	$^{2}_{1}\text{H}$ (deuterium)	. . .	0.015
	$^{3}_{1}\text{H}$ (tritium)	beta	. . .
Helium	$^{3}_{2}\text{He}$	. . .	0.00014
	$^{4}_{2}\text{He}$	. . .	99.9999
	$^{5}_{2}\text{He}$	alpha	. . .
	$^{6}_{2}\text{He}$	beta	. . .
	$^{8}_{2}\text{He}$	beta, gamma	. . .
Carbon	$^{12}_{6}\text{C}$	. . .	98.90
	$^{13}_{6}\text{C}$	. . .	1.10
	$^{14}_{6}\text{C}$	beta	trace
	$^{15}_{6}\text{C}$	beta	. . .
Silver	$^{107}_{47}\text{Ag}^{*}$	gamma, then stable	51.84
	$^{108}_{47}\text{Ag}$	beta, gamma	. . .
	$^{109}_{47}\text{Ag}^{*}$	gamma, then stable	48.16
	$^{110}_{47}\text{Ag}$	beta, gamma	. . .
Uranium	$^{232}_{92}\text{U}$	alpha, gamma	. . .
	$^{233}_{92}\text{U}$	alpha, gamma	. . .
	$^{234}_{92}\text{U}$	alpha, gamma	0.0055
	$^{235}_{92}\text{U}$	alpha, gamma	0.72
	$^{236}_{92}\text{U}$	alpha, gamma	. . .
	$^{237}_{92}\text{U}$	beta, gamma	. . .
	$^{238}_{92}\text{U}$	alpha, gamma	99.27
	$^{239}_{92}\text{U}$	beta, gamma	. . .

Isotopes in excited nuclear states are marked with asterisks (*) in the table. Not every known isotope for the elements carbon, silver, and uranium is included here.

Several different devices are used to detect radiation, the most common being the *Geiger counter*. It exploits the fact that nuclear radiation is ionizing radiation. A gas-filled cylinder is equipped with a fine wire running along its axis (Figure 11.4). A high voltage is maintained between the cylinder wall and the wire, producing a strong electric field. When an alpha, beta, or gamma ray enters the cylinder and ionizes some of the gas atoms, the freed electrons are accelerated by the electric field. These in turn ionize other atoms, and an avalanche of electrons reaches the wire and causes a current pulse. Most Geiger counters emit an audible click each time a ray is detected. They also keep track of how many are detected each second. In other words, they indicate the count rate—the number of alpha, beta, or gamma rays detected each second.

Alpha Decay

An alpha particle is really four particles tightly bound together: two protons and two neutrons. It is identical to a nucleus of helium-4. For this reason, an alpha particle can be represented as

$$\text{alpha particle:} \quad \alpha \quad \text{or} \quad ^{4}_{2}\text{He}$$

A nucleus that undergoes alpha decay *does not* contain a helium-4 nucleus. This just happens to be a particularly stable combination of nuclear particles that can be ejected as a group from the nucleus.

The emission of an alpha particle reduces both the atomic number Z and the neutron number N by 2. The mass number A is reduced by 4. Figure 11.5 is a diagram of the alpha decay of a plutonium-242 nucleus. The atomic number of the nucleus decreases from 94 to 92: The nucleus is transformed from plutonium to uranium. The original plutonium-242 nucleus is called the "parent," and the resulting uranium-238 nucleus is called the "daughter."

Figure 11.5 shows why the isotopic notation introduced earlier is so convenient. When used to represent a nuclear process, like alpha

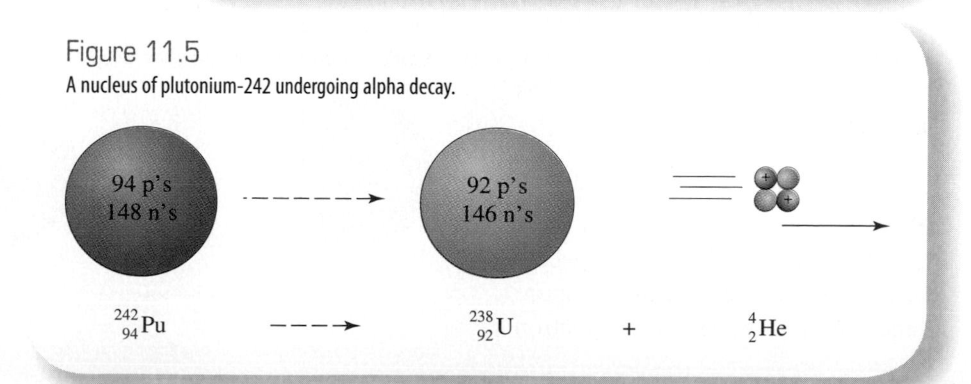

Figure 11.5
A nucleus of plutonium-242 undergoing alpha decay.

decay, it clearly shows how the mass and the charge of the nucleus are affected. Because both conservation of charge and conservation of mass must hold in nuclear transformations (more on this in Chapter 12), the total electric charge and the total number of protons and neutrons must be the same before and after the process. This means that the sum of the subscripts on the right side of the arrow must equal the sum of those on the left. The same is true for the superscripts. This allows one to determine what the decay product (the daughter) is when a given nucleus undergoes alpha decay.

EXAMPLE 11.1

The isotope radium-226 undergoes alpha decay. Write the reaction equation, and determine the identity of the daughter nucleus.

From the periodic table of the elements (see the periodic table tear-out card), we find that the atomic number of radium is 88 and its chemical symbol is Ra. So the reaction will appear as follows:

$$^{226}_{88}\text{Ra} \rightarrow ? + ^{4}_{2}\text{He}$$

The mass number A of the daughter nucleus must be 222 for the superscripts to agree on both sides of the arrow. By the same reasoning, the atomic number Z must be 86. From the periodic table, we find the element radon (Rn) has $Z = 86$. The daughter nucleus is radon-222.

$$^{226}_{88}\text{Ra} \rightarrow ^{222}_{86}\text{Rn} + ^{4}_{2}\text{He}$$

Because alpha decay causes such a drastic change in the mass of the nucleus, it generally occurs only in radioisotopes with high atomic numbers. The alpha particle is ejected with very high speed (typically around one-twentieth the speed of light). Alpha particles are quickly absorbed when they enter matter: Even something as thin as a sheet of paper can stop a significant fraction of them.

Beta Decay

Beta decay is easily the oddest of the three kinds of radioactivity. A beta particle is simply an electron ejected from a nucleus. This means that a beta particle has the same symbol as an electron.

Beta particle: β or $^{0}_{-1}\text{e}$

But wait a minute: There aren't any electrons in a nucleus. During beta decay, one of the neutrons is spontaneously converted into an electron (the beta particle) and a proton. The electron is ejected with very high speed, and the proton remains in the nucleus.* We can represent this process as follows:

$$\text{neutron} \rightarrow \text{proton} + \text{electron}$$

$$^{1}_{0}\text{n} \rightarrow ^{1}_{1}\text{p} + ^{0}_{-1}\text{e}$$

The total electric charge remains the same—zero.

A nucleus that undergoes beta decay loses one neutron and gains one proton. Figure 11.6 shows the beta decay of a carbon-14 nucleus. Note that the mass number A of the nucleus is unchanged.

EXAMPLE 11.2

The isotope iodine-131 undergoes beta decay. Write the reaction equation, and determine the identity of the daughter nucleus.

From the periodic table (see the periodic table tear-out card), we find that iodine's chemical symbol and atomic number are I and 53, respectively. Therefore:

$$^{131}_{53}\text{I} \rightarrow ? + ^{0}_{-1}\text{e}$$

The mass number stays the same, and the atomic number is increased by 1 to 54. Appealing again to the periodic table, we find that this is the element xenon (Xe). The daughter nucleus is xenon-131.

$$^{131}_{53}\text{I} \rightarrow ^{131}_{54}\text{Xe} + ^{0}_{-1}\text{e}$$

Like many nuclear processes, the decay of a neutron shown above is reversible. That is, it is possible for a proton to combine with an electron to produce a neutron. This is what happens during an *electron capture process*: An atomic electron penetrates the nucleus and

*Another particle, a neutrino (from the Italian for "little neutral one"), is also emitted in beta decay. Neutrinos are very strange little beasts—they have no charge, their mass is extremely small, and they rarely interact with matter. Neutrinos routinely pass through the entire Earth without being absorbed, deflected, or otherwise affected. For our purposes, the neutrino can be regarded merely as part of the energy that is released during beta decay. (Chapter 12 has more to say about neutrinos.)

Figure 11.6

A nucleus of carbon-14 undergoing beta decay.

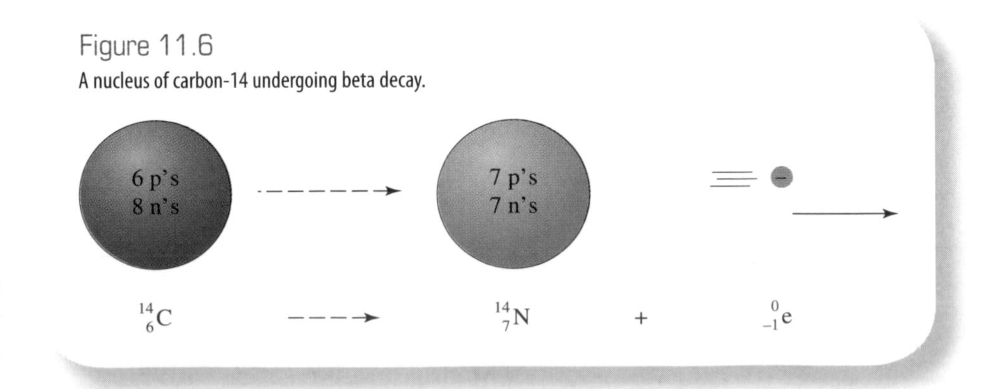

Gamma rays are the most penetrating.

© iStockphoto.com

interacts with a bound proton to form a bound neutron in what might be thought of as an *inverse beta decay.* In such circumstances, the nucleus loses a proton and gains a neutron. As with ordinary beta decay, the mass number of the nucleus remains the same, but now the atomic number is reduced by one unit.

Often the daughter nucleus in both alpha and beta decay is itself radioactive and decays into another isotope. Plutonium-242 undergoes alpha decay to uranium-238. This is a radioisotope that undergoes alpha decay into thorium-234. This process continues until, after a total of nine alpha decays and six beta decays, the stable isotope lead-206 is reached. This is called a **decay chain**. When the Earth was formed from the debris of exploded stars, hundreds of different radioisotopes were present in varying amounts. Geological formations that were originally rich in uranium-238 now contain large amounts of lead-206 as well.

Gamma Decay

Because a gamma ray has no mass or electric charge, gamma-ray emission has no effect on the mass number A or the atomic number Z of a nucleus. Nuclei can exist in excited states in much the same way that atomic electrons can. Gamma rays are emitted when the nucleus makes a transition to a lower energy state. Figure 11.7 shows the gamma decay of a strontium-87 nucleus. The identity of the nucleus is not changed during the process.

Gamma-ray photons, as indicated in Figure 10.4, have energies from about 100,000 electronvolts to more than a billion electronvolts. Most gamma-ray photons emitted in gamma decay are around 1 million electronvolts. (The unit of energy that is used most often in nuclear physics is the megaelectronvolt [MeV], which is 1 million electronvolts.)

One way to compare the different decay processes is to focus on three properties of the nucleus: mass, electric charge, and energy. Alpha decay alters all three: An alpha particle carries away considerable mass, two charged protons, and a great deal of kinetic energy. Beta decay has little effect on the mass of the nucleus, but it does increase the positive charge of the nucleus, and it takes away energy in the form of the beta particle's kinetic energy. Gamma decay only takes energy away from the nucleus.

The three types of nuclear radiation differ considerably in their ability to penetrate solid matter. All three are ionizing radiation. They ionize atoms as they pass through matter. Alpha particles are the least penetrating. Their positive charge causes them to interact strongly with atomic electrons and nuclei, and they are absorbed after traveling only a short distance. Beta particles are more penetrating than alpha particles. For comparison, a thin sheet of aluminum that will block essentially all alpha particles will stop only a fraction of beta particles. Gamma rays are the most penetrating of the three. Since they have no electric charge and travel at the speed of light, they interact with atoms much less frequently. It requires several centimeters of lead to block gamma rays (see Figure 11.8).

Radioactivity and Energy

What becomes of the energy of nuclear radiation as it is absorbed? Most of it goes to heat the material. Early experimenters with radioactivity noticed that highly radioactive samples were physically warmer than their surroundings. As long as the substance continued to emit radiation, it stayed warm. Because of this, radioactivity has been used as an energy source on spacecraft. Many interplanetary space probes, including the Cassini spacecraft sent to probe the planet Saturn

Figure 11.7

A nucleus of strontium-87 undergoing gamma decay. The asterisk (*) on the symbol for the parent nucleus indicates that it is in an excited state.

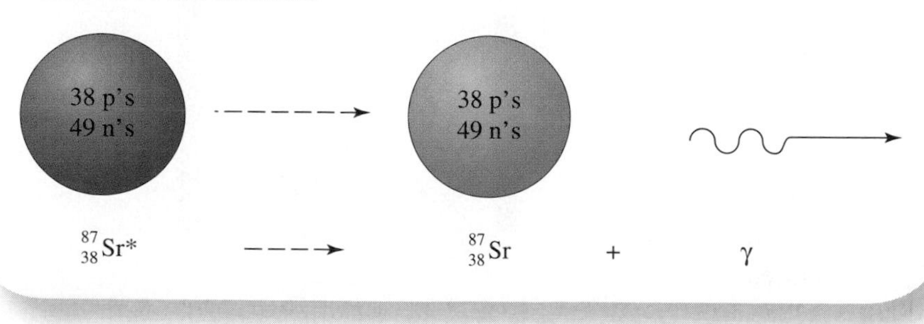

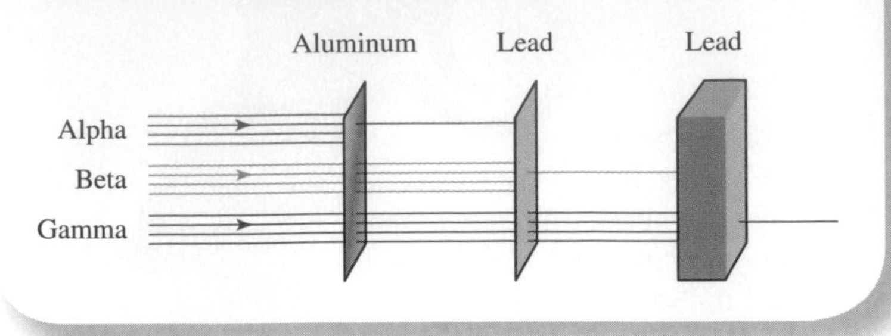

Figure 11.8
Alpha, beta, and gamma radiation differ a great deal in their ability to penetrate matter. Alpha rays are the least penetrating, while gamma rays are the most.

and its moons (see page 77),were equipped with *radioisotope thermoelectric generators* (RTGs). The heat from the radioactive decay of an isotope such as plutonium-238 is converted into electricity to operate cameras, radio transmitters, and other onboard equipment.

The interior of the Earth is so hot that much of it is molten. Some of this heat reaches the surface in volcanoes, in geysers, and through other geothermal processes. Geologists are not certain why the Earth's interior is so hot, but they know what keeps it from cooling off: heat from radioactive decay. In the Earth's interior, relatively small amounts of radioisotopes are still present. The energy released as these radioisotopes decay is enough to compensate for the transfer of heat to the Earth's surface. Since much of this radioactive material is concentrated in a layer just below the Earth's outer crust, this layer is kept in a partially liquid state by heat from the radioactivity. This in turn allows the crustal plates (continental land masses) to slowly slide around over the Earth in a process called *continental drift.*

Applications

Radiation from radioactive decay is used routinely for many purposes. Nuclear medicine makes use of radioisotopes for diagnosis and treatment (see the special feature article on page 306). Dozens of industrial facilities around the world use gamma radiation from cobalt-60 and other isotopes to sterilize disposable medical products such as syringes and physician gloves. The gamma rays can penetrate deep into containers of items and kill bacteria and other microorganisms. Many of these facilities also irradiate food items, from spices and herbs to fresh meat, with gamma rays and other ionizing radiation to prevent spoilage, prolong shelf life, and kill harmful bacteria like salmonella.

Because gamma rays are absorbed in a predictable way as they pass through solid material, they can be used to measure the thickness and integrity of manufactured items. Flaws in jet engine parts and cracks in large cables can be detected because more gamma rays will pass through items with gaps in them than through solid material. The thickness of metal sheets as they are manufactured can be measured by monitoring how much gamma radiation passes through them.

Another way in which radioactivity is put to good use is in the most common type of smoke detector. The alpha radiation from a radioisotope (americium-241) ionizes the air between two plates connected to the terminals of a battery (Figure 11.9). Consequently, an electric current flows through the circuit. When

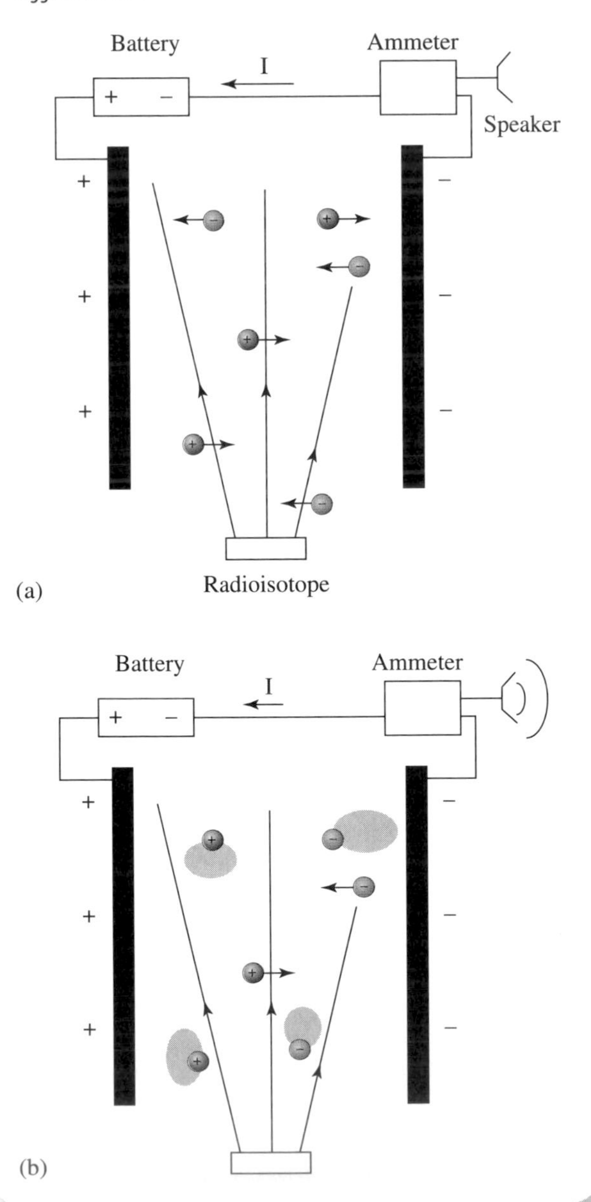

Figure 11.9
Simplified diagram of a smoke detector. (a) A current flows between the plates because the radiation ionizes the air. (b) Ions are attracted to the smoke particles and attach to them. This reduces the current, which triggers the alarm.

smoke enters the detection chamber, the ions are attracted to the smoke particles and attach to them. This reduces the flow of charge between the plates. An ammeter built into the electronic circuitry detects the resulting decrease in the current and triggers an audible alarm. Thousands of lives have been saved by these simple, inexpensive devices.

Analysis of radioactive decay is an important tool in nuclear physics. The type of radiation emitted by a particular radioisotope, along with the amount of energy released, provide clues to the structure of the nucleus. Since nuclei are much too small for us to examine with our eyes, we have to use indirect information, such as that carried by radiation when it leaves a nucleus, to learn about them. The Concept Map on your Review Card summarizes the concept of radioactive decay.

11.3 Half-Life

We now address an aspect of radioactive decay that we have avoided so far—time. If a nucleus is radioactive, when will it decay? In truth, there is no way of knowing exactly when a particular nucleus will decay: It may wait a billion years or only a millionth of a second. Radioactive decay is a random process, much like throwing dice. One can't predict exactly what will come up each throw, but one can analyze the results of hundreds of throws and establish how likely it is that each possible value will come up. Similarly, we can predict how much time will elapse, on the average, before a nucleus of a given radioisotope will decay. There is wide variation among the thousands of different radioisotopes. Nearly all nuclei of some isotopes will decay in less than a second, while only a fraction of the nuclei of other isotopes will decay in 4.5 billion years, the age of the Earth. This leads us to the concept of **half-life**.

The exact value of one roll of dice can't be predicted, but the approximate number of times that eleven will turn up in 1,000 rolls can be predicted quite closely. Similarly, the exact time at which a given nucleus will decay can't be predicted, but the fraction of 1 million nuclei that will decay during some time interval can be established very accurately.

© Jill Battaglia/iStockphoto.com / © C Squared Studios/Photodisc/Getty Images

DEFINITION

Half-Life *The time it takes for half the nuclei in a sample of a radioisotope to decay. The time interval during which each nucleus has a 50% probability of decaying.*

The half-lives of radioisotopes range from a tiny fraction of a second to billions of years (Table 11.3). During the span of one half-life, approximately half of the nuclei in a sample will decay—that is, emit their radiation. Half of the remaining nuclei will decay during the span of a second half-life, leaving only one-fourth of the original nuclei. After three half-lives, one-eighth remain, and so on. After n half-lives, one-half raised to the power n of the original nuclei will remain undecayed. That is,

$$N = N_0\left(\tfrac{1}{2}\right)^n$$

Here, N_0 is the original number of radioactive nuclei present in the sample, and N is the number of radioactive nuclei remaining after the elapse of n half-lives.

For example, let's say that we start with 8 million nuclei of a radioisotope with a half-life of 5 minutes. About 4 million of the nuclei will decay in the first 5 minutes. Half of the remaining 4 million will decay during the next 5 minutes, leaving 2 million. After 15 minutes, there will be 1 million left undecayed. After 50 minutes (10 half-lives), there will be about 7,800 nuclei left (8 million $\times$ $(\tfrac{1}{2})^{10} = 7{,}800$).

EXAMPLE 11.3

A pure sample of uranium-237 is prepared. As the uranium nuclei decay, the sample becomes "contaminated" with decay products. How much time will elapse before only one-fourth of the sample is uranium-237?

One-half of the uranium nuclei decay during 6.75 days, the half-life of uranium-237 (from Table 11.3). After another 6.75 days, one-half of the remaining nuclei decay, leaving one-fourth of the original amount. Therefore, after a total of *13.5 days*, only one-fourth of the sample will be uranium-237.

Table 11.3 Half-Lives of Isotopes in Table 11.2

Element	Isotope	Half-Life
Hydrogen	$^{1}_{1}H$	...
	$^{2}_{1}H$ deuterium	...
	$^{3}_{1}H$ tritium	12.3 yr
Helium	$^{3}_{2}He$	...
	$^{4}_{2}He$	...
	$^{5}_{2}He$	2×10^{-21} s
	$^{6}_{2}He$	0.805 s
	$^{8}_{2}He$	0.119 s
Carbon[a]	$^{12}_{6}C$	...
	$^{13}_{6}C$	...
	$^{14}_{6}C$	5,730 yr
	$^{15}_{6}C$	24 s
Silver[a]	$^{107}_{47}Ag^{*}$[b]	44.2 s
	$^{108}_{47}Ag$	2.42 min
	$^{109}_{47}Ag$	39.8 s
	$^{110}_{47}Ag$	24.6 s
Uranium[a]	$^{232}_{92}U$	70 yr
	$^{233}_{92}U$	159,000 yr
	$^{234}_{92}U$	245,000 yr
	$^{235}_{92}U$	704,000,000 yr
	$^{236}_{92}U$	23,400,000 yr
	$^{237}_{92}U$	6.75 d
	$^{238}_{92}U$	4,470,000,000 yr
	$^{239}_{92}U$	23.5 min

[a]Not every isotope of the elements C, Ag, and U is given.
[b]Asterisks (*) indicate that the nuclei are in an excited state.

In reality, it isn't possible to count how many nuclei are left undecayed every 5 minutes. But with a Geiger counter, one can keep track of how the rate of decay, the number of nuclei that decay each minute, decreases. In the example mentioned earlier, 4 million decay during the first 5 minutes, 2 million during the next 5 minutes, and so on. A Geiger counter placed nearby might show an initial count rate of 10,000 counts per minute. (The actual, observed count rate would depend on how close the counter is to the sample, as well as how efficient the counter is in detecting the type of nuclear radiation emitted.) Five minutes later, the count rate would be one-half that, about 5,000 counts per minute. In other words, the count rate also is halved during each half-life (Figure 11.10). Thus, by monitoring the count rate of the nuclear radiation emitted by a radioisotope, one can determine the isotope's half-life.

Is it useful to know the half-lives of radioisotopes? Yes, for a number of reasons. Knowing the half-life of a radioisotope allows us to estimate over what period of time a sample will emit appreciable amounts of radiation. The isotope americium-241 is used in smoke detectors partly because its long half-life (432 years) ensures that it will continue to produce the ions needed for the detector's operation throughout the device's lifetime.

Small amounts of certain radioisotopes are sometimes used medically in the human body (see the special feature on page 306). The flow of the isotope through the bloodstream (for example) can be monitored with a

Figure 11.10
(a) Graph of the number of remaining nuclei versus time for a radioisotope with a 5-minute half-life. (N_0 is the initial number of nuclei.) (b) Graph of the relative count rate versus time for the same radioisotope.

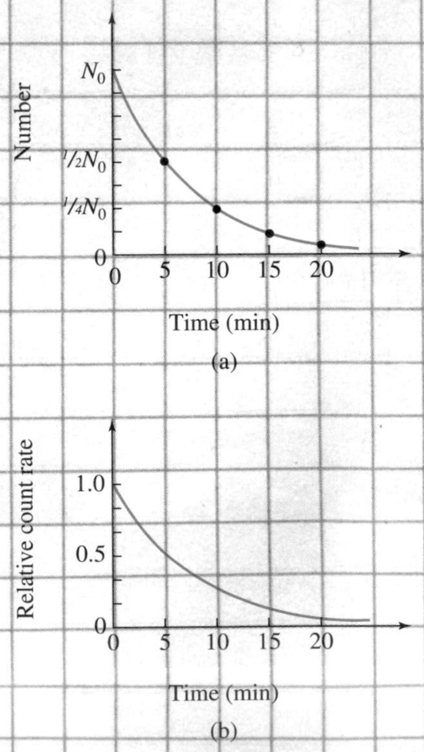

© Ivan Kmit/iStockphoto.com

Geiger counter. The radioisotope that is used must have a long enough half-life to remain detectable during the time that it takes to move through the body. Its half-life must not be too long, however. To minimize any possible harmful effects, the body should not be subjected to the radiation any longer than necessary. One of the most commonly used radioisotopes in nuclear medicine is technetium-99, a gamma emitter with a half-life of 6 hours.

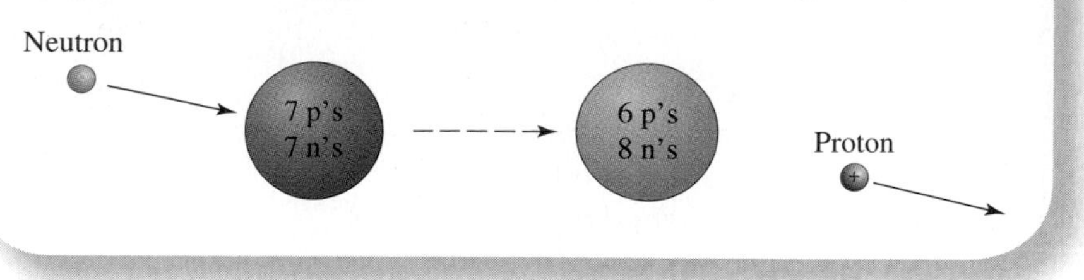

Figure 11.11
The formation of carbon-14 from nitrogen-14. This process occurs naturally in the upper atmosphere.

Dating

The regular rate of decay of radioisotopes can be used like a clock. **Carbon-14 dating** is a good example. Carbon is a key element in all living things on Earth. Carbon enters the food chain as plants take in carbon dioxide from the air. The complex carbon-based molecules formed by plants are ingested by animals and people. About 99% of this carbon is the stable isotope carbon-12. About one out of every trillion carbon atoms is radioactive carbon-14. In the upper atmosphere, carbon-14 is constantly being produced as cosmic rays from outer space collide with atoms in air molecules. Some of the fragments are free neutrons that cause the formation of carbon-14 when they collide with nitrogen-14. The reaction goes like this:

$$^{1}_{0}n + {}^{14}_{7}N \rightarrow {}^{14}_{6}C + {}^{1}_{1}p$$

(See Figure 11.11.) The carbon-14 atoms can combine with oxygen atoms to form carbon dioxide molecules, then mix in the atmosphere and enter the food chain. A small percentage of all the carbon atoms in plants, animals, and people is carbon-14. (You are slightly radioactive! The amount of carbon-14 is so small that the radiation—beta particles—is not a hazard.) The half-life of carbon-14 is about 5,700 years, so very little of it decays during the lifetime of most organisms. An exception would be certain trees that can live thousands of years.

After an organism dies, no new carbon-14 is added, so the percentage of carbon-14 decreases as the nuclei undergo radioactive decay. This can be used to measure the age of the remains. For example, if a tree is cut down and used to build a shelter, 5,700 years later there will be one-half as much carbon-14 in the wood as in a live tree (Figure 11.12). After 11,400 years, there would be one-fourth as much, and so on. Thus, by measuring the carbon-14 content in ancient logs, charcoal, bones, fabric, or other such artifacts, archaeologists can estimate their age. This process has become an invaluable tool to archaeologists and is quite accurate for material ranging up to about 40,000 years old.

Geologists use radioisotopes to estimate the ages of rock formations and the Earth itself. The ratio of the amount of the parent radioisotope to the amount of the daughter is used. For example, about 28% of the naturally occurring atoms of the element rubidium are the radioisotope rubidium-87. Its half-life is so long—49 billion

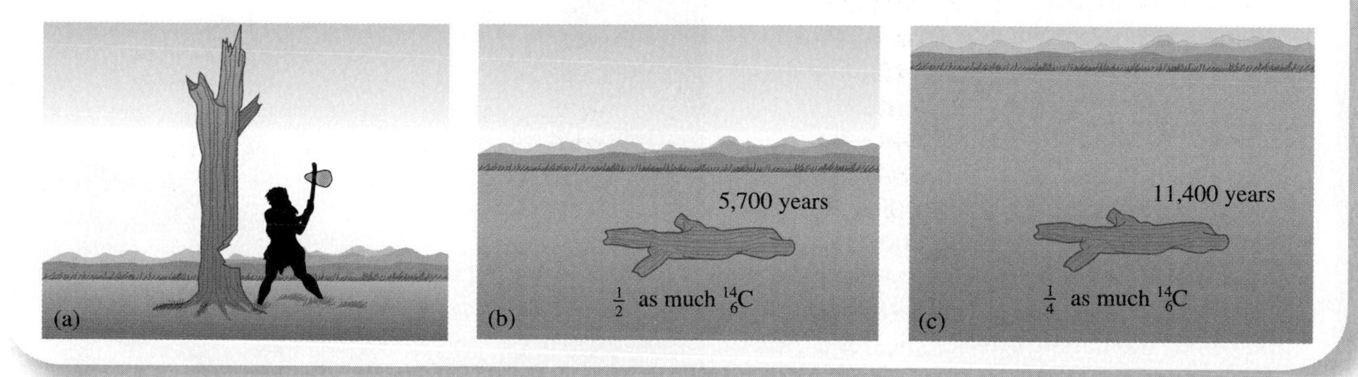

Figure 11.12
(a) As soon as a tree dies, the amount of carbon-14 in it begins to decrease. (b) After 5,700 years (one half-life), it contains one-half as much. (c) After 11,400 years, it contains one-fourth as much.

NUCLEAR MEDICINE

For the same reasons that radiation can be harmful to life, radiation can be a powerful tool in identifying and combating disease. Approximately one third of all hospitalized patients today benefit from the use of radioisotopes for diagnosis or treatment (i.e., nuclear medicine).

Diagnostics

Often, diagnosing an ailment involves finding abnormally high or low activity in some part of the body or determining if substances are circulating as they should. Radioactive nuclei make excellent probes for these tasks because they move through the body emitting radiation. A small quantity of a radioactive material can be placed in the bloodstream. Then, its movement through the circulatory system can be monitored with a Geiger counter placed near a vein or with a camera that responds to radiation instead of light.

Gamma-emitting radioisotopes are preferred for diagnostic procedures that require placing the material inside the body. Since gamma rays interact less with matter than some other forms of radiation, they do less damage to tissue and are more likely to make it out of the body where they can be detected. The half-life of an implanted radioisotope is a critical parameter for determining dosages, since it indicates over what period of time the irradiation will be appreciable.

Some chemical elements naturally accumulate in specific places. For example, the body uses calcium to "build" bones and teeth. The element iodine accumulates in the thyroid gland at a rate affected by such conditions as congestive heart failure or improper thyroid function. In one standard test, a patient ingests a pill that includes a radioactive isotope of iodine, such as iodine-123 (half-life = 13.1 hours). The iodine enters the bloodstream and begins to accumulate in the thyroid after a few hours. A radiation detector placed by the patient's neck indicates if the accumulation rate is normal or not.

Treatment

Radiation is used to fight cancer because radiation can kill the cancer cells while limiting the damage to normal tissue. External radiation therapy involves aiming a beam of radiation at the cancerous tissue in the body. Doctors limit exposure of healthy cells to the radiation by rotating the beam so that it hits the cancer from different directions. It is like shining a flashlight at a particular spot on the floor, while at the same time walking in a circle about that spot. X-rays and gamma rays are usually used for this purpose, but in some cases, a beam of high-speed neutrons or protons is more effective.

There are also therapies that place the source of radiation inside the body. In some cases, radioactive material is sealed in a small container and placed next to the cancer. This allows high doses to be administered to very specific locations. An oral dose of the radioisotope iodine-131 (half-life = 8 days) can be used to irradiate the thyroid selectively. The beta and gamma radiation it emits destroys cancerous tissue in the thyroid where the iodine accumulates. A more recent innovation is to incorporate radioactive nuclei in antibodies that naturally seek out cancer cells and attach to them. The radiation from the radioisotope can then kill the cells.

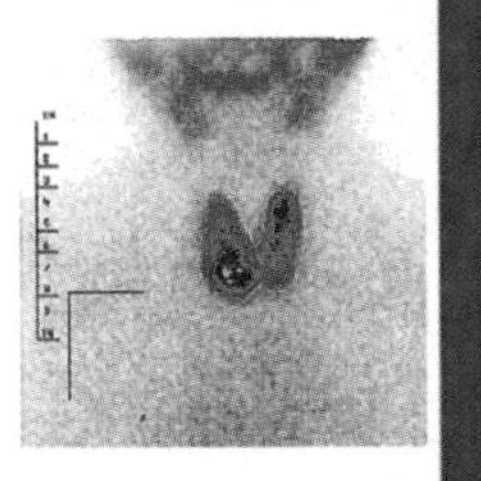

© Serghei Velusceac/iStockphoto.com / © Miriam Maslo/Photo Researchers, Inc.

years—that only a small percentage of it has decayed since the Earth was formed. Rubidium-87 undergoes beta decay into the stable isotope strontium-87. The age of rock formations that contain rubidium-87 can be estimated by measuring the relative amount of strontium-87 that is present. The more strontium-87 present, the older the rock.

Because of radioactive decay, most of the radioisotopes present when the Earth formed have long since decayed away. Only radioisotopes with very long half-lives, like rubidium-87 and uranium-238, have survived to this day. A few radioisotopes with short half-lives occur naturally because they are constantly being produced, carbon-14 being a good example. We humans have added a large number of radioisotopes to the environment through processes like nuclear explosions and nuclear power production. Concern over the amount of radioactive fallout in the atmosphere produced by nuclear weapons testing led to the limited Test Ban Treaty of 1963.

11.4 Artificial Nuclear Reactions

Radioactivity and the formation of carbon-14 from nitrogen-14 are examples of natural nuclear reactions. With the exception of gamma decay, each of these results in *transmutation*—the conversion of an atom of one element into an atom of another (the medieval alchemist's dream come true). Many other types of nuclear reactions can be induced artificially in laboratories. A common example of this is the bombardment of nuclei with alpha particles, beta particles, neutrons, protons, or other nuclei. If a nucleus "captures" the bombarding particle, it will become a different element or isotope. (This is called *artificial transmutation.* Note that here something is added to the nucleus, whereas in radioactive decay something leaves the nucleus.)

One example of a useful artificial reaction involves bombardment of uranium-238 nuclei with neutrons. The result is uranium-239.

$$^{238}_{92}\text{U} + {}^{1}_{0}\text{n} \rightarrow {}^{239}_{92}\text{U}$$

This new nucleus undergoes two beta decays, resulting in plutonium-239.

$$^{239}_{92}\text{U} \rightarrow {}^{239}_{93}\text{Np} + {}^{0}_{-1}\text{e}$$

$$^{239}_{93}\text{Np} \rightarrow {}^{239}_{94}\text{Pu} + {}^{0}_{-1}\text{e}$$

The plutonium-239 can be used directly in nuclear reactors, but the original uranium-238 can't. A breeder reactor is a nuclear reactor designed to use neutrons to produce, or "breed," reactor fuel: the plutonium-239. This process also shows how elements with Z greater than 92 can be produced artificially.

Similarly, neutron bombardment can be used to produce hundreds of other isotopes, most of which are radioactive. Some of the radioisotopes used in nuclear medicine (refer to page 306) are produced this way. **Neutron activation analysis** is an accurate method for determining what elements are present in a substance. The material to be tested is bombarded with neutrons. Many of the nuclei in the substance are transformed into radioisotopes. As shown in Figure 11.13, it is possible to determine which elements were originally present by monitoring the intensity and energy of the nuclear

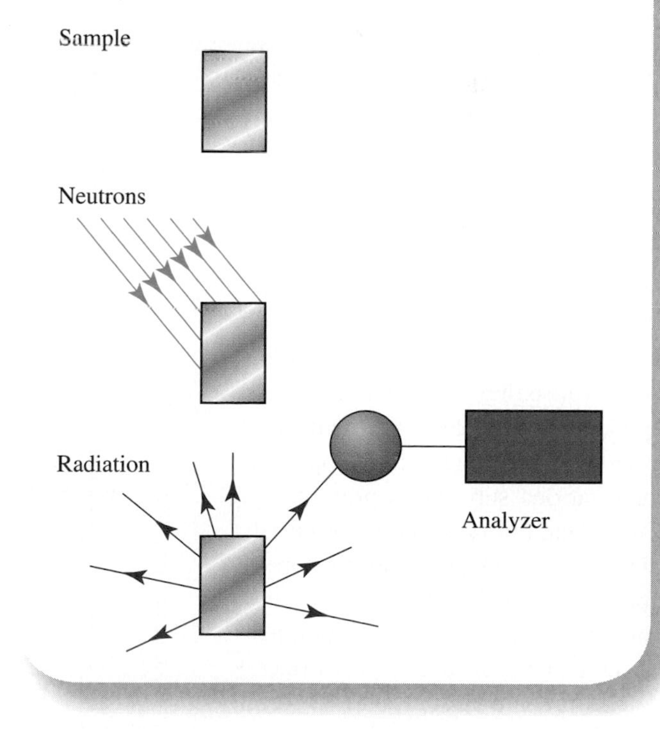

Figure 11.13
In neutron activation analysis, the sample is irradiated with neutrons. This produces radioisotopes in the sample. Analysis of the radiation that is emitted makes it possible to determine the original composition of the sample.

radiation that is then emitted. For example, the only naturally occurring isotope of sodium, sodium-23, becomes radioactive sodium-24 by neutron absorption.

$$^{23}_{11}\text{Na} + ^{1}_{0}\text{n} \rightarrow ^{24}_{11}\text{Na}$$

The sodium-24 emits both gamma and beta rays with specific energies. If sodium is present in an unknown substance, it can be detected by bombarding the substance with neutrons and then looking to see whether gamma and beta rays with the proper energies are emitted.

Neutron activation analysis played a key role in the discovery that a large asteroid or comet struck the Earth 65 million years ago, contributing to the extinction of dinosaurs. The element iridium is rare on Earth, but in the 1970s, neutron activation analysis revealed a high concentration of it in ancient sediments. Scientists concluded that a massive impact threw iridium-laden debris into the atmosphere, inducing adverse climatic conditions that eventually proved fatal for all but the smallest land animal species. In time, the iridium-rich material settled on the Earth's surface and became incorporated into the rock record.

Neutron activation analysis has a number of uses related to law enforcement. For example, it can reveal the presence of arsenic in a single hair from a victim who was poisoned. Likewise, an art forgery can be detected by using neutron activation analysis to determine if the paint or other materials conform to what was known to be in use when the piece was supposedly created.

This procedure can be readily used to find explosives and illegal drugs—even if they are inside a locked vehicle. Most chemical explosives contain nitrogen, and some illegal drugs, including cocaine, contain chlorine. A truck-mounted neutron activation analysis system can be parked next to a suspicious vehicle and scan it for the presence of either element. The vehicle is bombarded with neutrons, thereby producing radioisotopes in the various substances present inside. Gamma rays that are emitted in the decay of these radioisotopes pass into the detecting system and are analyzed to see if their energies match those of activated nitrogen or chlorine nuclei.

11.5 Nuclear Binding Energy

A very important and useful characteristic of a nucleus is its binding energy. Imagine the following experiment: A nucleus is dismantled by removing each proton and neutron one at a time, and the total amount of work done in the process of overcoming the strong nuclear force that binds the nucleons is measured. If the process is now reversed, and these protons and neutrons are allowed to reassemble under their mutual strong force attraction to form the original nucleus, an amount of energy equal to the previous work done would be released. This energy is called the **binding energy** of the nucleus. It indicates how tightly bound a nucleus is. A useful quantity for comparing different nuclei is the average binding energy for each proton and neutron, called the *binding energy per nucleon.* It is just the total binding energy divided by the total number of protons and neutrons in the nucleus—that is, the mass number. If we measure the binding energy per nucleon for all of the elements, we find that it varies considerably, from about 1 MeV to nearly 9 MeV (Figure 11.14).

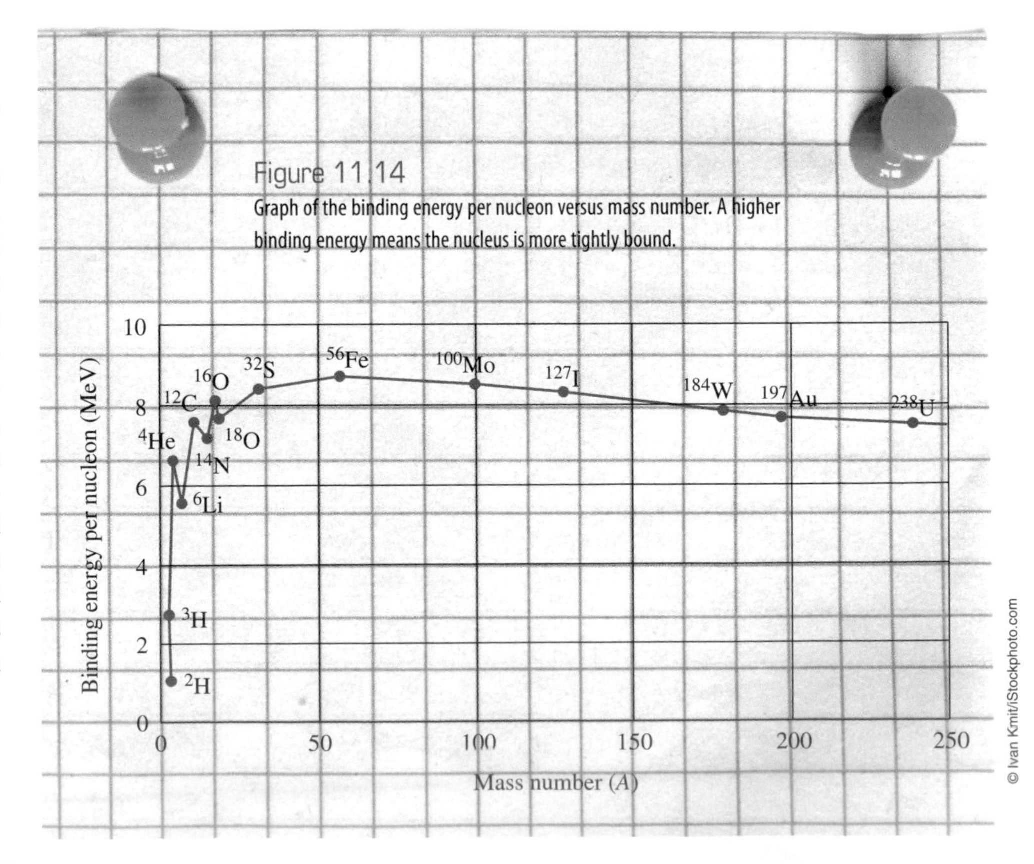

Figure 11.14
Graph of the binding energy per nucleon versus mass number. A higher binding energy means the nucleus is more tightly bound.

© Ivan Kmit/iStockphoto.com

Nuclei with mass numbers around 50 have the highest binding energies per nucleon: The protons and neutrons are more tightly bound to the nucleus than in larger or smaller nuclei, so these nuclei are the most stable. The relatively high stability of the helium-4 nucleus (alpha particle) is indicated by the small peak in the graph at $A = 4$.

A proton or neutron bound to a nucleus is similar to a ball resting in a hole in the ground; its total energy is negative. The ball's binding energy is the amount of energy that would have to be provided (the amount of work that would have to be done) to lift it out of the hole, thus releasing it from its confinement. A deeper hole means a higher binding energy.

The graph of binding energy per nucleon suggests how one can tap nuclear energy. Imagine taking a large nucleus with A around 200 and splitting it into two smaller nuclei. The graph shows that each of the smaller nuclei, say with A around 100, has a higher binding energy per nucleon than the original nucleus. All of the neutrons and protons have become more tightly bound together and have *released* energy in the process. (This is like moving the ball to a deeper part of the same hole. Its potential energy is decreased [made more negative], so it has given up some energy.) The act of splitting a large nucleus, referred to as *nuclear fission*, releases energy.

Similarly, we can combine two very small nuclei into one larger nucleus and release energy. If a hydrogen-1 nucleus and a hydrogen-2 nucleus are combined to form helium-3, the binding energy of each proton and neutron is increased, and energy is again released. This process is called *nuclear fusion.*

Clearly, the graph of binding energy per nucleon versus mass number is very important in nuclear physics because it shows why nuclear fission and nuclear fusion release energy. Fission is exploited in nuclear power plants and atomic bombs, and fusion is the source of energy in the Sun, the stars, and in hydrogen bombs. But how does one actually go about measuring the binding energy of a nucleus? It is not practical to actually dismantle a nucleus and to keep track of the amount of work done. (Remember that the largest nuclei contain over 200 nucleons.) The best way to measure binding energy is to use the equivalence of mass and energy.

One of the predictions of Einstein's special theory of relativity, presented in Section 12.1, is that the mass of

So, How *Do* You Weigh a Nucleus?

Well, you sure don't do it on a scale. One device that can determine the mass of a nucleus is a mass spectrometer. The particular kind of mass spectrometer shown to the right is used to analyze the surface of a sample. A pulsed ion beam removes molecules from the sample's surface. The molecules are ionised and accelerated towards a detector. The time it takes them to reach the detector depends on their mass.

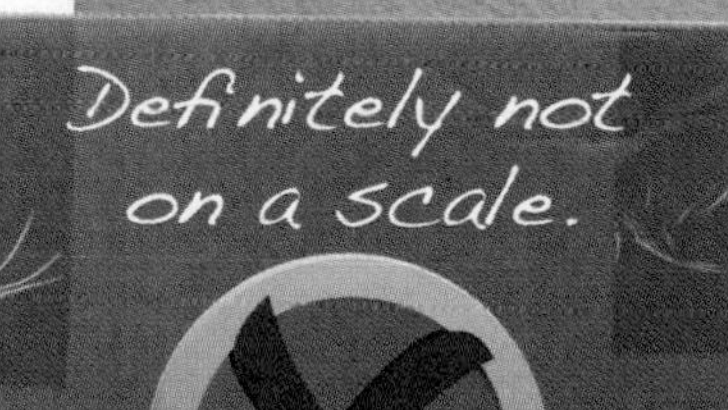

© Giorgio Fochesato/iStockphoto.com / © iStockphoto.com / © Tek Image/Photo Researchers, Inc. / © Filipp Bezlutskiy/iStockphoto.com

an object increases when it gains energy. The exact relationship between the energy E of a particle and its mass m is the famous equation

$$E = mc^2$$

When a particle is given energy, its mass increases; and when it loses energy, its mass decreases. In other words, mass can be converted into energy, and vice versa. The amount of mass "lost" (converted into energy) by things around us when we take energy from them is generally much too small to be measured. But the quantities of energy involved in nuclear processes are so great that mass-energy conversion can be measured.

For example, the mass of a hydrogen-1 atom (one proton and one electron) is 1.00785 u, and the mass of one neutron is 1.00869 u (see Table 11.1). The total mass of a hydrogen atom and a free neutron is 2.01654 u. But careful measurements of the mass of a hydrogen-2 (deuterium) atom give the value 2.01410 u. When the neutron and the proton are bound together in the hydrogen-2 nucleus, their combined mass is 0.00244 u *less* than when they are apart (Figure 11.15). This much mass is converted into energy when a proton and a neutron combine, and this energy is the binding energy of the hydrogen-2 nucleus.

In a similar way, we find that the masses of all nuclei are less than the combined masses of the individual protons and neutrons. The energy equivalent of this "mass defect" is the total binding energy of the nucleus. The binding energy divided by the mass number is the binding energy per nucleon of the nucleus. This is a good way to compute the binding energy per nucleon for the different isotopes.

Incidentally, the atomic mass unit, u, is defined to be exactly one-twelfth of the mass of one atom of carbon-12. In other words, u was defined so that the mass of an atom of carbon-12 is exactly 12 u. Other isotopes could have been used for establishing the size of u, but carbon-12 is convenient because it is very common and plays an important role in the life cycle of most organisms on Earth.

Fission and fusion are processes that convert matter into energy. In both cases, the total mass of all the nucleons afterwards is less than the total mass before. Some of the original mass, typically from 0.1 to 0.3%, is converted into energy. In the next two sections, we take a closer look at fission and at fusion. The Concept Map on your Review Card summarizes the concepts in Section 11.5.

Figure 11.15
A proton and a neutron bound together in a nucleus have *less mass* than a proton and a neutron separated. When the two combine, some of their mass is converted into energy—the binding energy of the nucleus. (The electron is not shown.)

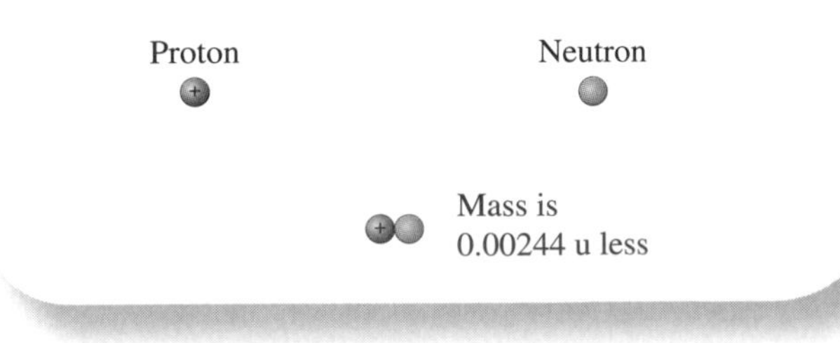

11.6 Nuclear Fission

In the 1930s, a discovery was made during neutron bombardment experiments that was one of the most fateful in human history: The nuclei of certain isotopes actually split when they absorb neutrons. Uranium-235 and plutonium-239 are the two most important nuclei that do this. Energy is released in the process along with several free neutrons. This process is called **nuclear fission**.

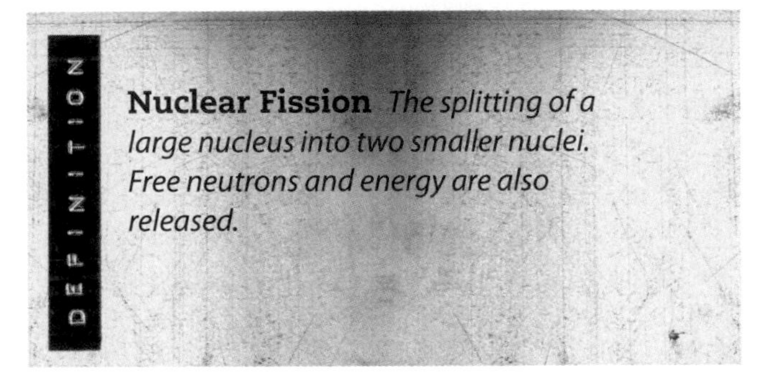

DEFINITION

Nuclear Fission *The splitting of a large nucleus into two smaller nuclei. Free neutrons and energy are also released.*

The two resulting nuclei are called *fission fragments.* For a particular type of fissioning isotope, the set of fission fragments produced in the decay of the activated species will not always be the same. There are dozens of different ways that a nucleus can split, and there are dozens of different possible fission fragments.

Fission is most commonly induced by bombarding nuclei with neutrons. Protons, alpha particles, and gamma rays also have been used. Upon absorption of a neutron, the nucleus becomes highly unstable and quickly divides (Figure 11.16). Two of the many possible fission reactions of uranium-235 are

$${}^{1}_{0}\text{n} + {}^{235}_{92}\text{U} \rightarrow {}^{236}_{92}\text{U}^{*} \rightarrow {}^{141}_{56}\text{Ba} + {}^{92}_{36}\text{Kr} + 3{}^{1}_{0}\text{n}$$

$${}^{1}_{0}\text{n} + {}^{235}_{92}\text{U} \rightarrow {}^{236}_{92}\text{U}^{*} \rightarrow {}^{140}_{54}\text{Xe} + {}^{94}_{38}\text{Sr} + 2{}^{1}_{0}\text{n}$$

Figure 11.16

A schematic representation of a nucleus of uranium undergoing fission. In this example, three neutrons are released.

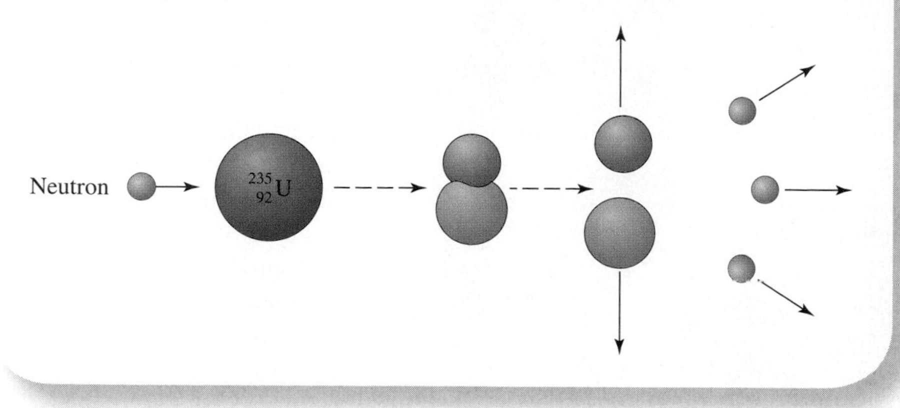

each fission induces another fission, a *stable* chain reaction results. The number of nuclei that split each second remains constant. Energy is therefore released at a steady rate. This type of reaction is used in nuclear power plants. If, on the other hand, more than one of the neutrons causes other fissions, an *unstable* chain reaction results—a nuclear explosion. One fission could trigger two, these two could trigger four more, then eight, sixteen, and so on. Energy is released at a rapidly increasing rate. This is the process used in atomic bombs.

The asterisk (*) indicates that the uranium-236 is unstable, so much so that it splits immediately. In both cases, energy is released during the fission event. The total mass of the fission fragments and the neutrons is less than the mass of the original uranium nucleus and the neutron. The missing mass is converted into energy. Most of this energy appears as kinetic energy of the fission fragments; they emerge from the interaction with high speeds.

The average amount of energy released by the fissioning of one uranium-235 nucleus is 215 million electronvolts (3.4×10^{-11} joules). By comparison, the amount of energy released during chemical processes like burning, metabolism in your body, and chemical explosions is typically about 10 electronvolts for each molecule involved. This is why nuclear-powered ships and submarines can go years without refueling, while ships that use diesel or oil must take on tons of fuel for each trip.

In addition to the energy released during fission, two other aspects of this process are extremely important.

1. Almost all of the possible fission fragments are themselves radioactive. The ratio of neutrons to protons in most fission fragments is too high for the nuclei to be stable, and they undergo beta decay. These radioactive fission fragments are important components of the radioactive fallout from nuclear explosions and of the nuclear waste produced in nuclear power plants.

2. The neutrons that are released during fission can strike other nuclei and cause them to split. This process leads to what is called a *chain reaction.*

For uranium-235, 2.5 neutrons are released on the average by each fission. If just one of the neutrons from

We can make a comparison between a nuclear chain reaction (fission) and a chemical chain reaction (burning). A stable chain reaction is similar to the burning of natural gas in a furnace or on a stovetop. The number of gas molecules that burn each second is kept constant, and the energy is released at a steady rate. An unstable chain reaction is like the explosion of gunpowder in a firecracker. The energy released at the start quickly causes the rest of the powder to burn rapidly until all of it is consumed in a violent, uncontrolled fashion.

In the remainder of this section, we look at some of the details involved in constructing atomic bombs and nuclear power plants. Keep in mind that there are two different types of nuclear bombs: atomic bombs, which use nuclear fission, and hydrogen bombs (also called *thermonuclear bombs*), which use both fission and nuclear fusion. The latter will be discussed in the next section.

The key raw material for both atomic bombs and nuclear power plants is uranium. Naturally occurring uranium is approximately 99.3% uranium-238 and 0.7% uranium-235. The uranium-235 fissions readily; the uranium-238 does not (unless irradiated with extremely high-speed neutrons), but it can be "bred" into fissionable plutonium-239.

Atomic Bombs

To produce an uncontrolled fission chain reaction, one must ensure that the fission of each nucleus leads to more than one additional fission. This is accomplished by using a high density of fissionable nuclei so that each neutron emitted in a fission is likely to encounter another nucleus and cause it to fission (Figure 11.17). In other words, nearly pure uranium-235 or plutonium-239 must be used. Uranium-235 can be extracted from uranium ore through a complicated and costly series of processes referred to as *enrichment*, or *isotope separation.* Since all uranium isotopes have essentially the same chemical properties, the separation processes rely on the difference in the masses of the nuclei (recall the special feature on p. 309).

Not only is nearly pure uranium-235 or plutonium-239 necessary, there must be a sufficient amount of it and it must be put into the proper configuration—a sphere, for example. This is so that each neutron is likely to collide with a nucleus before it escapes through the surface of the containment chamber holding the fissionable material. When these conditions are met, there is a *critical mass* of fissionable material. If there are too few fissionable nuclei present, or they are spread out too far (the density is low), too many of the neutrons will escape without causing other fissions. Also, the critical mass must be held together long enough for the chain reaction to cause an explosion. If not, the initial energy and heat from the first fissions will blow the critical mass apart and stop the chain reaction: a "fizzle." A firecracker is tightly wrapped with paper for a similar reason.

Another important consideration is timing: A premature explosion is highly undesirable, to say the least. This is prevented by keeping the fissionable material out of the critical mass configuration before the explosion. Two different techniques are used. In a gun-barrel atomic bomb, two subcritical lumps of uranium-235 are placed at opposite ends of a large tube (Figure 11.18). The explosion is triggered by forcing the two lumps together into a critical mass. This type of bomb was dropped on Hiroshima, Japan.

The other type of bomb, the implosion bomb, is used with plutonium-239. A subcritical sphere of the plutonium is surrounded by a shell of specially shaped conventional explosives (Figure 11.19). These explosives squeeze the plutonium-239 so much that it becomes a critical mass, and an explosion occurs. This type of bomb was dropped on Nagasaki, Japan.

A nuclear explosion releases an enormous amount of energy in the form of heat, light, other electromagnetic and nuclear radiation. The temperature at the center of the blast reaches millions of degrees. Even a mile away, the heat radiation is intense enough to instantly ignite wood and other combustible materials. The air near the explosion is heated and expands rapidly, producing a

© Popperfoto/Getty Images / © Phil Cardamone/iStockphoto.com

Figure 11.17

Fission chain reaction in pure uranium-235. Since all of the nuclei are fissionable, each neutron emitted by one fission event is likely to strike another nucleus and cause it to undergo fission. This leads to an explosive chain reaction.

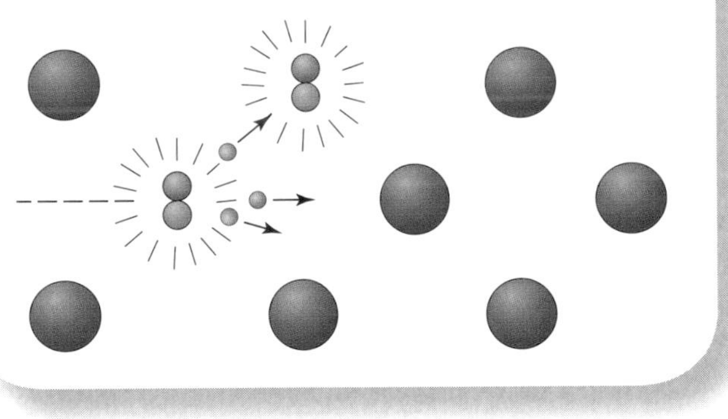

Figure 11.18

Gun-barrel atomic bomb. The two lumps of uranium are forced together to form a critical mass.

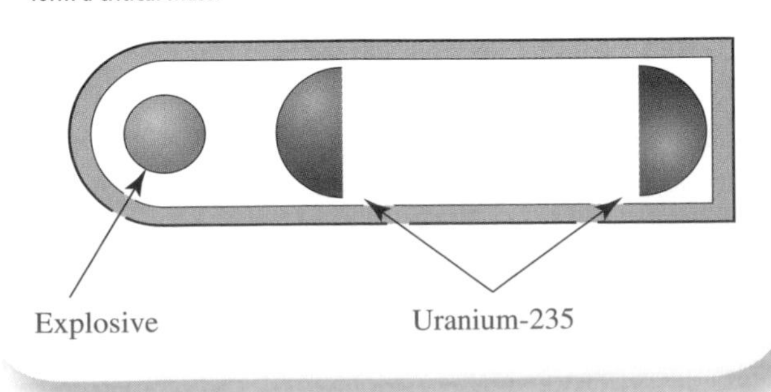

Figure 11.19

Implosion atomic bomb. The plutonium is subcritical until it implodes under the force of the explosives. This brings it to a critical mass.

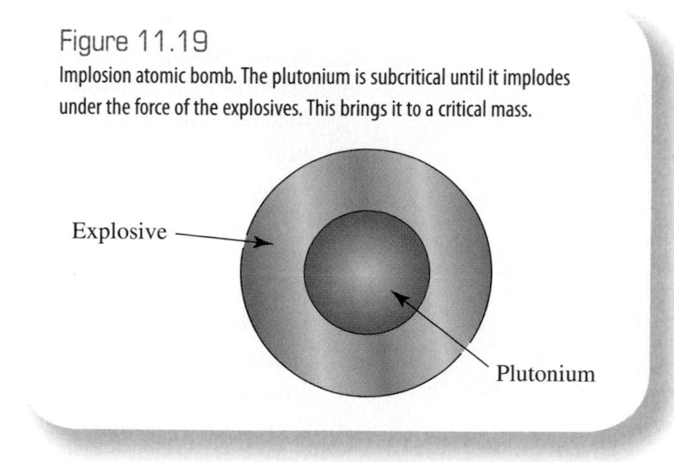

shock wave followed by hurricane-force winds. In addition to the radioactive fission fragments, large quantities of radioisotopes are produced by the neutrons and other radiation that bombard the air, dust, and debris. As this material settles back to the ground, it is collectively referred to as *fallout.*

The amount of energy released by an atomic bomb explosion is generally expressed in terms of the number of tons of TNT, a conventional high explosive, that would release the same amount of energy. One ton of TNT releases about 4.5 billion joules when it explodes. Most atomic bombs are in the 10–100 kiloton range: They are equivalent to between 10,000 and 100,000 tons of TNT. For comparison, an entire 100-unit freight train can carry only about 10,000 tons of commercial TNT. Some fission based nuclear weapons are small and lightweight enough to be carried by one person.

Nuclear Power Plants

About one-sixth of the world's electricity is generated by nuclear power plants. Over 435 of them are currently in operation around the globe, 104 of them in the United States alone. The reactors in these plants release the energy from nuclear fission in a controlled manner. The key to preventing a fission chain reaction from escalating to an explosion is controlling how the neutrons induce other fissions.

There are dozens of different designs for nuclear reactors. Most of them use only slightly enriched uranium fuel, typically about 3% uranium-235. This keeps the density of fissionable nuclei low so that not all neutrons from each fission are likely to strike other U-235 nuclei and to induce other fissions and makes it impossible for a critical mass configuration to arise: A nuclear reactor cannot explode like an atomic bomb (Figure 11.20). This is also economical because the enrichment process is very expensive.

Another important factor in producing and controlling a fission chain reaction in a nuclear reactor is the fact that slow neutrons are much better at causing uranium-235 nuclei to undergo fission than fast neutrons. Slowing down the neutrons released in the fissioning of uranium-235 makes it much more likely that they will be captured by other uranium-235 nuclei and cause them to split. This again makes it feasible to use a low concentration of uranium-235 in a reactor.

How are the neutrons slowed down? This is done through the use of a moderator, a substance that contains a large number of small nuclei. Small nuclei are more effective at taking kinetic energy away from the neutrons in the same way that a golf ball will lose more kinetic energy when it collides with a baseball at rest than with a bowling ball at rest. As neutrons pass through the moderator, they collide with the nuclei and lose some of their energy. They slow down. After many collisions, they have the same average kinetic energy as the surrounding nuclei. (For this reason, they are called *thermal neutrons* because their energies are determined by the temperature of the moderator.) Almost all nuclear power plant reactors in the United States use water as the moderator because of the hydrogen nuclei in the water molecules. Many reactors in England and the former Soviet Union use carbon in the form of graphite as the moderator.

The uranium fuel is shaped into long rods that are separated from each other. Large reactors have tens of thousands of individual fuel rods. Each rod relies on thermal neutrons from neighboring rods to sustain the chain reaction. This makes it possible to control the chain reaction by lowering *control rods* between the fuel rods (Figure 11.21). Control rods are made of materials that absorb neutrons effectively, such as the elements cadmium and boron. The rate of fissioning in a reactor

Figure 11.20

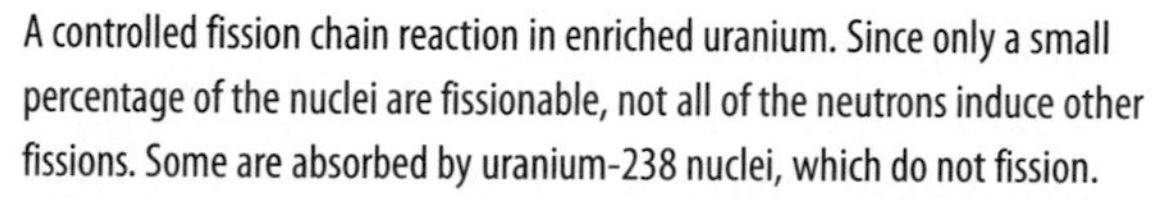

A controlled fission chain reaction in enriched uranium. Since only a small percentage of the nuclei are fissionable, not all of the neutrons induce other fissions. Some are absorbed by uranium-238 nuclei, which do not fission.

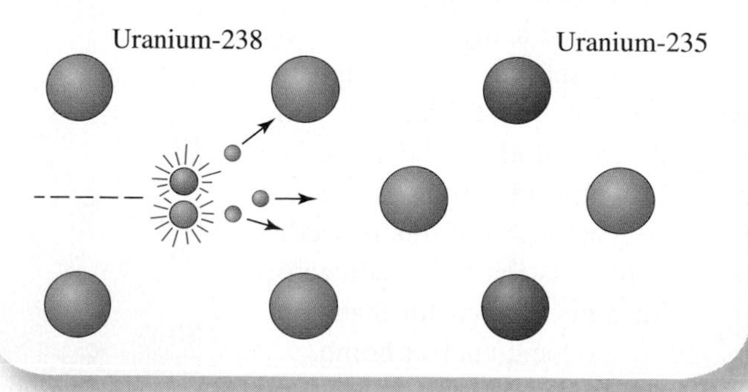

Figure 11.21

Simplified diagram of the core of a nuclear reactor. Neutron-absorbing control rods can be inserted between the fuel rods to control the chain reaction. The control rods prevent neutrons released in one fuel rod from inducing fissions in another fuel rod.

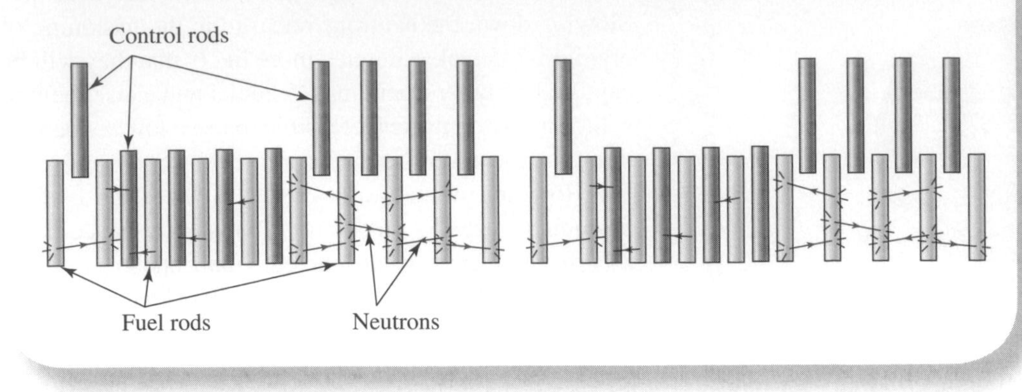

and, therefore, the rate of energy release, is regulated by the number of control rods that are withdrawn.

The energy released in a nuclear reactor is mainly in the form of heat. This heat is used to boil pressurized water into high-temperature steam. The rest of a nuclear power plant is much the same as that found in coal-fired power plants (refer to Figure 5.25). The steam turns a turbine, which turns a generator, which produces electricity. In most nuclear reactors, water flows around the fuel rods and serves both as a moderator slowing the released neutrons and as a coolant carrying away the heat in much the same way that the liquid in a car's radiator carries heat away from hot engine parts. If, for some reason, the reactor loses its water (called a "loss of coolant accident"), the reactor can become dangerously overheated.

Nuclear power plants are equipped with a sophisticated array of safety features and emergency backup systems. Millions of gallons of water are poised to flood the reactor if the temperature becomes too high or if the coolant system leaks. The control rods are designed to drop into place at the slightest sign of trouble. The entire nuclear reactor and supporting systems are housed inside a huge steel-reinforced concrete containment building. These buildings are typically about 200 feet high, dome shaped, with walls that are more than 1 foot thick. They are designed to seal airtight if there is a possibility that radiation has leaked from the reactor. All of this is needed because there is far more radioactivity inside a nuclear reactor than is released by a typical nuclear bomb.

Two notable accidents at nuclear power plants point out the potential for disaster that exists when human operators and designers make mistakes. In 1979, an unlikely series of errors and equipment malfunctions caused the core of a reactor at Three Mile Island, Pennsylvania, to lose most of its cooling water for 3 hours. Over half of the core melted. The containment building served its designed purpose however, and very little radiation escaped from the plant. But the costly reactor was ruined, and the core was reduced to a pile of intensely radioactive rubble. The billion-dollar cleanup lasted for more than a decade.

A major nuclear disaster occurred in April 1986 when a graphite-moderated power plant at Chernobyl (near Kiev, Ukraine) blew itself apart. Operators performing tests on the generator pushed the reactor beyond its limits. Since the plant did not have the kind of containment building used at Three Mile Island, tons of radioactive debris were released into the atmosphere and then carried by winds literally around the world. The huge mass of graphite was ignited by the heat and burned for 10 days, adding to the severity of the accident. Thirty-one people were killed by the immediate effects of the accident, and over 300,000 were required to move out of the contaminated area and resettle elsewhere. Early predictions that thousands might eventually die from radiation exposure

© Andrea Krause/iStockphoto.com

now appear to have been wrong. A United Nations report published in 2000 indicates that approximately 1,800 cases of thyroid cancer, mostly in individuals who were children at the time of exposure, were caused by the accident. Estimates of the total economic loss to the economies of Ukraine and neighboring countries run into hundreds of billions of dollars. Better reactor design and operator training should make a repeat of this accident unlikely.

Eventually, the fissionable isotopes in the fuel rods are consumed, and the spent rods have to be replaced. Spent fuel rods are highly radioactive because they contain large amounts of fission fragments, in the form of more than 200 different radioisotopes. So much radiation is emitted that the rods must be kept under water for months to remain cool. Even after the radioisotopes with short half-lives have decayed away and the rods have cooled, they remain dangerously radioactive for thousands of years. Finding a safe way to dispose of spent fuel rods and other nuclear waste is a major concern of government regulatory officials and nuclear power plant authorities.

Figure 11.22

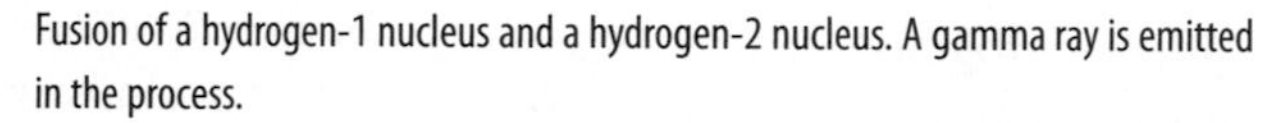

Fusion of a hydrogen-1 nucleus and a hydrogen-2 nucleus. A gamma ray is emitted in the process.

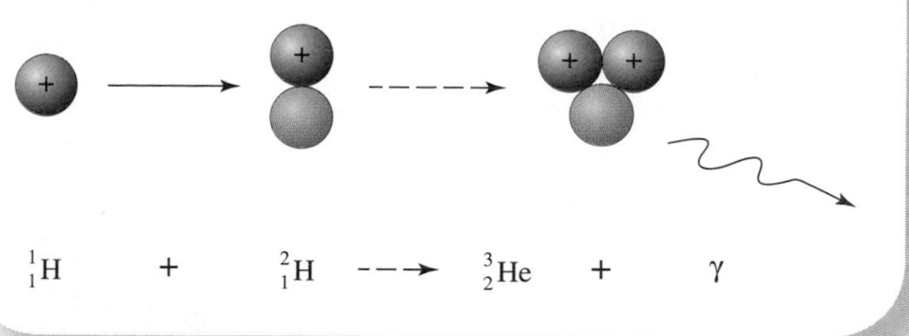

$^{1}_{1}H + ^{2}_{1}H \rightarrow ^{3}_{2}He + \gamma$

11.7 Nuclear Fusion

As incredible as it may seem, another source of nuclear energy exists that dwarfs nuclear fission: **nuclear fusion**. It is the source of energy for the Sun and the stars and for hydrogen bombs.

DEFINITION

Nuclear Fusion *The combining of two nuclei to form a larger nucleus with the accompanying release of energy.*

Fusion is like the reverse of fission, although it usually involves very small nuclei. One example of a fusion reaction is shown in Figure 11.22.

In this case, two hydrogen nuclei, one hydrogen-1 and the other hydrogen-2 (deuterium), fuse to form a nucleus of helium-3. There are many other possible fusion reactions that result in a release of energy. Some of these reactions, including the amount of energy released, are

$$^{2}_{1}H + ^{2}_{1}H \rightarrow ^{3}_{2}He + ^{1}_{0}n + 3.3 \text{ MeV}$$

$$^{2}_{1}H + ^{3}_{1}H \rightarrow ^{4}_{2}He + ^{1}_{0}n + 17.6 \text{ MeV}$$

$$^{2}_{1}H + ^{3}_{2}He \rightarrow ^{4}_{2}He + ^{1}_{1}p + 18.3 \text{ MeV}$$

In each case, energy is released because the total mass of the nucleons after the fusion is less than the total mass before. As with fission, the missing mass is converted into energy. This energy appears as kinetic energy of the fused nucleus and the energy of the proton, neutron, or gamma ray.

Fusion can occur only when the two nuclei are close enough for the short-range nuclear force to pull them together. This turns out to be a major problem in trying to induce a fusion reaction. Nuclei are positively charged. When brought close together, they exert strong repulsive forces on each other and resist fusion.

How can this difficulty be overcome? The most common way is with extremely high temperatures. If nuclei can be given enough average kinetic energy, their inertia will carry them close enough together to fuse when they collide (Figure 11.23). This process is called *thermonuclear fusion.*

Figure 11.23

In a high-speed collision of two nuclei, the repulsive force between them is overcome, and fusion occurs. At extremely high temperatures, nuclei possess such high speeds because of their thermal motions. This is thermonuclear fusion.

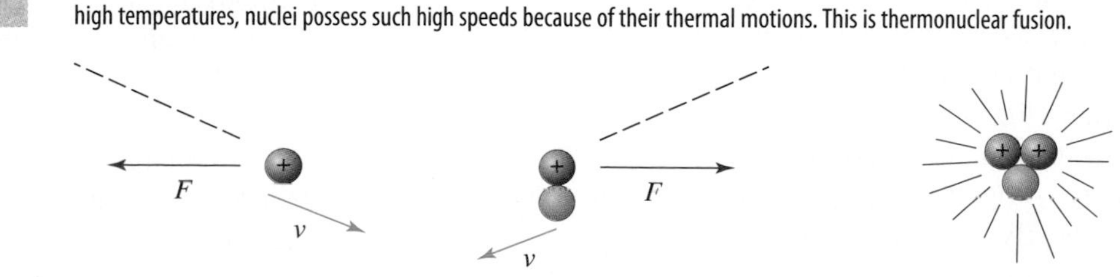

When we say that the temperatures are high, we really mean it: The hydrogen-1 plus hydrogen-2 fusion reaction requires a temperature of about *50 million* degrees Celsius. Some fusion reactions require several hundred million degrees Celsius. These unearthly temperatures do occur in the interiors of stars and at the centers of nuclear fission explosions.

Figure 11.24
Remnant of Tycho's supernova that erupted in November 1572 and was carefully observed and reported on by the famous Danish astronomer Tycho Brahe. This bubble of expanding gas and dust is located in the constellation of Cassiopeia at a distance of 7500 light years. The image is a composite made from observations taken in the x-ray region of the spectrum by the orbiting Chandra X-ray Observatory, in the infrared by the Spitzer Space Telescope, and in the optical using the 3.5-m Calar Alto telescope in Spain.

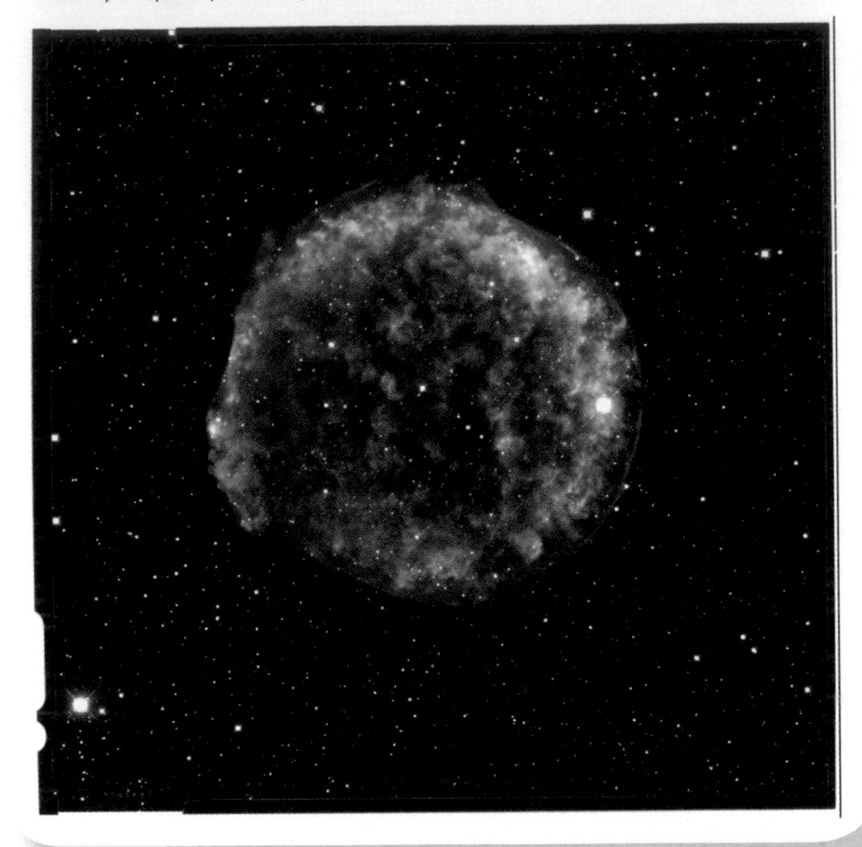

Courtesy of NASA/CXC/SAO, Infrared: NASA/JPL-Caltech; Optical: MPIA/CAHA / © Maria Toutoudaki/iStockphoto.com

Fusion in Stars

The Sun glows white hot and emits enormous amounts of energy. This energy originates in a natural fusion reaction in the Sun's interior. The Sun is composed mostly of hydrogen in a dense, high-temperature plasma. At the Sun's core, the temperature is around 15 million degrees Celsius, and the pressure is over one billion atmospheres. Under these conditions, the hydrogen undergoes a series of fusion reactions that results in the formation of helium. Each second, hundreds of millions of tons of hydrogen deep in the Sun's interior fuse, and more than 4 million tons of matter are converted into energy.

Stable stars get their energy from nuclear fusion. Some produce energy by reactions like those in the Sun, whereas others that are larger and have hotter cores use other fusion reactions. In large stars with core temperatures in excess of 100 million degrees Celsius, the helium fuses to form larger nuclei such as carbon and oxygen, which in turn may fuse to form silicon and, eventually, iron. In this way, the elements in the periodic table are built up from the basic raw material, hydrogen. The heaviest elements are produced during *supernova explosions*—the gigantic explosions of massive stars at the ends of their life cycles (Figure 11.24)—by successive neutron capture followed by beta decay. The majority of the elements contained in the Earth and everything on it (including you) were formed in the cataclysms of supernovas that occurred billions of years ago.

Life as we know it would be impossible on this planet without the Sun. Directly or indirectly, its energy supports the entire web of life. It is ironic that solar energy—that familiar, reliable, placid, and natural form of energy—originates from a violent nuclear process on a scale nearly beyond imagination.

Thermonuclear Weapons

The most destructive weapons in the world's arsenals are *thermonuclear warheads*, also called *hydrogen bombs.* These weapons get most of their explosive energy from the fusion of hydrogen. To produce the high temperatures necessary for the fusion reactions to occur, a nuclear fission explosion is used as a trigger. The fission explosion, an incredibly huge blast in itself, is but a primer for the monstrous fusion explosion. In terms of the relative amount of energy released, a hydrogen bomb is to an atomic bomb what a stick of dynamite is to a small firecracker.

Truly enormous amounts of energy are released. Many thermonuclear devices are in the *megaton* range; their energy output is given in terms of *millions* of tons of TNT. The largest weapon ever tested was rated at more than 50 megatons. To put this in perspective, this one blast released more energy than the total of all of the explosions in all of the wars in history, including the two atomic bomb blasts in World War II. However, it wasn't the most powerful blast to have ever occurred on Earth. Some volcanic eruptions, like the one on the island of Krakatoa in 1883, released more energy.

Currently, the world's arsenals contain thousands of nuclear warheads. Indications are that if even a fraction of these were used in a short "war," much of the life on this planet, including humankind, would be threatened with extinction.

Controlled Fusion

Soon after fusion was discovered, scientists looked for ways to harness it as an energy source. The initial success of nuclear fission reactors gave them hope. But fusion presents technical challenges that have resisted solutions for decades. For a thermonuclear fusion reaction to occur, two conditions must be met:

1. The nuclei must be raised to an extremely high temperature to ignite the fusion reaction.

   It is possible to produce temperatures in the millions of degrees, but the problem is containing (referred to as "confining") a plasma at this high temperature. Containers made of conventional materials would melt long before such temperatures would be reached.

2. There must be a high enough density of nuclei for the probability of collisions and, therefore, fusions, to remain high.

   If energy is to be released at a usable rate, a large number of fusions must occur each second. This means that the density of the plasma must be kept sufficiently high for the nuclei to collide often.

Research on controlled fusion is being pursued along a number of avenues. Several of these employ *magnetic confinement* of the plasma. Specially shaped magnetic fields are used to keep the plasma confined without letting it come into contact with other matter. This is possible because the nuclei are charged particles and consequently experience a force when moving in a magnetic field. A "magnetic bottle" is formed out of magnetic fields, and the plasma is injected into it. One of the most promising designs, called a *tokamak*, has the plasma confined inside a toroid—the shape of a doughnut (Figure 11.25). Fusion reactions have been produced, but so far, the amount of energy released has been less than the amount used to produce the reactions.

Another approach to controlled fusion is to use extremely intense bursts of laser light to produce miniature fusion explosions. A small capsule containing deuterium and tritium is blasted from several directions simultaneously. The enormous pressure inside the capsule, over 10^{11} atmospheres, squeezes the nuclei close enough together for fusion to take place. The National Ignition Facility (NIF) at Lawrence Livermore National Laboratory is the newest and largest such facility (Figure 11.26). At a cost of more than $4 billion, NIF, which was certified by the Department of Energy to begin testing in March 2009, will use short pulses from 192 laser beams to deliver 5×10^{14} watts of power to a fuel capsule. If good progress is made in testing the laser systems, ignition trials at the facility may begin as early as 2010.

A third technology has also achieved fusion and shows promise of one day producing a self-sustaining fusion reaction. The *pulsed power*, or *"Z," machine* makes use of an electromagnetic process in which hundreds of very thin wires are arranged symmetrically around, and parallel to, a central axis (the "z-axis"). A huge momentary current (thousands of amperes) is sent simultaneously through each wire. The wires are quickly vaporized by the ohmic heating and form a plasma. The huge magnetic field produced by the cumulative current exerts forces on the current-carrying plasma and compresses it violently;

Figure 11.25

Schematic of a tokamak fusion device. The plasma containing hydrogen nuclei is trapped by the magnetic field produced by the various electromagnets. [Adapted from *Scientific American* 249, no. 4 (October 1983):63.]

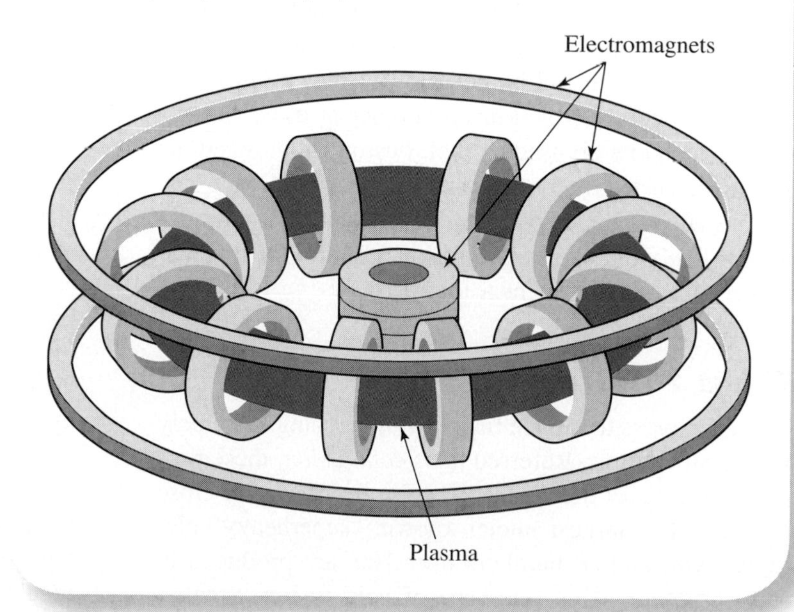

Figure 11.26
The 10-meter-diameter target chamber (blue) during construction of the National Ignition Facility. One hundred ninety-two high-powered laser beams will converge on a fuel capsule at the center of the chamber.

Courtesy Lawrence Livermore National Laboratory

the rapid, inward acceleration of the charged particles produces an intense pulse of x-rays. These x-rays trigger fusion in a BB-sized deuterium pellet. This route to sustained fusion is being pursued by scientists in France developing the *Laser Mégajoule* (LMJ) project slated for completion in 2012. It remains to be seen whether this avenue of research, or any of the others being pursued, will ever lead to commercial power from nuclear fusion.

There are good reasons why controlled fusion would make a good energy source. The main one is that the oceans contain an enormous supply of the fuel: hydrogen nuclei. Compared to fission reactors, much less radioactive waste would be generated. Chief by-products would be helium, which has a number of uses, and tritium, which can be used as fuel. Fusion chain reactions would be easier to control than fission chain reactions. But because of the huge technical challenges, it does not appear likely that controlled fusion will provide a viable source of energy in the near future.

Cold Fusion

There are ways to induce fusion without using extremely high temperatures. Referred to as *cold fusion*, these processes use other means to bring the fusing nuclei close together. The largest nuclei known, "superheavy" elements with atomic numbers over 100, are produced in the laboratory using one type of cold fusion. Smaller nuclei are accelerated to high speeds and collide with large nuclei. Under proper conditions, the nuclei fuse to form larger nuclei. Between 1958 and 1974, scientists in the United States and in the USSR accelerated very small nuclei to synthesize elements 102 through 106. In the 1980s, a team in Germany specialized in creating even larger nuclei by colliding not-so-small nuclei with large nuclei. For example, in 1996, they produced element 112 by bombarding lead-208 with zinc-70. The reaction equation is

$$^{70}_{30}\text{Zn} + {}^{208}_{82}\text{Pb} \rightarrow {}^{277}_{112}\text{Cp} + {}^{1}_{0}\text{n}$$

Cp is the symbol for copernicium, the proposed name for element 112. (See the footnote on page 86.)

The kinetic energy of the incoming nucleus must be just large enough to overcome the electrostatic repulsion between the two nuclei but not so large that the resulting nucleus has enough excess energy to undergo immediate fission. Researchers are optimistic that, using this technique and fine tuning the choices of the targets, projectiles, and their relative energies, prospects for creating increasingly larger and more massive nuclei will remain promising.

Another form of cold fusion seen in the laboratory is produced with the aid of an exotic elementary particle known as the *muon.* (Muons and other elementary particles are discussed in Chapter 12.) Hydrogen atoms normally combine in pairs to form H_2 molecules. If one of the two electrons is removed, the result is two hydrogen nuclei bound together by their mutual attraction to the electron. But, in this case, the nuclei are too far apart to undergo fusion. The negatively charged muon is basically an overweight electron: Its mass is about 200 times larger. Consequently, a "muonic atom" can exist that is just a muon in orbit about a hydrogen nucleus. Now, if the electron in the molecule described above is replaced by a muon, the two nuclei will be about 200 times closer together, and it is possible for them to fuse. This type of cold fusion has been observed, but it is not likely to be used as a source of energy. Muons are unstable, with a half-life of only 2.2×10^{-6} seconds. Too much energy would be needed to create a constant supply of muons,

questions&problems

Questions

(▶ Indicates a review question, which means it requires only a basic understanding of the material to answer. The others involve integrating or extending the concepts presented thus far.)

1. ▶ Why do different isotopes of an element have the same chemical properties?
2. The atomic number of one particular isotope is equal to its mass number. Which isotope is it?
3. A mixture of two common isotopes of oxygen, oxygen-16 and oxygen-18, is put in a chamber that is then spun around at a very high speed. It is found that one isotope is more concentrated near the axis of rotation of the chamber and the other is more concentrated near the outer part of the chamber. Why is that, and which isotope is where?
4. ▶ What is the name of the force that holds protons and neutrons together in the nucleus?
5. ▶ What aspects of the composition of a nucleus can cause it to be unstable?
6. ▶ Describe the common types of radioactive decay. What effect does each have on a nucleus?
7. A nuclear explosion far out in space releases a large amount of alpha, beta, and gamma radiation. Which of these would be detected first by a radiation detector on Earth?
8. The deflection of an alpha particle as it passes through a magnetic field is much less than the deflection of a beta particle (see Figure 11.3). Why?
9. A standard treatment for some cancers inside the body is to use nuclear radiation to kill cancer cells. If the radiation has to pass through normal tissue before reaching the cancer, why would alpha radiation not be a good choice?
10. A concrete wall in a building is found to contain a radioactive isotope that emits alpha radiation. What could be done to protect people from the radiation (short of razing the building)? What if it were gamma radiation that was being emitted?
11. ▶ Explain the concept of half-life.
12. ▶ One-half of the nuclei of a given radioisotope decays during one half-life. Why doesn't the remaining half decay during the next half-life?
13. A large number of regular six-sided dice are shaken together in a box, then dumped onto a table. Those showing "1" or "2" are removed, and the process is repeated with the remaining ones. Is the half-life of the dice greater than, equal to, or less than one throw?
14. ▶ How is carbon-14 used to determine the ages of wood, bones, and other artifacts?
15. One cause of uncertainty in carbon-14 dating is that the relative abundance of carbon-14 in atmospheric carbon dioxide is not always constant. If it is discovered that during some era in the past carbon-14 was more abundant than it is now, what effect would this have on the estimated ages of artifacts dated from that period?
16. The half-life of plutonium-238, the isotope used to generate electricity on the *Voyager* spacecraft, is about 88 years. What effect might this have on the spacecraft's anticipated useful lifetime?
17. ▶ The half-lives of most radioisotopes used in nuclear medicine range between a few hours and a few weeks. Why?
18. ▶ What are the principal steps in neutron activation analysis?
19. During the normal operation of nuclear power plants and nuclear processing facilities, machinery, building materials, and other things can become radioactive even if they never come into physical contact with radioactive material. What causes this?
20. If the binding energy per nucleon (see the graph in Figure 11.14) increased steadily with mass number instead of peaking around $A = 56$, would nuclear fission and nuclear fusion reactions work the same way they do now? Explain.
21. ▶ How can a nucleus of uranium-235 be induced to fission? Describe what happens to the nucleus.
22. ▶ What aspect of nuclear fission makes it possible for a chain reaction to occur? What is the difference between a chain reaction in a bomb and one in a nuclear power plant?
23. ▶ Explain how materials that absorb neutrons are used to control nuclear fission chain reactions.
24. ▶ What are fission fragments, and why are they so dangerous?
25. ▶ There is much more uranium-235 in a typical nuclear power plant than there was in the bomb that destroyed the city of Hiroshima. Why can't the reactor explode like an atomic bomb?
26. After a fuel rod in a fission reactor reaches the end of its life cycle (typically 3 years), most of the energy that it produces comes from the fissioning of plutonium-239. How can this be?
27. ▶ Why is a nuclear fusion reaction so difficult to induce?
28. If the strong nuclear force had a longer range than it does, what effect (if any) would that have on efforts to harness controlled fusion as an energy source?
29. ▶ Why are extremely high temperatures effective at causing fusion? What is used to produce such temperatures in a thermonuclear warhead?
30. ▶ Why is magnetic confinement being used in fusion research?
31. ▶ What is meant by the term "cold fusion"?

Problems

(Note: In some of these, you may need to use the Periodic Table of the Elements on your tear-out card.)

1. Determine the nuclear composition (number of protons and neutrons) of the following isotopes.
   a) carbon-14　　b) calcium-45
   c) silver-108　　d) radon-225
   e) plutonium-242
2. The isotope helium-6 undergoes beta decay. Write the reaction equation, and determine the identity of the daughter nucleus.
3. The isotope silver-110 undergoes beta decay. Write the reaction equation, and determine the identity of the daughter nucleus.
4. A nucleus of oxygen-15 undergoes electron capture. Write out the reaction equation, and determine the identity of the daughter nucleus.
5. The isotope polonium-210 undergoes alpha decay. Write the reaction equation, and determine the identity of the daughter nucleus.
6. The isotope plutonium-239 undergoes alpha decay. Write the reaction equation, and determine the identity of the daughter nucleus.
7. The isotope silver-107* undergoes gamma decay. Write the reaction equation, and determine the identity of the daughter nucleus.
8. The following is a possible fission reaction. Determine the identity of the missing nucleus.

   $${}^{1}_{0}n + {}^{235}_{92}U \longrightarrow {}^{236}_{92}U^{*} \longrightarrow {}^{95}_{39}Y + ? + 2{}^{1}_{0}n$$

9. The following is a possible fission reaction. Determine the identity of the missing nucleus.

   $${}^{1}_{0}n + {}^{235}_{92}U \longrightarrow {}^{236}_{92}U^{*} \longrightarrow {}^{143}_{57}La + ? + 3{}^{1}_{0}n$$

10. Two deuterium nuclei can undergo two different fusion reactions. One of them is given at the beginning of Section 11.7. In the second possible reaction, two deuterium nuclei fuse to form a new nucleus plus a lone proton. Write the reaction equation, and determine the identity of the resulting nucleus.
11. At the centers of some stars, nitrogen-15 can undergo a fusion reaction with a proton (a hydrogen-1 nucleus) to produce two different nuclei. One of the products is an alpha particle (a helium-4 nucleus). What is the other? Write down the reaction equation for this transformation.
12. Iron-58 and bismuth-209 fuse into a large nucleus plus a neutron. Write the reaction equation, and determine the identity of the resulting nucleus.
13. A Geiger counter registers a count rate of 4,000 counts per minute from a sample of a radioisotope. Twelve minutes later, the count rate is 1,000 counts per minute. What is the half-life of the radioisotope?
14. Iodine-131, a beta emitter, has a half-life of 8 days. A 2-gram sample of initially pure iodine-131 is stored for 32 days. How much iodine-131 remains in the sample afterwards?
15. An accident in a laboratory results in a room being contaminated by a radioisotope with a half-life of 3 days. If the radiation is measured to be eight times the maximum permissible level, how much time must elapse before the room is safe to enter?
16. As a general rule, the radioactivity from a particular radioisotope is considered to be reduced to a safe level after 10 half-lives have elapsed. (Obviously, the initial quantity of the isotope is also important.)
    a) By how much is the rate of emission of radiation reduced after 10 half-lives?
    b) Plutonium-239 is considered to be one of the most dangerous radioisotopes. Its half-life is about 25,000 years. How long would a sample of plutonium-239 have to be kept isolated before it could be considered safe?
17. The amount of carbon-14 in an ancient wooden bowl is found to be one-half that in a new piece of wood. How old is the bowl?
18. When the plutonium bomb was tested in New Mexico in 1945, approximately 1 gram of matter was converted into energy. How many joules of energy were released by the explosion?
19. A nucleus of element 112 is formed using the reaction equation given near the end of Section 11.7 on page 318. It then undergoes 6 successive alpha decays. Give the identity of the isotope that results after each step of this process.
20. A nucleus of element 114 is produced by fusing calcium-48 with plutonium-244. Write the reaction equation assuming three neutrons are also released.

There are more quizzes and study tools online at 4ltrpress.cengage.com/physics.

"They are written in **concise, down-to-earth language.** There are tons of pictures and interesting blurbs of information. It's very relevant to my life. It's nice to have a book/website that seems to **reach out to students and actually care** about how we learn and try to tailor to our needs as much as possible. Thank you for this."

– Alice Brent, Student at Arizona State University

SPEAK UP!

THEY DID

PHYSICS was built on a simple principle: to create a new teaching and learning solution that reflects the way today's faculty teach and the way you learn.

Through conversations, focus groups, surveys, and interviews, we collected data that drove the creation of the current version of PHYSICS that you are using today. But it doesn't stop there – in order to make PHYSICS an even better learning experience, we'd like you to SPEAK UP and tell us how PHYSICS worked for you.

What did you like about it?
What would you change?
Are there additional ideas you have that would help us build a better product for next semester's physics students?

At **4ltrpress.cengage.com/physics** you'll find all of the resources you need to succeed in physics – **simulations, interactive quizzes, flashcards,** and more.

Speak Up! Go to **4ltrpress.cengage.com/physics.**

Special Relativity and Elementary Particles

12.1 Special Relativity: The Physics of High Velocity

Imagine the following experiment: You are seated in the cargo area of a small pickup truck moving directly away from a companion at a constant velocity of 20 km/h. Your friend tosses a baseball to you with a horizontal velocity of 50 km/h (Figure 12.1). From your point of view, what is the ball's velocity? If you answered 30 km/h, you're right. Clearly, the velocity of the ball with respect to you depends upon *your own velocity*—that is, *the velocity of the observer.* This is common sense: Newton would have agreed with you completely.

Now consider a second hypothetical experiment: You enter a spacecraft and leave the Earth, traveling uniformly at a velocity of 200,000 km/s. After a time, a friend sends out a light ray, which moves at the speed of light—300,000 km/s—in your direction

Figure 12.1
If you are traveling at 20 km/h, and a friend throws a ball toward you at 50 km/h, you see the ball approaching at 30 km/h.

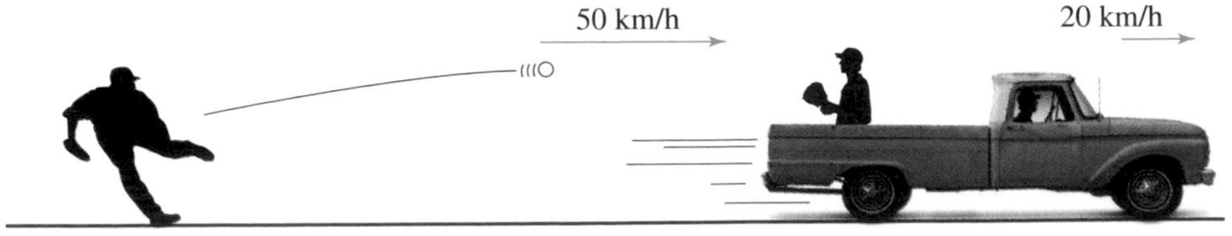

© Jon Patton/iStockphoto.com / © iStockphoto.com / © iStockphoto.com

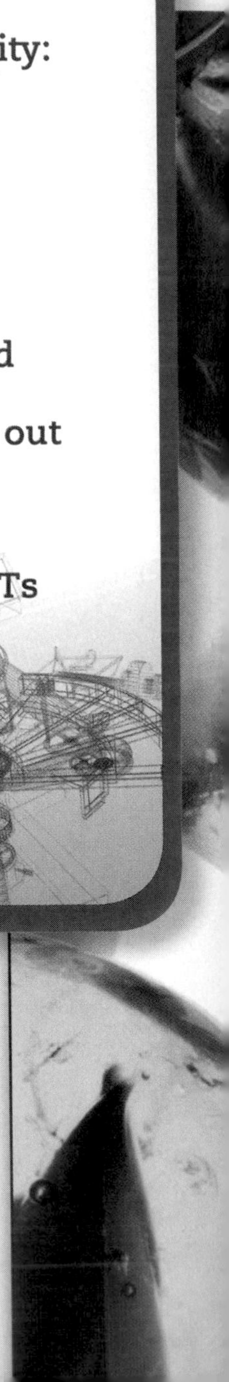

(Figure 12.2). When the light ray reaches you, what would you measure its speed to be? If you answered 100,000 km/s, you are wrong. Strange as it may seem, you would find the speed of the light to be 300,000 km/s, just what it is for your friend back on Earth. Evidently, *the speed of light does not depend upon the motion of the observer.* Light (and all other forms of electromagnetic radiation), behaving as Maxwell predicted, does *not* act as we would expect, based on the physics of Newton. The theory of Newtonian mechanics and the theory of electromagnetism appear to be in conflict.

Postulates of Special Relativity

Albert Einstein recognized the contradiction between the predictions of classical mechanics and those of electromagnetism as regards the propagation of light. He set about reconciling the two by adopting these two postulates.

1. The speed of light, c = 300,000 km/s, is the same for all observers, regardless of their motion.
   Like the gravitational constant G in Newton's law of universal gravitation or like Planck's constant, h, in the equation quantizing the energy of atomic oscillators, c is a fundamental constant

© Milos Luzanin/iStockphoto.com

Figure 12.2
The light approaches you at 300,000 km/s, *even if you are moving away* from the source at 200,000 km/s. (Drawing not to scale.)

300,000 km/s

200,000 km/s

© Jan Rysavy/iStockphoto.com / © Sergii Tsololo/iStockphoto.com / © iStockphoto.com / © Leon Bonaventura/iStockphoto.com

of Nature. The fact that the speed of light is constant was just starting to be accepted when Einstein began his studies in the early 1900s. Over the past one hundred years, precise experiments have demonstrated beyond doubt that the speed of light is invariant under all circumstances. For example, measurements of the speed of photons emitted in the decay of subatomic particles called *pions* (π mesons, see page 334) traveling at $0.9998c$, instead of producing $\sim 2c$ as expected from Newtonian kinematics, yield c to within 0.02%, in excellent agreement with predictions based on Einstein's first postulate.

2. The laws of physics are the same for all observers moving uniformly—that is, at a constant velocity. This is the *principle of relativity.*

   This means that if two observers traveling toward one another at a constant speed perform identical experiments, they will get identical results. Moreover, *no* experiment can be performed by either observer that will indicate whether they are moving or what their speed is. Two people playing air hockey in the lounge of a 747 jet plane cannot tell from the motion of the puck on the table whether they are aloft and traveling uniformly at 800 km/h or sitting at rest on the airport tarmac. You can play billiards or shuffleboard on a cruise ship and not be able to tell the ship is in motion so long as its velocity is constant. In each case, the results of the "experiments" (playing air hockey or shuffleboard) cannot be used to demonstrate uniform motion relative to the Earth because the laws of mechanics (and of physics more generally) are the same for you and an Earth-bound observer. But, you say, you can look out the window of the jet or the ship and see that you are in motion. True, but how can you *prove* that you are not really at rest and that, by some magic, the clouds, trees, mountains, and so forth are not moving uniformly past you in the opposite direction? In point of fact, you can't! The principle of relativity has been confirmed numerous times in particle-collision experiments where energy and momentum measurements show excellent agreement with predictions based on this statement.

Based on these two postulates, Einstein developed his **special theory of relativity**, which was published in 1905. It describes how two observers, in uniform relative motion, perceive space and time differently. One of the interesting aspects of this theory is that once you have accepted the experimentally verified postulates on which it is based, the fundamental predictions can be understood with only elementary algebra. The equations of special relativity are only a little more difficult than Newton's laws of motion or his law of universal gravitation.

Predictions of Special Relativity

Let's consider one of the predictions of special relativity, something called **time dilation**. Imagine that we construct two identical "clocks" consisting of a flashbulb, a mirror, and a light-sensitive detector (Figure 12.3). The flashbulb emits flashes of light at some preset rate. Each light pulse is reflected by the mirror into the detector. Each time the detector receives a pulse, an audible click is emitted like that of a standard wall clock. Let's now synchronize our two clocks and give one to some friends who are to travel in a spaceship with velocity v relative to you on Earth. The question is, Will the two clocks keep the same time? That is, will they continue to tick at the same rate? The answer seems obvious: Yes! This is the answer Newton would have provided. Unfortunately, given the postulates of special relativity, it is the wrong answer. Let's see why.

Figure 12.3

(a) A "light clock" at rest in a laboratory on Earth. (b) The light clock on board a spacecraft traveling uniformly with velocity v, as it would be seen by an observer on Earth.

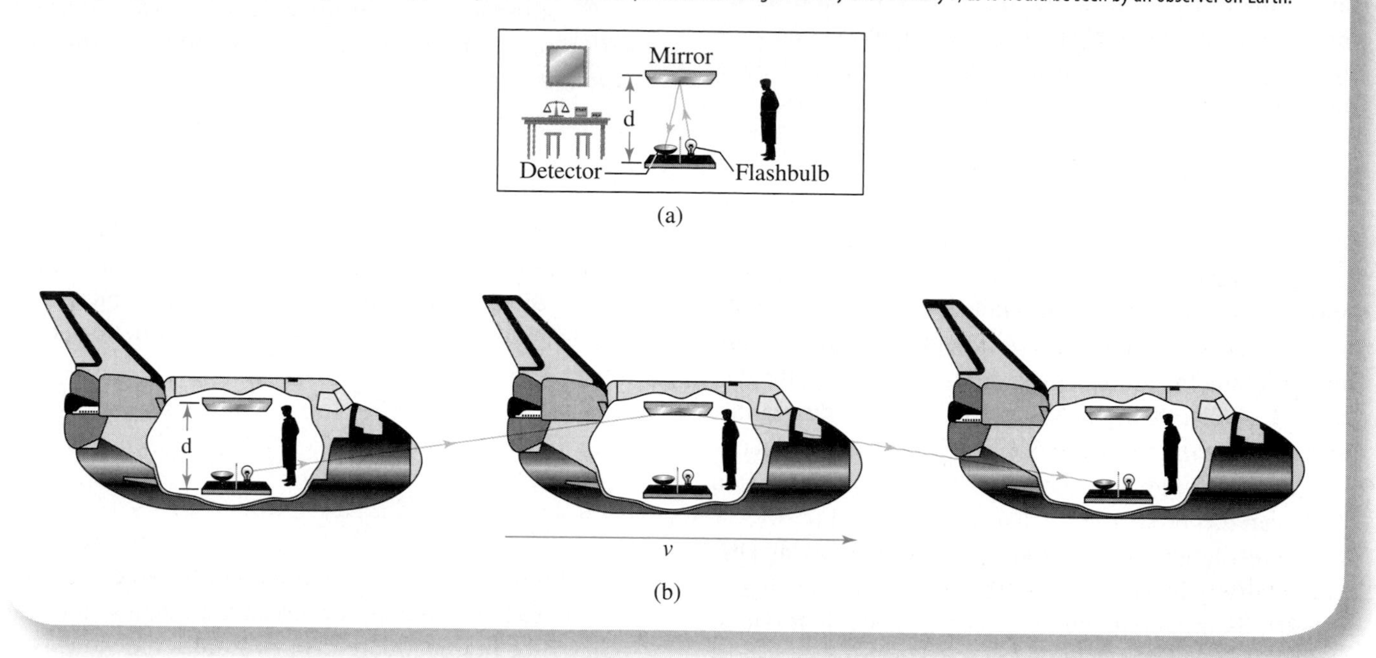

Consider the clock in the spaceship. When your friends took it on board, everyone agreed that it was a properly working "standard" clock. Consequently, they note nothing peculiar in its performance as they travel along. Indeed, they *cannot* identify anything different about the clock, because if they did, they could know they were moving. Such a circumstance would violate the principle of relativity, which says that physics is the same for all uniformly moving observers. The clock on the spacecraft, *as seen by your friends aboard the craft*, ticks along at the same rate it did when they first received it.

But what about the clock on board the spaceship as seen by you, an external observer? If you track the motion of the clock, say, with a powerful telescope, you see that the light, in going from flashbulb to detector, follows a zigzag path, since the pulse (moving with the spaceship) has a sideways component to its velocity in addition to a vertical component (see Figure 1.7). Evidently, the path that the light travels in the moving clock is longer than the path it follows in your laboratory clock. Consequently, since the speed of light is the same in both cases, you conclude that the time it takes the light to reflect back to the detector is longer for the moving clock than for your clock. In other words, the moving clock is running slow: The rate at which it ticks is slower than that for your clock. Or, put yet another way, the time *intervals* between ticks on the moving clock have been dilated or expanded.

Of course, if your friends read your clock from their spaceship, it is *your* clock that appears to be running slow, because from their vantage point, your laboratory appears to be moving with uniform velocity v in the *opposite direction.* The symmetry between the observations made on Earth and in the spaceship is guaranteed by the principle of relativity. But who is *really* right, you ask? Whose clock is *really* running slow? Both observers are right, and each clock is really running slow when compared to the other. The observers perceive the rate of flow of time differently because of their relative motion. But because no experiment can determine which observer is really in uniform motion, each observer's perception is as good or as true as the other's. *Time, then, is not an absolute:* It depends on the observer and their state of *relative* motion.

By how much will the interval between successive ticks differ for the two clocks discussed above? Not very much, it turns out, unless v is very close to the speed of light. If we let Δt be the time between ticks on the clock at rest with an observer and let $\Delta t'$ be the observed time between ticks on the clock moving with velocity v relative to the observer, then Δt and $\Delta t'$ are related by the following equation:

$$\Delta t' = \frac{\Delta t}{\sqrt{1 - v^2/c^2}}$$

Figure 12.4 is the graph of $\Delta t'$ for different velocities v. Even for velocities as high as one-half the speed of light,

150,000 km/s, $\Delta t'$ is not much larger than Δt; the clocks tick at very nearly the same rate. Effects of time dilation are virtually unknown in our daily lives, because we do not experience extremely high speeds. However, these effects are as real as any other phenomena in physics, and they have been observed in a very interesting manner.

A subatomic particle called the *muon* decays spontaneously into an electron plus some other particles in an average time of 0.000002 seconds. These muons are produced in large numbers by collisions between high-energy particles from space (*cosmic rays*) and atmospheric molecules some 10 or more kilometers above the Earth. Given their short lifetimes, we should find very few muons reaching the ground, even though they cover the distance between their place of production and the Earth's surface at nearly the speed of light. To the contrary, however, we detect great numbers of muons at ground level. How can this be?

A resolution to this paradox comes by applying special relativity theory. Because the muons are traveling so rapidly, their internal clocks, which regulate their rate of decay, appear to us to be running some ten times too slow. Consequently, from our perspective, there is ample time for them to reach the ground and to be detected—which is what happens. Of course, from the point of view of the muons, they do decay in 0.000002 seconds according to their own clocks, and, again from their perspective, it is our clocks that are running ten times too slow.

EXAMPLE 12.1

What is the mean lifetime of a muon as measured in the laboratory if it is traveling at 0.90c with respect to the laboratory? The mean lifetime of a muon at rest is 2.2 × 10⁻⁶ seconds.

If an observer were moving along with the muon, then the muon would appear to be at rest to such an observer. The muon would decay in an average time $\Delta t = 2.2 \times 10^{-6}$ seconds as seen by this observer. For an observer in the laboratory, the muon lives longer because of time dilation. Applying our equation, we find that the average muon lifetime, $\Delta t'$, as determined in the laboratory is

$$\Delta t' = \frac{\Delta t}{\sqrt{1 - v^2/c^2}} = \frac{2.2 \times 10^{-6}\ \text{s}}{\sqrt{1 - (0.90c)^2/c^2}}$$

$$= \frac{2.2 \times 10^{-6}\ \text{s}}{\sqrt{0.19}} = 5.0 \times 10^{-6}\ \text{s}$$

This is about 2.3 times the average lifetime of a muon at rest. To the laboratory observer, the muon's clock appears to be running more than two times too slow. How fast would the muon have to be traveling for its clock to appear to be running ten times too slow, as mentioned in the text?

Time dilation is one verified prediction of special relativity. There are two other important ones. The first is **length contraction**, in which moving rulers are shortened in the direction of motion. A convenient way to measure a distance is to time how long it takes light to traverse it. But if moving clocks run slow, so that the elapsed light-travel time is smaller, then moving rulers must be too short in the direction of motion. (Remember, distance = velocity × time.) The length of a meter stick moving relative to an observer is decreased by the same factor as the time between ticks in the two light clocks. Thus, moving observers disagree on issues involving *both* length *and* time.

Like time dilation, length contraction isn't ordinarily observed because the speeds at which we normally travel are

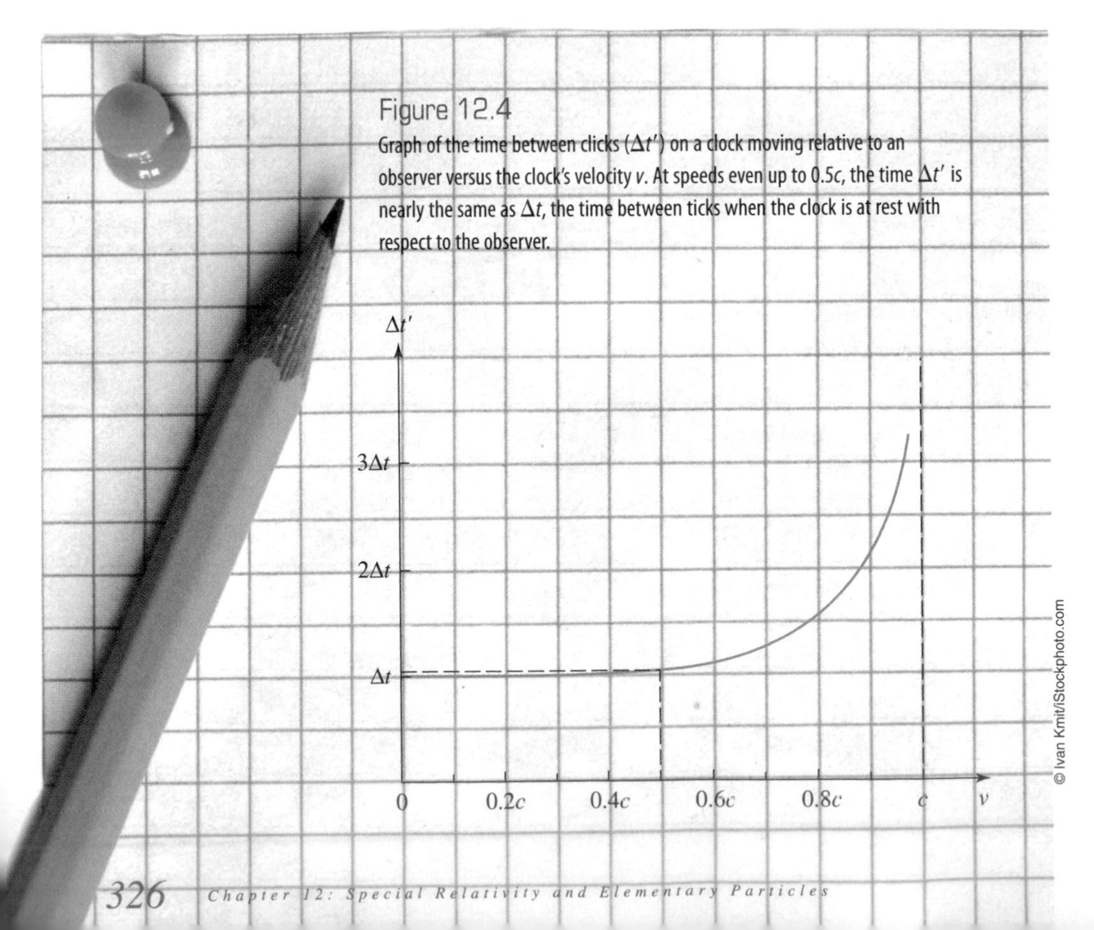

Figure 12.4
Graph of the time between clicks ($\Delta t'$) on a clock moving relative to an observer versus the clock's velocity v. At speeds even up to 0.5c, the time $\Delta t'$ is nearly the same as Δt, the time between ticks when the clock is at rest with respect to the observer.

© Ivan Kmit/iStockphoto.com

so very small compared to the speed of light. But it is a real effect. When high-speed electrons move through the Stanford Linear Accelerator in California, their electric field lines are compressed in the direction of motion by length contraction. As the relativistic electrons pass through coils of wire arrayed along the accelerator, they produce a brief signal that is demonstrably different from that of more slowly moving electrons. The observed difference is precisely accounted for in terms of the special relativistic predictions of length contraction.

The speed of light is not only a constant for all observers, it is also an absolute speed barrier that no object can cross.

Another consequence of special relativity, and the most important one for particle physics, is the equivalence of energy and mass that was introduced in Section 11.5. If moving observers disagree on matters involving length and time, then they will also disagree on the velocities of material particles. For example, if a collision between two electrons occurs in one laboratory setting, the initial and final velocities of the particles as determined by an observer in that laboratory will not agree in general with those determined by another observer moving uniformly relative to the first. Yet both observers *must* agree that the physics of the collision is the same. In particular, both must agree that momentum and energy are conserved during the collision.

It turns out that for the laws of conservation of momentum and energy to be preserved in such cases, the observers must each include in their calculations the **rest energy** E_0 of the particles, given by Einstein's famous equation

$$E_0 = mc^2$$

where m is the ordinary mass of each particle (sometimes called the *rest mass*) and c is the speed of light. Thus, as Einstein wrote in 1921, "Mass and energy are therefore essentially alike; they are only different expressions for the same thing."

© Isabelle Zacher-Finet/iStockphoto.com

$E_0 = mc^2$

In special relativity theory, then, the total **relativistic energy** of a particle as measured by an observer is comprised of two parts: the rest energy, mc^2, of the particle plus whatever additional energy the particle has due to its motion—that is, its kinetic energy. The rest energy of a particle is clearly the same for all observers, but the kinetic energy (and hence the total energy) of the particle is not. It depends upon the frame of reference of the observer. Specifically, the energy of a particle of mass m moving with velocity v relative to a particular observer is equal to

$$E_{rel} = KE_{rel} + mc^2 = \frac{mc^2}{\sqrt{1 - v^2/c^2}}$$

Solving for the relativistic kinetic energy, we find

$$KE_{rel} = \frac{mc^2}{\sqrt{1 - v^2/c^2}} - mc^2$$

For very low particle speeds, this equation can be shown to reduce to the familiar Newtonian form $\frac{1}{2}mv^2$. However, for velocities approaching the speed of light, the energy increases without limit. Thus, to accelerate a particle to the speed of light would require an infinite amount of energy. This is the reason why no material particle can ever travel at the speed of light. The energy and power demands for doing so simply cannot be met! The speed of light is not only a constant for all observers, it is also an absolute speed barrier that no object can cross.

The fact that the (rest) mass of a particle is the same (that is, *invariant*) for all observers, does *not* mean that the mass cannot change. Quite the contrary! In an inelastic collision, mass frequently changes and is transformed into energy. Conversely, in the course of such encounters, energy may be converted into mass. This state of affairs is the direct result of Einstein's recognition of the equivalence of these two things. Thus, when two balls of clay collide and stick together, some of the initial energy is converted to internal energy. Within the framework of special relativity, the energy that has gone into internal energy (and any other forms of internal excitation present in the final system) is measured exactly by the increase in the rest mass of the final system over that of the initial. Now in most practical situations in everyday life, the changes in mass that accompany interactions of this type are far too small to be detected. However, we have discussed some examples in which this type of conversion does lead to very dramatic effects in connection with nuclear reactions in Section 11.5. As we shall see, similar conversions taking place in inelastic collisions involving high-speed particles are the bread and butter of experimental high-energy physics and lead to the creation of exotic new species seldom seen in Nature.

EXAMPLE 12.2

In an x-ray tube (Figure 8.27), an electron with mass m = 9.1 × 10^−31 kilograms is accelerated to a speed of 1.8 × 10^8 m/s. How much energy does the electron possess? Give the answer in joules and in MeVs (million electronvolts).

The total relativistic energy of the electron, E_{rel}, is its relativistic kinetic energy plus its rest energy. From our equation, we see that

$$E_{rel} = KE_{rel} + mc^2 = \frac{mc^2}{\sqrt{1 - v^2/c^2}}$$

Let's first determine at what fraction of the speed of light the electron is moving.

$$\frac{v}{c} = \frac{1.8 \times 10^8 \text{ m/s}}{3.0 \times 10^8 \text{ m/s}} = 0.60$$

$$v = 0.60c$$

The electron travels at 60% of the speed of light.

Evaluating the square root in the equation for E_{rel} gives

$$\sqrt{1 - v^2/c^2} = \sqrt{1 - (0.60)^2} = 0.80$$

The energy of the electron is then given by

$$E_{rel} = \frac{(9.1 \times 10^{-31} \text{ kg})(3.0 \times 10^8 \text{ m/s})^2}{0.80}$$

$$= 1.02 \times 10^{-13} \text{ J}$$

But 1 joule = 6.25 × 10^18 electron volts (see the conversion table card), so

$$E_{rel} = (1.02 \times 10^{-13})(1 \text{ J})$$

$$= (1.02 \times 10^{-13})(6.25 \times 10^{18} \text{ eV})$$

or

$$E_{rel} = 637{,}500 \text{ eV}$$

Since 1 MeV = 1 × 10^6 eV, the energy of the electron is approximately 0.638 MeV.

Notice, with E in MeV, the *equivalent* mass of the electron could be given as 0.638 MeV/c^2. This is a frequently used and very convenient way of representing subatomic particle masses because it eliminates the small numbers that necessitate the cumbersome exponential notation.

Let's compare the relativistic kinetic energy of the electron to that given by classical physics. The rest energy of the electron, mc^2, may be easily shown to be 0.511 MeV following the model above. Then

$$KE_{rel} = E_{rel} - mc^2 = 0.638 \text{ MeV} - 0.511 \text{ MeV}$$

$$= 0.127 \text{ MeV}$$

According to Newtonian mechanics,

$$KE_{classical} = \frac{1}{2}mv^2$$

$$= \frac{1}{2}(9.1 \times 10^{-31} \text{ kg})(1.8 \times 10^8 \text{ m/s})^2$$

$$= 1.47 \times 10^{-14} \text{ J} = 92{,}100 \text{ eV} = 0.092 \text{ MeV}$$

The classical result *underestimates* the electron's kinetic energy by almost 30%.

If we reflect on the special theory of relativity, we see that it accomplishes a profound unification in physics: It reconciles the physics of low speeds with that of high speeds. It is a better, more comprehensive system of mechanics than that formulated by Newton because it works for all particles, regardless of their relative velocities. In the limit of small velocities, we recover the laws of classical mechanics as we specified them in Chapters 1–3; for high velocities, we find that Einstein's theory predicts new effects not contained in Newton's physics that are confirmed experimentally. We will return to this theme of unification in Section 12.5 after we consider elementary particles and the forces they mediate, since it has been, and remains today, one of the overriding goals of physical science.

The Concept Map on your Review Card summarizes the basic postulates and implications of Einstein's special theory of relativity.

12.2 Forces and Particles

The Four Forces: Natural Interactions among Particles

At the end of Section 2.8, we introduced the four fundamental forces of Nature. Table 12.1 lists these forces and includes some properties of each. It is important to acknowledge that *all* the interactions that occur in our environment are due to these forces. They produce the

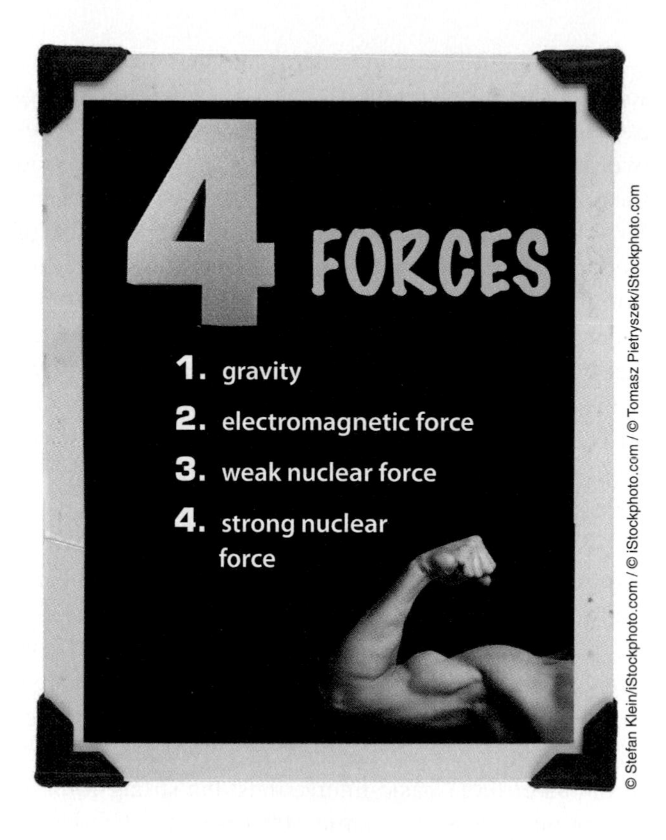

beauty, variety, and change that we witness daily in the world around us.

In Chapter 2, we defined a force as a push or pull acting on a body that usually causes a distortion or a change in velocity (or both). This is a perfectly good description of what we mean by a force in classical physics, but to investigate the realm of particle physics, we must broaden our definition to include every change, reaction, creation, annihilation, disintegration, and so on, that particles can undergo. Thus, when a radioactive nucleus spontaneously decays (see Section 11.2), we will describe this decay in terms of a force that acts between the parent nucleus and its decay products. Similarly, when two particles collide and undergo a nuclear reaction to create new particles (see Section 11.4), we say that there is a force responsible for this transformation.

Because the roles played by forces in particle physics are somewhat different from those traditionally assigned to them in classical physics, it is often the case that they are referred to as the *four basic interactions* of Nature instead of the four basic forces. In this context, we use the word "interaction" to mean the mutual action or influence of one or more particles on another. With this in mind, we return to Table 12.1 and discuss each of the four fundamental interactions briefly, beginning with the most familiar, the gravitational interaction.

Gravity, a very important force in our everyday lives, has been investigated at some length in Chapter 2. Several aspects of this interaction as it pertains to particle physics should be reviewed. First, although gravity affects *all* particles, its importance in particle physics is entirely negligible because its strength is so feeble when compared to the other interactions that can occur. To get a feel for just how inconsequential gravity is on a subatomic level, we can compare the strength of the gravitational attraction between two protons bound in a nucleus with the electrical repulsion between these charged particles. A simple calculation shows the electrical interaction to be more than 10^{36} times stronger than the gravitational interaction. Comparisons between the strength of gravity and the other interactions of Nature are given in Table 12.1. In each case, the effects of gravity are too small to be considered seriously.

Before moving to a discussion of the other forces, it is worth remarking upon two aspects of gravity that do have important consequences for *large-scale* interactions: First, gravitational interactions may dominate in circumstances where charge neutrality prevails. If many

Table 12.1 The Four Fundamental Forces or Interactions

Name (example)	Relative Strength	Range (m)	Carrier: Particle	Carrier: Mass (MeV/c^2)	Carrier: Spin
Strong (binds the nucleus)	1	$\approx 10^{-15}$	Meson[a]	$>10^2$	1
Electromagnetic (binds atoms and molecules)	10^{-2}	Infinite	Photon	0	1
Weak[b] (produces beta decay)	10^{-7}	$\lesssim 10^{-18}$	Z^0, $W^{\pm}$	$\lesssim 10^5$	1
Gravitational (binds planets to the Sun)	10^{-38}	Infinite	Graviton	0	2

[a]At the level of the nucleons, we may regard the messengers of the strong force to be the mesons, although, as discussed in Section 12.5, the true carriers of this interaction are the massless, chargeless *gluons*.

[b]All the messengers of the basic forces are uncharged, except for two of those of the weak force. The W^+ particle carries one unit of positive charge, while the W^- holds one unit of negative charge. The other carrier of the weak force, the Z^0, is electrically neutral.

particles interact together at once and the number of positive charges balances the number of negative ones, electrical forces may cancel out, leaving gravity the dominant interaction. Unlike the electrical interaction, gravity cannot be shielded out or eliminated because there is only one kind of mass. And second, gravity is a long-range interaction. The gravitational force varies inversely as the distance squared, and although it grows ever weaker with separation, it never completely disappears. Thus, gravitational effects may reach over vast regions of space, accumulating in such a fashion as to affect the structure and evolution of the entire universe.*

Aside from gravity, the next most familiar force or interaction is the **electromagnetic interaction**, discussed in Chapter 8. Unlike the gravitational force that is always attractive, the electromagnetic force can be either attractive or repulsive, depending on the signs of the interacting charges. But, like gravity, the electromagnetic force is a long-range force, becoming smaller as the distance separating the charges increases. The electromagnetic force also manifests itself in the magnetic forces associated with moving charges. This interaction is responsible for all the kinds of electromagnetic radiation, from gamma rays to radio waves, investigated in Chapter 8.

It is important to note that although the electric and magnetic forces act only between charged particles, the electromagnetic interaction can have an influence on uncharged particles as well. For example, a photon is not a charged particle, but the absorption or emission of a photon by an atom is an electromagnetic process.

Next among Nature's interactions is the **weak nuclear force**, which is responsible for beta decay (the conversion of a neutron to a proton within the nucleus; see Section 11.2). The term *weak* may be interpreted in a variety of ways. For example, this interaction is weak in the sense that it is effective only over very short distances: at least 100 times smaller than the range of the strong nuclear force and infinitesimal compared to the ranges of gravity and electromagnetism. The weak force is also "weak" because the probability is quite small that interactions involving this force occur. Indeed, particle interactions involving the weak force generally happen only as a last resort when all other interaction mechanisms are prohibited.

* To properly account for the interaction between massive objects like stars and galaxies or to develop models for the global structure of the universe that are consistent with observation, use must be made of Einstein's *general theory of relativity*. This is a more comprehensive and accurate theory of gravity than that originally proposed by Newton, and it permits scientists to interpret correctly such phenomena as the deviation in the path of starlight passing near the Sun. Additional discussion of general relativity is given in the Epilogue.

Although it is not apparent from what we have said so far, a very close relationship exists between the electromagnetic and weak interactions. The similarity between these two was first noted in the late 1950s, and further studies of the connections between the weak force and electromagnetism have led to a unification of these two interactions into one, the **electroweak interaction**. This is much like the unification between electricity and magnetism in Maxwell's theory. It is now recognized that the source of the electromagnetic and weak forces is the same but that their practical manifestations differ considerably, leading to a separate classification for each. More will be said about the issue of unification of forces in Section 12.5.

The last of the forces of Nature is the **strong nuclear force**. The strong force is responsible for holding the nuclei of atoms together and is involved in nuclear fusion reactions (see Section 11.7). It is a short-range, attractive interaction that does not depend upon electric charge. Considering the probability that two colliding particles will interact by the strong force as opposed to any one of the other three basic interactions, the strong force is indeed quite "strong," some 100 times as effective in bringing about a reaction as the electromagnetic force. Comparisons like these between the relative probability that a reaction will occur via a particular interaction have been used to establish the measures of the relative strengths of the four forces given in Table 12.1.

Having compared the properties of the four forces, let's now consider *how* we have acquired our knowledge about their characteristics. Einstein's theory of special relativity includes the postulate that the speed of light is finite; 300,000 km/s is the maximum speed attainable by particles in the universe. This is also the maximum speed at which information may be propagated through the universe. The fact that a star 150 million kilometers away suddenly explodes *cannot* be known to us until at least 500 seconds (about 8 minutes) later because the particles ejected from the event require that long to make their way to Earth. Information about this explosion thus comes to us through the intermediary of particles that race out from the interaction site carrying data about the nature of the event to our location.

In the same way that we come to understand the details of a stellar explosion by the particles emitted during the event, particle physicists come to know the characteristics of the four forces of Nature by the particles produced in these interactions. In fact, current theories associate with each force a carrier, or mediator, of the interaction. These carriers are exchanged between the particles experiencing the forces, and they communicate the interaction between the reactants and the products. For this reason, **carrier particles** are also known as exchange particles. For example, if we wiggle an electron, the change in its electric field will propagate outward as a wave traveling at the speed of light. The disturbance in the field produces forces on other charged particles in the neighborhood of the electron and communicates to them information about the electron's motion. The role of the wave is that of a messenger, and this has led physicists to conceive of the influence of the electric field as being conveyed or carried by a messenger particle—the photon (recall Section 10.5 and the discussion of wave-particle duality). In this view, all the effects of an electromagnetic field may be explained by the exchange of photons.The photons that mediate the electromagnetic interaction are not real photons like those produced in the emission of light by excited atoms but what are called *virtual photons.* These particles are undetectable in the traditional sense but are responsible for the transmission of forces between other ordinary, observable particles. Thus, while the virtual photons themselves cannot be directly observed, the *effects* of the virtual photons as carriers of the electromagnetic interaction can be directly seen.

Table 12.1 includes the carriers of the four basic interactions as well as their masses (measured in equivalent energy units; see Example 12.2). In the next section, we will explore the characteristics of these and other elementary particles in greater detail. Before doing so, we show Figure 12.5, which depicts how a particle physicist might represent the weak interaction that converts a neutron inside a nucleus into a proton (a beta decay) and how modern physics views the repulsion between two electrons.

Figure 12.5

(a) Modern representation of the beta decay of a neutron. The neutron transmutes into a proton after emitting a W^- particle, and the W^- subsequently decays into an electron (e^-) and an antineutrino ($\overline{\nu}_e$). (b) In particle physics, the electromagnetic repulsion between two electrons is viewed as being caused by the exchange of (virtual) photons (γ). This process is shown in what is called a "Feynman diagram," after Nobel prize winning theoretical physicist Richard P. Feynman (1912–1988) who pioneered its use.

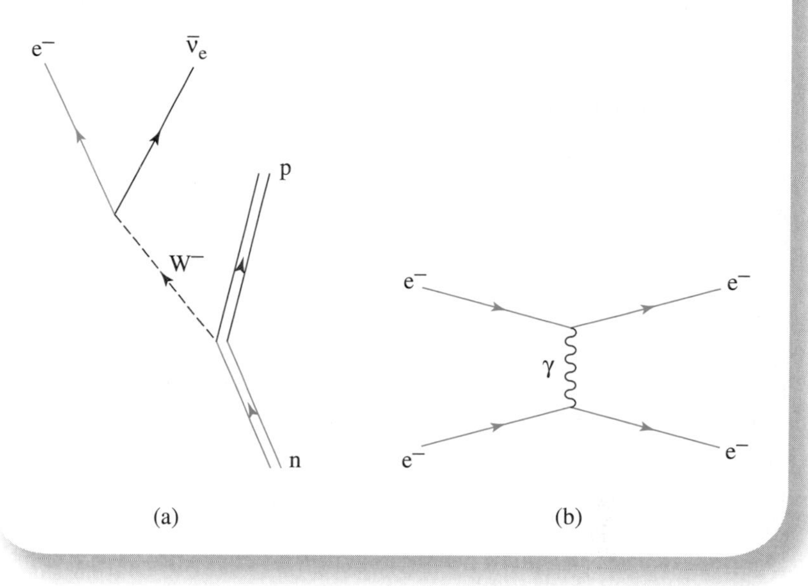

Classification Schemes for Particles

In Chapter 4, we classified matter into solid, liquid, gas, and plasma phases. We also classified matter according to the number and kinds of atoms that are present: elements, compounds, mixtures, and so forth. We were further able to categorize the properties of matter on the basis of the forces that acted between the constituents: large forces in solids, smaller forces in liquids, and so on.

Just as we could classify bulk matter in several different ways, it is possible to classify elementary particles using different schemes. In this section, we will consider three ways of doing so: on the basis of spin, on the basis of interaction, and on the basis of mass. Before going any further in this discussion, however, it is worth defining what we mean by an **elementary particle** and by an **antiparticle**.

DEFINITION

Elementary Particles *The basic, indivisible building blocks of the universe. The fundamental constituents from which all matter, antimatter, and their interactions derive. They are believed to be true "point" particles, devoid of substructure or measurable size.*

Antiparticle *A charge-reversed version of an ordinary particle. A particle of the same mass (and spin) as its cogener but of opposite electric charge (and certain other quantum mechanical "charges").*

Every known particle has a corresponding antiparticle. There are antielectrons (more commonly called **positrons**), antiprotons, antineutrons, and so on. Collections of antiparticles form antimatter, just as collections of ordinary particles form (ordinary) matter. The first antiparticle, the positron, was discovered in 1932 by Carl Anderson in cosmic rays. The antiproton was discovered in 1955, and the antineutron the year following. Particle physicists now routinely create and store small quantities of antimatter with high-energy accelerators.

When matter and antimatter meet, mutual annihilation results, accompanied by a burst of energy in the form of gamma rays (see Figure 12.6). For example, if an electron and positron collide, they will quickly annihilate to produce two high-energy gamma rays. This process has been harnessed by modern medicine in the diagnostic procedure called *positron emission tomography* or PET. PET is a medical imaging technique that uses positrons to monitor biochemical processes in the body. Small amounts of tracer compounds containing positron-emitting isotopes with short half-lives, like carbon-11, oxygen-15, and fluorine-18, are ingested or inhaled by the patient. As the radioactive nuclei decay, they release positrons that quickly annihilate with electrons in the surrounding tissue generating gamma rays. The gamma rays are observed with specialized detectors, and an image is created that reveals the location and concentration of the radioisotope in the body part scanned. PET has revolutionized medical research and treatment in several areas, especially neurology, where it has dramatically enhanced our understanding of brain chemistry and function.

The positron-emitting radioisotopes used in PET scans are frequently created in small reactors near or within the hospital facilities that employ these radioactive species. A case in point is fluorine-18 which may be created by colliding oxygen-18 nuclei with high-speed protons. In 1995, using similar high-energy physics techniques, scientists at the Conseil Europeén pour la Recherche Nucléaire (CERN) were successful in producing antihydrogen, an atom of antimatter composed of an antiproton and a positron.

In what follows, we shall have several occasions to examine the creation and annihilation of particles and antiparticles. In those reactions, we will distinguish antiparticles using the same symbol as that for the corresponding particle but with a "bar" over it. Thus, for example, if n designates a neutron, then $\bar{n}$ (pronounced "en-bar") represents an antineutron.*

The kinds of particles that have been termed "elementary" have gradually changed with time. Before 1890 and the discovery of the electron, atoms were regarded as the smallest units of matter, and they were believed to possess no internal structure of their own. In the 1930s, it was believed that the basic building blocks of Nature consisted of the proton, the neutron, the electron, the positron, the photon, and the neutrino (extremely low-mass particles, typically produced in beta decay reactions, that have no charge and only very weakly interact with ordinary matter). Circa 1934, these were the "atoms" (indivisible particles) sought by the ancient Greeks. In the last 75 years, particle physicists have discovered that not only are there more than six "elementary" particles in Nature but that some of the original six are not really elementary at all! They themselves are composed of still more basic and elusive particles. At the time of this writing, there exist well over 100 different subatomic particles, and it is well accepted that the proton and the neutron (among others) are not the ultimate constituents of matter. The majority of the remainder of this chapter will be spent developing the story of our changing perspective on what constitutes a truly "elementary" particle. But, first, a bit more about the properties of subatomic particles.

Figure 12.6

Pair annihilation. A particle and an antiparticle collide and annihilate one another to produce photons (gamma rays). For each reaction to occur, the total energy of the gamma rays must at least be equal to the mass-energy of the proton and antiproton.

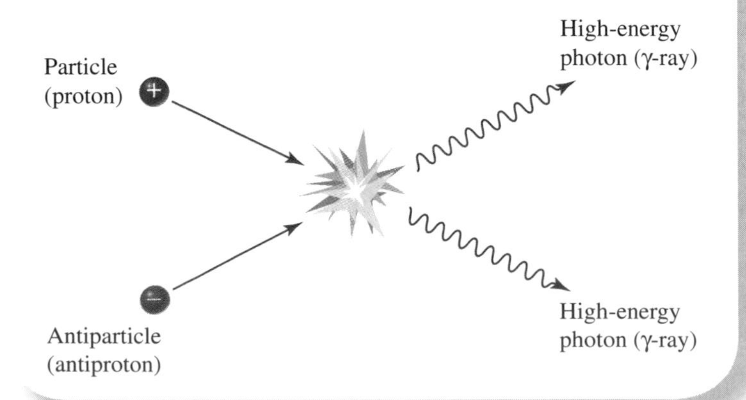

Spin

Spin, like mass and charge, is an intrinsic property of all elementary particles and measures the angular momentum carried by the particle. If we treat

* There are several noteworthy exceptions to this rule, however. For example, an antielectron (or positron) is symbolized e^+, not $\bar{e}^-$. Likewise, an antimuon is denoted μ^+, not $\bar{\mu}^-$. See Table 12.3 for some other cases in which the "bar" rule is not adhered to strictly.

an elementary particle as a simple hard sphere, then we can picture spin as due to the rotation of the particle about an axis through itself. Unlike, say, a basketball whirling at the end of one's finger, which can have any amount of spin, the spin of elementary particles is quantized (see Section 10.5) in units of $h/2\pi$, where h is Planck's constant. The spins of all known particles are either integral or half-integral multiples of this basic unit. In other words, spin can take on values of 0, $\frac{1}{2}$, 1, $\frac{3}{2}$, and so on, in units of $h/2\pi$. Experiments have shown, for example, that the spin of the electron and the proton is $\frac{1}{2}$, but that of the photon is 1.

Stephen Hawking has provided another view of spin that may be helpful in thinking about this property of elementary particles. It has the advantage of avoiding any conflicts with quantum mechanics that arise from thinking of particles as little marbles or ball bearings. In Hawking's approach, the spin of a particle tells us what the particle looks like from different directions. Thus, a particle of spin zero is like the period at the end of this sentence. It looks the same from all directions (Figure 12.7). A spin 1 particle is like an arrow. It looks different when seen from different directions, but a rotation through a complete circle (360°) restores the original view. A spin 2 particle is similar to a two-headed arrow; rotation by 180° gives the same appearance. In this picture, spin $\frac{1}{2}$ particles, like electrons, have the remarkable property that *two* complete rotations through 360° are required for the particle to "look" the same. Can you think of any common object or figure that exhibits this type of symmetry?

Figure 12.7

Illustrations of objects exhibiting symmetry properties analogous to those of elementary particles, as described by Stephen Hawking. (a) Dot (spin = 0); (b) ace of clubs (spin = 1); (c) jack of hearts (spin = 2).

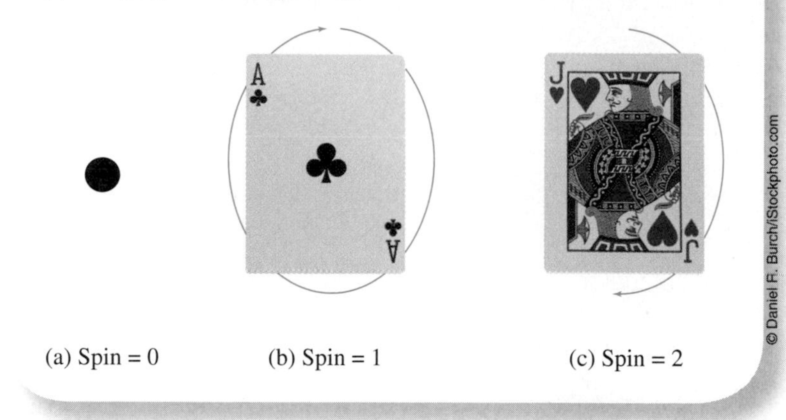

Particles possessing half-integral spins are called **fermions**, after Italian-American scientist Enrico Fermi, who carefully investigated the behavior of collections of such particles. Particles with integer spins are called **bosons**, after Indian physicist Satyendra Bose, who, with Einstein, developed the laws describing their collective behavior. The main difference between these two types of particles is that the former obey the Pauli exclusion principle (page 284) while the latter do not. This law, for which Wolfgang Pauli won the Nobel Prize in 1945, states that no two interacting fermions of the same type can be in exactly the same quantum state. They must be distinguishable in some manner. Thus, in a normal helium atom, when the two electrons are in the lowest atomic energy state (the ground state), the exclusion principle demands that these spin $\frac{1}{2}$ particles differ in some way. How can this be achieved? Isn't one electron just like any other? Same mass, same charge, same spin? Yes. But let's return to our rotating sphere model for a moment. Relative to the axis of rotation, the marble may spin either clockwise or counterclockwise. For a given total angular momentum, two distinct spin states associated with the directions of rotation exist (Figure 12.8). In the same way, one can associate with the electron two different spin configurations, called *spin-up* and *spin-down*, each with the same total amount of spin, $h/4\pi$. With this addition, it is now possible to satisfy the Pauli principle for helium by requiring one of the electrons to be in a spin-up state while the other is in a spin-down state. The state of every electron in every atom can be accounted for by this principle.

Bosons, by contrast, do not obey the Pauli principle. They are truly indistinguishable. An unlimited number of bosons can occupy a given energy state and so be concentrated in any given volume of space without violating any physical laws. This accounts for the fact that there is no restriction on the number of photons (spin 1 particles) that can be packed into a beam of light and hence no (theoretical) limit to the intensity of the beam. There is also no limit to the number of force-carrying particles that can be exchanged in a given reaction, because *all* the mediators of the fundamental interactions have integral spins (Table 12.1).

Elementary Particle Lexicon

Separating particles according to spin divides them into two groups. Establishing additional selection criteria further subdivides these two groups. A very useful way of doing so is to identify those particles that participate in strong force interactions and those that do not.

Figure 12.8

Spin-up, (a), and spin-down, (b), configurations for a rotating marble.

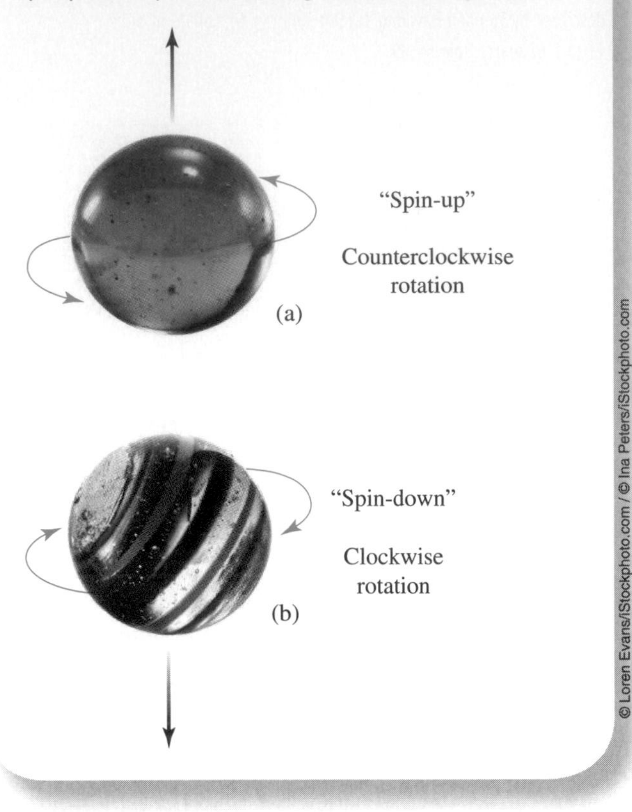

We thus distinguish the four groups of particles given in Table 12.2: the baryons, the leptons, the mesons, and the intermediate (or *gauge*) bosons. Examples of several particles of each type are included in the table. An electron is a lepton with spin $\frac{1}{2}$ that is unaffected by the strong force; a proton is a baryon with spin $\frac{1}{2}$ that does interact via the strong force. Photons are bosons that do not participate in strong interactions.

The names for these groups derive largely from earlier classification schemes based on experimentally determined masses for these particles. The word **baryon** comes from the Greek *barys* meaning heavy, while the word **lepton** means "light one" in Greek. **Mesons** refers to the "middle ones" with intermediate mass. At the time these groups were named, the heaviest particles known were found among the baryons and the lightest were included among the leptons. Recent discoveries, however, have revealed mesons and at least one lepton with masses larger than those of protons and neutrons. Thus, it is no longer possible to specify the correct class of an elementary particle by its mass alone, although for most species this is still a useful guide.

The force-carrying particles, the intermediate bosons, exhibit a wide variety of mass. Because they are bosons, there is no limit to the number that can be exchanged in any interaction, but there is a close correlation between the range of the force they mediate and their mass. If the carrier particles have a high mass, it will generally be difficult to produce them and to exchange them over long distances. Thus, a force carried by massive particles will have only a short range. This is the case for the W and Z particles that mediate the weak interaction. However, if the carrier particles have no mass of their own (like the photon and the graviton), the forces associated with them (in these cases, the electromagnetic and gravitational forces, respectively) will be of long range.

Table 12.2 Classification of Particles

Spin Group	Particles Interacting Via the Strong Force (Hadrons)	Particles Unaffected by the Strong Force
Fermions (Half-integer spin)	*Baryons* (Protons, neutrons, lambdas, sigmas, . . .)	*Leptons* (Electrons, muons, neutrinos, . . .)
Bosons (Integer spin)	*Mesons* (Pions, kaons, etas, . . .)	*Intermediate (or gauge) bosons* (Photons, Z^0, $W^{\pm}$, gravitons)

Before leaving this section, we introduce one additional bit of nomenclature that is used commonly in connection with elementary particles, the word *hadron.* This word is also of Greek origin, coming from *hadros*, meaning "thick" or "strong." Baryons and mesons are collectively referred to as hadrons (Table 12.3) because they interact by the strong force, a distinction not shared by the leptons or the gauge particles. The Concept Map on your Review Card shows ways elementary particles may be classified, as well as the connections among them.

12.3 Conservation Laws, Revisited

In Chapter 3, we emphasized the importance of conservation laws in physics. We used the laws of conservation of mass, energy, linear momentum, and angular momentum to derive information about physical systems "before-and-after." Such an approach permitted us to

Table 12.3 Properties of Some Long-Lived Hadrons

Class	Particle Name	Symbol[a] (Quark Content)	Antiparticle	Mass (MeV/c^2)	B[b]	S[c]	Lifetime (s)
Baryons	Proton	p (uud)	$\bar{p}$	938.3	+1	0	Stable(?)
	Neutron	n (udd)	$\bar{n}$	939.6	+1	0	886
	Lambda	Λ^0 (uds)[g]	$\overline{\Lambda}^0$	1,115.7	+1	−1	2.6×10^{-10}
Spin = $\frac{1}{2}$	Sigma	Σ^+ (uus)	$\overline{\Sigma}^-$	1,189.4	+1	−1	0.8×10^{-10}
		Σ^0 (uds)[g]	$\overline{\Sigma}^0$	1,192.6	+1	−1	7.4×10^{-20}
		Σ^- (dds)	$\overline{\Sigma}^+$	1,197.5	+1	−1	1.5×10^{-10}
	Xi	Ξ^0 (uss)	$\overline{\Xi}^0$	1,315	+1	−2	2.9×10^{-10}
		Ξ^- (dss)	$\overline{\Xi}^+$	1,321	+1	−2	1.6×10^{-10}
Spin = $\frac{3}{2}$	Omega	Ω^- (sss)	$\overline{\Omega}^+$	1,672	+1	−3	0.8×10^{-10}
Mesons	Pion	π^+ ($u\bar{d}$)	π^-	139.6	0	0	2.6×10^{-8}
		π^0 ($u\bar{u}$, $d\bar{d}$)[f]	Self[e]	135.0	0	0	8.4×10^{-17}
	Kaon[d]	K^+ ($u\bar{s}$)	K^-	493.7	0	+1	1.2×10^{-8}
		K^0 ($d\bar{s}$)	$\overline{K}^0$	497.7	0	+1	0.9×10^{-10}
Spin = 0							5.2×10^{-8}
	Eta	η^0 ($u\bar{u}$, $d\bar{d}$, $s\bar{s}$)[f]	Self	547.8	0	0	5.5×10^{-19}
	"Dee-plus"	D^+ ($\bar{d}c$)	D^-	1,869	0	0	1.0×10^{-12}
	"Bee-plus"	B^+ ($\bar{b}u$)	B^-	5,279	0	0	1.7×10^{-12}
Spin = 1	Psi	ψ^0 ($c\bar{c}$)	Self	3,097	0	0	7.6×10^{-21}
	Upsilon	Υ^0 ($b\bar{b}$)	Self	9,460	0	0	1.2×10^{-20}

[a] Superscripts to the right of the particle symbols indicate the charge carried by the particle in units of the proton charge (see Section 12.3). See Sections 12.4 and 12.5 for discussions of the quark composition of baryons and mesons.
[b] Baryon number (see Section 12.3).
[c] Strangeness (see Section 12.3).
[d] Although created in strong interactions, K^0 mesons decay by the weak force with two characteristic times, one some 575 times shorter than the other, due to what is called *neutral particle oscillations* (cf. a similar phenomenon for neutrinos on p. 349).
[e] Some neutral particles are their own antiparticles. Thus, when two π^0's meet, they annihilate one another to form γ rays.
[f] A few mesons are composed of admixtures of two or three quark - antiquark states.
[g] While having the same quark content, these two baryons differ in terms of an internal quantum number called *isospin* and so represent different states of the same quark system.

draw some general conclusions about the systems without knowing all the details of the physics of the interaction taking place. A good example of this is the application of the principle of conservation of linear momentum during a collision. Here, information about the initial speeds of the colliding objects may be extracted from that concerning the final speeds without knowing the details of the very complicated interactions occurring in the crash itself.

In elementary particle physics, the same technique may be applied because the same physical laws are at work. When two particles collide, linear momentum must be conserved: The net momentum present before the collision must equal that after the collision. Likewise, mass-energy must be conserved: The total energy present at the start of an interaction between particles, including that stored as mass, must be the same as that present after the interaction is completed. Together, these two laws mandate that no single isolated particle can spontaneously decay into particles whose mass exceeds its own, and that a particle cannot decay into a *single* particle lighter than itself. A high-energy photon (γ ray) may spontaneously break up into an electron and a positron, each of which has more (rest) mass than the photon, but this can happen only in the vicinity of another particle, usually an atomic nucleus, that absorbs the excess momentum to keep the total constant.

In particle physics, the law of conservation of angular momentum that we used to analyze spinning ice skaters and orbiting spacecraft in Chapter 3 sees application largely in the preservation of total spin in particle reactions. For example, when a photon turns into an electron-positron pair, spin must be conserved. This means that since the spin of the photon is 1, the spins of the departing electron and positron (each of magnitude $\frac{1}{2}$) must both lie along the same direction so that their sum equals 1 $\left(\frac{1}{2}+\frac{1}{2}\right)$. Similarly, if a particle of spin zero is spontaneously converted into two particles of spin $\frac{1}{2}$, their spins must be oppositely aligned so that they add up appropriately $\left[\frac{1}{2}+\left(-\frac{1}{2}\right)=0\right]$.

The conservation laws come from classical physics but work equally effectively at the level of subatomic particles.

The conservation laws discussed above come from classical physics but work equally effectively at the level of subatomic particles. Another classical conservation law also must be obeyed at the submicroscopic level: conservation of charge. In any reaction or interaction among particles, the total (or net) charge present before must equal the total (or net) charge present after. Charge can neither be created nor destroyed during an interaction between elementary particles. Continuing with our example of the spontaneous decay of a photon to produce an electron and a positron, we see that the net charge before the decay is zero and that the net charge afterward is also zero, since the electron possesses one unit of negative charge while an antielectron possesses one unit of positive charge. Clearly, the sum of these is zero.

It is interesting that the stability of the electron against spontaneous decay owes its origin to a classical conservation law: All the particles lighter than an electron that could be produced by its decay are uncharged. If the electron were to undergo a change to produce such particles, a violation of conservation of charge would occur. Consequently, such conversions are forbidden in Nature, and the electron is believed to be absolutely stable.

New Conservation Laws

The four conservation laws of classical physics were developed from the observed behavior of macroscopic systems, but they are found to apply equally well to submicroscopic systems, in particular to the interaction of elementary particles. However, when experiments involving collisions between particles are carried out, a curious thing is noticed: Some reactions that are *not* forbidden by the classical conservation laws are *never* observed. Currently, in particle physics, it is widely assumed that any reaction that is not strictly forbidden *will* occur, albeit sometimes with low probability or frequency. The fact that, regardless of the number of times a particular collision happens, a certain outcome is never seen has led physicists to suspect that there are *additional* conservation laws that operate *only* at the submicroscopic scale. These new conservation laws may be *ad hoc*, approximate rules, or they may be fundamental statements about Nature like that of charge conservation. In either case, the recognition, statement, and use of these additional conservation laws have led to remarkable progress in particle physics, and we take time now to consider several of them. To describe these new regulations on particle reactions, new "charges" or quantum numbers (see page 280) have had to be invented.

Conservation of Baryon and Lepton Numbers

Consider a collision between two protons. Of the many particles that can be produced in such an interaction, mesons are never seen as the *sole* products, even in cases like the one shown below, which does not violate any of the four classical conservation laws.*

$$\mathrm{p}+\mathrm{p} \not\rightarrow \pi^{+}+\pi^{+}+\pi^{0}$$

(You'll have to take our word for it that this reaction doesn't violate momentum or mass-energy conservation, but you can verify for yourself that it conserves charge and spin using the information in Table 12.3.) Alternatively, mesons *may* be the sole products of the collision between

* In this chapter, we dispense with the superscripts and subscripts giving the mass and atomic number, respectively, for the elementary particles as was done in Chapter 11. Thus, the proton will be designated p instead of $^{1}_{1}$p and the neutron will be symbolized n instead of $^{1}_{0}$n. We will denote the electron as e^{-}.

a proton and an antiproton. A possible reaction of this type is the following:

$$p + \bar{p} \rightarrow \pi^{+} + \pi^{-} + \pi^{0}$$

Evidently, there seems to be a hidden conservation law that prevents the first reaction from happening yet allows the second.

Investigations of instances similar to this have led particle physicists to formulate the **law of conservation of baryon number**.

LAWS

Law of Conservation of Baryon Number *In a particle interaction, the baryon number (B) must remain constant. The net baryon number at the start of a reaction must equal the net baryon number at the end.*

Like the law of conservation of charge, the law of conservation of baryon number is a simple counting rule. Each baryon is assigned baryon number $B = +1$; each antibaryon is assigned baryon number $B = -1$. Leptons and mesons have baryon number zero. To apply the conservation of baryon number, simply add up the baryon numbers of the reactants, and compare that value to the total baryon number of the products. If the two agree, then baryon number is conserved in the reaction, and the reaction may occur (unless forbidden by other conservation laws). If the values of B before and after the reaction are not equal, baryon conservation is violated, and the reaction cannot occur.

Looking back at the proton-proton reaction given above, we see immediately why it is not observed: The total baryon number at the start is 2, while that at the end is zero—mesons have no baryon number. However, in the interaction between a proton and an antiproton, the initial baryon number is zero. This fact permits the reaction to yield only mesons under certain circumstances.

Conservation of baryon number is obeyed in all interactions—strong, weak, or electromagnetic. It has never been observed to be violated, and it dictates that baryons are created and destroyed in pairs. When a baryon is born in a reaction, an antibaryon must also be created to conserve baryon number. For this reason, the number of baryons in the universe minus the number of antibaryons is constant.

Just as conservation of charge is responsible for the stability of the electron, conservation of baryon number is believed to be the cause of the stability of the proton. The proton is the lightest baryon known. It cannot decay to any other lighter baryons, because none exist. But conservation of baryon number prevents a proton from decaying to any other type of particle, like mesons or leptons, too. Thus it appears that the proton is absolutely stable *if* baryon conservation is a fundamental law and not an approximate one. Some recent theories that attempt to unify the strong force with the electroweak force require that the proton decay spontaneously to lighter particles, in violation of baryon conservation, in an average time somewhere between 10^{30} and 10^{32} years. This result, if true, may not seem to have much relevance for us here and now—we are in no immediate danger of spontaneously disappearing in a burst of

"We know in our bones that the proton's lifetime is very long if not infinite. If the lifetime were shorter than 10^{16} years, the radioactivity stemming from the decay of protons in our body would imperil our lives."

–Physicist Maurice Goldhaber

© Vasiliy Koval/iStockphoto.com / © Maria Toutoudaki/iStockphoto.com

radioactivity due to the decay of protons in our bodies. But the stability of the proton does have considerable impact on the fate of the universe, whose age is currently estimated to be around 10^{10} years.

Table 12.4 Lepton Families[a]

Family	Particle Name	Symbol[b]	Antiparticle	Mass (MeV/c^2)	Lepton Number L_e	L_μ	L_τ
Electron	Electron	e^-	e^+ (positron)	0.511	+1	0	0
	Neutrino	ν_e	$\overline{\nu}_e$	$\sim 0(\lesssim 2 \times 10^{-6})$	+1	0	0
Muon	Muon	μ^-	μ^+	105.7	0	+1	0
	Neutrino	ν_μ	$\overline{\nu}_\mu$	$\sim 0(<0.17)$	0	+1	0
Tau	Tau	τ^-	τ^+	1,777	0	0	+1
	Neutrino	ν_τ	$\overline{\nu}_\tau$	$\sim 0(<15.5)$	0	0	+1

[a]The spins of all the leptons, regardless of family, are $\frac{1}{2}$. For this reason, we have not separately listed this property for each particle.

[b]Superscripts to the right of the particle symbols indicate the charge carried by the particle in units of the proton charge. Thus, the electron possesses one unit of negative charge, while the antimuon carries one unit of positive charge. The neutrinos carry no charge.

A conservation law similar to that adhered to by baryons exists for leptons. Electrons, muons, and neutrinos all are created and annihilated in lepton-antilepton pairs. For example, a photon, a nonlepton with lepton number zero, decays into an electron and a positron to conserve lepton number. In beta decay, when a neutron decays into a proton and an electron, an antilepton—in this case, an antineutrino—*must* accompany the process to conserve lepton number (see Section 11.2):

$$n \rightarrow p + e^- + \overline{\nu}_e$$

There is, however, one additional complication to the application of this lepton counting rule that does not appear when using conservation of baryon number. Specifically, leptons belong to *families*, and within each family, lepton number is conserved separately. Thus the six known leptons (and their anti's) divide into three families: the "electron family," consisting of the electron and its neutrino; the "tau family," containing the tau and its neutrino;* and the "muon family," with the muon and the muon neutrino (Table 12.4). Within each family, the particles are assigned lepton number +1, and the anti's have lepton number −1. In elementary particle physics reactions, the electron, muon, and tau lepton numbers before an interaction must balance those after.

More on the Tau Neutrino

Although its existence was never in doubt, the first direct evidence of the tau neutrino was reported only in July 2000 by an international team of scientists in experiments conducted at Fermilab outside Chicago. High-energy protons were driven into a tungsten target to produce (among numerous other particles) tau neutrinos that were detected through their rare interactions with nuclei in a specially prepared photographic emulsion. Of the roughly 200 candidate interactions captured in the emulsion after 5 months of experimentation, 4 had all the characteristics to mark them as involving tau neutrinos.

© iStockphoto.com

LAWS

Law of Conservation of Lepton Number *In particle interactions, lepton number (L) is conserved; within each lepton family (electron, muon, and tau) the value of the lepton number at the start of a reaction must equal its value at the end.*

Conservation of lepton number in this form explains why the following reaction is not observed, even though it does not appear to violate any other conservation laws:

$$\mu^- \nrightarrow e^- + \gamma$$

If all leptons had the same kind of lepton number, then this reaction should be possible, since the lepton number at the start is +1, and that at the finish is also +1. However, experiments have clearly demonstrated that the properties of leptons differ according to family and that because of these differences, a reaction of this type is precluded: It violates lepton number conservation within each lepton family, since the muon lepton number on the left is +1, while that on the right is zero. Thus the muon cannot disintegrate as shown but may decay to an electron by the following route, which *does* satisfy all the relevant conservation laws:

$$\mu^- \rightarrow e^- + \overline{\nu}_e + \nu_\mu$$

Conservation of Strangeness

Beginning in the early 1950s, particle physicists began to detect new particles, which they labeled kaons (K), lambdas (Λ), and sigmas (Σ). These new members of the elementary particle zoo exhibited some very strange properties. First, they were always observed to be formed in pairs. The following two reactions are typical of the way these *strange particles*, as they came to be called, are produced:

$$\pi^- + p \rightarrow \Lambda^0 + K^0$$

$$\bar{p} + p \rightarrow K^- + K^0 + \pi^+ + \pi^0$$

Conversely, the following reaction, which did not seem to violate any then-known conservation laws, was found not to occur:

$$p + p \nrightarrow p + \Lambda^0 + \pi^+$$

Reactions like this that produced single, strange hadrons were never observed.

The second strange aspect of these particles involved their decay rates. The production of these strange species occurred with very high frequency or probability, provided enough energy was available in the collisions. This seemed to indicate that the strong interaction was their source. However, once produced, the lifetimes of these strange particles were much too long for strongly interacting entities: Unstable, strongly interacting particles typically decay to other strongly interacting particles on time scales of $\sim 10^{-23}$ seconds. The strange particles decayed to other hadrons only after enormously longer times on the order of 10^{-10} to 10^{-8} seconds. These times turn out to be more characteristic of particles that decay not by the strong interaction but by the weak one!

In 1953, Murray Gell-Mann and Kazuhiko Nishijima independently proposed that certain particles possess another type of "charge" or another quantum number termed **strangeness** (S). This quantity is conserved in strong and electromagnetic interactions, but is not conserved in weak interactions. Strangeness is a partially conserved quantity. Table 12.3 lists the strange charges for some of the better-known strange particles. Their anti's possess strangeness in equal magnitude but of opposite sign. Nonstrange particles, like protons and neutrons, have zero strangeness.

Nonstrange particles, like protons and neutrons, have zero strangeness.

Strangeness is a partially conserved quantity.

DEFINITION

Strangeness *In strong and electromagnetic interactions, strangeness (S) is conserved; in weak interactions, strangeness may change by ±1 unit. Strangeness is a partially conserved quantity.*

The impact of the Gell-Mann–Nishijima theory was immediate and revolutionary. It completely solved all the puzzles presented by the strange particles. When strange particles are produced in strong interactions, the total strangeness of the products must be zero to conserve strangeness. This guarantees that strange particles will be formed in strange-antistrange pairs. (Recall this same behavior is found in strong interactions that produce baryons to conserve baryon number.) Thus in the reaction

$$\pi^- + p \rightarrow \Lambda^0 + K^0$$

$S = 0$ initially, because the pion and proton are nonstrange, but $S = 0$ afterward also, because the strangeness of Λ^0 is -1 while that of K^0 is $+1$. Once created, a strange particle cannot decay via the strong or electromagnetic force to other particles with no strangeness or even to particles of lower strangeness than its own, since this would constitute a violation of the conservation of strangeness. Consequently, the only decay channel left to a strange particle is the weak interaction, which, because of its lower probability of occurrence, simply takes longer to happen. In weak processes, strangeness is not conserved, and there can be a net change in the strangeness in such reactions. An example of this is shown below:

$$\begin{array}{cccc} \Lambda^0 & \xrightarrow{\hspace{3em}} & p + \pi^- & \\ S = -1 & \text{(weak interaction)} & 0 \quad 0 & (\Delta S = +1) \end{array}$$

Table 12.5 summarizes our current knowledge of some of the conservation laws. Those listed were adequate to explain all the reactions between particles studied up to about 1970. After this date, as we shall discover in the next section, several new conservation laws and "charges" had to be invented by particle physicists to explain the observation of new species of elementary particles. The following examples demonstrate how conservation laws may be used to determine whether a given reaction can occur and to predict the identity of "missing" particles in simple reactions.

Table 12.5 Some Important Conservation Laws in Particle Physics

Physical Quantity	Interactions in Which Quantity Is Conserved
Mass-energy	Strong, electromagnetic, and weak
Linear momentum	Strong, electromagnetic, and weak
Angular momentum	Strong, electromagnetic, and weak
Electric charge	Strong, electromagnetic, and weak
Baryon number	Strong, electromagnetic, and weak
Electron lepton number	Electromagnetic and weak
Muon lepton number	Electromagnetic and weak
Strangeness	Strong and electromagnetic

EXAMPLE 12.3

Identify the conservation law(s) that would be violated in each of the following reactions: (a) $\pi^+ \rightarrow e^+ + \gamma$***; (b)*** $\pi^- + p \rightarrow K^0 + p + \pi^0$***; (c)*** $\Lambda^0 \rightarrow \pi^- + \pi^+$.

The conservation laws that can be checked easily are those involving charge, baryon number, lepton number, and strangeness. Let's examine each reaction in the light of these regulations.

(a) Since the photon (γ) is neutral, the decay does not violate charge conservation. None of the reactants is a strange particle or a baryon, so conservation laws involving these quantum numbers are irrelevant. But the reaction does violate conservation of lepton number.

Because mesons have lepton number 0, as do photons, lepton number is not conserved: $L_e = 0$ initially and $L_e = -1$ finally. (Remember, an antielectron has a lepton number opposite that of an electron.)

(b) This interaction clearly violates charge conservation because the total charge at the start of the reaction is zero, but the net charge at the finish is $+1$, the K^0 and π^0 being neutral.

Because none of the particles in this reaction are leptons, lepton number conservation need not be considered. Although baryon number is conserved in this interaction (mesons have $B = 0$ while protons have $B = +1$), strangeness is not. Protons are strongly interacting particles, so this represents a strong interaction that must conserve strangeness. The π's and the p's are nonstrange particles (see Table 12.3); the K^0 has one unit of strangeness. Before the reaction $S = 0$; afterward, $S = +1$. Strangeness is not constant in this *strong* interaction as it must be. (Put another way, since strange particles must be produced in pairs, this reaction obviously violates conservation of strangeness as there is only one strange particle, the K^0, created in this interaction.)

(c) In this decay, a baryon ($B = +1$ for Λ^0) is converted into two mesons ($B = 0$ for $\pi^\pm$). This is forbidden by conservation of baryon number. Again, because this reaction involves no leptons, conservation of lepton number is irrelevant. However, charge obviously is conserved: $Q_{\text{initial}} = 0$ (the Λ^0 is uncharged); $Q_{\text{final}} = -1 + 1 = 0$.

This reaction also *would not* violate conservation of strangeness because the decay of the strange particle Λ^0 occurs by the *weak* interaction for which S may change by one unit, as it does here: $S_{\text{initial}} = -1$; $S_{\text{after}} = 0$ (since the mesons are nonstrange); and $\Delta S = +1$, which is permitted in weak interactions (see Table 12.5).

In this chapter, we use Q to represent charge, reserving the symbol q for generic reference to quarks beginning in Section 12.4. This notation differs from what we adopted in Chapter 7 and should not be confused with the symbol for *heat* used in Chapter 5.

© iStockphoto.com

EXAMPLE 12.4

Each of the reactions below is missing a single particle. Figure out what it must be if these interactions are permitted.

(a) $p + \overline{p} \rightarrow n +$ ________

(b) $\overline{\nu}_\mu + p \rightarrow n +$ ________

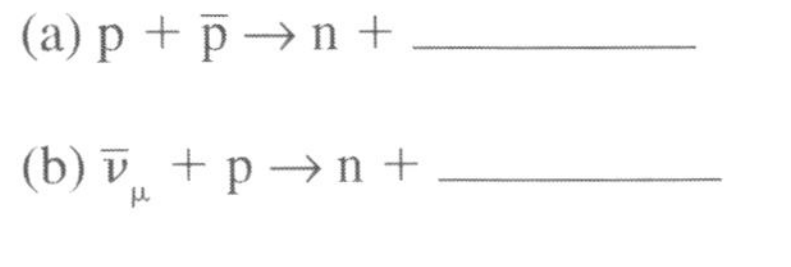

(c) $p + p \rightarrow p + \Lambda^0 +$ ________

(a) Here we have a proton encountering an antiproton and annihilating. The annihilation energy is used to create two new particles. Since the initial charge is zero (the antiproton has one unit of *negative* charge), the missing particle must be neutral like the neutron. But the starting baryon number is also zero ($B = +1$ for p and $B = -1$ for $\bar{p}$). Thus, the missing particle must be a baryon with $B = -1$ to cancel the baryon number (+1) of the neutron.

As this is a strong interaction, strangeness must be conserved. Since $S = 0$ at the beginning and n itself is nonstrange (see Table 12.3), the absent particle must also have strangeness zero. The only particle satisfying *all* these criteria is $\bar{n}$, the antineutron.

(b) The key conservation law to use to identify the missing particle is conservation of muon lepton number. The $\bar{\nu}_\mu$ is a muon antineutrino with $L_\mu = -1$. The p and the n are clearly baryons with $L = 0$. The absent species must belong to the muon family and have $L_\mu = -1$.

To conserve charge, the missing lepton must also possess one unit of positive charge to balance the initial charge of the proton. (Remember, neutrinos have no charge.) The only particle that meets these requirements is the μ^+, an antimuon.

(c) To fill in the missing entity in this reaction, we rely upon conservation of strangeness. This is a strong interaction, so we know strangeness must remain constant: $S_{initial} = 0$; therefore S_{final} must be zero as well. As it stands, the sum of the strangeness of the products, *excluding* the missing particle is $0 + (-1) = -1$. The particle that is lacking *must* have $S = +1$.

Conservation of charge demands that the missing particle have one unit of positive charge because $Q_{initial} = +2$, while $Q_{final} = +1$ *without* the additional contribution of the missing particle.

The missing particle *cannot* be a baryon, however, because adding another baryon to the products destroys the equality that already exists in baryon number. We are obviously searching for a singly charged, positive meson having one unit of positive strangeness. Examining Table 12.3 shows that we need a K^+, a kaon.

12.4 Quarks: Order out of Chaos*

The rapid proliferation of subatomic particles during the 1960s and early 1970s caused many physicists to wonder if "elementary particles" were really so "elementary" after all. Maybe they themselves were composites of still smaller entities, the "*really* elementary" particles. But even before this, in 1956, a Japanese physicist named Shoichi Sakata had proposed a model in which all hadrons were made out of just six: the proton, the neutron, the lambda, and their anti's. Although there were some peculiar aspects to this model concerning its predictions for the masses and binding energies of some hadrons, it did a pretty good job of accounting for the properties of the then-known baryons and mesons, and it violated no then-recognized laws of physics.

In 1961, Murray Gell-Mann and Yuval Ne'eman independently gave a new way to classify hadrons. They introduced an additional property of subatomic particles called *unitary spin*. This quantum characteristic is conserved in strong interactions and has eight components, each of which is a combination of the quantum numbers we have seen before (plus a few others that are too esoteric for us to consider at this level). Because of the eight-part structure of the basic "charge" in this theory and the theory's potential for leading to a deeper understanding of Nature, Gell-Mann christened this the "eightfold way," in analogy with "the noble eightfold way" of Buddhism that leads to nirvana.

Many of the particles listed in Table 12.3 had not yet been discovered at the time these two competing theories for the composition of hadrons were proposed. In the best tradition of the scientific method, a test was soon devised to distinguish which of the two, if either, gave the better description of Nature. Before 1963, the spin of the sigma particles was not known. The Sakata model predicted that these particles should have spin $\frac{3}{2}$. The eightfold way predicted a spin of only $\frac{1}{2}$. When the spins of the Σ^0 and $\Sigma^\pm$ were finally measured, they were found to be $\frac{1}{2}$, in agreement with the eightfold way.

Additional support for the Gell-Mann–Ne'eman model came a little later, with the discovery of the omega minus, Ω^-. Based on the eightfold way, the existence of a single, massive particle with charge -1, strangeness -3, and spin $\frac{3}{2}$, which had never been seen before, was predicted. Particle physicists began to search for this species in new, higher-energy collision experiments, and in February 1964, the successful detection of a particle having all the predicted characteristics was announced by scientists at the Brookhaven National Laboratory. Figures 12.9 and 12.10 show the original photograph of the particle tracks in the discovery

* As used here, the word *chaos* is synonymous with confusion: It does not refer to that branch of dynamics wherein experimental outcomes become highly unpredictable due to their extreme sensitivity to the precision of the starting parameters.

Figure 12.9

The first photograph of the Ω^- particle, taken at the Brookhaven National Laboratory in 1964. The path of the Ω^- is marked with an arrow at the lower left of the picture.

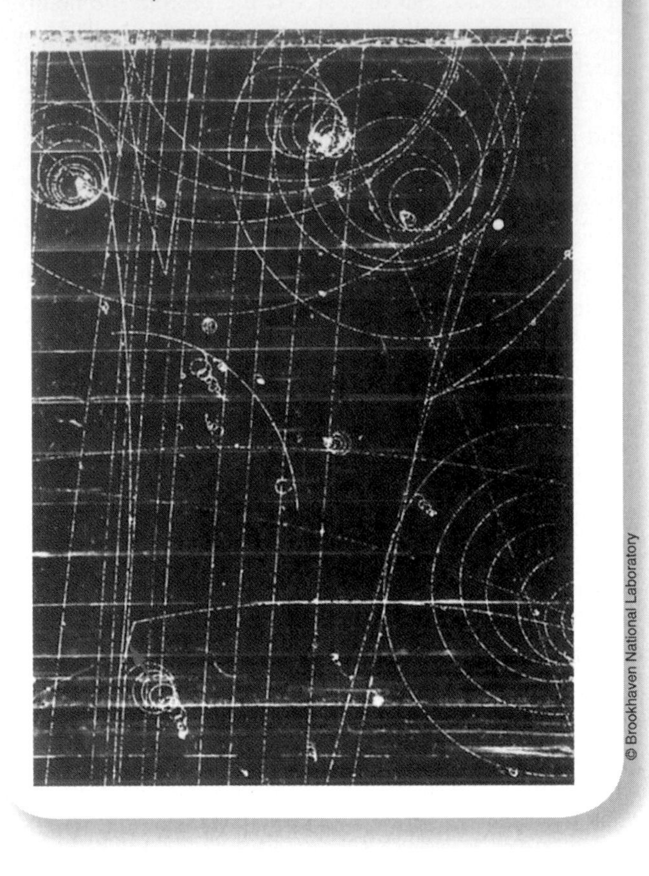

Figure 12.10

A schematic reconstruction of some of the particle tracks shown in Figure 12.9. Solid lines indicate the trajectories of charged particles, while dashed lines give the paths of neutral particles (which do not show on the original photograph). The formation and decay of the Ω^- involve the following reactions: (1) $K^- + p \rightarrow \Omega^- + K^+ + K^0$; (2) $\Omega^- \rightarrow \Xi^0 + \pi^-$.

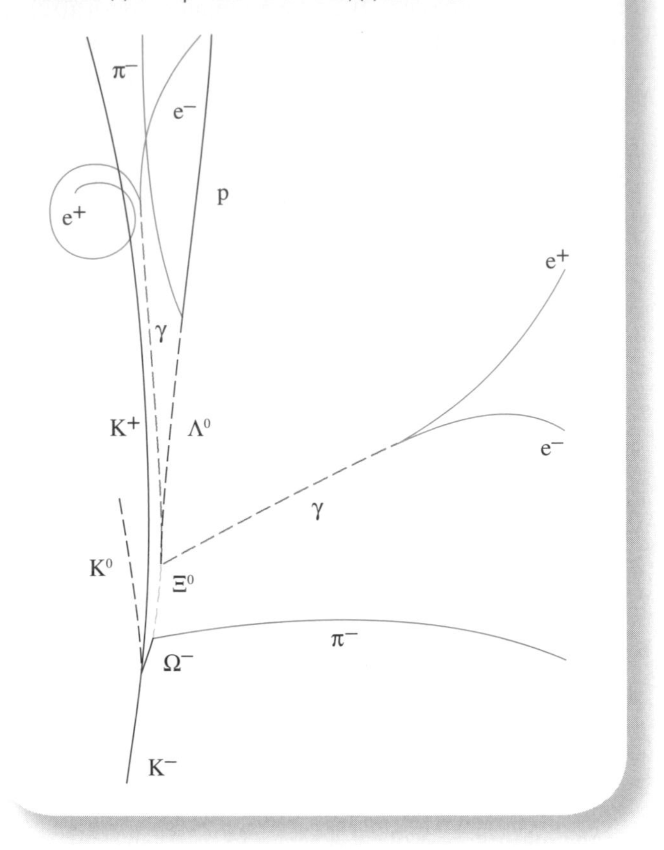

experiment and the analyses that led to its interpretation in terms of the Ω^-.

Quarks

Further investigation of the implications of the eightfold way led to a refinement of the theory in 1964. At that time Gell-Mann and George Zweig postulated that all hadrons were formed from three fundamental particles, which Gell-Mann called **quarks**, and their anti's. The quarks were designated u (for "up"), d (for "down"), and s (for "strange"). Table 12.6 gives the properties of these three particles and their anti's. Notice that all the quarks have spin $\frac{1}{2}$, baryon number $\pm\frac{1}{3}$, and charge $\pm\frac{1}{3}$ or $\pm\frac{2}{3}$ (in units of the proton charge). These are fractionally charged particles, unlike anything we have dealt with before!

Surprising as this may appear, the introduction of these noninteger charged particles enabled physicists to describe perfectly all the hadrons discovered prior to about 1970 and, with some extensions, all the heavy particles found since. The two rules governing the formation of hadrons from quarks are simple:*

1. Mesons are composed of quark-antiquark pairs, like $u\bar{d}$ and $d\bar{s}$.
2. Baryons are constructed out of three-quark combinations; antibaryons are made up of three antiquarks.

For example, a π^+ meson is equivalent to a $u\bar{d}$ pair with their spins oppositely directed. This combination gives a particle of spin 0, baryon number $0 \left(= \frac{1}{3} - \frac{1}{3}\right)$, and charge

* In 2003, a new kind of elementary particle—an *exotic baryon*—consisting of five quarks was reported by four experimental groups in Japan, the United States, Germany, and Russia. The Θ^+, as the new particle is now generally called, has a mass of 1,540 MeV/c^2, and is believed to be composed of two ud quark pairs plus an $\bar{s}$ quark. Theorists, extending the original model of Gell-Mann and Zweig, had predicted the existence of such exotic hadrons more than a decade earlier. Although some experiments through 2005 seem to confirm the existence of the Omega-plus, other large-scale studies, including ones at Fermilab, Brookhaven, SLAC, and DESY in Germany, have yielded only null results.

Table 12.6 Some Properties of the Originally Proposed Quarks[a]

Quark	Electric Charge (Q)[b]	Baryon Number (B)	Strangeness (S)
u	$+\frac{2}{3}$	$+\frac{1}{3}$	0
d	$-\frac{1}{3}$	$+\frac{1}{3}$	0
s	$-\frac{1}{3}$	$+\frac{1}{3}$	-1
$\bar{u}$	$-\frac{2}{3}$	$-\frac{1}{3}$	0
$\bar{d}$	$+\frac{1}{3}$	$-\frac{1}{3}$	0
$\bar{s}$	$+\frac{1}{3}$	$-\frac{1}{3}$	$+1$

[a]The quarks are all spin $\frac{1}{2}$ particles.
[b]These values are in units of the proton charge. Thus the charge of an up quark in SI units would be $\left(\frac{2}{3}\right)(1.6 \times 10^{-19}\text{ C}) = 1.07 \times 10^{-19}\text{ C}$.

$+1$ $\left(= \frac{2}{3} + \frac{1}{3}\right.$; see Table 12.6$\left.\right)$ as required. A K^0 meson may be shown to consist of a combination of a d quark and an $\bar{s}$ quark.

Similarly, a proton is a collection of uud quarks, while the neutron is a udd quark combination. (You should take a few minutes to assure yourself that the addition of quarks as indicated gives the usual properties of the proton and the neutron that you are familiar with. The data in Tables 12.3 and 12.6 will be helpful in this regard.) Other three-quark arrangements give other baryons: $\Sigma^+ =$ uus, $\Lambda^0 =$ uds, and $\Omega^- =$ sss.

EXAMPLE 12.5

To what hadron does the combination of a d and a $\bar{u}$ quark correspond? Assume the spins of the quarks are antiparallel.

A quark-antiquark pair gives a meson, so that is the type of hadron we're looking for. Now a d quark has charge $-\frac{1}{3}$ and a $\bar{u}$ quark has charge $-\frac{2}{3}$. The total charge of our mystery particle is thus -1. Since neither the d nor the $\bar{u}$ quark is strange, their combination must yield a nonstrange particle. If the spins of the d and $\bar{u}$ quarks are oppositely aligned, the net spin will be zero for the meson. Evidently, we seek a spin zero meson with strangeness 0 and charge -1. Consulting Table 12.3, we find that the particle in question is the π^- meson.

EXAMPLE 12.6

Give the quark combination associated with the xi minus, Ξ^-, baryon.

Baryons are composed of three quarks. The Ξ^- has charge -1, strangeness -2, and spin $\frac{1}{2}$. Because the quarks each are spin $\frac{1}{2}$ particles, we see immediately that the spins of two of the quarks making up the Ξ^- must be paired off (spin-up plus spin-down) so that the net spin is only $\frac{1}{2}$. Since the Ξ^- is a strange baryon, it must contain s quarks; they are the only quarks that carry this property. Given that an s quark possesses -1 unit of strangeness, the Ξ^- must contain two of these quarks to have net strangeness -2. If the Ξ^- has two s's, then together they contribute a total charge of $-\frac{2}{3}\left(= -\frac{1}{3} - \frac{1}{3}\right)$. To make up the additional $-\frac{1}{3}$ charge needed to obtain the total charge of -1 for the Ξ^- requires a nonstrange quark of charge $-\frac{1}{3}$. The only candidate is the d quark. Thus a Ξ^- is a dss quark combination.

It is important to emphasize at this point that the quark model *applies only* to hadrons. Leptons, like the electron, the muon, and the neutrinos, are *not* made of quarks. Indeed, the status of the leptons and the quarks is much the same within the realm of particle physics: Both are groups consisting of fundamental, irreducible spin $\frac{1}{2}$ fermions that together make up the matter in the universe.

The quark model has several distinct advantages over any competing theories of subatomic particles. First, it explains why mesons all have integral spins—that is, why mesons are bosons. Because mesons are two-quark combinations, the mesons, as a class, can only have total spin 0 (when the spins of the two constituent quarks point in opposite directions) or 1 (when the two spins are aligned). Second, the model also accounts for the fact that baryons all are fermions, half-integral spin particles: Any arrangement of three quarks will always yield either a spin $\frac{1}{2}$ or a spin $\frac{3}{2}$ particle, depending on whether the spins of two of the three are paired (spin $\frac{1}{2}$) or whether all three spins are parallel (spin $\frac{3}{2}$). By the same token, the quark model permits a new interpretation of strangeness and its conservation. Strangeness is just the difference between the number of strange antiquarks and the number of strange quarks making up the particle. Conservation of strangeness in strong interactions may now be seen as the prohibition of the conversion of an s quark to a d or a u quark during a reaction.

Given the abundant success of the quark model, it was not too long before particle physicists began looking for evidence of the existence of quarks. Despite

Remarkably, the force that binds the quarks actually gets weaker as the separation between the quarks in the hadrons diminishes. This property, called "asymptotic freedom," was discovered by David J. Gross, H. David Politzer, and Frank Wilczek, who shared the 2004 Nobel Prize for their work.

© Snezana Negovanovic/iStockphoto.com / © Andrew Johnson/iStockphoto.com

many careful searches among many different types of particle interactions, no free quarks have ever been found.* According to prevailing theories of quark confinement, quarks are inescapably bound within hadrons by what is called the *color force.* Within each hadron, the color force is mediated by exchange particles called *gluons*, and the "strings" that bind the individual quarks comprising the hadrons are called *gluon tubes.* (See Section 12.5 for more on gluons and the color force.) Thus, the strong force that exists between hadrons is now revealed to be just a shadow of the "*really* strong force" (the color force) that binds the quarks. But, if all this is true, we are still left with our original question: What evidence is there for the existence of quarks?

The experimental evidence for quarks comes from two sources: first, the scattering of high-energy electrons off protons and, second, the observation of jets of hadrons coming from collisions between electrons and positrons or protons and antiprotons. If the proton were a uniform spherical distribution of positive charge, then the deflection suffered by a high-speed electron penetrating the interior of a proton would be dependent in large part on how much charge the electron "saw" as it passed through the proton. On the other hand, if the proton were made up of three small, fractionally charged quarks, the deflection of the incoming electron would be slight, unless it happened to hit one of the quarks "head on." Then the electrical force between the electron and the quark would be substantial, and the electron might be scattered through a very large angle. This would seldom happen, of course, since the proton in the quark model is mostly empty space. And what is observed? Just what is expected from the quark picture! In particular, the number of electrons deflected by large amounts is in good agreement with the predictions of the quark model.

Figure 12.11

This track is an example of real data collected from the DELPHI detector on the Large Electron-Positron (LEP) collider at CERN, which ran between 1989 and 2000. Here a Z^0 particle is produced in the collision between an electron and positron, which then decays into a quark-antiquark pair. The quark pair is seen as a pair of hadron jets in the detector.

© CERN

A second type of evidence for quarks involves the products of high-energy collisions of beams of electrons and positrons. In these reactions, the energy derived from the e^--e^+ annihilation goes into the creation of q-$\bar{q}$ pairs. As these particles move off in opposite directions (as required to conserve linear momentum), they produce a stream of new q-$\bar{q}$ pairs. The quarks in this bidirectional jet quickly combine to give various hadrons, so that at the level of the experiment, what is seen is two trains of heavy particles, oriented at 180°, traveling away from one another (Figure 12.11). Similar

© Catherine dée Auvil/iStockphoto.com / © CERN/Photo Researchers, Inc.

* Tantalizing evidence for the existence of free quarks in what is called a quark-gluon plasma (QGP) is beginning to emerge, however, from the Brookhaven National Laboratory's Relativistic Heavy Ion Collider (RHIC) experiments conducted in 2003. By colliding opposing beams of high-speed gold ions, scientists have succeeded in creating, for a brief instant, conditions like those that existed in the universe less than a second after the Big Bang, when the temperature was about 2×10^{12} K. In such circumstances, the protons and neutrons in the gold nuclei "melt," releasing the quarks and gluons that compose them to produce a tiny sample of "soup"—the QGP. Like bound quarks, evidence for a QGP comes from studying the particle jets that emerge from the head-on collisions of the gold ions, and further experimentation, like some planned with the LHC (see page 204), will be necessary to confirm these early results.

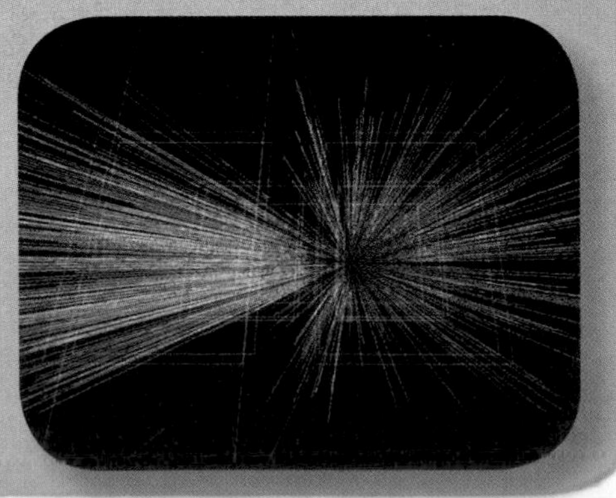

jets of hadrons are observed in high-energy collisions between protons and antiprotons. Coupled with the data from the electron-scattering experiments described in the previous paragraph, these observations provide persuasive evidence for the overall correctness of the quark model and the existence of these elusive particles.

The quark model also provides a handy way of understanding and analyzing decays and reactions in particle physics. Only two rules need to be remembered:

1. Quark-antiquark pairs can be created from energy in the form of gamma rays or collisional kinetic energy.
2. The weak interaction alone can change one type of quark into another. The strong and electromagnetic interactions cannot cause such changes. Thus we see quark types are conserved by the strong and electromagnetic forces but not by the weak force.

EXAMPLE 12.7

Analyze the decay $n \rightarrow p + e^- + \bar{\nu}_e$ in terms of the quark model.

The beta decay of a neutron may be rewritten in terms of the quarks making up the various hadrons present in the reaction as follows. (*Remember*: The electron and its antineutrino, being leptons, are not composed of quarks.)

$$udd \rightarrow uud$$

Figure 12.12
An alternative representation of the beta decay of a neutron emphasizing the quark content of the neutron and the proton. The weak interaction transforms a d quark into a u quark, thereby converting the original neutron (udd) into a proton (udu). Compare this description of the reaction to that depicted in Figure 12.5a.

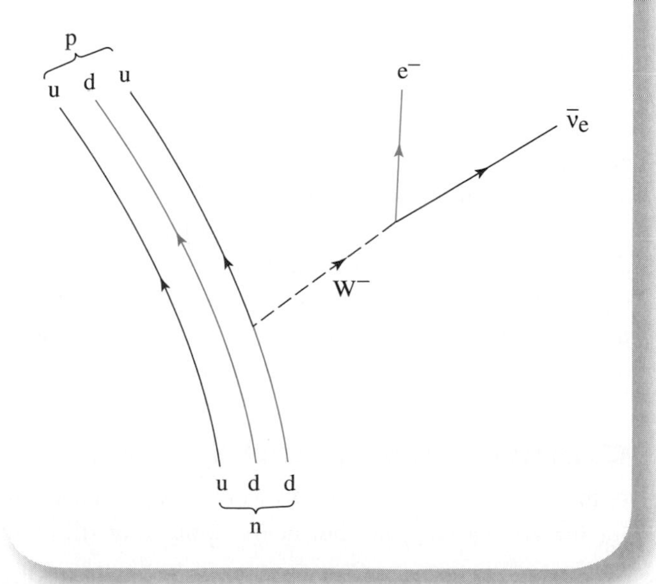

Canceling a u and a d quark on both sides of the arrow yields:

$$d \rightarrow u$$

The fundamental decay is the conversion of a d quark to a u quark by the weak force. A W^- particle is created in the process from the mass deficit between the d and u quarks; this carrier particle subsequently decays into an electron and an antineutrino (see Figure 12.12).

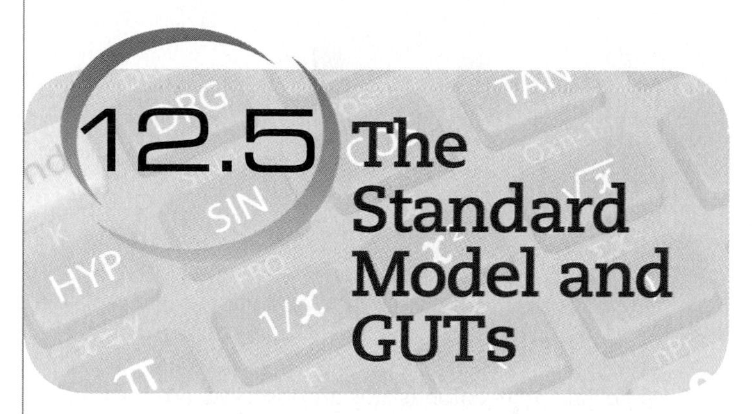

12.5 The Standard Model and GUTs

Quarks are fermions. They have half-integral spins, and they should therefore obey the Pauli exclusion principle. But if this is true, how do we explain the existence of the Ω^-? This particle is composed of three strange quarks, all of which share the same mass, spin, charge, and so on—that is, the same quantum numbers. It would appear that the Ω^- is made up of three identical, interacting s quarks. Doesn't this violate the exclusion principle? How can we reconcile these circumstances?

One way would be to argue that quarks are somehow different from other fermions and, consequently, don't have to conform to the Pauli principle. When particles like the Ω^- were discovered, some theoretical physicists did suggest this explanation to resolve the dilemma. However, other scientists were reluctant to make exceptions to the exclusion principle and instead proposed that quarks carry an additional property that makes them distinguishable within hadrons like the Ω^-. For this scheme to work, this new characteristic had to come in three varieties to permit, for example, the three s quarks in the Ω^- to be different from each other. The name that particle theorists gave to this new quantum property or number was *color*, although it has absolutely nothing to do with what we commonly refer to as color—that is, our subjective perception of certain frequencies in the electromagnetic spectrum. The three quark colors were labeled after the three primary colors of the artist's palette: red, blue, and green. Antiquarks are colored antired, antiblue, and antigreen (Figure 12.13).

Figure 12.13
Each quark (and antiquark) comes in three colors (or anticolors).

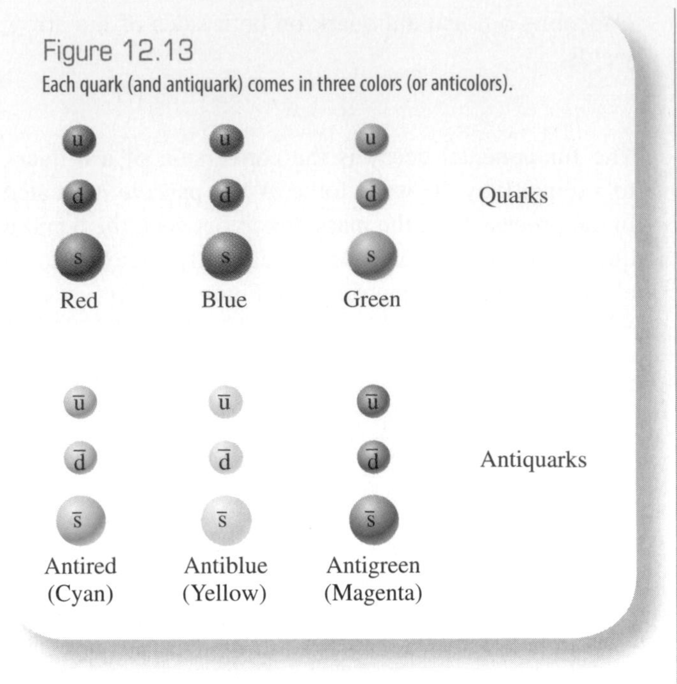

The fact that color is not an observed property of hadrons indicates that these particles are "colorless." If they are composed of colored quarks, then the way the colors come together within the hadron must be such as to produce something with no net color, something that is color neutral. For this to be true, the three quarks that make up baryons must each possess a color different from their companions, one red, one blue, and one green. The addition of the primary colors produces the result we call "white," so baryons containing three different colored quarks are considered to be white or neutral as concerns the color charge. In an analogous manner, for mesons to be color-neutral requires them to be made up of a quark of one color and an antiquark possessing the corresponding anticolor. Returning to our previous example of the π^+ meson, we see in the light of this new color physics that if the u quark is red, the $\bar{d}$ quark must be antired.

You have probably noticed that the language of particle physics is rather whimsical in comparison with the other areas of physics that we have studied. To carry this whimsy one step further, we note that often the different types of quarks, u, d, s (and others that we will shortly introduce), are designated quark *flavors.* Quarks come in six flavors (not counting the antiflavors), and each flavor comes in three colors. Complicated as this may seem to you now, its still less troublesome than making a choice from among 31 varieties at a Baskin-Robbins® ice cream store.

Color is another example of an internal quantum number, like unitary spin. It is not directly observed in hadrons, and it cannot be used to classify them. Moreover, it does not influence the interactions among hadrons. So what is the significance of the color charge? What possible role does it play in particle physics?

Current theories now suggest that the color charge is the source of the interquark force, much as ordinary electric charge is the source of the electrical force. And just as the electrical force between charged particles is carried by the photon, the color force between the bound quarks is carried by what are called *gluons.* Gluons are electrically neutral, zero mass, spin 1 particles that are exchanged between the bound quarks in hadrons. There are eight gluons in all, and the emission or absorption of a gluon by a quark can lead to a change in the color of the quark. Such changes occur very rapidly and continuously among the quarks forming a given hadron, but always in such a way that the overall color of the hadron remains neutral. The decay of a single meson ($q\bar{q}$) into two may also be interpreted in terms of the emission of a gluon by one quark bound in the original meson. The radiated gluon is here quickly converted into matter and antimatter—namely, a new quark and antiquark, which subsequently bind to the existing pair to form two mesons. Unlike virtual photons, which mediate the electrical interaction but carry no electric charge themselves, gluons, the carriers of the color force, are themselves colored, that is, they possess color charge. This fact permits the gluons to interact among themselves by the attractive color force and form what are known as *glueballs.*

The color force is sometimes referred to as the *chromodynamic force*, from the Greek word *chroma*, meaning "color." The theory of color and its connection to gluons is called *quantum chromodynamics.* As is the case for the quarks, evidence for the existence of gluons may be found in high-energy particle-collision experiments, particularly ones in which three or four hadron jets are seen (Figure 12.14). Data to support the existence of glueballs are much less compelling, although some physicists believe that the iota (ι) particle, produced in the decay of the ψ (see the following subsection), may represent a candidate for a glueball. Other glueball candidates include the short-lived f_0 (1500) and f_J (1710) particles (the numbers in parentheses refer to the mass of the particle in MeV/c^2) produced at CERN, although again the evidence is inconclusive. Several new experiments are being planned involving proton-antiproton beam collisions that hold great promise for providing more definitive data on the existence and properties of glueballs.

Charm, Truth, and Beauty

In the early 1960s, three quarks were enough to account for the known hadrons, but at the time, four different

Figure 12.14

This image is an electronic display of an electron-positron collision in the ALEPH detector at CERN, the European particle physics laboratory near Geneva . The display shows a cross-section of the detector, with the beam tube in the center (blue) surrounded by various detector components (blue and red). The electron and positron, accelerated to high energy in CERN's LEP collider, annihilate at center to create a quark and an antiquark, one of which radiates a gluon (carrier of the strong nuclear force). Although too short-lived to be detected directly, the quark, antiquark, and gluon each decays into jets of hadrons that spread out through the detector.

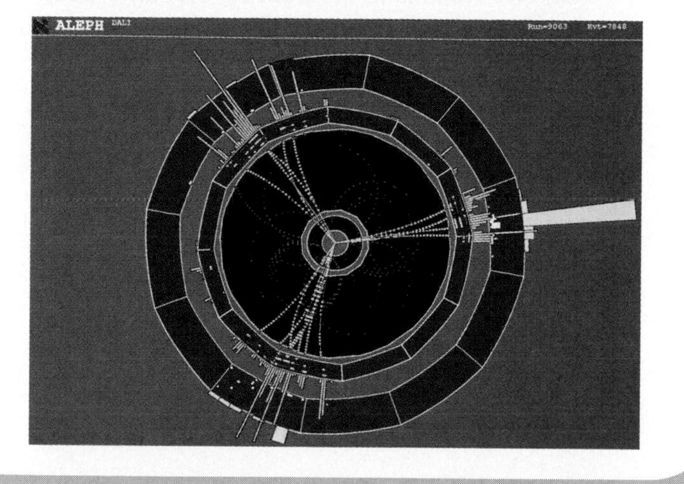

© CERN/SPL/Photo Researchers, Inc.

leptons were known: the electron and its neutrino and the muon and its neutrino. (The tau was not discovered until 1975.) Because of this asymmetry in the numbers of quarks and leptons, it was suggested that there should exist a fourth quark to balance out the leptons. Other, "harder" evidence for the existence of a fourth quark came in the form of certain rare, weak interactions involving hadrons in which their charges remained constant but their strangeness changed. Together, these bits of data convinced particle physicists that another quark, the c quark, was present in Nature and that it carried a new quantum "charge" called *charm*. Theory indicated that the charmed quark should have electric charge $+\frac{2}{3}$, strangeness 0, and a mass greater than any of the previously identified quarks.

The search for particles containing the c quark—so-called charmed particles—began in earnest in 1974 at Brookhaven and at the Stanford Linear Accelerator Center (SLAC) in California. In November 1974, Samuel Ting, leader of the Brookhaven group, and Burton Richter, head of the SLAC team, jointly reported the discovery of a short-lived particle produced in e^--e^+ annihilation events that seemed to "fit the bill." The particle, called J by Ting's group and ψ (psi) by Richter's, is now believed to be a $c\bar{c}$ combination—one state of several belonging to what has been referred to as "charmonium"—having a mass of 3.1 GeV/c^2. In 1976, Ting and Richter shared the Nobel Prize in physics for their work.

The existence of charmonium clearly indicated the presence of a new quark and its anti, but to prove beyond doubt that this special type of quark existed required the identification of a particle, say, a meson, composed of one charmed quark (or its anti) plus another, different quark, like a u or a $\bar{d}$. This was necessary so that the property of charm could be explicitly manifested and not cancel out as it does in the ψ. After two years of searching, in 1976, a charmed meson, labeled D^0, was found; it appears to be made up of a $c\bar{u}$ quark combination. By the end of the 1970s, other mesons possessing both charm and strangeness were discovered, and even a charmed baryon—the Λ_c = udc! The existence of charm in the universe was secure.

But even more surprises were in store for particle physicists. In 1977, in experiments conducted at Fermilab outside Chicago, a very heavy meson, called the Υ (upsilon), was discovered. It could not be accounted for in terms of the four previously postulated quarks. So to explain the properties of this hadron, researchers proposed that a fifth quark, the b or "bottom" quark (early on referred to as *beauty*), existed. The Υ was taken to be a $b\bar{b}$ pair.

The b quark then carries yet another special quantum charge. The discovery of the Υ meson spurred the search for additional mesons composed of a b (or a $\bar{b}$) quark and a lighter u or d quark, much as the identification of ψ led to the search for charmed mesons. Because the masses of these mesons were expected to be very large and their lifetimes extremely short, many, many ultrahigh-energy collisions had to be carried out and carefully studied before enough evidence to corroborate the existence of such particles could be gathered. Finally, in 1982, after examining over 140,000 e^--e^+ collisions, investigators at Cornell University identified 18 events that they concluded corresponded to decays involving B mesons, hadrons with masses near 5.3 GeV/c^2 composed of b quarks: $B^+ = \bar{b}u$; $B^- = b\bar{u}$; and $B^0 = \bar{b}d$. Here was strong confirmation of the reality of the fifth quark.

In 1975, the announcement of the discovery of the tau and its (presumed) neutrino (recall this particle was not observed in collision reactions until 2003) brought the number of leptons to six. With the identification of the b quark in 1982, the quarks once again lagged the leptons by one. To maintain the symmetry between the two groups, the need to have a sixth quark was pointed out. The presence of this newest quark, the t or "top" quark (sometimes called *truth*), permits the quarks to be paired off into three families much as the leptons are: (u,d); (s,c); and (b,t). In 1985, in experiments done at CERN, some preliminary evidence for a meson containing the top quark was found.

In March 1995, based on data from proton-antiproton collision experiments carried out using the Fermilab Tevatron accelerator, representatives of a consortium of over 400 scientists and engineers announced the discovery of more than 100 events out of several trillion associated with the production of the top quark with a probable error of less than 1 in 500,000. Figure 12.15 shows a schematic representation of a typical "top event" resulting from a p-$\bar{p}$ collision and the subsequent decay of the top and antitop quarks into other particles. Beginning in 2007, "single" top quarks in t$\bar{b}$ meson combinations have been reported in experiments at Fermilab, offering unequivocal proof of the reality of the top quark. Based on the analyses of these events, the current value of the top quark mass is 174.3 GeV/c^2, a value greater than the mass of the nucleus of an atom of gold!

If current plans are followed, the nearly $4 billion Large Hadron Collider (LHC, near Geneva, Switzerland, see page 204) at CERN will begin full operation in late 2009 with proton beams having collision energies up to 14 TeV (*tera* $= 10^{12}$), and will produce about 8×10^6 t$\bar{t}$ pairs per experiment. Within a very short time, physicists will be able to use the top quark to answer many questions that still remain about matter and the forces that govern the physical universe. Table 12.7 summarizes the properties of the six quarks.

Are there more than six quarks in Nature? More than six leptons? Should we expect to see an expansion in the numbers of families of these particles beyond three in the future? And if so, does this provide evidence that these "elementary" particles may themselves be composites of still more fundamental species? The answers to all of these questions cannot be given at this time, but the results of experiments done within the last few years seem to indicate that there cannot be more than three distinct families of quarks and leptons. With the confirmation of the existence of the top quark, it appears that we have found *all* of the fundamental building blocks of Nature. Figure 12.16 summarizes the properties of the three quark and lepton families, as well as those of their anti's, in what is now referred to as the *Standard Model* of elementary particle physics.

Although highly successful at accounting for much of what is known about elementary particles, the Standard Model has limitations. For example, it predicts that the masses of the neutrinos should all be precisely zero. Laboratory experiments have placed upper limits on these masses (see Table 12.4), but the most compelling evidence that neutrinos must have mass comes from recent observations at the Sudbury Neutrino Observatory (SNO). Operated by a consortium of scientists from 15

Figure 12.15
Candidate event for top-antitop production. Each top quark decays at the p$\bar{p}$ primary collision vertex into a W boson plus a bottom quark. The W^+ decays into a positron e^+ plus an invisible neutrino (ν_e), and the W^- decays into a quark and an antiquark, which appear as two jets of hadrons (jets 1 and 2). Each bottom (b) quark becomes a neutral B meson that travels a few millimeters from the production vertex before its decay creates a hadron jet. (From "Top-ology" by Chris Quigg, *Physics Today* 50, no. 5 (May 1997), p. 24.)

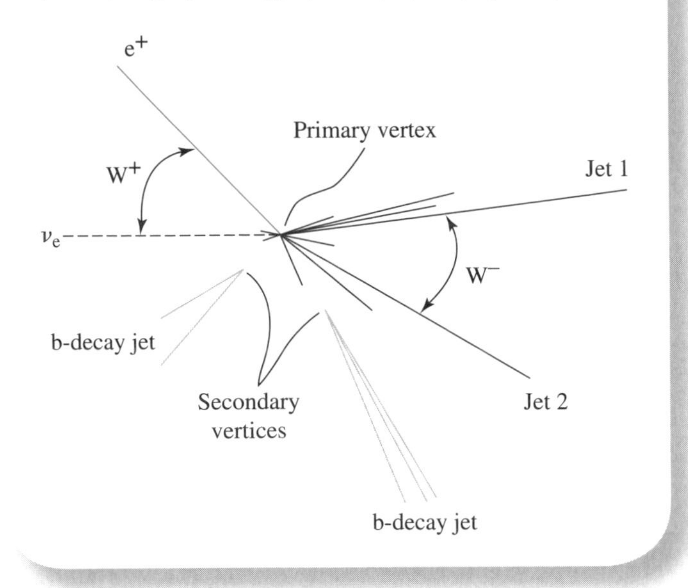

Table 12.7 Summary of Basic Characteristics of u, d, s, c, b, and t Quarks

	Quantum Number or Charge[a]					
Quark	Electric Charge	Strangeness	Charm	Bottomness	Topness	Mass[b] (MeV/c^2)
u	$+\frac{2}{3}$	0	0	0	0	2.5
d	$-\frac{1}{3}$	0	0	0	0	4.8
c	$+\frac{2}{3}$	0	1	0	0	1,270
s	$-\frac{1}{3}$	−1	0	0	0	104
t	$+\frac{2}{3}$	0	0	0	1	171,200
b	$-\frac{1}{3}$	0	0	1	0	4,250

[a]All the quarks have baryon number $\frac{1}{3}$ and spin $\frac{1}{2}$.
[b]Values given are approximate upper limits to quark mass ranges.

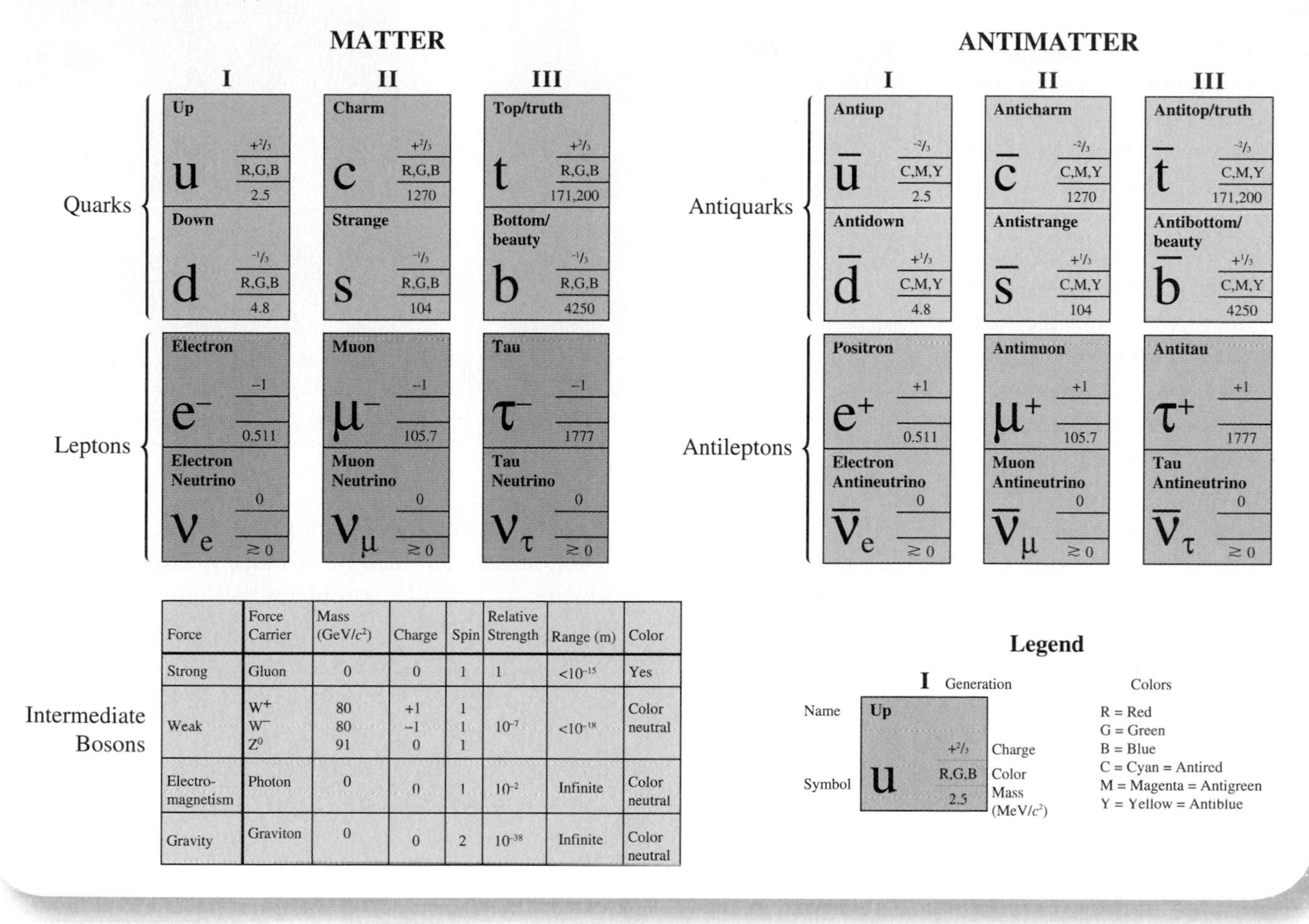

Force	Force Carrier	Mass (GeV/c^2)	Charge	Spin	Relative Strength	Range (m)	Color
Strong	Gluon	0	0	1	1	$<10^{-15}$	Yes
Weak	W^+ W^- Z^0	80 80 91	+1 −1 0	1 1 1	10^{-7}	$<10^{-18}$	Color neutral
Electro-magnetism	Photon	0	0	1	10^{-2}	Infinite	Color neutral
Gravity	Graviton	0	0	2	10^{-38}	Infinite	Color neutral

Figure 12.16
Standard Model of elementary particles.

institutions from the United States, Great Britain, and Canada, the heart of this facility consists of 1,000 tons of "heavy" water (D_2O) in a 12-m-diameter containment vessel buried 2 km underground. Nearly 9,500 light-sensitive detectors placed inside the vessel look for faint bursts of light that accompany the interaction of neutrinos (mainly emitted by nuclear reactions in the core of the Sun) with the heavy water. Unlike other neutrino detectors, SNO has the capacity to distinguish the type of neutrino (electron, muon, or tau) involved in the interaction, and recent results indicate that only about one-third of the total number of neutrinos reaching the Earth are of the electron variety.*

* This confirms results reported by Raymond Davis Jr. and collaborators more than forty years ago showing that the number of electron neutrinos detected from the Sun is smaller than what is expected on the basis of our best models for the solar interior. For his pioneering work in what was called the "solar neutrino problem," Davis shared the 2002 Nobel Prize in physics.

Almost *all* the neutrinos produced by the Sun are electron neutrinos, so two-thirds of them must have changed flavor during their approximately eight-minute flight to Earth. Since massless neutrinos cannot spontaneously change type, the implication of this discovery is that neutrinos must have non-zero mass. Theoretical models to explain these *neutrino oscillations*, as they are called, do not provide explicit values for the individual neutrino masses, only their mass differences. Current values for the mass differences, although highly model dependent, are generally small, ranging from about 10^{-1} to as little as 10^{-4} eV/c^2. If the scale of the neutrino masses is set by the electron neutrino ($\sim$1 eV/c^2), then all the neutrino masses are very small. But the fact that these elusive particles have any mass at all remains something that the Standard Model cannot explain. At present, much of the effort in theoretical particle physics is devoted to revising the Standard Model to remedy this and other deficiencies.

The Electroweak Interaction and GUTs

In Section 12.2, we mentioned that similarities between the electromagnetic and the weak nuclear forces first noted in the 1950s have led scientists to the conclusion that these two interactions are really just two different manifestations of a more basic interaction called the **electroweak interaction**. The theory describing this interaction was developed independently by Steven Weinberg (1967) and Abdus Salam (1968), and is often referred to as the Weinberg-Salam theory. Weinberg and Salam, together with Sheldon Glashow, who had also made contributions to the unification of these two forces, shared the Nobel Prize in physics for 1979.

High energy state at the roulette table. When the wheel spins rapidly, there exists only one state of the system—with the ball whirling around in the groove.

© Ken Whitmore/Getty Images

When the wheel slows down (low energy state), there appear to be 38 possible states for the system corresponding to the 38 slots into which the ball can lodge.

© iStockphoto.com / © Slavoljub Pantelic/iStockphoto.com

According to the Weinberg-Salam model, at moderate to high energies (of the order of 100 billion electron-volts—100 GeV), the electroweak interaction is carried by four massless bosons and is described by equations that are symmetric (i.e., they remain unchanged under certain mathematical operations on some specific characteristics of these particles). As the energy of the system is lowered, however, the symmetry is broken spontaneously, with the result that the original family of four massless bosons separates into two subfamilies, the first consisting of the massless photon, which mediates what we call the electromagnetic interaction, and the second including the very massive intermediate (or gauge) bosons ($W^{\pm}$, Z^0), which carry the weak interaction.

The reasons why some of the originally massless carriers of the electroweak interaction acquire mass as the symmetry of the system is broken involves what are called *gauge fields* and *Higgs mechanisms* and are too complicated for us to explore in any detail, but an example involving broken symmetry in another context might help remove some of the mystery associated with this concept.*

* In the simplest model, the initial symmetry of the system is broken by a new particle, the *Higgs*, a massive spin 0 boson, which interacts with the intermediate gauge bosons, as well as the leptons and the quarks. These latter particles acquire mass in the process, the values being determined by how strongly each "couples" to or interacts with the Higgs. For the recently confirmed top quark whose mass is very large, the coupling with the Higgs is relatively strong. The LHC's ability to produce large numbers of top quarks makes it ideal for exploring possible connections between the t and the eagerly sought-after Higgs.

This example has been given by Stephen Hawking and connects the idea of symmetry within a system to its energy. Consider a roulette ball on a roulette wheel (Figure 12.17). At high energies when the wheel is spun rapidly, the ball behaves in basically one way: it rolls around and around in one direction in the groove of the wheel. We might say that at high energies there is only one state in which the ball can be found, and all rapidly spinning roulette wheels display the same state: They're all symmetrical with respect to one another. However, as the wheel slows and the ball's energy decreases, it eventually drops into one of the 38 slots molded into the wheel; at low energies, it appears that there are 38 different states in which the ball can exist. If we surveyed roulette wheels solely at low energies, we might be led to the conclusion that these 38 possibilities were the only ones allowed to the ball. We would miss the fact that at high energies these different states merge into one. This is analogous to what happens to the carrier family for the electroweak interaction. At high energy, these carriers are all massless and behave like one another. But at low energy when the symmetry is broken, this group appears as two families of completely different particles.

At the time Weinberg and Salam first published their theory, particle accelerators did not exist that could produce the energies required to test their predictions, particularly those concerning the existence of the massive $W^{\pm}$ and Z^0 particles that carried the weak nuclear force. By the early 1980s, however, machines were beginning to come on line that could achieve the needed energies, and, in January 1983, Carlo Rubbia of CERN reported the detection of the W particles in nine events out of over a million recorded p-$\bar{p}$ collisions. In July 1984, the CERN group also announced the discovery of the Z^0 particle together with additional evidence to support their earlier detection of the W particles. The 1984 Nobel Prize in physics was awarded to Rubbia and his collaborator Simon van der Meer for their search and discovery of the carriers of the weak interaction. These results completely vindicated the Weinberg-Salam model and left little doubt about its correctness.

The success of the unification of the electromagnetic and the weak nuclear forces caused many physicists to wonder whether it might not also be possible to unite the strong nuclear force with the electroweak force to produce what has come to be called a **grand unified theory**, or **GUT**. As Hawking has pointed out, these theories (there are actually several competing ones) are neither all that grand nor fully unified. For example, they do not include gravity. Nevertheless, they represent the first steps toward a true unification of the interactions of Nature, and they do present some interesting implications for the long-term future of the universe.

The basic idea of GUTs is the following: Just as the electromagnetic and weak forces represent different aspects of the same basic force that coalesce at high energies, the electroweak and strong nuclear forces are thought to be different manifestations of yet another even more fundamental force that emerges at still higher energies. The energy at which all three of these forces become fused into a single force is called the *grand unification energy*. Its value is not well known but would probably have to be at least 10^{15} billion electronvolts, far above the energies ever likely to be attainable with terrestrial laboratory particle accelerators. The GUTs also predict that at this energy the different spin $\frac{1}{2}$ particles, such as quarks and leptons, would all behave in the same manner. This is similar to the formation of a single family of electroweak carrier particles from the photons and the intermediate bosons that mediate the separate electromagnetic and weak forces.

Because of the impossibility of reaching the energies needed to test GUTs directly in the laboratory, scientists have begun searching for indirect, low-energy consequences of these theories. One search pertains to the possible decay of the proton. In Section 12.3, we argued that conservation of baryon number in strong interactions demands that the proton be absolutely stable, since there exist no lighter baryons to which it can decay. But at grand unification energies, the quarks that make up the proton are indistinguishable from leptons, so that the spontaneous decay of a proton (through quark conversion) to lighter particles such as positrons becomes possible. The three quarks bound inside a proton do not normally have sufficient energy to make this transition. We live in an environment in which the available energies are typically far below the threshold energy for grand unification. However, like all interactions governed by quantum mechanics, a very, very small, but non-zero, probability exists that such a transition can occur, in which case, the proton would decay. The chances of this happening, as you might guess, are very small. If you could watch a single proton, hoping to catch it in the act of decaying as described, you'd have to wait and watch for something like a million million million million million (or 10^{30}) years! Clearly, this is not practical. The entire universe has been in existence for only about 14 billion years or so.

To increase your chances of witnessing the decay of the proton and, hence, of validating one version of GUTs, instead of watching only one proton, you could watch lots of them. You could monitor a large quantity of matter containing an enormous number of protons—maybe 10^{31} of them. If this were done for a year, you would expect, on average, to see as many as 10 proton decays. Increasing the number of protons in the sample obviously improves your chances of seeing a decay.

This is precisely the approach taken by researchers running the Super-Kamiokande experiment in Japan. Using an underground tank filled with 50 kilotons of water (each water molecule contributes 10 protons) and equipped with thousands of light-sensitive devices, experimenters look for the flashes of radiation emitted by the high-speed positrons believed to accompany the decay of the protons. The results of this experiment have been discouraging: No clearly defined proton-decay events have been recorded

© Bruno Vincent/Getty Images / © Shelly Perry/iStockphoto.com

since data began being collected in 1996. In order to be consistent with these data, the average lifetime of the proton must be greater than 6×10^{33} years. This value is larger than that predicted by the simplest GUTs, but there are more elaborate versions of these theories that give longer proton lifetimes, up to as much as 10^{36} years.

One thing is clear. If the proton is found to be unstable, even on such long time frames, then eventually all the matter in the universe will evaporate. This is so because the positrons, as products of the proton decay, will annihilate with electrons to produce gamma rays. Assuming the universe survives for the requisite amount of time, it will ultimately be stripped of all matter and reduced to a cold, dark space filled with photons (which steadily lose energy) and a few isolated electrons and neutrinos (which escaped destruction)—a grim fate indeed, but one that may well await our now-glorious universe. More so than most, this example demonstrates how the physics of the subatomic domain can influence that of the largest scale known. But the reverse is also true.

We have seen that it is not feasible to expect that terrestrial laboratory experiments will ever achieve the energies required to verify directly the predictions of GUTs. Indeed, one might well ask, "Have such energies ever been available in the entire history of the universe?" And the answer is, "Yes, in the very first few moments after the birth of the universe in the Big Bang!"

Thus we begin to see that the ultimate tests of GUTs may rest in the field of cosmology, the study of the structure and evolution of the universe on the largest scale possible (see the Epilogue). The justification of theories involving the smallest of physical entities, quarks and electrons, may depend on our understanding of the largest of physical structures, galaxies and clusters of galaxies.

questions&problems

Questions

(▶ Indicates a review question, which means it requires only a basic understanding of the material to answer. The others involve integrating or extending the concepts presented thus far.)

1. ▶ Describe the two fundamental postulates underlying Einstein's special theory of relativity.
2. Suppose you were traveling toward the Sun at a constant velocity of $0.25c$. With what speed does the light streaming out from the Sun go past you? Explain your reasoning.
3. Light travels in water at a speed of 2.25×10^8 m/s. Is it possible for particles to travel through water at a speed $v > 2.25 \times 10^8$ m/s? Why or why not? Explain.
4. ▶ In your own words, define what is meant by *time dilation* in special relativity theory. Provide a similar definition for *length contraction*. Give an example in which the effects of time dilation are actually observed.
5. Galileo used his pulse like a clock to measure time intervals by counting the number of heartbeats. If Galileo were traveling in a spaceship, moving uniformly at a speed near that of light, would he notice any change in his heart rate, assuming the circumstances of his travel produced no significant physiological stress on him? If someone on Earth were observing Galileo with a powerful telescope, would he or she detect any change in Galileo's heart rate relative to the resting rate on Earth? Explain your answers.
6. Newton wrote: "Absolute, true, and mathematical time, of itself, and from its own nature, flows equally without relation to anything external." Comment on the significance of this statement for two timekeepers in relative motion. In the light of special relativity, is Newton's statement valid? Explain.
7. Why don't we generally notice the effects of special relativity in our daily lives? Be specific.
8. Does $E_0 = mc^2$ apply only to objects traveling at the speed of light? Why or why not?
9. If a horseshoe is heated in a blacksmith's furnace until it glows red hot, does the mass of the horseshoe change? If a spring is stretched to twice its equilibrium length, has its mass been altered in the process? If so, explain how and why in each case.
10. ▶ List the four fundamental interactions of Nature, and discuss their relative strengths and effective ranges.
11. ▶ What common feature of the electromagnetic and gravitational interactions requires that their carrier (or exchange) particles be massless?
12. ▶ What is an antiparticle? What may happen when a particle and its anti collide?
13. ▶ What does the acronym PET stand for? Why is PET a good example of the relevance of elementary particle physics to our everyday lives?
14. Some neutral particles, like the π^0, are their own antiparticles, but not the neutron. In what ways are n and $\bar{n}$ the same? Speculate on how they might be different.

15. According to Table 12.4, the rest mass of an electron is 0.511 MeV/c^2. What is the rest mass of a positron?

16. ▶ Distinguish between fermions and bosons in as many different ways as you can.

17. ▶ Give some ways by which physicists classify elementary particles.

18. ▶ In which of the four basic interactions does an electron participate? A neutrino? A proton? A photon?

19. ▶ Baryon conservation and lepton conservation are laws used frequently by particle physicists to decide whether a reaction involving elementary particles is possible or not. Explain how these laws are applied to make such determinations.

20. ▶ In your own words, describe what a physicist means when using the term *strangeness*.

21. ▶ What is a quark? How many different types of quarks are now known? What are some of the basic properties that distinguish these quarks?

22. ▶ Describe the kinds of evidence that have led scientists to conclude that quarks exist.

23. ▶ How many quarks form a baryon? A meson? What is the relationship (if any) between a quark and a lepton (e.g., an electron)?

24. In the quark model, is it possible to have a baryon with strangeness −1 and electric charge +2? Explain.

25. What kind of a particle (baryon, meson, or lepton) corresponds to a $t\bar{t}$, that is, to a top-antitop quark combination? Describe some of the properties such a particle would have.

26. Although doubly charged baryons have been found (that is, baryons with net charge +2; see Problem 26), no doubly charged mesons have yet been identified. What effect on the quark model would there be if a meson of charge +2 were discovered? How could such a meson be interpreted within the quark model?

27. ▶ Quarks are said to possess "color." What does this mean? Are physicists really suggesting that quarks look red like ripe strawberries or blue like the cloudless sky? Explain.

28. ▶ Describe the *Standard Model* of elementary particle physics.

29. A bumper "snicker" on a car belonging to the chairperson of a physics department reads: "Particle physicists have GUTs!" Explain in your own words the meaning of this little joke or "play on words."

30. ▶ Unification of its basic laws and theories has long been a goal in physics. Describe some ways in which physicists have been successful in unifying certain forces and theories. In what area(s) of physics is (are) the process(es) of unification still ongoing?

31. If a proton can decay, then its lifetime is of the order of 10^{33} years, far longer than the current age of the universe. Does this necessarily imply that a proton decay has not yet occurred in the entire history of the universe? Explain.

Problems

[For many of the exercises, referring to Table 12.3 and Figure 12.16 will be very helpful.]

1. The lifetime of a certain type of elementary particle is 2.6×10^{-8} s. If this particle were traveling at 95% the speed of light relative to a laboratory observer, what would this observer measure the particle's lifetime to be?

2. How fast would a muon have to be traveling relative to an observer for its lifetime as measured by this observer to be 10 times longer than its lifetime when at rest relative to the observer?

3. The lifetime of a free neutron is 886 s. If a neutron moves with a speed of 2.9×10^8 m/s relative to an observer in the lab, what does the observer measure the neutron's lifetime to be?

4. A computer in a laboratory requires 2.50 μs to make a certain calculation, as measured by a scientist in the lab. To someone moving past the lab at a relative speed of 0.995c, how long will the same calculation take?

5. The formula for length contraction gives the length of an interval on a ruler moving with velocity v relative to an observer as $\sqrt{1 - v^2/c^2}$ times the length of the same interval on a ruler at rest with respect to the observer. By what fraction is the length of a meter stick reduced if its velocity relative to you is measured to be 95% the speed of light?

6. If an electron is speeding down the two-mile-long Stanford Linear Accelerator at 99.98% the speed of light, how many meters long is the trip as seen from the perspective of the electron? (See Problem 5.)

7. Calculate the rest energy of a proton in joules and MeVs. What is the mass of a proton in MeV/c^2?

8. The tau is the heaviest of all the known leptons, having a mass of 1,777 MeV/c^2. Find the rest energy of a tau in MeVs and joules. What is the mass of the tau in kilograms? Compare your result with the mass in kilograms of an electron.

9. If a 1.0-kg mass is completely converted into energy, how much energy, in joules, would be released? Compare this value to the amount of energy released when 1.0 kg of liquid water at 0°C freezes.

10. In a particular beam of protons, each particle moves with an average speed of 0.8c. Determine the total relativistic energy of each proton in joules and MeVs.

11. A particle of rest energy 140 MeV moves at a sufficiently high speed that its total relativistic energy is 280 MeV. How fast is it traveling?

12. If the relativistic kinetic energy of a particle is 9 times its rest energy, at what fraction of the speed of light must the particle be traveling?

13. If a proton and an antiproton, both at rest, were to completely annihilate each other, how much energy would be liberated?

14. Indicate which of the following decays are possible. For any that are forbidden, give the conservation law(s) that is/are violated.

 a) $\Sigma^+ \rightarrow n + \pi^0$
 b) $n \rightarrow p + \pi^-$
 c) $\pi^0 \rightarrow e^+ + e^- + \nu_e$
 d) $\pi^- \rightarrow \mu^+ + \nu_e$

15. Indicate which of the following reactions may occur, assuming sufficient energy is available. For any that are forbidden, give the conservation law(s) that is/are violated.

 a) $p + p \rightarrow p + n + K^+$
 b) $\gamma + n \rightarrow \pi^+ + p$
 c) $K^- + p \rightarrow n + \Lambda^0$
 d) $p + p \rightarrow p + \pi^+ + \Lambda^0 + K^0$

16. Supply the missing particle in each of the following *strong* interactions, assuming sufficient energy is present to cause the reaction:

 a) $n + p \rightarrow p + p +$ _____
 b) $p + \pi^+ \rightarrow \Sigma^+ +$ _____
 c) $K^- + p \rightarrow \Lambda^0 +$ _____
 d) $\pi^- + p \rightarrow \Xi^0 + K^0 +$ _____

17. Identify the missing particle(s) in each of the following *weak* decays:

 a) $\Omega^- \rightarrow \Xi^0 +$ _____
 b) $K^0 \rightarrow \pi^+ +$ _____
 c) $\pi^+ \rightarrow \mu^+ +$ _____
 d) $\tau^+ \rightarrow \mu^+ +$ _____ $+$ _____

18. A 0.1-kg ball connected to a fixed point by a taut string whirls around a circular path of radius 0.5 m at a speed of 3 m/s. Find the angular momentum of the orbiting ball. Compare this with the intrinsic angular momentum of an electron. How many times larger is the former than the latter?

19. What is the quark combination corresponding to an antineutron? An antiproton?

20. Give three (3) combinations of quarks (not antiquarks!) that will give baryons with charge: (a) +1; (b) −1; and (c) 0.

21. Give all the quark-antiquark pairs that result in mesons that have no charge, no strangeness, no charm, and no bottomness. How do you think such particles might be distinguished from one another?

22. Distinguish a particle composed of a combination of $\overline{dus}$ quarks from one composed of dus quarks in as many ways as you can. Do the same for a dss combination and a $\overline{dss}$ combination.

23. The Δ^- (delta minus) is a short-lived baryon with charge −1, strangeness 0, and spin $\frac{3}{2}$. To what quark combination does this subatomic particle correspond?

24. The F^- meson possesses −1 unit of charge, −1 unit of strangeness, and −1 unit of charm. Identify the quark combination making up this rare subatomic particle.

25. The D^+ meson is a charmed particle with charge +1, charm +1, and strangeness 0. Work out the quark combination for this species of elementary particle.

26. What is the quark composition of the Δ^{++} baryon? It has no strangeness, no charm, and no topness or bottomness. Its spin is $\frac{3}{2}$.

27. Analyze the following reactions in terms of their constituent quarks:

 a) $\pi^+ + p \rightarrow \Sigma^+ + K^+$
 b) $\gamma + n \rightarrow \pi^- + p$
 c) $p + p \rightarrow p + p + p + \bar{p}$
 d) $K^- + p \rightarrow K^+ + K^0 + \Omega^-$

28. Analyze the following decays in terms of the quark contents of the particles:

 a) $\Sigma^- \rightarrow n + \pi^-$
 b) $\Lambda^0 \rightarrow p + \pi^-$
 c) $K^{*+} \rightarrow K^0 + \pi^+$
 d) $\Omega^- \rightarrow \Lambda^0 + K^-$

29. The properties of the top quark have been confirmed with the 2007 discovery of mesons composed of a *t* quark and a $\bar{b}$ quark. What charge does such a particle carry? What is its spin? How many units of topness and bottomness does this particle have?

30. In the electroweak theory, symmetry breaking occurs on a length scale of about 10^{-17} m. Compute the frequency of particles whose de Broglie wavelength (see Section 10.5) equals this value, and, using this result, determine their energies by the Planck formula (page 267). Show that the mass of particles having such energies is on the order of the mass of the W particles (see Figure 12.16).

31. If the average lifetime of a proton was 10^{33} years, about how many protons would you have to assemble together and observe simultaneously to witness a total of 100 proton decays in one year? Explain the reasoning that led to your conclusion.

***remember**

There are more quizzes and study tools online at 4ltrpress.cengage.com/physics.

Discover your **PHYSICS** online experience at **4ltrpress.cengage.com/physics**.

You'll find everything you need to succeed in your class.

- Interactive Quizzes
- Interactive and Printable Flashcards
- Simulations
- Glossary
- And more

4ltrpress.cengage.com/physics

Epilogue

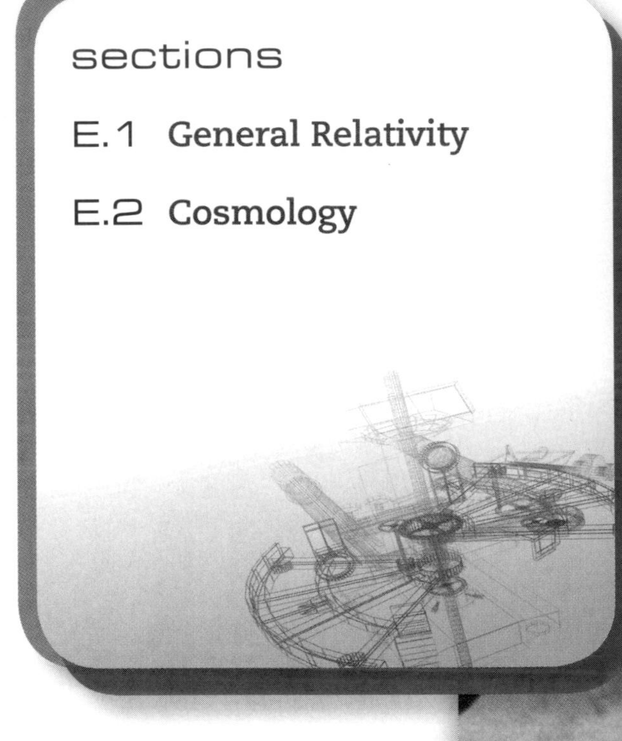

sections

Albert Einstein called it "the greatest blunder of my life." But was it? The *cosmological constant*, Λ, has a long and checkered history in physics. Introduced by Einstein in 1917 as part of his *general theory of relativity*, it provided a repulsive force between galaxies, keeping them from collapsing together under their mutual gravitational attraction and thereby stabilizing the universe on the largest possible scale. Ten years later, when Edwin Hubble discovered that the galaxies were all rushing apart and that the universe was expanding, Einstein discarded his cosmological constant as being inconsistent with observation. It was reintroduced in 1948 by Fred Hoyle (who coined the term "Big Bang" to describe the cataclysmic event that produced our universe) and his collaborators, only to be rejected again. Now, this cosmological parameter has once more taken center stage as a result of work by two independent research groups seeking to understand the nature of the expansion of the universe by studying supernova explosions.

The groups, the High-Z Supernova Search Team and the Supernova Cosmology Project (SCP), use complementary but independent analysis techniques to determine the distances and redshifts,* z, for particular types of supernovae. If gravity were the only influence on the expansion of the universe, then we'd expect to see galaxies gradually slowing down as time goes on due to the attractive forces between them. The greater the mass-energy density in the universe, the greater the attraction and the more rapid the slowing. Because of the constancy of the speed of light, looking farther out in space is equivalent to looking further back in time to earlier epochs in the universe's history. Thus, we should see the most remote galaxies moving faster—that is, having higher redshifts—than the nearby ones. A plot of velocity (y-axis) versus distance (x-axis)—a so-called *Hubble plot*—should curve upward. The greater the mass-energy density, the greater the upward curvature."

* Since the galaxies are all receding from us, the frequencies of their spectral lines are all reduced from what they would be if the galaxies were stationary. The wavelengths of the lines are thus increased and shifted to the red end of the spectrum. The amount of the redshift is given by $z = (\lambda_{\text{observed}} - \lambda_{\text{emitted}})/\lambda_{\text{emitted}}$. A redshift of $z = 1$ implies that the observed wavelength is twice as large as the emitted wavelength.

But look at Figure E.1! It shows a Hubble plot that includes data for the so-called type Ia supernovae, investigated by the High-Z and SCP teams, and extends out to redshifts near $z = 1.0$. (For reference, $z = 0.5$ corresponds to a look-back time extending to about one-third the present age of the universe, or ~4.5 billion years.) Clearly, the data appear to be turning unmistakably *downward*, not upward, suggesting that the cosmic expansion has been *accelerating*, not decelerating, at least in the epoch since $z \leq 1$. Even for unreasonably small values of the universe's mass-energy density, the data seem to demand that a *repulsive* force be included in our cosmological models. This is precisely the kind of term Einstein envisioned when he introduced the cosmological constant.

Although the negative cosmic pressure seemingly required by the type Ia supernova observations is probably not constant over all space and time like Einstein's Λ, the fact of its existence, now confirmed by supernova observations out to $z > 1.7$, will continue to fuel what Washington University physicist Clifford Will has called a "renaissance" in general relativity. Among the most interesting questions spawned by this work concerns the source of this cosmic repulsion. And, while there are even now many speculative answers to this question, at this juncture in our inquiry into physics,

© J.W. Burkey/Getty Images

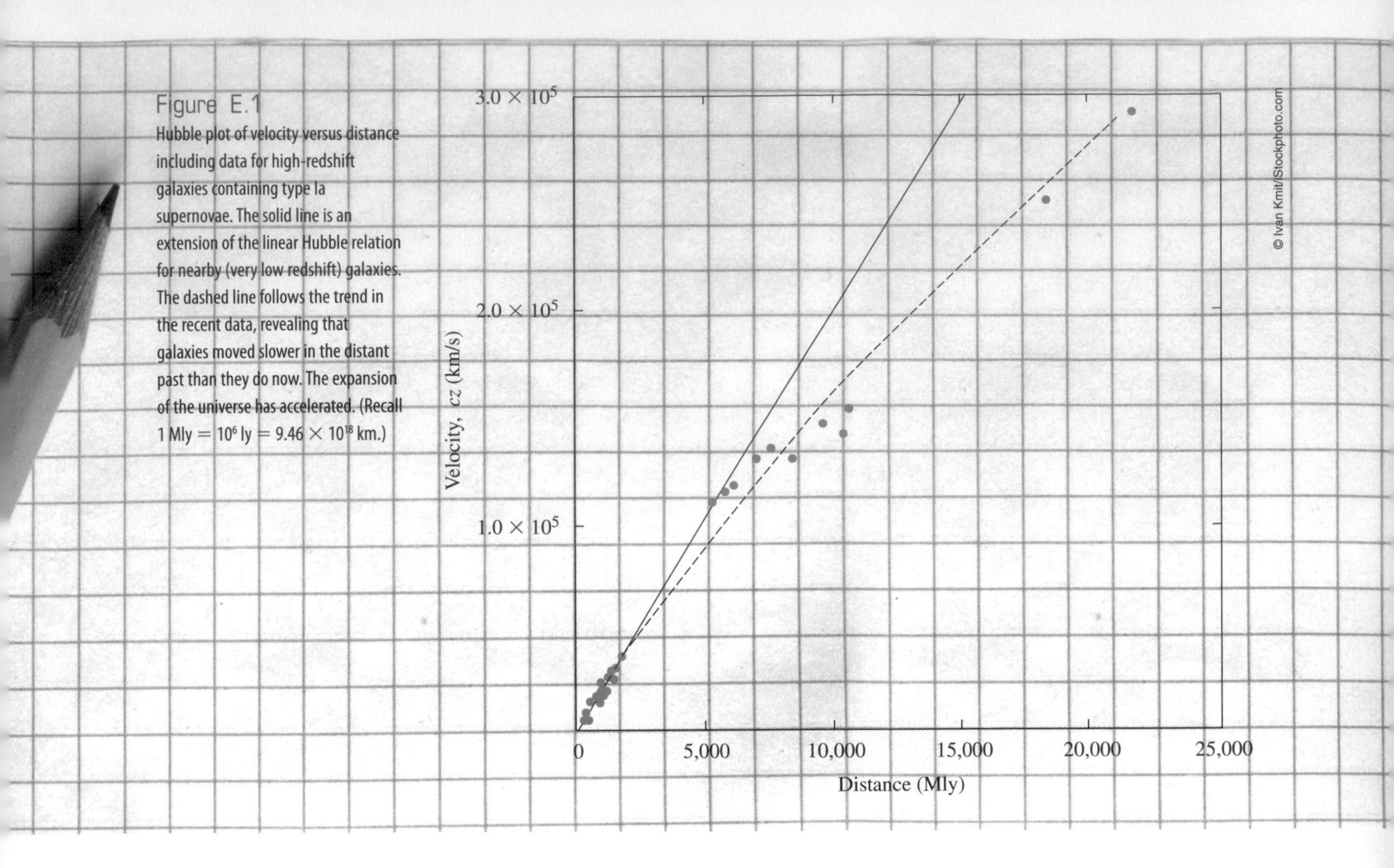

Figure E.1
Hubble plot of velocity versus distance including data for high-redshift galaxies containing type Ia supernovae. The solid line is an extension of the linear Hubble relation for nearby (very low redshift) galaxies. The dashed line follows the trend in the recent data, revealing that galaxies moved slower in the distant past than they do now. The expansion of the universe has accelerated. (Recall 1 Mly = 10^6 ly = 9.46×10^{18} km.)

the bigger question is what exactly is this general theory of relativity that forms the basis for any discussion of cosmology?

In what follows, we will consider several important features of general relativity, as well as some implications of this theory for the structure and evolution of the universe. In so doing, we will discover how the various branches of physics (mechanics, thermodynamics, electromagnetism, quantum mechanics, etc.) must be drawn together in order to even begin to address the most fundamental and interesting questions about the cosmos.

E.1 General Relativity

The **general theory of relativity**, developed by Einstein beginning in 1915, is basically a theory of gravity. It incorporates special relativity theory (see Section 12.1), and permits us to understand the motion of material particles and photons traveling in strong gravitational fields where Newton's law of universal gravitation (compare Section 2.8) gives only approximately correct answers. There exist several well-documented examples wherein the superiority of Einstein's theory of gravity to that of Newton's has been clearly demonstrated. We will consider some of these in due course. Let's first consider some ways in which Einstein's conception of gravity differs fundamentally from Newton's.

A basic postulate of general relativity is the *principle of equivalence.* It asserts that in a uniform gravitational field, all objects, regardless of their size, shape, or composition, accelerate at the same rate. Einstein recognized that this condition would make it impossible to physically distinguish the motion of a freely falling object near the Earth's surface from the motion of the same object released in a laboratory that was accelerating upward at a rate of 9.8 m/s^2 ($1g$) *in the absence of a local gravitational field.* As far as the laws of free fall are concerned, an accelerated laboratory is equivalent to one that is unaccelerated but possesses gravity. Einstein even went one step further and argued that not only are the laws of freely falling bodies the same for two such laboratories, but *all* the laws of physics are the same in such circumstances. The adoption of this so-called *strong* principle of equivalence leads to a very striking consequence: If the

Figure E.2

Two views of the same experiment involving the propagation of a pulse of light. Light, which travels in a straight line according to an observer at rest outside an elevator rising rapidly with an acceleration of 1*g*, will follow a curved trajectory according to an observer in the elevator. Both observers agree that the light enters the elevator near the top ledge of the left-hand window (1), and exits near the bottom ledge of the right-hand window (4′). The external observer (left) associates this fact with the upward motion of the elevator during the time required for the light pulse to cross the elevator car. The observer in the elevator (right), believing himself to be in an Earthlike gravitational field, interprets the observation to mean that gravity has caused the light path to bend downward. Both descriptions are equally valid. (This image is not drawn to scale and greatly exaggerates the amount of bending an actual light ray would suffer under these circumstances.)

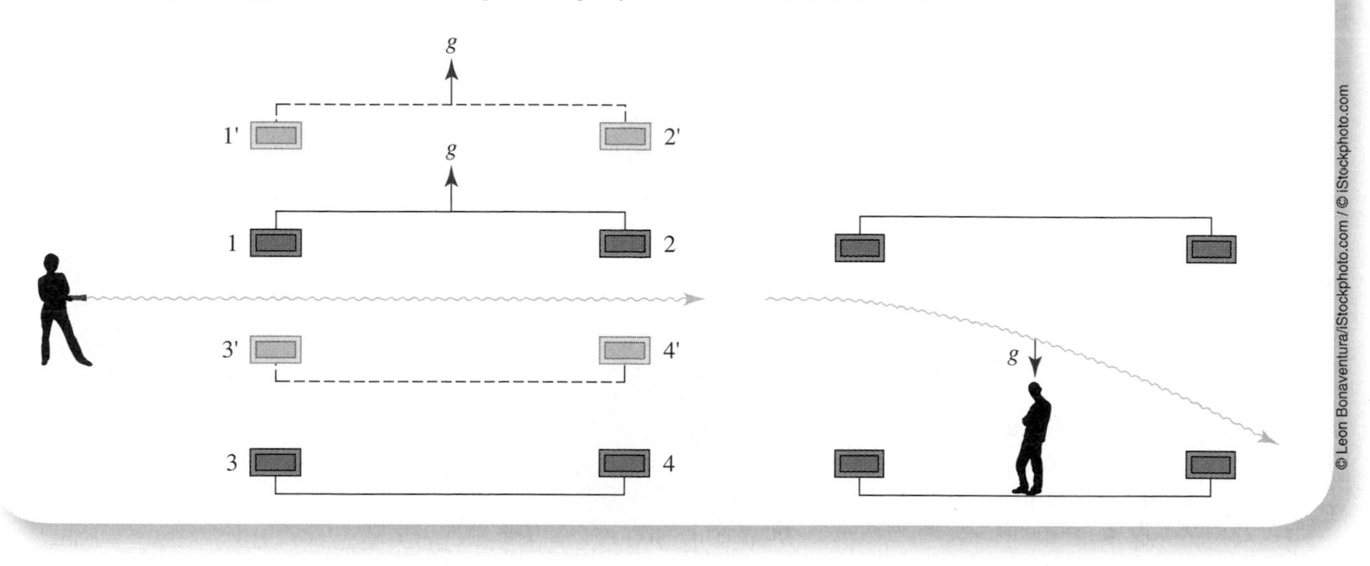

© Leon Bonaventura/iStockphoto.com / © iStockphoto.com

path of a light ray is observed to deviate from a straight line in an accelerating "laboratory" like the elevator in Figure E.2, then the path of a light ray will be similarly deflected in a gravitational field produced, say, by a massive object like the Sun.

Einstein's interpretation of the cause of the deflection of a light ray in a gravitational field was quite different from that which Newton might give. Newton would likely speak of the attractive force acting between the mass of the gravitating body and the effective mass of the photons making up the light (compare Section 10.2). (Recall that photons have no rest mass. Their mass derives from the energy they possess according to Einstein's equation $E_0 = mc^2$; see Section 12.1.) This force produces a centripetal acceleration of the photon and a subsequent deviation in its motion from the straight line predicted by Newton's first law (see Section 2.2). Indeed, one can calculate, using Newtonian methods, the

Force tells mass how to accelerate. Mass tells gravity how to exert force.

Curved spacetime tells mass-energy how to move.

Mass-energy tells spacetime how to curve.

© Bart Coenders/iStockphoto.com / © Popperfoto/Getty Images / © U.P. Images/iStockphoto.com

expected deflection for such a light ray, but the result, sadly, turns out to be equal to only one-half the observed deflection.

The remaining amount of deviation, and the correct explanation for the effect, belong to Einstein, who viewed the situation in a revolutionary way. In particular, he argued that there was no need to describe the deflection in terms of forces, but noted instead that the paths followed by the photons are merely their natural or *geodesic* trajectories in a curved space (or, more properly, *spacetime*, because in relativity space and time are inextricably linked) that has been distorted by the presence of mass (and/or energy). Princeton physicist John A. Wheeler has contrasted the views of Einstein and Newton on this subject as shown on the bottom of the previous page.

Perhaps a fable will help distinguish these two viewpoints more clearly. Imagine you are covering the final hole of the Master's Golf Championship from a helicopter above the eighteenth green. Sharing the reporting responsibilities with you are none other than Isaac Newton and Albert Einstein. As the three of you watch, Tiger Woods prepares to putt out for a birdie. After lining up his shot, he holes the putt and waves triumphantly to the cheering crowd. The trajectory followed by the ball to the cup, as seen from your aerial vantage point, is shown in Figure E.3a.

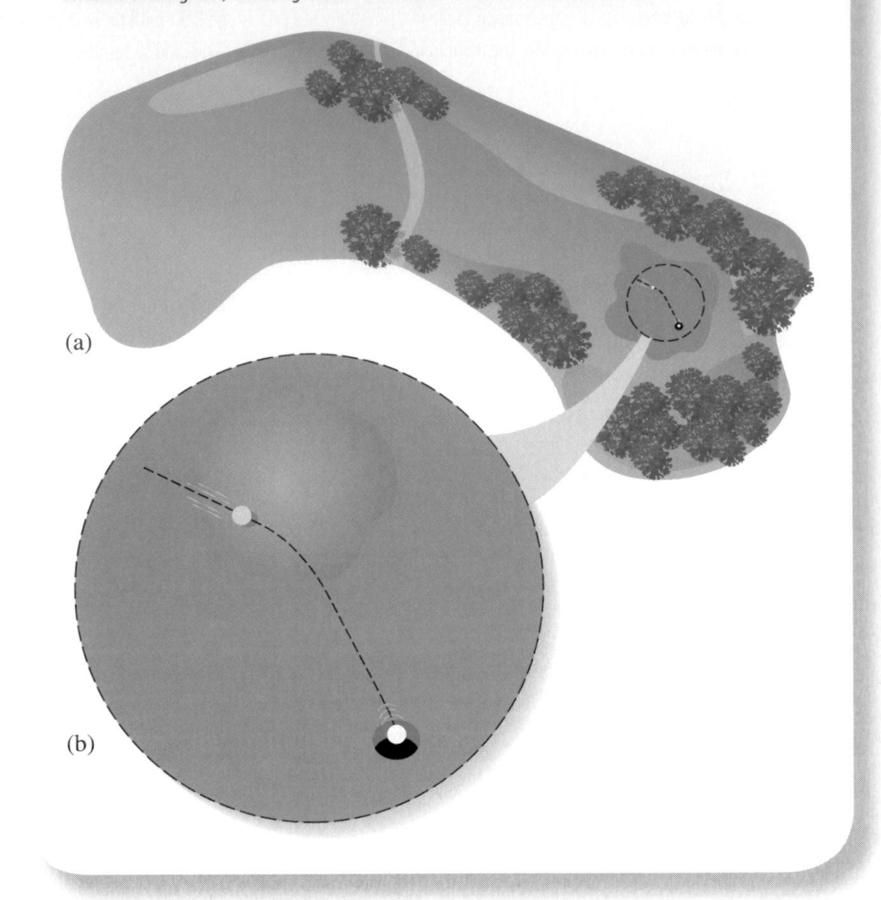

Figure E.3

(a) Path of a golf ball as seen from high above the surface of the green. The deviation from straight-line motion appears to require the action of some (unknown) force. (b) Close-up view of the surface of the green along which the ball moved. In rolling into the cup, the ball followed the natural contours of the green, including a small mound that deflected it toward the hole.

Noting that the path deviated from a straight line, Newton remarks that there must have been a force acting on the ball to accelerate it along the curved trajectory it followed to the cup. He speculates that perhaps there was a strong cross wind, which diverted the ball or that Woods was using a steel-core ball and there was an ore deposit beneath the surface of the green that provided a magnetic force on the ball; or . . . At this juncture, we can imagine Einstein interrupting to point out that all this talk of forces is unnecessary because the ball was just following the natural contours of the curved surface of the eighteenth green on which it moved. To prove this, Einstein suggests that the helicopter land and the green be inspected close up. The result is depicted in Figure E.3b.

Einstein knew what all good golfers know: The surfaces of most greens are not flat planes but are instead rolling contours designed to challenge the skill of the players at "reading the green" and putting the ball so as to take advantage of the dips and curves. In making his final shot, Tiger Woods relied on no mysterious forces to move the ball but simply recognized that the ball would be traveling along a warped surface. He accurately assessed the natural path the ball should take in this space to reach the hole. This is how Einstein would have us view the effects of what we call "gravity."

As in all parables, there are limitations to the fictional story just told, but it serves to emphasize the fundamental differences in the world views of Einstein and Newton. Thus enlightened, we may now return to the issue of the deflection of light and address the source of the curvature: mass-energy. In Einstein's theory of general relativity, matter and energy warp space (and alter time). The greater the density of matter and energy present, the greater the warping—that is, the stronger the curvature. Again, an analogy in two dimensions may be helpful. Consider a large rubber sheet, stretched taut between supporting posts. If we place a marble near the center of the sheet, the weight of the marble will deform the surface only very slightly in its immediate vicinity (Figure E.4a). The space surrounding the marble will remain flat, and the path of a small ball bearing rolled past the marble will be deviated by just the tiniest amount from a straight line due to the bending of the surface of the sheet. However, if we replace the marble with a bowling ball, the rubber will be stretched to a much greater degree, and the deformation of the space surrounding the bowling ball from a flat plane will be far larger and extend more widely around the ball. A ball bearing now rolled past the bowling ball will experience a space exhibiting considerable curvature, and its path will deviate significantly from a "straight" line (Figure E.4b).

A similar deflection effect happens in three dimensions for the light from distant stars that passes near the Sun or the light from remote galaxies that passes near a foreground cluster of galaxies. The first detection of this phenomenon was made in 1919 when the positions of stars close to the Sun on the sky were measured during the relative darkness provided by a solar eclipse. The observed differences in the positions of the stars during and outside the eclipse (when the Sun was far removed from them on the sky) agreed extremely well with the predictions of general relativity but only poorly with those of Newtonian gravitation.

More recently, astronomers have turned the deflection of light by massive objects from a test of the validity of general relativity to a tool to assess mass. Figure E.5a shows a Hubble Space Telescope (HST) image of a yellow cluster of galaxies acting as what is called a *gravitational lens.* The mass of this cluster has distorted the "fabric" of space in its vicinity and, like a lens, bends the light from a more distant blue galaxy to produce multiple images of it arrayed along circular arcs surrounding the cluster (Figure E.5b). Careful analyses of the images of a lensed galaxy in pictures like this using general relativity permit astronomers to establish the mass of the foreground cluster. In cases where comparisons can be made between this method and others used to determine cluster masses, the agreement has been found to be highly satisfactory. This lends credence to the view that we are correctly interpreting the nature of the observations, and it gives us confidence in the masses established via gravitational lensing alone in circumstances where other methods cannot be employed.

Another prediction of Einstein's general theory is that, because of the warping of space by mass, the orbital path of one spherical object about another will not be a closed ellipse as predicted by the inverse square law of Newton. Instead, it will be an open curve for which the point of closest approach between the two bodies slowly advances in space (an effect called **precession**). This is observed in the solar system in the motion of Mercury about the Sun (Figure E.6): The *perihelion* point

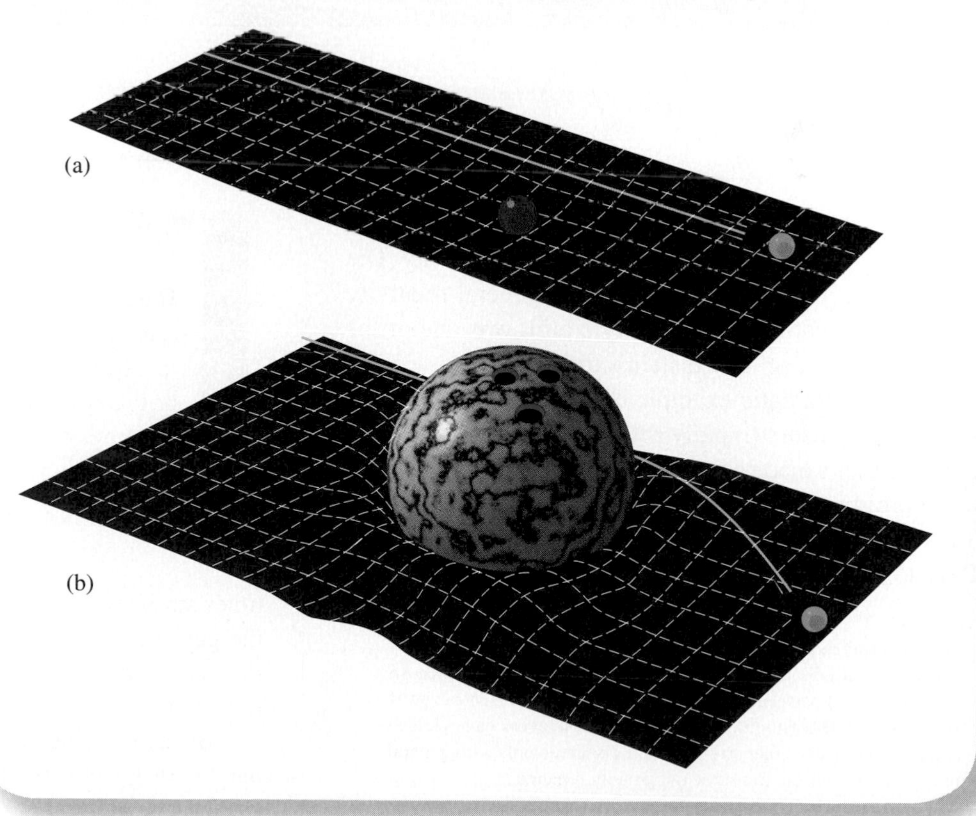

Figure E.4
The warping of space by massive objects causes deviations in the paths of particles moving nearby. The greater the mass, the larger the deformation of the space surrounding it. (a) The marble (red) produces only a small warp in the two-dimensional surface of a rubber sheet, and consequently a negligible alteration in the motion of the ball bearing (green) from a straight line. (b) The bowling ball creates a considerable depression in the surface of the sheet so that the path of the ball bearing deviates significantly from that of a straight line. In each case, the ball bearing is following a *geodesic* path along the surface of the sheet, but in (b), the surface has been warped from a flat plane by the action of a massive object.

Figure E.5

(a) HST picture of a cluster of yellow, elliptical galaxies acting like a gravitational lens. The blue ovals surrounding the cluster are distorted images of a more distant galaxy lying almost directly behind the cluster. (b) The large mass of the cluster warps the space in its vicinity and bends the paths of light rays passing nearby from a background galaxy. Light arriving at Earth from different directions produces multiple, distorted images of the more distant galaxy.

Courtesy of Hubble

(a)

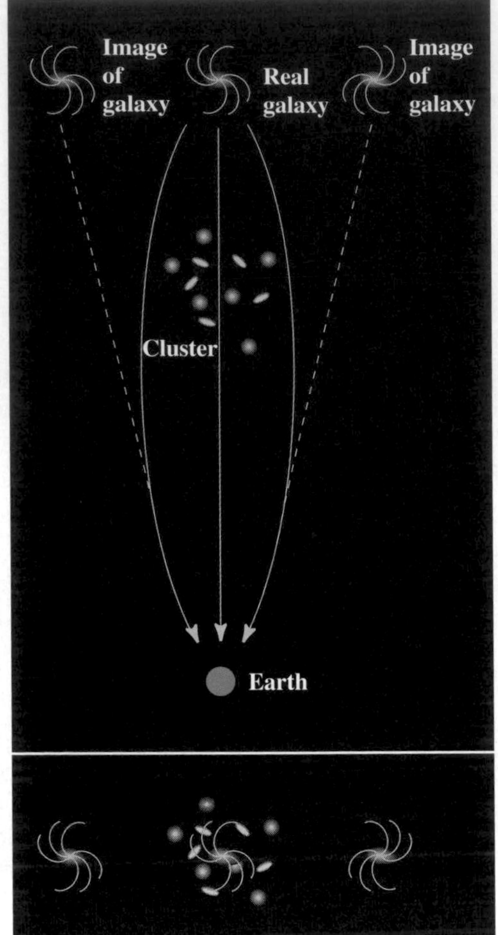

View as seen from Earth: Multiple images of same galaxy. (Diagram not to scale.)

(b)

From *The Cosmic Perspective*, 3rd ed. by Jeffrey Bennett et al., p. 687. Copyright © 2004 Pearson Education, Inc. Reprinted by permission.

of Mercury's orbit precesses at a rate of 43 arcseconds (0.012 degrees) per century due to general relativity. The agreement between the predictions of general relativity and the observations for Mercury's orbit was one of the first successful tests of Einstein's theory.*

A more dramatic example of this is seen in the motion of two pulsars (rapidly rotating, highly magnetized, collapsed stars made up primarily of neutrons) discovered to be orbiting one another by Russell Hulse and Joseph Taylor in 1974. The objects in this system (catalogued as PSR 1913 + 16) experience gravitational fields 10,000 times stronger than the Sun's field at Mercury, rendering the general relativistic effects far more significant than they are in the solar system. Indeed, the precession rate in PSR 1913 + 16 is a bit more than 4° *per year*!

DEFINITION

Perihelion (or Orbital) Precession
The slow rotation in space of the orbit of a celestial body. In the solar system, the gradual rotation in space of the line joining the points of closest and furthest approach, the perihelion *and* aphelion, *respectively, of an orbiting body to the Sun.*

By associating the observed precession in this binary with general relativistic effects, Hulse and Taylor

* Some controversy surrounds the interpretation of the precession of Mercury's perihelion wholly in terms of general relativity, because deviations from perfect sphericity in the shape of the Sun can produce similar effects. Measurements of the Sun's oblateness have yielded small enough values that no significant disagreements with general relativity have arisen, but some uncertainty remains.

For a system like the binary pulsar, general relativity predicts that, because of the gravitational radiation emitted from the centripetally accelerated masses, the two stars should gradually spiral inward toward one another, their orbital period getting shorter by a minuscule 75 *microseconds* per year. Astonishingly, by 1983, data on this highly unusual double star had accumulated to yield a measured orbital decay rate of 76 ± 2 microseconds per year in near perfect agreement with general relativity. Thus, although the direct detection of gravitational waves remains an elusive goal,* their existence, based on the precise agreement between the predictions of general relativity and the observations of PSR 1913 + 16, can scarcely be in doubt. For these achievements, Taylor and Hulse were selected as the 1993 recipients of the Nobel Prize in physics.

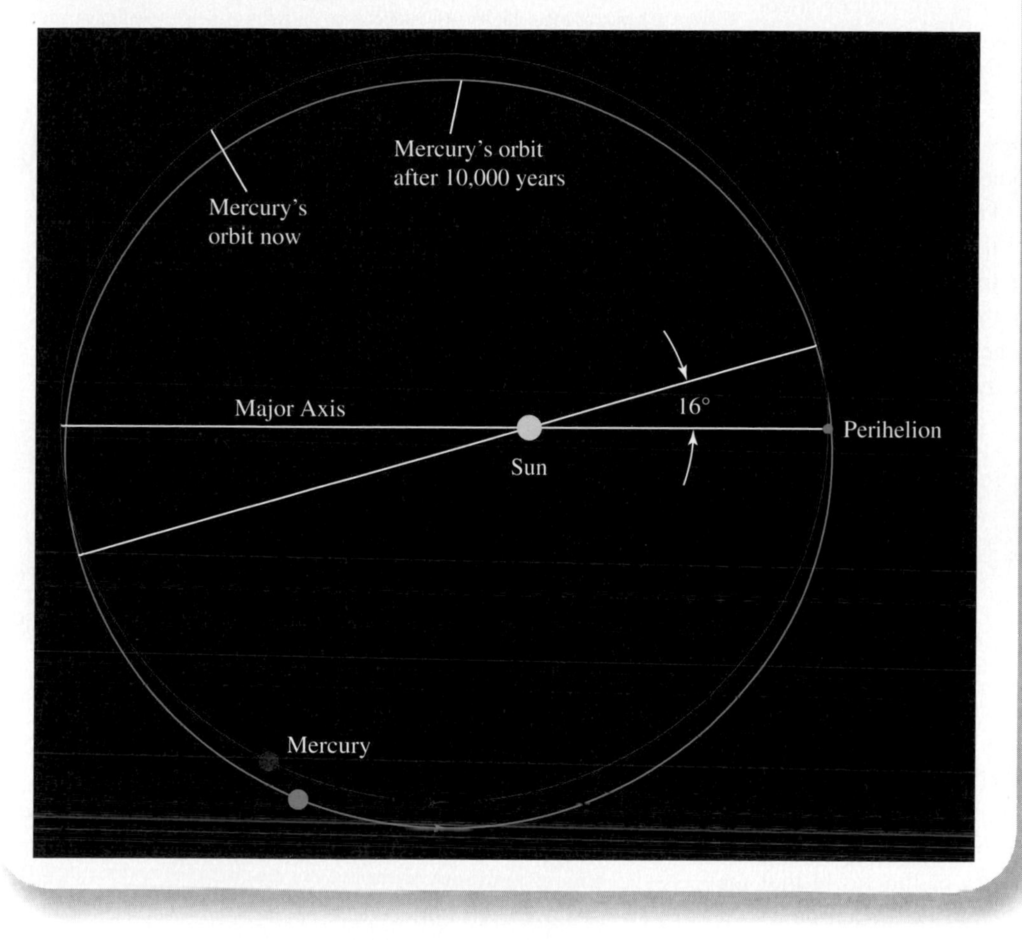

Figure E.6

Display of the precession of the perihelion of Mercury. This effect arises because the gravitational attraction of the Sun for Mercury deviates from a strict $1/r^2$ force law due, in part, to general relativistic effects. The actual precession of Mercury is small, accumulating to only about 16° every 10,000 years from all sources (including contributions caused by gravitational perturbations associated with nearby planets).

were able to deduce the masses of the two stars and then conduct a very sensitive test of another aspect of general relativity theory: the prediction of a gradual decrease in the orbital period of the system due to energy losses in the form of *gravitational radiation.* This is something that has no analog in Newtonian theory.

Recall from Section 8.5 that when an electric charge is accelerated, it radiates electromagnetic waves, which propagate outward at the speed of light. Similarly, according to general relativity, if a massive object is accelerated, it, too, radiates energy in the form of *gravitational waves* that also travel at the speed of light. Compared to more common electromagnetic waves, gravitational waves are much less intense and very difficult to detect directly. Some very sophisticated experiments have been conducted in laboratories around the world to detect gravitational waves from celestial sources, but none has succeeded unambiguously.

Once we acknowledge the validity of general relativity and that mass and energy deform space(time), it becomes possible to inquire about the effect that *all* the matter and energy contained in the universe have on its global structure, evolution, and geometry. After all, the

* In 2002, the U.S.-based Laser Interferometer Gravitational Wave Observatory (LIGO) began operation. Its purpose is to directly detect gravitational radiation from space. Using two installations 2,000 miles apart, each with 1.2-meter-diameter vacuum pipes arranged in an L with 4-kilometer-long arms, small test masses fitted with mirrors that reflect ultrastable laser beams are monitored for motion attributable to passing gravitational waves. The use of widely separated stations is necessary to rule out test-mass motion due to local effects (vibrations and the like) that mimic the response to gravitational radiation. Through the end of the latest scientific run in 2007, no gravitational wave detections could be claimed, although analysis of more than 700 terabytes of raw data is continuing. A new LIGO run began in July 2009 and will collect data for 12 months at the highest instrumental sensitivities available.

largest possible concentration of matter and energy is that which is present in the *entire* universe! Is the total amount of "stuff" in the universe large enough to significantly alter the "shape" or geometry of the universe from the flat space of Euclid, much as the presence of a heavy bowling ball deforms a flat rubber sheet? Questions like this one bring us into the realm of cosmology—the search for an understanding of what our universe is like on the grandest of scales, how it evolved to this state from its beginnings in the Big Bang, and what the future holds for us some 5 or 10 or 100 billion years from now. The answers to all these questions are not known with certainty, and it is not possible in the space that remains here for us to provide more than a glimpse of some of the current thinking about these issues. We shall thus end with a beginning, and focus on what is called the *flatness problem.*

Figure E.7

Unlike the individual shown in this wood engraving that appeared in an 1888 publication by the French astronomer Camille Flammarion, it is impossible to encounter, much less poke through, a bounding surface to the universe. Modern cosmology teaches that any model universe, regardless of its specific geometrical properties, is unbounded in the sense of having no edges. The universe encompasses the entire spacetime volume that exists. There is nothing outside the universe to be experienced as suggested by this image.

© Bettman/Corbis / © Bakaleev Aleksey/iStockphoto.com

E.2 Cosmology

Before the voyages of exploration made by Columbus and others more than 500 years ago, many otherwise educated and urbane people believed that the world was flat. They shunned traveling too far from home for fear that it would lead them to fall off the edge of the world. The view of the global geometry of the world that they adhered to was strictly a bounded Euclidean one.

According to general relativity, the curvature of the universe is dependent on its total mass-energy density, symbolized here as Ω_T, including the energy contained in the field associated with the cosmic repulsion, now commonly called *dark energy.* If, in the dimensionless units employed in general relativity, this value equals unity, space is flat on a cosmological scale. Any deviation from $\Omega_T = 1$ means that the geometry of the universe will not be that of a three-dimensional Euclidean space(time). Besides the geometry of Euclid, two other general types of geometry exist that could describe the global structure of the universe. They differ from Euclidean geometry only in terms of the parallel-line postulate. One is called *Riemannian geometry*, after mathematician Georg F. B. Riemann, and it adopts the postulate that through any point not on a line, *no* parallels may be drawn. This leads to geometrical characteristics in three dimensions analogous to those in two dimensions on the surface of a sphere. The other type of geometry demands that through any point not on a line, an *infinite* number of parallel (non-intersecting) lines may be constructed. The characteristics of this geometry were developed by Nikolai Lobachevski and, in three dimensions, are similar to the properties of two-dimensional surfaces shaped like the bells of trombones.

You may recall from your high school math courses that the geometry of Euclid is based on five postulates, one of which states that through a point not on a line, one and only one line may be drawn that is parallel to the original line. Although our appreciation for the vastness and complexity of the cosmos has increased during the last 500 years, it is remarkable that humankind is still struggling to understand the geometry of the universe, and that the early savants may have been correct after all: On the largest of scales, the universe may well be Euclidean, although it is clearly without boundaries or edges (Figure E.7).

© Susan Campbell/iStockphoto.com / © Matthew Rambo/iStockphoto.com

A variety of observations in the last few years suggest that the contribution to Ω_T from matter (particles), Ω_M, is about 30% of the critical value, that is, $\Omega_M \sim 0.3$. Of this, most (perhaps 99%) of the matter is "dark" (that is, nonluminous), and a significant fraction (more than about 85%) is likely to be nonbaryonic (compare with the discussion in Section 12.2). Particle physics offers three candidates for such material: (1) *axions*, theoretical particles whose masses range from ~ 1 eV/c^2 to as low as 10^{-16} eV/c^2 depending on the model; (2) *neutralinos*, hypothetical so-called *weakly interacting massive particles* (WIMPs), with masses between about 10 and 500 GeV/c^2; and (3) ordinary neutrinos of mass ≤ 1 eV/c^2. Current estimates place the neutrino contribution to the mass-energy density of the universe at between 0.1 and 5%, far too small to account for the bulk of the nonbaryonic contribution to Ω_T. Neutralinos are electrically neutral particles that interact with normal matter only through gravitation. Such particles, if they exist, would likely attach themselves to galaxies and form halos around them. All searches for WIMPs, including neutralinos, have failed to produce any conclusive results to date. Similarly, searches for axions with masses of up to 0.02 eV/c^2 with the CERN Axion Solar Telescope have also yielded no definitive detections. The dark matter thus remains elusive.

Matter then provides only about a third of the critical mass-energy density required for a flat universe. What about the dark-energy contribution, Ω_Λ? Based on observations of type Ia supernovae, the High-Z and SCP groups independently arrive at a value for Ω_Λ of about 70%. Remarkably, the sum $(\Omega_M + \Omega_\Lambda)$ is very close to unity (Figure E.8). Consequently, the universe is very close to being Euclidean in structure. If the present mass-energy density of the universe is even within a factor of 2 of the critical density after $\sim$14 billion years of expansion, then, extrapolating back to the time of the Big Bang, the original mass-energy density must have agreed with the critical value to better than a few parts in 10^{61}! Thus, it would seem that the universe appears flat now because it was flat at the beginning.

The precise nature of the dark-energy field that appears to be responsible for the observed acceleration in the expansion of the universe is not yet known. Clues to the nature of the dark-energy source will likely be found in the fine details of the history of the cosmos, because different dark-energy models provide different expansion rates. The differences, however, are extremely small and will require measurements of supernovae brightnesses that are more accurate by up to a factor of ten, as well as the extension of the observations to larger redshifts. This is a big challenge but one that may well be met in the next decade with the advent of new dedicated space missions like the Supernova Acceleration Probe (SNAP) and the Dark Energy Space Telescope (DESTiny), and the launch in 2014 of NASA's \$3.5-billion replacement for HST, the 6.5-meter James Webb Space Telescope.

But why should the universe have started out with just exactly the right amount of matter and energy to render it flat instead of some other value that would have caused it to be curved in a different manner? How are we to interpret the fact that the universe seems to have begun with this carefully selected initial condition? The original Big Bang scenarios had no good answers to these questions. And physicists are always suspicious of specially arranged circumstances that do not themselves emerge naturally from the physics of the situation but have to be imposed from outside the theory. This is the essence of the *flatness problem*.

The currently proposed solution to this problem avoids the necessity of prescribing special initial conditions by introducing some new physics. It provides the missing link that led to the deficiency in the explanatory power of the initial Big Bang models of the universe. Elementary particle physicists have come to the rescue of cosmologists and have developed a new approach that has solved the flatness problem (and several other difficulties inherent in the original Big Bang scenarios). The model for the universe that has emerged from the

Figure E.8

Pie chart showing the composition of the universe. Most of the mass-energy density of the cosmos is tied up in dark energy, the source of the negative pressure responsible for the acceleration of the universe. The remainder is mostly composed of non-baryonic dark matter. Only about 4% of the mass-energy of the universe is made up of ordinary matter, and most of it also is "dark," consisting of gas distributed around and between galaxies. The luminous material in the universe (stars, nebulae, and galaxies) constitutes only about 1% of its total mass-energy density.

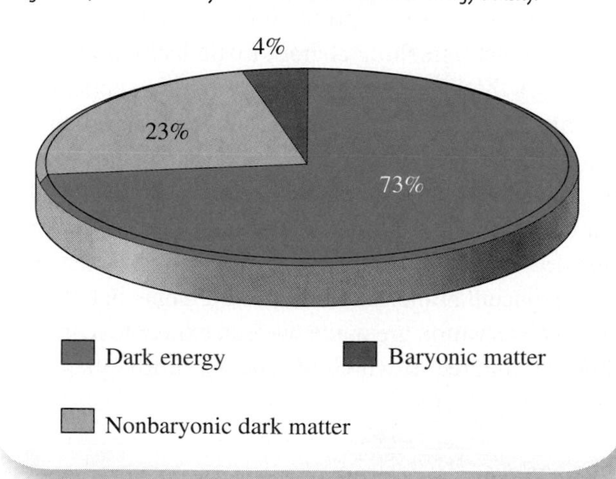

marriage of the ultrasmall and the ultralarge is known as the *inflationary universe.*

As discussed in Section 12.5, grand unified theories (GUTs) seek to amalgamate the fundamental interactions of nature into a unified whole. Such a unification, in which the four forces become indistinguishable, can occur only at very high energies, about 10^{16} to 10^{19} GeV. Such energies are unattainable in terrestrial laboratories but were amply present at the creation of the universe in the Big Bang. The era in which these enormous energies prevailed, some 10^{-43} s after the Big Bang, is known as the *TOE* (or Theory of Everything) *epoch.* Initial attempts to develop a TOE focused on *superstring theory*, in which the fundamental particles correspond to different vibrational frequencies of 10-dimensional strings. Since space and time together provide only four dimensions, the remaining six dimensions in these theories (there are several variants) remain curled up into knots with sizes of the order of 10^{-33} meters, much too tiny to detect.

More recent efforts have succeeded in merging the competing string theories into a grander, 11-dimensional theory called *M-Theory* (variously for membrane, master, magic, or mystery). According to this model, our universe is "glued" to a three-dimensional "brane" (short for membrane) embedded in a higher-dimensional hyperspace. Many other parallel universes may exist on other branes, but they remain invisible to us because photons, in this conception, cannot cross brane boundaries. Only gravitons (Table 12.1) can do so, indicating that we may be able to "feel" the influence of these other universes (and the matter in them) through the gravitational force. This has been a suggested solution for the dark-matter problem.

Fantastic as these predictions are, M-Theory does have one distinguishing characteristic lacking in the earlier string theories: Some of its predictions appear to be testable. For example, if strong sources of gravitational radiation are ever detected with LIGO (see the footnote on page 363) that cannot be identified optically, these sources could be candidates for concentrations of gravitons transmitted to our brane from normal matter on another adjacent brane. As further refinements in both theory and observation are made, we can expect to learn more about the degree to which M-Theory fulfills its promise to unify the realm of microphysics (quantum mechanics) with that of macrophysics (general relativity).

© Slavoljub Pantelic/iStockphoto.com

At the very beginning of time, then, we believe there was complete symmetry among the forces of nature. As the universe evolved, the energy density declined, and a spontaneous break in the symmetry occurred, as one after the other of the forces was separated out and assumed its own identity. These episodes of spontaneously broken symmetry corresponded to phase changes in the universe. Several things happened to the structure and evolution of the early universe, assuming that such symmetries existed but then were broken. One of them was that the universe suffered a period of rapid and enormous inflation in size, which effectively rendered it flat, regardless of its initial curvature. Let's see how this developed.

As the universe expanded and cooled, the first phase transition that occurred was the one that split off the gravitational force. The next one, and the one that is important for our discussion, was that which pared the strong nuclear force from the remaining electroweak force. This phase change may be compared to the solidification of water (refer to Sections 4.1 and 5.6). If a vessel of water is slowly cooled and the container is not disturbed during the process, it is possible to *supercool* the water below 0°C without it freezing. The system gets stuck in a metastable liquid phase. Eventually, some perturbation will occur, and nucleation will begin. Very quickly, ice crystals will form and spread through the volume, converting the symmetric liquid phase (in which the molecules show no preferred orientation with respect to any particular direction) to a distinctly asymmetric solid phase (which exhibits characteristic anisotropies along different crystalline directions). We believe the same kind of thing happened in the universe when the strong nuclear force was "frozen out."

Like the water in the example above, we think the universe got stuck in a metastable state, called a *false vacuum state*, as it expanded and cooled. If this phase was prolonged, the energy density (there was no mass in the universe at this very early period) consisted principally of that associated with the vacuum state. Only a small part of the energy density was tied up in radiation. The nature of a vacuum energy density may be likened to that of the latent heat of fusion in a liquid (see page 132). The

© David Joyner/iStockphoto.com

Figure E.9

By inflating a balloon quickly to very large dimensions, the curvature of the surface can be reduced to a point where it is difficult, by local measurements, to distinguish it from a flat, Euclidean space. The same effect may have occurred in the early universe if the inflationary hypothesis is correct. If so, then the flatness problem that plagued early versions of the Big Bang can be solved.

© iStockphoto.com

latent heat of fusion is the amount of energy that must be present at the melting point to maintain the molecules of a substance in the liquid phase. That is, it is the amount of internal energy required to prevent the molecules from binding together to form a solid. Similarly, the vacuum energy density was the energy needed to maintain the universe in a symmetric phase wherein the strong and electroweak forces were joined. It was this vacuum energy that ultimately drove the inflation.

When the universe eventually underwent the phase transition that severed the strong interaction from its electroweak sibling, enormous amounts of energy were released, exactly as happens when water freezes and heat is removed to the surroundings. The energy thus provided produced a rapid and extraordinary expansion of the scale of the universe. In fact, the equations of general relativity show that the inflation occurred in an exponential fashion causing the universe to grow by a factor of at least 10^{25} (and perhaps as much as 10^{50}) in the period when the universe was about 10^{-32} seconds old. The era of inflation was very short (only about 10^{-34} seconds or so long) and ended when the transition to the asymmetric state was fully achieved and the vacuum energy was completely depleted. After this, the universe resumed its normal expansion characteristics, but the effects of the inflationary period were profound.

In particular, if regions of spacetime in the very early universe were highly curved before the phase change, the inflation reduced their curvature, rendering them flat. A vivid two-dimensional analogue that helps explain how this could happen involves the surface of a balloon that is being inflated (Figure E.9). At the beginning of the process, the balloon's surface is quite small and highly curved. As air gradually enters the balloon, the surface expands and, as perceived by a local observer who samples only a small portion of the space, becomes less highly curved. Imagine now blowing into the balloon very hard so that it inflates rapidly to extremely large dimensions. After the period of inflation is over, the surface of the balloon is so distended that *all* local observers *anywhere* in the two-dimensional space of the surface of the fabric see a flat space. Regardless of what the curvature of the balloon's surface was at the start, once the inflation is complete, it looks flat. A similar circumstance is believed to have occurred in the early universe, and it is for this reason that in the present epoch we see the mass-energy density so close to the critical density for a flat, Euclidean world. A nice aspect about this explanation for the flatness problem is that it requires no special initial conditions for its success. *Whatever* the initial state of the universe, after the inflation, it always arrives at a final state that is flat in terms of its spacetime curvature. Put another way, Ω_T naturally approaches unity due to the inflation with no need for *ad hoc* assumptions about what conditions in the Big Bang were like prior to an age of 10^{-34} seconds.

The inflationary cosmology, developed in the 1980s by Alan Guth, Paul Steinhardt, and Andre Linde, has also provided solutions to several other nagging problems associated with the original Big Bang picture. The success of this model derives partly from its reliance on physical theories designed to explain the fundamental structure of matter on the tiniest of dimensions. The physics of the smallest scale has thus led to the elucidation of the physics of the largest scale. Elementary particle physics and cosmology have been inextricably linked, and advances in one area have fostered advances in the other. This type of synergy is at the heart of modern scientific exploration, whether in physics, chemistry, biology, medicine, environmental science, or any other field.

Appendix

Solutions to Odd-Numbered Problems

Chapter 1 Problems

1. $d = 20$ m
From conversion table: 1 m = 3.28 ft
$d = 20 \text{ m} = 20 \times 1 \text{ m} = 20 \times 3.28 \text{ ft}$
$d = 65.6$ ft

3. milli- means 1/1000 or 0.001; so: 1 second = 1000 milliseconds
$0.0452 \text{ s} = 0.0452 \times 1 \text{ s} = 0.0452 \times 1000 \text{ ms}$
$0.0452 \text{ s} = 45.2 \text{ ms}$

5. $T = 0.8$ s (period)
$$f = \frac{1}{T} = \frac{1}{0.8 \text{ s}}$$
$f = 1.25$ Hz

7. $\text{average speed} = \dfrac{\text{distance}}{\text{time}}$
$$v = \frac{d}{t} = \frac{1{,}200 \text{ mi}}{2.5 \text{ h}}$$
$v = 480$ mph

9. In Figure 1.8c, the length of the resultant velocity arrow is 0.66 times the length of the arrow representing 8 m/s (v_1). Therefore the resultant velocity is $0.66 \times 8 \text{ m/s} = 5.3 \text{ m/s}$.
In Figure 1.8d, the resultant velocity is 12.5 m/s.

11. $d = vt$
$d = 25 \text{ m/s} \times 5 \text{ s}$
$d = 125$ m
for $v = 250$ m/s: $d = vt$
$d = 250 \text{ m/s} \times 5 \text{ s}$
$d = 1{,}250$ m

13.

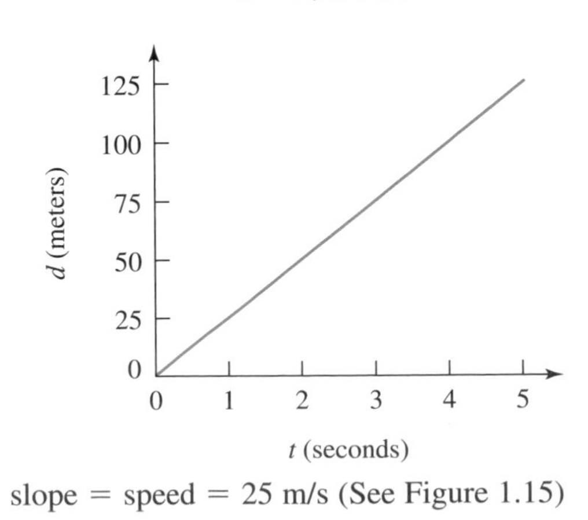

slope = speed = 25 m/s (See Figure 1.15)

15. (a) $a = 5.66 \text{ m/s}^2$ (See Example 1.3)
(b) $a = -9.4 \text{ m/s}^2$ (See Example 1.4)

17. $a = 200 \text{ m/s}^2$ (See Example 1.5)

19. $a = 2.86 \text{ m/s}^2$ (See Example 1.5)

21. (a) $v = at \qquad a = 60 \text{ m/s}^2 \qquad t = 40 \text{ s}$
$v = 60 \text{ m/s}^2 \times 40 \text{ s}$
$v = 2{,}400 \text{ m/s}$

(b) $v = at$
$7{,}500 \text{ m/s} = 60 \text{ m/s}^2\, t$
$$\frac{7{,}500 \text{ m/s}}{60 \text{ m/s}^2} = t \qquad t = 125 \text{ s}$$

23. (a)

(b)

25. (a) $v = at \qquad a = g \qquad t = 3 \text{ s}$
$v = gt = 9.8 \text{ m/s}^2 \times 3 \text{ s}$
$v = 29.4 \text{ m/s}$

(b) $d = \frac{1}{2}at^2 \qquad d = \frac{1}{2}gt^2$
$$d = \frac{1}{2}(9.8 \text{ m/s}^2) \times (3 \text{ s})^2$$
$d = 4.9 \text{ m/s}^2 \times 9 \text{ s}^2$
$d = 44.1$ m

27. $a = 4.9 \text{ m/s}^2$

(a) $v = at \qquad t = 3 \text{ s}$

$v = 4.9 \text{ m/s}^2 \times 3 \text{ s}$

$v = 14.7 \text{ m/s}$

(b) $d = \frac{1}{2}at^2$

$d = \frac{1}{2}(4.9 \text{ m/s}^2) \times (3 \text{ s})^2$

$d = 2.45 \text{ m/s}^2 \times 9 \text{ s}^2$

$d = 22.05 \text{ m}$

29. at "a": $a = \text{slope} = \dfrac{500 \text{ m/s}}{0.001 \text{ s}}$

$a = 500{,}000 \text{ m/s}^2$

at "b": $a = 0 \text{ m/s}^2$

at "c": $a = -2{,}500{,}000 \text{ m/s}^2$

31. $a = \dfrac{\Delta v}{\Delta t} = \dfrac{300 \text{ mph}}{5 \text{ s}} = 60 \text{ mph/s} = 2.7\ g$

Chapter 2 Problems

1. One example: $W = 150 \text{ lb} \qquad 1 \text{ lb} = 4.45 \text{ N}$

$W = 150 \times 1 \text{ lb} = 150 \times 4.45 \text{ N}$

$W = 667.5 \text{ N}$

$W = mg$

$667.5 \text{ N} = m\, 9.8 \text{ m/s}$

$\dfrac{667.5 \text{ N}}{9.8 \text{ m/s}^2} = m$

$m = 68.1 \text{ kg}$

3. (a) $W = mg = 30 \text{ kg} \times 9.8 \text{ m/s}^2$

$W = 294 \text{ N}$

(b) $W = 294 \text{ N} \qquad 1 \text{ N} = 0.225 \text{ lb}$

$W = 294 \times 1 \text{ N} = 294 \times 0.225 \text{ lb}$

$W = 66.15 \text{ lb}$

5. $F = 36{,}000 \text{ N}$ (See Example 2.1)

7. $F = ma$

$10 \text{ N} = 2 \text{ kg} \times a$

$10 \text{ N}/2 \text{ kg} = a$

$a = 5 \text{ m/s}^2$

9. $F = ma$

$20{,}000{,}000 \text{ N} = m \times 0.1 \text{ m/s}^2$

$\dfrac{20{,}000{,}000 \text{ N}}{0.1 \text{ m/s}^2} = m$

$m = 200{,}000{,}000 \text{ kg}$

11. $m = 4{,}500 \text{ kg} \qquad F = 60{,}000 \text{ N}$

(a) $F = ma$

$60{,}000 \text{ N} = 4{,}500 \text{ kg} \times a$

$\dfrac{60{,}000 \text{ N}}{4{,}500 \text{ kg}} = a$

$a = 13.3 \text{ m/s}^2$

(b) $v = at$

$v = 13.3 \text{ m/s}^2 \times 8 \text{ s}$

$v = 106.4 \text{ m/s}$

(c) $d = \frac{1}{2}at^2$

$d = \frac{1}{2}(13.3 \text{ m/s}^2) \times (8 \text{ s})^2$

$d = 6.65 \text{ m/s}^2 \times 64 \text{ s}^2$

$d = 425.6 \text{ m}$

13. (a) $a = 3 \text{ m/s}^2$ (See Example 1.3)

(b) $F = 240 \text{ N}$ (See Example 2.1)

(c) $d = \frac{1}{2}at^2 = \frac{1}{2}(3 \text{m/s}^2) \times (3 \text{ s})^2$

$d = 1.5 \text{ m/s}^2 \times 9 \text{ s}^2$

$d = 13.5 \text{ m}$

15. (a) $a = 28 \text{ m/s}^2 = 2.86\ g$ (See Example 1.3)

(b) $d = 87.5 \text{ m}$ (See solution of 13(c))

(c) $F = 504{,}000 \text{ N}$ (See Example 2.1)

17. $a = 6\ g \qquad g = 9.8 \text{ m/s}^2$

$= 6 \times 9.8 \text{ m/s}^2$

$a = 58.8 \text{ m/s}^2$

$F = ma \qquad m = 1{,}200 \text{ kg}$

$= 1{,}200 \text{ kg} \times 58.8 \text{ m/s}^2$

$F = 70{,}560 \text{ N}$

19. $v = 60 \text{ m/s} \qquad r = 400 \text{ m} \qquad m = 600 \text{ kg}$

(a) $a = \dfrac{v^2}{r} = \dfrac{(60 \text{ m/s})^2}{400 \text{ m}}$

$a = \dfrac{3{,}600 \text{ m}^2/\text{s}^2}{400 \text{ m}}$

$a = 9 \text{ m/s}^2$

$g = 9.8 \text{ m/s}^2 \qquad 1 \text{ m/s}^2 = \dfrac{g}{9.8}$

$a = 9 \times 1 \text{ m/s}^2 = 9 \times \dfrac{g}{9.8}$

$a = \dfrac{9}{9.8}\, g$

$a = 0.918\ g$

(b) $F = ma$

$= 600 \text{ kg} \times 9 \text{ m/s}^2$

$F = 5{,}400 \text{ N}$

21. $F = m\dfrac{v^2}{r}$

$60 \text{ N} = 0.1 \text{ kg} \times \dfrac{v^2}{1 \text{ m}}$

$v^2 = \dfrac{60 \text{ N-m}}{0.1 \text{ kg}} = 600 \text{ m}^2/\text{s}^2$

$v = 24.5 \text{ m/s}$

23. $F = m\dfrac{v^2}{r}$

$200 \text{ N} = 1{,}000 \text{ kg} \times \dfrac{(5{,}000 \text{ m/s})^2}{r} = \dfrac{2.5 \times 10^{10}}{r}$

$r = \dfrac{2.5 \times 10^{10}}{200} = 1.25 \times 10^8 \text{ m}$

25. (a) At *twice* as far away the force is *one-fourth* as large.

$F = 150 \text{ lb}$

(b) $F = 66.7$ lb (one-ninth as large)
(c) $F = 6$ lb (one one-hundredth as large)

Chapter 3 Problems

1. $mv = 650$ kg m/s (See Section 3.2)

3.
$$F = \frac{\Delta mv}{\Delta t} = \frac{(mv)_f - (mv)_i}{\Delta t}$$
$$mv_i = 1{,}000 \text{ kg} \times 0 \text{ m/s}$$
$$mv_i = 0 \text{ kg m/s}$$
$$mv_f = 1{,}000 \text{ kg} \times 27 \text{ m/s}$$
$$mv_f = 27{,}000 \text{ kg m/s}$$
$$\Delta t = 10 \text{ s}$$
$$F = \frac{(mv)_f - (mv)_i}{\Delta t} = \frac{27{,}000 \text{ kg m/s} - 0 \text{ kg m/s}}{10 \text{ s}}$$
$$F = \frac{27{,}000 \text{ kg m/s}}{10 \text{ s}}$$
$$F = 2{,}700 \text{ N}$$

5. $v = 39$ m/s (See Example 3.2)

7.
$$mv_{\text{before}} = 50 \text{ kg} \times 5 \text{ m/s}$$
$$mv_{\text{before}} = 250 \text{ kg m/s}$$
$$mv_{\text{after}} = mv = (40 \text{ kg} + 50 \text{ kg}) \times v$$
$$mv_{\text{after}} = 90 \text{ kg} \times v$$
$$mv_{\text{before}} = mv_{\text{after}}$$
$$250 \text{ kg m/s} = 90 \text{ kg} \times v$$
$$\frac{250 \text{ kg m/s}}{90 \text{ kg}} = v$$
$$v = 2.78 \text{ m/s}$$

9.
$$mv_{\text{before}} = 0 \text{ kg m/s}$$
$$mv_{\text{after}} = (mv)\text{gun} + (mv)\text{bullet}$$
$$mv_{\text{after}} = 1.2 \text{ kg} \times v + 0.02 \text{ kg} \times 300 \text{ m/s}$$
$$mv_{\text{after}} = 1.2 \text{ kg} \times v + 6 \text{ kg m/s}$$
$$mv_{\text{before}} = mv_{\text{after}}$$
$$0 \text{ kg m/s} = 1.2 \text{ kg} \times v + 6 \text{ kg m/s}$$
$$-6 \text{ kg m/s} = 1.2 \text{ kg} \times v$$
$$\frac{-6 \text{ kg m/s}}{1.2 \text{ kg}} = v$$
$$v = -5 \text{ m/s} \quad \text{(opposite direction of bullet)}$$
or use: $$\frac{v \text{ gun}}{v \text{ bullet}} = \frac{-m \text{ bullet}}{m \text{ gun}}$$

11. Work = 30,000 J (See Example 3.3)

13. $PE = 2{,}156$ J (See Example 3.7)

15. $KE = 45{,}000$ J (See Example 3.6)

17.
$$KE = \frac{1}{2}mv^2$$
$$60{,}000 \text{ J} = \frac{1}{2} \times 300 \text{ kg} \times v^2$$
$$\frac{60{,}000 \text{ J}}{150 \text{ kg}} = v^2$$
$$400 \text{ J/kg} = v^2$$
$$v = 20 \text{ m/s}$$

19. $v = 111$ m/s (See Example 3.8)

21.
$$v^2 = 2gd$$
$$(7.7 \text{ m/s})^2 = 2 \times 9.8 \text{ m/s}^2 \times d$$
$$59.3 \text{ m}^2\text{/s}^2 = 19.6 \text{ m/s}^2 \times d$$
$$\frac{59.3 \text{ m}^2\text{/s}^2}{19.6 \text{ m/s}^2} = d$$
$$d = 3 \text{ m} \quad \text{(or use Table 3.2)}$$

23. (a) $PE = 323{,}000$ J (See Example 3.7)
(b) $v = 46.4$ m/s (104 mph; See Example 3.8)

25. (a) $KE = 4{,}000$ J (See Example 3.6)
(b)

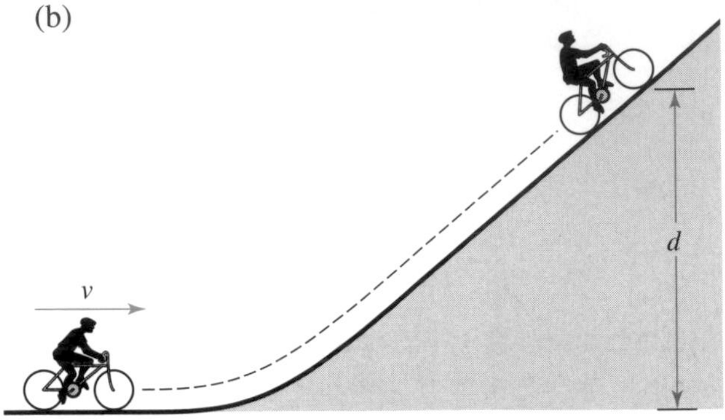

$$KE \text{ at bottom} = PE \text{ when stopped on hill}$$
$$4{,}000 \text{ J} = mgd = 80 \text{ kg} \times 9.8 \text{ m/s}^2 \times d$$
$$\frac{4{,}000 \text{ J}}{784 \text{ kg m/s}^2} = d$$
$$d = 5.1 \text{ m} \quad \text{or use: } d = \frac{v^2}{2g}$$

27. It would have to be thrown vertically, with just enough speed that it would slow to a stop just as it reached the ceiling: $v = \sqrt{2gd}$
$= \sqrt{2 \times 9.8 \text{ m/s}^2 \times 20 \text{ m}} = 19.8$ m/s (= 44 mph)

29.
$$KE_{\text{before}} = \frac{1}{2}mv^2 = \frac{1}{2} \times 50 \text{ kg} \times (5 \text{ m/s})^2$$
$$KE_{\text{before}} = 25 \text{ kg} \times 25 \text{ m}^2\text{/s}^2$$
$$KE_{\text{before}} = 625 \text{ J}$$
$$KE_{\text{after}} = \frac{1}{2}mv^2 = \frac{1}{2} \times 90 \text{ kg} \times (2.78 \text{ m/s})^2$$
$$KE_{\text{after}} = 45 \text{ kg} \times 7.73 \text{ m}^2\text{/s}^2$$
$$KE_{\text{after}} = 348 \text{ J}$$
$$KE_{\text{lost}} = 625 \text{ J} - 348 \text{ J}$$
$$KE_{\text{lost}} = 277 \text{ J}$$

31.
$$P = \frac{\text{work}}{t}$$
$$200 \text{ W} = \frac{10{,}000 \text{ J}}{t}$$
$$t = \frac{10{,}000 \text{ J}}{200 \text{ W}}$$
$$t = 50 \, s$$

33. $P = \frac{\text{work}}{t} = \frac{PE}{t}$

$PE = mgd = 1{,}000 \text{ kg} \times 9.8 \text{ m/s}^2 \times 30 \text{ m}$

$PE = 294{,}000 \text{ J}$

$P = \frac{PE}{t} = \frac{294{,}000 \text{ J}}{10 \text{ s}}$

$P = 29{,}400 \text{ W}$

35. $P = 155 \text{ W}$ (See Problem 33)

Chapter 4 Problems

1. $p = \frac{F}{A} = \frac{2{,}000{,}000 \text{ lb}}{400 \text{ ft}^2}$ (see Example 4.1)

$p = 5{,}000 \frac{\text{lb}}{\text{ft}^2}$

$1 \text{ ft}^2 = 144 \text{ in.}^2 \quad p = 5{,}000 \frac{\text{lb}}{\text{ft}^2} = 5{,}000 \frac{\text{lb}}{144 \text{ in.}^2}$

$p = \frac{5{,}000 \text{ lb}}{144 \text{ in.}^2}$

$p = 34.7 \text{ psi}$

3. $F = pA$

$F = 80 \text{ psi} \times 1{,}200 \text{ in.}^2$

$F = 96{,}000 \text{ lb}$

5. $F = 1{,}176 \text{ lb}$ (See Example 4.2)

7. (a) $D = \frac{m}{V} = \frac{393 \text{ kg}}{0.05 \text{ m}^3}$ (see Example 4.3)

$D = 7{,}860 \text{ kg/m}^3$

$D_w = D \times g = 7{,}860 \text{ kg/m}^3 \times 9.8 \text{ m/s}^2$

$D_w = 77{,}028 \text{ N/m}^3$

(b) Iron (From Table 4.4)

9. (a) $D_w = \frac{W}{V}$

water: $D_w = 62.4 \text{ lb/ft}^3$ (See Table 4.4)

$62.4 \text{ lb/ft}^3 = \frac{40{,}000 \text{ lb}}{V}$

$V = \frac{40{,}000 \text{ lb}}{62.4 \text{ lb/ft}^3}$

$V = 641 \text{ ft}^3$

(b) gasoline: $D_w = 42 \text{ lb/ft}^3$

$D_w = 42 \text{ lb/ft}^3 = \frac{40{,}000 \text{ lb}}{V}$

$V = \frac{40{,}000 \text{ lb}}{42 \text{ lb/ft}^3}$

$V = 952 \text{ ft}^3$

11. $m = 162 \text{ kg}$ (See Example 4.4 and Table 4.4)

13. (a) $m = 2{,}190 \text{ kg}$ (See Example 4.4 and Table 4.4)

(b) $W = mg = 2{,}190 \text{ kg} \times 9.8 \text{ m/s}^2$

$W = 21{,}500 \text{ N} = 4{,}830 \text{ lb}$

15. $p = 0.433\, d = 0.433 \times 12 \text{ ft}$ (See Example 4.6)

$p = 5.196 \text{ psi}$

17. $p = 3{,}030{,}000 \text{ Pa}$ (See Example 4.6)

19. About 6 psi

21. $F_b = W_{\text{water displaced}} \quad W = (D_w)_{\text{water}} \times V$

$F_b = 62.4 \text{ lb/ft}^3 \times 12 \text{ ft}^3$

$F_b = 749 \text{ lb}$

23. (a) $W_{He} = D_W \times V = 0.011 \text{ lb/ft}^3 \times 200{,}000 \text{ ft}^3$

$= 2{,}200 \text{ lb}$ (D_W from Table 4.4)

(b) $F_b = W_{\text{air}} = D_W \times V = 0.08 \text{ lb/ft}^3 \times 200{,}000 \text{ ft}^3$

$= 16{,}000 \text{ lb}$

(c) $F_{\text{net}} = F_b - W_{He} = 16{,}000 \text{ lb} - 2{,}200 \text{ lb}$

$= 13{,}800 \text{ lb}$

25. (a) $W = 468 \text{ lb}$ (See Example 4.5 and Table 4.4)

(b) $F_b = W_{\text{water}} = (D_w)_{\text{water}} \times V$

$= 62.4 \text{ lb/ft}^3 \times 3 \text{ ft}^3$

$F_b = 187.2 \text{ lb}$

(c)

$F_b = 187.2$ lb

W = 468 lb

$F_{\text{net}} = 468 \text{ lb} - 187.2 \text{ lb}$

$F_{\text{net}} = 280.8 \text{ lb}$ (downward)

27. (a) $W = 5{,}720{,}000 \text{ lb}$

(See Example 4.5 and Table 4.4)

(b) $W_{\text{seawater}} = F_b = W_{\text{ice}}$ (since it floats)

$W_{\text{seawater}} = D_w V \quad D_w(\text{seawater}) = 64.3 \text{ lb/ft}^3$

$5{,}720{,}000 \text{ lb} = 64.3 \text{ lb/ft}^3 \times V$

$\frac{5{,}720{,}000 \text{ lb}}{64.3 \text{ lb/ft}^3} = V$

$V_{\text{seawater}} = 88{,}960 \text{ ft}^3$

(c) $V_{\text{out}} = V_{\text{total}} - V_{\text{underwater}}$

$= 100{,}000 \text{ ft}^3 - 88{,}960 \text{ ft}^3$

$V_{\text{out}} = 11{,}040 \text{ ft}^3$

29. $W_{\text{Al}} = 100 \text{ N}$ Scale reading $= 100 \text{ N} - F_b$

$F_b = W_{\text{water}} = D_{\text{water}} \times g \times V_{\text{Al}}$

$W_{\text{Al}} = D_{\text{Al}} \times g \times V_{\text{Al}} \quad V_{\text{Al}} = 0.00378 \text{ m}^3$

$F_b = 37.0 \text{ N}$

Scale reading $= 63.0 \text{ N}$

Chapter 5 Problems

1. $30°\text{C} = 86°\text{F}$ (no jacket needed)

3. $\Delta l = -0.336 \text{ ft}$ means it is shorter

(See Example 5.1 and Table 5.2)

$l = 699.664 \text{ ft}$

5. Need to shorten diameter from 5 mm to 4.997 mm
$\Delta l = -0.003$ mm
$\Delta l = \alpha l \Delta T \qquad \alpha = 12 \times 10^{-6}/°C$
$-0.003 \text{ mm} = 12 \times 10^{-6}/°C \times 5 \text{ mm} \times \Delta T$

$$\frac{-0.003 \text{ mm}}{60 \times 10^{-6} \text{ mm/°C}} = \Delta T$$

$$\frac{-0.003}{60} \times 10^{6}°C = \Delta T$$

$$-0.00005 \times 10^{6}°C = \Delta T$$

$$\Delta T = -50°C$$

7. $\Delta U = \text{work} + Q \qquad Q = -2\ J$ (heat lost)
Work $= Fd = 50 \text{ N} \times 0.1$ m
Work = 5 J (Work done *on* gas)
$\Delta U = \text{Work} + Q = 5 \text{ J} + (-2 \text{ J})$
$\Delta U = 3$ J

9. $Q = 1{,}081{,}000$ J
(See Example 5.2 and Table 5.3)

11. (a) $Q = Cm\Delta T$
$C = 4{,}180$ J/kg-°C for water
$\Delta T = 100°C$
$Q = Cm\,\Delta T = 4{,}180 \text{ J/kg-°C} \times 1 \text{ kg} \times 100°C$
$Q = 418{,}000$ J
(b) $Q = 2{,}260{,}000$ J (See page 132)

13. (a) $KE = 375{,}000$ J (See Example 3.6)
(b) $Q = Cm\Delta T \qquad Q = KE$
$KE = Cm\Delta T \qquad C = 460$ J/kg-°C for iron
$375{,}000 \text{ J} = 460 \text{ J/kg-°C} \times 20 \text{ kg} \times \Delta T$

$$\frac{375{,}000 \text{ J}}{9{,}200 \text{ J/°C}} = \Delta T$$

$$\Delta T = 40.8°C$$

15. $\Delta T = 47.4°C$ (See Example 5.3 and Table 5.3)

17. (a) Rel. Hum. = 62.5%
(See Example 5.6 and Table 5.5)
(b) Rel. Hum. = 5.78%
(See Example 5.6 and Table 5.5)

19. The amount of water vapor in the air is given by the water vapor density, which is the humidity.

$$\text{Rel. Hum.} = \frac{\text{Humidity}}{\text{Sat. Den.}} \times 100\%$$

At 20°C, Sat. Den. = 0.0173 kg/m^3

$$40\% = \frac{\text{Humidity}}{0.0173 \text{ kg/m}^3} \times 100\%$$

$$\frac{40\% \times 0.0173 \text{ kg}}{\text{m}^3/\ 100\%} = \text{Humidity}$$

Humidity = 0.00692 kg/m^3
There is 0.00692 kg of water vapor in each m^3 of air.

21. $m = DV \qquad V = 150 \text{ m}^3$

$$\text{relative humidity} = \frac{\text{humidity}}{\text{saturation density}} \times 100\%$$

$$60\% = \frac{D}{0.0228} \times 100\% \qquad D = 0.0137 \text{ kg/m}^3$$

$m = 0.0137 \text{ kg/m}^3 \times 150 \text{ m}^3$
$m = 2.1$ kg

23. Humidity = 0.0094 kg/m^3
(See the approach in Problem 19)
From Table 5.5, air with this humidity will be saturated when cooled to 10°C. Therefore the air must be cooled 20°C, which means it must rise 2,000 m.

25. Carnot eff. = 34.9% (See Example 5.7)

27. (a) $\text{Eff.} = \dfrac{\text{work output}}{\text{energy input}} \times 100\%$

$$\text{Eff.} = \frac{3{,}000 \text{ J}}{15{,}000 \text{ J}} \times 100\%$$

Eff. = 20%
(b) Carnot Eff. = 88% (See Example 5.7)

Chapter 6 Problems

1. (a) $\rho = 0.167$ kg/m (See Example 6.1)
(b) $v = 15.5$ m/s (See Example 6.1)

3. $v = 388$ m/s (See Example 6.2)

5. $v = f\lambda$
$v = 4 \text{ Hz} \times 0.5$ m
$v = 2$ m/s

7. $v = f\lambda$
$80 \text{ m/s} = f \times 3.2$ m

$$\frac{80 \text{ m/s}}{3.2 \text{ m}} = f$$

$$f = 25 \text{ Hz}$$

9. For $f = 20$ Hz:
$\lambda = 17.2$ m
$\lambda = 56.4$ ft (See Example 6.3)
For $f = 20{,}000$ Hz
$\lambda = 0.0172$ m
$\lambda = 0.677$ in. (See Example 6.3)

11. (a) $\lambda = 1.315$ m (See Example 6.3)
(b) $v = f\lambda \quad v = 1{,}440$ m/s (water, from Table 6.1)
$1{,}440 \text{ m/s} = 261.6 \text{ Hz} \times \lambda$

$$\frac{1{,}440 \text{ m/s}}{261.6 \text{ Hz}} = \lambda$$

$$\lambda = 5.505 \text{ m}$$

13. $v = 347$ m/s (See the approach in Problem 5)
$v = 20.1 \times \sqrt{T} = 347$ m/s
$(20.1)^2 \times T = (347 \text{ m/s})^2$
$T = 298 \text{ K} = 25°C$

15. This is destructive interference, so the difference in the two distances equals one half the wavelength of the sound.

$$7.2 \text{ m} - 7 \text{ m} = 0.2 \text{ m} = \frac{1}{2}\lambda$$

$\lambda = 0.4$ m
$f = 860$ Hz (See the approach in Problem 7)

17. (a) Total distance sound travels: $d = vt =$ $320 \text{ m/s} \times 0.03 \text{ s} = 9.6 \text{ m}$ (v from Table 6.1)

Distance to snow $= \frac{1}{2} d = 4.8 \text{ m}$

(b) Depth of snow $= 5 \text{ m} - 4.8 \text{ m} = 0.2 \text{ m}$

19. $d = vt \qquad t = \frac{d}{v} = \frac{4{,}782{,}000 \text{ m}}{344 \text{ m/s}}$

$t = 13{,}900 \text{ s} = 3.86 \text{ h}$

21. (a) $d = vt \qquad v = 344 \text{ m/s}$ (air)

$8{,}000 \text{ m} = 344 \text{ m/s} \times t$

$\frac{8{,}000 \text{ m}}{344 \text{ m/s}} = t$

$t = 23.2 \text{ s}$

(b) $d = vt \qquad v = 4{,}000 \text{ m/s}$ (granite, Table 6.1)

$8{,}000 \text{ m} = 4{,}000 \text{ m/s} \times t$

$\frac{8{,}000 \text{ m}}{4{,}000 \text{ m/s}} = t$

$t = 2 \text{ s}$

23. Distance sound travels underwater in 0.01 second:

$d = vt \qquad v = 1{,}440 \text{ m/s}$ (See Table 6.1)

$d = 1{,}440 \text{ m/s} \times 0.01 \text{ s}$

$d = 14.4 \text{ m}$

Sound travels to fish, reflects, and then returns. Distance to fish is $\frac{1}{2}$ the distance sound traveled.

$d = 7.2 \text{ m}$

25. For each 10 dB, sound is about 2 times louder.

$40 \text{ dB} = 4 \times 10 \text{ dB}$

$2 \times 2 \times 2 \times 2 = 16$

Sound is 16 times louder.

27. (a) Harmonics of note have frequencies equal to 2, 3, 4, . . . times frequency of note. Harmonics of 4,186 Hz note have frequencies: 8,372 Hz, 12,558 Hz, 16,744 Hz, 20,930 Hz, etc. We can't hear frequencies over 20,000 Hz. So, highest harmonic we can hear is 16,744 Hz. Including the note itself (4,186 Hz), we can hear 4 harmonics.

(b) The note one octave below has $\frac{1}{2}$ the frequency: $f = 2{,}093$ Hz.

Harmonics: 4,186 Hz, 6,279 Hz, 8,372 Hz, 10,465 Hz, 12,558 Hz, 14,651 Hz, 16,744 Hz, 18,837 Hz, 20,930 Hz. The highest harmonic we can hear is 18,837 Hz. Including the note itself (2,093 Hz), we can hear 9 harmonics.

Chapter 7 Problems

1. $I = \frac{q}{t} = \frac{250 \text{ C}}{30 \text{ s}}$

$I = 8.33 \text{ A}$

3. $I = \frac{q}{t} \qquad t = 1 \text{ min} = 60 \text{ s}$

$0.7 \text{ A} = \frac{q}{60 \text{ s}}$

$0.7 \text{ A} \times 60 \text{ s} = q$

$q = 42 \text{ C}$

5. $V = IR$

$120 \text{ V} = 12 \text{ A} \times R$

$\frac{120 \text{ V}}{12 \text{ A}} = R$

$R = 10\ \Omega$

7. $I = 1.8 \text{ A}$ (See Example 7.1)

9. $V = 20 \text{ V}$ (See Example 7.2)

11. $P = 1{,}440 \text{ W}$ (See Example 7.4)

13. $P = 1{,}500{,}000 \text{ W}$ (See Example 7.4)

15. (a) $P = 40 \text{ W} \qquad I = 3.33 \text{ A}$ (See Example 7.5)

$P = 50 \text{ W} \qquad I = 4.17 \text{ A}$ (See Example 7.5)

(b) $P = 40 \text{ W}: \qquad R = 3.6\ \Omega$

(See the approach in Problem 5)

$P = 50 \text{ W}: \qquad R = 2.88\ \Omega$

(See the approach in Problem 5)

17. $E = 9{,}600{,}000 \text{ J}$ (See Example 7.6)

or: $\qquad P = 4 \text{ kW} \qquad t = 40 \text{ min} = \frac{2}{3} \text{ h}$

$E = 4 \text{ kW} \times \frac{2}{3} \text{ h}$

$E = 2.67 \text{ kWh}$

19. Hair dryer: $E = 360{,}000 \text{ J}$ (See Example 7.6)

Lamp: $E = 2{,}160{,}000 \text{ J}$ (See Example 7.6)

The lamp costs more to run because it uses more energy.

21. (a) $P = 1{,}080 \text{ W}$ (See Example 7.4)

(b) $E = 64{,}800 \text{ J}$ (See Example 7.6)

or: $\qquad P = 1.08 \text{ kW} \qquad t = 1 \text{ min} = \frac{1}{60} \text{ h}$

$E = Pt = 1.08 \text{ kW} \times \frac{1}{60} \text{ h}$

$E = 0.018 \text{ kWh}$

23. (a) $P = IV$

$1{,}000{,}000{,}000 \text{ W} = I \times 24{,}000 \text{ V}$

$\frac{1{,}000{,}000{,}000 \text{ W}}{24{,}000 \text{ V}} = I$

$I = 41{,}670 \text{ A}$

(b) $E = Pt \qquad t = 24 \text{ h} = 24 \times 3{,}600 \text{ s}$

$t = 86{,}400 \text{ s}$

$E = 1{,}000{,}000{,}000 \text{ W} \times 86{,}400 \text{ s}$

$E = 8.64 \times 10^{13} \text{ J}$

In kWh: $\qquad E = 1{,}000{,}000 \text{ kW} \times 24 \text{ h}$

$E = 24{,}000{,}000 \text{ kWh}$

(c) 10¢ = \$0.1 $\qquad$ Revenue = \$2,400,000

25. (a) $E = 28{,}800{,}000 \text{ J}$ (See Example 7.6) or:

$E = Pt = 4 \text{ kW} \times 2\text{h} = 8 \text{ kWh}$

(b) $I = 133$ A (See Example 7.5)

(c) $E = Pt \quad P = \frac{E}{t} = \frac{8 \text{ kWh}}{1 \text{ h}}$

$P = 8 \text{ kW} = 8{,}000 \text{ W}$

27. $P = IV$

$I = \frac{P}{V} = \frac{400 \times 10^{-3} \text{ W}}{3.6 \text{ V}}$

$I = 0.11$ A

This is about 3 times the current in a clock radio. (See Table 7.1)

Chapter 8 Problems

1. $N_o = 41.7$ (See Example 8.1)

3. Example: $f = 92.5 \text{ MHz} = 92{,}500{,}000 \text{ Hz}$

$$c = f\lambda$$
$$3 \times 10^8 \text{ m/s} = 9.25 \times 10^7 \text{ Hz} \times \lambda$$
$$\frac{3 \times 10^8 \text{ m/s}}{9.25 \times 10^7 \text{ Hz}} = \lambda$$
$$\lambda = 3.24 \text{ m}$$

5. $c = f\lambda$

$$3 \times 10^8 \text{ m/s} = f \times 0.0254 \text{ m}$$
$$\frac{3 \times 10^8 \text{ m/s}}{0.0254 \text{ m}} = f$$
$$f = 1.18 \times 10^{10} \text{ Hz} = 11{,}800 \text{ MHz}$$

7. UV band: $f = 7.5 \times 10^{14}$ Hz to $f = 10^{18}$ Hz

λ's:

(See Figure 8.25)

$$c = f\lambda$$
$$3 \times 10^8 \text{ m/s} = 7.5 \times 10^{14} \text{ Hz} \times \lambda$$
$$\frac{3 \times 10^8 \text{ m/s}}{7.5 \times 10^{14} \text{ Hz}} = \lambda$$
$$\lambda = 4 \times 10^{-7} \text{ m}$$
$$c = f\lambda$$
$$3 \times 10^8 \text{ m/s} = 10^{18} \text{ Hz} \times \lambda$$
$$\frac{3 \times 10^8 \text{ m/s}}{1 \times 10^{18} \text{ Hz}} = \lambda$$
$$\lambda = 3 \times 10^{-10} \text{ m}$$

9. Energy emitted $\propto T^4$. From 300 K to 3,000 K, T increases 10 times.

Energy output increases $10^4 = 10{,}000$ times

11. $\lambda_{max} = 2.9 \times 10^{-10}$ m X-ray (See Example 8.3 and Figure 8.25)

13. $\lambda = \frac{0.0029}{T}$

$T \times \lambda = T \times 0.0000012 \text{ m} = 0.0029$

$T = 2{,}420$ K

Chapter 9 Problems

1. Review the figure below. By the Law of Reflection, the angle of reflection equals the angle of incidence, 50°. Therefore, the reflected ray makes an angle of 50° with respect to the normal.

For the refracted ray, the light slows upon entering the glass, so the ray is bent toward the normal—the angle of refraction is smaller than the angle of incidence. The actual angle is found using Figure 9.27. The refracted ray makes an angle of about 32° with respect to the normal.

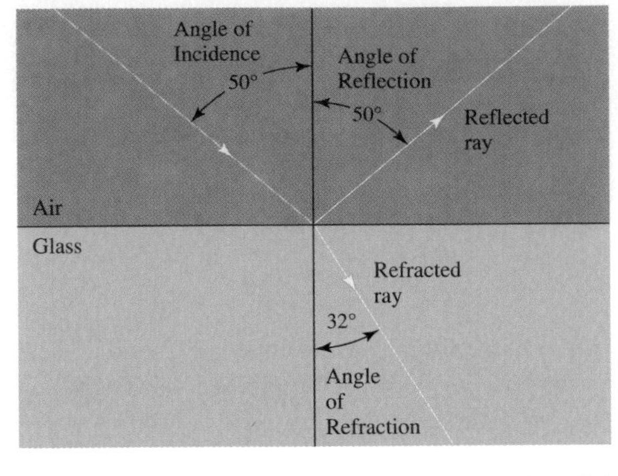

3.

Angle of Incidence (°)	Angle of Refraction (°)*
5	3
10	7
20	14

*From Figure 9.27

Doubling the angle of incidence approximately doubles the angle of refraction. Doubling the angle of incidence from 20° to 40° should result in an angle of refraction of about 2 × 14° or 28°. Figure 9.27 gives an angle of refraction of about 27° for this result, in good agreement with this estimate.

5. Rays skimming the water's surface are refracted to the fish's eye at the critical angle for an air-water interface, about 49° (see Example 9.2). The opening angle of the cone of light thus seen by the fish is twice this angle, or about 98°. (See the accompanying diagram.)

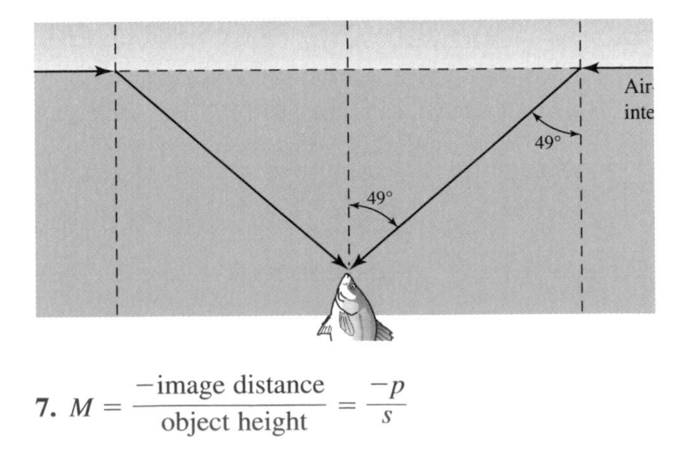

7. $M = \frac{-\text{image distance}}{\text{object height}} = \frac{-p}{s}$

The virtual image formed by the lens is upright, so $M = +2.5$. For a virtual image, the image is on the same side of the lens as the object:
$p = -15$ cm. (See Example 9.4)

$$2.5 = \frac{-(-15 \text{ cm})}{s} = \frac{15 \text{ cm}}{s}$$

$$s = \frac{15 \text{ cm}}{2.5} = 6.0 \text{ cm}$$

9. $$M = \frac{\text{image height}}{\text{object height}} = \frac{-0.005 \text{ m}}{2 \text{ m}}$$

$M = -0.0025$

11. $p = 15$ cm (to the right of the lens)
(See Example 9.3)
Image is real and inverted.

13. (a) $f = 50$ mm $\quad p = 60$ mm

$$p = \frac{sf}{s - f}$$

$$p^{-1} = \frac{(s - f)}{sf} = f^{-1} - s^{-1}$$

$$s^{-1} = f^{-1} - p^{-1} = \left(\frac{1}{50} \text{ mm}\right) - \left(\frac{1}{60} \text{ mm}\right)$$

$$s^{-1} = 0.0200 \text{ mm}^{-1} - 0.0167 \text{ mm}^{-1}$$
$$= 0.0033 \text{ mm}^{-1}$$
$$s = 300 \text{ mm}$$

$$M = -\frac{p}{s} = \frac{-60 \text{ mm}}{300 \text{ mm}}$$

$M = -0.20$ (The object is reduced in size by a factor of 5 and is inverted)

(b) Following the approach of part (a),
$s = 100$ mm
$M - -1$ (The object is actual size and inverted)

15. $$p = \frac{sf}{(s - f)} \qquad f = 8 \text{ cm} \qquad s = 10 \text{ cm}$$

$$p = (10 \text{ cm} \times 8 \text{ cm}) \div (10 \text{ cm} - 8 \text{ cm})$$
$$= \left(\frac{80 \text{ cm}}{2}\right)$$

$p = 40$ cm (p is positive implying the image is real)

17. Light scattered per second $\propto \lambda^{-4}$. (See p. 258.)
If the wavelength, λ, is doubled, the amount of scattered light is *reduced* by a factor of $(2)^4 = 16$.

Chapter 10 Problems

1. $E = 41.4$ eV (See Example 10.1)

3. First find the frequency; then use Figure 10.4.

$E = hf \qquad h = 6.63 \times 10^{-34}$ J/Hz

$9.5 \times 10^{-25} \text{ J} = 6.63 \times 10^{-34} \text{ J/Hz} \times f$

$$\frac{9.5 \times 10^{-25} \text{ J}}{6.63 \times 10^{-34} \text{ J/Hz}} = f$$

$$\frac{9.5}{6.63} \times \frac{10^{-25}}{10^{-34}} \text{ Hz} = f$$

$f = 1.43 \times 10^{9}$ Hz $\qquad$ low-frequency microwave

$$1 \text{ eV} = 1.6 \times 10^{-19} \text{ J} \qquad 1 \text{ J} = \frac{1 \text{ eV}}{1.6 \times 10^{-19}}$$

$E = 9.5 \times 10^{-25}$ J

$$E = 9.5 \times 10^{-25} \times \left(\frac{1 \text{ eV}}{1.6 \times 10^{-19}}\right)$$

$$E = \frac{9.5 \times 10^{-25}}{1.6 \times 10^{-19}} = \frac{9.5}{1.6} \times \frac{10^{-25}}{10^{-19}}$$

$E = 5.94 \times 10^{-6}$ eV

5. $\lambda = 9.1 \times 10^{-12}$ m $\qquad$ (See Example 10.2)
$\lambda = 0.0091$ nm

7. $$\lambda = \frac{h}{mv}$$

$$v = \frac{h}{m\lambda} = \frac{6.63 \times 10^{-34} \text{ J/Hz}}{(9.11 \times 10^{-31} \text{ kg}) \times (6.7 \times 10^{-7} \text{ m})}$$

$v = 0.109 \times 10^{(-34 + 31 + 7)}$ m/s
$v = 1{,}090$ m/s

9. (a) $n = 4 \qquad r = 0.847 \times 10^{-9}$ m

$$r = \frac{\lambda}{2\pi}$$

$\lambda = 2\pi r = 2 \times 3.14 \times (0.847 \times 10^{-9} \text{ m})$
$\lambda = 5.32 \times 10^{-9} \text{ m} = 5.32$ nm

(b) $$\lambda = \frac{h}{mv}$$

$$mv = \frac{h}{\lambda} = \frac{6.63 \times 10^{-34} \text{ J-s}}{5.32 \times 10^{-9} \text{ m}}$$

$mv = 1.25 \times 10^{-25}$ kg-m/s

11. (a) To be ionized, the atom's energy must be $E_\infty = 0$ since the electron goes to the $n = \infty$ level. If it starts in the $n = 2$ level, the energy it must gain, ΔE, is:
$\Delta E = E_\infty - E_2$
From Figure 10.21:
$E_2 = -3.40$ eV
$\Delta E = 0 - (-3.40 \text{ eV})$
$\Delta E = 3.40$ eV

(b) The energy of a photon that would ionize the atom equals ΔE.
$E_{photon} = \Delta E = 3.4$ eV
To find the frequency:
$E_{photon} = hf$
$3.40 \text{ eV} = 4.136 \times 10^{-15} \text{ eV/Hz} \times f$

$$\frac{3.40 \text{ eV}}{4.136 \times 10^{-15} \text{ eV/Hz}} = f$$

$f = 8.22 \times 10^{14}$ Hz

13. Using Figure 10.23 as a model, the possible energy level transitions are shown in the figure.

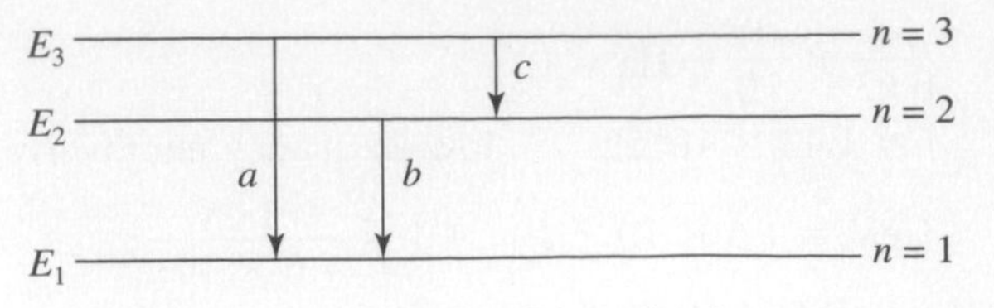

Only *downward* transitions lead to *emission* of photons.

The frequency of each photon is proportional to the change in energy of the electron during the transition. On the energy level diagram, this energy is proportional to the length of the arrow showing the transition. Since a's arrow is longest and c's is shortest,

$E_a > E_b > E_c$

$f_a > f_b > f_c$

The photon emitted in the $n = 3$ to $n = 1$ level transition has the highest frequency, followed by the $n = 2$ to $n = 1$ photon, and the $n = 3$ to $n = 2$ photon.

15. The number of electrons able to be accommodated in each atomic energy level equals $2n^2$, where n is the principal quantum number. Thus, the $n = 1$ level can have two electrons, the $n = 2$ level 8 electrons, and the $n = 3$ level 18 electrons. The total number of electrons able to be placed in the first three energy levels is $(2 + 8 + 18) = 28$. Zinc, with $Z = 30$, contains 30 electrons in its normal state. Thus, the minimum number of levels required to accommodate all 30 of the electrons in the zinc atom is $n = 4$.

17. For K_α(Cu): $\lambda_\alpha = 0.154$ nm ($n = 2 \Rightarrow n = 1$)
For K_β(Cu): $\lambda_\beta = 0.139$ nm ($n = 3 \Rightarrow n = 1$)

$$\frac{E_3 - E_1}{E_3 - E_1} = \frac{hf_\beta}{hf_\alpha} = \frac{f_\beta}{f_\alpha} = \frac{(c/\lambda_\beta)}{(c/\lambda_\alpha)}$$

$$\frac{E_3 - E_1}{E_2 - E_1} = (\lambda_\alpha/\lambda_\beta) = \frac{0.154 \text{ nm}}{0.139 \text{ nm}} = 1.11$$

The energy interval between the $n = 1$ level and the $n = 3$ level is about 10% larger than the corresponding interval between the $n = 1$ level and the $n = 2$ level.

19. $E_{photon} = hf = (hc/\lambda)$

$\lambda = 0.072 \text{ nm} = 0.072 \times 10^{-9}$ m

$$E_{photon} = \frac{(6.63 \times 10^{-34} \text{ J-s}) \times (3 \times 10^8 \text{ m/s})}{(0.072 \times 10^{-9} \text{ m})}$$

$E_{photon} = 2.76 \times 10^{-15}$ J

$$E_{photon} = \frac{(2.76 \times 10^{-15} \text{ J})}{(1.6 \times 10^{-19} \text{ J/eV})}$$

$E_{photon} = 17{,}266 \text{ eV} = 17.3$ keV
(x-ray photon; see Figure 10.4)

21. (a) $\Delta E = hf$ (See plan for Problem 11)

$0.117 \text{ eV} = (4.136 \times 10^{-15} \text{ eV/Hz}) \times f$

$$\frac{0.117 \text{ eV}}{(4.136 \times 10^{-15} \text{ eV/Hz})} = f$$

$f = 28.29 \times 10^{12}$ Hz

(b) This is an IR photon; see Figure 10.4.

23. (a) Because $\sqrt{f} \propto Z$, we see that $f \propto Z^2$. Thus, the highest frequency x-ray photon belongs to the element with the largest value of Z. This is iridium ($Z = 77$).

(b) The lowest frequency x-ray photon belongs to the element with the smallest atomic number. This is calcium ($Z = 20$).

Chapter 11 Problems

1. (a) carbon-14
$A = 14$ $Z = 6$ $N = 8$
6 protons and 8 neutrons

(b) calcium-45
$A = 45$ $Z = 20$ $N = 25$
20 protons and 25 neutrons

(c) silver-108
$A = 108$ $Z = 47$ $N = 61$
47 protons and 61 neutrons

(d) radon-225
$A = 225$ $Z = 86$ $N = 139$
86 protons and 139 neutrons

(e) plutonium-242
$A = 242$ $Z = 94$ $N = 148$
94 protons and 148 neutrons

3.

$${}^{110}_{47}\text{Ag} \rightarrow \underset{\substack{\uparrow \\ \text{daughter}}}{{}^{110}_{48}\text{Cd}} + {}^{0}_{-1}\text{e}$$

5. Alpha decay of ^{210}Po:

$${}^{210}_{84}\text{Po} \rightarrow {}^{4}_{2}\text{He} + {}^{A}_{Z}$$

A: $210 = 4 + A$
$210 - 4 = A$
$A = 206$

Z: $84 = 2 + Z$
$84 - 2 = Z$
$Z = 82$

Daughter nucleus: ${}^{206}_{82}$Pb (lead)

7. Gamma decay of ^{107}Ag*:
Because gamma decay does not alter the charge or the mass of the nucleus, but only removes extra energy from the excited nucleus, the daughter nuclide is ${}^{107}_{47}$Ag.

9. $^{1}_{0}\text{n} + {}^{235}_{92}\text{U} \rightarrow {}^{236}_{92}\text{U}^* \rightarrow {}^{143}_{57}\text{La} + {}^{A}_{Z} + 3\,{}^{1}_{0}\text{n}$

A: $236 = 143 + A + 3$
$236 - 143 - 3 = A$
$A = 90$

Z: $92 = 57 + Z + 0$
$92 - 57 = Z$
$Z = 35$

Additional nuclide is $^{90}_{35}\text{Br}$ (bromine)

11. $^{15}_{7}\text{N} + {}^{1}_{1}\text{p} \rightarrow {}^{4}_{2}\text{He} + {}^{A}_{Z}$

A: $15 + 1 = 16 = 4 + A$
$16 - 4 = A$
$A = 12$

Z: $7 + 1 = 8 = 2 + Z$
$8 - 2 = Z$
$Z = 6$

The nucleus formed is $^{12}_{6}\text{C}$ (carbon)

13. 4,000 cts/min to 1,000 cts/min in 12 min
$\frac{1}{4}$ as large   2 half lives $12 \text{ min} = 2t_{1/2}$
$t_{1/2} = 6$ min.

15. After 1 half-life, count rate is reduced to $\frac{1}{2}$.
After 2 half-lives, count rate is reduced to $\frac{1}{4}$.
After 3 half-lives, count rate is reduced to $\frac{1}{8}$.
Three half-lives must elapse for count rate to drop to a safe level.
$t_{1/2} = 3$ days
$t_{1/2} = 3 \times 3 \text{ days} = 9$ days

17. For each half-life, the amount of C-14 is reduced by one-half. Thus, a wooden sample with half as much C-14 as that in a new piece of wood must have existed for a time equal to one half-life of C-14 or about 5730 years (see Table 11.3).

19. $^{273}_{110}\text{Ds}$, $^{269}_{108}\text{Hs}$, $^{265}_{106}\text{Sg}$, $^{261}_{104}\text{Rf}$, $^{257}_{102}\text{No}$, $^{253}_{100}\text{Fm}$
(See Example 11.1)

Chapter 12 Problems

1. $\Delta t' = 8.3 \times 10^{-8}$ s   (See Example 12.1)

3. $\Delta t' = 3{,}460 \text{ s} = 57.7$ min.   (See Example 12.1)

5. $L' = L\sqrt{(1 - v^2/c^2)} = 1.0 \text{ m}\sqrt{(1 - (0.95c)^2/c^2)}$
$= 1.0 \text{ m}\sqrt{(1 - 0.9025)} = 0.312 \text{ m}$
The meter stick's length is reduced to 31.2 cm, or shortened by a factor of about 3.2.

7. $E_0 = mc^2 = (1.673 \times 10^{-27} \text{ kg}) \times (3 \times 10^8 \text{ m/s})^2$
$= 1.506 \times 10^{-10}$ J
$E_0 = 1.506 \times 10^{-10} \text{ J} \times (6.25 \times 10^{18} \text{ eV/J})$
$= 9.41 \times 10^8$ eV
$= 9.41 \times 10^8 \text{ eV} \times 10^{-6} \text{ MeV/eV}$
$= 9.41 \times 10^2 \text{ MeV} = 941$ MeV
$m = 941 \text{ MeV}/c^2$ (See Table 12.3; the small difference in this answer from what appears in Table 12.3 is due to the use of a convenient, but approximate, value for the speed of light.)

9. $E_0 = mc^2 = 1.0 \text{ kg} \times (3 \times 10^8 \text{ m/s})^2$
$= 9.0 \times 10^{16}$ J
$E = (\text{mass of water}) \times (\text{latent heat of fusion})$
$= 1.0 \text{ kg} \times (334{,}000 \text{ J/kg}) = 3.34 \times 10^5$ J
$\dfrac{E_0}{E} = \dfrac{9.0 \times 10^{16} \text{ J}}{3.34 \times 10^5 \text{ J}} = 2.69 \times 10^{11}$ (269 billion times larger)

11. $E_{rel} = \dfrac{mc^2}{\sqrt{(1 - v^2/c^2)}} = \dfrac{140 \text{ MeV}}{\sqrt{(1 - v^2/c^2)}} = 280 \text{ MeV}$
$\sqrt{(1 - v^2/c^2)} = 0.5$
$1 - v^2/c^2 = 0.25$
$v^2/c^2 = 0.75$
$v = 0.87\,c$

13. $E_0 = mc^2 = (2m_pc^2) = 2(1.673 \times 10^{-27} \text{ kg}) \times (3 \times 10^8 \text{ m/s})^2 = 3.01 \times 10^{-10}$ J
$= 3.01 \times 10^{-10} \text{ J} \times (6.25 \times 10^{18} \text{ eV/J})$
$= 1.88 \times 10^9$ eV
$= 1.88 \times 10^9 \text{ eV} \times 10^{-6} \text{ MeV/eV}$
$= 1.88 \times 10^3$ MeV
$= 1{,}880$ MeV

15. (a) Not possible; violates strangeness conservation
(b) Not possible; violates charge conservation
(c) Not possible; violates baryon number conservation and spin conservation
(d) Possible

17. (a) π^-
(b) π^-
(c) ν_μ
(d) ν_μ; $\bar{\nu}_\tau$

19. $\bar{\text{n}} = \bar{\text{d}}\,\bar{\text{d}}\,\bar{\text{u}}$; $\bar{\text{p}} = \bar{\text{u}}\,\bar{\text{u}}\,\bar{\text{d}}$

21. $\text{u}\bar{\text{u}}$, $\text{d}\bar{\text{d}}$, $\text{t}\bar{\text{t}}$, $\text{c}\bar{\text{c}}$, $\text{s}\bar{\text{s}}$, $\text{b}\bar{\text{b}}$, $\text{u}\bar{\text{t}}$, and $\bar{\text{u}}\text{t}$

23. d d d; all three quarks must have their spins aligned to yield a net spin of 3/2

25. D^+ mesons must be formed from a quark and an antiquark. With charge +1, charm +1, and strangeness 0, the only possible quark-antiquark combination to yield these characteristics is the ($\text{c}\,\bar{\text{d}}$): A charmed quark and an anti-$\bar{\text{d}}$ quark. (See Table 12.7)

27. (a) $\pi^+ + \text{p} \rightarrow \Sigma^+ + \text{K}^+$ (Use Table 12.3)
$(\text{u}\bar{\text{d}}) + (\text{uud}) \rightarrow (\text{uus}) + (\text{u}\bar{\text{s}})$
Canceling the common quarks on both sides the arrow yields:
$(\text{d} + \bar{\text{d}}) \rightarrow (\text{s} + \bar{\text{s}})$

The annihilation of the d and anti-$\bar{d}$ quarks produces energy that is used to create a new $(s + \bar{s})$ combination.

(b) $\gamma + n \rightarrow \pi^- + p$

$\gamma + (udd) \rightarrow (d\bar{u}) + (uud)$

$\gamma \rightarrow (u + \bar{u})$

The energy of the gamma ray is converted into mass as a $(u + \bar{u})$ pair.

(c) $p + p \rightarrow p + p + p + \bar{p}$

$(uud) + (uud) \rightarrow (uud) + (uud) + (uud) + (\bar{u}\bar{u}\bar{d})$

KE (of colliding protons) $\rightarrow (uud) + (\bar{u}\bar{u}\bar{d})$

$KE \rightarrow 2(u + \bar{u}) + (d + \bar{d})$

The original KE is used to create two $(u + \bar{u})$ pairs plus a $(d + \bar{d})$ pair.

(d) $K^- + p \rightarrow K^+ + K^0 + \Omega^-$

$(s\bar{u}) + (uud) \rightarrow (u\bar{s}) + (d\bar{s}) + (sss)$

$(u + \bar{u}) \rightarrow 2(s + \bar{s})$

The annihilation energy of the $(u + \bar{u})$ pair is used to create two $(s + \bar{s})$ pairs.

29. The $(t\bar{b})$ meson has $Q = (+2/3) + (+1/3) = +1$, topness equal to $+1$, and bottomness equal to -1. The spin of the particle will be either 0 if the spins of the quark and anti-quark are opposed or $+1$ if the spins are parallel. Either case is consistent with the spins of the mesons having integer values.

31. If every proton lasts about 10^{33} years, then in a collection of 10^{33} protons, one proton in the group would, on average, decay each year. For 100 decays to occur each year, then 100 times as many protons would be needed: $100 \times 10^{33} = 10^2 \times 10^{33} = 10^{35}$ protons.

Index

Tables are indicated by *t* and footnotes by *n* following the page number.

A

B

C

D

E

F

G

N

O

P

prepcard

CHAPTER 1
THE STUDY OF MOTION

What's Inside

In this first chapter, we build on the groundwork for the study of physics that was begun in the Prologue. Most of this chapter is an introduction to the branch of physics called *mechanics*—the study of motion and its causes. The main topic of this chapter is motion and how it is described using the concepts of speed, velocity, and acceleration. Both examples and graphs are used to illustrate and define these three concepts. An important example treated in detail is the motion of a body undergoing free fall. We conclude with a brief account of the work of two men who made important contributions to the development of mechanics.

Sections and Objectives

1.1 Fundamental Physical Quantities

- Explain what is meant by the term *fundamental physical quantity*, giving examples.
- Remember the factors of ten associated with the common metric prefixes milli-, centi-, kilo-, mega-, and giga-.
- Convert distances and speeds from metric to English units and vice versa.
- Explain what area and volume measure.
- Define *period* and *frequency*, and explain their relationship to each other.
- Remember what the hertz (Hz) measures.
- Try to explain the physical meaning of mass.

1.2 Speed and Velocity

- Make the distinction between average speed and instantaneous speed.
- Calculate average speeds from position vs. time data.
- Distinguish between speed and velocity.
- Explain how positive and negative velocities can represent movement in opposite directions.
- Do vector addition graphically as well as resolve a vector into its components.

1.3 Acceleration

- Explain carefully the concept of acceleration and distinguish the physics usage of the term from its everyday meaning.
- Compute accelerations from velocity vs. time data.
- Understand centripetal acceleration and do computations using the expression v^2/r.

1.4 Simple Types of Motion

- Analyze simple types of motion (e.g. uniform motion, free fall, other motion with constant acceleration) using tables, graphs, and equations relating position, velocity, and acceleration as functions of time.
- Explain the physical meaning of the slope of *d-t* and *v-t* graphs.

Key Terms

* Terms with an asterisk appear in definition boxes in the chapter.

Fundamental Physical Quantities

Distance

Time

Period*

Frequency*

Mass

Speed*

Velocity*

Vector

Scalar

Acceleration*

Centripetal Acceleration

Equations Covered

$$T = \frac{1}{f} \qquad d = vt \qquad a = \frac{v^2}{r}$$

$$f = \frac{1}{T} \qquad a = \frac{\Delta v}{\Delta t} \qquad v = at$$

$$v = \frac{\Delta d}{\Delta t} \qquad a = g \qquad d = \frac{1}{2}at^2$$

Common Misconceptions

My students often express worries like, "Are we going to have to memorize all the conversion factors?" I reassure them that the answer is certainly no. The only conversion I think worth remembering is 1 inch = *exactly* 2.54 cm. Others simply stick with you if you end up using them a lot, but it is not worth trying to memorize them. Just look them up. Practicing *how to do* unit conversions is the important thing.

Since velocity is a vector, the idea of an increase in velocity doesn't really make sense—you can speak of increases or decreases in *speed* without difficulty, but since velocity incorporates direction as well as speed there is no way to characterize a direction change as an increase or a decrease—it is simply a *change* in velocity. Of course you do encounter phrases like "increasing velocity" frequently, but in this case the word *velocity* is being used in the everyday fashion as a synonym for speed, not in the rigorous physical sense we are trying to learn.

The concept of a negative acceleration can cause confusion because it is very tempting to think that a negative acceleration is a deceleration. Example 1.4 on page 14 points out that the negative sign represents an *opposite direction, not a decrease in magnitude.* This is a very important point worth emphasizing. Plus and minus are used in vector component notation to represent opposite *directions,* not increasing vs. decreasing. This is the source of much trouble in more advanced courses (and in textbooks!).

Teaching Tips and Demonstrations

1.1 Fundamental Physical Quantities

The method of unit conversion presented is not the most common one, and you may not want to use it. Have the students attempt a unit conversion during class as soon as you've introduced the material. Spending just a few minutes looking over their shoulders as they work will give you plenty to fix.

Bring a pendulum for discussion of period, frequency, and time keeping. Set it swinging and have the students count the number of oscillations between different start and stop times. Students counting the swings of a pendulum can suffer from the notorious "off by one" error (they start from *one* instead of from *zero*), and from doubling the count because of the symmetry of "back" vs. "forth". You might try to trip them up by picking a start or stop time in the middle of a swing.

1.2 Speed and Velocity

The distinction between speed and velocity has to be stressed, and I point out that changes in direction of motion are equally as important as changes in speed as far as mechanics is concerned. Vector addition and the resolution of a vector into components occur only a few times in the text and are presented geometrically.

DEMONSTRATION: Look up the record times for Olympic racing events in running, swimming, ice skating, or bicycling. (These can be found in a book of world records or online.) For several events, calculate the average speed, and notice how it is lower for longer races. Why is this so? For comparison, compute the average speeds for the same distance in two different events, such as 1,500 meters running and swimming. Why are they so different?

You can "add" vertical and horizontal velocities by drawing a vertical line on the blackboard while standing still and then doing the same thing as you walk parallel to the board. The direction of the resultant velocity of the chalk is given by the line on the board. By varying the two speeds, you can show that the direction changes. With a stop watch and meter stick, you can measure the horizontal, vertical, and resultant speeds.

DEMONSTRATION: When lightning strikes, the flash of light reaches us in a fraction of a second, but the sound (thunder) is delayed. Why is that? The information in Table 1.2 should help you answer that. The time delay between the flash and the sound can be used to estimate how far away a lightning strike is. The sound travels about 340 meters in 1 second or 1 mile in 5 seconds. So what is the simple rule that relates the distance to the lightning to the number of seconds between the light and the sound? (Note: Follow the safety guidelines whenever there is a potential for lightning: Go into a building or a vehicle with a metal roof, and stay away from windows, plumbing, telephones with cords, and so on.)

1.3 Acceleration

The precise concept of acceleration is new to most students, and it deserves a lot of attention because of its importance in Chapter 2. The fact that a change in direction of motion is an acceleration is foreign to students and has to be explained. If you are in a colder climate, you might describe how a perfectly slick sheet of ice keeps a car from speeding up, slowing down, or going around a curve. All three would be accelerations.

You might bring up the "cornering acceleration" of cars in *g*'s as presented in automotive magazines, or the use of "*g*-suits" by pilots of high-performance aircraft.

DEMONSTRATION: It's easy to compute the acceleration of a car (in *g*'s) as it speeds up or slows down. Simply time how long (in seconds) it takes for the speed to change by 22 mph. Do this when you are a passenger and can see the speedometer. For example, it may take 4 seconds for a car to speed up from 20 mph to 42 mph. How do you find the acceleration in *g*'s using these numbers? (Hint: If it took 1 second for the speed to increase by 22 mph, the acceleration would be 22 mph/s = 1 *g*.)

1.4 Simple Types of Motion

The two main thrusts of section 1.4 are to show how distance, speed, and time are related to each other in simple types of motion, and to show how relationships can be expressed in different ways—using words, equations, tables, and graphs. Try constructing motion graphs by moving around (back and forth along a line) in front of the class and having the students try to draw the *d-t* or *v-t* graph corresponding to your movement. Also try acting out graphs dreamed up by the students. (Sometimes this requires infinite quickness or the ability to split in two!) You may not want to stress the graphing, particularly the calculation of slope, as much as I have unless you use it in labs as I do.

I usually point out, but do not dwell on, the fact that air resistance eventually affects the acceleration of a falling body. This is treated in Section 2.6.

A Demonstration Handbook for Physics (*DHP*), page M-6, item Mb-12 (a long string with balls tied along it with appropriate spacing so that when the whole thing is hung vertically and released the balls hit at equal time intervals) is an interesting way to show how distance traveled increases during free fall.

The graphs in Figures 1.22 (car acceleration) and 1.23 (karate chop) can lead to some interesting discussions. Encourage (make) the students to go to the library and find the whole story on the karate blow in the September 1983 *American Journal of Physics*. You might jump ahead a bit and mention that the force on the fist during the deceleration is on the order of 500 lbs. At about 25 ms the position of the fist drops below zero, and the velocity of the fist goes more negative (downward faster), so the concrete block must have cracked and given way.

You can find *A Demonstration Handbook for Physics* (*DHP*) at www.aapt.org.

What's Inside

In this chapter we present Newton's three laws of motion and his law of universal gravitation. These laws form the basis of mechanics. Force, a central concept in physics, is introduced, and several common examples are given. Each law is used to extend your understanding of motion, of how forces affect motion, and of gravity. The cause of an object's acceleration, the principle behind rockets, and the nature of orbits are some of the applications of these laws that are described. The chapter concludes with a close look at the life of Sir Isaac Newton.

Sections and Objectives

2.1 Force

- Develop a better understanding of the specific meaning of *force* in mechanics.
- Understand that gravity causes weight.
- Explain the mechanisms underlying friction, the laws of friction, and the difference between static and kinetic friction.

2.2 Newton's First Law of Motion

- State Newton's first law and explain the meaning of *net* force.

2.3 Mass

- Compare mass, inertia, and weight and explain the differences.

2.4 Newton's Second Law of Motion

- State Newton's second law in words as well as mathematically, and be able to perform calculations using it.

2.5 The International System of Units (SI)

- Be comfortable with the SI system of units and give some examples.

2.6 Different Forces, Different Motions

- Give examples of free fall.
- Explain projectile motion, especially the fact that the vertical and horizontal components are independent.
- Explain simple harmonic motion.
- Explain the changes that result from adding air resistance to a free fall situation, and explain terminal speed.

2.7 Newton's Third Law of Motion

- State Newton's third law, and give several examples of action-reaction force pairs.

2.8 The Law of Universal Gravitation

- State Newton's law of universal gravitation, in words and mathematical notation.
- Describe the Cavendish experiment.
- Explain how gravity is involved in orbits, and relate Newton's "cannon" idea for a satellite launch.
- Make an ellipse with tacks and a loop of string.
- Picture a gravitational field, and explain the benefit of this concept over "action at a distance."
- Explain how tides are created.

Key Terms

* Terms with an asterisk appear in definition boxes in the chapter.

Force*

Weight*

Friction*

Static Friction

Kinetic Friction

Centripetal Force

Mass*

International System (SI)

Simple Harmonic Motion

Laws

Newton's First Law of Motion

Newton's Second Law of Motion

Newton's Third Law of Motion

Newton's Law of Universal Gravitation

Equations Covered

$F = ma$

$W = mg$

$F = \frac{mv^2}{r}$

$F = \frac{Gm_1m_2}{d^2}$

Common Misconceptions

Newton's first law can lead to confusion if students place it in an unintended context. They might be misled by the need to exert a force on an object to bring it up to speed from rest. But Newton's first law does not refer to an object's *history*, only what is going on right now. If no forces are acting, the object simply *coasts* at constant velocity. They also may be unwittingly thinking of a concept closer to that of *energy* or *momentum* when they mistakenly use the term *force*.

Check the teaching resources online at 4ltrpress.cengage.com/physics for more common misconceptions that students have about physics in the real world.

Teaching Tips and Demonstrations

2.1 Force

The idea of weight being a force—a pull—is often new to students and must be stressed. You may not wish to dwell on friction to the extent presented in Section 2.1, although kinetic friction is mentioned by name in later chapters.

Question 5 about pressing and pushing on a book resting on a table is a "must do" in class. Have everyone try this, pressing lightly at first, then harder and harder. Then try it again with a sheet of paper between your hand and the book. (When I tried this variation on my computer table using a different book, I couldn't get the book to slip no matter how hard I pressed!)

2.2 Newton's First Law of Motion

The key idea to present in connection with Newton's first law is that a force is needed to cause any change in motion—in speed *or* direction. The concept of a *net* force and an *external* force may have to be explained. Describe several situations in which a centripetal force keeps something moving along a circular path. Figure 2.9 and the following demonstration with the sock should be pointed out and used to dispense with the common misconception that the object would move radially rather than tangentially. If the students don't get around to trying it right away, tie a rubber stopper to a weak string (strong thread) and swing it in a circle overhead. Hold a sharp knife in front of you and ask the class to predict where the stopper will go after you raise the knife so it will cut the string.

DEMONSTRATION: For this you need a sock, a piece of thread about an arm's length or more long, a reasonably sharp knife (be careful!), and a place where a flying, rolled-up sock isn't going to hurt anything. Roll up the sock and tie it to one end of the thread. Grasp the other end and whirl the sock in a horizontal circle above your head. With your free hand, or with the aid of an assistant, quickly (but carefully!) move the knife into the path of the thread so that the knife cuts the thread near its middle at the moment the sock is passing directly in front of you. Stow the knife, and then find the sock. In which direction did it go after the force on it was removed (the string was cut)? Not sure? Tie the two pieces of thread together and do it again.

2.3 Mass

The distinction between mass and weight should be stressed again, and the fact that weight is caused by something outside of an object. The concept of weightlessness in an orbiting spacecraft is very tricky to explain to students at this level.

2.4 Newton's Second Law of Motion

I usually work out several examples and/or problems on the board to illustrate Newton's Second Law. Incidentally, the momentum form of the second law is presented in Section 3.2.

2.5 The International System of Units (SI)

I use the term SI a great deal and try to get the students to automatically use SI units.

2.6 Different Forces, Different Motions

The fact that the states of no motion and of uniform motion are equivalent as far as forces are concerned needs to be pointed out again. Projectile motion, simple harmonic motion, and falling body with air resistance can be treated lightly if desired. The sinusoidal graphs for simple harmonic motion do appear in later chapters.

Use a mass on a spring, a simple pendulum, a metronome, a glider on an air track attached to two springs, or other systems to show simple harmonic motion.

2.7 Newton's Third Law of Motion

The fact that a wall or other passive object can exert a force is a new and important concept for students.

DEMONSTRATION: Stand facing a wall. Push on the wall so hard that you have to take a step backward to keep from falling over. What was the direction of the force you exerted? What was the direction in which your body accelerated? Is there a contradiction here?

2.8 The Law of Universal Gravitation

The concepts of an inverse square force and a field are important and will be seen again in Chapter 7. The text relates Newton's reasoning about gravity being an inverse square law early in Section 2.8 (the moon is 60 times farther away from the center of the Earth than something on the Earth's surface is from the center of the Earth, and the acceleration of the moon is 1/3600th—1/60th squared—of the acceleration of, say, a falling *apple*) but does not go into the more difficult problem of proving that it does not matter that the mass of the Earth is spread throughout the interior of a huge planet. (It might be interesting to have students chase down people who can do calculus and are willing to show them the details of the proof.)

DEMONSTRATION: You can draw the correct shape of an ellipse using two tacks or pins, some string or thread, a ruler, a piece of paper, and a surface into which you can push the tacks (a bulletin board works).

1. Place the tacks into the paper 10 centimeters apart. Wrap the string around the tacks, and tie it to form a tight loop around them. Move the tacks a few centimeters closer together. Insert a pen into the loop, move it outward until the string is tight, and draw the complete path around the tacks while holding the pen against the string. The resulting figure is an ellipse with each tack at one focus of the ellipse.
2. Place the tacks 4 centimeters apart. The ellipse you draw has the shape of Pluto's orbit. (The Sun would be at one of the pins.) Note that it is hard to distinguish it from a circle. How should you place the pins to draw a true circle?
3. Place the tacks 8.6 centimeters apart to draw the shape of the orbit of Nereid, a moon of Neptune that has a highly elliptical orbit.

What's Inside

In this chapter, we introduce the use of conservation laws in the study of how objects move. The laws of conservation of linear momentum, energy, and angular momentum are shown to be simple but powerful tools for analyzing processes—such as collisions—that we could not handle using only the concepts from Chapter 2. Work and power, two important physical quantities related to energy, are illustrated as well. The concept of energy is one of the most important in physics, and its usefulness extends well beyond the area of mechanics.

Sections and Objectives

3.1 Conservation Laws

- Explain what it means for a physical quantity to be conserved.

3.2 Linear Momentum

- Explain the concept of momentum.
- Give some examples of collisions demonstrating momentum conservation.

3.3 Work: The Key to Energy

- Explain the meaning of work, giving examples of work done *by* and *against* forces, as well as forces doing *no* work.
- Describe how levers and inclined planes work as simple machines.

3.4 Energy

- Define *kinetic energy.*
- Define *potential energy.*
- Give examples of gravitational potential energy, elastic potential energy, and internal energy.

3.5 The Conservation of Energy

- Solve problems using the principles of conservation of energy and conservation of momentum.

3.6 Collisions: An Energy Point of View

- Understand how momentum and energy are involved in elastic and inelastic collisions.

3.7 Power

- Define *power* and compute it.

3.8 Rotation and Angular Momentum

- Understand torque as a rotational analog of force.
- Define *angular momentum.*
- Use the principle of conservation of angular momentum.

Key Terms

* Terms with an asterisk appear in definition boxes in the chapter.

Conservation Laws
Linear Momentum*
Work*
Energy*
Kinetic Energy*
Potential Energy*
Gravitational Potential Energy
Elastic Potential Energy
Internal Energy
Elastic Collision*
Inelastic Collision*
Power*
Angular Momentum

Principle

Total mass in an isolated system is constant.

Laws

Newton's Second Law of Motion (alternate form)

Law of Conservation of Linear Momentum

Law of Conservation of Energy

Law of Conservation of Angular Momentum

Equations Covered

$$\text{linear momentum} = mv$$

$$\text{force} = \frac{\text{change in momentum}}{\text{change in time}}$$

$$F = \frac{\Delta\,(mv)}{\Delta t}$$

$$\text{total } mv \text{ before} = \text{total } mv \text{ after}$$

$$\text{work} = Fd$$

$$KE = \frac{1}{2}mv^2$$

$$PE = Wd = mgd$$

$$v = \sqrt{2gd}$$

$$d = \frac{v^2}{2g}$$

$$P = \frac{\text{work}}{t}; \quad P = \frac{E}{t}$$

$$\text{angular momentum} = mvr$$

Teaching Tips and Demonstrations

3.1 Conservation Laws

As a motivation for learning about conservation laws, explain how conservation laws make it simple to calculate the speed of a roller coaster at the bottom of a hill, whereas it would be very difficult to do so using Newton's second law.

3.2 Linear Momentum

For Section 3.2, Linear Momentum, review Newton's second law in Section 2.4 and Newton's third law in Section 2.7. When covering Section 3.2, the momentum form of the second law and the concept of impulse can be omitted. Demonstrate several examples of problems using linear momentum conservation. Newton's third law is never very far away.

3.3 Work: The Key to Energy

Use a "blackboard mechanics" kit or similar setup to show levers and ramps, or use some homemade apparatus.

3.4 Energy

When covering Section 3.4, point out that in the previous section no work is done as the box is carried or as the ball moves in a circle because the energy of the object doesn't change.

Accelerate a cart or ball with a shove while explaining how work is being done to give it kinetic energy. Lift an object to show work being done to give the object gravitational potential energy. Compress a spring, perhaps in a dynamics cart or a toy dart gun, to show work done to give the object elastic potential energy.

DEMONSTRATION: Battery race. You will need two identical cylindrical batteries (AAs or AAAs work well), a ramp for the batteries to roll on (a piece of cardboard or a thin hardcover book), and a book or some other kind of backstop.

1. Place the backstop in front of the ramp a distance from the ramp's lower end equal to its length (see figure below). Release a battery from the top of the ramp and then from places lower down the ramp. Is it traveling the same speed at the bottom each time? Why?

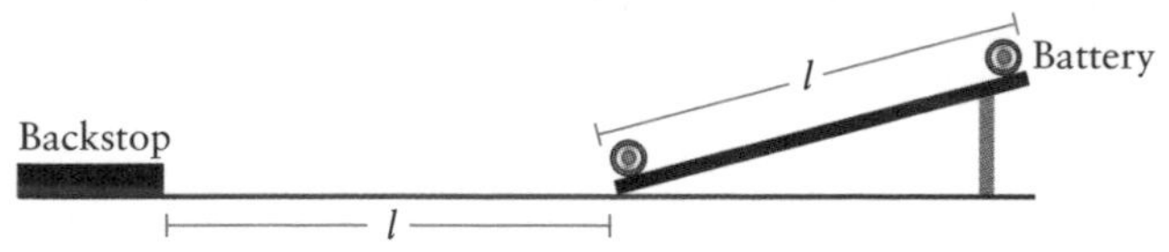

2. Hold one battery at the top of the ramp and the other one one-fourth of the way up from the bottom of the ramp. Release them at the same time. Which one wins the race to the backstop? Repeat with the lower battery starting at different places on the ramp. What happens in these races?

3.5 Conservation of Energy

For the law of conservation of energy presented in Section 3.5, the list of devices in Table 3.1 can be expanded easily. Spend some time relating the various types of energy back to the concepts of work and mechanical energy by describing a mechanism that can utilize each form of energy to do work lifting an object. The discussion of negative total energy and bound states may be too much for your students. It can be skipped, but it is referred to in later chapters.

Attach a bowling ball or other large, heavy object to a cable attached to the ceiling. Pull it to the side and up to your nose. It will swing away and back, stopping just in front of your nose because energy is conserved. Show conversion of elastic potential energy into kinetic using the dynamics carts shown in Figure 2.24, or with a dart gun.

3.6 Collisions: An Energy Point of View

When explaining where the energy goes in inelastic collisions, it might be useful to jump ahead to Section 5.5 where the temperature change of a concrete block hitting the ground is given (Example 5.3, p. 129–130).

Show the two collisions in Figure 3.19 using dynamics carts or an air track. Figure 3.20 shows a simple exoergic inelastic collision.

The slingshot effect shown on page 77 can be a starting point for discussions about deep space exploration.

3.7 Power

When covering Section 3.7 on power and its role in mechanics, point out the "full circle" that the units have gone through from familiar to unfamiliar to familiar—meters, newtons, joules, to watts.

Human-powered flight is an intriguing topic to discuss and a timely one because of human-powered projects since 1970, including helicopters. You might consider working out problem 34 at the end of the chapter in class.

3.8 Rotation and Angular Momentum

Sport Science has some interesting estimates of the relative sizes of the moments of inertia of the human body rotating in various ways to use for Section 3.8.

It is our duty to do the "spinning ice skater" demonstration. It is possible to replicate this in the classroom, but practice first. You might check ahead of time to see if any of your students has studied ballet and would do a pirouette in class. Purchase two pipes that have the same length and mass but different moments of inertia. (Or make your own by securing two 2 or 5 kg masses at the center of one pipe and fastening identical weights at the ends of another pipe.) Have a volunteer hold each pipe in turn and twist it back and forth as quickly as possible.

DEMONSTRATION: This is the "spinning ice skater" demonstration. You will need a chair that can spin in circles easily. (Caution: Don't try this if you are prone to dizziness.) Move it to a place at least 5 feet from walls, furniture, and other objects. Ask for a volunteer to sit in it with both arms straight out from her sides; then she should use her feet to make the chair spin. Have her lift her feet, and then pull her arms tight against her body. Ask the volunteer to extend her arms out to the side again. (You can enhance the effect by having the volunteer hold a weight in each hand.)

prepcard

CHAPTER 4
PHYSICS OF MATTER

What's Inside

In this chapter, we look into the physics of extended matter, particularly liquids and gases. The first section is an overview of the types, properties, and submicroscopic compositions of matter. Then the crucial physical quantities, *pressure* and *density*, are presented. The remainder of the chapter covers several laws and principles that explain a variety of phenomena involving fluids. The ideas presented in this chapter—particularly the law of fluid pressure, Archimedes' principle, and the concept of density—make it easy to understand how things like airships and altimeters work.

Sections and Objectives

4.1 Matter: Phases, Forms, and Forces

- List the four phases of matter and their properties.
- Give several examples of chemical elements.
- Explain the difference between elements and compounds.
- Explain why a physical mixture is not the same as a chemical compound.
- Link the properties of solids, liquids and gases to the strengths of the forces between their constituent particles.

4.2 Pressure

- Do calculations based upon the relationship of pressure, force, and area.
- Explain the relationship between gauge pressure and absolute pressure.
- Explain how a tire pressure gauge works.
- Do the "pop can crushing" demonstration.

4.3 Density

- Do computations based upon the definitions of mass density and weight density.
- Explain what specific gravity means.

4.4 Fluid Pressure and Gravity

- Be able to calculate pressures at various depths in fluids.
- Explain how barometers work, and tell what the phrase "millimeters (or inches) of mercury" means.

4.5 Archimedes' Principle

- Explain the origin of buoyant forces and discuss the cases of an object floating, sinking, or rising in a fluid.
- State Archimedes' principle.
- Explain how to experimentally measure specific gravities.
- Explain how an antifreeze tester works.

4.6 Pascal's Principle

- Explain Pascal's principle and how it is applied in hydraulic brake systems.

4.7 Bernoulli's Principle

- Explain Bernoulli's principle and how it relates to the operation of a perfume atomizer.

Key Terms

* Terms with an asterisk appear in definition boxes in the chapter.

Solids*	Nucleus	Solution
Liquids*	Electrons	Pressure*
Gases*	Protons	Mass Density*
Plasmas*	Neutrons	Weight Density*
Phases	Compound	Specific Gravity
Elements	Molecule	Buoyant Force*
Atoms	Mixture	

Principles

In a liquid, the absolute pressure at a depth h is greater than the pressure at the surface by an amount equal to the weight density of the liquid times the depth.

$$p = D_W h = Dgh \text{ (gauge pressure in a liquid)}$$

Archimedes' Principle The buoyant force acting on a substance in a fluid at rest is equal to the weight of the fluid displaced by the substance.

$$F_b = \text{weight of displaced fluid}$$

Pascal's Principle Pressure applied to an enclosed fluid is transmitted undiminished to all parts of the fluid and to the walls of the container.

Bernoulli's Principle For a fluid undergoing steady flow, the pressure is lower where the fluid is flowing faster.

Law

Law of Fluid Pressure

Equations Covered

$$p = \frac{F}{A}$$

$$D = \frac{m}{V}$$

$$D_W = \frac{W}{V}$$

$$p = D_W h = Dgh$$

$$p = 0.433 \text{ psi/ft} \times h$$

$$F_b = \text{weight of displaced fluid}$$

Teaching Tips and Demonstrations

4.1 Matter: Phases, Forms, and Forces

Most students have seen most of the material on the description and classification of matter in its various forms in prior physical science classes, but don't count on them remembering it. The nature of the forces between atoms and molecules, and how those forces affect the properties of a substance, are important because they are referred to in later chapters. Figure 4.4 represents just one of many different ways to emphasize the extremely small sizes of atoms.

4.2 Pressure

We chose to use the English unit of pressure (psi) quite a bit because it is so much more common than the metric units. The benefit of having a familiar unit of measure outweighs the break from the metric system, in our opinion.

DEMONSTRATION: This is the "pop can crushing" demonstration. Put an ounce or so of water into an empty aluminum can. Heat the can until the water is boiling vigorously and you can clearly see mist coming out of the opening. Let the water boil for about a minute. With a gloved hand, quickly but carefully turn the (hot!) can upside down over a pan or sink with cold water in it and plunge the top of it an inch or so into the water. What happens? How is the atmosphere involved in this?

4.3 Density

Both weight density and mass density are used in Section 4.3, the former mainly because of its use in fluid pressures in Section 4.4. You may need to stress that the density of a given substance is the same, regardless of the quantity under consideration.

Measure the volume and mass (or weight) of a block of aluminum, copper, or other material listed in Table 4.4, and compute its mass (or weight) density.

4.4 Fluid Pressure and Gravity

Ask students to estimate the weight of all of the air in your lecture room, and then compute it with dimensions you measured earlier. Afterwards you may want to give the figure for the weight that is corrected for air temperature and elevation.

Figure 4.18 shows the nonlinear (exponential) decrease of atmospheric pressure versus height above sea level. When discussing the principles behind barometers and altimeters, it should be stressed that it is the surrounding air that pushes mercury up a column and a drink up a soda straw.

4.5 Archimedes' Principle

When covering Section 4.5, it should be stressed that the buoyant force depends on the surrounding fluid and is independent of the composition of the object under consideration.

Perform the demonstration depicted in Figure 4.24. With a Styrofoam cup on the smaller scale you don't have to re-zero it.

4.6 Pascal's Principle

*DHP**, page F-4, item Fb-3 shows a sort of "push of war" with two hypodermic syringes of different sizes connected together. Students pushing on the syringe with the smaller plunger easily win.

4.7 Bernoulli's Principle

*DHP**, pages F-15 to F-19, has several demonstrations of Bernoulli's principle.

DEMONSTRATION: Hold the top of a piece of paper horizontally just below your lips, so that the paper hangs limp (as seen below). Blow hard over the top of the paper. What does the paper do? What causes this?

© Vern Ostdeik

Common Misconceptions

Al Bartlett, in his masterful lecture on the dangers to society of exponential growth, used to tell a true story about a former high-level government advisor who proposed the following solution to a forecasted shortage of copper: "We'll just make more from other materials!" See if your students recognize what is horribly wrong about this statement. Copper simply cannot be made from other materials. It is an element. Trusting his advice would be disastrous. This story should remind all of us how knowledge of the physical world is important to policy decisions that affect society at large.

It is easy to mistakenly think that a buoyant force *greater* than an object's weight is needed in order to make the object float. After all, it seems logical—if objects sink when the buoyant force is less than the weight, shouldn't they float when the reverse is true? The trouble is that floating is not the opposite of sinking. *Rising* is the opposite of sinking. It is true that an object will rise if the buoyant force on it is greater than its weight.

Check the teaching resources online at 4ltrpress.cengage.com/physics for more common misconceptions that students have about physics in the real world.

* Remember, you can find *A Demonstration Handbook for Physics* (*DHP*) at www.aapt.org.

What's Inside

In this chapter we take a formal look at the concepts of temperature and heat. First we consider temperature—what it is, how it is measured, and how matter is affected when it changes. Then we refine the concepts of internal energy and heat by showing how they are involved in changing the temperature or phase of matter. The chapter concludes with a discussion of heat engines and heat movers.

Sections and Objectives

5.1 Temperature
- Explain the physical meaning of temperature.
- Relate the Fahrenheit, Celsius, and Kelvin temperature scales.
- Understand the idea of absolute zero.
- Explain how thermometers work.

5.2 Thermal Expansion
- Compute amounts of thermal expansion.
- Remember some of the noteworthy properties of water (like its high specific heat capacity and unusual thermal expansion properties).
- Use the ideal gas law.

5.3 The First Law of Thermodynamics
- State the first law of thermodynamics.
- Explain the terms *heat* and *internal energy*, and give historical units of heat and relate some of the early confusion regarding these concepts.

5.4 Heat Transfer
- Distinguish the three types of heat transfer and give examples of each.

5.5 Specific Heat Capacity
- Perform calculations of heat transfer using specific heat capacity.

5.6 Phase Transitions
- Explain what a phase transition is, and tell why the temperature stays constant during a phase transition.
- Explain the difference between humidity and relative humidity, and do computations.
- Explain what *dew point* means.

5.7 Heat Engines and the Second Law of Thermodynamics
- State the second law of thermodynamics.
- Explain how refrigerators and heat pumps work.
- Begin to relate the concepts of Carnot efficiency, usable energy, and entropy.

Key Terms

* Terms with an asterisk appear in definition boxes in the chapter.

Temperature

Absolute Zero

Thermal Expansion

Coefficient of Linear Expansion

Internal Energy*

Heat*

Conduction*

Convection*

Radiation*

Latent Heat of Fusion

Latent Heat of Vaporization

Evaporation

Humidity*

Relative Humidity*

Heat Engine*

Heat Movers

Entropy

Principle

The Kelvin temperature of matter is proportional to the average kinetic energy of the constituent particles.

$$\text{Kelvin scale temperature} \propto \text{average } KE \text{ of atoms and molecules}$$

Laws

Ideal Gas Law

First Law of Thermodynamics

Second Law of Thermodynamics

Second Law of Thermodynamics (alternate form)

Equations Covered

$$\Delta l = \alpha\, l\, \Delta T$$

$$pV = (\text{constant})\, T$$

$$\Delta U = \text{work} + Q$$

$$Q = C\, m\, \Delta T$$

$$\text{relative humidity} = \frac{\text{humidity}}{\text{saturation density}} \times 100\%$$

$$\text{efficiency} = \frac{\text{energy or work output}}{\text{energy or work input}} \times 100\%$$

$$\text{Carnot efficiency} = \frac{T_h - T_l}{T_h} \times 100\%$$

Teaching Tips and Demonstrations

5.1 Temperature

It is a good idea to spend time distinguishing the concepts of temperature and heat.

DEMONSTRATION: Dip a racquetball or a flower in liquid nitrogen and drop it to show the effect of very low temperatures. A frozen onion will shatter spectacularly, too.

5.2 Thermal Expansion

The anomalous temperature dependence of the density of water below 4 degrees Celsius discussed on pages 120–121 can be omitted.

DEMONSTRATION: Use a brass ball and ring with the same diameters to show thermal expansion. Heat the ball until it can't pass through the ring, then heat the ring until the ball can fit through. This also shows that a hole in a solid expands when the solid's temperature rises.

*DHP**, pages H-3 to H-5, has several thermal expansion demonstrations.

5.3 The First Law of Thermodynamics

Internal energy is a rather abstract concept, and I stress that it is most useful when considering phase transitions, the topic of Section 5.6.

DEMONSTRATION: For this, you need a bicycle tire pump and a flat bicycle tire that you can inflate. First touch the lower end of the pump (or the metal fittings on the hose if it has one). Pump up the tire to the recommended pressure, and then touch the lower end (or metal fittings) again. What is different? What has happened?

A variation of this demonstration is to take off the hose or fittings on a hand-held bicycle pump, place this end next to your leg just above the knee (for example) and pump it once or twice. The fabric allows the compressed heated air to pass through, and you can quickly feel the heat.

5.4 Heat Transfer

Heat transfer by radiation is a rather tough topic because, at this point, most students don't really know what electromagnetic waves are. Campfires and heat lamps are good examples to use to convey the idea.

5.5 Specific Heat Capacity

In the satellite reentry example (Example 5.4) the simplifying assumptions should be pointed out.

5.6 Phase Transitions

The actual meaning of humidity has to be stressed since the word is often incorrectly used in place of relative humidity. I work out a couple of examples concerning relative humidity and how it changes when the temperature changes. You may want to compute the mass of water in the air in your room, or how much water condenses out of air when it is cooled. The cooling of a liquid by evaporation is analogous to a baseball team's batting average going down when the best hitters are traded.

5.7 Heat Engines and the Second Law of Thermodynamics

The fact that heat engines always have to reject heat is an important but somewhat counter-intuitive concept for students. The concept of entropy is also a tough one and may be omitted.

*DHP**, pages H-32 to H-34, shows several examples of heat engine demonstrations. A solar-powered Stirling engine is another interesting demonstration.

Common Misconceptions

Careless usage of the term *heat* continues to cause trouble to this day. It is very important to convey the narrow definition—that heat is energy transferred due to a temperature difference. The discussion about the hot pizza in Section 5.3 should be helpful in clarifying this issue. Any reference to an object "containing heat" is an error. Be careful to use the term *internal energy* in such a case. Sloppy usage coupled with terms that have a messy history to begin with will cause great student frustration.

Other books often speak of specific heat and heat capacity separately. Here, the term *specific heat capacity* is used and is the same thing as what the others call *specific heat*.

It's true that stirring a liquid can warm it—this is Joule's experiment—but here that effect is swamped by the greater effect of continually bringing new hot liquid to the surface where it cools most quickly. Evaporative cooling plays a big role here. As in many other "real-life" situations, more than one mechanism is operating simultaneously. The process can't be adequately modeled by considering just one of them.

Check the teaching resources online at 4ltrpress.cengage.com/physics for more common misconceptions that students have about physics in the real world.

* Remember, you can find *A Demonstration Handbook for Physics* (*DHP*) at www.aapt.org.

prepcard

CHAPTER 6
WAVES AND SOUND

What's Inside

In the first part of this chapter, we look at simple waves and examine some of their general properties. The remainder of the chapter is about sound—how it is produced, how it travels in matter, and how it is perceived by humans.

Sections and Objectives

6.1 Waves—Types and Properties

- Explain what a wave is.
- Explain what a wave medium is, and give examples of waves in different media.
- Compare a wave pulse and a continuous wave.
- Demonstrate both transverse and longitudinal waves on a Slinky, and give other examples of these types of waves.
- Compute the speed of a wave on a rope from its mass density and the tension applied.
- Compute the speed of sound in air given the temperature.
- Identify wavelength and amplitude on a sketch of a continuous wave.
- Remember the equation relating the velocity, frequency, and wavelength of a continuous wave.

6.2 Aspects of Wave Propagation

- Understand how waves can be represented by both wavefronts and rays.
- Explain why the amplitude of a wave gets smaller farther from the source.
- Define a *plane wave*.
- Give concrete examples of reflection of waves.
- Explain the consequences of the Doppler effect.
- Explain what causes a sonic boom.
- Explain what diffraction is, and give examples of situations where diffraction can be observed.
- Give an explanation of how the phase relationship of superposed waves determines whether they interfere constructively or destructively.

6.3 Sound

- Give an explanation of what sound is.
- Discuss the differences between pure tones, complex tones, and noise.
- Explain how ultrasound is different from audible sound, and give examples of some of its applications.

6.4 Production of Sound

- Recognize some differences in the ways various musical instruments produce sound.

6.5 Propagation of Sound

- Measure the reverberation time of a room and relate some of the factors that determine it.

6.6 Perception of Sound

- Describe the subjective impressions of pitch, loudness, and tone quality, and relate each of them to the physical quantity upon which it most depends.
- Appreciate some of the remarkable aspects of human hearing.
- Explain how the decibel scale measures loudness quantitatively.

Key Terms

* Terms with an asterisk appear in definition boxes in the chapter.

Wave*
Transverse Wave*
Longitudinal Wave*
Linear Mass Density
Amplitude*
Wavelength*
Frequency*
Compressions
Expansions
Wave Speed
Wavefront Model
Ray Model
Echolocation
Shock Wave
Diffraction
Interference
Constructive Interference
Destructive Interference
Waveform
Pure Tone
Complex Tone
Noise
Infrasound
Ultrasound
Sonoluminescence
Reverberation
Pitch
Loudness
Tone Quality
Sound Level
Harmonics

Principle

Any complex waveform is equivalent to a combination of two or more sinusoidal waveforms with definite amplitudes. These component waveforms are called *harmonics*. The frequencies of the harmonics are whole-number multiples of the frequency of the complex waveform.

Equations Covered

$$v = \sqrt{\frac{F}{\rho}}$$

$$v = 20.1 \times \sqrt{T}$$

$$\rho = \frac{m}{l}$$

$$v = f\lambda$$

Teaching Tips and Demonstrations

6.1 Waves—Types and Properties

The concept of a wave is a bit abstract to the student, so I try to spend time talking about the very common waves one encounters regularly.

DEMONSTRATION: For this, you need a Slinky, preferably the metal kind. (If you don't already have one, get one. You won't regret it.) On a smooth table, large desk, or bare floor, stretch the Slinky out about 5 feet, with a partner holding the other end—or attach it to something heavy so it stays put.

1. Send a transverse pulse down the Slinky by quickly moving your hand to the side and back. Send a longitudinal pulse by quickly moving your hand toward the other end and back. Do the two pulses seem to travel at the same speed? (Do they take about the same amount of time to get to the other end?)
2. Again send a transverse pulse down the Slinky. Watch what happens when it reaches the other end. Does it reflect? If so, is the reflected pulse identical to the original pulse?

A hand-cranked "wave box" showing a row of little balls moving up and down as they form a traveling continuous wave is very useful for talking about frequency, wavelength, and wave speed.

6.2 Aspects of Wave Propagation

Reflection, the Doppler effect, echolocation, shock waves from a moving wave source, diffraction, and interference are rather difficult for the students to grasp the first time around and should be demonstrated. They could be skipped now and also when they appear in Chapters 8 and 9 in connection with electromagnetic waves and light.

A ripple tank is a good way to demonstrate much of the material in this section, particularly if it is set up for a moving wave source. I prefer to use film loops of ripple tank demonstrations because they are reliable and take up a lot less class time.

6.3 Sound

A good way to illustrate waveforms is to use a sound level meter with its output jack connected to an oscilloscope that has good, stable triggering. The range setting on the meter makes it easy to adjust the size of the trace. If you have a lot of low-frequency noise from lights and ventilation systems, set the meter on the A-weighted scale. This reduces the meter's response to low frequencies. A tuning fork, steady whistle, and a beer bottle blown across the top make good pure tones. Your own spoken or sung vowel sounds and a note played on a harmonica are good complex tones. The latter is convenient because it is easy to get a good sound without being musically inclined and it has nice complicated waveforms. You probably have several students who are musicians—have them bring in their own instruments to test.

6.4 Production of Sound

Use the long wave demonstration spring to present the model of sound production in string instruments. Changes in tension and length alter the frequency of the pulse's back-and-forth oscillation, just like tuning a string or pressing a guitar string against a fret does. It may be useful to work out Problem 2.

6.5 Propagation of Sound

Use a sound level meter or microphone connected to a storage oscilloscope to show the decay of reverberant sound. Set the sweep on about 0.1 s/cm, depending on the room, and have a student clap their hands once or burst a balloon just as you trigger a single sweep.

6.6 Perception of Sound

If your class is mathematically astute, you might point out that the notes in the equally tempered musical scale are based on the twelfth root of 2: the frequency of each note is $\sqrt[12]{2}$ (= 1.059463094 . . .) times the frequency of the adjacent lower note. Sound amplitude instead of intensity is used when discussing loudness because it is much easier for students to grasp.

The problem with high sound levels is a fertile area of discussion: physiological effects, psychological effects, noise pollution, and noise control ordinances are some possible topics.

Common Misconceptions

As with the term "acceleration" back in Chapter 1, we run into trouble when a word is used differently in everyday speech than in physics. Students often take "interference" to mean something like "messing up," as if two waves interfering with each other get wrecked. Their "courses"—if that means directions—are certainly not affected at all. All that happens is that they superpose, and you observe the combined *net* effect of the two waves together. What is interesting is that this net effect can in certain cases be zero—no wave at all (but only in certain locations).

Check the teaching resources online at 4ltrpress.cengage.com/physics for more common misconceptions that students have about physics in the real world.

What's Inside

In this chapter, we consider some of the basic aspects of electricity, starting with electrostatics—phenomena involving electric charges that are not moving. In the remainder of the chapter, we discuss the physics of moving charges—electric current. We introduce the important quantities voltage, resistance, and current and show how they are involved in many devices around us and in living things. The last two sections deal with power and energy in electric circuits and the two types of electric current—AC and DC.

Sections and Objectives

7.1 Electric Charge

- Attempt to define *electric charge*, explain positive vs. negative charge, and tell what a coulomb is.
- Explain the difference between neutral atoms and ions.

7.2 Electric Force and Coulomb's Law

- Explain how the electrostatic force between charges depends on their +/− signs, magnitudes, and distance apart.
- Explain how a charged object is able to attract an *un*charged object.
- Explain what an electric field is, and show how field lines are used to represent one.
- Explain the mechanism behind "static electricity" sparks.

7.3 Electric Currents—Superconductivity

- Relate a few examples of modern uses of semiconductor technology.
- Define *electric current* and the ampere.
- Compare properties of conductors, insulators, semiconductors, and superconductors, and give examples of each.
- Define *electrical resistance* and the ohm.
- Explain what superconductivity is, relate recent developments in the "high-T_C" field, and list some applications.

7.4 Electric Circuits and Ohm's Law

- Define *voltage* and the volt.
- Make the analogy between electric circuits and water flowing in pipes.
- Remember Ohm's law and apply the relation $V = IR$ in solving circuit problems.
- Recognize series and parallel circuits, and draw circuit diagrams.

7.5 Power and Energy in Electric Currents

- Compute electric power from $P = VI$ and explain ohmic heating.
- Tell why commercial power distribution systems use very high voltages.
- Explain what a kilowatt-hour measures.

7.6 AC and DC

- Explain the difference between DC and AC, and tell why home electric outlets provide AC.

Key Terms

* Terms with an asterisk appear in definition boxes in the chapter.

Electric Charge*

Negative Ion

Positive Ion

Current*

Resistance*

Voltage*

Power Output

Ohmic Heating

Direct Current (DC)

Alternating Current (AC)

Laws

Coulomb's Law

Ohm's Law

Equations Covered

$$F = \frac{(9 \times 10^9) q_1 q_2}{d^2}$$

$$I = \frac{q}{t}$$

$$V = \frac{E}{q} = \frac{\text{work}}{q}$$

$$I = \frac{V}{R}$$

$$V = IR$$

$$P = VI$$

$$E = Pt$$

Teaching Tips and Demonstrations

7.1 Electric Charge

Show electrostatic attraction using a plastic rod rubbed with fur, attracting bits of paper or Styrofoam packing peanuts. Even better is a black plastic golf club tube rubbed with an oven roasting bag. Have some fun trying other materials. (I found some white plastic dinner plates at a local catering shop that charged wonderfully well when rubbed with oven roasting bags. And we found that photocopier transparency sheets charged oppositely when also rubbed with the oven roasting bags.)

DEMONSTRATION: Note: Exercises like this sometimes do not work well when the relative humidity is high. You need a Styrofoam cup (or inflated balloon), short pieces of thread, and fingernail-sized pieces of paper. "Charge" the cup by rubbing its sides back and forth in your hair or on fur-like material. Lower the cup to a few centimeters above the pieces of paper and thread. What happens? Does electricity appear to be stronger than gravity? (The cup can be "recharged" as needed. To "discharge" a cup, you can rub its outside on a metal water faucet.)

7.2 Electric Force and Coulomb's Law

For Section 7.2, I usually compare and contrast the electrostatic force and the gravitational force and write down the law of universal gravitation next to Coulomb's Law. The concept of electric field is very abstract to students and must be presented carefully. I try to get them to focus on the field itself in preparation for electromagnetic waves and other situations where an electric field can exist even when there are no electric charges around.

DEMONSTRATION: Note: Exercises like this sometimes do not work when the relative humidity is high. Attach a thread (roughly an arm's length) to a Styrofoam cup. Charge it and then let it dangle from the thread. Charge a second Styrofoam cup, and then bring it close to the hanging one. What happens? The size of the force on the cup is related to how far the thread is angled from vertical. Does the force on the cup seem to be affected by how far apart the two cups are?

DEMONSTRATION: Note: Exercises like this sometimes do not work well when the relative humidity is high. Turn on a water faucet just enough to have a very small but steady stream trickling out. Bring a charged Styrofoam cup near the stream. What happens?

*DHP**, pages E-3 to E-20, has many neat electrostatic demonstrations, many of them utilizing a Wimshurst machine.

7.3 Electric Currents—Superconductivity

For Section 7.3, the positive current convention is used and should be explained.

If you have a high-T_C superconductor and access to liquid nitrogen, you can show the Meissner effect now or wait until the discussion of magnetism in Chapter 8.

Semiconductor technology and its impact on society in recent decades and superconductivity are good topics.

7.4 Electric Circuits and Ohm's Law

Connect light bulbs in series and in parallel with a power supply. Disconnect one bulb in each circuit to make the point shown in Figures 7.20 and 7.21.

7.5 Power and Energy in Electric Currents

The paragraphs following Example 7.6 can lead to a discussion of how inexpensive electrical energy is. Ask the class what would have to be done if a law was passed that banned lethal voltages (over, say, 30 volts) in our society.

7.6 AC and DC

DEMONSTRATION: You will have to be in a room lit only by tube-type fluorescent lights (compact fluorescent lights don't work as well), and have a pen, pencil, or other thin object that is white or light colored. Position yourself so that the fluorescent light is behind or above you. Hold the pen in such a way that the light shines on it and there is a dark background behind it. Move the pen rapidly back and forth (sideways) with your hand. You should see faint lines parallel to the pen. What causes this? (Hint: AC is involved.) Why doesn't this work nearly as well with incandescent light bulbs?

Common Misconceptions

Some people put up quite a fuss over confusion of Ohm's law and the definition of resistance. Even though I think most people using the phrase "Ohm's law" are making a reference to the equation $V = IR$, this relationship is *not* Ohm's law in the strict sense. Ohm's law is this equation combined with the fact that the resistance R does not depend upon V or I. For materials that obey Ohm's law, the voltage and current are directly proportional—graphs of V vs. I are straight lines, as shown in Figure 7.22. Devices such as diodes or the light bulb of Figure 7.23 that have nonlinear V–I graphs do *not* obey Ohm's law because the resistance is different at different applied voltages.

Students have to spend some time with the idea of electric field lines to comprehend their true meaning. Just like the concept of rays in Chapter 5, they are abstractions, and students might attach an unwarranted reality to them. Make sure they realize the field exists in between the field lines even though no field lines are drawn there. The three-dimensional nature of the field could be explored, too.

Check the teaching resources online at 4ltrpress.cengage.com/physics for more common misconceptions that students have about physics in the real world.

* Remember, you can find *A Demonstration Handbook for Physics* (*DHP*) at www.aapt.org.

prepcard

CHAPTER 8 ELECTROMAGNETISM AND EM WAVES

What's Inside

Magnetism and its useful interrelationship with electricity are the subjects of this chapter. First, the properties of permanent magnets and the Earth's magnetic field are described. Next, we demonstrate how electric fields and magnetic fields intertwine whenever motion or change is involved. These concepts are used to explain how many common electrical devices operate. They also suggest the existence of electromagnetic (EM) waves. The properties and uses of the different types of EM waves are the main topics of the latter half of this chapter.

Sections and Objectives

8.1 Magnetism

- Discuss permanent magnets and north and south magnetic poles, and sketch the magnetic field of a bar magnet.
- Explain why compass needles don't point exactly to the Earth's geographic poles.
- Clear up any confusion about the Earth's north geographic pole being its *south* magnetic pole and vice versa.
- Tell what the Meissner effect is.
- Compare and contrast electrostatics and magnetism.

8.2 Interactions between Electricity and Magnetism

- Remember that moving electric charges produce magnetic fields.
- Remember that magnetic fields exert forces on moving electric charges, and explain the direction of the force.
- Remember that changing magnetic fields produce electric fields and vice versa.
- Discuss solenoids and electromagnets.
- Explain what particle accelerators are used for.
- Explain how electromagnetic induction is used in generators and how motors and generators are related.

8.3 Principles of Electromagnetism

- Explain how transformers work and why they only work with AC electricity.

8.4 Applications to Sound Reproduction

- Explain how microphones, speakers, and magnetic tape work.
- Explain how digital compact disc technology leads to better quality sound reproduction.

8.5 Electromagnetic Waves

- Explain and sketch the electric and magnetic field structure of an electromagnetic wave.
- Realize that all the types of electromagnetic waves have the same basic structure, differing only in frequency.
- List the various types of EM waves, and have a feel for their frequency and wavelength ranges.
- Explain how a microwave oven heats food.
- Relate the color of visible light to its wavelength.
- Explain what *ionizing radiation* means.

8.6 Blackbody Radiation

- Explain what blackbody radiation is, and explain how features of its spectrum depend upon temperature.

8.7 EM Waves and the Earth's Atmosphere

- Discuss the greenhouse effect and how it relates to global warming.
- Explain how the ionosphere plays a role in radio communication.
- Explain why better astronomical observations can be made from Earth's orbit than from the ground.

Key Terms

* Terms with an asterisk appear in definition boxes in the chapter.

North Pole

South Pole

Electromagnetic Induction

Electromagnetic (EM) Wave

Radio Waves

Microwaves

Infrared Radiation

Visible Light

Ultraviolet Radiation

X-Rays

Gamma Rays (γ Rays)

Electromagnetic Spectrum

Ionizing Radiation

Blackbody Radiation (BBR)

Greenhouse Effect

Principles

1. An electric current or a changing electric field induces a magnetic field.
2. A changing magnetic field induces an electric field.

Observations

Observation 1: A moving electric charge produces a magnetic field in the space around it. An electric current produces a magnetic field around it.

Observation 2: A magnetic field exerts a force on a moving electric charge. Therefore, a magnetic field exerts a force on a current-carrying wire.

Observation 3: A moving magnet produces an electric field in the space around it. A coil of wire in motion relative to a magnet has a current induced in it.

Equations Covered

$$\frac{V_o}{V_i} = \frac{N_o}{N_i} \qquad c = f\lambda \qquad \lambda_{max} = \frac{0.0029}{T}$$

Teaching Tips and Demonstrations

8.1 Magnetism

Show as many different shapes and sizes of permanent magnets as you can. Have a large demonstration compass available for showing polarity. Place a pair of small, powerful bar magnets on an overhead projector to show attraction and repulsion. To show the shape of a magnetic field, place a magnet on an overhead projector, place a glass plate over it, and sprinkle iron filings on it. Tap on the glass to help align the filings.

Show the Meissner effect by levitating a magnet over a superconductor.

The Earth's magnetic field and the fact that its poles move around is an interesting topic. This is used in archaeology to estimate the ages of kilns and ovens. Magnetic ores align with the Earth's field upon being heated and then cooled. The precise orientation of such materials in a kiln or oven indicates the location of the Earth's magnetic poles at the time it was used.

8.2 Interactions between Electricity and Magnetism

In preparation for Section 8.2, review electric fields in Section 7.2.

DEMONSTRATION: You will need a car or similar vehicle and a compass. Open the hood of the car (when the engine is not running), and locate the battery and the large cables that carry current from it to the starter. Close the hood, and hold a compass over the hood just above where a cable is located. Have a friend start the car, and watch the compass while the starter is engaged. What happens? What causes this?

Cyclotrons and other particle accelerators are good topics for discussion. Learning about the termination of the Superconducting Super Collider project would be appropriate.

8.3 Principles of Electromagnetism

Review the information on alternating current in Section 7.6 in preparation for study of the transformer.

Remove the bottom cover from a high intensity desk lamp and connect AC voltmeters to both the input and the output of the transformer. Switch on the lamp and read the voltages.

8.4 Applications to Sound Reproduction

Sound can be reviewed in Section 6.3. In Section 8.4, the operations of a dynamic microphone, speaker, and tape head are described, as well as the process of digital sound recording and playback. This section may be skipped, but you might suggest it to students who are audiophiles.

You can make a speaker from scratch by gluing several turns of fine wire to a piece of stiff paper and driving the coil with a power amplifier while holding it in front of a large permanent magnet. I used wire-wrap wire (slid off the end of the spool, still coiled) glued to a piece of post card weight paper with white glue (really "gooped" on and thoroughly dried). It took a hefty power amplifier and a big magnet to get much volume, but the audio quality was astoundingly good. We were all amazed.

8.5 Electromagnetic Waves

Review general aspects of waves in Section 6.1 to be set for electromagnetic waves in particular.

Use a microwave demonstration setup to show reflection and diffraction, and to measure the wavelength of microwaves.

8.6 Blackbody Radiation

Review the information on heat transfer in Section 5.4 to be ready for the deeper treatment of heat radiation here. Blackbody radiation does appear at the beginning of Chapter 10, so most of Section 8.6 can be skipped if it is done carefully. The temperature dependence of the blackbody radiation emitted by an object is described, followed by several applications.

8.7 EM Waves and the Earth's Atmosphere

Some of the important ways that EM waves interact with the atmosphere are presented in Section 8.7. Ozone depletion, global warming, and the greenhouse effect are important topics that are likely to become increasingly critical throughout your students' lives. Several astronomical satellites have been launched in recent years and should provide current discussion topics.

Common Misconceptions

"Electromagnetic waves" is not simply a label that is used artificially to encompass radio, microwaves, *IR*, light, UV, x-rays and gamma rays. It is not an arbitrary classification scheme. The whole point is that all of these are examples of the exact same specific phenomenon. They all go at the speed of light. They all have the "crossed" electric and magnetic field structure shown in Figure 8.24. The *only* difference is the frequency. It is astounding that such seemingly diverse phenomena are all so closely connected. The discovery that they can all be brought together in such a grand synthesis is one of the greatest triumphs of physics.

Another troublesome word is "radiation." Ionizing radiation or some vague concept of something invisible yet dangerous is often what comes to mind upon mention of this word. Many students say radiation sounds pretty scary. The word "radiation" is simply too vague to be useful. All electromagnetic waves can be referred to as radiation. In later chapters we learn that particles like electrons, neutrons, and alpha particles can also be called radiation. Just take a look at the number of page references in the index under "Radiation"! We must be more specific.

Check the teaching resources online at 4ltrpress.cengage.com/physics for more common misconceptions that students have about physics in the real world.

What's Inside

In this chapter, we will investigate the law of refraction together with its counterpart, the law of reflection, to discover that they provide the basis for many practical devices, such as cameras, telescopes, and liquid-crystal displays, as well as for many of the most striking natural phenomena, like rainbows, and soap-bubble iridescence. The subject of this chapter is *optics*, the study of light and its interaction with matter.

Sections and Objectives

9.1 Light Waves

- Define what a nanometer (nm) is, and give typical wavelengths of visible light in nm.
- Explain the difference between specular and diffuse reflection.
- Explain how selective absorption gives rise to the colors of objects.
- Discuss two-slit and thin-film interference.
- Describe what it means for light to be polarized.
- Explain how Polaroid sunglasses and liquid crystal displays (LCDs) work.

9.2 Mirrors: Plane and Not So Simple

- Explain how concave, convex, and "one-way" mirrors work.
- Define what the *normal* to a surface is.
- Explain the cause of and cure for spherical aberration.
- Relate some technical innovations in telescope mirror design.

9.3 Refraction

- Explain what occurs when a light ray refracts at an interface, as well as why it happens.
- Do Snell's Law calculations graphically using Figure 9.27.
- Describe total internal reflection and what is meant by the critical angle of incidence.
- Explain how optical fibers work.

9.4 Lenses and Images

- Recognize converging and diverging lenses from their shapes.
- Identify focal points and focal lengths of lenses.
- Locate images via ray diagrams and the thin lens equation.
- Explain the difference between real and virtual images.
- Compute magnification factors for simple optical systems.
- Explain what chromatic aberration is and how it can be corrected.

9.5 The Human Eye

- Describe the basic structure of the human eye, and relate how eyeglasses and surgical techniques can correct nearsightedness and farsightedness.

9.6 Dispersion and Color

- Explain how dispersion relates to prisms forming beautiful spectra.

9.7 Atmospheric Optics: Rainbows, Halos, and Blue Skies

- Explain what makes rainbows.
- Identify and explain lunar and solar halos.
- Tell why the sky is blue.

Key Terms

* Terms with an asterisk appear in definition boxes in the chapter.

Specular Reflection
Normal
Angle of Incidence
Angle of Reflection
Diffuse Reflection
Constructive Interference
Destructive Interference
Polarization
Focal Point
Spherical Aberration
Angle of Refraction
Critical Angle
Total Internal Reflection
Focal Length
Principal Rays
Object Distance
Image Distance
Real Image
Virtual Image
Magnification

Laws

Law of Reflection

Law of Refraction

Equations Covered

$$p = \frac{sf}{s - f}$$

$$M = \frac{\text{image height}}{\text{object height}}$$

$$M = \frac{-p}{s}$$

Teaching Tips and Demonstrations

9.1 Light Waves

Review the basic properties of waves in Sections 6.1 and 6.2, and electromagnetic waves in Section 8.5.

A blackboard optics kit is very useful for a host of demonstrations throughout Chapter 9.

Looking at a color TV screen with a magnifier is a must. Instead of searching for a channel with a black-and-white image, you can hunt for the color adjustment knob on the set and kill the color. It is amazing to see that a *black-and-white* picture is made entirely of red, green, and blue dots.

9.2 Mirrors: Plane and Not So Simple

Use blackboard optics to show the effect of concave and convex mirrors on parallel light rays.

DEMONSTRATION: Stand a small, rectangular plane mirror vertically on a table. Now place another similar plane mirror adjacent to the first with one of its vertical edges running alongside that of the other. Adjust the angle between the mirrors to be about 45°. Place a small object (a coin or a die will do nicely) midway between the mirrors. How many images of the object do you see reflected in the mirrors? Make the angle between the mirrors roughly 60°. How many reflected images do you see now? Set the mirror angle to 30°. Count the number of images in this case. Do you see a pattern developing between the total number of objects (actual plus images) arrayed around the cluster and the angle separating the mirrors? How do you think your observations might be applied to the construction of a kaleidoscope?

The mirror problem with the Hubble Space Telescope is an interesting example of how mirror shapes are so critical. Incidentally, its primary mirror is hyperbolic, not parabolic. (See, for example, *Physics Today*, November, 1990, pages 19–20.) Make sure the students realize the telescope has been successfully repaired.

9.3 Refraction

Use blackboard optics to show refraction, and perhaps even measure the angles of incidence and refraction. To show total internal reflection, shine a beam of light into the semicircular lens so that it enters the curved surface along a normal. The ray will strike the middle of the straight side of the lens. Vary the angle of incidence to show increasing angle of refraction and total internal reflection.

Figure 9.24 shows an interesting optical illusion that fish hunters have to be aware of.

Fiber optics is always a fertile topic because of the many interesting applications, such as using it to look inside the human body.

9.4 Lenses and Images

For Section 9.4, the difference between a real image and a virtual image needs to be emphasized and demonstrated.

The blackboard optics setup works very well for showing image formation of a single point (the light source). Image distance, object distance, focal length, and magnification can all be measured at the board. To show that the image is inverted, use two light sources, one directly above the other, with a red filter in front of one of them.

9.5 The Human Eye

If you have students who are camera buffs you may want to go into things like *f*-stop, depth of field, macro lenses, and so on. Correction of myopia by changing the shape of the cornea is a good topic. Recent innovations in eye surgery using lasers are noteworthy.

9.6 Dispersion and Color

Shine a bright, narrow beam of white light into a 60° flint glass prism and project the spectrum on a screen. Send a laser beam through the same prism to show that it consists of only one color.

9.7 Atmospheric Optics: Rainbows, Halos, and Blue Skies

An important step in the explanation of the primary rainbow is showing how red, even though it emerges from drops lower than the other colors, appears at the top of the rainbow.

Look at a rainbow through Polaroid sunglasses. Take them off and rotate them 90° and look through them—the rainbow disappears.

Common Misconceptions

It is easy to be misled by the term "focal point." It sounds like that is where the light comes to a focus, and it is—for light coming from an *infinite distance* from the lens. For rays originating somewhere closer than infinity, the place where the rays converge to a focus is not at the focal point—it is at a place we call the *image location. Focal point* is a name reserved for the special image location in the case where the object is infinitely far away.

Even though the wavelength of visible light is small, it isn't *that* small. For example, if you look at the millimeter divisions on a metric ruler and imagine them each being divided into ten parts, the wavelength of visible light is only about 200 times smaller than that.

Check the teaching resources online at 4ltrpress.cengage.com/physics for more common misconceptions that students have about physics in the real world.

* Remember, you can find *A Demonstration Handbook for Physics (DHP)* at www.aapt.org.

What's Inside

The two main topics of this chapter are modern atomic physics and the quantum nature of light. The first part of the chapter describes how the efforts of early twentieth-century physicists to understand three puzzling physical processes led to the concept of photons and the Bohr model of the atom. This is followed by a description of the wave nature of atomic particles and the emergence of the revolutionary new physics known as quantum mechanics. The rest of the chapter shows how the quantum-mechanical model of the atom successfully explains the production of atomic spectra and x-rays and the operation of lasers.

Sections and Objectives

10.1 The Quantum Hypothesis

- Explain what it means for something to be *quantized.*
- Sketch how Planck discovered that a model of oscillating atoms having quantized energy levels could explain the spectral features of blackbody radiation.
- Explain the role played by Planck's constant, *h.*

10.2 The Photoelectric Effect and Photons

- Tell how Einstein explained the photoelectric effect by inventing the idea of photons.
- Define the electronvolt (eV).
- Explain how photocopiers work.

10.3 Atomic Spectra

- Know the difference between continuous spectra and line spectra, and explain how elements can be identified from their spectra.

10.4 The Bohr Model of the Atom

- Relate the basic features of the Bohr model of the atom, and tell how it could account for elements having their observed line spectra.
- Criticize the assumptions of the Bohr model.

10.5 Quantum Mechanics

- Compute the de Broglie wavelengths of material particles.
- Tell why electron microscopes and scanning tunneling microscopes can outperform optical microscopes.
- Explain how de Broglie's wave hypothesis can be used to help defend Bohr's assumption of quantized angular momenta for the electron orbits in an atom.
- Associate Heisenberg and Schrödinger with the origin of quantum mechanics.
- Recall what the uncertainty principle says about position and momentum measurements.

10.6 Atomic Structure

- Distinguish the modern quantum mechanical model of the atom from the old Bohr atom.
- Compute the energies and corresponding wavelengths and frequencies of photons emitted or absorbed when atomic electrons undergo energy level transitions.
- Interpret an energy-level diagram—in particular, read off an atom's ionization energy, the quantum number *n* for various states, and identify sets of transitions that generate spectral series.
- Explain what is meant by the term *ground state.*
- Explain how neon signs, fluorescent lights, and plasma displays work.
- Explain the *Pauli exclusion principle* and its consequences.

10.7 X-Ray Spectra

- Explain how both continuous and characteristic x-ray spectra are produced.

10.8 Lasers

- Explain how *stimulated emission* and *population inversion* are involved in laser operation.
- Give several examples of types of lasers and some of their uses.

Key Terms

* Terms with an asterisk appear in definition boxes in the chapter.

Blackbody Radiation	Bohr Model
Photoelectric Effect	Ionization Energy
Atomic Spectra	Absorption Spectrum
Quantized	de Broglie Wavelength
Photon	Quantum Mechanics
Continuous Spectrum	Energy Level
Emission-Line Spectrum	Quantum Number

Principles

If we let Δx represent the uncertainty in the position of a particle, and Δmv represent the uncertainty in the momentum of the particle, then:

$$\Delta x \, \Delta mv \geq h \qquad \text{(uncertainty principle)}$$

Two electrons cannot occupy precisely the same quantum state at the same time. (Pauli exclusion principle)

Equations Covered

$E = hf$ $\lambda = \frac{h}{mv}$

$\Delta E = hf$ $\Delta x \, \Delta mv \geq h$

Teaching Tips and Demonstrations

10.1 The Quantum Hypothesis

Review blackbody radiation in Section 8.6.

The concept of quantization is quite important and should be emphasized. There are a number of analogies in addition to the one depicted in Figure 10.2. For example, a linked chain is quantized in length whereas a rope or cable is not. You might have a set of steps and a ramp handy when discussing quantization of energy.

10.2 The Photoelectric Effect and Photons

The concepts of photon and wave-particle duality are difficult for students who are usually still thinking in classical physics terms. Figure 10.4 is the same electromagnetic wave spectrum seen in Chapter 8, but with the photon energy scale added at the top.

10.3 Atomic Spectra

The importance of spectroscopy in physics, chemistry, and astronomy can't be stressed too much.

DEMONSTRATION: Pass around individual spectroscopes or just diffraction gratings to each student and have them look at different spectrum tubes in a darkened classroom. With spectroscopes you can also have them look at the continuous spectrum from an incandescent light bulb and the combination continuous and line spectrum from a fluorescent lamp.

10.4 The Bohr Model of the Atom

You may want to emphasize that Bohr's model is not quite the final version of the theory of electronic transitions in atoms and that the results of quantum mechanics alter the nature of the discrete orbits depicted in the drawings in Section 10.4.

The radical nature of Bohr's model, particularly in light of the unexplained assumptions described at the end of Section 10.4, is an interesting topic for discussion.

10.5 Quantum Mechanics

Quantum mechanics in general and Heisenberg's uncertainly principle in particular pose many philosophical questions. You may want to describe Einstein's reluctance to accept the indeterminacy in quantum mechanics.

EXPLORING ON THE WEB: You can have students investigate in more detail the diffraction and interference of electrons passing through slits using a suite of programs developed at Kansas State University by Professor Dean Zollman and collaborators. Have them visit the Visual Quantum Mechanics Web site at http://phys.educ.ksu.edu/vqm/ and select one of the available simulations (for example, Double Slit Diffraction). Follow the instructions and explore the dependence of the resulting electron interference patterns on such parameters as slit width, slit spacing, and electron-beam intensity. Compare your findings with the observed characteristics of light-wave diffraction and interference discussed in Chapters 6 and 9.

10.6 Atomic Structure

Compare the energy level diagram of the hydrogen atom to stairs with steps that get shorter at higher levels. The photon energies in the emission spectrum can be compared to the potential energy lost by a weight dropped from one level to another. You may want to show the emission spectrum of hydrogen again and point out the specific energy level transition associated with each line, as well as the energy, frequency, and wavelength of the photons.

10.7 X-Ray Spectra

Moseley's short career and his death in World War I can lead to a general discussion of the impact of wars and other historical events on science and scientists.

10.8 Lasers

Show a helium-neon laser and any other type of laser you may have access to (pocket-sized "laser pointers" are very inexpensive). To see the path of the beam, turn down the lights and put chalk dust or smoke along the beam. Send the beam through a prism or diffraction grating to show that it is indeed monochromatic.

The incredible number and variety of uses of lasers is a wide open topic, well worth exploring.

Common Misconceptions

We shouldn't lose sight of how difficult this material really is. What makes it especially tough is the way the chapter touches only briefly on several complex subjects. The students need to be reassured when they don't get it that this is completely normal. Beginners have no way to see how much more there is to, say, quantum mechanics than just the taste to be had here. They often try to read and reread a treatment like that in Section 10.5 in order to understand it completely. Their inevitable failure can leave some of them very discouraged.

Check the teaching resources online at 4ltrpress.cengage.com/physics for more common misconceptions that students have about physics in the real world.

prepcard

CHAPTER 11
NUCLEAR PHYSICS

What's Inside

The basic properties of nuclei, the many applications of nuclear processes, and the promises and perils of nuclear energy are the main topics of this chapter. The first two sections describe the composition of the nucleus and the process of *radioactive decay*. This leads to the concept of half-life and the use of radioisotopes as natural clocks. Most of the remainder of the chapter deals with the role of energy in nuclear physics, particularly the two principal means of tapping that energy—nuclear fission and nuclear fusion.

Sections and Objectives

11.1 The Nucleus

- Appreciate of how small the nucleus of an atom really is.
- Define the atomic mass unit, u.
- Define and relate the atomic number *Z*, the mass number *A*, and the neutron number *N*, for an atom.
- Tell how isotopes of a given element differ.
- Read and interpret isotopic notation such as $^{4}_{2}\text{He}$.
- Realize how nuclear reactions differ from chemical and other physical processes that are the results of atomic electron interactions.
- Tell why large nuclei tend to have more neutrons than protons in them.

11.2 Radioactivity

- Explain what radioactivity is.
- Explain how a Geiger counter works.
- Compare the features and effects of alpha, beta, and gamma decay.

11.3 Half-life

- Clearly define the concept of half-life and use it in calculations.
- Explain how archaeologists are able to use carbon-14 dating.

11.4 Artificial Nuclear Reactions

- Tell how a breeder reactor can manufacture reactor fuel.
- Explain what neutron activation analysis is and how it can be used.

11.5 Nuclear Binding Energy

- Explain how the variation in nuclear binding energy per nucleon for various elements makes it possible for both fission and fusion reactions to release energy.
- Explain why the masses of all nuclei are less than the combined masses of the protons and neutrons in them.

11.6 Nuclear Fission

- Relate details of the process of nuclear fission.
- Compare stable and unstable chain reactions.
- Explain the differences between atomic fission bombs and thermonuclear fusion bombs.
- Explain the role of the moderator in a nuclear reactor.
- Tell what safety features are incorporated into nuclear power plants.
- Tell what happened in the accidents at Three Mile Island and Chernobyl.

11.7 Nuclear Fusion

- Relate details of the process of nuclear fusion.
- Be alert for news of progress in fusion reactor research.

Key Terms

* Terms with an asterisk appear in definition boxes in the chapter.

Atomic Number (*Z*)*

Neutron Number (*N*)*

Mass Number (*A*)*

Nucleons

Isotopes*

Radioactivity

Radioisotopes

Alpha Radiation (α)

Beta Radiation (β)

Gamma Radiation (γ)

Decay Chain

Half-Life*

Carbon-14 Dating

Neutron Activation Analysis

Binding Energy

Nuclear Fission*

Nuclear Fusion*

Equations Covered

$N = N_0\left(\frac{1}{2}\right)^n$

$E = mc^2$

Teaching Tips and Demonstrations

11.1 The Nucleus

Learning the isotopic notation should be stressed since it is used throughout the chapter.

Figure 11.2 shows how the neutron to proton ratio of stable nuclei rises for larger nuclei. You may want to refer back to this when talking about radioactive fission fragments.

11.2 Radioactivity

DEMONSTRATION: Use a Geiger counter with a rate meter to show radioactivity. Use standard radioactive samples, a radioisotope-type smoke detector, a "mantle" used in white-gasoline camping lanterns, an old luminous dial wristwatch, or other sources of radiation.

DEMONSTRATION: Use a large magnet to deflect beta rays, as shown in Figure 11.3. Use a lead pipe or channel to collimate the beta rays. Place the GM tube to the side of the beam and bring the magnet into place so that the beta particles are deflected into the GM tube and the count rate goes up. Reverse the polarity of the magnet to reduce the count rate. You may want to go through the right-hand-rule to show that the beta rays are negatively charged.

11.3 Half-life

A common misconception you may have to dispel is that all of the nuclei of a radioisotope will decay during two half-lives. Carbon-14 dating is a nice example of a very useful, non-dangerous application of nuclear physics.

DEMONSTRATION: Students can simulate the decay of radioactive nuclei with a large number of pennies or similar flat objects. It's best to use 50 or more. You might perform this demonstration as a community effort with several hundred pennies. If you use only 50 or so, do it two or more times and point out the statistical fluctuation in the data.

1. Place the pennies in a small box or bag and shake them thoroughly. Dump the pennies out on a flat surface like a desktop.
2. Treat each penny with "tails" showing as a nucleus that has decayed. Push these off to the side, and record how many pennies are left. Collect these "undecayed" pennies, and repeat the procedure several times.
3. Each penny has a 50% chance of turning up "tails" after each throw. Thus each throw represents one half-life for the pennies. On the average, half of them "decay" during each throw.
4. Notice how the number of pennies left undecayed decreases. Make a simple graph of these numbers. You should get graphs like those in Figure 11.10.
5. Think about the probabilities: After, say, the fifth throw, any surviving penny has turned up "heads" five times in a row. What is its chance of turning up "heads" on the next throw?

11.4 Artificial Nuclear Reactions

When covering Section 11.4, the conversion of uranium-238 to plutonium-239 should be stressed since it is important in nuclear fission technology.

11.6 Nuclear Fission

Nuclear power and its potential hazards has been a hot debate topic for decades. You may want to go into more detail about reactor design and describe what happened at Three Mile Island and at Chernobyl. The report of the President's Commission on the Accident at Three Mile Island, often referred to as the Kemeny Commission, is a thorough account of the accident and its causes. It's been almost 65 years since atomic bombs were developed during the Manhattan Project in World War II. There is potential for discussion from several angles: technology, politics, the personalities of the scientists involved, and so on.

11.7 Nuclear Fusion

Controlled fusion is a good topic for discussion for many reasons. The technology is so demanding that a huge effort during several decades has failed to bring about commercial fusion power plants. At this time the results are more promising than before. The "table-top" cold fusion uproar of March 1989 (not mentioned in the text) is a fascinating story with many different facets. The enormous destructive power of megaton-range thermonuclear weapons, both individually and in huge stockpiles, is always a timely topic, though it is sobering as well. The possible effects on the Earth's climate of a large nuclear exchange, particularly the "nuclear winter" hypothesis, could be mentioned.

Common Misconceptions

Radioactivity seems rather mysterious and scary. Bad sci-fi movies and other misinformation do not help. Some students may need to have their conceptions about radioactivity straightened out. They need to see that the effects of ionizing radiation are indeed understood and can be predicted. Ionizing radiation can be readily detected at intensity levels *way* below those that are cause for concern. This is a very good thing. A Geiger counter can immediately reveal a radiation hazard, and action can be taken right away. A dangerous situation involving toxic chemicals rather than radioactive materials would be harder to detect and evaluate.

Questions about the safety of nuclear reactors and disposal of nuclear waste are important, but they should be weighed against the truly astonishing amount of energy that nuclear power can produce from a very small amount of fuel. Society's demand for energy is surely going to increase in the future, and nuclear power will be called upon to satisfy that demand when other sources aren't sufficient. People will need to make big decisions, and if these are to be wise decisions, all facets of the issue need to be considered.

Check the teaching resources online at 4ltrpress.cengage.com/physics for more common misconceptions that students have about physics in the real world.

prepcard

CHAPTER 12
SPECIAL RELATIVITY AND ELEMENTARY PARTICLES

What's Inside

This chapter will consider questions related to the properties of elementary particles and their interactions. Our goal is to develop an appreciation for what particle physicists call the Standard Model and to forge links between submicroscopic physics and cosmology. As we shall see, the holy grail of high-energy physicists and cosmologists is the unification of physics, including all the known forces of nature, into one grand theory of everything. We start with one of the seminal theories in this unification process developed by Albert Einstein—the special theory of relativity.

Sections and Objectives

12.1 Special Relativity: The Physics of High Velocity

- State the two postulates of the special theory of relativity.
- Explain how relativistic time dilation allows muons created by cosmic rays high in the atmosphere to get all the way to the ground before decaying.
- Relate the concepts of relativistic time dilation and length contraction.
- Demonstrate an understanding of Einstein's famous equation $E_0 = mc^2$ and link it with the correct modern interpretation of mass in relativity.
- Express masses of particles in energy units (e.g., MeV/c^2).

12.2 Forces and Particles

- Identify the four forces of nature and their "carrier" particles.
- Explain the similarities and differences between elementary particles and their antiparticles, and be familiar with the "bar" notation used for antiparticles.
- Know what is meant by the phrases "integral spin" and "half-integral spin", and how these relate to fermions and bosons.
- Distinguish between fermions and bosons (e.g., the former obey the Pauli exclusion principle, while the latter do not).
- Classify particles as baryons, leptons, mesons, or gauge bosons.

12.3 Conservation Laws, Revisited

- Apply the conservation of mass-energy, linear momentum, angular momentum, and electric charge in analyzing particle interactions.
- Use the conservation laws for baryon number, lepton number (for all three families), and strangeness in analyzing particle interactions.

12.4 Quarks: Order out of Chaos

- Outline the quark model for hadrons.
- Analyze particle interactions in terms of their quark constituents.

12.5 The Standard Model and GUTs

- Describe what is meant by "quark colors" and what the role of gluons is in mediating the color force between quarks.
- Present accurately the major features of the standard model of elementary particles, including the six quark flavors and six leptons.
- Demonstrate an understanding of the significance of the achievement of Weinberg and Salam in successfully unifying the electromagnetic and weak nuclear interactions.
- Explain what is meant by "spontaneous symmetry breaking."
- Describe the goal of grand unified theories, or GUTs.

Key Terms

* Terms with an asterisk appear in definition boxes in the chapter.

Special Theory of Relativity	Antiparticle*
Time Dilation	Positrons
Length Contraction	Spin
Rest Energy	Fermion
Relativistic Energy	Boson
Gravity	Baryon
Electromagnetic Interaction	Lepton
Weak Nuclear Force	Meson
Electroweak Interaction	Strangeness*
Strong Nuclear Force	Quarks
Carrier (or Exchange) Particle	Grand Unified Theory (GUT)
Elementary Particles*	

Laws

Law of Conservation of Baryon Number

Law of Conservation of Lepton Number

Equations Covered

$$\Delta t' = \frac{\Delta t}{\sqrt{1 - v^2/c^2}}$$

$$E_0 = mc^2$$

$$E_{rel} = \frac{mc^2}{\sqrt{1 - v^2/c^2}}$$

$$KE_{rel} = \frac{mc^2}{\sqrt{1 - v^2/c^2}} - mc^2$$

Teaching Tips and Demonstrations

12.1 Special Relativity: The Physics of High Velocity

When discussing the experimental basis for the constancy of the speed of light, you may wish to elaborate upon the Michelson-Morley experiment and its "null" result that led to the demise of the aether theory. In connection with time dilation, you could present the twin paradox and use it as a springboard to a discussion of "space-time" travel. Emphasizing the inviolability of the speed of light often engenders questions about "faster-than-light" particles (*tachyons*). A good reference if you wish to pursue such issues is L. M. Feldman's article "Short Bibliography on Faster-Than-Light Particles (Tachyons)" in the *American Journal of Physics*, 42(3), March 1974, pp. 179–182.

12.2 Forces and Particles

When starting Section 12.2 it may be helpful to review gravity in Section 2.9, the electromagnetic interaction in Sections 8.1 and 8.2, and the elements of nuclear physics covered in Sections 11.1, 11.2 and 11.4. A reminder to students about the concept of quantization covered in Section 10.1 is also helpful.

In connection with the four fundamental forces of nature, it is important to emphasize the need to broaden the definition of "force" to include all types of changes that elementary particles may undergo (spontaneous disintegrations, annihilations, creations, etc.) and to acknowledge the modern view that these forces are mediated by *carrier* particles. An example of the mediation of the electrical repulsion between two electrons by the exchange of (virtual) photons is given. The basic ways of classifying elementary particles (via spin, mass, and participation in the strong interaction) are presented. Although spin is a purely quantum mechanical property, at least for spin $\frac{1}{2}$ particles the classical analogy of a BB rotating clockwise or counterclockwise is successful at establishing their quantization in either the spin-up or spin-down configuration.

12.3 Conservation Laws, Revisited

The discovery and use of new conservation laws and conserved quantities to analyze subatomic particle reactions are discussed; here it may be helpful to review conservation laws in classical mechanics in Chapter 3. In this context, it is important to emphasize the probabilistic aspects of these interactions and the fact that frequently there is as much to learn from what is *not* observed as there is from what *is* observed.

12.4 Quarks: Order out of Chaos

Focusing attention on the bubble-chamber discovery photograph of the Ω^- particle shown in Figures 12.9 and 12.10 is a convenient excuse to digress briefly into the "hardware" aspects of high-energy physics, that is, the production and detection of elementary particles.

EXPLORING ON THE WEB: Want to build your own baryons? Play particle pinball? Find out what life might be like without some of the fundamental forces? All this and more awaits the curious student in the Fermilabyrinth at the online Lederman Science Center (http://www-ed.fnal.gov/ed_lsc.html). Have students pay a visit to this site and play some of the many educational games provided to learn more about elementary particle physics and especially about the detection and analysis techniques used to probe the details of particle-collision reactions.

DEMONSTRATION: Analysis of bubble chamber photographs to determine the mass, momentum, and lifetime of particles from direct measurements of their tracks provide good opportunities to apply many of the concepts and principles introduced in Chapters 11 and 12. The equipment requirements are minimal (rulers, compasses, and, in some cases, map measurers), and worksheets and bubble chamber photos may be ordered from the Education Office at Fermilab.

12.5 The Standard Model and GUTs

Other analogies may be used to help students get a handle on spontaneously broken symmetry and its connection with energy. For example, a pencil standing vertically on a tabletop, delicately balanced on its point, represents a single, initial, unstable, high (potential) energy state relative to the table's surface. A small perturbation may cause the pencil to "decay" (tip over) into any one of a large number of distinguishable lower energy states corresponding to the various compass directions it may point while lying on the tabletop. See H. Pagel's book *Perfect Symmetry* (pp. 203–205) for an additional example involving ferromagnets.

Common Misconceptions

It is worth reiterating the earlier comments about the phrase "relativistic mass." Such usage is very entrenched, and attempts to dislodge it are rather recent. Many books you and your students will encounter may still contain this terminology, and there will be great confusion. The most modern view is that there is no such thing as relativistic mass or a change in mass as objects accelerate to relativistic speeds. "Mass is mass is mass," says Edwin F. Taylor in *Spacetime Physics.*

Check the teaching resources online at 4ltrpress.cengage.com/physics for more common misconceptions that students have about physics in the real world.